The Soil Nexus

The Soil Nexus

Understanding Ecosystem Dynamics and Digital Innovations

Edited by

Hamid Reza Pourghasemi

Professor, Department of Soil Science, College of Agriculture, Shiraz university, Shiraz, Iran

Sedigheh Maleki

Assistant Professor, Department of Soil Science, College of Agriculture, Shiraz university, Shiraz, Iran

ELSEVIER

Elsevier
Radarweg 29, PO Box 211, 1000 AE Amsterdam, Netherlands
125 London Wall, London EC2Y 5AS, United Kingdom
50 Hampshire Street, 5th Floor, Cambridge, MA 02139, United States

ISBN: 978-0-443-45286-4

For information on all Elsevier publications visit our website at
https://www.elsevier.com/books-and-journals

Publisher: Peter Llewellyn
Acquisitions Editor: Peter Llewellyn
Editorial Project Manager: Arushi Rawat
Production Project Manager: Jenela Priscy
Cover Designer: Mark Rogers

Typeset by TNQ Tech

Dedication

To my son, Yara, and my wife, Kimia.

—Hamid Reza Pourghasemi

To Pouriyar, my beloved son.

—Sedigheh Maleki

Contents

PART III Economic valuation of soil and policy frameworks for soil management and sustainable development

PART V Soil and food security

Contributors

Narges Abdar
Department of Soil Science, College of Agriculture, Shiraz University, Shiraz, Iran

Morteza Akbari
Department of Arid and Desert Region Management, Faculty of Natural Resources and Environment, Ferdowsi University of Mashhad, Mashhad, Iran

Abdel Razzaq Al-Tawaha
National Agricultural Research Center (NARC), Baq'a, Jordan

Seyed Kazem Alavipanah
Department of Remote Sensing and GIS, Faculty of Geography, University of Tehran, Tehran, Iran

Majid Baghernejad
Department of Soil Sciences, School of Agriculture, Shiraz University, Shiraz, Iran

Abdolhossein Boali
Department of Arid and Desert Region Management, Faculty of Natural Resources and Environment, Ferdowsi University of Mashhad, Mashhad, Iran; Department of Arid Zone Management, Gorgan University of Agricultural Sciences and Natural Resources, Gorgan, Iran

Michele Delchiaro
Department of Earth Sciences, Sapienza University of Rome, Rome, Italy

Sayyed Mahmoud Enjavinezhad
Department of Soil Sciences, School of Agriculture, Shiraz University, Shiraz, Iran

Foroogh Golkar
Department of Water Engineering & Oceanic and Atmospheric Research Centre, College of Agriculture, Shiraz University, Shiraz, Iran

Yousef Hasheminejhad
Khorasan Razavi Agricultural and Natural Resources Research and Education Center, Agricultural Research, Education and Extension Organization (AREEO), Mashhad, Iran

Mahvan Hassanzadeh Bashtian
Department of Soil Science, College of Agriculture, Ferdowsi University of Mashhad, Mashhad, Iran

Alireza Karimi
Department of Soil Science, College of Agriculture, Ferdowsi University of Mashhad, Mashhad, Iran

Narges Kariminejad
Department of Natural Resources and Environmental Engineering, College of Agriculture, Shiraz University, Shiraz, Iran

Arun Karnwal
Department of Microbiology, Graphic Era (Deemed to be University), Dehradun, Uttarakhand, India

Farhad Khormali
Department of Earth Sciences, Southern Methodist University, Dallas, TX, United States

Anupam Kumar
Department of Biotechnology, Faculty of Applied, Pure and Health Sciences, Parul University, Quepem, Goa, India

Deepak Kumar
Department of Chemistry, Manipal University Jaipur, Jaipur, Rajasthan, India

Gaurav Kumar
Department of Microbiology, Amity Institute of Microbial Technology, Amity University Rajasthan, Jaipur, Rajasthan, India

Sedigheh Maleki
Department of Soil Science, College of Agriculture, Shiraz University, Shiraz, Iran

Mahsa Mardani
Department of Water Resources Engineering and Management, University of Birjand, Birjand, Iran

Najmeh Mazloom
Khorasan Razavi Agricultural and Natural Resources Research and Education Center, Agricultural Research, Education and Extension Organization (AREEO), Mashhad, Iran

Abdel Rahman Mohammad Said Al-Tawaha
Department of Biological Sciences, Al Hussein Bin Talal University, Ma'an, Jordan

Ali Akbar Moosavi
Department of Soil Science, College of Agriculture, Shiraz University, Shiraz, Iran

Mohammad Moumeni
Department of Science, University of Roma Tre, Rome, Italy

Amin Mousavi
Department of Remote Sensing and GIS, Faculty of Geography, University of Tehran, Tehran, Iran

Seyed Javad Naghibi
Department of Soil Sciences, School of Agriculture, Shiraz University, Shiraz, Iran; Landscape & Green Spaces Organization of Shiraz Municipality, Shiraz, Iran

Ali Nasirian
Department of Water Resources Engineering and Management, University of Birjand, Birjand, Iran

Natalia Nesterova
National University of Life and Environmental Sciences of Ukraine, Kyiv, Ukraine

Hamid Reza Pourghasemi
Department of Soil Science, College of Agriculture, Shiraz University, Shiraz, Iran

Parviz Rezvani Moghaddam
Faculty of Agriculture, Ferdowsi University of Mashhad, Mashhad, Iran

Amin Sabbagh
Department of Civil Engineering, Faculty of Engineering, Shahid Bahonar University of Kerman, Kerman, Iran

Niloofar Sadri
Department of Soil Science, College of Agriculture, Shiraz University, Shiraz, Iran

Tayebeh Safari
Department of Soil Science, College of Agriculture, Ferdowsi University of Mashhad, Mashhad, Iran

Neha Saklani
Department of Microbiology, Graphic Era (Deemed to be University), Dehradun, Uttarakhand, India

Raziyeh Shamshirgaran
Department of Water Resources Engineering and Management, University of Birjand, Birjand, Iran

Afshin Shayeghi
Department of Geography & Environmental Sustainability, University of Oklahoma, Norman, OK, United States

Morteza Poozesh Shirazi
Department of Soil and Water Research, Fars Agricultural and Natural Resources, Research and Education Center, AREEO, Shiraz, Iran

Elham Sirjani
Department of Soil Science, Ferdowsi University of Mashhad, Mashhad, Iran

Moein Tosan
Department of Water Science and Engineering, Ferdowsi University of Mashhad, Mashhad, Iran

Morteza Yaqubi
Saffron Institute, University of Torbat Heydarieh, Torbat-e Heydarieh, Iran

Maryam Zahedifar
Department of Range and Watershed Management (Nature Engineering), Faculty of Agriculture, Fasa University, Fasa, Iran

Leila Zare
Department of Soil Science, College of Agriculture, Shiraz University, Shiraz, Iran

Mehdi Zarei
Department of Soil Science, College of Agriculture, Shiraz University, Shiraz, Iran; Department of Agriculture and Natural Resources, Higher Education Center of Eghlid, Eghlid, Iran

PART I

Defining soil and its role in ecosystems: Introduction to soil science and its contemporary relevance

CHAPTER 1

Soil's contribution to biogeochemical cycles (water, carbon, nitrogen)

Leila Zare
Department of Soil Science, College of Agriculture, Shiraz University, Shiraz, Iran

1. Introduction: Soil as a critical biogeochemical interface

Soil is far more than a passive medium for plant growth, it is a dynamic, living system that serves as a central hub linking the atmosphere, biosphere, hydrosphere, and lithosphere. Biogeochemical cycles of water, carbon (C), and nitrogen (N) are fundamental to life on Earth, governing nutrient availability, climate regulation, and ecosystem productivity (Shah et al., 2019; Saraswathi, 2023). These cycles are intrinsically coupled within soils, where physical structure, chemical composition, and biological activity interact across scales to mediate the continuous exchange of energy and matter. Soil organic matter (SOM) formed through microbial transformation of plant residues, constitutes the largest terrestrial C pool (storing 2000–3000 Pg C) (Shahbaz et al., 2017; Zhang, Cai, et al., 2022, Zhang, Chen, Jin, et al., 2022, Zhang, Chen, Yao, et al., 2022). This vast reservoir is highly sensitive to land use and climate change, with even small losses potentially amplifying atmospheric C dioxide (CO_2) and methane (CH_4) concentrations. Similarly, soil drives the N cycle through processes such as mineralization, nitrification, and denitrification, regulating nutrient supply while influencing emissions of nitrous oxide (N_2O), a potent greenhouse gas (GHG) (Robertson & Groffman, 2024). Crucially, C and N cycling are tightly linked, with microbial activity often co-limited by the availability of both elements. The hydrological cycle is equally dependent on soil function. Through its capacity to absorb, store, and transmit water, soil regulates infiltration, runoff, groundwater recharge, and evapotranspiration (Yang et al., 2021). Soil moisture, in turn, controls oxygen availability and microbial metabolism, acting as a master variable that integrates water, C, and N dynamics (Amador et al., 2005).

Yet, human activities and climate change are disrupting these interconnected processes, accelerating soil degradation and weakening its regulatory capacity (Raimi et al., 2021), thereby threatening both soil health and global environmental stability. This chapter explores the pivotal role of soil in mediating the water, C, and N cycles, emphasizing the interdependence of physical, chemical, and biological processes. We examine the mechanisms underlying these cycles, their interactions, and the impacts of anthropogenic disturbances, concluding with pathways for sustainable soil management in an era of global change.

The Soil Nexus. https://doi.org/10.1016/B978-0-443-45286-4.00002-X

2. Carbon cycle

2.1 Soil as a central component of the global carbon cycle

The global C cycle is a fundamental biogeochemical process encompassing a series of transformations such as photosynthesis, respiration, decomposition, and sequestration which regulate atmospheric CO_2 concentrations and influence Earth's climate system. Among terrestrial ecosystems, soil plays a pivotal role in the global C cycle, functioning as both the largest terrestrial C reservoir and a dynamic regulator of C fluxes (Crowther et al., 2019; Zaman et al., 2025). The upper 1 m of soil contains approximately 1500 Pg (Pg) of organic C (OC) more than the atmosphere (850 Pg) and aboveground vegetation (500 Pg) combined (Azevedo et al., 2024; Huys et al., 2022). This vast pool is not static and is maintained through a continuous balance between inputs and losses driven by biogeochemical processes (Hussain et al., 2021; Xu et al., 2021).

C enters the soil primarily through plant litter, root exudates, and microbial byproducts, and is processed via complex biological and chemical transformations. Simultaneously, C is released back into the atmosphere mainly through soil respiration -the second-largest C flux between the terrestrial biosphere and the atmosphere, amounting to 98 Pg C yr^{-1}. This flux consists of heterotrophic and autotrophic respiration (Feng et al., 2024). Despite being a net source of CO_2, soil also acts as a vital C sink through sequestration processes. These include the conversion of atmospheric CO_2 into stable soil organic C (SOC) pools via humification, the formation of secondary carbonates, and bicarbonate leaching into groundwater, all of which contribute to long-term C storage and climate change mitigation (Hussain et al., 2021). However, human activities such as deforestation, intensive agriculture, and land degradation have led to the loss of an estimated 133 Pg of SOC since the dawn of agriculture (Ewing et al., 2023). These disturbances disrupt the natural balance of the C cycle and trigger feedbacks to climate change. Therefore, understanding soil's dual role as both a C reservoir and a dynamic processor of C fluxes is essential for predicting future climate trajectories and developing sustainable land-use practices that enhance C sequestration and resilience. The following sections examine the sources and stabilization of SOC, microbial mediation of transformations, sequestration dynamics, soil–atmosphere gas exchange, anthropogenic impacts, and the role of SOC in climate mitigation policy.

2.2 Sources of soil carbon

Soil C occurs in inorganic, elemental, and organic forms. Inorganic C, primarily as carbonates, is found in arid and semiarid soils derived from carbonate-rich parent material, where limited leaching allows accumulation (Yang et al., 2025) and is about one-third proportion of the total C (Singh, 2024). SOC is the most dynamic and biologically active fraction, sustaining soil fertility and agricultural productivity through its influence on soil physical, chemical, and biological properties (Wu et al., 2024). SOC originates from diverse biological sources and is distributed across functional pools based on turnover and stabilization mechanisms. The main input of SOC is derived from plant-litter including aboveground litter (e.g., senescent leaves and crop residues) and root-derived input (e.g., root litter and mycorrhizal hyphae), root exudates and microbial necromass -the cellular remains of dead microorganisms (Feng et al., 2024; Zhang, Cai, et al., 2022, Zhang, Chen, Jin, et al., 2022, Zhang, Chen, Yao, et al., 2022). Notably, anthropogenic inputs such as compost, manure, and biochar must also be considered significant sources of SOC, as they introduce both labile and recalcitrant C, stimulating microbial activity while contributing to long-term C storage (Fig. 1.1).

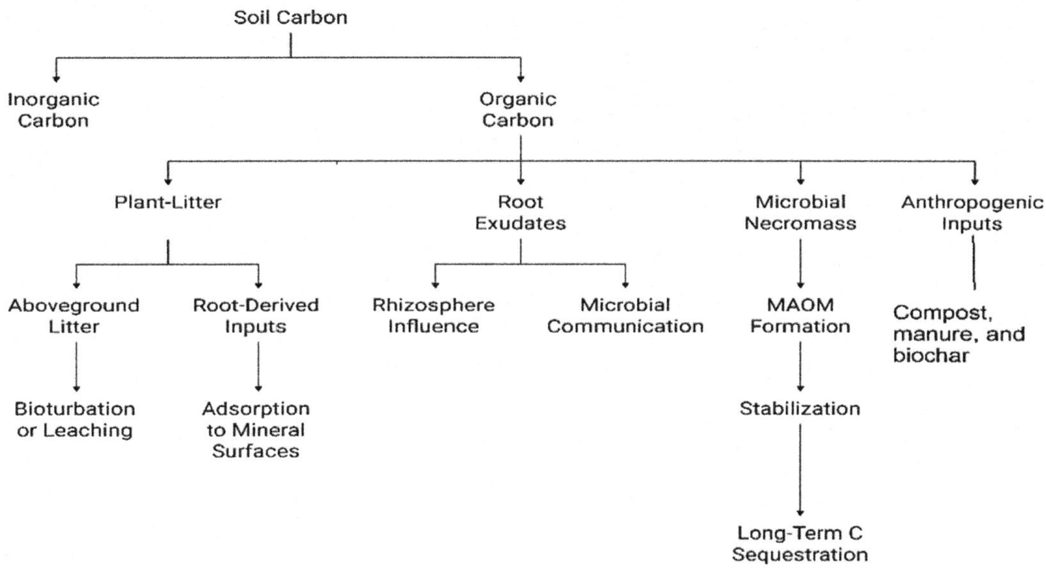

FIGURE 1.1

Soil organic carbon sources.

Plant-litter is a major source of SOC, returning approximately 50 Gt of OC annually to soils (Zhang, Cai, et al., 2022, Zhang, Chen, Jin, et al., 2022, Zhang, Chen, Yao, et al., 2022) and mainly consist of complex polymers such as cellulose, hemicellulose, lignin, cutin, and suberin. The chemical properties of these tissues influence not only how quickly they decompose but also how they affect key soil properties such as aggregate stability, soil pH, and cation exchange capacity (CEC) (Feng et al., 2024; Xu et al., 2021). Aboveground litter accumulates at the soil surface and is incorporated via bioturbation or leaching, whereas root-derived inputs is deposited directly at depth. While aboveground litter releases soluble OC, root-derived inputs particularly from fine roots have more aliphatic compounds which are quickly adsorbed to mineral surfaces and play a disproportionately large role in SOC dynamics due to their intimate interaction with minerals, microorganisms, and aggregates in soils.

Root exudates contain low molecular weight compounds (LMW) such as sugars and organic acids (OA) which directly release into the rhizosphere, influencing bio-physicochemical properties of rhizosphere, serving roles in microbial communication and nutrient mobilization (Villarino et al., 2021). Zhang, Cai, et al. (2022), Zhang, Chen, Jin, et al. (2022), Zhang, Chen, Yao, et al. (2022) reported quantified C inputs from litter, fine roots, and root exudates, showing they contributed 85.34%, 10.39%, and 4.27%, respectively, to total plant-derived C input. Although the root exudation represents a smaller fraction, their contribution to SOC is significant. Soil microorganisms easily metabolize LMW compounds in root exudates and have high mineral-associated organic matter (MAOM) formation efficiency. Thus, they should be highly considered in C cycle models (Carvalho et al., 2023).

Contrary to traditional views that emphasize plant-derived materials as the primary contributors to SOC, emerging evidence underscores the central role of microbial anabolic processes in SOC formation and stabilization. Microbes transform plant-derived C into cellular components and metabolic byproducts, which accumulate as microbial necromass now recognized as a dominant source of persistent SOC, particularly within MAOM fractions (Kögel-Knabner et al., 2022; Wu et al., 2024). This microbial-derived C is stabilized through close physicochemical interactions with soil minerals, enhancing its resistance to decomposition and contributing significantly to long-term C sequestration (Wu et al., 2024).

2.3 Forms of soil organic carbon

SOC stability arises from interacting soil and environmental controls, including biochemical resistance, physical protection, chemisorption, and microbial activity. Among these, physical protection is widely regarded as the dominant stabilizing mechanism (Qi et al., 2023). Based on physical protection and turnover dynamics, SOC is commonly categorized into three functional pools: dissolved organic C (DOC), particulate organic C (POC) and mineral-associated organic C (MAOC) (Lei et al., 2023). DOC consists of labile, soluble compounds including fulvic acids, LMWOA, peptides, carbohydrates, and aromatic substances and serves as a key microbial substrate, playing a critical role in C sequestration, transport, and stabilization in mineral soils (Aumtong et al., 2023). It can leach into subsoil, contributing to deep C inputs, but may also induce priming effects (PE) that accelerate decomposition of existing SOC, potentially offsetting C gains by 14.8%–59.4% (Huys et al., 2022; Nakhavali et al., 2021; Xu et al., 2021; Zhang, Cai, et al., 2022, Zhang, Chen, Jin, et al., 2022, Zhang, Chen, Yao, et al., 2022). In SOC formation, simpler compounds such as root exudates tend to promote MAOC formation, while more complex plant residues contribute to POC accumulation (Lei et al., 2023; Shi et al., 2024; Villarino et al., 2021). POC is relatively labile, with a turnover time of years to decades. It is protected by aggregate entrapment and chemical recalcitrance, is sensitive to disturbance, and has high accumulation potential. MAOC is more stable, accounting for 60%–90% of total SOC, and persists for decades to centuries. It is stabilized through adsorption to mineral surface, occlusion within nano to micrometer sized microaggregates and coprecipitation with iron (Fe) and aluminum (Al) oxides (Hu et al., 2025; Qi et al., 2023; Yang et al., 2025).

2.4 Microbial and faunal drivers of soil organic carbon dynamics

2.4.1 Macro and mesofauna in carbon cycle

Soil fauna strongly regulate the rate, pathway, and outcome of plant-litter decomposition by directly consuming and fragmenting it which increasing surface area for microbial colonization. They also influence microbial community size and composition on litter, with effects that can be either stimulatory or inhibitory, as they both graze on microbes and supply substrates for microbial growth (Jackson et al., 2017). Therefore, greater soil fauna diversity can enhance SOC cycling, as different species contribute uniquely or synergistically some performing specialized functions essential for C processing and stabilization (Lipton et al., 2025). Despite the critical roles of macrofauna (e.g., earthworms, termites) and mesofauna (e.g., mites, springtails) in soil processes, their contributions to SOC dynamics remain understudied. Macrofauna act as ecosystem engineers, driving SOC redistribution and stabilization through bioturbation, a process that often exceeds the vertical mixing capacity of roots and leaching. They incorporate organic litter into mineral soil layers as POC via mucus mediated aggregation, enzyme sorption, gut processing, and fecal pellet deposition, thereby promoting organomineral associations and creating biostructures that physically protect OC. Mesofauna accelerate decomposition through litter fragmentation and stimulation of microbial activity; their absence has been shown to reduce global litter decay rates by approximately 35% (Azevedo et al., 2024; Carvalho et al., 2023). Recent work by Jílková et al. (2024) further demonstrates that macrofauna significantly enhance the decomposition of low-quality shrub litter, leading to greater C stabilization as POC, whereas micro and mesofauna primarily drive the decomposition of high-quality grass litter, favoring the formation of MAOC. These effects are amplified through synergies with fungi (in shrub litter decomposition) and bacteria (in grass litter decomposition), supporting the hypothesis that functional faunal groups of soil differentially regulate decomposition pathways and C stabilization depending on litter quality.

2.4.2 Microorganism in carbon cycle

The soil microbiome represents one of the most diverse biological communities on Earth, harboring at least 25% of global biodiversity (Wu et al., 2024), engaging in complex interactions and actively contribute to multiple biogeochemical cycles (Azevedo et al., 2024; Liu et al., 2024). Bacteria are the most abundant soil microorganisms, representing 15% of global living biomass, while fungi and archaea account for 2% and 1%, respectively. In comparison, Funal contribute only 0.3% to the planet's total biomass, highlighting the dominant role of soil microorganisms in global biogeochemical cycles (Daunoras et al., 2024). Bacteria and fungi are central to C cycling, and their necromass serves as the primary source of microbial-derived C that underpins the long-term stabilization of SOC (Wu et al., 2024). The following sections detail how microbes regulate C fate through enzymatic catalysis (2.4.2.1), environmental and substrate controls (2.4.2.2), and physiological efficiency (2.4.2.3).

2.4.2.1 Enzymatic breakdown of organic carbon

Extracellular enzymes secreted with microorganism initiate SOM breakdown through depolymerization, releasing labile organic compounds that microbes assimilate for metabolic energy and biomass synthesis, with a portion respired as CO_2 (Manzoni et al., 2019). As noted in Section 2.2, C enters the soil from diverse sources, each containing distinct types of organic compounds that are degraded by specific extracellular enzymes. Polysaccharides including cellulose and hemicellulose are degraded by hydrolases enzymes such as β-glucosidase (BG), xylanase (XYL), chitinase, and amylase into DOC. Aromatic compounds like lignin are broken down by phenoloxidases (PO) and peroxidases. Aliphatic compounds, including fatty acid esters, require lipolytic enzymes (LIP) (Daunoras et al., 2024; Heitkötter et al., 2017). This enzymatic transformation is not only essential for C cycling but also reflects microbial metabolic and stoichiometric demands. As such, the production of specific enzymes indicates microbial investment in acquiring particular resources, aligning with ecological homeostasis (Daunoras et al., 2024). Notably, despite having a C/N ratio similar to leaves, roots decompose more slowly due to their biochemical composition particularly higher levels of recalcitrant compounds such as lignin, suberin, phenols, and tannins. Soil microorganisms degrading these resistant components produce enzymes capable of cometabolizing analogous compounds in SOM, thereby influencing overall SOM turnover (Shahbaz et al., 2017).

Microbial enzymatic activity is also subject to substrate hierarchy. When labile, energy rich substrates such as root exudates and fresh litter are abundant, microbes favorably utilize them, suppressing investment in energetically costly enzymes targeting recalcitrant SOM. This phenomenon sometimes termed priming inhibition (PI) contrasts with the classical PE, wherein fresh C inputs stimulate SOM decomposition. In this context, high rates of residue mineralization often reflect substrate abundance rather than enhanced SOM destabilization, highlighting that microbial enzyme expression is dynamically regulated by substrate quality and availability (Shahbaz et al., 2017). Ultimately, enzymatic breakdown is not only a gateway to C mineralization, it is a key regulatory node in the SOC cycle, determining the partitioning of organic inputs between microbial biomass, stabilized SOC, and atmospheric CO_2.

2.4.2.2 Substrate quality and environmental conditions as key drivers of microbial activity in soil

Soil microorganisms' activity and thus the fate of SOC is governed by two main controls including substrate quality and environmental conditions. Soil microbial turnover is strongly influenced by substrate quality and availability, which are defined by chemical complexity, nutrient stoichiometry and physical accessibility. Labile, nutrient-rich substrates promote rapid microbial growth and higher C use efficiency (CUE). In contrast, when substrates are limited, microbial activity declines, leading to dormancy or cell death and often greater respiratory losses per unit C processed. This results in reduced microbial biomass, a response consistently observed in laboratory incubation studies (Shahbaz et al., 2017). Simultaneously, the activity of soil microorganisms tightly regulated by environmental factors. Temperature and moisture directly influence microbial metabolic rates, which typically rise from 0 °C to an optimum at 25–39 °C, declining sharply above 65 °C (Marschner, 2021). Water availability affects microbial physiology, nutrient diffusion, and redox conditions, shaping the availability of electron acceptors. However, limited O_2 diffusion promotes anaerobic metabolism, reducing C processing efficiency (Azevedo et al., 2024). Consequently, extreme or prolonged changes in temperature and moisture can disrupt microbial community stability and reduce C use efficiency (Crowther et al., 2019). Microbial functioning is also modulated by pH and nutrient availability, which are governed by soil chemistry and organic substrate quality. Bacteria tend to dominate in neutral soils, whereas fungi prevail in acidic conditions (Malik et al., 2018). In the case of soil texture, OM is protected in clay-rich soils through the formation of MAOC, limiting microbial access and enhancing C sequestration (Crowther et al., 2019). Indeed, favorable conditions can enhance microbial turnover and necromass production, while stressors may trigger dormancy, reduce biomass, and accelerate PE or C loss. Thus, substrate and environment jointly determine whether soil acts as a net C sink or source and converge in a central microbial physiological trait, CUE, which determines the ultimate partitioning of C between biomass and respiration (Azevedo et al., 2024).

2.4.2.3 Microbial carbon use efficiency (CUE)

Microbial CUE refers to the fraction of assimilated C allocated to biomass rather than respiration and governs microbial growth yield and critically influences the SOC stock. Even small changes in microbial CUE can profoundly influence SOC fate (Manzoni et al., 2012; Wang & Kuzyakov, 2023). Higher CUE reduces respiratory CO_2 loss, promoting C retention in microbial biomass; upon cell death, this biomass becomes microbial necromass, a chemically complex and mineral associated residue that contributes to long term SOC storage, particularly as MAOC. Conversely, low CUE induced by nutrient limitation, substrate recalcitrance, or environmental stress shifts C flux toward respiration, accelerating CO_2 release (Manzoni et al., 2021). Therefore, microbial CUE is not a constant value and depends on environmental conditions, the microbial necromass and physiology and chemical composition of plant-litter as a substrate and nutrient resource (Jackson et al., 2017; Kästner et al., 2021).

Soil water availability strongly influences microbial CUE. Drought stress limits substrate diffusion and growth, induces stress responses, and can lower CUE especially upon rewetting due to osmolyte release and increased C excretion. Conversely, waterlogged (anaerobic) conditions shift metabolism toward incomplete oxidation, producing acetate and CH_4 instead of CO_2, which also reduces CUE despite maintained C uptake (Manzoni et al., 2012). Substrate quality influences microbial CUE through degradability and energy content. Complex substrates such as aboveground litter of plants requiring extensive enzymatic processing, lower biomass conversion efficiency and increase respiratory losses (Manzoni et al., 2012). In contrast, roots rich in aliphatic compounds that readily adsorb to minerals, along with their exudates, enhance microbial CUE more effectively than aboveground litter and on average, 45% of belowground C inputs are stabilized, compared to just 8% of aboveground litter, a more than fivefold difference (Carvalho et al., 2023; Jackson et al., 2017). Microbial community composition influences CUE and suggested higher CUE in fungal-dominated systems. Greater differences arise between functional groups: zymogenous (opportunistic) microbes exhibit low CUE under high substrate availability, whereas autochthonous (slow growing) microbes achieve high CUE under low C conditions. Soil fauna also affect CUE, grazing by protozoa can stimulate microbial growth and increase CUE (Manzoni et al., 2012). CUE links short-term microbial physiology to long-term C sequestration, positioning soil microbes not only as decomposers but as key architects of SOC persistence. Understanding and potentially managing CUE offers a biologically grounded strategy to enhance soil's role as a C sink in the global biogeochemical cycle. As explored in the next section, the stabilization of microbial-derived C particularly through mineral association represents the final step in transforming transient inputs into persistent SOC stocks.

2.5 Mechanisms of carbon sequestration

SOC sequestration refers to the process by which atmospheric CO_2 is fixed into plant biomass via photosynthesis and subsequently transferred into soil as OM. This occurs in three stages: (1) CO_2 capture from the atmosphere, (2) incorporation of C into plant biomass, and (3) transfer of plant-derived C to soil as SOC. According to the physicochemical and microbial controls discussed in Sections 2.3 and 2.4, C sequestration represents the net outcome of these processes which transforme plant-litter and microbial inputs into stabilized, long-term SOC pools (Kögel-Knabner et al., 2022; Zaman et al., 2025; Shah et al., 2019). SOC is stabilized through multiple interacting mechanisms including chemically, via adsorption onto clay minerals; biochemically, through transformation into recalcitrant compounds such as humic substances and physically, by occlusion within soil aggregates that limit microbial access (Shah et al., 2019; Xu et al., 2021). These stabilization pathways are governed by biotic and abiotic factors which collectively determine SOC persistence. Anthropogenic land-use practices further modulate these dynamics (Carvalho et al., 2023) (Fig. 1.2).

2.5.1 Abiotic factors

Soil properties such as mineralogy, pH, texture, porosity and soil depth critically influence C stabilization pathways. In soils, OM primarily associates with different soil's mineral. Phyllosilicates exhibit permanent negative charge on their siloxane surfaces due to isomorphic substitution, enabling OM binding, while their edge sites display variable charge and facilitate inner and outer sphere complexation and H bonding with organic ligands. Pedogenic Fe and Al occur as monomers complexed with organic ligands or as polymeric species such as oxides, hydroxides, oxyhydroxides, and

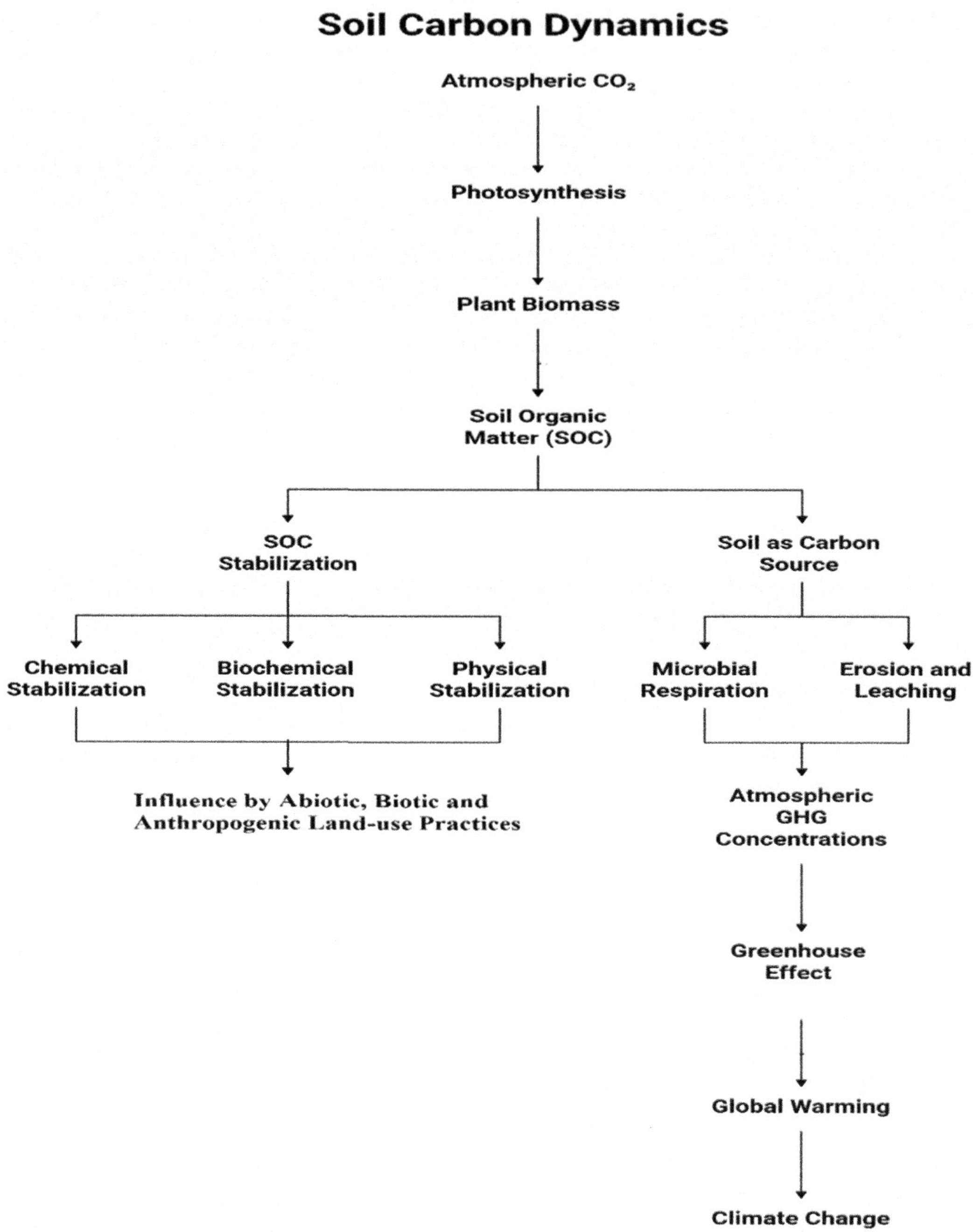

FIGURE 1.2

Soil carbon dynamic as either a net source or sink of C.

aluminosilicates which exhibit high OM sorption capacity due to abundant surface hydroxyl groups and large specific surface area (Kögel-Knabner et al., 2022). The organomineral associations form through diverse mechanisms including specific and nonspecific adsorption, hydrogen bonding, van der Waals forces, and hydrophobic interactions (Rasmussen et al., 2018). These associations markedly enhance SOC stability due to higher thermal decomposition temperatures, slower decomposition rates, and turnover times up to four times longer than free OC (Kögel-Knabner et al., 2022). Soil pH is a master regulator of C sequestration, primarily through its control of the PE and SOC dynamics. It influences SOC chemistry, stabilization, and availability by governing organomineral interactions, enzyme activities, microbial community structure, and nutrient cycling. Lower pH enhances OM sorption to Fe and Al oxides via increased positive surface charge, promoting physical protection, while higher pH favors calcium (Ca) stabilization. Microbial activity and enzyme function exhibit pH optima (typically pH 6–8), with maximal PE intensities, whereas acidic ($pH < 4.5$) or alkaline ($pH > 7$) extremes often suppress PE (Wang & Kuzyakov, 2024). Li et al. (2023) reported that under alkaline conditions, pH primarily modulates microbial activity and mineral stability, which in turn govern SOC stabilization. In acidic soils, OAs dissolve silicate and metal minerals, promoting metal organic complexation that enhances SOC stability; however, this dissolution can also release obstructed OC.

Soil texture determines whether plant residues are stabilized primarily through physical protection via aggregation or through chemical stabilization as MAOC. Clay- and silt-sized particles enhance long-term SOC storage by strongly adsorbing organic compounds through electrostatic interactions, ligand exchange, and surface complexation (Aumtong et al., 2023). These mineral organic bonds are often recalcitrant to microbial enzymatic degradation, rendering MAOC resistant to decomposition and enabling C stabilization over decades to centuries. As a result, MAOC formation is increasingly recognized as a critical mechanism for mitigating erosion-induced C losses and enhancing long-term C sequestration in agricultural soils (Hussain et al., 2021; Shi et al., 2024). Aumtong et al. (2023) found that the clay- and silt-rich soils have greater DOC retention via adsorption to mineral surfaces particularly Fe and Al oxides, enhancing MAOC formation and long-term SOC stabilization. In contrast, sandy soils with low clay, low CEC, and acidic pH exhibited reduced MAOC formation due to limited adsorption capacity and potential DOC leaching. They also revealed that water regime and soil texture together control DOC availability and its conversion to MAOC. Soils with higher residue inputs and clay content retained more DOC, raising MAOC formation and greater SOC storage. Thus, DOC acts not only as a transient C pool but as a key mediator linking hydrological dynamics, microbial processing, and long-term C stabilization via MAOC with clay content and moisture regime as dominant environmental filters. Hussain et al. (2021) reported that kaolinite-dominated soils lose SOC more rapidly due to their low-activity clay content, poor fertility and water-holding capacity, uncontrolled internal drainage, weak soil structure and limited aggregation.

Soil porosity controls SOC decomposition by regulating microbial access and O_2 and nutrient diffusion. Micropores (<13 μm) isolate OM and microbes, suppressing mineralization, especially under anaerobic, while larger pores enhance decomposition. Ultimately, the soil porosity governs physical distance between decomposers and substrates at the microscale thereby controlling C persistence, making soil structure a key regulator of SOC stability (Carvalho et al., 2023; Kögel-Knabner et al., 2022). Soil depth is another vital issue. The relative influence of biological versus physical controls on SOC turnover varies across soil depths. For example, surface soils in temperate grasslands are often microbially driven with rapid turnover, whereas subsoil C persistence is more strongly influenced by mineral protection and slower diffusion rates (Manzoni et al., 2019). Thus, with increasing soil depth, temperature and moisture regimes stabilize, reducing environmental variability and amplifying the influence of mineral chemistry on OM dynamics and C persistence (Jackson et al., 2017). Hu et al. (2025) found that soil depth does not directly control SOC storage or turnover; rather, it indirectly affects decomposition rates by influencing plant C inputs, moisture availability, and mineral composition. They revealed that across global forests, POC increases with latitude but declines with depth, while MAOC dominates in subsoils and exhibits stronger microbial processing particularly in low-latitude systems.

2.5.2 Biotic factors

Microbial processing mediates the transformation of labile C into stable pools. POC can initially degraded by microbial extracellular enzymes, with a portion assimilated into microbial biomass. Upon cell death, microbial necromass and metabolic byproducts can be stabilized through associations with clay minerals and metal oxides, forming MAOC. These pathways underscore microbial activity as a key mediator in transforming labile OC into stable SOC pools (Murray, 2024). The composition of microbial community is highly affected the C sequestration. Due to microbial community complexity, bioindicators like the fungus to bacteria (F:B) ratio are used to infer C sequestration potential. As fungi, especially mycorrhizal types (associated with >90% of plants), efficiently convert substrates into biomass (up to 50%), transport plant-derived C into soil, stimulate decomposers, contribute necromass, and stabilize aggregates which is collectively enhancing C sequestration, the higher F:B ratios correlate with greater SOC accumulation and stability. So that promoting fungal symbioses thus offers a promising strategy for SOC storage (Carvalho et al., 2023). Jackson et al. (2017) further note that soils dominated by fungi often accumulate more C due to higher microbial CUE and the production of recalcitrant compounds. Soils dominated by ectomycorrhizal (ECM) typically store more SOC than those dominated by arbuscular mycorrhizal (AM), likely because ECM fungi access organic N, limiting its availability to decomposers. In addition to microbial processes, the quantity and quality of plant inputs which discussed in Section 2.2, critically influence SOC sequestration. Root-derived inputs contribute 18%–58% more to SOC storage than aboveground residues, enhancing both C stabilization and turnover (Jackson et al., 2017; Xu et al., 2021; Zhang, Cai, et al., 2022, Zhang, Chen, Jin, et al., 2022, Zhang, Chen, Yao, et al., 2022). C from dead roots exhibits a residence time approximately 2–4 times longer than that from aboveground litter at a certain spatial and temporal scale (Carvalho et al., 2023; Lei et al., 2023), underscoring the importance of belowground allocation in long-term C persistence. Shrublands and grasslands typically allocate more C to roots than forests or croplands, leading to higher SOC stocks even after accounting for climate, soil texture, and other factors (Jackson et al., 2017). Furthermore, microbial necromass has been shown to constitute 24%–60% of the total SOC pool across terrestrial ecosystems, playing a far greater role in C sequestration than previously recognized and highlighting its centrality in C stabilization (Kästner et al., 2021). While increased SOC inputs are often assumed to enhance C sequestration, growing evidence shows they can also trigger the priming effect. Root exudates, as a major source of labile C, are key drivers of this phenomenon via two mechanisms: (1) the microbial metabolism theory, wherein exudates serve as an energy source that alters microbial community structure, composition, and activity, thereby enhancing SOC decomposition; and (2) a physicochemical mechanism, whereby exudates such as oxalic acid destabilize MAOC complexes, releasing protected C for microbial utilization (Lei et al., 2023).

2.6 Pathways of carbon loss from soil

Soil can function as either a net source or sink of C (Fig. 1.2), depending on environmental conditions, microbial activity, substrate quality, and anthropogenic influences. When soils acting as a source of C, contribute to atmospheric GHG concentrations, mainly as CO_2, with a smaller fraction as CH_4 (Shah et al., 2019; Wu et al., 2024). Both gases increasing the greenhouse effect, trap heat in the Earth's atmosphere and drive global warming, thereby accelerating climate change (Singh, 2024). It seems that the global warming, with Earth's temperature projected to exceed 1.5 °C above by the end of this century. This warming is altering large-scale weather patterns which may accelerate land-use change and influencing terrestrial C storage (Hussain et al., 2021). As the largest terrestrial C pool, soils play a critical role in climate change mitigation through the sequestration of atmospheric CO_2. However, SOC can be lost through several pathways: primarily via microbial respiration during decomposition, but also through erosion and leaching of DOC.

2.6.1 Respiration

The majority of soil CO_2 efflux arises from soil respiration, which comprises two main components including autotrophic respiration (from plant roots) and heterotrophic respiration (from microbial decomposition of OM). Heterotrophic respiration dominates total CO_2 emissions, accounting for over two-thirds of the flux, and is particularly responsive to labile organic inputs such as plant litter and root exudates (Feng et al., 2024; Wang et al., 2020; Zhuang et al., 2023). In contrast, CH_4 emissions occur under anaerobic conditions through the activity of methanogenic microorganisms, primarily in waterlogged environments such as wetlands, rice paddies, and peatlands. Although CH_4 has a global warming potential approximately 28 times greater than CO_2 over a 100 year horizon, its atmospheric lifetime is shorter (12 years), and its overall contribution to radiative forcing remains smaller due to lower atmospheric concentrations (Singh, 2024).

Soil respiration estimates remain highly uncertain due to the complex interplay of factors including soil microclimate, microbial activity, plant growth, temperature, and moisture that drive its variability (Feng et al., 2024). Soil respiration is strongly regulated by temperature and moisture. Temperature governs the rates of biochemical reactions, while moisture controls the diffusion of solutes and oxygen to microbial cells, thereby modulating metabolic activity. SOC decomposition responds non linearly to temperature. While rising global temperatures driven by anthropogenic GHG emissions are expected to enhance respiration, they may simultaneously intensify evapotranspiration and precipitation variability, leading to reduced soil moisture. This moisture limitation can constrain enzyme-substrate diffusion and suppress microbial activity, ultimately inhibiting SOC decomposition (Murray, 2024). Litter inputs influence soil respiration both indirectly by moderating soil temperature and moisture through surface insulation and directly by altering microbial biomass and activity through changes in C availability (Feng et al., 2024). However, root exudates particularly labile compounds such as sugars and OAS play a disproportionately large role in stimulating microbial metabolism and enhancing CO_2 release, often exceeding the effect of litter inputs and underscoring the primacy of root-derived C in driving belowground carbon cycling (Lei et al., 2023; Zhuang et al., 2023). Feng et al. (2024) found that microorganisms preferentially metabolize root-exudate C, resulting in a stronger stimulation of soil respiration compared to litter-derived C. Notably, root-exudate C decomposes at a rate comparable to bulk SOC, contributing 7%–31% of total belowground respiration in grassland ecosystems and representing a significant, yet often overlooked, source of soil CO_2 (Lei et al., 2023).

2.6.2 Erosion and leaching

Soil DOC serves as a key energy source for soil microorganisms and plays a central role in the cycling, transport, and stabilization of SOC. Due to its high lability, DOC is particularly sensitive to erosion and hydrological processes. Rainfall intensity and surface runoff drive DOC leaching and vertical migration through the soil profile, with higher DOC concentrations typically observed at eroding sites compared to depositional areas (Zheng et al., 2025). Globally, DOC production from litter decomposition and SOC mineralization in mineral soils is estimated at 1.4 Gt C yr^{-1}. Of this, approximately 40% is mineralized to CO_2, 40% is reincorporated into stable SOC pools, and 20% (0.28 Gt C yr^{-1}) is leached into aquatic systems. Both DOC production and export are highest in tropical regions accounting for 60% of global leaching fluxes, which aligns with elevated riverine CO_2 emissions in these areas due to high biological activity and rapid decomposition rates (Nakhavali et al., 2021).

Erosion, driven by mountain uplift, water, wind, and human activities, plays a significant role in soil–atmosphere C exchange. Water-driven physical and chemical erosion are the dominant contributors to erosion-induced C fluxes, while wind erosion contributes relatively little (Zheng et al., 2025). Globally, erosion displaces approximately 5.7 Pg C yr^{-1} of SOC, resulting in the redistribution and redeposition of 4.0 Pg C, the emission of 1.14 Pg C to the atmosphere (via accelerated mineralization), and the lateral transfer of 0.57 Pg C to the ocean (Shi et al., 2024). Agricultural soils are especially vulnerable to erosion due to vegetation removal and soil disturbance associated with conventional tillage practices. Intensive tillage has contributed to the loss of 60–90 Pg C from SOC stocks over recent decades. Broader land-use changes since 1750 including mechanization and deforestation have further depleted global SOC pools by an additional 66–90 Pg C and reduced them by 22%, respectively (Hussain et al., 2021). Conversion of natural ecosystems to agricultural systems leads to rapid SOC depletion: losses of 25%–50% within 20–50 years in temperate zones, and 50%–75% within just 5–20 years following tropical deforestation. Consequently, agricultural erosion not only degrades soil fertility but also disrupts the global C cycle by exporting SOC-rich sediments via surface runoff, effectively transferring C from terrestrial sinks to atmospheric and aquatic reservoirs (Hussain et al., 2021).

2.7 Anthropogenic impacts and climate feedbacks

The disturbances driven by human such as land-use change, intensive agriculture, and ecosystem degradation disrupt soil biogeochemical cycles, shifting soils from C sinks to net C sources. Deforestation reduces SOC sequestration by removing aboveground biomass, while agricultural practices such as tillage accelerate OM decomposition, enhancing CO_2 emissions. Urbanization further exacerbates these effects by diminishing vegetative cover and compacting soils, thereby limiting their C storage capacity. Collectively, these anthropogenic activities disrupt C storage in plant biomass, with significant consequences for SOC sequestration and increased atmospheric CO_2 concentration (Zaman et al., 2025). Globally, agricultural soils have lost an estimated 133 Pg of SOC since the dawn of farming, weakening a major terrestrial C sink and amplifying atmospheric CO_2. Climate change intensifies this loss, warming accelerates microbial respiration, while altered precipitation patterns increase erosion and leaching (Rumpel & Chabbi, 2021). Improving the management of just 4% of the world's agricultural soils could sequester 2–3 gigatons of CO_2 annually, potentially offsetting 20%–35% of global anthropogenic GHG emissions (Pan et al., 2021). Key strategies to enhance SOC storage include conservation tillage, agroforestry, organic amendments and crop straw recycling practices that minimize soil disturbance, increase OC inputs, and stimulate microbial activity that stabilizes C in soil (Rumpel & Chabbi, 2021).

Straw incorporation, in particular, is a well-documented approach for SOC restoration. Dai et al. (2021) found that deep tillage in the North China Plain increased the relative abundance of genes involved in carbohydrate transport and metabolism, potentially enhancing the conversion of straw-derived C into SOC. In contrast, no tillage (NT) reduced straw decomposition capacity by decreasing the abundance of fungi and carbohydrate-binding module genes. Therefore, the tillage management is closely interrelated with residue management and these two practices should always be investigated simultaneously. Notably, straw recycling has emerged as a promising approach to replenish plant-available silicon (H_4SiO_4) and promote long-term C sequestration. A portion of this C is stored in phytoliths -silica bodies formed in plants, that are returned to soil via litter. Given their chemical stability and potential persistence in soils for centuries to millennia, phytolith-occluded C is increasingly recognized as a durable long-term C sink (Zaman et al., 2025). Global initiatives such as the 4 per 1000 Initiative aim to scale such practices to simultaneously advance climate mitigation, adaptation, and food security (Rumpel & Chabbi, 2021). The 4 per 1000 Initiative, launched at COP21 by UNFCC in Paris in 2015, proposes that increasing global SOC stocks by 0.4% annually in the top 30–40 cm of non-permafrost soils could offset the annual rise in atmospheric CO_2. This target is feasible primarily in actively managed soils such as croplands, pastures, and agroforestry systems and would require sequestering approximately 4.7 Gt C yr^{-1} in agricultural soils alone. Realizing this potential depends on widespread adoption of improved practices, including sustainable cropland and grazing management, better livestock and manure handling, restoration of degraded lands, and agroforestry (Rodrigo-Comino et al., 2020).

3. Nitrogen cycle

N is the fourth most abundant element in biomass which is critical for plant growth and agricultural productivity. Unlike most nutrients, which are primarily derived from rock weathering, N enters ecosystems mainly through biological processes particularly N fixation and is predominantly removed via microbial denitrification rather than sedimentation (Zhang et al., 2020). Soil is a place where various biotic and abiotic N processes occur and plays a central role in N cycle by mediating key biogeochemical transformations that regulate N availability, retention, and loss. Through microbial processes such as mineralization, nitrification and denitrification, soil converts organic N into inorganic forms (ammonium (NH_4^+) and nitrate (NO_3^-)) usable by plants. However, environmental losses of N occurring via processes such as leaching, volatilization, and denitrification, can lead to elevated concentrations of inorganic anions, particularly NO_3^-, in groundwater, and contribute to increased emissions of GHGs, notably N_2O (Pan et al., 2021; Hayashi, 2022). Soil enzymes such as urease and nitrate reductase and physicochemical properties of soil (pH, moisture, C availability) critically influence these transformations. Therefore, understanding soil's role in the N cycle is essential for optimizing agricultural productivity while minimizing environmental impacts (Robertson & Groffman, 2024). The following sections examine the key biogeochemical processes governing N dynamics in soil, the factors influencing N retention and loss, and the implications for sustainable nutrient management.

3.1 Nitrogen pools in soil: Forms and bioavailability

N is a fundamental biogenic element, essential to a wide range of organic and inorganic compounds (Popoviciu, 2024), and its pools vary widely in size and turnover rate across terrestrial ecosystems (Myrold, 2021). Dinitrogen (N_2), the dominant gaseous form, constitutes approximately 78% of the

Earth's atmosphere by volume which equivalent to an estimated 5.148×10^{18} kg and represents the largest global N pool. However, due to its strong triple bond, N_2 is largely inert, with a residence time on the order of millions of years. Trace atmospheric N oxides (NO, NO_2, N_2O) are far less abundant and relatively unstable. In soils, organic N constitutes the second-largest pool, accounting for over 90% of total soil N and turning over on decadal timescales. Its quantity and composition vary with climate, vegetation, and soil properties, and it encompasses a diverse array of molecules, including amino acids, peptides, proteins, nucleobases, nucleotides, nucleic acids, as well as amines, amides, alkaloids, and vitamins. In contrast, inorganic N (primarily NH_4^+ and NO_3^-) represents the smallest yet most dynamic soil N pool, often turning over multiple times per day despite concentrations typically ranging only a few mg N kg^{-1} in natural systems (Myrold, 2021; Popoviciu, 2024). Soil N whether organic or inorganic derives from multiple sources, including synthetic fertilizers, manure applications, irrigation water, atmospheric deposition, and biological nitrogen fixation (BNF) (Hayashi, 2022; Ramond et al., 2022).

3.2 Nitrogen transformation in soil

N exhibits more chemical forms than any other essential element, all formed by microbial activity. Even when reactive N (Nr) enters the biosphere via abiotic pathways such as synthetic fertilizers, atmospheric deposition, or lightning their subsequent transformations, retention, and loss in soil are predominantly governed by microbial metabolism. Microorganisms play a central role in the N cycle because they possess genes encoding key enzymes involved in N metabolism. Thus, understanding the cycling and fate of N at any scale requires knowledge of the microorganisms that drive each process (Grzyb et al., 2021; Robertson & Groffman, 2024). Soil microbes are functionally specialized, mediating key transformations such as the oxidation of NH_4^+ to NO_3^- (nitrification), the reduction of NO_3^- to N_2 (denitrification), and the mineralization of OM to NH_4^+ (ammonification). Through these activities, they regulate the availability of N to plants and influence critical ecosystem outcomes, including crop productivity, GHG emissions, and water quality (Noor et al., 2019) (Fig. 1.3). Their activities are highly sensitive to environmental conditions such as oxygen concentration, redox potential, C availability, and pH, which together determine the dominant transformation pathway in any given soil ecosystem. The following subsections detail the major microbial N transformation processes in soil including BNF, ammonification, nitrification, immobilization, denitrification, and NH_3 volatilization along with the soil, climatic, and management factors that regulate their rates and environmental impacts.

3.2.1 Nitrogen fixation

Atmospheric N enters soils via three processes including gaseous deposition through dry process (NO_x, HNO_3, and NH_3), precipitation through wet process, and N fixation by lightning and BNF (Ramond et al., 2022). Wet N deposition refers to the gaseous and particulate Nr compounds in the atmosphere scoured to the Earth's surface by precipitation (rain or snow), while dry deposition involves the direct uptake of gaseous and particulate Nr onto surfaces without precipitation which strongly influenced by soil surface properties and chemical reactivity (Zhang et al., 2021). Atmospheric N deposition has been risen steadily since the Industrial Revolution due to agricultural intensification and industrial emissions. While it supplies exogenous N that can enhance plant growth, it also intensifies N leaching and environmental degradation. Global N deposition rates vary widely,

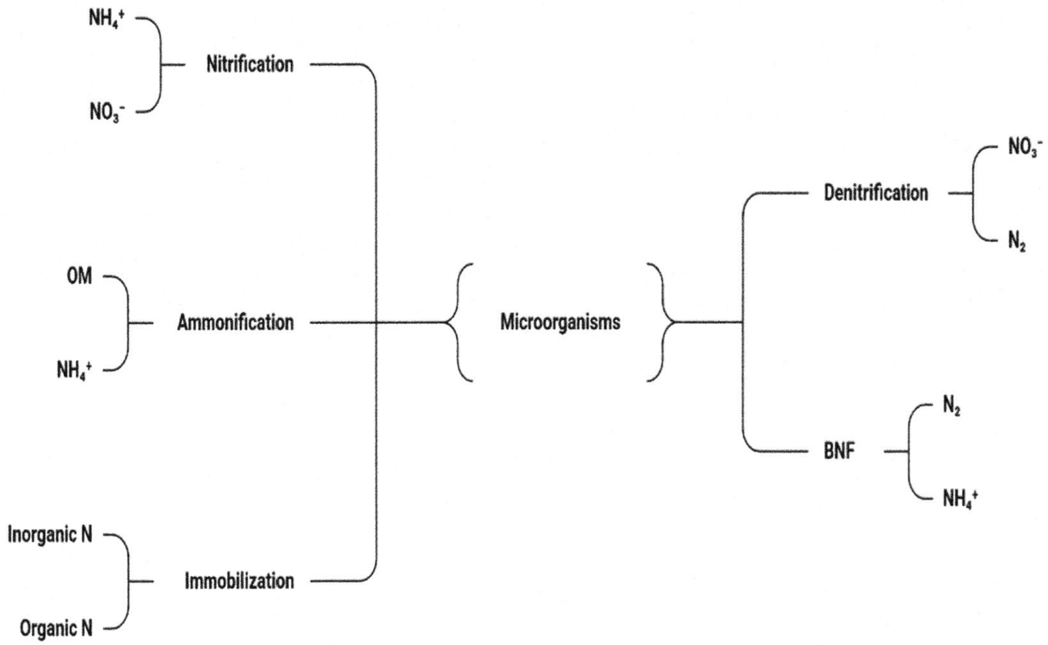

FIGURE 1.3

Role of microorganisms in the N cycle.

ranging from less than one to nearly 60 kg N ha^{-1} yr^{-1}. Under high N deposition, N immobilization becomes critical for retaining mineral N in soil, thereby mitigating excessive N losses and preventing phytotoxicity (Song et al., 2021).

While atmospheric deposition delivers Nr to ecosystems, the conversion of N_2 into bioavailable forms occurs primarily through natural and anthropogenic fixation processes which require substantial energy input and are central to the global N cycle (Robertson & Groffman, 2024). Globally, N fixation by lightning, primarily as NO_x is estimated at approximately 7 Tg N yr^{-1}. Anthropogenic Nr production primarily through the Haber-Bosch process (industrial N_2 fixation to NH_3) and fossil fuel combustion reached 226 Tg N yr^{-1} in 2020, representing 3.9 times the natural terrestrial BNF input (Hayashi, 2022). BNF represents the largest natural source of Nr to terrestrial and aquatic ecosystems, second only to anthropogenic inputs (Hayashi, 2022; Popoviciu, 2024) and contributes an estimated 58 Tg N yr^{-1} (Ramond et al., 2022). BNF is mediated exclusively by prokaryotes that employ a large, structurally complex, iron-rich metalloenzyme called nitrogenase composed of dinitrogenase reductase (NifH) and dinitrogenase (NifDK) along with substantial amounts of ATP and reducing equivalents to catalyze the energetically demanding reduction of N_2 to NH_3 under ambient conditions (Zhang et al., 2020). These diazotrophs are commonly grouped into three functional categories including symbiotic (e.g., rhizobia–legume associations), associative (e.g., *Azospirillum* colonizing root surfaces), and free-living (e.g., *Clostridium*, *Cyanobacteria*) (Robertson & Groffman, 2024). BNF is tightly constrained by ecological and biogeochemical factors. Soil water directly influences the growth and activity of rhizobia and therefor the BNF by altering water activity and diffusion of nutrients, and indirectly by modifying plant root exudates. Dry conditions reduce rhizobial populations and nodulation, especially in arid soils. Soil texture and pore structure also important as smaller pores in stable aggregates retain more water, creating favorable microhabitats for rhizobia. Fine-textured soils typically support higher BNF due to greater CEC and buffering capacity, whereas sandy soils often limit symbiosis through leaching and poor resource retention (Robertson & Groffman, 2024). BNF is further regulated by nutrient availability and soil chemistry. High concentrations of inorganic N suppress nitrogenase synthesis via feedback inhibition, while deficiencies in phosphorus (P), Fe, molybdenum (Mo), potassium (K), and Ca impair both microbial metabolism and symbiotic establishment. Ca, in particular, is critical for rhizobial attachment to root hairs. Near-neutral pH (6.0–7.5) is optimal for BNF. Acidic conditions reduce Mo solubility and increase aluminum (Al^{3+}) toxicity, whereas highly alkaline soils limit Fe and P availability. Additionally, DOC derived from root exudates or SOM decomposition provides the energy required to fuel the ATP-intensive nitrogenase reaction, thereby linking BNF activity closely to SOC status (Mohammadi et al., 2012). Temperature also exerts strong control, with optimal N fixation generally occurring between 20 and 25 °C (Grzyb et al., 2021).

3.2.2 Ammonification

Ammonification is the process by which soil organic N is converted into NH_4^+ and it is recognized as a first step of N mineralization. Ammonification is a microbially mediated process in which heterotrophic bacteria and fungi including *Bacillus, Pseudomonas, Aspergillus,* and *Penicillium* decompose organic residues, converting amino acids and proteins into NH_4^+ via proteases, ureases, and deaminases (Myrold, 2021; Zhang et al., 2020). Indeed, these enzymes break down complex organic polymers into LMW compounds which are taken up by microbial cells and further metabolized for energy, releasing NH_4^+ into the soil solution as a byproduct (Myrold, 2021). In ammonification process NH_4^+ can be retained with soil CEC or fixed within the lattice of certain clay minerals such as illite and vermiculite, chemically bound to organic compounds or lost as NH_3 through volatilization and biologically may be taken up by plants, assimilated by microbes, or oxidized to NO_3^- by nitrifying microorganisms (Myrold, 2021; Zhu et al., 2018). Ammonification process is primarily controlled by SOM quantity and chemistry, temperature, and moisture, soil properties, topography, and management which is carried out by diverse soil biota under both aerobic and anaerobic conditions (Yao et al., 2019; Zhu et al., 2015, 2018). The critical determinant of whether N is mineralized or immobilized is the C:N ratio of the SOM. Substrates with C:N < 25:1 typically yield net N mineralization, enhancing plant-available N, whereas high C:N materials induce net N immobilization as microbes scavenge NH_4^+ to meet biosynthetic demands (Robertson & Groffman, 2024). Kuśmierz et al. (2023) reported that OM contributed significantly to the availability of mineral N (NH_4^+ and NO_3^-) in the soils which the maximum content of mineral N (18.24–18.31 mg kg^{-1}) was noted in soils containing above 2% of C.

Soil moisture and temperature are key hydrothermal drivers of N transformation, regulate microbial activity and substrate dynamics. Increasing in soil temperature (optimal efficiency between 20 and 37 °C) increased N ammonification through enhancing the activity of ammonia-oxidizing microbes and soil enzymes (Fu et al., 2025). This process is highest at 80% of field water capacity of soil (Yao et al., 2019) while under waterlogged or freezing conditions low microbial activity leads to organic N accumulation due to suppressed mineralization (Robertson & Groffman, 2024). As reported by Fu et al. (2025) ammonification decreases with increasing pH. Acidic soils show positive net ammonification (NH_4^+ accumulation) because nitrification is suppressed attributed to inhibition of autotrophic nitrifiers, dominance of less-efficient heterotrophic nitrifiers, and feedback inhibition by accumulated NH_4^+. In contrast, neutral and calcareous soils exhibit negative net ammonification (NH_4^+ depletion) as microbial uptake for nitrification and assimilation outpaces replenishment. Topographically, N ammonification is generally higher in lower slopes due to greater SOM and moisture, which tend to decline with increasing slope position. Liu et al. (2017) reported that N ammonification declines with increasing latitude and altitude, driven by cooler temperatures, reduced microbial biomass, lower soil clay content, and higher soil C:N ratios, all of which limit substrate quality and microbial activity. However, in some cases, upper slopes may exhibit elevated mineral N concentrations due to denser vegetation cover and higher soil temperatures (Yao et al., 2019). Regarding soil texture, Skeletal soils (sandy loam, with relatively higher clay content) showed greater N ammonification than Aeolian soils (loamy sand), likely owing to higher SOM, lower bulk density (BD), better aeration, and more favorable moisture dynamics (Yao et al., 2019). Tillage profoundly influences N mineralization process, as conventional tillage (CT) accelerates OM decomposition, leading to rapid N ammonification and high NO_3^- accumulation, thereby increasing leaching risk. In contrast, NT systems exhibit slower early-season mineralization due to cooler, moister, and less aerated soils, which can cause temporary N deficiency but enhance long-term N retention through greater surface residue, higher organic N pools, and more efficient N cycling (Martens, 2001).

3.2.3 Nitrification

Nitrification is a key process in the N cycle, carried out commonly under aerobic conditions by specialized microorganisms namely ammonia-oxidizing bacteria (AOB) and ammonia-oxidizing archaea (AOA) (e.g., *Nitrosomonas, Nitrosopumilus*), which employ ammonia monooxygenase (AMO) and hydroxylamine oxidoreductase (HAO) to convert NH_3 to NO_2^-. This is followed by the oxidation of NO_2^- to NO_3^- by nitrite-oxidizing bacteria (NOB) (e.g., *Nitrobacter, Nitrospira*), which use nitrite oxidoreductase (NXR) to produce NO_3^- and completing the two-step process (Zhang et al., 2020; Zhu et al., 2018). AOB and AOA derive substantial energy from NH_3 oxidation (up to 440 kJ mol^{-1} NH_3) and constitute the primary biological source of NO_3^- in most ecosystems, including agricultural systems where NH_3-based fertilizers serve as substrates (Hayashi, 2022; Robertson & Groffman, 2024).

Nitrification is tightly controlled by NH_4^+ availability, soil pH, temperature, and other environmental factors. The primary regulator of nitrification in most soils is NH_4^+ availability. Nitrification rates are low when NH_4^+ supply is limited by slow mineralization or high immobilization by plants and microbes, and increase when disturbances such as tillage, fire, fertilization, or N deposition elevate NH_4^+ pools beyond biological demand (Robertson & Groffman, 2024). Temperature critically influences nitrification as AOA exhibit peak activity at 30–37°C, while AOB are most active at 16–23°C, with AOA generally increasing and AOB declining under warming conditions (Ayiti & Babalola, 2022). Soil pH strongly influences nitrification. As reported by Ayiti & Babalola (2022) optimal pH ranges are 7.5 for AOB and 7.0 for nitrite-oxidizing bacteria (NOB). Kuśmierz et al. (2023) reported that NO_3^- concentrations declined markedly as pH decreased (from 15.43 mg kg^{-1} at pH > 7.2 to 5.22 mg kg^{-1} at pH < 4.5), reflecting reduced NH_3 availability and suppressed nitrifier activity under acidic conditions. While higher pH promotes nitrification and enhances plant NO_3^- uptake, it also increases the risks of NO_3^- leaching and NH_3 volatilization, suggesting that controlled soil acidification could help mitigate NO_3^- losses in certain agroecosystems. They also found that soil texture affects N dynamics. NH_4^+accumulated in both clay-rich soils (due to higher CEC and SOM content) and sandy soils (where acidic conditions and rapid mineralization limit nitrification). In contrast, NO_3^- concentrations were lowest in sandy soils, owing to limited nitrification and high leaching potential. Additionally, nitrification is modulated by tillage practices and N availability. Tilled soils typically favor NO_3^- due to higher nitrification, while undisturbed native systems retain more NH_4^+, indicating suppressed nitrification. In NT systems, nitrification is generally slower under low N input due to substrate limitation and poor competitiveness of nitrifiers, but can accelerate after NH_4^+ fertilization due to higher surface moisture and nitrifier populations. Thus, NT's effect on nitrification depends on N and moisture context (Martens, 2001). Oxygen of soil is also critical, as obligate aerobes, nitrifiers are largely inactive in waterlogged soils, persisting only in thin oxic microzones (Robertson & Groffman, 2024). Under oxygen-limited conditions such as high soil water-filled pore space (WFPS) nitrifying bacteria can use NO_2^- as an alternative electron acceptor, reducing it to NO and N_2O. Consequently, nitrification rates, N_2O production, and the N_2O to NO_3^- ratio all increase with rising WFPS. NO is typically produced more rapidly than N_2O during nitrification, resulting in higher NO:N_2O emission ratios from well-aerated soils (Smith et al., 2018). Elevated soil salinity has a similarly inhibitory effect. At salinity levels of 17–27 dS m^{-1}, the second stage of nitrification is impeded, leading to NO_2^- accumulation, while extreme salinity (27 dS m^{-1}) stops the first stage, resulting in NH_4^+ buildup and subsequent volatilization (Grzyb et al., 2021).

Like NH_4^+, NO_3^- has many competing fates in the soil, with leaching being a major loss pathway due to its anionic nature and low soil retention. NO_3^- Leaching not only depletes plant-available N but also carries away essential cations to maintain charge balance, reducing base saturation and increasing soil acidity. Once in groundwater or surface waters, NO_3^- can cause eutrophication, trigger algal blooms, and pose human health risks such as methemoglobinemia. Biologically, NO_3^- is assimilated by plants and microbes or used as an electron acceptor in anaerobic microbial processes like denitrification and dissimilatory nitrate reduction (Myrold, 2021). Furthermore, by influencing the relative availability of NH_4^+ versus NO_3^-, nitrifiers can shape plant and microbial community composition, as many organisms exhibit N-form preferences (Robertson & Groffman, 2024). Plants vary in N-form preference. NO_3^- assimilation requires energy to reduce it to NH_4^+ before incorporation in plant metabolism and NH_4^+ is energetically cheaper to use, but many plants readily take up NO_3^- especially when energy from photosynthesis is abundant and may favor NO_3^- in heterogeneous soils where its higher mobility reduces the need for extensive root growth (Myrold, 2021).

3.2.4 Immobilization

N immobilization is a process of converting inorganic-N into organic forms, for N retention in soil and it is influenced by both biotic and abiotic processes (Cao et al., 2021). Biotic immobilization is driven by microorganisms through assimilating either small organic N compounds or inorganic N to synthesize essential cellular components. Fungi, due to their filamentous growth, are well-adapted to utilize the less mobile NH_4^+, whereas bacteria and archaea are more efficient at using NO_3^-, especially under low SOC conditions. Abiotic N immobilization affected by several mechanisms, including NH_4^+ fixation by 2:1 clay minerals, condensation of NH_3 with phenolic or quinone compounds, and nitrosation of NO_2^- with phenolic substances derived from SOM decomposition (Wang et al., 2020). The complex interplay of soil properties and residue quality makes it challenging to quantify the relative contributions of biotic and abiotic processes to N immobilization.

The biotic immobilization is strongly influenced by the availability of C and the C:N ratio of organic inputs. Globally, soil microbial biomass has a C:N ratio of 7.6–8.6 and the critical substrate C/

N threshold is at 20–25, which above it, microbes become N-limited and immobilize inorganic N to process extra C and balance their stoichiometric demands. resulting in net immobilization (Cao et al., 2021; Wang et al., 2020). Cao et al. (2021) reported that labile compounds like glucose are rapidly utilized by fast-growing microbes, strongly stimulating microbial N immobilization. In contrast, recalcitrant compounds like lignin decompose slowly via oxidative enzymes and provide little readily available C, resulting in abiotic N immobilization through chemical reactions with inorganic N. Soil pH further modulates immobilization pathways. Near-neutral to slightly alkaline conditions (pH 7–8.4) enhance microbial NH_4^+ uptake but suppress abiotic NO_3^- binding due to increased negative surface charges on soil particles. In contrast, acidic soils (pH < 6.2) promote abiotic retention of both NH_4^+ (via fixation in 2:1 clays) and NO_3^- (through reactions with SOM), while limiting microbial NO_3^- assimilation due to energetic constraints (Yansheng et al., 2020). Soil N availability exerts contrasting effects on biotic and abiotic N immobilization. Microbial immobilization (biotic) of both NH_4^+ and NO_3^- typically increases under N-limited conditions as microbes assimilate inorganic N to support growth, but declines under sustained high N inputs. In contrast, abiotic NH_4^+ immobilization is largely insensitive to N availability, while abiotic NO_3^- immobilization tends to decrease as N levels rise. Indeed, in N-rich ecosystems, microbial activity is often suppressed due to inhibited enzyme function, reduced F:B ratios, and slower decomposition of recalcitrant OM, shifting the balance toward abiotic retention mechanisms (Yansheng et al., 2020). Wang et al. (2020) reported that microbial assimilation is the dominant pathway for NH_4^+ retention following long-term application of straw and animal manure, whereas abiotic processes play a larger role in NO_3^- retention. Soil texture and management also play critical roles. High clay content enhances both biotic (via microbial habitat preservation) and abiotic (via surface adsorption) immobilization, leading to greater N retention in fine-textured soils. Tillage practices further modulate this balance. NT systems promote NH_4^+ immobilization through surface residue accumulation and elevated microbial N demand, whereas CT accelerates residue decomposition, often creating C-limited conditions that restrict immobilization (Martens, 2001; Wang et al., 2020).

3.2.5 Denitrification

Denitrification is the anaerobic microbial reduction of NO_3^- and NO_2^- to gaseous NO, N_2O, and N_2 and represents a major pathway for Nr loss from terrestrial ecosystems. This process is environmentally significant as the gaseous intermediates (NO and N_2O) contribute to air pollution and climate forcing, while incomplete denitrification or residual NO_3^- can contribute to groundwater contamination and eutrophication (Van Groenigen et al., 2015; Zhu et al., 2018). Denitrification initiates when microbial and root respiration deplete soil oxygen faster than atmospheric diffusion can replenish it, creating anaerobic microsites where NO_3^- acts as a terminal electron acceptor (Daunoras et al., 2024; Hayashi, 2022). Denitrifying bacteria such as *Pseudomonas, Paracoccus*, and B*acillus* respire NO_3^- instead of oxygen, sequentially reducing it to gaseous N forms through the activity of NO_3^- reductase (Nar), NO_2^- reductase (Nir), NO reductase (Nor), and N_2O reductase (Nos), a pathway responsible for the production and emission of nitrous oxide (N_2O) and the ultimate return of N_2 to the atmosphere. In contrast, some anaerobic bacteria, including *Clostridium*, *Desulfovibrio*, and *Shewanella*, perform dissimilatory NO_3^- reduction to NH_4^+ (DNRA), where NO_3^- is reduced to NH_4^+ rather than lost as gas, mediated by cytochrome *c* NO_2^- reductase (NrfA), thereby conserving N within the soil system. Another significant anaerobic pathway is anammox (anaerobic NH_4^+ oxidation), carried out by *Planctomycetota* such as *Candidatus Brocadia* and *Candidatus Kuenenia*. These microorganisms use hydrazine synthase (HZS) and hydrazine dehydrogenase (HDH) to couple NH_4^+ oxidation with NO_2^- reduction, producing N_2 through the unique intermediate hydrazine (N_2H_4) (Zhang et al., 2020).

The N_2O:N_2 emission ratio is strongly influenced by soil moisture and structure. In well-aerated soils, N_2O diffuses rapidly to the atmosphere before further reduction, whereas in water-saturated aggregates, prolonged anoxia favors complete reduction to N_2. Although NO is an obligate intermediate, it is rarely emitted from wet soils due to rapid microbial consumption. Consequently, the NO:N_2O ratio declines sharply with increasing WFPS, falling below 1 at >80% WFPS, and measurable NO emissions typically occur only under controlled experimental conditions that inhibit its re-assimilation (Smith et al., 2018). During microbial NO_3^- reduction, NO_2^- accumulates transiently and may follow divergent pathways. It can be fully denitrified to N_2O/N_2 or diverted to NH_4^+ via dissimilatory NO_3^- reduction to NH_4^+ (DNRA), depending on environmental conditions. The rates and end-product stoichiometry of these processes are controlled by a suite of interacting factors, including oxygen availability, NO_3^- and OC supply, temperature, pH, soil texture, and the abundance and composition of NO_3^--reducing microbial communities. Li et al. (2022) identify substrate availability particularly NO_3^- and labile OC and microbial community composition as the primary direct drivers of denitrification, with soil and climatic variables acting indirectly by modulating denitrifier activity and habitat conditions. Denitrification rates increase with rising soil NO_3^- concentrations, as elevated NO_3^- stimulates the proliferation and metabolic activity of denitrifying microorganisms, enhancing the expression of Nir. Mean annual temperature further regulates process rates, with denitrification approximately doubling per 10°C increase driven by accelerated microbial metabolism, greater substrate availability from enhanced mineralization, and increased nitrification-derived NO_3^- pools.

Soil pH exhibits a strong negative correlation with the N_2O ratio. Lower pH generally leads to higher N_2O:N_2 emissions due to its chemical and biological effects, particularly the inhibition of Nos activity under acidic conditions (Khalifah & Foltz, 2024). Hanif (2023) and Myrold (2021) demonstrated that acidic conditions suppress denitrifier activity, whereas near-neutral to alkaline conditions enhance it, altering the N_2O:N_2 emission ratio. Notably, although most studies focus on pH 5–8 (the typical range for active denitrification), denitrification can still occur at pH < 4 or >10 if sufficient C, NO_3^-, and adapted microbial communities are present. Soil texture modulates denitrification through its influence on WFPS and aeration. Fine-textured soils retain more water, sustain longer anaerobic periods, and often exhibit higher N_2O emissions than sandy soils, which drain rapidly and limit anoxia (Grzyb et al., 2021; Khalifah & Foltz, 2024). Malique et al. (2019) reported that N_2O emissions tend to be higher in clay loam than sandy soils. Cui et al. (2023) demonstrated that finer-textured soils typically harbor more mineral-associated OM, exhibit higher N mineralization rates, and support greater denitrification potential due to microaerophilic micropores, often leading to higher N_2O emissions than sandy soils. Elevated soil water content can restrict oxygen diffusion to microbial communities, thereby fostering anaerobic conditions that promote denitrification. Wang and Kuzyakov (2023) demonstrated that as moisture increases and oxygen becomes limited, denitrification surpasses nitrification as the primary N_2O source, with the transition typically occurring between 60% and 70% WFPS, though this threshold varies with soil properties. Smith et al. (2018) reported that drainage of organic soils by lowering the water table shifts redox conditions from anaerobic to aerobic and significantly increasing N_2O emissions. In contrast, flooded conditions typically result in negligible N_2O emissions, as any N_2O produced during denitrification is fully reduced to N_2. Management practices such as NT can enhance denitrification potential near the soil surface by maintaining higher moisture and organic carbon levels compared to conventional tillage (Martens, 2001).

3.2.6 Ammonia volatilization

NH_3 volatilization from croplands is a significant concern as it results in the loss of Nr from soils and contributes to atmospheric N pollution. This process is strongly influenced by soil properties including pH, CEC, moisture, and temperature as well as meteorological factors (e.g., wind speed) and agronomic practices, particularly the type and method of fertilizer application. Many of these factors are amenable to management (Hayashi, 2022; Zhu et al., 2015). High CEC and moist conditions retain more NH_4^+ on exchange sites or in soil solution, reducing volatilization, but pH exerts the strongest control. At pH 7, less than 0.5% of total NH_4^+ exists as volatile NH_3, rising to over 5% at pH 8. This effect is particularly pronounced with surface-applied urea, which undergoes hydrolysis catalyzed by the enzyme urease, releasing NH_3 and locally elevating pH by 2–3 units and creating hotspots for gaseous N loss (Myrold, 2021) which enhance microbial activity and urease expression; for instance, a temperature increase from 7°C to 32°C can elevate NH_3 losses by up to fivefold. Regional soil conditions further modulate volatilization risk. Soils in semiarid and arid regions are typically alkaline, and their inorganic N is predominantly present as NO_3^- because high pH promotes rapid nitrification, which depletes NH_4^+ and consequently limits NH_3 volatilization. In contrast, soils in humid subtropical and tropical regions are typically acidic, leading to NH_4^+ dominance in inorganic N pools due to suppressed nitrification at low pH. This reduction in NO_3^- production minimizes NO_3^- leaching and runoff, while the acidic conditions also effectively inhibit NH_3 volatilization (Xie et al., 2015). Nevertheless, NH_3 losses can still be substantial in dry, hot, high-pH soils where surface-accumulated NH_4^+ from fertilizer is exposed to favorable volatilization conditions (Zhu et al., 2018). Best management practices to mitigate NH_3 loss include applying fertilizers to cool, moist soils; incorporating them below the surface; and using urease inhibitors (e.g., NBPT) or controlled-release formulations, which can reduce emissions by 30%–50% (Myrold, 2021). These strategies are especially critical in NT systems, where surface residue accumulation increases soil moisture and harbors elevated populations of urease-producing microbes, boosting urease activity 3–4 times in the top 10 cm compared to conventional tillage (Martens, 2001) thereby exacerbating volatilization when urea is surface-applied.

It is also important to recognize that plants actively participate in NH_3 exchange. Plant stomata serve as primary pathways for gas exchange, facilitating the movement of NH_3, CO_2, and water vapor. When the NH_3 concentration in the leaf intercellular air space exceeds that of the ambient atmosphere, NH_3 is emitted. Conversely, when ambient NH_3 levels are higher, plants absorb NH_3 via dry deposition. Therefore, the relative roles of soil and vegetation should be assessed independently, as both can act as NH_3 sources, and NH_3 emitted from the soil may be partially recaptured by the plant canopy (Hayashi, 2022). NH_3 volatilization is a major cause of low N use efficiency (NUE) when NH_4^+-based fertilizers especially urea is surface-applied, particularly in NT systems.

3.3 Environmental and human influences on soil nitrogen dynamics

Anthropogenic activities have doubled global Nr inputs over the past century, profoundly altering N cycle in agricultural ecosystems (Kuśmierz et al., 2023). Although synthetic N fertilizers have been crucial for sustaining crop production and can indirectly boost SOC through increased residue inputs, low NUE often leads to losses of inorganic N in the soil profile (Cao et al., 2021; Zaman et al., 2025). Approximately 25% of applied N is lost to the atmosphere and 20% leached as NO_3^-, reducing crop productivity and degrading environmental quality (Kuśmierz et al., 2023). These losses contribute to degraded groundwater and surface water quality, elevated atmospheric GHGs, loss of plant biodiversity, and deteriorated rural air quality (Myrold, 2021; Robertson & Groffman, 2024). Ayiti & Babalola (2022) reported that excess NO_3^- from fertilizers leaches into waterways or volatilizes as NO_x and NH_3 and contributes to soil acidification particularly from urea-based fertilizers which impairs nutrient availability (e.g., Ca, Mg, K, P, Mo), reduces soil biodiversity, and degrades long-term soil health. In light-textured soils, unbalanced mineral fertilization may even accelerate SOC mineralization. Similarly, liming to adjust soil pH can enhance C and N retention, yet it may also promote gaseous losses (CO_2, N_2O, NO_x, NH_3) and dissolved C and N leaching (Kuśmierz et al., 2023). Soil microorganisms mediate the conversion of N-rich inputs, such as synthetic fertilizers and manure, into N_2O and excessive fertilizer application triggers soil microbes to emit disproportionately high levels of N_2O (Noor et al., 2019). Ayiti & Babalola (2022) reported that chemical fertilizers stimulate nitrifying and denitrifying microbes, increasing emissions of NO, NO_2 and N_2O, while reducing plant NUE. Notably, while CO_2 is the dominant GHG, N_2O makes substantial contributions to global warming despite its lower atmospheric concentrations. Over a 100-year timeframe, N_2O have global warming potentials approximately 300 times greater than CO_2. Soils are a major source of it potent gases, contributing roughly two-thirds of N_2O emissions (Smith et al., 2018). Global emissions of atmospheric Nr primarily NH_3 and NO_x were estimated at approximately 103 Tg N yr^{-1} in the early 1990s and are projected to double by 2050. A significant portion of this anthropogenic Nr returns to terrestrial ecosystems via atmospheric N deposition (Zhang et al., 2021), thereby substantially perturbing the global N biogeochemical cycle (Ramond et al., 2022).

Soil critically influences NUE through its pH, OM contents, texture, structure, depth, and water holding capacity (Noor et al., 2019). For instance, acidification impairs plant N uptake and reduces NUE. Urea application temporarily elevates soil pH, promoting the conversion of NH_4^+ to NH_3, thereby increasing volatilization losses. Soil texture governs water and nutrient retention, aeration, and drainage, all of which affect NUE and loss pathways. Coarse-textured soils exhibit higher N leaching due to rapid drainage through macropores, whereas fine-textured soils retain more water and nutrients, supporting higher crop yields and NUE. However, fine-textured soils are prone to poor aeration under high moisture, raising anaerobic conditions that promote denitrification and associated N_2O emissions (Hanif, 2023) . Effective N management in soils is therefore essential for sustaining food and feed productivity while minimizing Nr losses to the environment. This is achieved by improving NUE through strategies such as reducing synthetic fertilizer inputs and recycling available N sources, particularly manure and crop residues (Hayashi, 2022; Noor et al., 2019; Zaman et al., 2025). Applying N in split doses aligned with crop demand, using a mix of NH_4^+ and NO_3^-, precision instead of surface broadcasting is critical to obtain optimal NUE. Tillage effects on NUE are context-dependent, varying with soil type, climate, and residue management. CT can enhance soil quality and water use efficiency but may increase denitrification due to reduced aeration and higher moisture. In sandy soils, NT can exacerbate NO_3^- leaching through continuous macropores and straw incorporation further fuels denitrification by supplying DOC (Hanif, 2023). Critically, optimizing NUE extends beyond incremental agronomic adjustments. As Zhang et al. (2020) emphasize, effective N stewardship requires integrated soil–crop–climate–policy strategies that recognize soil as a living, biogeochemically active system. Soils with high OM, stable aggregates, and diverse microbial communities exhibit greater N retention through immobilization and reduced gaseous or leaching losses, underscoring that NUE is fundamentally linked to soil health. Moreover, the anthropogenic N problem differs qualitatively from the CO_2 problem. Nr perturbations are more local, more immediate, and more reversible, offering tangible opportunities for mitigation through precision agriculture, circular nutrient economies, and policy instruments such as fertilizer taxation or nutrient trading. Thus, optimizing NUE is not merely a technical challenge but a systems-level imperative, balancing food security.

3.4 Soil-plant interactions in nitrogen utilization

Soil–plant interactions play a central role in N utilization and cycling within terrestrial ecosystems. Plants affect N dynamics through root morphology, exudate chemistry, and symbiotic relationships with rhizosphere microorganisms. Root systems create preferential flow paths that alter soil structure and facilitate N migration, which can enhance N uptake under optimal conditions but may accelerate NO_3^- leaching in coarse-textured soils with high drainage. Sandy loam is more likely to form preferential paths for plant roots in comparison to clay loam. Plant metabolism governs N assimilation via enzymatic pathways, converting inorganic N (NO_3^- and NH_4^+) into amino acids and proteins essential for growth. Plants residues further contribute to N transformation by returning organic N to the soil, stimulating microbial activity, and affects soil temperature, moisture, DOC, and N_2O emissions depending on their C:N ratios. High C:N residues (>25) promoting NH_4^+ immobilization and reducing N_2O losses by up to 80%. However, the net effect on N cycling is highly varying with plant species, residue chemistry, soil properties, and environmental conditions (Henneron et al., 2020; Zhang, Cai, et al., 2022, Zhang, Chen, Jin, et al., 2022, Zhang, Chen, Yao, et al., 2022). Chen et al. (2019) showed that plant-derived inputs are fundamental drivers of soil biological activity and biogeochemical cycling. In a long-term N enrichment experiment on semiarid grassland, plant removal dramatically reduced soil C and N mineralization rates and suppressed the abundance of most soil organisms particularly bacteria, fungi, and bacterial- and plant-feeding nematodes underscoring their reliance on recent plant C. Root exudates can inhibit nitrification via biological nitrification inhibitors or stimulate N mineralization and fixation by supplying labile C that fuels microbial activity. While some compounds suppress denitrification, others enhance it, leading to complex effects on N_2O emissions depend on plant species, age, and environmental conditions like elevated CO_2 or temperature, and is also influenced by soil N availability, creating a dynamic feedback loop between plants and microbes (Zhang, Cai, et al., 2022, Zhang, Chen, Jin, et al., 2022, Zhang, Chen, Yao, et al., 2022). Malique et al. (2019) found that denitrification was enhanced in planted arable soil early in growth due to root-exuded labile C but declined as plants matured and competed with microbes for inorganic N. Ryegrass stimulated denitrification more than barley or wheat, highlighting species-specific effects. While some plants promote denitrification, others suppress it through root-exuded inhibitors. These findings suggest that strategic plant selection could help manage N losses. Thus, the plant–soil interface functions as an integrated biogeochemical unit where plant-driven physical, metabolic, and biochemical processes collectively determine N retention, loss, and bioavailability, highlighting the need for plant-centered strategies in sustainable N management.

4. Water cycle

The water cycle, often called the hydrologic cycle involves the continuous circulation of water among atmosphere, hydrosphere, lithosphere, and biosphere. The hydrologic cycle includes physical processes such as evaporation, precipitation, infiltration, surface run-off, and subsurface flow (Cui et al., 2018). Within this dynamic system, soil serves as a critical interface that mediates exchanges between the atmosphere, biosphere, and lithosphere and playing a critical role in the hydrological cycle by storing water for plant use, filtering contaminants, and cycling nutrients. These soil's functions are influenced by its chemical, physical and biological properties (Zewide, 2021). However, due to its inherent complexity, alterations in one soil parameter often trigger cascading effects across multiple hydrological and agronomic processes. Consequently, soil-water interactions fundamentally shape both the quantity and quality of freshwater resources, with direct implications for ecosystem health, agricultural productivity, and water security in the face of global environmental change (Lal et a., 2017). Given soil's central role in mediating hydrological processes, the following subsections examine key aspects of soil-water interactions.

4.1 Water infiltration and percolation through soil

4.1.1 Infiltration

Precipitation is the primary source of freshwater on Earth. A substantial portion of rainfall infiltrates the soil, the process by which water at the surface enters the soil profile. From a hydrologic perspective, rain-driven infiltration is the most important pathway for replenishing soil moisture and groundwater. However, if rainfall intensity exceeds the soil's infiltration capacity, water accumulates on the surface, forming temporary ponds or generating surface runoff (Novák & Hlaváčiková, 2018; Zewide, 2021). Thus, infiltration is indeed a key soil function that governs leaching, runoff, and crop water availability, and it is influenced by a range of physical and chemical soil characteristics that exhibit both spatial variability and temporal dynamics due to agricultural practices, water management, and biological activity (Seidel, 2018). Soil texture is an inherent property, which highly affected soil infiltration. Sandy soils generally exhibit high infiltration rates due to their large pore spaces, whereas clayey soils typically have low infiltration rates because of their small pores especially when compacted or lacking structure. However, under dry conditions, certain clayey soils develop shrinkage cracks, which act as preferential pathways and can significantly increase infiltration (Haghnazari et al., 2015). Over the last 40 years, soil structure has received high interest as a dynamic soil factor affecting water movement in soil at multiple scales. Basset et al, (2023) demonstrate that soil structure is a dominant control on infiltration dynamics with strong empirical correlations between steady infiltration rate and key structural indicators including mean weight diameter (MWD), water-stable aggregates (WSA), total porosity, SOC, and bulk density (inversely related). Management practices profoundly shape these relationships as soil amendments and crop management strategies consistently enhance soil structure, increasing MWD, WSA, SOC, and porosity while reducing bulk density, thereby boosting infiltration. Their findings show that soil texture alone is insufficient to predict infiltration and well-structured clayey soils can infiltrate water more effectively than poorly structured sandy soils, underscoring structure's primacy over texture.

Optimal water infiltration depends not just on the quantity of pores but on their continuity, shape, distribution, and stability all of which are enhanced by SOM, biological activity, appropriate land cover, and conservation-oriented management such as NT, crop rotation, mulching, and agroforestry (Seidel, 2018). Ju et al. (2024) demonstrated that water infiltration in red soils region of southern China is primarily governed by the connectivity and continuity of macropores, not just total porosity. Zewide (2021) reported that porosity and soil structure, deeply influenced by SOM content which promotes stable aggregates and supports soil biota like earthworms, whose burrowing creates continuous pores that facilitate water entry. Lal et al. (2017) reported that poor soil aggregation degrades macropore structure, reducing size and connectivity, which restricts root growth, increases surface runoff and erosion, and can cause waterlogging in flat areas. While tillage may temporarily increase porosity in compacted soils, long-term CT generally reduces infiltration especially in fine- and medium-textured soils by disrupting soil structure, decreasing surface SOM, and often increasing bulk density (Seidel, 2018). In contrast, surface SOC accumulation under NT fosters a favorable soil architecture that enhances infiltration, water retention, and resistance to compaction, underscoring the importance of SOC stratification as an indicator of soil health and hydrological function (Franzluebbers, 2002). In the Ouriyori watershed of northern Benin, saturated hydraulic conductivity (K_S) was significantly lower in cropland than in fallow or natural vegetation. This reduction in infiltration is due to land management practices, particularly continuous tillage without residue incorporation, which compacts soil and disrupts macropore networks. Fallow lands also retained higher soil moisture during the rainy season, despite overall low moisture levels (Togbévi et al., 2022). Soil organisms including microbes, fauna, and plant roots profoundly influence hydraulic conductivity by shaping aggregate stability and pore structure. Roots and burrowing animals create macropores that enhance preferential flow, especially near saturation, while dead roots form persistent channels that boost late-season infiltration. However, microbial activity can also reduce conductivity when metabolic byproducts clog pores (Haghnazari et al., 2015).

The hydraulic properties of the most restrictive layer often govern infiltration capacity. As reported by Haghnazari et al. (2015) in layered soils, infiltration is often limited by the most restrictive horizon. Paradoxically, a coarse sand layer within fine-textured soil can impede flow due to insufficient matric suction. Anthropogenic practices like land leveling for irrigation further alter hydraulic behavior by redistributing soil, increasing bulk density, exposing subsoil, and modifying spatial patterns of texture and structure. Thus, both inherent texture and management-induced changes shape field-scale infiltration dynamics (Zewide, 2021).

4.1.2 Percolation

Water that infiltrates at the land surface may percolate through the unsaturated zone and eventually recharge groundwater reservoirs. The groundwater serves as a freshwater storage that can be utilized during droughts when surface water supplies are reduced and constitutes 22% of Earth's freshwater (Deng & Makinde, 2020). It can also discharge into wetlands, or particularly in coastal settings mix with saline groundwater, potentially contributing to saltwater intrusion. Ultimately, it may discharge into the ocean, where it becomes available for evaporation, thereby closing the loop of the hydrologic cycle. However, not all infiltrated water reaches the water table to constitute true aquifer recharge. As emphasized by Shanafield & Cook (2014), factors such as deep water tables, high evapotranspiration demand, short-duration flood events, and the presence of low-permeability layers (clay pans or silty deposits) can intercept or delay percolation, sometimes preventing recharge entirely. In such cases, infiltrated water may be lost to evaporation or plant uptake, or become temporarily stored in perched aquifers above restrictive horizons. Critically, antecedent soil moisture exerts strong control over recharge efficiency. Wetter initial conditions reduce capillary-driven resistance, shorten travel times through the vadose zone, and increase the fraction of infiltration that successfully reaches the aquifer. Thus, the transformation of infiltration into groundwater recharge is neither guaranteed nor instantaneous, it is a highly conditional process governed by interacting hydrologic, geologic, and climatic factors.

4.2 Soil water retention and storage

Soil water retention is a key hydraulic property that governs soil water storage, redistribution, availability, and evaporation, thereby influencing hydrological and ecological processes in terrestrial ecosystems. Soil particle size distribution, SOM content, and bulk density are the primary factors controlling water retention and are widely used in pedotransfer functions (PTFs) for prediction. Additional properties such as soil structure, CEC, carbonate content, plastic limit, and specific surface area (SSA) are also correlated with water retention and often included as secondary predictors in PTFs to improve estimation accuracy (Yang et al., 2014). Soil water retention is closely linked to SOM content and composition, depending on how SOM influences soil structure particularly aggregation and pore architecture. SOM enhances porosity, infiltration, and aggregate stability through hydrophilic compounds such as polysaccharides, although certain organic components can impart hydrophobicity, thereby affecting soil wettability and water distribution (Lal, 2020). As reported by Minasny & McBratney (2018) SOM enhances soil water retention, with the greatest effect observed at saturation (0 kPa), followed by field capacity (−10 to −33 kPa), and the smallest impact at the wilting point (−1500 kPa), a pattern observed across diverse soil types though the magnitude varies. This improvement is primarily driven by a reduction in bulk density and an increase in total porosity, as SOM being less dense than mineral particles dilutes the soil matrix, and by the promotion of soil aggregation, which fosters a more complex pore system comprising macropores, mesopores, and micropores. They reported that the benefits of SOM are most pronounced in coarse-textured soils (clay $<20\%$), with available water capacity (AWC) increasing significantly in sandy to sandy clay loam soils, while effects diminish or become negligible in soils with $>40\%$ clay. Similar results reported by Rawls et al. (2003) who demonstrated that SOM content is a key soil property that significantly improves the prediction of soil water retention beyond what can be achieved using texture alone. Including SOM in predictive models such as regression trees and Group Method of Data Handling (GMDH) neural networks reduced root mean square error (RMSE) by 15% at −33 kPa and 10% at −1500 kPa, indicating enhanced accuracy in estimating both AWC and water retention at wilting point. Yang et al. (2014) demonstrated that SOM exhibits significant correlations with soil water content across the entire matric potential range (0–1500 kPa), indicating its dominant control over soil water retention. Two mechanisms are proposed: 1) At higher matric potentials, SOM enhances water retention primarily by improving soil structure, reflected in reduced bulk density, 2) At lower matric potentials, SOM increases water retention through enhanced adsorption capacity due to its hydrophilic properties. Additionally, the SOM-rich A horizons show the highest water retention across all moisture conditions, underscoring their hydrological significance as key components of the critical zone in alpine ecosystems.

Traditional agricultural practices involving repeated tillage and residue removal are increasingly being replaced by NT systems due to concerns over soil degradation, and erosion. NT offers multiple benefits, including improved soil structural stability, enhanced water infiltration, reduced runoff and erosion, greater soil water storage, increased SOC, improved nutrient retention, and enhanced soil biological diversity. Fifty years of data from the Hermitage long-term experiment demonstrate that NT combined with stubble retention significantly enhances soil water storage during fallow compared to CT with stubble burning. This improved water accumulation is attributed to better soil structure in NT systems, including the maintenance of large, continuous pores even under moist conditions, which facilitate deeper water infiltration and reduced runoff (Page et al., 2019).

4.3 Evaporation and transpiration from soils

Evapotranspiration which includes evaporation from the soil surface and transpiration from plants is one of the most important fluxes in the terrestrial water cycle (Cui et al., 2018). As the second-largest hydrologic flux after precipitation, evapotranspiration has been significantly affected by global climate change over recent decades. However, estimates of evapotranspiration remain highly uncertain, particularly with respect to the partitioning between evaporation and transpiration. Transpiration is estimated to account for 25%–64% of total evapotranspiration, but may contribute as much as 80%–90% in certain ecosystems, highlighting significant variability across biomes (Vereecken et al., 2015). The evapotranspiration process is governed by two categories of factors: external (atmospheric) and internal (soil-related). External factors encompass atmospheric conditions that drive the vapor transfer from the soil and plant surface, including solar radiation, wind speed, air temperature, vapor pressure deficit, and relative humidity. Internal factors are primarily linked to intrinsic soil properties (An et al., 2018), with soil playing a central role in regulating both evaporation and transpiration by modulating water availability. Key soil characteristics including texture, structure, SOM content, and hydraulic conductivity control infiltration, water retention, root-zone storage, and ultimately influence the partitioning of evapotranspiration into evaporation and transpiration. Soils rich in OM and with well-developed structure exhibit enhanced porosity and aggregate stability, increasing plant-AWC and prolonging moisture availability during dry periods, thereby supporting sustained transpiration. In contrast, coarse-textured soils allow rapid drainage, promoting shallow water storage and higher evaporative losses, while compacted or degraded soils restrict root development and reduce water uptake efficiency, limiting transpirative performance (Suliman et al., 2024). A study by An et al. (2018) identified soil texture (specifically sand content), sample thickness, and initial compactive state as critical factors governing evaporation. Coarser, thicker, and loosely compacted soils generally promote faster and more sustained evaporation, whereas finer, thinner, and highly compacted soils tend to reduce it.

Climate change is altering the hydrological cycle and soil water balance globally, with significant implications for agriculture. Rising temperatures and longer sunshine duration have increased potential evapotranspiration, while shifting precipitation patterns wetter winters and drier springs and summers are disrupting water supply during key growing seasons. A key strategy to mitigate these impacts is the intentional build-up of SOC through the OM incorporation such as crop residues, green manure and biochar. SOC improves soil structure and increases water retention, enhancing plant-AWC, especially in coarse-textured soils. However, the depth at which SOC accumulates plays a critical role. Under CT, SOC tends to concentrate near the surface, leading to shallow, organic-rich layers that are prone to high evaporative losses. This can reduce the effective water availability for crops over time and limit deep percolation, thereby decreasing groundwater recharge (Feifel et al., 2024).

While increased vegetation density can significantly reduce sediment yield by enhancing ground cover and mitigating erosion, excessive vegetation also elevates water consumption via elevated evapotranspiration. In arid and semi-arid regions, where soil water is the primary constraint on plant growth, determining the maximum sustainable vegetation density is essential to balance ecological benefits with water availability (Xia & Shao, 2008).

Thus, soil acts not only as a reservoir but also as a regulator of evapotranspiration, with its physical and chemical properties mediating the exchange of water between the land surface, plants, and atmosphere. Managing soil health through conservation practices, organic amendments, and reduced compaction is essential for optimizing evapotranspiration dynamics, improving water use efficiency, and sustaining ecosystem productivity under changing climate conditions.

4.4 Soil as a regulator of water quality

Groundwater quality is closely linked to human health and highly vulnerable to anthropogenic disturbances of the hydrological system, including leaching of nutrients and pesticides from agricultural lands, salinization in irrigated areas, and contamination by NO_3^-, heavy metals, and wastewater application in agriculture (Keesstra et al., 2012). Soil acts as a natural filter and plays a critical role in regulating water quality (Cheng et al., 2021), with its filtering efficiency depending on three main factors: (1) pollutant properties (e.g., solubility, sorption affinity, degradability), (2) soil characteristics (e.g., texture, SOM content, mineralogy, and structure), and (3) hydrological dynamics particularly the occurrence of preferential flow, which can bypass the soil matrix and reduce filtration effectiveness. Pollutants may be retained through adsorption onto clay minerals and SOM, immobilized via precipitation or complexation, or degraded by microbial activity (Keesstra et al., 2012). However, the rapid expansion of industrial and agricultural activities has led to a significant influx of pollutants into soils, often exceeding their natural self-purification capacity and resulting in soil degradation and pollution. In 2018, global agricultural inputs reached approximately 108.7 Mt of N, 40.6 Mt of P, and 4.1 Mt of pesticides (Cheng et al., 2021).

Results from a study by Singer & Brown (2018) highlight that while soils can act as effective natural biofilters improving water quality through enhanced sorption, microbial degradation, and immobilization, soil properties critically determine whether filtration mitigates or exacerbates contamination risks. Two contrasting soils an acidic, sandy forest soil and an organically amended silt-loam farm soil were evaluated for their capacity to filter Class A reclaimed water. Soil filtration significantly reduced estrogenic activity by 86%–99.6% and lowered NH_3 and phosphate (PO_4^{3-}) concentrations, with the forest soil demonstrating superior PO_4^{3-} retention. However, NO_3^- concentrations increased in leachate due to net mineralization, particularly in the forest soil with 21:1 of C:N ratio. Metal dynamics varied between soils. Arsenic (As) and Ca were mobilized in both, while copper (Cu) and manganese (Mn) levels increased in the farm soil leachate.

Sometimes degradation products can exhibit higher solubility than parent compounds. For instance, in conventional banana production, frequent and intensive use of the fungicide Mancozeb to control *Black Sigatoka* leads to significant environmental contamination. Although Mancozeb itself has low persistence in soil (1–7 days), it rapidly degrades into ethylene thiourea (ETU), a highly soluble and persistent metabolite with carcinogenic and mutagenic potential. A study in a tropical banana-growing region of Mexico found that Mn and zinc (Zn) from Mancozeb accumulated in topsoil, ETU was barely detectable in soil due to its weak adsorption and high mobility. Instead, ETU leached into surface and subsurface waters, contaminating drinking water sources with concentrations as high as 22.5 mg/L in surface water and 4.3 mg/L in shallow groundwater which is far exceeding safety thresholds. Despite the region's clay-rich soils, which typically enhance pollutant retention, the high solubility of ETU overcame soil filtering capacity, leading to widespread aquatic contamination. This highlights that even in soils with high adsorption potential, highly mobile pesticide metabolites can bypass natural filtration, posing serious risks to human and ecosystem health (Geissen et al., 2010). Earthworm burrows act as preferential flow paths that significantly influence water infiltration and solute transport in soils. While these macropores enhance rapid water movement, their walls also serve as sorption sites and microbiological hotspots, promoting the retention and biodegradation of pollutants (Keesstra et al., 2012). Land use practices profoundly influence the soil filter function. Studies in southeastern Australia show that after 30 years of winery wastewater application, concentrations of potentially toxic elements such as Zn, Cu, B, Al, and Fe, as well as macronutrients like K and Mg, significantly increased in the soil and were exceeding established threshold values and posing risks to plant growth. Additionally, winery wastewater disrupts microbial processes involved in C and N cycling. These findings indicate that while soil can initially buffer winery wastewater inputs, prolonged or unmanaged application compromises its filtering capacity and biogeochemical function (Mosse et al., 2012). Regarding the critical role of subsurface soil depth and texture in filtration, studies have shown that thin soils limit the soil's filtering capacity, making groundwater vulnerable to anthropogenic pollution. As a result, forest management guidelines discourage fertilization and intensive tillage near aquifers. Shallow phreatic aquifers in agricultural areas with humic-sandy and loamy-sandy soils are important sources of drinking water, but are vulnerable to contamination due to preferential flow processes such as fingered and funneled flow. These pathways can facilitate rapid transport of pesticides like bentazone through sandy soils, bypassing the soil matrix and reducing filtration efficiency (Keesstra et al., 2012).

4.5 Human impacts on soil's role in the water cycle

Soil and water are essential inputs into a food production strategy of agroecosystems. Global water demand has increased six fold over the past century and is projected to rise further, with irrigated agriculture consuming 60%–80% of global freshwater, yet the water use efficiency (WUE) is low. With a potential 40% global water deficit by 2030 under current trends, improving WUE through reduced losses (runoff, evaporation, deep percolation) is critical (Lal et al., 2017). Human activities profoundly influence groundwater recharge. Urban development introduces impervious surfaces such as roads and rooftops, which reduce infiltration by redirecting storm water into drainage systems that discharge directly to surface waters, bypassing aquifers and increasing flood and pollution risks. As urbanization intensifies or storms grow larger, the remaining pervious areas may become overwhelmed, leading to significantly reduced groundwater recharge. Vegetation management also plays a critical role. Agricultural practices, logging, and the introduction of non-native species alter evapotranspiration rates and soil permeability. Most significantly, anthropogenic climate change is altering recharge patterns globally. Warming increases evapotranspiration, reduces snowpack and glacial storage, and intensifies hydrological extremes leading to more droughts and heavier rainfall events (Owen, 2021). There are four factors changing in the climate that will affect WUE including rising CO_2 levels, increasing temperatures, more variable precipitation, and higher atmospheric humidity. These changes elevate atmospheric evaporative demand, intensify extreme rainfall events, and expand drought-prone areas, thus reducing soil water availability, especially in regions with poor water holding soils (Hatfield & Dold, 2019). Effective soil and water management practices can improve water and nutrient availability to plants, potentially increasing WUE defined as the ratio of plant biomass produced per unit of water used by 15%–25%. Soil fertility plays a key role in WUE, supporting plant growth and regulating transpiration. Practices that increase SOM improve water retention, root-zone moisture, and nutrient uptake, thereby boosting crop yields and WUE (Gavrilescu, 2021). Enhancing soil health particularly through SOC sequestration offers a nexus-based solution, as it improves soil structure, macroporosity, hydraulic conductivity, moisture retention, and mitigates hardpan formation. Mulching using plastic or crop residues reduces soil water loss, enhances soil moisture retention, and increases WUE, particularly in semi-arid regions with less than 400 mm annual precipitation. Irrigation method strongly affects WUE. Micro-irrigation systems such as drip or furrow reduce non-beneficial evaporation, allowing substantial water savings up to 37% in cotton and 23% in wheat while maintaining or even increasing yields. Strategic deficit irrigation can optimize WUE, especially when maximizing efficiency is prioritized over maximum yield (Hatfield & Dold, 2019).

5. Soil water-carbon- nitrogen

Soil health is fundamentally governed by the concentration, quality, and dynamics of SOC in the root zone. A functional threshold of 1.5%–2.0% SOC is often associated with the maintenance of key soil ecosystem services and below this level, physical, chemical, and biological functions deteriorate significantly. Maintaining SOC above this threshold is essential for stabilizing soil structure including porosity, pore-size distribution, and pore connectivity which in turn enhances water infiltration, reduces surface runoff, increases plant-AWC, and moderates soil temperature (Lal et al., 2017). SOC supports vital soil chemical functions including improving CEC, buffers soil pH, and availability of nutrients, especially N and as the primary energy source for soil biota, fuels the activity and diversity of microbial and faunal communities, which in turn regulate biogeochemical cycling and elemental transformations. Critically, the linkage between C and N cycles is mediated by soil moisture dynamics (Dutta et al., 2024). As Porporato et al. (2003) demonstrate stochastic rainfall patterns drive fluctuations in soil water content, which directly control microbial decomposition rates, the balance between N mineralization and immobilization, nitrification, plant N uptake, and leaching losses. For instance, microbial activity peaks near field capacity, whereas both excess moisture (anoxia) and drought suppress decomposition and N transformations. Moreover, the C:N ratio of organic inputs determines whether microbes mineralize excess N or immobilize inorganic N, directly affecting N availability and loss potential. This tight C–N coupling is often disrupted in intensively managed agroecosystems. Inorganic N fertilization without concurrent organic C inputs leads to accumulation of mineral N, elevating risks of NO_3^-leaching and N_2O losses processes exacerbated under high soil moisture. In contrast, extensively managed or natural ecosystems maintain a balanced C–N stoichiometry through continuous SOM returns, enhancing N retention and use efficiency (Yansheng et al., 2020). Re-establishing this C–N linkage in agricultural systems through practices such as crop residue retention, cover cropping, compost application, or agroforestry not only mitigates N losses but also builds SOC stocks. Enhanced SOC further contributes to climate regulation by sequestering atmospheric CO_2, suppressing soil erosion, and reducing non point source pollution through improved infiltration and nutrient buffering. Thus, integrated management of soil water, carbon, and nitrogen is central to sustainable land and water management, climate resilience, and food security. Long-term field studies (2015–2022) in cereal-based systems (rice–wheat–sesbania and rice–barley–green gram) demonstrate that integrated and organic nutrient management strategies (50% organic + 50% inorganic and 100% organic amendments) significantly enhance soil water retention, SOC sequestration, and N availability compared to conventional inorganic fertilization. Organic inputs such as farmyard manure, green manures, crop residues, and root exudates not only increase total C and total N but also improve soil structure, porosity, and aggregation, thereby enhancing water-holding capacity and reducing runoff. The build-up of labile C fractions and biologically active N pools in organically managed soils fosters a favorable microenvironment for root growth, microbial activity, and nutrient mineralization which all critical for efficient water and nutrient use (Dutta et al., 2024). Residue management directly modulates the soil water–C–N nexus, with significant implications for nutrient supply and SOM stabilization. The placement of crop residues whether as surface mulch or incorporated into the soil profoundly shapes the interplay among soil water, C, and N dynamics. Surface mulch conserves soil moisture by reducing evaporation, creating a more uniform and favorable water environment in the upper soil layers that enhances microbial activity and stimulates the mineralization of native organic matter. In contrast, incorporated residues decompose more rapidly due to greater contact with soil microbes and optimal moisture in the 5–10 cm depth, leading to higher short-term C mineralization but faster loss of DOC (Coppens et al., 2006).

6. Conclusion

Soil is far more than a passive growing medium, it is a dynamic, living biogeochemical interface that integrates the water, C, and N cycles, sustaining ecosystem function and human well-being. As demonstrated throughout this chapter, these three cycles are tightly interwoven. Soil water governs microbial activity and redox conditions, which in turn regulate the SOC sequestration, and N transformations such as mineralization, nitrification, and denitrification. SOC serves as the linchpin of this nexus, influencing water retention, nutrient availability, aggregate stability, and microbial diversity and below its critical thresholds, key soil functions deteriorate, undermining agricultural productivity, water quality, and climate resilience. Human activities especially intensive agriculture, deforestation, and urbanization have profoundly disrupted these coupled cycles, leading to widespread soil degradation, GHGS emissions, nutrient losses, and diminished water regulation capacity. Integrated management strategies that simultaneously address water, C, and N such as organic amendments (manure, compost, crop residues, biochar), diversified cropping systems, and residue placement optimization can restore soil health by enhancing SOC stocks, improving water infiltration and retention, and balancing N cycling to minimize losses while maximizing plant uptake. In essence, the sustainable management of the soil water–C–N nexus is not merely an agronomic objective, it is a foundational strategy for planetary health. By restoring and protecting soils, we secure the very basis of life on Earth for generations to come.

References

Amador, J. A., Görres, J. H., & Savin, M. C. (2005). Role of soil water content in the carbon and nitrogen dynamics of lumbricus terrestris L. burrow soil. *Applied Soil Ecology, 28*(1), 15–22.

An, N., Tang, C. S., Xu, S. K., Gong, X. P., Shi, B., & Inyang, H. I. (2018). Effects of soil characteristics on moisture evaporation. *Engineering geology, 239*, 126–135.

Aumtong, S., Chotamonsak, C., & Glomchinda, T. (2023). Study of the interaction of dissolved organic carbon, available nutrients, and clay content driving soil carbon storage in the rice rotation cropping system in Northern Thailand. *Agronomy, 13*(1), 142.

Ayiti, O. E., & Babalola, O. O. (2022). Factors influencing soil nitrification process and the effect on environment and health. *Frontiers in Sustainable Food Systems, 6*, Article 821994.

Azevedo, L. C. B., Bertini, S. C. B., Ferreira, A. S., Rodovalho, N. S., Ferreira, L. F. R., & Kumar, A. (2024). Microbial contribution to the carbon flux in the soil: A literature review. *Revista Brasileira de Ciência do Solo, 48*, Article e0230065.

Basset, C., Abou Najm, M., Ghezzehei, T., Hao, X., & Daccache, A. (2023). How does soil structure affect water infiltration? A meta-data systematic review. *Soil and Tillage Research, 226*, 105577.

Cao, Y., He, Z., Zhu, T., & Zhao, F. (2021). Organic-C quality as a key driver of microbial nitrogen immobilization in soil: A meta-analysis. *Geoderma, 383*, Article 114784.

Carvalho, M. L., Maciel, V. F., Bordonal, R. D. O., Carvalho, J. L. N., Ferreira, T. O., Cerri, C. E. P., & Cherubin, M. R. (2023). Stabilization of organic matter in soils: Drivers, mechanisms, and analytical Tools–Aliterature review. *Revista Brasileira de Ciência do Solo, 47*, Article e0230130.

Chen, D., Xing, W., Lan, Z., Saleem, M., Wu, Y., Hu, S., & Bai, Y. (2019). Direct and indirect effects of nitrogen enrichment on soil organisms and carbon and nitrogen mineralization in a semi-arid grassland. *Functional Ecology, 33*(1), 175–187.

Cheng, K., Xu, X., Cui, L., Li, Y., Zheng, J., Wu, W., ... & Pan, G. (2021). The role of soils in regulation of freshwater and coastal water quality. *Philosophical Transactions of the Royal Society B: Biological Sciences, 376*(1834).

Coppens, F., Garnier, P., De Gryze, S., Merckx, R., & Recous, S. (2006). Soil moisture, carbon and nitrogen dynamics following incorporation and surface application of labelled crop residues in soil columns. *European Journal of Soil Science, 57*(6), 894–905.

Crowther, T. W., Van den Hoogen, J., Wan, J., Mayes, M. A., Keiser, A. D., Mo, L., … Maynard, D. S. (2019). The global soil community and its influence on biogeochemistry. *Science, 365*(6455), Article eaav0550.

Cui, P., Chen, Z., Fan, F., Yin, C., Song, A., Li, T., … Liang, Y. (2023). Soil texture is an easily overlooked factor affecting the temperature sensitivity of N2O emissions. *Science of the Total Environment, 862*, Article 160648.

Cui, Y., Chen, X., Gao, J., Yan, B., Tang, G., & Hong, Y. (2018). Global water cycle and remote sensing big data: Overview, challenge, and opportunities. *Big Earth Data, 2*(3), 282–297.

Dai, H., Zhang, H., Li, Z., Liu, K., & Zamanian, K. (2021). Tillage practice impacts on the carbon sequestration potential of topsoil microbial communities in an agricultural field. *Agronomy, 11*(1), 60. https://doi.org/10.3390/agronomy11010060

Daunoras, J., Kačergius, A., & Gudiukaitė, R. (2024). Role of soil microbiota enzymes in soil health and activity changes depending on climate change and the type of soil ecosystem. *Biology, 13*(2), 85.

Deng, J. M. M., & Makinde, O. D. (2020). Biomechanics of surface runoff and soil water percolation. In *Defect and diffusion forum* (Vol 401, pp. 36–46). Trans Tech Publications Ltd.

Dutta, D., Meena, A. L., Bhanu, C., Ghasal, P. C., Choudhary, J., Kumar, S., … Yadav, P. (2024). Sustainable soil management for climate resilience: Long-term management effects on soil carbon sequestration and nitrogen dynamics in a semi-arid tropical inceptisol of India. *Journal of Soil Science and Plant Nutrition, 24*(3), 4407–4426.

Ewing, P. M., Tu, X., Runck, B. C., Nord, A., Chikowo, R., & Snapp, S. S. (2023). Smallholder farms have and can store more carbon than previously estimated. *Global Change Biology, 29*(6), 1471–1483.

Feifel, M., Durner, W., Hohenbrink, T. L., & Peters, A. (2024). Effects of improved water retention by increased soil organic matter on the water balance of arable soils: A numerical analysis. *Vadose Zone Journal, 23*(1), Article e20302.

Feng, J., Wang, C., Gao, J., Ma, H., Li, Z., Hao, Y., … Wan, S. (2024). Changes in plant litter and root carbon inputs alter soil respiration in three different forests of a climate transitional region. *Agricultural and Forest Meteorology, 358*, Article 110212.

Franzluebbers, A. J. (2002). Water infiltration and soil structure related to organic matter and its stratification with depth. *Soil and Tillage Research, 66*(2), 197–205.

Fu, Y., Ren, X., & Zhu, B. (2025). Responses of soil nitrogen transformation and N2O emission to soil pH and hydrothermal changes. *Agronomy, 15*(5), 1005.

Gavrilescu, M. (2021). Water, soil, and plants interactions in a threatened environment. *Water, 13*(19), 2746.

Geissen, V., Ramos, F. Q., de, J., Bastidas-Bastidas, P., Díaz-González, G., Bello-Mendoza, R., Huerta-Lwanga, E., & Ruiz-Suárez, L. E. (2010). Soil and water pollution in a banana production region in tropical Mexico. *Bulletin of Environmental Contamination and Toxicology, 85*(4), 407–413.

Grzyb, A., Wolna-Maruwka, A., & Niewiadomska, A. (2021). The significance of microbial transformation of nitrogen compounds in the light of integrated crop management. *Agronomy, 11*(7), 1415.

Haghnazari, F., Shahgholi, H., & Feizi, M. (2015). Factors affecting the infiltration of agricultural soils. *International Journal of Agronomy and Agricultural Research, 6*(5), 21–35.

Hanif, M. N. (2023). Factors affecting nitrogen use efficiency (NUE): Meta analysis. *Türkiye Tarımsal Arastırmalar Dergisi, 10*(2), 231–242.

Hatfield, J. L., & Dold, C. (2019). Water-use efficiency: Advances and challenges in a changing climate. *Frontiers in plant science, 10*, 103.

Hayashi, K. (2022). Nitrogen cycling and management focusing on the central role of soils: A review. *Soil Science & Plant Nutrition, 68*(5–6), 514–525.

Heitkötter, J., Heinze, S., & Marschner, B. (2017). Relevance of substrate quality and nutrients for microbial C-turnover in top-and subsoil of a dystric cambisol. *Geoderma, 302*, 89–99.

Henneron, L., Kardol, P., Wardle, D. A., Cros, C., & Fontaine, S. (2020). Rhizosphere control of soil nitrogen cycling: A key component of plant economic strategies. *New Phytologist, 228*(4), 1269–1282.

Hu, Y., Ao, G., Feng, J., Chen, X., & Zhu, B. (2025). The patterns of forest soil particulate and mineral associated organic carbon characteristics with latitude and soil depth across eastern China. *Forest Ecosystems, 12*, Article 100291.

Hussain, S., Hussain, S., Guo, R., Sarwar, M., Ren, X., Krstic, D., … El-Esawi, M. A. (2021). Carbon sequestration to avoid soil degradation: A review on the role of conservation tillage. *Plants, 10*(10), 2001.

Huys, R., Poirier, V., Bourget, M. Y., Roumet, C., Hättenschwiler, S., Fromin, N., … Freschet, G. T. (2022). Plant litter chemistry controls coarse-textured soil carbon dynamics. *Journal of Ecology, 110*(12), 2911–2928.

Jílková, V., Macek, P., Angst, G., Bartuška, M., Starý, J., Šustr, V., & Devetter, M. (2024). Macrofauna amplify plant litter decomposition and stabilization in arctic soils in a warming climate. *Soil Biology and Biochemistry, 188*, Article 109245.

Jackson, R. B., Lajtha, K., Crow, S. E., Hugelius, G., Kramer, M. G., & Piñeiro, G. (2017). The ecology of soil carbon: Pools, vulnerabilities, and biotic and abiotic controls. *Annual Review of Ecology, Evolution, and Systematics, 48*(1), 419–445.

Ju, X., Gao, L., She, D., Jia, Y., Pang, Z., & Wang, Y. (2024). Impacts of the soil pore structure on infiltration characteristics at the profile scale in the red soil region. *Soil and Tillage Research, 236*, Article 105922.

Kästner, M., Miltner, A., Thiele-Bruhn, S., & Liang, C. (2021). Microbial necromass in Soils—Linking microbes to soil processes and carbon turnover. *Frontiers in Environmental Science, 9*, Article 756378.

Kögel-Knabner, I., Wiesmeier, M., & Mayer, S. (2022). Mechanisms of soil organic carbon sequestration and implications for management. *Nature Reviews Earth & Environment, 4*(3), 188–204.

Keesstra, S. D., Geissen, V., Mosse, K., Piiranen, S., Scudiero, E., Leistra, M., & van Schaik, L. (2012). Soil as a filter for groundwater quality. *Current Opinion in Environmental Sustainability, 4*(5), 507–516.

Khalifah, S., & Foltz, M. E. (2024). The ratio of denitrification end-products were influenced by soil pH and clay content across different texture classes in Oklahoma soils. *Frontiers in Soil Science, 4*, Article 1342986.

Kuśmierz, S., Skowrońska, M., Tkaczyk, P., Lipiński, W., & Mielniczuk, J. (2023). Soil organic carbon and mineral nitrogen contents in soils as affected by their pH, texture and fertilization. *Agronomy, 13*(1), 267.

Lal, R. (2020). Soil organic matter and water retention. *Agronomy Journal, 112*(5), 3265–3277.

Lal, R., Mohtar, R. H., Assi, A. T., Ray, R., Baybil, H., & Jahn, M. (2017). Soil as a basic nexus tool: Soils at the center of the food–energy–water nexus. *Current Sustainable/Renewable Energy Reports, 4*(3), 117–129.

Lei, X., Shen, Y., Zhao, J., Huang, J., Wang, H., Yu, Y., & Xiao, C. (2023). Root exudates mediate the processes of soil organic carbon input and efflux. *Plants, 12*(3), 630.

Li, Z., Tang, Z., Song, Z., Chen, W., Tian, D., Tang, S., … Niu, S. (2022). Variations and controlling factors of soil denitrification rate. *Global Change Biology, 28*(6), 2133–2145.

Li, Q., Wang, L., Fu, Y., Lin, D., Hou, M., Li, X., … Wang, Z. (2023). Transformation of soil organic matter subjected to environmental disturbance and preservation of organic matter bound to soil minerals: A review. *Journal of Soils and Sediments, 23*(3), 1485–1500.

Lipton, S., Ceja-Navarro, J. A., Philpott, S. M., Bowles, T. M., & Franco, A. L. (2025). Soil fauna impacts on carbon cycling in managed grasslands: A review. *Advances in Agronomy, 192*, 251–291.

Liu, Z., Gu, H., Yao, Q., Jiao, F., Hu, X., Liu, J., … Wang, G. (2024). Soil pH and carbon quality index regulate the biogeochemical cycle couplings of carbon, nitrogen and phosphorus in the profiles of isohumosols. *Science of the Total Environment, 922*, Article 171269.

Liu, Y., Wang, C., He, N., Wen, X., Gao, Y., Li, S., … Yu, G. (2017). A global synthesis of the rate and temperature sensitivity of soil nitrogen mineralization: Latitudinal patterns and mechanisms. *Global Change Biology, 23* (1), 455–464.

Malik, A. A., Puissant, J., Buckeridge, K. M., Goodall, T., Jehmlich, N., Chowdhury, S., … Griffiths, R. I. (2018). Land use driven change in soil pH affects microbial carbon cycling processes. *Nature Communications, 9*(1), 3591.

Malique, F., Ke, P., Boettcher, J., Dannenmann, M., & Butterbach-Bahl, K. (2019). Plant and soil effects on denitrification potential in agricultural soils. *Plant and Soil, 439*(1), 459–474.

Manzoni, S., Ahmed, M. H., & Porporato, A. (2019). Ecohydrological and stoichiometric controls on soil carbon and nitrogen dynamics in drylands. In P. D'Odorico, A. Porporato, & C. Wilkinson Runyan (Eds.), *Dryland ecohydrology.* Cham: Springer. https://doi.org/10.1007/978-3-030-23269-6_11

Manzoni, S., Chakrawal, A., Spohn, M., & Lindahl, B. D. (2021). Modeling microbial adaptations to nutrient limitation during litter decomposition. *Frontiers in Forests and Global Change, 4*, Article 686945. https://doi.org/10.3389/ffgc.2021.686945

Manzoni, S., Taylor, P., Richter, A., Porporato, A., & Ågren, G. I. (2012). Environmental and stoichiometric controls on microbial carbon-use efficiency in soils. *New Phytologist, 196*(1), 79–91.

Marschner, P. (2021). Processes in submerged soils–linking redox potential, soil organic matter turnover and plants to nutrient cycling. *Plant and Soil, 464*(1), 1–12.

Martens, D. A. (2001). *Nitrogen cycling under different soil management systems.*

Minasny, B., & McBratney, A. B. (2018). Limited effect of organic matter on soil available water capacity. *European Journal of Soil Science, 69*(1), 39–47.

Mohammadi, K., Sohrabi, Y., Heidari, G., Khalesro, S., & Majidi, M. (2012). Effective factors on biological nitrogen fixation. *African Journal of Agricultural Research, 7*(12), 1782–1788.

Mosse, K. P. M., Patti, A. F., Smernik, R. J., Christen, E. W., & Cavagnaro, T. R. (2012). Physicochemical and microbiological effects of long-and short-term winery wastewater application to soils. *Journal of Hazardous Materials, 201*, 219–228.

Murray, J. G. (2024). *Soil organic matter processes from the treetops to the clay particle: The role of climate, minerals, and organic matter chemistry in soil carbon accumulation and stability.*

Myrold, D. D. (2021). Transformations of nitrogen. In *Principles and applications of soil microbiology* (pp. 385–421). Elsevier.

Nakhavali, M., Lauerwald, R., Regnier, P., Guenet, B., Chadburn, S., & Friedlingstein, P. (2021). Leaching of dissolved organic carbon from mineral soils plays a significant role in the terrestrial carbon balance. *Global Change Biology, 27*(5), 1083–1096.

Noor, M. A., Nawaz, M. M., Hassan, M. U., Sher, A., Shah, T., Abrar, M. M., … Ma, W. (2019). Small farmers and sustainable N and P management: Implications and potential under changing climate. In *Carbon and nitrogen cycling in soil* (pp. 185–219). Singapore: Springer Singapore.

Novák, V., & Hlaváčiková, H. (2018). Infiltration of water into soil. In *Applied soil hydrology* (pp. 137–164). Cham: Springer International Publishing.

Owen, D. (2021). Law, land use, and groundwater recharge. *Stanford Law Review, 73*, 1163.

Page, K. L., Dang, Y. P., Dalal, R. C., Reeves, S., Thomas, G., Wang, W., & Thompson, J. P. (2019). Changes in soil water storage with no-tillage and crop residue retention on a vertisol: Impact on productivity and profitability over a 50 year period. *Soil and Tillage Research, 194*, Article 104319.

Pan, S. Y., Dong, C. D., Su, J. F., Wang, P. Y., Chen, C. W., Chang, J. S., ... Hung, C. M. (2021). The role of biochar in regulating the carbon, phosphorus, and nitrogen cycles exemplified by soil systems. *Sustainability, 13*(10), 5612.

Popoviciu, D. R. (2024). *Soil biology & ecology: The basics*. Bentham Science Publishers.

Porporato, A., D'odorico, P., Laio, F., & Rodriguez-Iturbe, I. (2003). Hydrologic controls on soil carbon and nitrogen cycles. I. modeling scheme. *Advances in Water Resources, 26*(1), 45–58.

Qi, P., Chen, J., Wang, X., Zhang, R., Cai, L., Jiao, Y., ... Han, G. (2023). Changes in soil particulate and mineral-associated organic carbon concentrations under nitrogen addition in China—A meta-analysis. *Plant and Soil, 489*(1), 439–452.

Raimi, M. O., Abiola, I., & Omini, D. E. (2021). Exploring how human activities disturb the balance of biogeochemical cycles: Evidence from the carbon, nitrogen and hydrologic cycles. *Nitrogen and Hydrologic Cycles. Research on World Agricultural Economy, 2*(3), 23–44.

Ramond, J. B., Jordaan, K., Díez, B., Heinzelmann, S. M., & Cowan, D. A. (2022). Microbial biogeochemical cycling of nitrogen in arid ecosystems. *Microbiology and Molecular Biology Reviews, 86*(2), e00109–e00121.

Rasmussen, C., Heckman, K., Wieder, W. R., Keiluweit, M., Lawrence, C. R., Berhe, A. A., ... Wagai, R. (2018). Beyond clay: Towards an improved set of variables for predicting soil organic matter content. *Biogeochemistry, 137*(3), 297–306.

Rawls, W. J., Pachepsky, Y. A., Ritchie, J. C., Sobecki, T. M., & Bloodworth, H. (2003). Effect of soil organic carbon on soil water retention. *Geoderma, 116*(1–2), 61–76.

Robertson, G. P., & Groffman, P. M. (2024). Nitrogen transformations. In *Soil microbiology, ecology and biochemistry* (pp. 407–438). Elsevier.

Rodrigo-Comino, J., López-Vicente, M., Kumar, V., Rodríguez-Seijo, A., Valkó, O., Rojas, C., ... Panagos, P. (2020). Soil science challenges in a new era: A transdisciplinary overview of relevant topics. *Air, Soil and Water Research, 13*, Article 1178622120977491.

Rumpel, C., & Chabbi, A. (2021). Managing soil organic carbon for mitigating climate change and increasing food security. *Agronomy, 11*(8), 1553.

Saraswathi, S. G. (2023). *Soil: The dynamic nexus of life, science and sustainability: Specially crafted concise guide for students of higher education*. Shanlax Publications.

Seidel, E. (2018). Processes of soil infiltration and water retention and strategies to increase their capacity. *Journal of Experimental Agriculture International, 20*(2), 1–14.

Shah, T., Lateef, S., & Noor, M. A. (2019). Carbon and nitrogen cycling in agroecosystems: An overview. *Carbon and Nitrogen Cycling in Soil*, 1–15.

Shahbaz, M., Kuzyakov, Y., Sanaullah, M., Heitkamp, F., Zelenev, V., Kumar, A., & Blagodatskaya, E. (2017). Microbial decomposition of soil organic matter is mediated by quality and quantity of crop residues: Mechanisms and thresholds. *Biology and Fertility of Soils, 53*(3), 287–301.

Shanafield, M., & Cook, P. G. (2014). Transmission losses, infiltration and groundwater recharge through ephemeral and intermittent streambeds: A review of applied methods. *Journal of Hydrology, 511*, 518–529.

Shi, J., Lv, J., Peng, Y., Yao, Y., Wei, X., & Wang, X. (2024). Mechanisms controlling the stability and sequestration of mineral associated organic carbon upon erosion and deposition. *Catena, 242*, Article 108119.

Singer, R., & Brown, S. (2018). Impact of soil filtration on metals, nutrients, and estrogenic activity of reclaimed water. *Journal of Environmental Quality, 47*(6), 1504–1512.

Singh, V. (2024). Biogeochemical cycles. In *Textbook of environment and ecology*. Singapore: Springer. https://doi.org/10.1007/978-981-99-8846-4_5

Smith, K. A., Ball, T., Conen, F., Dobbie, K. E., Massheder, J., & Rey, A. (2018). Exchange of greenhouse gases between soil and atmosphere: Interactions of soil physical factors and biological processes. *European Journal of Soil Science, 69*(1), 10–20.

Song, L., Li, Z., & Niu, S. (2021). Global soil gross nitrogen transformation under increasing nitrogen deposition. *Global Biogeochemical Cycles, 35*(1), Article e2020GB006711.

Suliman, M., Scaini, A., Manzoni, S., & Vico, G. (2024). Soil properties modulate actual evapotranspiration and precipitation impacts on crop yields in the USA. *Science of the Total Environment, 949*, Article 175172.

Togbévi, Q. F., Van Der Ploeg, M., Tohoun, K. A., Agodzo, S. K., & Preko, K. (2022). Assessing the effects of anthropogenic land use on soil infiltration rate in a tropical West African watershed (ouriyori, benin). *Applied and Environmental Soil Science, 2022*(1), Article 8565571.

Van Groenigen, J. W., Huygens, D., Boeckx, P., Kuyper, T. W., Lubbers, I. M., Rütting, T., & Groffman, P. M. (2015). The soil N cycle: New insights and key challenges. *Soil, 1*(1), 235–256.

Vereecken, H., Huisman, J. A., Hendricks Franssen, H. J., Brüggemann, N., Bogena, H. R., Kollet, S., ... Vanderborght, J. (2015). Soil hydrology: Recent methodological advances, challenges, and perspectives. *Water Resources Research, 51*(4), 2616–2633.

Villarino, S. H., Pinto, P., Jackson, R. B., & Piñeiro, G. (2021). Plant rhizodeposition: A key factor for soil organic matter formation in stable fractions. *Science Advances, 7*(16), Article eabd3176.

Wang, C., & Kuzyakov, Y. (2023). Energy use efficiency of soil microorganisms: Driven by carbon recycling and reduction. *Global Change Biology, 29*(22), 6170–6187.

Wang, C., & Kuzyakov, Y. (2024). Soil organic matter priming: The pH effects. *Global Change Biology, 30*(6), Article e17349.

Wang, J., Liu, Y., Cao, W., Li, W., Wang, X., Zhang, D., Shi, S., Pan, D., & Liu, W. (2020). Effects of grazing exclusion on soil respiration components in an alpine meadow on the north-eastern qinghai-tibet plateau. *Catena, 194*, Article 104750.

Wu, H., Cui, H., Fu, C., Li, R., Qi, F., Liu, Z., ... Qiao, M. (2024). Unveiling the crucial role of soil microorganisms in carbon cycling: A review. *Science of the Total Environment, 909*, Article 168627.

Xia, Y. Q., & Shao, M. A. (2008). Soil water carrying capacity for vegetation: A hydrologic and biogeochemical process model solution. *Ecological Modelling, 214*(2–4), 112–124.

Xie, Y., Zhang, J., Meng, L., Müller, C., & Cai, Z. (2015). Variations of soil N transformation and N2O emissions in tropical secondary forests along an aridity gradient. *Journal of Soils and Sediments, 15*(7), 1538–1548.

Xu, S., Sayer, E. J., Eisenhauer, N., Lu, X., Wang, J., & Liu, C. (2021). Aboveground litter inputs determine carbon storage across soil profiles: A meta-analysis. *Plant and Soil, 462*(1), 429–444.

Yang, D., Yang, Y., & Xia, J. (2021). Hydrological cycle and water resources in a changing world: A review. *Geography and Sustainability, 2*(2), 115–122.

Yang, N., Zhang, M., Pan, X., Dang, P., Wang, S., Han, X., ... Qin, X. (2025). Characteristics of changes to POC and MAOC after straw returning in China: A meta-analysis. *Land Degradation and Development, 36*(15), 5199–5211.

Yang, F., Zhang, G. L., Yang, J. L., Li, D. C., Zhao, Y. G., Liu, F., ... Yang, F. (2014). Organic matter controls of soil water retention in an alpine grassland and its significance for hydrological processes. *Journal of Hydrology, 519*, 3086–3093.

Yansheng, C., Fengliang, Z., Zhongyi, Z., Tongbin, Z., & Huayun, X. (2020). Biotic and abiotic nitrogen immobilization in soil incorporated with crop residue. *Soil and Tillage Research, 202*, Article 104664.

Yao, Y., Shao, M., Fu, X., Wang, X., & Wei, X. (2019). Effect of grassland afforestation on soil N mineralization and its response to soil texture and slope position. *Agriculture, Ecosystems and Environment, 276*, 64–72.

Zaman, W., Ayaz, A., & Puppe, D. (2025). Biogeochemical cycles in plant–soil systems: Significance for agriculture, interconnections, and anthropogenic disruptions. *Biology, 14*(4), 433.

Zewide, I. (2021). Review paper on effect of natural condition on soil infiltration. *Internation Journal of Green Chemistry, 7*(1), 34–41.

Zhang, C., Cai, Y., Zhang, T., He, T., Li, J., Li, X., & Zhao, Q. (2022a). Litter removal increases the plant carbon input to soil in a Pinus massoniana plantation. *European Journal of Forest Research, 141*(5), 833–843.

Zhang, B., Chen, L., Jin, S., Guo, Q., & Hou, J. (2022b). The influence of plants on the migration and transformation of nitrogen in plant-soil systems: A review. *Journal of Soil Science and Plant Nutrition, 22*(4), 4084–4102.

Zhang, F., Chen, X., Yao, S., Ye, Y., & Zhang, B. (2022c). Responses of soil mineral-associated and particulate organic carbon to carbon input: A meta-analysis. *Science of the Total Environment, 829*, Article 154626.

Zhang, Q., Li, Y., Wang, M., Wang, K., Meng, F., Liu, L., … Zhang, F. (2021). Atmospheric nitrogen deposition: A review of quantification methods and its spatial pattern derived from the global monitoring networks. *Ecotoxicology and Environmental Safety, 216*, Article 112180.

Zhang, X., Ward, B. B., & Sigman, D. M. (2020). Global nitrogen cycle: Critical enzymes, organisms, and processes for nitrogen budgets and dynamics. *Chemical Reviews, 120*(12), 5308–5351.

Zheng, H., Miao, C., Huntingford, C., Tarolli, P., Li, D., Panagos, P., … Van Oost, K. (2025). The impacts of erosion on the carbon cycle. *Reviews of Geophysics, 63*(1), Article e2023RG000829.

Zhu, Q., Castellano, M. J., & Yang, G. (2018). Coupling soil water processes and the nitrogen cycle across spatial scales: Potentials, bottlenecks and solutions. *Earth-Science Reviews, 187*, 248–258.

Zhu, X., Zhang, W., Chen, H., & Mo, J. (2015). Impacts of nitrogen deposition on soil nitrogen cycle in forest ecosystems: A review. *Acta Ecologica Sinica, 35*(3), 35–43.

Zhuang, W., Liu, M., Wu, Y., Ma, J., Zhang, Y., Su, L., … Fu, S. (2023). Litter inputs exert greater influence over soil respiration and its temperature sensitivity than roots in a coniferous forest in North-South Transition Zone. *Science of the Total Environment, 886*, Article 164009.

CHAPTER

Microbial drivers of nutrient cycling (N, P, C) and soil formation in arid and semiarid ecosystems

2

Mahvan Hassanzadeh Bashtian and Alireza Karimi
Department of Soil Science, College of Agriculture, Ferdowsi University of Mashhad, Mashhad, Iran

1. Introduction

Approximately 40% of the total surface area of the world is classified as arid or semiarid, which is considered as having a lack of rain and high evaporation from the land and large temperature changes (Reynolds et al., 2007). Environments that have these constraints result in a unique collection of microorganisms, including the soil, that have adapted to exist with limited access to water and nutrients. In spite of these extreme conditions, the microorganisms in these areas are of vital importance in supporting the plant's survival and productivity, providing the necessary nutrients to the plants, and facilitating the formation of soil (Belnap, 2003).

Nutrient cycling in dryland regions is typically pulse-driven as short-term rainy (wet) periods catalyze the high level of microbial activity, followed by long-term periods when there is little if any microbial activity (Austin et al., 2004). The rapid pulse-driven cycling of nutrients has implications for the development and stabilization of soil organic carbon (SOC) and SOC's availability (Maestre et al., 2016). Therefore, understanding how these processes occur is critical for determining how dryland soils will react with regard to land use and climate changes, as there are predictions of increased aridity in many countries around the world (Maestre et al., 2016).

Microorganisms present in the soil ecosystems play a pivotal role as drivers of the biogeochemistry of these dry ecosystems. Microorganisms involved in nitrogen cycling include a variety of non-symbiotic N2-fixing bacteria (e.g., Azotobacter) and cyanobacteria in biological soil crusts (e.g., Nostoc), as well as obligate symbiotic rhizobia that utilize atmospheric nitrogen during the moisture pulses of event-based precipitation (Pérez-García et al., 2016; Barger et al., 2016). The availability of phosphorus from the soil is mediated by microorganisms through their ability to solubilize mineral phosphates through the secretion of Organic Acids and Phosphatases, and through mycorrhizal fungi, which allow Polar routes for the roots to obtain phosphorus from soils with restricted diffusion of Phosphorus and Water (Barea et al., 2005; Rodríguez & Fraga, 1999). The microbial decomposition of organic substrates generates carbon cycling and is limited by low soil moisture. However, specialized decomposer microorganisms such as Actinobacteria and Saprotrophic fungi that have adaptations to withstand drought stress facilitate the movement of carbon nutrients and contribute to soil organic matter to store additional carbon in the soil system (Collins et al., 2008).

The Soil Nexus. https://doi.org/10.1016/B978-0-443-45286-4.00019-5

In addition to nutrient cycling, microorganisms directly affect the weathering of rocks to create a medium for soil development through Bioweathering, production of aggregate-stabilizing extracellular polymers, and production of erosion-resistant biological soil crusts (Mager & Thomas, 2011; Rossi & De Philippis, 2015). In dry ecosystems, biological soil crusts serve as a critical factor for nutrient cycling and soil stabilization. Biological soil crusts (biocrusts) are made of a complex assemblage of organisms containing cyanobacteria, algae, microfungi, lichens, and mosses (Weber et al., 2016). The findings from Iranian drylands indicate that biocrusts are important to improving the quality of soils in the semiarid regions of Iran through enhancing three of the main attributes of soil quality, nutrient availability, organic matter content, and aggregate stability (Bashtian et al., 2024; Hassanzadeh Bashtian et al., 2024). Recent advances in technology, specifically hyperspectral remote sensing and drone imagery, have enabled us to monitor the distribution and physiological condition of biocrust over large areas. In addition, the use of in situ sensor networks for real-time measurements of soil moisture, temperature, and CO2 fluxes has made it possible for us to accurately track the dynamics of "pulse" events of biocrust microbial communities (Blanco-Sacristán et al., 2021; Ramson et al., 2021). In view of their critical importance to the function of ecosystems, soil microbial communities should be considered as central components of restoration strategies for drylands, climate adaptation, and sustainable farming in arid and semiarid environments worldwide.

2. Importance of microbial communities in ecosystem function

2.1 Nitrogen (N) cycling

The maximum amount of energy available to produce plant biomass and productivity within a dryland ecosystem will always be limited by nitrogen. The movement of nitrogen within the dryland ecosystem is achieved through several biological processes (transformations) by the vegetative states of multiple communities of microbes, beginning with the biological fixation of atmospheric gas and ending with the activity of hydrogenotrophic microbes removing nitrogen from the dryland ecosystem. The transformation and movement of nitrogen will occur in niche areas within the larger ecosystems of drylands. Niche areas in dryland ecosystems consist of biological crusts, root zones in the subsoil, and colonized rock outcrops, and possibly a small number of other unique microhabitats (Elbert et al., 2012; Ramond et al., 2022).

2.1.1 Key microbial processes and players

Biological nitrogen fixation (BNF): The majority of the nitrogen fixed biologically within dryland ecosystems is performed by cyanobacteria (Nostoc spp. and Microcoleus spp.) and free-living diazotrophs (Azotobacter, Azospirillum and Clostridia), all of which can perform BNF within biological crusts. In dryland BNF systems, the timing and amount of BNF are determined by available water due to either rainfall or groundwater recharge (Elbert et al., 2012; Ramond et al., 2022). A small amount of BNF occurs in drylands through associated organisms of plants (plant-microbial symbionts), such as through legumes by symbiotic bacteria that form nodules on plant roots and through nodulation at root base locations (resource islands) (Ramond et al., 2022).

Nitrification: The process of ammonia oxidation is attributed to both ammonia-oxidizing bacteria (AOB), such as Nitrosomonas and Nitrosospira, and archaea (AOA), such as Nitrososphaera. However, organisms known as comammox (complete ammonia oxidizers from the genus Nitrospira) have provided evidence for the ability of these organisms to completely oxidize ammonia (NH_4^+) to nitrate

(NO_3^-) through nitrite (NO_2^-) via a single organism. In oligotrophic and low-moisture soil environments, AOA (and in some cases comammox) (Fig. 2.1) dominate nitrification due to the high substrate affinity and ability of AOA to withstand low-nutrient environments (Kits et al., 2017). The relative contributions of AOA, AOB, and comammox to nitrification change in response to soil pH, availability of substrates, and pulses of moisture (Chisholm et al., 2025; Li et al., 2025).

Denitrification and Nitrogen Loss: Heterotrophic denitrifiers (e.g., Pseudomonas, Paracoccus, members of Alcaligenes, and various Firmicutes) eventually reduce NO_3^-/NO_2^- to N_2, with N_2O as a byproduct. Denitrification tends to be spatially distributed and contingent upon the presence of anaerobic microsites stimulated by rewetting events or developed via the formation of soil aggregates or rhizosphere hotspots (Paracoccus and Pseudomonas are common denitrifying genera). Other forms of denitrification, such as fungal denitrification and chemodenitrification (abiotic), may also be of some importance, especially after a soil is re-wetted (Ramond et al., 2022).

Ammonification/mineralization: The organic nitrogen is mineralized by heterotrophic bacteria and fungi through their enzymes (proteases and peptidases), which in turn release NH_4^+. The

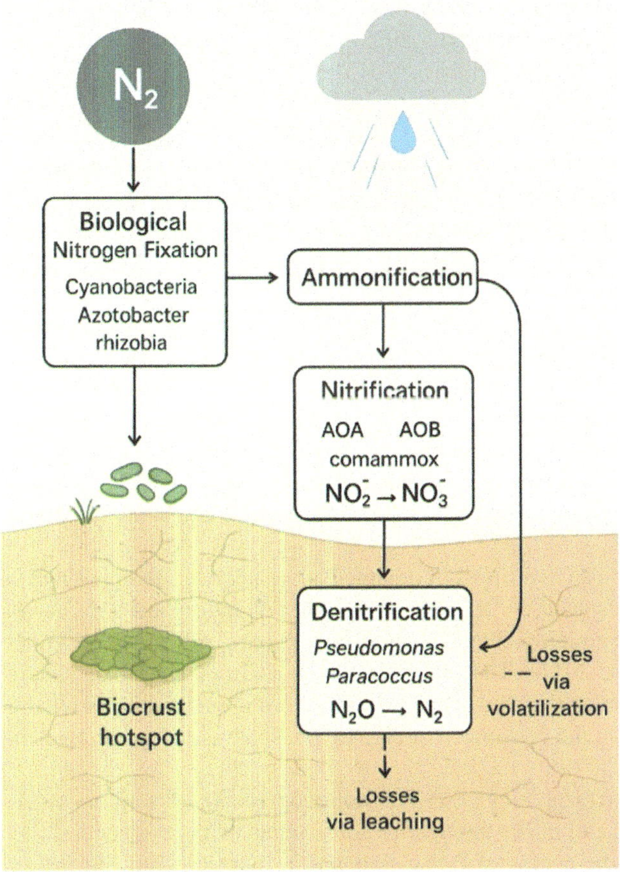

FIGURE 2.1

Conceptual model of microbial-mediated nitrogen cycling in arid and semiarid soils, highlighting biological nitrogen fixation (BNF) by cyanobacteria and free-living diazotrophs, nitrification by ammonia-oxidizing archaea (AOA), bacteria (AOB), and comammox Nitrospira, and denitrification by heterotrophic bacteria. Arrows indicate transformation pathways; dashed lines represent gaseous or leaching losses.

Adapted from Elbert, W., Weber, B., Burrows, S., Steinkamp, J., Büdel, B., Andreae, M. O., & Pöschl, U. (2012). Contribution of cryptogamic covers to the global cycles of carbon and nitrogen. Nature Geoscience, 5(7), 459-462. https://doi.org/10.1038/ngeo1486 and Ramond, J. B., Jordaan, K., Díez, B., Heinzelmann, S. M., & Cowan, D. A. (2022). Microbial biogeochemical cycling of nitrogen in arid ecosystems. Microbiology and Molecular Biology Reviews, 86(2), e00109-21. https://doi.org/10.1128/mmbr.00109-21.

microorganisms involved are sensitive to the prevailing environmental conditions and substrate quality, which determines their activities (Collins et al., 2008; Ramond et al., 2022).

Pulse dynamics and coupling/decoupling: In the case of dryland, the nitrogen cycling process is extremely pulsed. Small rainfall events frequently facilitate mineralization and nitrification through microbial action but may not support the uptake of plants, while larger pulses can lead to the temporary connection between microbial transformations and plant demand. There is a threshold behavior in the aridity gradients (decoupling of plant and microbial dynamics in nitrogen below certain aridity indices) (Maestre et al., 2015).

2.1.2 Ecological and management implications

Microbial nitrogen fixation and retention by biocrusts, as well as plant-microbe interactions, are key processes that support nitrogen inputs in arid regions, and restoration practices usually focus on either crust restoration or planting of legumes to enhance nitrogen availability (Elbert et al., 2012; Ramond et al., 2022).

Nitrous oxide (N_2O) emissions from rewetting induced denitrification are an important but spatiotemporally variable source of greenhouse gases in arid regions (pulse events) that have been misunderstood in the past. Microbial gene abundance (amoA, nirK/nirS, nosZ) studies help to improve predictions of N transformations and losses (Ramond et al., 2022).

2.2 Phosphorus (P) cycling

The amount of phosphorus that can be utilized in arid areas is influenced more by the geochemical processes, such as parent materials (calcium-bound P in calcareous soils) and dust input, than by leaching. Thus, microbial processes that liberate and recycle P are of paramount importance for plant nutrition in places where the liberation is slow (Delgado-Baquerizo et al., 2013; Richardson & Simpson, 2011; Rodríguez & Fraga, 1999).

2.2.1 Key microbial processes and players

Phosphate-solubilizing microorganisms (PSM): Several genera of bacteria, for instance, Pseudomonas, Bacillus, and certain Rhizobium and Burkholderia, along with fungi and yeasts, for example, Penicillium and Aspergillus, have the ability to solubilize the mineral P by the secretion of low-molecular-weight organic acids (gluconic and oxalic), proton extrusion, siderophore production, and via extracellular phosphatases (Richardson & Simpson, 2011; Rodríguez & Fraga, 1999; Sharma et al., 2013).

Phosphatases and organic P turnover: The hydrolysis of organic P (e.g., phytate, phospholipids) is done by microbial alkaline and acid phosphatases, and the release of orthophosphate (Pi) occurs. Phosphatase activity has a strong relation to microbial biomass, moisture pulses, and plant P demand in resource islands (Richardson & Simpson, 2011). The molecular level study of microbial alkaline phosphatase activity is often done by targeting the phoD gene, which is a common marker for alkaline phosphatase–coding microbes that facilitate the conversion of organic phosphorus into plant-available inorganic P in soils (Ragot et al., 2017).

Mycorrhizal fungi: Arbuscular mycorrhizal fungi (AMF; Glomeromycota) as well as several ectomycorrhizal types increase the P uptake of plants through the soil volume that is extended by their presence (hyphal foraging) and through the access of P that is bound in both organic and mineral fractions (Fig. 2.2). AMF can play a very important role in drylands since water around the root's limits plant access to soil microsites and then the diffusion of Pi is also a limiting factor (Smith & Read, 2010; Van Der Heijden et al., 2016).

Biocrusts and dust cycling: P is immobilized, concentrated, and released by biocrust communities in a later stage; atmospheric dust deposition is considered an important P source for drylands, and microbial activity can help to mobilize it (Delgado-Baquerizo et al., 2013). In semiarid and arid regions with alkaline soils, such as the Iranian plateaus and the American Southwest, the predominant form of phosphorus is calcium-bound minerals, which include hydroxyapatite and fluorapatite. In these environments, the organic acids produced by biofilm-forming microorganisms and by phosphate-solubilizing microorganisms not only decrease pH but also take a critical role through chelating Ca^{2+} ions, thus making Ca–P complexes unstable and freeing the inorganic phosphate (Pi) that is bioavailable (Richardson & Simpson, 2011; Rodríguez & Fraga, 1999).

Environmental controls: Soil pH, Ca and Fe content (which bind phosphate), soil moisture pulses, and root exudates control the success of microbial P-mobilization. Calcareous soils, which are prevalent in many semiarid areas, retain phosphorus as Ca-phosphates, which usually require either acidification or organic ligand production for solubilization.

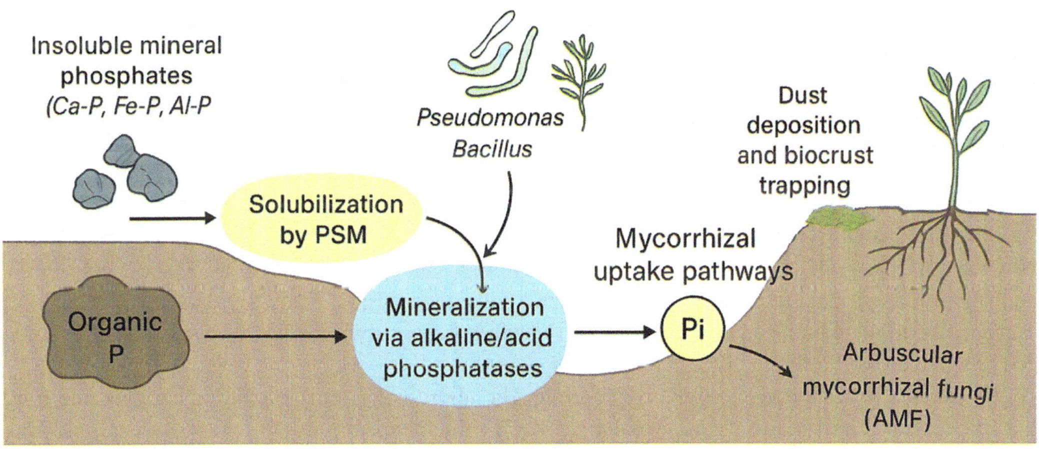

FIGURE 2.2

Conceptual model of microbial phosphorus cycling in arid and semiarid ecosystems. Phosphate-solubilizing microorganisms (PSM) such as Pseudomonas and Bacillus release inorganic phosphate (Pi) from mineral complexes via organic acids and enzymes. Organic P is mineralized by phosphatases. Arbuscular mycorrhizal fungi (AMF) extend root access to P beyond depletion zones. Biocrusts trap aeolian dust, providing additional P inputs.

Adapted from Richardson, A. E., & Simpson, R. J. (2011). Soil microorganisms mediating phosphorus availability update on microbial phosphorus. Plant physiology, 156(3), 989-996. https://doi.org/10.1104/pp.111.175448, and Delgado-Baquerizo, M., Maestre, F. T., Gallardo, A., Bowker, M. A., Wallenstein, M. D., Quero, J. L., ... & Zaady, E. (2013). Decoupling of soil nutrient cycles as a function of aridity in global drylands. Nature, 502(7473), 672-676. https://doi.org/10.1038/nature12670.

2.2.2 Ecological and management implications

Inoculation with P-solubilizing strains that are effective and management that encourages the development of mycorrhizal communities (such as reducing tillage and maintaining plant diversity) are potential P-limited drylands restoration tools (Richardson & Simpson, 2011; Sharma et al., 2013). Long-term N enrichment (through either fertilization or deposition) can have an impact on P cycling by encouraging the production of microbial phosphatase or altering plant distribution, thus changing P limitation dynamics.

Biocrust communities can also help to regulate P supply by getting phosphorus from dust and keeping it in the surface layers. Research conducted in semiarid regions shows that biofilms contribute positively to soil quality, as they help to stabilize soil structure, increase P availability, and promote microbial biomass. They are crucial for the P cycle to occur amid extreme environmental conditions, which is in line with Bashtian et al. (2024) findings.

2.3 Carbon (C) cycling

The C cycle in drylands is an intricate process. It involves inputs from autotrophs (primarily vascular plants) and is regulated by microbial fixation and decomposition, which ultimately govern soil organic carbon (SOC) dynamics. The soils in arid regions generally have lower SOC than mesic regions, but they still contribute significantly to the total carbon stored in the soil, and the presence of cryptogamic covers can increase the amount of C fixed (Elbert et al., 2012; Garcia-Pichel, 2023). Respiration of microorganisms in arid and calcareous soils has a double effect on soil inorganic carbon dynamics since it raises the partial pressure of the CO_2 in the soil pores, leading to the precipitation of carbonates such as calcite during drying phases. This process links carbon cycling directly to soil formation and long-term carbon storage in the form of soil inorganic carbon (SIC) (Monger, 1999; Zamanian et al., 2016).

2.3.1 Key microbial processes and players

Autotrophic fixation by biocrusts: The cyanobacteria, together with phototrophic algae present in the biocrusts, fix CO_2 and then release some of the labile C onto the surface soils. The cyanobacteria's contribution to producing extracellular polymeric substances (EPS) leads to the deposition of polysaccharide carbon, which can then be temporarily stabilized or subsequently decomposed (Elbert et al., 2012; Rossi & De Philippis, 2015).

Decomposition and heterotrophic respiration: The microbial groups, including bacteria and fungi, that process the dead matter (e.g., Actinobacteria, Firmicutes, saprotrophic Ascomycota, and Basidiomycota) are the main contributors of organic matter to the soil through their mediating effects on litterfall and necromass. Soil decomposition in drylands is severely limited by moisture and takes place mostly during the wetting events, which explains the existence of microbial communities in arid soils that are often made up of oligotrophic and desiccation-tolerant taxa (Bastida et al., 2017; Fierer et al., 2009).

Soil organic carbon stabilization: According to the Microbial Carbon Pump theory, microbial remains (necromass) account for the bulk of long-term soil organic carbon (SOC) stocks, and they are sometimes even more significant than the undecomposed plant litter contribution. The role of EPS in dry and semi-dry soils can be characterized as "living" in the sense that they are dynamic and thus support the microbial community in the root zone, specifically biocrust-associated cyanobacteria and bacteria communities, through the processes of binding mineral particles, which leads to aggregate formation and eventually to organic matter being physically protected during the moisture pulses. On the contrary, the microbial necromass of the past consists of cell wall fragments and intracellular biomolecules that can be grouped into organo-mineral associations in such a way that they last beyond wet-dry cycles and thus act as a "permanent" stabilizer of SOC in dryland soils (Kallenbach et al., 2016; Schmidt et al., 2011).

Microbial carbon use efficiency (CUE) and low-moisture dynamics: Microorganisms in dryland areas frequently have very rapid turnover rates. For instance, they can quickly uptake water and have low CUE after pulses, or they can have slow metabolism and higher CUE in oligotrophic niches. The ratio of differences in microbial CUE and biomass formation will be the factor deciding whether the SOC will be increased or lost after the pulses (Manzoni et al., 2019).

Biocrusts & aggregate formation: The surface particles become agglomerated into stable aggregates through the activities of EPS and filamentous cyanobacteria, and thus the soil is less prone to erosion and more capable of holding organic matter; this process not only causes an increase in C inputs but also leads to SOC stabilization (Fig. 2.3) (Rossi & De Philippis, 2015).

2.3.2 Ecological and management implications

The measures of protecting and restoring biocrusts, keeping plant cover, and controlling land use to minimize disturbance can be very influential in enhancing the factors of C sequestration and surface stability (Elbert et al., 2012; Garcia-Pichel, 2023; Butterbach-Bahl).

The impacts of climate change (changing the timing and intensity of precipitation) may alter the dynamics of the pulse and thus alter whether dryland soils act as net carbon sources or sinks.

In summary, the cycling of nitrogen, phosphorus, and carbon in arid and semi-arid ecosystems is governed by a diverse consortium of microorganisms, each performing specialized functions. Microbial groups with their main roles and important examples are shown in Table 2.1. Such functional diversity guarantees the resilience of the ecosystems and also highlights the soil microorganisms' indispensable role in nutrient balancing even under extreme environmental conditions.

3. Unique challenges and microbial adaptations in arid and semi-arid ecosystems

Arid and semiarid ecosystems require special physicochemical soil conditions that have the presence of high salinity in some areas, low organic matter content and poor mobility of nutrients, which directly affect the number and type of microorganisms as well as their functions (Pointing & Belnap, 2012). One of the main problems for microbial life in such regions is the intermittent resource availability. Nutrient and water pulses are rare and of short duration, which controls the cycles of microbial activity (Austin et al., 2004). During dry periods, a lot of microorganisms go into a dormant state or form highly resistant structures such as spores, cysts, and akinetes (Schimel, 2018). When moisture is provided, the metabolic activity quickly resumes, resulting in the occurrence of intense nutrient mineralization and transformation events. Heat is another factor that modifies the adaptability of microbes. Heat-resistant bacteria and fungi synthesize heat shock proteins to ensure proper functioning of cells during the daily heat stress (Alsharif et al., 2020; Parasar et al., 2024), whereas cold-adaptation methods, such as the production of antifreeze proteins, allow one to survive through the cold nights or during the freeze-thaw cycle (Pastore et al., 2023; D'Amico et al., 2006). Another

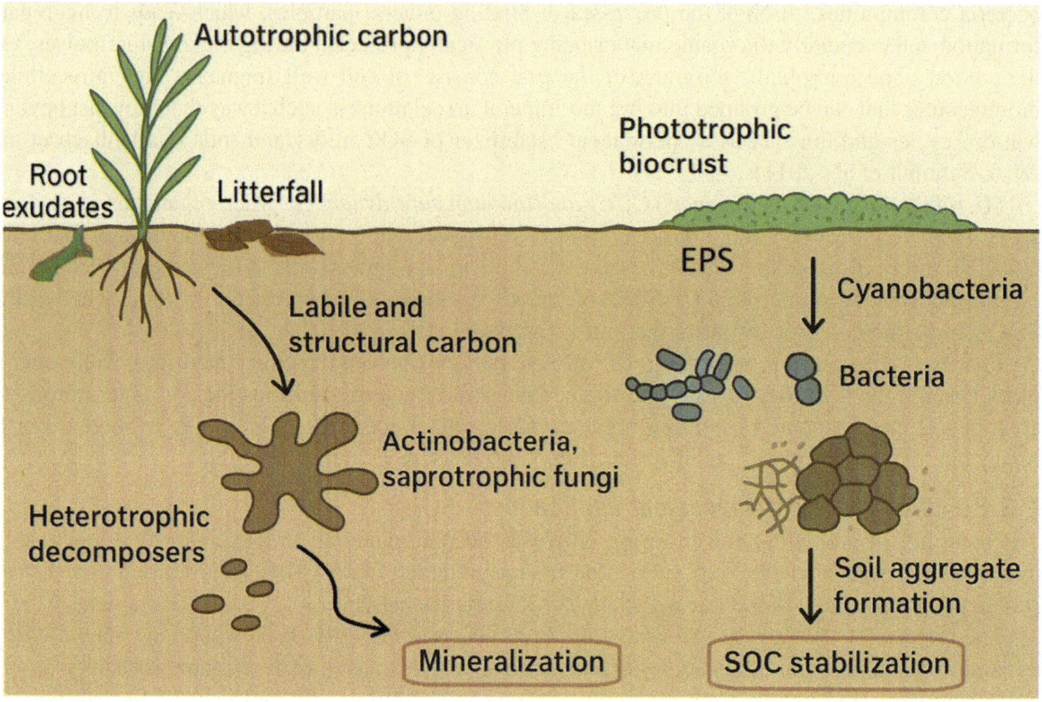

FIGURE 2.3

Conceptual model of microbial carbon cycling in arid and semiarid ecosystems. Autotrophic carbon inputs from vascular plants and phototrophic biocrusts contribute labile and structural carbon to soils. Heterotrophic decomposers (e.g., Actinobacteria, saprotrophic fungi) mineralize organic matter during moisture pulses. Extracellular polymeric substances (EPS) from cyanobacteria and bacteria promote soil aggregate formation, enhancing soil organic carbon (SOC) stabilization.

Adapted from Elbert, W., Weber, B., Burrows, S., Steinkamp, J., Büdel, B., Andreae, M. O., & Pöschl, U. (2012). Contribution of cryptogamic covers to the global cycles of carbon and nitrogen. Nature Geoscience, 5(7), 459-462. https://doi.org/10.1038/ngeo1486 and Rossi, F., & De Philippis, R. (2015). Role of cyanobacterial exopolysaccharides in phototrophic biofilms and in complex microbial mats. Life, 5(2), 1218-1238. https://doi.org/10.3390/life5021218, and Garcia-Pichel, F. (2023). The microbiology of biological soil crusts. Annual Review of Microbiology, 77(1), 149-171. https://doi.org/10.1146/annurev-micro-032521-015202.

crucial difficulty is the lack of nutrients, especially nitrogen and phosphorus. The presence of low organic matter in the soils is a major cause of limited nitrogen input in arid soils, whereas strong phosphate adsorption on calcium carbonate and iron oxide minerals renders phosphorus bioavailability very low (Uygur, 2009; Von Wandruszka, 2006).

Microorganisms have evolved adaptive strategies such as:

- Producing exopolysaccharides to maintain their moisture and to capture particles of dust that are rich in nutrients (Mager & Thomas, 2011).

Table 2.1 Key microbial groups involved in nutrient cycling (N, P, C) in arid and semiarid ecosystems.

Nutrient	Microbial groups (example genera/ functional groups)	Main function(s)	References
Nitrogen (N)	*Nostoc, Microcoleus* (cyanobacteria); *Azotobacter, Azospirillum* (free-living diazotrophs); rhizobia (legume symbionts)	Biological N fixation (BNF) — introduce new N to soils, especially via biocrusts and root nodules; support plant N in resource islands.	Elbert et al., 2012; Ramond et al., 2022
	Ammonia-oxidizing archaea (AOA; *Nitrososphaera*), ammonia-oxidizing bacteria (AOB; *Nitrosomonas, Nitrosospira*), comammox (*Nitrospira*)	Nitrification ($NH_4^+ \rightarrow NO_2^- \rightarrow NO_3^-$); control NO_3^- availability and legacy N speciation.	Lehtovirta-Morley et al., 2011; Kits et al., 2017
	Pseudomonas, Paracoccus, diverse denitrifying taxa	Denitrification ($NO_3^- \rightarrow N_2O \rightarrow N_2$) — gaseous N losses; N_2O source during wetting events.	Butterbach-Bahl et al., 2013
Phosphorus (P)	*Pseudomonas, Bacillus, Rhizobium, Burkholderia* (PSB); *Penicillium, Aspergillus* (fungal PSMs)	Phosphate solubilization via organic acid production; mineral P mobilization; enzyme-mediated hydrolysis of organic P.	Rodríguez & Fraga, 1999; Richardson & Simpson, 2011; Sharma et al., 2013
	Arbuscular mycorrhizal fungi (AMF; Glomeromycota) and ectomycorrhizal fungi	Hyphal foraging and direct plant P transfer; enhance plant P uptake in diffusion-limited soils.	Smith & Read, 2010; Lekberg et al., 2013; Van Der Heijden et al., 2016
	Biocrust phototrophs and heterotrophs (cyanobacteria, microalgae, heterotrophic bacteria)	Dust and biocrust P capture, short-term immobilization and later release; stabilization of P in surface horizon.	Delgado-Baquerizo et al., 2013; Garcia-Pichel, 2023
Carbon (C)	Cyanobacteria (e.g., *Microcoleus, Nostoc*), cryptogams (lichens, mosses)	Autotrophic C fixation at soil surface; EPS production that contributes labile and structural C to soils.	Elbert et al., 2012; Rossi & De Philippis, 2015; Garcia-Pichel, 2023
	Actinobacteria (e.g., *Streptomyces* spp.), diverse saprotrophic bacteria	Decomposition of complex polymers (cellulose, chitin, lignin precursors) — early	Fierer et al., 2009; Mohammadipanah & Wink, 2016

Continued

Table 2.1 Key microbial groups involved in nutrient cycling (N, P, C) in arid and semiarid ecosystems.—cont'd

Nutrient	Microbial groups (example genera/ functional groups)	Main function(s)	References
		stages of decomposition; extracellular enzyme production.	
	Saprotrophic fungi (Ascomycota, Basidiomycota)	Decomposition of recalcitrant organic matter; formation of stable microbial necromass contributing to SOC.	Smith & Read, 2010; Bardgett & van der Putten, 2014
	EPS-producing bacteria & cyanobacteria	Aggregate formation and SOC physical protection; increase water retention and stabilize surface soils.	Rossi & De Philippis, 2015; Mager & Thomas, 2011

- Increasing their secretion of extracellular enzymes (such as alkaline phosphatases and proteases) to break down tough organic material and make nutrients, especially phosphorus and nitrogen, available to them under conditions of chronic nutrient deficiency (Richardson & Simpson, 2011; Vranová et al., 2013).
- Forming partnerships with plants through mycorrhiza or rhizobia to obtain nutrients that would be otherwise inaccessible.
- Biocrusts are very adaptable to these harsh conditions. They develop surface layers that are cohesive and thus reduce erosion, improve infiltration, and significantly contribute to nutrient inputs through biological nitrogen fixation, phosphorus trapping, and carbon sequestration (Belnap, 2003; Weber et al., 2016).

These characteristics present an instance of a pulsed storage pattern in dryland ecology, where microorganisms store resources in wet periods, although drought is still prolonged, thus sustaining life (Austin et al., 2004). Getting to know these challenges and adaptations is vital for making predictions of the microbial nutrient cycling changes in the future due to global change drivers such as increased aridity, warming, and land use intensification.

3.1 Exopolysaccharide production

In arid and semiarid soils, exopolysaccharide production is a widely used microbial strategy, especially among cyanobacteria, Actinobacteria, and some Proteobacteria. The high-molecular-weight polymers are secreted out of the cell and create a hydrated matrix that effectively shields the cells from desiccation, UV radiation, and temperature fluctuations (Mager & Thomas, 2011). In biocrusts, EPS plays a dual role:

1) Water retention: exopolysaccharide has the capacity to hold water equivalent to several times its own weight, which prolongs the period of moisture after rain or dew (Rossi & De Philippis, 2015). 2) Soil aggregation: exopolysaccharide attaches to mineral particles, which makes aggregates more stable, prevents erosion, and provides microsites that are safe for organic matter from decomposition (Chenu, 1995). Additionally, exopolysaccharide captures and holds airborne dust that is rich in mineral nutrients (P and micronutrients), effectively acting as a nutrient capture mechanism in dust-dominated arid regions such as Iran (Karimi et al., 2024; Svirčev et al., 2019).

3.2 Enzyme secretion

Enzyme activity in drylands is mostly dependent on moisture and its highest level is reached right after wetting (Collins et al., 2008). Certain microbes produce enzymes that are firmly attached to soil particles and thus are capable of surviving through extremely dry conditions (Kelley et al., 2011; Olagoke et al., 2020). Even in drylands with limited nutrients, microorganisms still offer a range of extracellular enzymes to get the nutrients from otherwise inaccessible sources:

- Alkaline phosphatases release inorganic phosphates from organic P compounds in soils with high pH.
- Proteases break down proteins into amino acids, thus they play a part in nitrogen mineralization.
- Fungi and Actinobacteria are the main ones producing cellulases and ligninases that aid in decomposition of plant litter, thereby releasing both carbon and nitrogen.

3.3 Symbiosis

Interactions with plants are beneficial and essential in the adaptation process of living in conditions where there are little moisture and lack of nutrients, such as:

- The association between rhizobia and legumes makes nitrogen more accessible through the colonization of root nodules and the fixation of N_2 (Sprent, 2009; Udvardi & Poole, 2013).
- Mycorrhizal fungi increase both phosphorus and water uptake by extending the effective root system into nutrient-depleted zones (Barea et al., 2005).
- The presence of endophytic bacteria and fungi can increase the capacity of a plant to withstand environmental stresses through the production of phytohormones, root architecture modification, and enhancement of drought resistance (Numan et al., 2018). The process of nutrient absorption is not the sole purpose of many of the endophytic and rhizosphere-associated microorganisms in arid and semi-arid soils that produce 1-aminocyclopropane-1-carboxylate (ACC) deaminase, which is the enzyme that breaks down ACC—the direct precursor of ethylene—thereby lessening the amount of ethylene in the plants that are under drought stress. Microbes that produce ACC-deaminase not only suppress signaling through the stress-induced ethylene pathways associated with aging and death of the plant but also help to keep the plant active physiologically with the help of small droughts, thus, the plant is dead and can recover when water becomes available in Fazal et al. (2025); Hamza et al. (2025); Jain and Saraf (2023).

Such symbiotic relations are especially useful in arid agroecosystems, where the enhancement of microbial-plant interactions is a strategy that can be followed to improve soil fertility and the resilience of crops without being heavily reliant on chemical fertilizers.

4. Microbial contributions to soil formation

Microorganisms in arid and semiarid ecosystems are not merely nutrient cyclers; they are fundamental agents of pedogenesis, directly contributing to aggregate stability, mineral transformation, and the development of surface soil structure. In environments where physical weathering predominates due to extreme temperature fluctuations and wind erosion, microbial processes are critical for initiating and sustaining soil formation.

4.1 Biological soil crusts (biocrusts)

Biocrusts function as ecosystem engineers in drylands, forming complex surface communities that significantly modify their environment. Their roles include:

- Stabilizing soil surfaces through the production of exopolysaccharides and the filamentous networks of cyanobacteria, such as Microcoleus vaginatus.
- Enhancing nutrient inputs via biological nitrogen fixation by genera such as Nostoc and Scytonema, and through the trapping of nutrient-rich dust particles.
- Increasing water infiltration and reducing erosion by forming cohesive surface layers. Studies in Iranian drylands have demonstrated that biocrusts can double topsoil aggregate stability and significantly mitigate wind erosion (Bashtian et al., 2024; Karimi et al., 2024).

In arid and semiarid regions, microorganisms are not just nutrient recyclers; they are the main players of pedogenesis, with a direct impact on the size of the particles, mineral change, and the formation of soil structure at the surface. In such places where physical weathering is the main process due to high temperature variation and wind erosion, the microorganisms' activities are indispensable for soil creation and maintenance.

Biocrusts play the role of ecosystem engineers in arid areas by creating the complex surface communities that greatly alter the environment. Their functions are as follows:

- Exopolysaccharides production and the filamentous networks of cyanobacteria, such as Microcoleus vaginatus, stabilize soil surfaces.
- Generating nutrient supplies through the activity of nitrogen-fixing bacteria such as Nostoc and Scytonema, and the collection of nutrient-laden dust particles.
- Strong Surface layers will not only let water pass but also will keep erosion down. Research conducted in Iranian deserts has indicated that biocrusts are capable of increasing the stability of topsoil aggregates twice over and significantly lessening the effect of wind erosion (Bashtian et al., 2024; Karimi et al., 2024).

4.2 Microbial-induced mineral bioweathering

Microorganisms trigger the bioweathering of the parent material through various mechanisms:

- The talk of organic acids being secreted by the microbes is a crucial one. For instance, Pseudomonas and Bacillus, two bacterial genera, and Aspergillus, a fungal genus, are some of the organisms involved in this process. Organic acids dissolve the mineral silicates and, along with them, release key nutrients (e.g., K, Ca, and P).

- Microorganisms can also produce siderophores, which are compounds that can bind iron and aluminum, which helps in the formation of secondary minerals.
- One of the key processes involved in this is the biochemical redox reactions. A good example of the kind of microbial activity that changes the mineral content and thus the nutrients available for plant uptake is the reduction of Fe^{3+} by the species Shewanella. In calcareous soils, the bacterial population associated with the plant roots causes a decrease in pH, and primary mineral carbonate is dissolved, thus increasing the availability of the nutrient for plants (Hambo et al., 2025; Tamilselvi et al., 2016).
- While bacterial respiration and release of certain materials by them can help in the precipitation of calcareous soils, the microbial alteration of silicate minerals is a major factor in the formation of clay minerals such as palygorskite and other fibrous phyllosilicates. Thus, these actions by microorganisms are directly associated with the processes of soil horizon differentiation, carbonate accumulation, and long-term pedogenic evolution in dryland landscapes (Fomina & Skorochod, 2020; Perelomov, 2023).

4.3 EPS-mediated soil aggregation

EPS play a paramount role in the aggregation of soil minerals into stable soil blocks. This activity not only assists in the formation of protective microhabitats that microorganisms can tolerate with respect to moisture and heat (Chenu, 1995) but also prevents the rapid breakdown of organic matter. The involvement of these mechanisms consists of:

- Initially, the cyanobacterial EPS in the form of biocrusts creates the surface layer of soil that is most cohesive in early soil development.
- Fungi take care of soil by linking it with their hyphae, and this layered structure enhances the stability of the main aggregate.
- The bacterial EPS not only keeps the clay and organic matter together but also, particularly in sandy soils, supports the process that requires low natural aggregation due to its nature.

Studies conducted in the semiarid areas have discovered that the use of inoculants consisting of EPS-producing cyanobacteria (e.g., Nostoc commune) results in a substantial increase in micro-aggregate stability and a decrease in soil penetration resistance during wetting-drying cycles (Guida et al., 2023; Karimi et al., 2023).

5. Impacts of climate change on microbial functions in drylands

Climate change, particularly the rise in temperature and changes in precipitation patterns, will have a strong impact on microbial driving forces of nutrient cycling and soil formation in desert and semi-desert ecosystems. Empirical and modeling studies of the last few years have pointed out the following main effects.

5.1 Altered pulse timing and magnitude

The modification of the precipitation pulses in terms of their frequency and intensity leads to the changing of the timing and the magnitude of the microbial responses (the dynamics of the Birch effect), which, in turn, may lead to the enhancement of periodic nitrogen losses through denitrification (N_2O pulses) and to the shift in the net carbon balance (Ramond et al., 2022).

5.2 Increased aridity and threshold effects

Research along aridity gradients has indicated the existence of non-linear thresholds (e.g., AI $\approx$ 0.32) beyond which the microbial cycling is separated from the plant uptake and the gaseous N losses change direction (Berdugo et al., 2020; Delgado-Baquerizo et al., 2013; Wang et al., 2014). Passage through such thresholds might cause permanent drops in microbial biomass and enzyme activity.

5.3 Biocrust vulnerability

Cyanobacterial crusts are less vulnerable than lichen crusts; however, they are still affected by the changes in temperature, UV radiation, and moisture regimes. The decrease in biocrust cover leads to reduced nitrogen fixation, carbon inputs, and surface stability (Ferrenberg et al., 2015; Reed et al., 2016).

5.4 Impacts of soil salinization and irrigation

In several semiarid agricultural areas, the poor management of irrigation leads to the accumulation of salts, which selects for halotolerant microorganisms, changes enzyme expression, and decreases the diversity and activity of the soil microorganisms overall (Rengasamy, 2006).

5.5 Community reassembly and functional changes

Thermal enhancement can favor thermotolerant or drought-tolerant taxa (e.g., specific Actinobacteria), which will affect the rates of decomposition and CUE (carbon use efficiency), leading to different consequences for SOC stocks (Bastida et al., 2017).

5.6 Land use and N deposition interactions

N deposition and fertilization can induce P limitation and influence microbial community structure (higher copiotrophs), thus altering enzymatic stoichiometry and possibly causing SOC losses to be accelerated in some scenarios (Leff et al., 2015; Ren et al., 2024).

5.7 Management and research implications

In order to reliably predict ecosystem development, models should include the pulse dynamics, the limit of microbial function along the gradients of aridity, and biocrust contributions to the ecosystem. Restoration methods (biocrust inoculation, reintroduction of native shrubs, no tillage) should be the priority to strengthen microbial buffers that are susceptible to the extremes of climate change (Maestre et al., 2013; Maggioli et al., 2025; Schultz et al., 2022).

The unevenness of the biocrust development across the landforms due to different factors has a very strong effect on the soil microbial biomass and the potential of the site for nutrient cycling (Couradeau et al., 2019; Havrilla & Villarreal, 2024). The presence of small hills and mounds in the case of alluvial fans and similar sites in Iran is shown to affect the extent of biocrust cover and, consequently, the size of the microbial population, the level of enzyme activity, and the soil structure development (Hassanzadeh Bashtian et al., 2024).

6. Monitoring the microbial soil nexus

The intricate relationship between soil microorganisms and their environment is a key factor in the global biogeochemical cycles and the ecosystem's well-being. Technological advancements are progressively boosting our capacities to observe, simulate, and foretell the working and endurance of this vital system over a variety of space and time scales.

6.1 Remote sensing of biocrusts

Biocrusts have become the subject of monitoring through remote sensing techniques, which are now indispensable for monitoring the distribution and health of biocrusts over extensive spatial scales:

- Hyperspectral Imagery: The said technique involves the capture of data by being swept through a very narrow and continuous part of the electromagnetic spectrum. Thus, it gives a detailed spectral signature for the diverse biocrusts (Cyanobacteria-dominated, Lichens-dominated, etc.) along with their states of health (Intact, disturbed). Some differences in the photosynthetic pigments (Chlorophyll and Carotenoids) and moisture content can be identified with a high degree of accuracy, thus allowing the formulation of biocrust distribution maps and the assessment of their vitality (Lehnert et al., 2018).
- Drones (Unmanned Aerial Vehicles - UAVs): Drones that are fitted with hyperspectral or multispectral sensors provide data at the landscape level that has a high spatial and temporal resolution (Blanco-Sacristán et al., 2021). This is beneficial in the monitoring of the ecosystem recovery and its stability through the tracking of very slight changes in the biocrust cover and the color that occurs following disturbance events, thus supplying critical data for conservation and management activities.

6.2 Sensor networks for pulse dynamics

Soil microbial activity is often described as a series of extremely rapid, very short-lived reactions pulse dynamics and these reactions are influenced by changes in the environment. One of the phenomena is the "Birch effect," which refers to the sudden and very quick release of CO_2 that takes place during the rewetting of dry soil. It is the most important sign of the melting of microbial dormancy and carbon decay.

In-Situ IoT Sensor Networks: The deployment of Internet of Things (IoT) sensor networks enables real-time and high-frequency data collection on important soil parameters. The sensors in these networks are used to detect:

- Soil Moisture and Temperature: These are the main environmental factors affecting the activation and dormancy of microorganisms.
- CO_2 Fluxes: These are automatically measured in chambers, thus giving a direct indication of microbial respiration and decay rates.
- Precise Measurement of Pulse Responses: The integration of real-time moisture fluctuations and instantaneous changes in CO_2 flux reveals the very fine time resolution that is important for the accurate measurement of the size and duration of the Birch effect and other microbial pulse responses to rainfall or freezing/thawing events, and this is ensuring that our understanding of

carbon cycling vulnerability in the face of erratic weather is greatly improved (Ramson et al., 2021).

7. Case studies and key research findings

7.1 Case study: Plant species and seasonal drivers of microbial function in the zagros woodlands (Iran)

In semiarid woodlands, the plant identity and seasonal pulses have a great effect on the soils, which in turn facilitate the microbial nutrient cycling. A remarkable piece of work done in the Zagros semiarid forests of Iran by Eslaminejad et al. (2020) laid bare the issue quantitatively. The researchers stated that tree species had much more microbial biomass C (MBC), microbial respiration, and enzyme activities than shrub or open-canopy areas. A large part of the difference is explained by the litter quality and root exudates that are higher under trees.

The study continued to track the important temporal dynamic, with enzyme activities (chiefly urease and alkaline phosphatase) reaching their highest point in spring when the temperature rose, and the moisture was lower compared to winter, thus showing a clear pulse-driven nutrient mineralization response. The modeling of statistics pointed out bulk density, soil organic carbon (SOC), and total nitrogen as the most important factors determining microbial function, where bulk density alone accounted for a large portion of the variation in microbial parameters.

The vegetation composition shapes the resource islands, which in turn create the distinct areas of microbial activity and nutrient turnover, with the tree species particularly encouraging the high microbial-mediated nitrogen and phosphorus cycling to improve the local fertility. At the same time, the findings point to the modulating effect of seasonal moisture and temperature changes on the enzymatic activity and microbial turnover through the pulse-driven responses.

These findings lead to direct management applications: among the methods to be applied, restoration and afforestation with the use of native tree species or a mixture of them can lead to an increase in microbial biomass and enzyme activity, which, in turn, will support soil formation and nutrient retention in semiarid woodlands. Besides, the monitoring of urease and phosphatase activities offers a practical, economical, and effective bioindicator framework for the evaluation of soil recovery progress in conservation and reforestation projects.

7.2 Case study: Shrub-induced enhancement of microbial hubs in mountainous rangelands (Iran)

The type of vegetation is the main factor that determines the microbial community in the soil fertility of semiarid rangelands, with the case of the presence of trees being the main reason for the formation of hotspots of microbial life and activity. The study of Kooch and Noghre (2020) conducted in the mountainous Diakooh region of Iran gives the best example of comparing shrubland and grassland ecosystems. The analysis indicated that the patches of bushes, mainly Crataegus and Berberis, were the ones to have the largest amount of microbial matter (MBC, MBN, MBP), the highest enzyme activities, the largest fractions of particulate and dissolved organic matter, and the most soil moisture when compared to the pasture sites.

Through the analysis of enzyme stoichiometry, the researchers were able to get deeper functional insights by revealing microbial nitrogen limitation being reduced in shrublands and carbon and phosphorus processing capacity being higher than in the case of plains. That said, the situation with degraded pastures, where the conversion of land drastically decreased the microbial services of the ecosystem point out how fragile these ecological processes are and how easily they can be disturbed by humans.

On the other hand, enzyme assays and SOM fractionation have become accepted as very useful diagnostic tools for the purpose of evaluating the success of restoration and also as a means of monitoring the recovery of microbial functions in rangelands that have become degraded.

7.3 Case study: Biocrust inoculation as a microbial tool for rapid nutrient cycling and soil stabilization (Mojave Desert, USA)

The regeneration of biocrusts in arid regions is a strong and clear indication of microbial communities' ecological functions, such as soil formation and nutrient cycling. Conducted by Chiquoine et al. (2016), a field trial on the severely degraded gypsum-rich soils of the Mojave Desert in the USA proved that using native biocrust materials can remarkably restore microbial functions. These functions include nitrogen transformation, carbon sequestration, and soil stabilization within just 18 months. The research unveiled that biocrust-inoculated plots exhibited significant advancements in lichen and moss cover, cyanobacteria density, soil chlorophyll *a* content, and ammonium (NH_4–N) concentrations to levels of undisturbed reference sites. In contrast, the soils of the disturbed area that were not inoculated did not support colonization by cyanobacteria, exhibited reduced chlorophyll, increased nitrate (NO_3–N), and had poor aggregate stability. Moreover, biocrust inoculation was the only process in which macroorganisms such as lichens and mosses were able to be reestablished, thus emphasizing the necessity of total community rather than functional group reintroduction.

Filamentous cyanobacteria restoration, which are the main soil binding agents, was an essential step for erosion control and soil structure improvement. The proofs obtained strongly state that the biocrusts are microbial hotspots that take up and manage nutrient fluxes in the arid landscape. Their coming back means not only that the cycle of nitrogen and carbon is fully restored, but also that the disturbance effects are lessened since they make the soil surfaces stable, and hence, dust emissions are reduced. It may be possible to use monitoring of cyanobacteria density and soil chlorophyll as bioindicators for measuring the success of restoration.

7.4 Case study: The fungal loop: A microbial engine for nutrient cycling in aridlands (New Mexico, USA)

Arid and semiarid ecosystems are a major concern for the traditional carbon (C) and nitrogen (N) cycling models, which have been developed mainly from the data of temperate and mesic climates (Collins et al., 2008). The Chihuahuan Desert and similar ecosystems provide evidence that soil fungi are the agents of the fungal loop. This microbial network connects plants such as the dominant perennial C_4 grasses Bouteloua eriopoda and B. gracilis (blue grama) with biocrusts consisting of cyanobacteria, lichens, and mosses. Among the fungi, the arbuscular mycorrhizal fungi (Glomeromycota), dark septate endophytes (especially the Pleosporales), and many saprotrophic fungi have taken over the critical processes because of their excellent tolerance to high temperature and low water potential, which is much better than that of bacteria. They are the ones who break down resistant organic compounds, carry out heterotrophic nitrification and assimilatory denitrification, and serve as networks for the movement of C and N between vegetation patches and biocrusts. The carbon loop is energized by biotrophic carbon from both plants and cyanobacteria, which means that nutrient transformations are closely connected with primary producers, even when decomposition and plant production are temporally disconnected.

Extracellular enzyme activity (EEA) profiles in these alkaline arid soils are different from those in mesic systems. The potentials for proteolytic enzymes (e.g., aminopeptidase) and oxidative enzymes (e.g., peroxidase) are remarkably high, which assists in the breakdown of the resistant C and proteins, thus impairing the accumulation of soil organic matter (SOM) and directly bringing N back to microbes and plants. This scenario indicates that the fungal-assisted protein mineralization is the major source of bioavailable N in these nutrient-poor soils.

The impact of this fungal-centered model is very significant as it relates to our knowledge of soil formation and nutrient retention. By the ailing of limited nutrients and stabilizing soil structure through hyphal networks, fungi not only increase the productivity of the resource islands below the plants but also minimize the nutrient losses. This microbial agent is very important in the formation and preservation of soil capital in the dryland regions.

Global change phenomena such as changes in precipitation patterns and nitrogen deposition call for the comprehension of microbial loop dynamics in order to predict the future of these large and increasingly covered areas.

7.5 Case study: Remote sensing of biocrusts to assess microbial controls on nutrient cycling and soil formation (Spain)

The incorporation of unmanned aerial vehicle (UAV) based remote sensing technology into the field of dryland ecology is a great way to demonstrate the role of microbial biocrust communities in nutrient cycling and soil formation. A UAV multisensor study in the semiarid gypsiferous landscapes of central Spain showed that high-resolution RGB, multispectral, and thermal imagery can map the composition of the biocrust and at the same time, infer the microbial processes that drive the nitrogen, phosphorus, and carbon dynamics in the heterogeneous terrain (Blanco-Sacristán et al., 2021). It was found that the functional types of biocrusts, which are remotely sensed, especially lichen versus moss-dominated communities, could indicate soil moisture, nutrient status, and microtopographic stability very convincingly. Through the means of strong chlorophyll absorption in the red band and high apparent thermal inertia (ATI), moss-dominated biocrusts were detected, and they were always linked with cool, moisture-retaining microenvironments. These zones were distinguished by higher organic carbon content, total nitrogen and phosphorus, reflecting more active processes of microbial respiration, nitrogen fixation, and biochemical weathering. On the other hand, lichen-dominated crusts were located on hot, dry, and Erosion-prone slopes with strong sunlight, which were identified by their high reflectance and low pigment absorption. These areas were marked by slight soil organic carbon, slow microbial nutrient turnover, and high soluble salts, all of which indicate that the microbial cycling and soil development are still in the early stages.

Importantly, the study demonstrated that thermal imaging derived ATI strongly predicted soil moisture ($r^2 = 0.83$), enabling moisture-driven microbial processes to be inferred at the landscape scale, an advance not achievable through field sampling alone. Spatial patterns of ATI, in combination with terrain indices such as slope and potential solar radiation, revealed the fine-scale hydrological redistribution that structures microbial activity and nutrient accumulation in biocrust covered soils.

The remote sensing approach also highlighted distinct functional trajectories within biocrust communities. Mixed lichen–moss crusts (BLM), representing later successional stages, developed preferentially in stable low slope areas where nutrient concentrations were higher. In contrast, pioneering lichen species such as Diploschistes diacapsis dominated exposed gypsum outcrops with extremely limited microbial nutrient cycling. These findings demonstrate that biocrust spatial distribution, when mapped via UAV sensors, provides an accurate proxy for microbial controls on N, P, C cycling and pedogenesis in drylands.

8. Synthesis and conclusions

The main creators of the nutrient cycling and soil formation processes in arid and semiarid ecosystems are microbial communities. They can execute extremely difficult environmental conditions by performing different activities such as biological nitrogen fixation, phosphorus solubilization, and carbon stabilization. For example, microbial adaptations such as the production of exopolysaccharides, the secretion of extracellular enzymes, and the establishment of symbiotic relationships with plants are among the most important tools they have to endure and overcome such constraints. The mentioned adaptations are acting not only as survival mechanisms but also as main factors that drive soil functions, such as aggregation, mineral weathering, and nutrient retention. Currently, climate change and intense land use are the factors that most influence the relationships between microbial communities and the ecosystem as a whole. However, it is noteworthy that our knowledge and management skills of these microbial influencers are being tremendously improved by the latest technologies. Remote sensing methods such as UAVs and hyperspectral imagery, in addition to IoT sensor networks deployed on-site, are allowing us to observe the microbial processes at the microsite scale and then extrapolate them to the landscape scale, incorporate non-rainfall moisture inputs into our models, and affiliate microbial activity with observable soil formation rates in real-time.

Monitoring, predictive modeling, and decision-making processes will then encompass the integration of the role of microbial communities with the sensor and remote sensing data in the future management and restoration strategies.

References

Alsharif, W., Saad, M. M., & Hirt, H. (2020). Desert microbes for boosting sustainable agriculture in extreme environments. *Frontiers in Microbiology, 11*, 1666. https://doi.org/10.3389/fmicb.2020.01666

Austin, A. T., Yahdjian, L., Stark, J. M., Belnap, J., Porporato, A., Norton, U., … Schaeffer, S. M. (2004). Water pulses and biogeochemical cycles in arid and semiarid ecosystems. *Oecologia, 141*(2), 221–235. https://doi.org/10.1007/s00442-004-1519-1

Bardgett, R. D., & Van Der Putten, W. H. (2014). Belowground biodiversity and ecosystem functioning. *Nature, 515*(7528), 505–511. https://doi.org/10.1038/nature13855

Barea, J. M., Pozo, M. J., Azcon, R., & Azcón-Aguilar, C. (2005). Microbial co-operation in the rhizosphere. *Journal of Experimental Botany, 56*(417), 1761–1778. https://doi.org/10.1093/jxb/eri197

Barger, N. N., Weber, B., Garcia-Pichel, F., Zaady, E., & Belnap, J. (2016). Patterns and controls on nitrogen cycling of biological soil crusts. In *Biological soil crusts: An organizing principle in drylands* (pp. 257–285). Cham: Springer International Publishing. https://doi.org/10.1007/978-3-319-30214-0_14

Bashtian, M. H., Karimi, A., Sepehr, A., Lakzian, A., & Bol, R. (2024). Functional roles of biocrusts in enhancing soil quality in a semi-arid environment. *Geoderma Regional, 37*, Article e00787.

Bastida, F., Torres, I. F., Romero-Trigueros, C., Baldrian, P., Větrovský, T., Bayona, J. M., ... Nicolás, E. (2017). Combined effects of reduced irrigation and water quality on the soil microbial community of a citrus orchard under semi-arid conditions. *Soil Biology and Biochemistry, 104*, 226–237.

Belnap, J. (2003). The world at your feet: Desert biological soil crusts. *Frontiers in Ecology and the Environment, 1*(4), 181–189. https://doi.org/10.1890/1540-9295(2003)001[0181:TWAYFD]2.0.CO;2

Berdugo, M., Delgado-Baquerizo, M., Soliveres, S., Hernández-Clemente, R., Zhao, Y., Gaitán, J. J., ... Maestre, F. T. (2020). Global ecosystem thresholds driven by aridity. *Science, 367*(6479), 787–790. https://doi.org/10.1126/science.aay5958

Blanco-Sacristán, J., Panigada, C., Gentili, R., Tagliabue, G., Garzonio, R., Martín, M. P., ... Rossini, M. (2021). UAV RGB, thermal infrared and multispectral imagery used to investigate the control of terrain on the spatial distribution of dryland biocrust. *Earth Surface Processes and Landforms, 46*(12), 2466–2484. https://doi.org/10.1002/esp.5189

Butterbach-Bahl, K., Baggs, E. M., Dannenmann, M., Kiese, R., & Zechmeister-Boltenstern, S. (2013). Nitrous oxide emissions from soils: how well do we understand the processes and their controls? *Philosophical Transactions of the Royal Society B: Biological Sciences, 368* (1621).

Chenu, C. (1995). Extracellular polysaccharides: An interface between microorganisms and soil constituents. *Environmental impact of soil component interactions, 1*, 217–233.

Chiquoine, L. P., Abella, S. R., & Bowker, M. A. (2016). Rapidly restoring biological soil crusts and ecosystem functions in a severely disturbed desert ecosystem. *Ecological Applications, 26*(4), 1260–1272. https://doi.org/10.1002/15-0973

Chisholm, C., Di, H., Cameron, K., Podolyan, A., Shen, J., Zhang, L., ... Che, X. (2025). Transcriptional activity of ammonia oxidisers in response to soil temperature, moisture and nitrogen amendment. *Frontiers in Microbiology, 15*, Article 1466991. https://doi.org/10.3389/fmicb.2024.1466991

Collins, S. L., Sinsabaugh, R. L., Crenshaw, C., Green, L., Porras-Alfaro, A., Stursova, M., & Zeglin, L. H. (2008). Pulse dynamics and microbial processes in aridland ecosystems. *Journal of Ecology, 96*(3), 413–420. https://doi.org/10.1111/j.1365-2745.2008.01362.x

Couradeau, E., Giraldo-Silva, A., De Martini, F., & Garcia-Pichel, F. (2019). Spatial segregation of the biological soil crust microbiome around its foundational cyanobacterium, microcoleus vaginatus, and the formation of a nitrogen-fixing cyanosphere. *Microbiome, 7*(1), 55. https://doi.org/10.1186/s40168-019-0661-2

D'Amico, S., Collins, T., Marx, J. C., Feller, G., Gerday, C., & Gerday, C. (2006). Psychrophilic microorganisms: Challenges for life. *EMBO Reports, 7*(4), 385–389. https://doi.org/10.1038/sj.embor.7400662

Delgado-Baquerizo, M., Maestre, F. T., Gallardo, A., Bowker, M. A., Wallenstein, M. D., Quero, J. L., ... Zaady, E. (2013). Decoupling of soil nutrient cycles as a function of aridity in global drylands. *Nature, 502* (7473), 672–676. https://doi.org/10.1038/nature12670

Elbert, W., Weber, B., Burrows, S., Steinkamp, J., Büdel, B., Andreae, M. O., & Pöschl, U. (2012). Contribution of cryptogamic covers to the global cycles of carbon and nitrogen. *Nature Geoscience, 5*(7), 459–462. https://doi.org/10.1038/ngeo1486

Eslaminejad, P., Heydari, M., Kakhki, F. V., Mirab-Balou, M., Omidipour, R., Muñoz-Rojas, M., & Lucas-Borja, M. E. (2020). Plant species and season influence soil physicochemical properties and microbial function in a semi-arid woodland ecosystem. *Plant and Soil, 456*(1), 43–59. https://doi.org/10.1007/s11104-020-04691-1

Fazal, F., Amin, A., & Barq, M. G. (2025). Microbial resilience in arid soils: Ecological responses to drought and salinity stress. *Current Microbiology, 82*(12), 576. https://doi.org/10.1007/s00284-025-04559-6

Ferrenberg, S., Reed, S. C., & Belnap, J. (2015). Climate change and physical disturbance cause similar community shifts in biological soil crusts. *Proceedings of the National Academy of Sciences, 112*(39), 12116–12121. https://doi.org/10.1073/pnas.1509150112

Fierer, N., Strickland, M. S., Liptzin, D., Bradford, M. A., & Cleveland, C. C. (2009). Global patterns in belowground communities. *Ecology Letters, 12*(11), 1238–1249. https://doi.org/10.1111/j.1461-0248.2009.01360.x

Fomina, M., & Skorochod, I. (2020). Microbial interaction with clay minerals and its environmental and biotechnological implications. *Minerals, 10*(10), 861. https://doi.org/10.3390/min10100861

Garcia-Pichel, F. (2023). The microbiology of biological soil crusts. *Annual Review of Microbiology, 77*(1), 149–171. https://doi.org/10.1146/annurev-micro-032521-015202

Guida, G., Nicosia, A., Settanni, L., & Ferro, V. (2023). A review on effects of biological soil crusts on hydrological processes. *Earth-Science Reviews, 243*, Article 104516. https://doi.org/10.1016/j.earscirev.2023.104516

Hambo, C. L., Shitindi, M. J., Sibuga, K. P., & Nzogela, Y. B. (2025). Calcite dissolving bacteria: Promising approach as biofertilizer, a review. *Frontiers in Microbiology, 16*, Article 1533127. https://doi.org/10.3389/fmicb.2025.1533127

Hamza, M. A., Abbas, M., Youssef, H. H., Nour, E. H., Abdelwahab, M. S., Tchakounté, G. V., … Hegazi, N. A. (2025). Desert aridity recruits xerophyte microbiota of diverse structure and function. *Discover Plants, 2*(1), 186. https://doi.org/10.1007/s44372-025-00268-3

Hassanzadeh Bashtian, M., Karimi, A., Sepehr, A., Lakzian, A., & Caballero, E. R. (2024). Spatial relationship of landform surface features and biocrusts, and their effect on soil microbial biomass on an alluvial fan. *Earth Surface Processes and Landforms, 49*(4), 1348–1360. https://doi.org/10.1002/esp.5771

Havrilla, C. A., & Villarreal, M. L. (2024). Soil cover heterogeneity associated with biocrusts predicts patch-level plant diversity patterns. *Landscape Ecology, 39*(11), 187. https://doi.org/10.1007/s10980-024-01986-x

Jain, R., & Saraf, M. (2023). ACC deaminase producing PGPR modulates nutrients uptake, soil properties and growth of cluster bean (Cyamopsis tetragonoloba L.) under deficit irrigation. *Biologia, 78*(9), 2303–2316. https://doi.org/10.1007/s11756-023-01376-9

Kallenbach, C. M., Frey, S. D., & Grandy, A. S. (2016). Direct evidence for microbial-derived soil organic matter formation and its ecophysiological controls. *Nature Communications, 7*(1), Article 13630. https://doi.org/10.1038/ncomms13630

Karimi, A., Tahmourespour, A., & Hoodaji, M. (2023). The formation of biocrust and improvement of soil properties by the exopolysaccharide-producing cyanobacterium: A biogeotechnological study. *Biomass Conversion and Biorefinery, 13*(17), 15489–15499. https://doi.org/10.1007/s13399-022-02336-0

Karimi, A., Tahmourespour, A., & Hoodaji, M. (2024). Cyanobacterial biocrust alters soil physical properties reducing soil erosion and aerosol production. *Brazilian Journal of Microbiology, 55*(3), 2453–2461. https://doi.org/10.1007/s42770-024-01377-z

Kelley, A. M., Fay, P. A., Polley, H. W., Gill, R. A., & Jackson, R. B. (2011). Atmospheric CO2 and soil extracellular enzyme activity: A meta-analysis and CO2 gradient experiment. *Ecosphere, 2*(8), 1–20. https://doi.org/10.1890/ES11-00117.1

Kits, K. D., Sedlacek, C. J., Lebedeva, E. V., Han, P., Bulaev, A., Pjevac, P., … Wagner, M. (2017). Kinetic analysis of a complete nitrifier reveals an oligotrophic lifestyle. *Nature, 549*(7671), 269–272. https://doi.org/10.1038/nature23679

Kooch, Y., & Noghre, N. (2020). Nutrient cycling and soil-related processes under different land covers of semi-arid rangeland ecosystems in northern Iran. Catena, 193, 104621. https://doi.org/10.1016/j.catena.2020.104621

Leff, J. W., Jones, S. E., Prober, S. M., Barberán, A., Borer, E. T., Firn, J. L., … Fierer, N. (2015). Consistent responses of soil microbial communities to elevated nutrient inputs in grasslands across the globe. *Proceedings of the National Academy of Sciences, 112*(35), 10967–10972. https://doi.org/10.1073/pnas.1508382112

Lehnert, L. W., Jung, P., Obermeier, W. A., Büdel, B., & Bendix, J. (2018). Estimating net photosynthesis of biological soil crusts in the atacama using hyperspectral remote sensing. *Remote Sensing, 10*(6), 891. https://doi.org/10.3390/rs10060891

Lehtovirta-Morley, L. E., Stoecker, K., Vilcinskas, A., Prosser, J. I., & Nicol, G. W. (2011). Cultivation of an obligate acidophilic ammonia oxidizer from a nitrifying acid soil. *Proceedings of the National Academy of Sciences, 108*(38), 15892–15897.

Lekberg, Y., Gibbons, S. M., Rosendahl, S., & Ramsey, P. W. (2013). Severe plant invasions can increase mycorrhizal fungal abundance and diversity. *The ISME journal, 7*(7), 1424–1433. https://doi.org/10.1038/ismej.2013.41

Li, L., Luo, Z., Nian, L., Li, L., Niu, Y., Zhang, Y., ... Liu, J. (2025). Regulation of soil ammonia-oxidizing microbial community assembly by alfalfa (Medicago sativa) planting duration in the loess plateau. *Frontiers in Microbiology, 16*, Article 1517296. https://doi.org/10.3389/fmicb.2025.1517296

Maestre, F. T., Escolar, C., de Guevara, M. L., Quero, J. L., Lázaro, R., Delgado-Baquerizo, M., ... Gallardo, A. (2013). Changes in biocrust cover drive carbon cycle responses to climate change in drylands. *Global Change Biology, 19*(12), 3835–3847. https://doi.org/10.1111/gcb.12306

Maestre, F. T., Delgado-Baquerizo, M., Jeffries, T. C., Eldridge, D. J., Ochoa, V., Gozalo, B., ... Singh, B. K. (2015). Increasing aridity reduces soil microbial diversity and abundance in global drylands. *Proceedings of the National Academy of Sciences, 112*(51), 15684–15689. https://doi.org/10.1073/pnas.1516684112

Maestre, F. T., Eldridge, D. J., Soliveres, S., Kéfi, S., Delgado-Baquerizo, M., Bowker, M. A., ... Berdugo, M. (2016). Structure and functioning of dryland ecosystems in a changing world. *Annual review of ecology, evolution, and systematics, 47*(1), 215–237. https://doi.org/10.1146/annurev-ecolsys-121415-032311

Mager, D. M., & Thomas, A. D. (2011). Extracellular polysaccharides from cyanobacterial soil crusts: A review of their role in dryland soil processes. *Journal of Arid Environments, 75*(2), 91–97.

Maggioli, L., Chamizo, S., Román, R., Rodríguez-Caballero, E., Roncero-Ramos, B., & Cantón, Y. (2025). Restoring dryland soils functionality through synergism between biocrust-forming cyanobacteria and paper waste amendment. *Soil Biology and Biochemistry.* , Article 109997. https://doi.org/10.1016/j.soilbio.2025.109997

Manzoni, S., Ahmed, M. H., & Porporato, A. (2019). Ecohydrological and stoichiometric controls on soil carbon and nitrogen dynamics in drylands. In *Dryland ecohydrology* (pp. 279–307). Cham: Springer International Publishing. https://doi.org/10.1007/978-3-030-23269-6_11

Mohammadipanah, F., & Wink, J. (2016). Actinobacteria from arid and desert habitats: Diversity and biological activity. *Frontiers in Microbiology, 6*, 1541. https://doi.org/10.3389/fmicb.2015.01541

Monger, H. C. (1999). Biotic and abiotic processes and rates of pedogenic carbonate accumulation in the Southwestern United States-relationship to atmospheric CO_2 sequestration. In *Global climate change and pedogenic carbonates* (1st ed., pp. 273–289). CRC Press.

Numan, M., Bashir, S., Khan, Y., Mumtaz, R., Shinwari, Z. K., Khan, A. L., ... Al-Harrasi, A. (2018). Plant growth promoting bacteria as an alternative strategy for salt tolerance in plants: A review. *Microbiological Research, 209*, 21–32. https://doi.org/10.1016/j.micres.2018.02.003

Olagoke, F. K., Kaiser, K., Mikutta, R., Kalbitz, K., & Vogel, C. (2020). Persistent activities of extracellular enzymes adsorbed to soil minerals. *Microorganisms, 8*(11), 1796. https://doi.org/10.3390/microorganisms8111796

Parasar, B. J., Kashyap, S., Sharma, I., Marme, S. D., Das, P., & Agarwala, N. (2024). Microbe mediated alleviation of drought and heat stress in plants-current understanding and future prospects. *Discover Plants, 1*(1), 26. https://doi.org/10.1007/s44372-024-00022-1

Pastore, M. A., Classen, A. T., English, M. E., Frey, S. D., Knorr, M. A., Rand, K., & Adair, E. C. (2023). Soil microbial legacies influence freeze–thaw responses of soil. *Functional Ecology, 37*(4), 1055–1066. https://doi.org/10.1111/1365-2435.14273

Perelomov, L. V. (2023). The role of interactions between bacteria and clay minerals in pedochemical processes. *Geochemistry International, 61*(10), 1026–1035. https://doi.org/10.1134/S0016702923100105

Perez-Garcia, O., Chandran, K., Villas-Boas, S. G., & Singhal, N. (2016). Assessment of nitric oxide (NO) redox reactions contribution to nitrous oxide (N2O) formation during nitrification using a multispecies metabolic network model. *Biotechnology and Bioengineering, 113*(5), 1124–1136. https://doi.org/10.1002/bit.25880

Pointing, S. B., & Belnap, J. (2012). Microbial colonization and controls in dryland systems. *Nature Reviews Microbiology, 10*(8), 551–562. https://doi.org/10.1038/nrmicro2831

Ragot, S. A., Kertesz, M. A., Meszaros, E., Frossard, E., & Bünemann, E. K. (2017). Soil phoD and phoX alkaline phosphatase gene diversity responds to multiple environmental factors. *FEMS Microbiology Ecology, 93*(1), Article fiw212. https://doi.org/10.1093/femsec/fiw212

Ramond, J. B., Jordaan, K., Díez, B., Heinzelmann, S. M., & Cowan, D. A. (2022). Microbial biogeochemical cycling of nitrogen in arid ecosystems. *Microbiology and Molecular Biology Reviews, 86*(2), Article e00109-21. https://doi.org/10.1128/mmbr.00109-21

Ramson, S. J., León-Salas, W. D., Brecheisen, Z., Foster, E. J., Johnston, C. T., Schulze, D. G., ... Malaga, M. P. (2021). A self-powered, real-time, LoRaWAN IoT-based soil health monitoring system. *IEEE Internet of Things Journal, 8*(11), 9278–9293. https://doi.org/10.1109/JIOT.2021.3056586

Reed, S. C., Maestre, F. T., Ochoa-Hueso, R., Kuske, C. R., Darrouzet-Nardi, A., Oliver, M., ... Belnap, J. (2016). Biocrusts in the context of global change. In *Biological soil crusts: An organizing principle in drylands* (pp. 451–476). Cham: Springer International Publishing. https://doi.org/10.1007/978-3-319-30214-0_22

Ren, Y., Wang, Y., Zhang, X., Liu, X., Liu, P., & Chen, L. (2024). Enzymatic stoichiometry reveals the metabolic limitations of soil microbes under nitrogen and phosphorus addition in Chinese fir plantations. *Microorganisms, 12*(8), 1716. https://doi.org/10.3390/microorganisms12081716

Rengasamy, P. (2006). World salinization with emphasis on Australia. *Journal of Experimental Botany, 57*(5), 1017–1023.

Reynolds, J. F., Maestre, F. T., Kemp, P. R., Stafford-Smith, D. M., & Lambin, E. (2007). Natural and human dimensions of land degradation in drylands: Causes and consequences. In *Terrestrial ecosystems in a changing world* (pp. 247–257). Berlin, Heidelberg: Springer Berlin Heidelberg. https://doi.org/10.1007/978-3-540-32730-1_20

Richardson, A. E., & Simpson, R. J. (2011). Soil microorganisms mediating phosphorus availability update on microbial phosphorus. *Plant physiology, 156*(3), 989–996. https://doi.org/10.1104/pp.111.175448

Rodríguez, A., & Fraga, R. (1999). Phosphate solubilizing bacteria and their role in plant growth promotion. *Biotechnology Advances, 17*(4–5), 319–339.

Rossi, F., & De Philippis, R. (2015). Role of cyanobacterial exopolysaccharides in phototrophic biofilms and in complex microbial mats. *Life, 5*(2), 1218–1238. https://doi.org/10.3390/life5021218

Schimel, J. P. (2018). Life in dry soils: Effects of drought on soil microbial communities and processes. *Annual review of ecology, evolution, and systematics, 49*(1), 409–432. https://doi.org/10.1146/annurev-ecolsys-110617-062614

Schmidt, M. W., Torn, M. S., Abiven, S., Dittmar, T., Guggenberger, G., Janssens, I. A., ... Trumbore, S. E. (2011). Persistence of soil organic matter as an ecosystem property. *Nature, 478*(7367), 49–56. https://doi.org/10.1038/nature10386

Schultz, N. L., Sluiter, I. R., Allen, G. G., Machado-de-Lima, N. M., & Muñoz-Rojas, M. (2022). Biocrust amendments to topsoils facilitate biocrust restoration in a post-mining arid environment. *Frontiers in Microbiology, 13*, Article 882673. https://doi.org/10.3389/fmicb.2022.882673

Sharma, S. B., Sayyed, R. Z., Trivedi, M. H., & Gobi, T. A. (2013). Phosphate solubilizing microbes: Sustainable approach for managing phosphorus deficiency in agricultural soils. *SpringerPlus, 2*(1), 587. https://doi.org/10.1186/2193-1801-2-587

Smith, S. E., & Read, D. J. (2010). *Mycorrhizal symbiosis*. Academic press.

Sprent, J. I. (2009). *Legume nodulation: A global perspective*.

Svirčev, Z., Dulić, T., Obreht, I., Codd, G. A., Lehmkuhl, F., Marković, S. B., ... Meriluoto, J. (2019). Cyanobacteria and loess—an underestimated interaction. *Plant and Soil, 439*(1), 293–308. https://doi.org/10.1007/s11104-019-04048-3

Tamilselvi, S. M., Thiyagarajan, C., & Uthandi, S. (2016). Calcite dissolution by brevibacterium sp. SOTI06: A futuristic approach for the reclamation of calcareous sodic soils. *Frontiers in Plant Science, 7*, 1828. https://doi.org/10.3389/fpls.2016.01828

Udvardi, M., & Poole, P. S. (2013). Transport and metabolism in legume-rhizobia symbioses. *Annual Review of Plant Biology, 64*, 781–805. https://doi.org/10.1146/annurev-arplant-050312-120235

Uygur, V. (2009). Phosphate sorption in calcareous soils: The role of iron oxide and carbonates. *Asian Journal of Chemistry, 21*(4), 3001.

Van Der Heijden, M. G., De Bruin, S., Luckerhoff, L., Van Logtestijn, R. S., & Schlaeppi, K. (2016). A widespread plant-fungal-bacterial symbiosis promotes plant biodiversity, plant nutrition and seedling recruitment. *The ISME Journal, 10*(2), 389–399. https://doi.org/10.1038/ismej.2015.120

Von Wandruszka, R. (2006). Phosphorus retention in calcareous soils and the effect of organic matter on its mobility. *Geochemical Transactions, 7*(1), 6. https://doi.org/10.1186/1467-4866-7-6

Vranova, V., Rejsek, K., & Formanek, P. (2013). Proteolytic activity in soil: A review. *Applied Soil Ecology, 70*, 23–32. https://doi.org/10.1016/j.apsoil.2013.04.003

Wang, C., Wang, X., Liu, D., Wu, H., Lu, X., Fang, Y., … Bai, E. (2014). Aridity threshold in controlling ecosystem nitrogen cycling in arid and semi-arid grasslands. *Nature Communications, 5*(1), 4799. https://doi.org/10.1038/ncomms5799

Weber, B., Budel, B., & Belnap, J. (Eds.). (2016). *Biological soil crusts: An organizing principle in drylands*. Springer. https://doi.org/10.1007/978-3-319-30214-0

Zamanian, K., Pustovoytov, K., & Kuzyakov, Y. (2016). Pedogenic carbonates: Forms and formation processes. *Earth-Science Reviews, 157*, 1–17. https://doi.org/10.1016/j.earscirev.2016.03.003

CHAPTER

Soil-plant interactions: Water retention, its availability to plants, and their influential factors

3

Ali Akbar Moosavi and Niloofar Sadri

Department of Soil Science, College of Agriculture, Shiraz University, Shiraz, Iran

1. Introduction

1.1 Soil-plant interaction: An overview

Soil is one of the most important and non-renewable environmental resources. It significantly affects human health, and its importance and roles are clear to everyone. In recent years, the growth of the population and the consequent increase in food demand have resulted in significant degradation of soils globally. Soil degradation decreases fertility, biodiversity, and water-holding capacity of the soils. It also results in remarkable disruption in nutrient cycles. Interactions among soil, plants, and microbes strongly affect soil quality and health (Baldi, 2021). Based on particle size distribution (PSD), soils are commonly classified into three main textural groups of coarse, medium-, and fine-textured soils, which directly affects their water-holding capacity and ability to reserve nutrient elements. Researchers reported direct and significant relationships between the soil texture and the amount of nutrients required for plant growth (Adomako et al., 2022; Brown et al., 2019; Xue et al., 2021). In other words, soil texture plays a critical role in assessing and absorbing nutrients by plants, which is one of the most important factors affecting the optimal growth of the plants. Frequent researchers have highlighted the significant role of soil physical attributes in ecological and environmental processes, emphasizing their highly significant effects on the growth and development of plant roots across various conditions. Changes in soil PSD result in remarkable changes in various processes, such as air, water, and nutrient translocation within the soil profile as well as toward plant roots, thermal, physical, mechanical, and hydraulic properties of the soils that directly influence plant growth (Jiang et al., 2020, 2022; Mohseni et al., 2017). Furthermore, changes in soil PSD have been reported as one of the most important factors resulting in the nonuniform distribution of nutrients in plants (Mohammadi et al., 2019). Fine particles of soil, clay minerals, play key roles in the retention of water and nutrients, which are essential for plant growth. Whereas coarse-textured soils have a lesser amount of available nutrients due to lower organic matter content. The issues confirm the important role of PSD in water and nutrient dynamics within soil and their uptake by plants (Dong et al., 2022).

The Soil Nexus. https://doi.org/10.1016/B978-0-443-45286-4.00009-2

1.2 Water retention

Water retention in soil is an important agricultural and environmental physical process (Kumar et al., 2023). It was significantly influenced by a wide range of soil attributes, including soil texture, soil structure, compaction, and organic matter content, as well as managerial activities such as tillage, human and animal traffic, land leveling, fertilization, and liming (Panagea et al., 2021). Retention of water in soil can affect the chemical fraction of nutrient elements, consequently their availability to plants and soil fertility and productivity. In addition, it decreases soil erosion and degradation, which both depend on the soil type (Vereecken et al., 1989). Therefore, studying water retention and its improvement via application of soil conditioners or applying appropriate managerial practices is essential due to its remarkable impacts on water and nutrient availability to plants, land productivity, and soil biological health, particularly in arid/semiarid conditions (Kumar et al., 2023; Panagea et al., 2021). In other words, water retention is very important in plant growth and development.

1.3 Role of roots in ecosystem functioning

Plants, as an important kind of living organisms, are connection bridges between the lithosphere and the hydrosphere through their roots and shoots (Schenk & Jackson, 2005). Root-soil interactions comprise physical, chemical, and biological types of interactions. Plant roots, along with their related microorganisms, affect ecosystem and environmental functions (Bardgett et al., 2014; Freschet & Roumet, 2017). For instance, roots uptake nutrient elements and water for plant survival, stabilize plants and soil, support soilborne microorganisms and biodiversity, and reduce soil erosion and land degradation (Dignac et al., 2017; Martin & Isaac, 2015).

2. Soil properties affecting water retention, dynamics, and its availability to plants

2.1 Soil texture and structure

Among the physical properties of soils, soil texture and structure are among the most important ones affecting the availability of water and nutrients to plants, and consequently, plant yield and soil productivity (Fig. 3.1). Soil texture represents the relative percentage of soil primary particles in the mineral fraction of the soil (i.e., sand, silt, and clay). Based on the percentages of these primary particles, soils are classified into 12 textural classes ranging from fine (clay) to coarse (sand) textured soils according to a standard soil texture triangle (Fig. 3.2).

The interaction of the mentioned components (i.e., sand, silt, and clay) results in a wide range of soil physical, mechanical, and hydraulic attributes, for example, water movement and retention (Gupta & Singh, 2022). The relationship between soil texture and water retention is crucial in managing environmental challenges, preserving land, and enhancing the quantity and quality of agricultural products (Gupta & Singh, 2022). For instance, sandy soils cannot provide the water requirements of plants under water shortage conditions that are common in arid and semiarid regions of the world. Whereas the water retention capacity of the fine-textured soils is high, and these soils are prone to

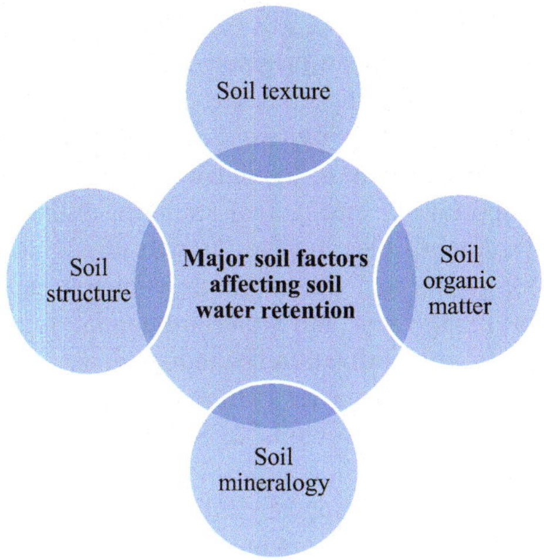

FIGURE 3.1

Major soil factors affecting soil water retention.

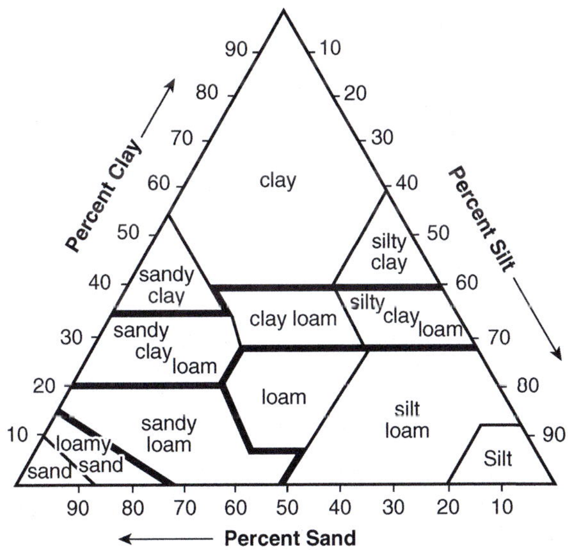

FIGURE 3.2

The USDA Soil Texture Triangle and 12 soil textural classes.

reduction conditions and waterlogging, they need to be drained through artificial drainage systems. On the other hand, loamy (medium-textured) soils with a balanced proportion of soil textural components (i.e., sand, silt, and clay) make an ideal medium and environmental conditions for plant growth. Consequently, identifying soil texture is important for sustainable agriculture and achieving economic plant production.

2.2 Soi organic matter

Organic matter (OM) is the living organism-originated materials containing CHO compounds of varying rates of decomposition. OM plays important direct/indirect roles in different soil functions and properties, for example, water and nutrient dynamics, adsorption-desorption process of materials, chemical forms of elements and heavy metals, aggregation and structural stability, water infiltration and movement, pollutant transportation, mechanical strength, drainage, soil temperature, and soil aeration. Furthermore, it plays crucial roles in plant growth through supplying some of the needed essential nutrient elements for plants and the supporting microorganisms, and also providing an optimal growing medium for plant roots and other influential living organisms. Soil OM can remarkably affect soil fertility via immobilization (Zahedifar et al., 2012), mineralization (Zahedifar et al., 2011), increasing water holding capacity (Minasny & McBratney, 2018), and providing an optimal habitat for soil microorganisms (Bouajila & Sanaa, 2011). Investigators reported positive influences of OM on the growth and chemical composition of plants and soil properties. For instance, Gavili et al. (2019a) stated that the application of cattle manure-derived biochar could increase the growing parameters and yield of soybean at drought-stressed conditions (Table 3.1). They concluded that application of biochar up to 2.5%wt could have positive impacts; whereas application of higher doses imposed negative influences due to increasing soil salinity (Table 3.2). The investigators also stated that application of cattle manure-derived biochar could positively affect various soil physical (Gavili et al., 2018) and chemical properties (Gavili et al., 2019b), as well as soil quality indices (Zahedifar et al., 2025).

The water cycle and its global balance are negatively affected by climate change and global warming, resulting from increased greenhouse gas emissions from fuel combustion and industrial and agricultural activities. On one hand, reduced rainfall and increased evapotranspiration and the resultant drought periods, and on the other hand, floods and waterlogging are direct consequences of climate change (Friedrich et al., 2022). The mentioned climate change-induced consequences, particularly drought and water shortage crisis is one of the most significant challenges for agricultural activities, especially in arid/semiarid regions of the world. If the soil and growing media cannot store sufficient moisture across the wet seasons (fall and winter), the agricultural activities will likely face serious matters during the following dry seasons (Marx, 2022; Schindler et al., 2007). Therefore, it is necessary to improve the water-holding capacity of soils, particularly in arid region soils. Addition of organic matter such as agricultural wastes, crop residues, animal manures, biochar, and sewage sludge is one of the effective ways for improving soil attributes as well as their water retention capacity (Rawls et al., 2003; Olness & Archer, 2005; Minasny & McBratney, 2018).

Furthermore, the addition of OM to soils can improve the plant root growth and development within soil through improving soil structure, increasing soil porosity, supplying nutrients, and providing the needed nutrients and habitat for microorganisms. Overall, incorporating OM into the soil can result in beneficial changes in various soil properties and plant growth as reported in previous studies.

2.3 Soil moisture dynamics

Soil moisture status significantly influences plant growth and development, nutrient transportation within soil and their uptake by plant roots, and the thermal properties of the soil media, as well as

Table 3.1 Effect of cattle manure-derived biochar and soil moisture status (SMS, %field capacity, FC) on some agronomic traits and yield components of soybean.

		Agronomic traits and yield components										
	Biochar	Stomatal conductance	Leaf area	WUE[b]	SFMY	SDMY	1000-GW	GY	PLH	PN	GN	HI
SMS	(%wt)	(μmol m^{-2} s^{-1})	(mm^2 pot^{-1})	(g L^{-1})	(g pot^{-1})	(g pot^{-1})	(g)	(g pot^{-1})	(cm)	(pot^{-1})	(pot^{-1})	(%)
FC	0	0.16[a] *b*	3465 *a*	0.26 *b*	19.6 *b*	5.9 *bc*	204 *ab*	10.1 *a*	134 *a*	25 *b*	50 *b*	28 *ab*
	1.25	0.22 *a*	3364 *ab*	0.40 *a*	28.3 *a*	8.5 *a*	221 *ab*	12.2 *a*	147 *a*	28 *a*	55 *a*	27 *abc*
	2.5	0.21 *ab*	3667 *a*	0.34 *ab*	17.7 *bc*	6.9 *ab*	237 *a*	11.1 *a*	105 *b*	23 *b*	47 *b*	27 *abc*
	5	0.17 *b*	2732 *cd*	0.24 b	11.2 ce	4.3 *ce*	177 *bc*	5.8 *cd*	94 *b*	18 *c*	32 *c*	23 *cd*
0.7FC	0	0.08 *f*	2329 *de*	0.28 *b*	10.3 *de*	4.5 *ce*	215 *ab*	6.6 *bc*	99 *b*	15 *cd*	31 *c*	30 *a*
	1.25	0.18 *b*	2936 *bc*	0.33 *ab*	14.8 *bd*	5.5 *bd*	223 *ab*	7.7 *b*	108 *b*	17 *c*	34 *c*	26 *abc*
	2.5	0.17 *b*	2320 *de*	0.24 *b*	10.4 *de*	3.5 *ef*	223 *ab*	5.5 *be*	66 *c*	14 *de*	25 *d*	27 *ab*
	5	0.14 *c*	2182 *e*	0.25 *b*	9.2 *de*	3.1 *ef*	173 *bc*	3.3 *e*	50 *cd*	10 *f*	19 *d*	22 *d*
0.55FC	0	0.07 *f*	2116 *e*	0.26 *b*	3.3 *de*	3.5 *ef*	210 *ab*	4.6 *ce*	47 *cd*	14 *de*	22 *d*	27 *ab*
	1.25	0.13 *cd*	2217 *de*	0.30 *ab*	3.3 *de*	3.8 *df*	226 *ab*	4.6 *ce*	67 *c*	13 *df*	21 *d*	26 *bc*
	2.5	0.12 *de*	1872 *e*	0.23 *b*	5.1 *e*	2.8 *ef*	192 *ac*	3.9 *de*	44 *cd*	11 *ef*	20 *d*	25 *bc*
	5	0.10 *e*	2237 *de*	0.24 *b*	5.0 *e*	2.2 *f*	138 *c*	0.9 *f*	32 *d*	5 *g*	7 *e*	16 *e*

[a] *Means of each attribute correspond to biochar levels at every SMS, followed by the same lowercase letter, are not significantly different based on Duncan's Test at* $P < .05$.
[b] *WUE, SFMY, SDMY, 1000-GW, GY, PLH, PN, GN, and HI are water use efficiency, straw fresh matter yield, straw dry matter yield, 1000- grain weight, grain yield, plant height, pod number, grain number, and harvest index, respectively.*
After Gavili, E., Moosavi, A. A., & Kamkar Haghighi, A. A. (2019a). Does biochar mitigate the adverse effects of drought on the agronomic traits and yield components of soybean? Industrial Crops and Products, 128, 445–454. https://doi.org/10.1016/j.indcrop.2018.11.047.

Table 3.2 Effect of cattle manure-derived biochar and soil moisture status (SMS, %field capacity, FC) on electrical conductivity (EC, dS/m) of soil saturated extract after soybean harvesting.

	Biochar (%wt)				
SMS	**0**	**1.25**	**2.5**	**5**	**Mean**
FC	1.16[a] *i*	3.15 *gi*	6.16 *df*	9.17 *ac*	**4.91 *B***
0.7FC	2.18 *hi*	4.39 *fh*	7.00 *ce*	9.43 *ab*	**5.75 *AB***
0.55FC	2.22 *hi*	5.01 *eg*	7.44 *bd*	10.20 *a*	**6.22 *A***
Mean	**1.86 *D***	**4.19 *C***	**6.87 *B***	**9.60 *A***	

[a]Means in each row or column followed by the same capital letter and the values in the table body followed by the same lowercase letters are not significantly different based on Duncan's Multiple Range Test at P <.05.
After Gavili, E., Moosavi, A. A., & Kamkar Haghighi, A. A. (2019a). Does biochar mitigate the adverse effects of drought on the agronomic traits and yield components of soybean? Industrial Crops and Products, 128, 445–454. https://doi.org/10.1016/j.indcrop.2018.11.047.

playing an important role in the irrigation of agricultural fields (Babaeian et al., 2019; Mohanty et al., 2017). Furthermore, it affects microorganism activity, soil compaction, and the quantity and quality of agricultural products. Soil water dynamics also influence the rate of water and nutrient uptake and consequently crop yield (Kumar et al., 2023).

Soil moisture dynamics show the changes of water across wetting or drying cycles, which is called the hysteresis phenomenon (Lu & Khorshidi, 2015; Nuth & Laloui, 2008). In other words, it depends on the initial soil moisture. Soil moisture retention curve (SMRC), which is a curve showing the relation of water content and matric potential, is used to show the state of soil water and its dynamics across drying-wetting cycles and provides valuable information about soil water storage (Hartge et al., 2016). SMRC is highly dependent on soil physical (soil texture, structure, porosity, etc.) and chemical (clay types, soluble salts, exchangeable sodium percentage, and organic matter) characteristics and managerial practices, including tillage, liming, and cropping (Huang et al., 2021; Pečan et al., 2023). For instance, Moradi-Choghamarani et al. (2024) reported that application of 3%wt sugarcane bagasse-derived biochar significantly increased stability ratio (SR) and structural stability index (SSI) of their three texturally different studied soils including silty clay in Inceptisols order, silty clay loam in Alfisols order, and loamy in Aridisols order (Khormali & Abtahi, 2003) by 35%–100%, as compared to those of the control. They stated that with increasing biochar doses from 0 to 3%wt, the plant available water (PAW), least limiting water range (LLWR) significantly decreased in the studied silty clay (18%–61%, 8%–44%, and 6%–35%) and silty clay loam (8%–44%, 18%–35%, and 20%–47%) soils; whereas, increased in loamy soil (4%–54%, 3%–61%, and 24%–111%).

2.4 Availability of soil water to plants

The soil's available water for plant uptake has been variously defined in the literature. Commonly, plant available water (PAW) is introduced as the difference between water content at field capacity (FC) and permanent wilting point (PWP), which themselves are measured using a pressure plate apparatus at average matric potential of 330 and 15,000 cm, respectively (Moradi-Choghamarani et al., 2024). This approach is based on a uniform water absorption pattern by plants in a cut-off form from FC to PWP and only considers water potential as a limitation factor for water uptake by plant roots (de Lima et al., 2016). In recent years, the term least limiting water range (LLWR) has been introduced (Benjamin et al., 2003). LLWR considers and integrates water potential, aeration, and penetration resistance as the three major limiting factors to plant growth that are obtained from SMRC and penetration resistance (PR) curves. An R package entitled *soilphysics* has been provided by de Lima et al. (2016) for calculating plant available water, in which the exponential SMRC (Ross et al., 1991) is considered for water potential restriction.

A power model (Eq. 3.1) can be used to consider penetration resistance (*PR*) limitation and create a continuous function to the relation of soil PR versus matric potential (*h*):

$$PR = bh^{c} \tag{3.1}$$

where *b* (MPa/cm) and *c* are fitting parameters (Groenevelt et al., 2001).

The water content ($\theta h_{PR\text{-}2}$, cm) and equivalent h ($h_{PR\text{-}2}$, cm) at which PR reaches the 2 MPa critical point are then calculated (Eqs. 3.1 and 3.2).

$$\theta = \theta_r + [(\theta_s - \theta_r)(1 + (\alpha h)^n)^{-m}] \tag{3.2}$$

where θ_r and θ_s are the residual and saturated water content, respectively. α and *n* are shape parameters that control the position and steepness of SMRC, respectively. *m* is another parameter of SMRC (m = 1-(1/*n*)), and its small values are obtained in soils having a wide PSD.

Considering the aeration restriction, water content at a 10% air-filled porosity ($\theta_{A\text{-}10}$, cm^3/cm^3) is calculated (Eq. 3.3).

$$\theta_{A-10} = \left(1 - \frac{\text{Bulk density}}{\text{Particle density}}\right) - 0.1 \tag{3.3}$$

The water content at FC or at $\theta_{A\text{-}10}$, whichever is smaller, is considered the upper limit of LLWR. Water content at PWP or $\theta_{PR\text{-}2}$, whichever is larger, is considered the lower limit of LLWR. Groenevelt et al. (2001) introduced the integral water capacity (IWC) term (Eq. 3.4) as the sharp cut-off between PAW and LLWR procedures to determine the available water as a continuous weighting function of its limiting factors.

$$\text{IWC} = \int_0^{\infty} \left(\prod_{i=1}^{p} \omega_i(h) \right) C(h) dh \tag{3.4}$$

where $\omega_i(h)$ are multiplicative weighting functions of the first to pth limiting soil factors as a function of h; *C*(*h*) is the absolute specific water capacity function (i.e., $|d\theta/dh|$) based on the van Genuchten (1980) equation. High penetration resistance and low soil hydraulic conductivity at the dry range, and lack of oxygen and high soil hydraulic conductivity at the wet range are considered as restriction factors to water uptake. The relative hydraulic conductivity ($K_r(h)$) equation (Mualem, 1976) is used to calculate IWC (Eq. 3.5).

$$K_r(h) = \left[1 - (\alpha h)^{n-1}[1 + (\alpha h)^n]^{\left(\frac{1}{n}-1\right)}\right]^2 [1 + (\alpha h)^n]^{\left(\frac{1-n}{2n}\right)} \tag{3.5}$$

The method proposed by Minasny & McBratney (2003) is used calculate the amount of energy required to remove the available water value defined by IWC (EI_{IWC}) as follows:

$$EI_{\text{IWC}} = \frac{1}{10\text{IWC}} \int_{h_i}^{h_f} \left(\prod_{i=1}^{m} \omega_i(h) \right) hC(h)dh \tag{3.6}$$

where h_i and h_f are boundary h values at the wet and dry ends, respectively. The detailed descriptions on the procedures and equations have been brought in Groenevelt et al. (2001), Minasny & McBratney (2003), Asgarzadeh et al. (2010, 2011), and de Lima et al. (2016).

Soil water availability-related attributes are influenced by various soil and managerial factors. For instance, Moradi-Choghamarani et al. (2024) stated that application of 1%, 2%, and 3% wt sugarcane bagasse-derived biochar produced at 300 and 600 °C significantly decreased PAW, LLWR, and IWC in their studied Inceptisols and Alfisols (Table 3.3). However, they reported opposite results in their studied Aridisols (Table 3.3).

3. Root architecture and its importance

3.1 Types of root systems

The root system of plants is the primary structure that fixes them in the ground and absorbs water and nutrients (Fig. 3.3). Plants commonly face various environmental stresses that harmfully affect their survival, yield, and efficiency. Drought is one of the most significant challenges for plant growth, which firstly and negatively affects the root system of the plants. It has been estimated that the plant's water requirement is going to be exceeded two-fold by 2050, while available fresh water will probably decrease by nearly 50% due to climate change (Amtmann et al., 2022; Gupta et al., 2020; Maqbool et al., 2022).

Different plant species employ various strategies to overcome the drought-induced adverse consequences. For example, leaves can regulate transpiration through opening and closing their stomata and producing a waxy film on their leaf surfaces to minimize water loss. Furthermore, the roots of some plants apply specific adaptations, such as physiological changes to overcome drought stress and the formation of the root exodermis, which reduces the root's radial hydraulic conductivity, resulting in more water retention. Besides, plants may decrease water loss by increasing the salt concentration of the root cells (Naveed et al., 2018, 2019).

Root depth and volume, diameter, and growth angle are defined as the root architecture indices (Pandey & Bennett, 2019). In water shortage conditions, roots can adjust the mentioned architectural indices to improve water retention. For instance, Thidar et al. (2020) reported that plants decreased both root density and length under drought conditions. They also concluded that the root length density distribution over various dimensions plays a vital role in water and nutrient uptake efficiency.

3.2 Adaptations to soil conditions

The roots' growth is remarkably influenced by gravity, soil aeration, soil composition, soil moisture, soil structure, density, mechanical resistance, soil porosity, and their related attributes, which consequently affect crop yield (Colombi et al., 2018; Rich & Watt, 2013). Root growth is limited in soils of high mechanical resistance (i.e., fine-textured soils) but improved in coarse-textured soils having a

Table 3.3 Effect of different levels of two sugarcane bagasse-derived biochars produced at 300°C (BC300) and 600°C (BC600) on the mean values of some water availability-related attributes in three different soil types.

Soil type	Biochar (BC)	Biochar level (%wt)	Plant available water (PAW) ($cm^3\ cm^{-3}$)	Least limiting water range (LLWR) ($cm^3\ cm^{-3}$)	Integral water capacity (IWC) ($cm^3\ cm^{-3}$)
Inceptisols	BC300	0	0.143§ a	0.100 a	0.136 a
		1	0.108 b	0.086 b	0.127 b
		2	0.081 c	0.076 c	0.120 c
		3	0.056 c	0.056 d	0.107 d
	BC600	0	0.143 a	0.100 a	0.136 a
		1	0.116 b	0.092 b	0.107 b
		2	0.081 c	0.071 c	0.090 c
		3	0.057 d	0.057 d	0.088 c
Alfisols	BC300	0	0.113 a	0.095 a	0.157 a
		1	0.089 b	0.078 b	0.126 b
		2	0.071 bc	0.068 c	0.112 c
		3	0.063 c	0.063 d	0.084 d
	BC600	0	0.113 a	0.095 a	0.157 a
		1	0.104 a	0.062 b	0.104 b
		2	0.065 b	0.063 b	0.092 b
		3	0.065 b	0.065 b	0.088 b
Aridisols	BC300	0	0.090 b	0.079 c	0.107 a
		1	0.086 b	0.077 c	0.133 b
		2	0.133 a	0.125 a	0.175 a
		3	0.104 b	0.104 b	0.194 a
	BC600	0	0.090 b	0.079 b	0.107 a
		1	0.133 a	0.115 a	0.152 b
		2	0.139 a	0.126 a	0.199 a
		3	0.127 a	0.127 a	0.226 a

§*Means of each attribute correspond to biochar levels for each studied soil type followed by the same lowercase letter, are not significantly different on the LSD test at* $P < .05$.
After Moradi-Choghamarani, F., Moosavi, A.A., & Sepaskhah, A.R. (2024). Sugarcane bagasse derived biochar potential to improve soil structure and water availability in texturally different soils. Scientific Reports, 14, Article 26541. 10.1038/s41598-024-77700-3.

high proportion of macropores (large pores) (Islam et al., 2021). On the other hand, high soil penetration resistance often results in root twisting and shortening that adversely affects plant growth and yield (Moraes & Gusmão, 2021; Popova et al., 2016).

Water availability is one of the most influential factors in root growth. In other words, the limited available water in drought conditions often results in a remarkable decrease in root elongation. However, the roots of certain plant species may grow and develop, but typically with lower diameters,

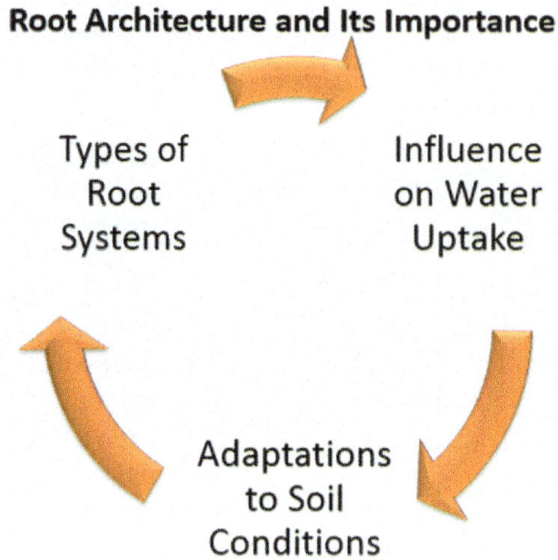

FIGURE 3.3

Interaction of root systems, soil, and water uptake.

under drought conditions. Overall, while roots can grow in various moisture conditions, their growth and development are limited under drought conditions (Gregory, 2006; Sharp et al., 1988).

3.3 Influence on water uptake

The root architecture strongly affects its ability to uptake water, especially under drought conditions, as well as nutrient uptake (Bouda & Saiers, 2017). The presence and distribution of water within soil pores significantly affect water and nutrient availability to plant roots and, consequently, influence plant growth and crop yield. Therefore, effective management of soil resources that facilitate root growth and enhance water and nutrient supply is crucial to achieve sustainable agriculture. Typically, nutrients move toward roots through mass flow (convection) and diffusion (Zarebanadkouki et al., 2019). The mentioned mechanisms depend on water flow within the soil and the nutrient concentration gradient, both of which depend on soil moisture content and are negatively influenced under drought conditions. As a consequence, global warming and climate change that affect soil moisture will adversely affect the degree of water and nutrient availability to plant roots (Griffiths & York, 2020).

4. Mechanisms of water retention in soil

4.1 Capillary action and soil moisture

The soil moisture retention curve (SMRC) shows the relationship between soil matric potential (suction) and water content (Fig. 3.4), and overall, the changes in available water as a function of matric potential (Lu et al., 2019). This curve directly depends on pore size distribution and the geometry of the soil particle (Montzka et al., 2017). The capillary water refers to the water exist in the capillary pores and can move by capillary movement through the soil profile. The soil water not only

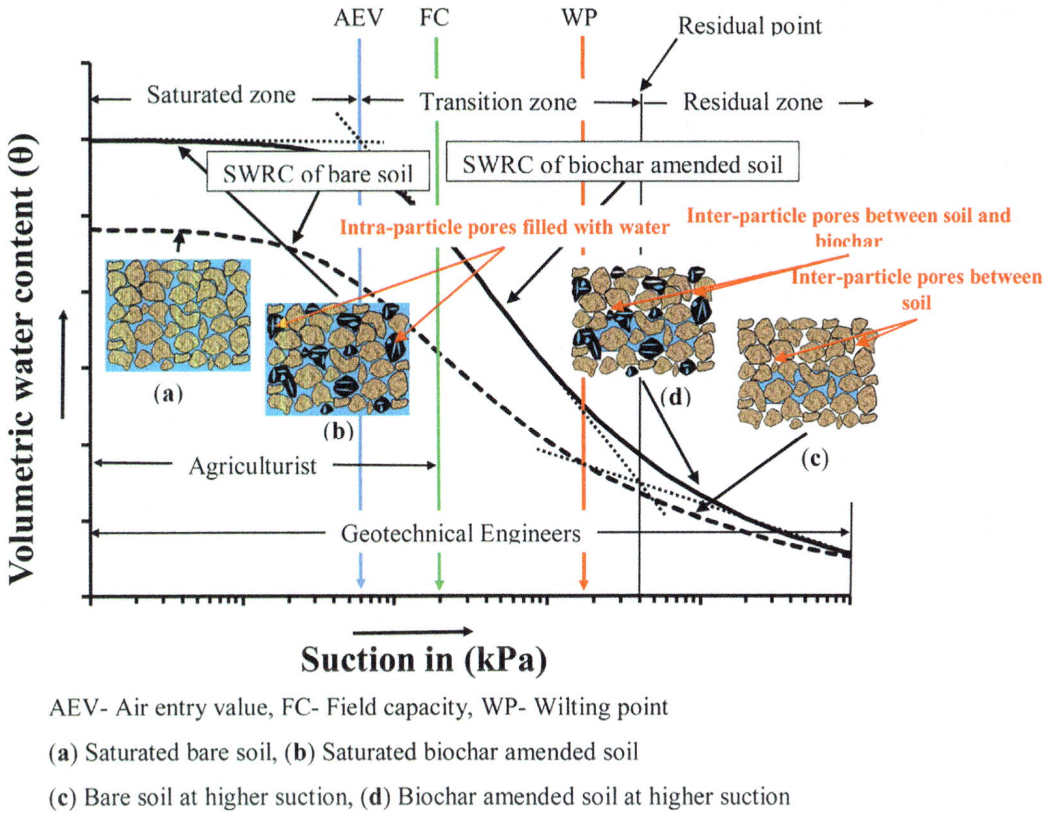

FIGURE 3.4

Schematic description of the soil moisture characteristics curve (SMRC) of a biochar-amended soil compared to a bare soil.

After Hussain, R., Ravi, K., & Garg, A. (2020). Influence of biochar on the soil water retention characteristics (SWRC): Potential application in geotechnical engineering structures. Soil and Tillage Research, 204, 104713.

depends on the matric potential, but also depends on hysteresis (Lu & Khorshidi, 2015; Nuth & Laloui, 2008). The difference between the SMRC obtained from desorption and sorption approaches is called the hysteresis phenomena than result in variation in water retention and release over wetting and drying cycles (Fig. 3.5). Different soil factors (including soil texture, organic matter, etc.) affect SMRC and hysteresis. For instance, fine-textured soils and soils rich in organic matter retain more water than coarse-textured soils and soils poor in organic matter (Konyai et al., 2009; Sun et al., 2020).

4.2 Role of soil aggregates

Soil structure and its stability meaningfully affect SMRC, water movement, and storage, as well as its availability to plants. In other words, the proportion and arrangement of soil micropores and

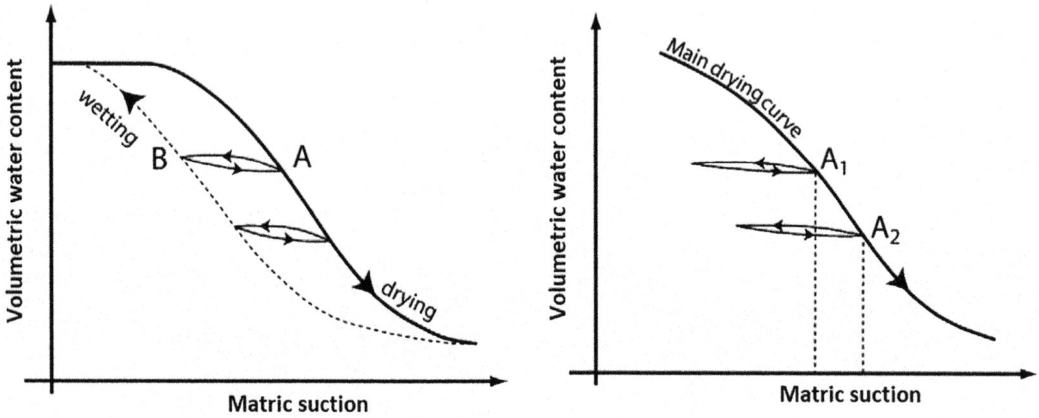

FIGURE 3.5

Hysteresis phenomena of the soil moisture characteristics curve.

After Nuth, M., & Laloui, L. (2008). Advances in modelling hysteretic water retention curve in deformable soils. Computers and Geotechnics, 35(6), 835-844.

macropores affect its potential for water retention and its movement through soil. Indeed, micropores contribute to water retention, while macropores facilitate water movement within the soil profile. Therefore, water infiltrates more easily and effectively into well-structured soils and consequently improves moisture content in the rooting depth. In contrast, poorly structured soils face high runoff and erosion problems as well as low water penetration into the soils, which may result in drought stress (Mishra & Singh, 2025). Several indicators, such as bulk density, total porosity, aggregate stability, and infiltration rate, can be applied to evaluate soil structure and the resultant attributes and processes, for example, water infiltration into the soil. Various strategies, including the application of soil amendments such as biochar, compost, and cover crops, are commonly used to improve soil structure and aggregation. These practices remarkably improve soil structure, porosity, and hydraulic conductivity, which consequently improve plant growth and crop yield (Gavili et al., 2018; Chen et al., 2020). Other investigators reported that incorporating 2% biochar into the soil significantly increased permeability (Wong et al., 2018). They also stated that incorporating biochar into soils enhanced soil water retention capacity.

5. Conclusion

The chapter discussed the soil-plant root systems interactions with respect to water availability and uptake. The most influential soil attributes related to water dynamics and uptake by plant roots, along with some common adaptation strategies used by plants to overcome the unfavorable conditions, were also discovered. Furthermore, some new quantitative indices of the availability of water to plants, including LLWR and IWC, were presented. Soil texture and organic matter content were recognized as two important influential factors to root growth and water availability to plants. So that, medium-textured (loamy) soils provide optimal conditions; whereas, coarse and fine-textured soils present unfavorable conditions (e.g., water and nutrient limitation and aeration challenges, respectively). Organic amendments, particularly biochar and at moderate application rates ($\leq$2.5% wt), showed significant ability to improve soil water retention and structural-related attributes. Although higher rates of the applied biochar restrict plant growth due to increasing salinity. Climate change and human-induced drought conditions necessitate an understanding of soil-plant-water dynamics and the implementation of integrated and appropriate managerial approaches, including the application of organic amendments to overcome drought conditions and achieve sustainable agriculture.

References

Adomako, M. O., Roiloa, S., & Yu, F. H. (2022). Potential roles of soil microorganisms in regulating the effect of soil nutrient heterogeneity on plant performance. *Microorganisms, 10*(12), 1–17. https://doi.org/10.3390/microorganisms10122399

Amtmann, A., Bennett, M. J., & Henry, A. (2022). Root phenotypes for the future. *Plant, Cell and Environment, 45* (3), 595–601. https://doi.org/10.1111/pce.14269

Asgarzadeh, H., Mosaddeghi, M. R., Mahboubi, A. A., Nosrati, A., & Dexter, A. R. (2010). Soil water availability for plants as quantified by conventional available water, least limiting water range and integral water capacity. *Plant and Soil, 335*, 229–244. https://doi.org/10.1007/s11104-010-0410-6

Asgarzadeh, H., Mosaddeghi, M. R., Mahboubi, A. A., Nosrati, A., & Dexter, A. R. (2011). Integral energy of conventional available water, least limiting water range and integral water capacity for better characterization of water availability and soil physical quality. *Geoderma, 166*, 34–42. https://doi.org/10.1016/j.geoderma.2011.06.009

Babaeian, E., Sadeghi, M., Jones, S. B., Montzka, C., Vereecken, H., & Tuller, M. (2019). Ground, proximal, and satellite remote sensing of soil moisture. *Reviews of Geophysics, 57*(2), 530–616.

Baldi, E. (2021). Soil–plant interaction: Effects on plant growth and soil biodiversity. *Agronomy, 11*(12), 10–12. https://doi.org/10.3390/agronomy11122378

Bardgett, R. D., Mommer, L., & De Vries, F. T. (2014). Going underground: Root traits as drivers of ecosystem processes. *Trends in Ecology and Evolution, 29*(12), 692–699. https://doi.org/10.1016/j.tree.2014.10.006

Benjamin, J. G., Nielsen, D. C., & Vigil, M. F. (2003). Quantifying effects of soil conditions on plant growth and crop production. *Geoderma, 116*(1–2), 137–148.

Bouajila, K., & Sanaa, M. (2011). Effects of organic amendments on soil physico-chemical and biological properties. *Journal of Materials and Environmental Science, 2*, 485–490.

Bouda, M., & Saiers, J. E. (2017). Dynamic effects of root system architecture improve root water uptake in 1-D process-based soil-root hydrodynamics. *Advances in Water Resources, 110*, 319–334. https://doi.org/10.1016/j.advwatres.2017.10.018

Brown, C., Oppon, K. J., & Cahill, J. F. (2019). Species-specific size vulnerabilities in a competitive arena: Nutrient heterogeneity and soil fertility alter plant competitive size asymmetries. *Functional Ecology, 33*(8), 1491–1503. https://doi.org/10.1111/1365-2435.13340

Chen, Q., Miao, F., Wang, H., Xu, Z., Tang, Z., Yang, L., & Qi, S. (2020). Downscaling of satellite remote sensing soil moisture products over the Tibetan Plateau based on the random forest algorithm: Preliminary results. *Earth and Space Science, 7*.

Colombi, T., Torres, L. C., Walter, A., & Keller, T. (2018). Feedbacks between soil penetration resistance, root architecture and water uptake limit water accessibility and crop growth – a vicious circle. *Science of the Total Environment, 626*, 1026–1035. https://doi.org/10.1016/j.scitotenv.2018.01.129

de Lima, R. P., da Silva, A. R., da Silva, A. P., Leao, T. P., & Mosaddeghi, M. R. (2016). Soilphysics: An R package for calculating soil water availability to plants by different soil physical indices. *Computers and Electronics in Agriculture, 120*, 63–71. https://doi.org/10.1016/j.compag.2015.11.003

Dignac, M. F., Derrien, D., Barré, P., Barot, S., Cécillon, L., Chenu, C., Chevallier, T., Freschet, G. T., Garnier, P., Guenet, B., Hedde, M., Klumpp, K., Lashermes, G., Maron, P. A., Nunan, N., Roumet, C., & Basile-Doelsch, I. (2017). Increasing soil carbon storage: Mechanisms, effects of agricultural practices and proxies. A review. *Agronomy for Sustainable Development, 37*(2). https://doi.org/10.1007/s13593-017-0421-2

Dong, L., Wang, J., Li, J., Wu, Y., Zheng, Y., Zhang, J., Li, Z., Yin, R., & Liang, C. (2022). Assessing the impact of grazing management on wind erosion risk in grasslands: A case study on how grazing affects aboveground biomass and soil particle composition in Inner Mongolia. *Global Ecology and Conservation, 40*, Article e02344. https://doi.org/10.1016/j.gecco.2022.e02344

Freschet, G. T., & Roumet, C. (2017). Sampling roots to capture plant and soil functions. *Functional Ecology, 31* (8), 1506–1518. https://doi.org/10.1111/1365-2435.12883

Friedrich, K., Imbery, F., & Klasen, U. (2022). *Klimastatusbericht Deutschland Jahr 2021*. Deutscher Wetterdienst. https://dwd.de/klimaatlas.

Gavili, E., Moosavi, A. A., & Kamkar Haghighi, A. A. (2019a). Does biochar mitigate the adverse effects of drought on the agronomic traits and yield components of soybean? *Industrial Crops and Products, 128*, 445–454. https://doi.org/10.1016/j.indcrop.2018.11.047

Gavili, E., Moosavi, A. A., & Moradi Choghamarani, F. (2018). Cattle manure biochar potential for ameliorating soil physical characteristics and spinach response under drought. *Archives of Agronomy and Soil Science, 64*, 1714–1727. https://doi.org/10.1080/03650340.2018.1453925

Gavili, E., Moosavi, A. A., & Zahedifar, M. (2019b). Integrated effects of cattle manure-derived biochar and soil moisture conditions on soil chemical characteristics and soybean yield. *Archives of Agronomy and Soil Science, 65*, 1758–1774. https://doi.org/10.1080/03650340.2019.1576864

Gregory, P. (2006). *Plant roots*. Blackwell Publishing Ltd.

Griffiths, M., & York, L. M. (2020). Targeting root ion uptake kinetics to increase plant productivity and nutrient use efficiency. *Plant Physiology, 182*(4), 1854–1868. https://doi.org/10.1104/pp.19.01496

Groenevelt, P. H., Grant, C. D., & Semetsa, S. (2001). A new procedure to determine soil water availability. *Australian Journal of Soil Research, 39*, 577–598. https://doi.org/10.1071/SR99084

Gupta, A., Rico-Medina, A., & Cano-Delgado, A. I. (2020). The physiology of plant responses to drought. *Science, 368*(6488), 266–269. https://doi.org/10.1126/science.aaz7614

Gupta, A., & Singh, S. (2022). Understanding soil texture and its impact on water retention. *International Journal for Advanced Research in Science and Technology, 12*(10), 196–201.

Hartge, K. H., Horn, R., Horton, R., Bachmann, J., & Peth, S. (2016). *Essential soil physics: An introduction to soil processes, structure, and mechanics*. Schweizerbart Science Publishers.

Huang, X., Wang, H., Zhang, M., Horn, R., & Ren, T. (2021). Soil water retention dynamics in a mollisol during a maize growing season under contrasting tillage systems. *Soil and Tillage Research, 209*, Article 104953. https://doi.org/10.1016/j.still.2021.104953

Hussain, R., Ravi, K., & Garg, A. (2020). Influence of biochar on the soil water retention characteristics (SWRC): Potential application in geotechnical engineering structures. *Soil and Tillage Research, 204*, 104713.

Islam, D. M., Price, A. H., & Hallett, P. D. (2021). Contrasting ability of deep and shallow rooting rice genotypes to grow through plough pans containing simulated biopores and cracks. *Plant and Soil, 466*(1), 515–530. https://doi.org/10.1007/s11104-021-05131-4

Jiang, P., Ding, W., Yuan, Y., & Ye, W. (2020). Diverse response of vegetation growth to multi-time-scale drought undedifferent soil textures in China's pastoral areas. *Journal of Environmental Management, 274*, Article 110992. https://doi.org/10.1016/j.jenvman.2020.110992

Jiang, L., Wu, L., Liu, H., He, W., Shi, L., Xu, C., & Xiang, C. (2022). Coarsened soil reduces drought resistance of fibrous-rooted species on degraded steppe. *Ecological Indicators, 145*, Article 109644. https://doi.org/10.1016/j.ecolind.2022.109644

Khormali, F., & Abtahi, A. (2003). Origin and distribution of clay minerals in calcareous arid and semi-arid soils of Fars province, Southern Iran. *Clay Minerals, 38*, Article 511e527. https://doi.org/10.1180/0009855023740112

Konyai, S., Sriboonlue, V., & Trelo-Ges, V. (2009). The effect of air entry values on the hysteresis of water retention curve in saline soil. *American Journal of Environmental Sciences, 5*(3), 341–345.

Kumar, S., Shri, D., Baba, S., & Ambikapur, A. M. (2023). A review of soil water retention and its influence on soil physical properties. *Research Zone: A Journal of Science and Humanities, 14*, 31–45.

Lu, N., & Khorshidi, M. (2015). Mechanisms for soil-water retention and hysteresis at high suction range. *Journal of Geotechnical and Geoenvironmental Engineering, 141*(8), 1–10. https://doi.org/10.1061/(asce)gt.1943-5606.0001325

Lu, S., Lu, Y., Peng, W., Ju, Z., & Ren, T. (2019). A generalized relationship between thermal conductivity and matric suction of soils. *Geoderma, 337*, 491–497. https://doi.org/10.1016/j.geoderma.2018.09.057

Maqbool, S., Hassan, M. A., Xia, X., York, L. M., Rasheed, A., & He, Z. (2022). Root system architecture in cereals: Progress, challenges and perspective. *The Plant Journal, 110*(1), 23–42. https://doi.org/10.1111/tpj.15669

Martin, A. R., & Isaac, M. E. (2015). Plant functional traits in agroecosystems: A blueprint for research. *Journal of Applied Ecology, 52*(6), 1425–1435. https://doi.org/10.1111/1365-2664.12526

Marx, A. (2022). Dürremonitor Deutschland. In *Helmholtz-Zentrum für Umweltforschung*. https://www.ufz.de/index.php?de=37937.

Minasny, B., & McBratney, A. B. (2003). Integral energy as a measure of soil-water availability. *Plant and Soil, 249*, 253–262. https://doi.org/10.1023/A:1022825732324

Minasny, B., & McBratney, A. B. (2018). Limited effect of organic matter on soil available water capacity. *European Journal of Soil Science, 69*(1), 39–47. https://doi.org/10.1111/ejss.12475

Mishra, R., & Singh, D. (2025). Role of soil texture and structure in water retention and crop productivity. *International Journal of Creative Research Thoughts, 13*(6), 2320–2882.

Mohammadi, M., Shabanpour, M., Mohammadi, M. H., & Davatgar, N. (2019). Characterizing spatial variability of soil textural fractions and fractal parameters derived from particle size distributions. *Pedosphere, 29*(2), 224–234. https://doi.org/10.1016/S1002-0160(17)60425-9

Mohanty, B. P., Cosh, M. H., Lakshmi, V., & Montzka, C. (2017). Soil moisture remote sensing: State-of-the-science. *Vadose Zone Journal, 16*(1), 1–9. https://doi.org/10.2136/vzj2016.10.0105

Mohseni, N., Hosseinzadeh, S. R., Sepehr, A., Golzarian, M. R., & Shabani, F. (2017). Variations in spatial patterns of soil-vegetation properties and the emergence of multiple resilience thresholds within different debris flow fan positions. *Geomorphology, 290*, 365–375. https://doi.org/10.1016/j.geomorph.2017.04.023

Montzka, C., Herbst, M., Weihermüller, L., Verhoef, A., & Vereecken, H. (2017). A global data set of soil hydraulic properties and sub-grid variability of soil water retention and hydraulic conductivity curves. *Earth System Science Data, 9*, 529–543. https://doi.org/10.5194/essd-9-529-2017

Moradi-Choghamarani, F., Moosavi, A. A., & Sepaskhah, A. R. (2024). Sugarcane bagasse derived biochar potential to improve soil structure and water availability in texturally different soils. *Scientific Reports, 14*, Article 26541. https://doi.org/10.1038/s41598-024-77700-3

Moraes, M. T.d., & Gusmão, A. G. (2021). How do water, compaction and heat stresses affect soybean root elongation? A review. *Rhizosphere, 19*, Article 100403. https://doi.org/10.1016/j.rhisph.2021.100403

Mualem, Y. (1976). A new model for predicting the hydraulic conductivity of unsaturated porous media. *Water Resources Research, 12*, 513–522. https://doi.org/10.1029/WR012i003p00513

Naveed, M., Ahmed, M. A., Benard, P., Brown, L. K., George, T., Bengough, A., Roose, T., Koebernick, N., & Hallett, P. (2019). Surface tension, rheology and hydrophobicity of rhizodeposits and seed mucilage influence soil water retention and hysteresis. *Plant and Soil, 437*, 65–81.

Naveed, M., Brown, L. K., Raffan, A., George, T. S., Bengough, A., Roose, T., Sinclair, I., Koebernick, N., Cooper, L., & Hallett, P. D. (2018). Rhizosphere-scale quantification of hydraulic and mechanical properties of soil impacted by root and seed exudates. *Vadose Zone Journal, 17*(1), 1–13.

Nuth, M., & Laloui, L. (2008). Advances in modelling hysteretic water retention curve in deformable soils. *Computers and Geotechnics, 35*(6), 835–844.

Olness, A., & Archer, D. (2005). Effect of organic carbon on available water in soil. *Soil Science, 170*(2), 90–101. https://doi.org/10.1097/00010694-200502000-00002

Panagea, I. S., Berti, A., Čermak, P., Diels, J., Elsen, A., Kusá, H., Piccoli, I., Poesen, J., Stoate, C., Tits, M., Toth, Z., & Wyseure, G. (2021). Soil water retention as affected by management induced changes of soil organic carbon: Analysis of long-term experiments in Europe. *Land, 10*(12), Article 1362. https://doi.org/10.3390/land10121362

Pandey, B. K., & Bennett, M. J. (2019). A new angle on how roots acclimate to sporadic rainfall. *Cell, 178*(2), 269–271. https://doi.org/10.1016/j.cell.2019.06.018

Pečan, U., Pintar, M., Mihelič, R., & Kastelec, D. (2023). Variability of in situ soil water retention curves under different tillage systems and growing seasons. *Soil and Tillage Research, 233*, Article 105779. https://doi.org/10.1016/j.still.2023.105779

Popova, L., Van Dusschoten, D., Nagel, K. A., Fiorani, F., & Mazzolai, B. (2016). Plant root tortuosity: An indicator of root path formation in soil with different composition and density. *Annals of Botany, 118*(4), 685–698. https://doi.org/10.1093/aob/mcw057

Rawls, W., Pachepsky, Y., Ritchie, J., Sobecki, T., & Bloodworth, H. (2003). Effect of soil organic carbon on soil water retention. *Geoderma, 116*(1–2), 61–76. https://doi.org/10.1016/s0016-7061(03)00094-6

Rich, S. M., & Watt, M. (2013). Soil conditions and cereal root system architecture: Review and considerations for linking Darwin and Weaver. *Journal of Experimental Botany, 64*(5), 1193–1208. https://doi.org/10.1093/jxb/ert043

Ross, P. J., Willians, J., & Bristow, K. L. (1991). Equations for extending water retention curves to dryness. *Soil Science Society of America Journal, 55*, 923–927. https://doi.org/10.2136/sssaj1991.03615995005500040004x

Schenk, H. J., & Jackson, R. B. (2005). Mapping the global distribution of deep roots in relation to climate and soil characteristics. *Geoderma, 126*(1–2), 129–140. https://doi.org/10.1016/j.geoderma.2004.11.018

Schindler, U., Steidl, J., Müller, L., Eulenstein, F., & Thiere, J. (2007). Drought risk to agricultural land in Northeast and central Germany. *Journal of Plant Nutrition and Soil Science, 170*(3), 357–362. https://doi.org/10.1002/jpln.200622045

Sharp, R., Silk, W., & Hsiao, T. (1988). Growth of the maize primary root at low water potentials. I. Spatial distribution of expansive growth. *Plant Physiology, 87*, 50–57.

Sun, T., Wang, Y., Hui, D., Jing, X., & Feng, W. (2020). Soil properties rather than climate and ecosystem type control the vertical variations of soil organic carbon, microbial carbon, and microbial quotient. *Soil Biology and Biochemistry, 148*, Article 107905. https://doi.org/10.1016/j.soilbio.2020.107905

Thidar, M., Gong, D., Mei, X., Gao, L., Li, H., Hao, W., et al. (2020). Mulching improved soil water, root distribution and yield of maize in the Loess plateau of Northwest China. *Agricultural Water Management, 241*, Article 106340. https://doi.org/10.1016/j.agwat.2020.106340

van Genuchten, M. T. (1980). A closed-form equation for predicting the hydraulic conductivity of unsaturated soils. *Soil Science Society of America Journal, 44*, 892–898. https://doi.org/10.2136/sssaj1980.03615995004400050002x

Vereecken, H., Maes, J., Feyen, J., & Darius, P. (1989). Estimating the soil moisture retention characteristic from texture, bulk density, and carbon content. *Soil Science, 148*(6), 389–403. https://doi.org/10.1097/00010694-198912000-00001

Wong, J. T. F., Chen, Z., Wong, A. Y. Y., Ng, C. W. W., & Wong, M. H. (2018). Effects of biochar on hydraulic conductivity of compacted kaolin clay. *Environmental Pollution, 234*, 468–472. https://doi.org/10.1016/j.envpol.2017.11.079

Xue, W., Huang, L., & Yu, F.-H. (2021). Increasing soil configurational heterogeneity promotes plant community evenness through equalizing differences in competitive ability. *Science of The Total Environment, 750*, Article 142308. https://doi.org/10.1016/j.scitotenv.2020.142308

Zahedifar, M., Karimian, N., Ronaghi, A., Yasrebi, J., & Emam, Y. (2011). Soil–plant nutrient relationship at different growth stages of spinach as affected by phosphorus and manure applications. *Communications in Soil Science and Plant Analysis, 42*, 1765–1781. https://doi.org/10.1080/00103624.2011.587567

Zahedifar, M., Karimian, N., & Yasrebi, J. (2012). Influence of applied zinc and organic matter on zinc desorption kinetics in calcareous soils. *Archives of Agronomy and Soil Science, 58*, 169–178. https://doi.org/10.1080/03650340.2010.507195

Zahedifar, M., Moosavi, A. A., & Gavili, E. (2025). Monitoring soil quality indices and soybean yield as influenced by integrated biochar and drought stress. *Environment, Development and Sustainability, 27*(1), 1877–1903.

Zarebanadkouki, M., Fink, T., Benard, P., & Banfield, C. C. (2019). Mucilage facilitates nutrient diffusion in the drying rhizosphere. *Vadose Zone Journal, 18*, Article 190021. https://doi.org/10.2136/vzj2019.02.0021

CHAPTER

4

Calcareous urban soils challenges: Exploring land use's role in mitigating heavy metal contamination in Southern Iran

Sayyed Mahmoud Enjavinezhad[1], Seyed Javad Naghibi[1,2], Morteza Poozesh Shirazi[3], Majid Baghernejad[1] and Ali Akbar Moosavi[4]

[1]*Department of Soil Sciences, School of Agriculture, Shiraz University, Shiraz, Iran;* [2]*Landscape & Green Spaces Organization of Shiraz Municipality, Shiraz, Iran;* [3]*Department of Soil and Water Research, Fars Agricultural and Natural Resources, Research and Education Center, AREEO, Shiraz, Iran;* [4]*Department of Soil Science, College of Agriculture, Shiraz University, Shiraz, Iran*

1. Introduction

Soil has a direct and indirect relationship with the overall life quality of urban residents as an essential element of urban ecosystems (Mihailović et al., 2015). Currently, 55% of people on Earth reside in cities or metropolitan areas, and by 2050, that number is predicted to rise to 70%, especially in low- and medium-income countries where rapid urbanization can also lead to potential inequalities and health problems (Kuddus et al., 2020). Urban areas offer several opportunities for employment and access to better services (health, education, social protection) that are necessary for good health and human development (Wan et al., 2022). Therefore, the state of the urban environment could impact a significant portion of the world's population, and adequate public health measures are crucial for health. Therefore, adequate healthy urban soils are needed for sustainable urban health (Ramírez-Rubio et al., 2019).

Urbanization is responsible for some negative environmental consequences, including soil contamination with heavy metals and metalloids (Romzaykina et al., 2021). Generally, man-made damage and anomalies in water and soil systems cause serious risks to human health and the living environment. Examples of such impact are mining activities, excess use of chemical fertilizers, pesticides, landfill leaks, land use changes, deforestation, and soil erosion (Dehkordi et al., 2024). Heavy metal pollution is one of the most common environmental problems due to its potential toxicity, non-degradability, and physiological effects on living organisms (Abd Elnabi et al., 2023; Sager, 2020). These metals generally belong to a group of intermediate elements of the periodic table with densities exceeding 5 $Mg\ m^{-3}$ (Wu et al., 2018). The most important heavy metals include cadmium (Cd), arsenic (As), cobalt (Co), vanadium (V), zinc (Zn), mercury (Hg), iron (Fe), manganese (Mn), nickel (Ni), lead (Pb), chromium (Cr), and copper (Cu) (Kishe & Machiwa, 2003). These metals originate from natural and man-made sources with toxicity problems for agricultural produce and potential risks

The Soil Nexus. https://doi.org/10.1016/B978-0-443-45286-4.00012-2

to ecosystem and human health (Wang et al., 2018). Important sources of soil pollutants, especially heavy metals, are human urban and industrial activities. These metals are transported over long distances as particles in the atmosphere. Some of these metals may end up in soil by atmospheric deposition, so the soils around industrial and urban areas are important places for their accumulation (Wu et al., 2015). Human and ecological risks of metals in soils under different land use types in an urban environment of Bangladesh showed metal hazard quotients are decreased in the order of As < Cr < Pb < Cd < Ni < Cu (Islam et al., 2020). Despite commonly formation of heavy metals carbonate precipitates in calcareous soils reduce their mobility within soil system and decrease their potential transport to groundwater, changes in pH and type of minerals may disrupt heavy metal sorption-desorption behaviors (Alghamdi & Alasmary, 2022).

Shiraz urban watershed with calcareous origin soils is located in Southwest of Islamic Republic of Iran. Lack of information on the investigation of the effect of land use and soil properties on controlling heavy metal contaminations (especially Pb and Cd) in this led to the present study. The objectives of current study were: 1) to study physical (soil texture), chemical (pH, electrical conductivity, organic matter and carbon, exchangeable cations, Cd, and Pb) properties from 150 soils collected from different land uses (agriculture, barren lands, garden, park and range lands), 2) to identify the possible origin (natural and/or anthropogenic) of the metals in the soils for obtaining the spatial distribution of the possible pollutants.

2. Methods and materials

2.1 Sampling and sample preparation

The present study was conducted in the Shiraz urban watershed, Fars province, southern Iran, with an area of 411,133 ha and in the metric position of the northern latitude from 662,390 to 632,643 and eastern longitude from 3,262,310 to 3,301,466 in UTM zone 39R (Fig. 4.1). The soil moisture and temperature regimes of the study area were xeric and thermic, respectively (Enjavinezhad et al., 2025). Soils of study area are classified as Entisols and Inceptisols (Naghibi, 2023, p. 203). A total of 150 topsoil samples (0–20 cm) under different land uses, including irrigated agriculture, gardens, green space of parks, barren lands, and range lands were collected in a systematic random method by a Garmin 62s GPS device. After transferring the samples to the laboratory, the soil samples were air-dried and then passed through a 2 mm sieve for further analysis.

2.2 Soil physicochemical analysis

Soil physical and chemical experiments were as follows. The soil particle size distribution was estimated by a hydrometer method (Bouyoucos, 1962). Calcium carbonate equivalent (CCE) was measured by neutralization with HCl (Loeppert & Suarez, 1996), and organic carbon (OC) was measured by wet oxidation (Nelson & Sommers, 1996). Soil pH and electrical conductivity (EC) were measured in saturated paste, and its extract, respectively (Page et al., 1982); and cation exchange capacity (CEC) was measured by the substitution of exchangeable cations with sodium acetate and replacing sodium with ammonium acetate (Chapman, 1965).

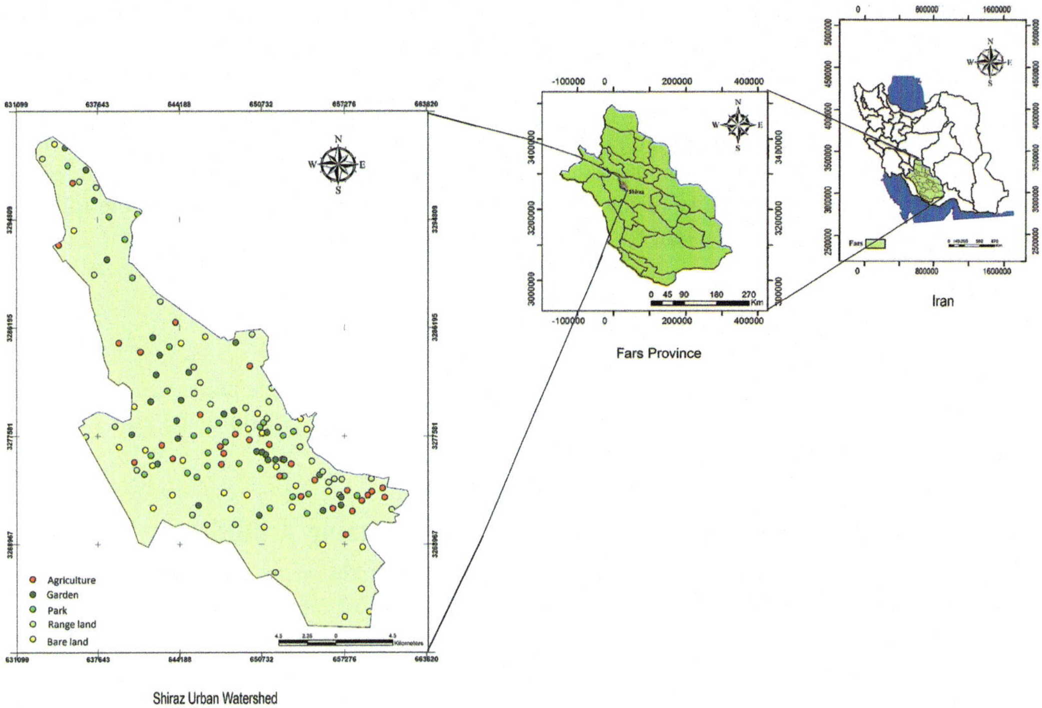

FIGURE 4.1

Study area and sampling points within different land uses, fars province, Iran. ***Map lines delineate study areas and do not necessarily depict accepted national boundaries.***

2.3 Measuring the amount of heavy metal pollution

Concentrations of heavy metals (Fe, Cd, and Pb) in soil extracts after digestion using aqua regia (mixture of concentrated hydrochloric acid and nitric acid in a ratio of 3: 1) at 18.

0°C for 2 h in a bain-marie and then measured by the Shimadzu AA-670 model atomic absorption spectrometer (Golia et al., 2007).

2.4 Data analysis

Using a One-Way ANOVA statistical test in SPSS23 software, the amount of different heavy metals was compared between different land uses.

2.5 Statistical analysis using GS$^+$ software

A summary of the descriptive statistics of selected soil properties and heavy metal contamination, including central statistics, data distribution, skewness, and kurtosis of data, was performed using the GS + geostatistical program. After performing the Kolmogorov–Smirnov test to check the normality of the data in SPSS23 software, the GS + software was applied for normalizing the data. The GS + geostatistical program is commonly used in geostatistical surveys based on the complex mathematical relationships obtained from an empirical semivariogram (Design, 2004). In the present study, to prepare maps of soils' physicochemical properties and also heavy metal contamination using the GS + software, a spatial correlation was determined between the studied properties at the measured points. For this purpose, the area semivariogram for each property was determined, and the anisotropy in the spatial continuity of the relevant data was investigated. Then, using the following equation, the experimental variogram for each property was calculated (Eq. 4.1):

$$\gamma(h) = \frac{1}{2N(h)} \sum_{i=1}^{n(h)} [z(xi) - z(xi + h)]^2 \tag{4.1}$$

where γ (h) is the semivariogram at h distance, N(h) is the number of sample pairs used to calculate γ (h) for each h interval, z(xi) is the variable value in the xi sampling position, and z(xi + h) is the variable value in the xi + h sampling position.

2.6 Soil pollution and selected properties mapping using ArcGIS

To determine the best estimation method in this program, in addition to using the results of geo-statistical calculations performed by the GS + program, the lowest mean root square error (RMSE) can be used as an important parameter. Subsequently, heavy metals pollution and selected soil properties mapping were implemented in ArcGIS 10.5 software.

3. Results and discussion

3.1 Descriptive statistics of data

The summary of descriptive statistics for soil properties and heavy metal contents is shown in Table 4.1. The Kolmogorov–Smirnov test was also used to ensure the normal distribution of the studied properties, and the same results were found by skewness and kurtosis of data distributions. The normalization test results showed that all the selected soil properties except EC and clay percentage had a normal distribution. Also, all the metal concentrations except Cd had a normal distribution.

The soil reaction of the study area was neutral to alkaline (pH range of 7.02–8.77 with a mean value of 7.69), which is possibly related to the geological formations (parent rocks) involved in soil-forming processes. According to the geological formation map, the northern parts of the study area with a high pH were mostly surrounded by upper Gachsaran and upper Bakhtiari formations which include limestone and marl, while the other parts were mostly surrounded by clay, silt, and sand deposits or lower Gachsaran formation and flood basin deposits which led to an observed uniform neutral pH in larger scales (Fig. 4.2). Also, the transport of soil from different places should not be ignored.

Although the EC of large parts of the study area was moderate (0.41–6.32 dS/m with a mean value of 2.74 dS/m) with no sign of a salinity problem, the southeastern parts of the area faced salinity problems due to the effect of the Maharloo Lake deposits. In addition, the observed local high EC amounts all over the area might be due to the transfer of soils from different places to build urban green spaces. The percentage of OM showed a moderate amount (0.02%–6.72% with a mean value of

Table 4.1 Descriptive statistics of soil properties and metal concentrations.

Soil properties (unit)	Minimum	Maximum	Mean	Standard deviation	Coefficient of variation (CV)[a]	Skewness[b]	Kurtosis[b]	Kol-Smi test[c]
pH of saturated paste	7.02	8.77	7.69	0.21	0.02	0.49	3.75	0.58
Electrical conductivity of saturated extract, EC (dS m^{-1})	0.41	15.32	2.74	2.02	0.73	3.02	14.39	0.001
Sand (%)	21.90	72.60	44.31	10.06	0.22	0.13	−0.11	0.566
Silt (%)	18.90	54.10	38.20	5.58	0.14	−0.48	0.86	0.478
Clay (%)	7.60	33.50	17.47	5.51	0.31	0.43	−0.59	0.003
Calcium carbonate equivalent, CCE (%)	31.15	85.82	52.67	7.18	0.13	0.47	2.59	0.183
Organic matter, OM (%)	0.02	6.72	3.10	2.11	0.68	−0.005	−1.45	0.23
Cation exchange capacity, CEC ($cmol_{(+)}$ kg^{-1})	4.58	20.91	12.31	3.86	0.31	0.01	−0.68	0.636
Iron, Fe (mg kg^{-1})	400	21,700	8442	2608	0.30	0.51	5.54	0.05
Lead, Pb (mg kg^{-1})	5.12	17.4	7.78	1.21	0.15	3.31	25.23	0.05
Cadmium, Cd (mg kg^{-1})	0.83	7.93	2.06	0.61	0.29	5.54	54.91	0.00

[a] *(very low variability <0.1; low variability 0.1–0.2; medium variability 0.2–0.3; and high variability >0.3).*
[b] *−2< Normal<2.*
[c] *Normal, sig>0.05.*

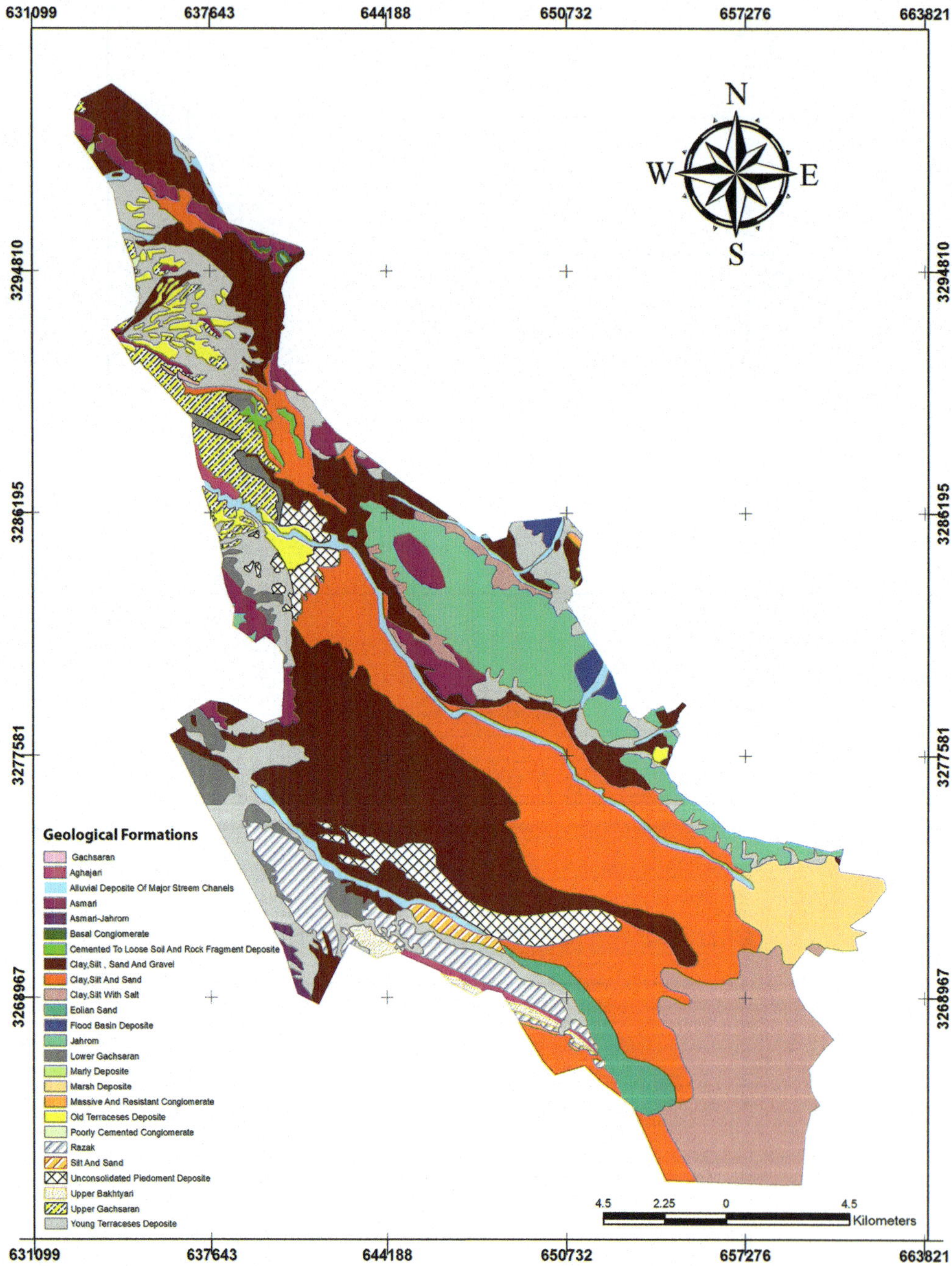

FIGURE 4.2

Geological formations of the study area. ***Map lines delineate study areas and do not necessarily depict accepted national boundaries.***

3.10%) over the study area, as expected for an arid and semi-arid region. The CEC of the study area was low (4.58–2.91 $cmol_{(+)}\ kg^{-1}$ with a mean value of 12.31 $cmol_{(+)}\ kg^{-1}$) and had an overlap with the OM and clay percentage. The percentage of sand, silt, and clay in the study area ranged from 21.90% to 72.60% with a mean of 44.31%; 18.90%–45.10% with a mean of 38.20%; and 7.60%–33.50% with a mean of 17.47%, respectively. Consequently, loam was the predominant soil texture class of the studied soils. The percentage of CCE ranged from 31.15% to 85.82% with a mean value of 52.67%, which is common in calcareous soils of arid and semiarid regions. The Fe content of the soils ranged from 400 to 21,700 (mean value of 2608) mg/kg; while Pb content ranged from 5.12 to 17.4 (mean value of 7.78) mg/kg, and Cd from 0.83 to 7.93 (mean value of 2.06) mg/kg.

According to Table 4.1 and based on the coefficient of variation, pH had very low variations all over the area, while the silt and CCE had low variations. Also, the clay, sand, and CEC had moderate variations, while the OM and EC had high variations. These findings showed that the variations of soil properties might be controlled by the nature of the parent material and also soil management. Meanwhile, soil management will usually have more impact in urban areas, causing extensive changes in soil properties.

The potential toxicity, non-degradability, and negative physiological effects on living organisms are heavy metal pollution characteristics that make it an important environmental feature to investigate (Islam et al., 2015). According to Table 4.2, the mean value of Pb content was in a safe health range determined by four different standards, while the mean value of Cd content exceeded the UK and FAO thresholds. According to Table 4.1, the variations of heavy metal pollution in the study area were high due to high CV and might be controlled by soil management. To perform a geostatistical investigation of selected soil properties and heavy metal contaminations, the obtained data distribution should be normal. Thus, logarithmic functions were used to normalize the data distribution deviating from normal (Ersahin, 2003; Rezaee et al., 2020).

3.2 Investigation of spatial variability

The model fitted to the experimental semi-variogram, and a summary of the geostatistical analysis of the studied features are shown in Table 4.3. The obtained models were of two types: spherical and exponential. Results showed that soil properties variogram models, such as pH, EC, silt, and CEC, were best fitted with the anisotropic models; while the others were isotropic. Regarding metal contaminations, only the Fe variogram model was anisotropic. The effective range of soil properties had such a trend: pH < CCE = OM < Silt < EC < Sand = Clay < CEC. The observed trend for the effective range of metals was as Fe < Pb=Cd. The spatially structured (C) to total (C_0+C) variations (C/C_0+C) ratio was in the range of 0.50–0.65 for soil properties and 0.50 to 0.89 for the studied metal

Table 4.2 Maximum permissible limits for heavy metals in soil based on some common global standards (STD) (after Nangia, 1991; Asemave & Anhwange, 2012; Hong et al., 2014).

Heavy metal	EU STD	UK STD	US STD	WHO
Cadmium, Cd (mg/kg)	3.0	1.4	4.0	0.002–0.5
Lead, Pb (mg/kg)	300	70	300	0.3–10

EU, *Europe;* UK, *United Kingdom;* US, *United States;* WHO, *World Health Organization.*

Table 4.3 The fitted models to the experimental semi-variograms of the studied soil attributes, along with their parameters.

Soil properties (unit)	Variogram direction	Model	Nugget (C_0)	Sill (C_0+C)	Effective range (m)	C/C_0+C	C_0/C_0+C	Determination coefficient (R^2)	Residual sum of squares (RSS)
pH of saturated paste	Anisotropic	Exponential	7×10^{-4}	2.01×10^{-3}	9087	0.65	0.35	0.22	1.41×10^{-4}
Electrical conductivity of saturated extract, EC (dS/m)	Anisotropic	Exponential	0.36	0.87	770	0.59	0.41	0.26	0.22
Sand (%)	Isotropic	Exponential	82.50	165.10	433	0.50	0.50	0.51	2389
Silt (%)	Anisotropic	Spherical	28.08	79.58	774	0.65	0.35	0.25	1320.55
Clay (%)	Isotropic	Exponential	24.80	49.61	433	0.50	0.50	0.59	186
Calcium carbonate equivalent, CCE (%)	Isotropic	Spherical	0.18	0.36	811	0.50	0.50	0.68	0.03
Organic matter, OM (%)	Isotropic	Spherical	3.37	6.74	811	0.50	0.50	0.82	7.02
Cation exchange capacity, CEC ($cmol_{(+)}$ kg^{-1})	Anisotropic	Exponential	15.01	37.43	376	0.60	0.40	0.56	238.33
Iron, Fe (mg/kg)	Anisotropic	Spherical	2,970,000	26,003,437	1061	0.89	0.11	0.59	6.41×10^{16}
Lead, Pb (mg/kg)	Isotropic	Exponential	0.02	0.03	433	0.50	0.67	0.86	1.80×10^{-4}
Cadmium, Cd (mg/kg)	Isotropic	Exponential	0.05	0.10	433	0.50	0.50	0.42	1.74×10^{-3}

contaminants. In other words, the C_0/C_0+C ratio was in the range of 0.35–0.50 for soil properties and 0.11 to 0.67 for the studied metal contaminants. Overall, the results indicated that the studied metals show stronger spatial structure than those of the other physicochemical attributes, probably due to their inherent variabilities across the study area or their different susceptibility to managerial and human activities.

The summary of the geostatistical analysis of the studied features and the fitted model of the experimental semivariograms are shown in Table 4.3. The direction of the variogram indicates the homogeneity of variation in the study area. The term anisotropic is used when heterogeneity of a variable is observed in different directions. In this case, the difference is observed in an effective range, and the variogram sill in a different direction. Among the investigated soil properties, including EC, pH, silt, and CEC, an anisotropic variogram was acceptable due to the different management of urban soils over different periods. Also, Fe had an anisotropic variogram, which was inconsistent with the % CCE of the study area. Generally, in calcareous soils, Fe deficiency is widespread (Taalab et al., 2019). The isotropic variogram of CCE and the application of Fe-fertilizers in some land uses of urban soils seemed an acceptable reason for this finding. The spatial isotropic ratio ($C/C_0 + C$) indicated the structure strength of spatial variables, so the stronger the spatial structure, the closer it is to 1. The spatial variability class ($C_0/C_0 + C$) is also used to affirm the degree of spatial variables' dependency. If the spatial dependency is very strong for a variable, this ratio is less than 0.25, while if the spatial dependency is moderate, the ratio is between 0.25 and 0.75. Also, if the spatial dependency is considered weak, this ratio is greater than 0.75 (Design, 2004). As presented in Table 4.3, soil physicochemical properties had strong spatial structure and moderate spatial dependency. For metal pollutants, Fe had a strong spatial structure and spatial dependency, while the other two contaminants had moderate spatial structure and dependency. The very strong spatial dependency is attributed to soil intrinsic factors such as soil parent material, soil texture, topography, and vegetation. Conversely, the moderately weak and very weak spatial dependency is related to soil extrinsic factors, such as fertilization and cultivation practices (Wang et al., 2009; Ye et al., 2018). In the present study, the observed moderate dependency of soil physicochemical properties and heavy metal contamination confirmed urban soils' characteristics change with the control of extrinsic factors. The strong dependency of Fe was mainly related to intrinsic factors such as soil pH and the abundance of calcium carbonate in calcareous soils. The effective range of soil properties in the study area showed that in semi-arid climates pH, CCE, OM, and EC could be investigated by sampling at longer distances. The observed low effective range for CEC or soil texture might be due to the soil disruption caused by transferring from other sites, and the effects of both natural and artificial land use in urban areas resulted in severe variations.

3.3 Digital mapping of spatial variations of soil properties and metal contaminants using ArcGIS

To map the spatial variations of soil properties and metal contaminants, after determining the best-selected model for each soil property and metal contaminant, two different estimator methods available in ArcGIS 10.5 software, the inverse distance weighting (IDW) method and ordinary kriging (Ord. K) method, were performed (Table 4.4). The best model for mapping spatial variations in soil properties and metal contaminants is a model that has the least absolute mean error and the root mean square error. Thus, the results of Table 4.4 showed that the IDW showed better results for EC, sand, and

Table 4.4 Results of prediction methods used for the prediction of soil properties.

	Inverse distance weighting (IDW)		Kriging		Chosen model
Soil properties (unit)	RMSE	MAE	RMSE	MAE	
pH of saturated paste	0.24	0.004	0.23	0.005	Kriging
Electrical conductivity of saturated extract, EC (dS/m)	16.89	−0.37	17.11	0.25	IDW
Sand (%)	10.63	0.04	10.70	0.15	IDW
Silt (%)	6.04	−0.01	6.07	−0.01	IDW
Clay (%)	5.71	−0.03	5.68	−0.08	Kriging
Calcium carbonate equivalent, CCE (%)	8.30	−0.38	8.28	−0.38	Kriging
Organic matter, OM (%)	2.38	0.18	2.20	0.03	Kriging
Cation exchange capacity, CEC ($cmol_{(+)}$ kg^{-1})	4.20	0.19	3.98	−0.08	Kriging
Iron, Fe (mg/kg)	2831	98.06	2734	45.89	Kriging
Lead, Pb (mg/kg)	1.36	0.02	39.48	−0.29	IDW
Cadmium, Cd (mg/kg)	0.67	0.008	0.63	−0.01	Kriging

MAE, *mean absolute error;* RMSE, *root mean square error.*

silt, similar to Pb. However, the other properties and metal contaminants were well-estimated by the kriging method. Overall, the kriging method had a wider application in the estimation of the properties and metal contaminants, but this does not mean which applied estimator was superior to another because different results might be obtained in other study areas with similar parameters. Some previous studies used kriging at unsampled locations and produced risk maps of soil contamination by these metals (Appiah et al., 2018; Reza et al., 2015).

3.4 Digital mapping of spatial variations of soil properties

According to the maps drawn with the ArcGIS 10.5 geostatistical program, soil pH in the study area was relatively alkaline and had a uniform distribution, but slight non-uniformity was observed in the northern and central parts (Fig. 4.3a). The relatively alkaline nature with a uniform trend all over the area is mainly due to soil parent materials that originated from calcareous parent rocks. The observed high pH values in the northern parts were due to the existence of highly calcareous parent rocks.

The EC amounts had a distinct trend in the study area, with higher and lower amounts observed in the southeastern and northern parts, respectively. The distribution of this property showed slight non-uniformity in the central parts (Fig. 4.3b). The regular changes in soil EC are mainly due to the origin of the soil parent material, the quality of irrigation water or groundwater, and the disturbance of the soil's soluble salt balance, which is the result of improper fertilization and soil management. In the present study, the higher values in the southern parts were observed mainly because of the origin of the soil parent material, which is affected by the location of Maharloo Lake and the recent occurrence of floods.

The final composition ArcGIS map of soil particle size distribution classes (soil texture) is shown in Fig. 4.3c. The predominant soil texture in the study area was loam, and some other loamy textures were found in small parts of the map. The presence of a loam-based texture with a relatively uniform trend in the area may be related to the lithology of the parent rocks.

The percentage of CCE was relatively high and was found mostly in the central parts of the study area, although a relatively low percentage was observed in the western, northern, and southeastern parts (Fig. 4.3d). The percentage of CCE in the study area was mainly due to lithology and geological formations. Thus, it showed a relatively uniform distribution all over the area. Despite the coarse-grained, sedimentary rocks such as conglomerates and sandstones, Shiraz soils were affected; the soil was mainly controlled by alluvial deposits.

The mean percentage of OM was relatively low and in the range of 1.86%–3.07% in the study area (Fig. 4.3e). Also, the lower and higher percentages of OM had non-uniformity all over the area. The observed non-uniform OM map may be related to the different land uses, transferring soils within urban areas, and recent plant cultivations to increase green spaces in the study area.

Despite the mapping of spatial variation of CEC showing that it had a non-uniform trend all over the area, the northern and southeastern parts, and also the central to eastern parts showed higher CEC values (Fig. 4.3f). According to the CEC map, the values of this property were affected mainly by the land use of sampling points and their clay and OM contents.

Generally, urban soil formation is affected by both direct and indirect anthropogenic factors, which lead to special characteristics and processes (Yang & Zhang, 2015). Indirect anthropogenic effects on soil-forming factors include changing carbon and nitrogen stocks and soil temperature and moisture regimes (Lorenz & Lal, 2015; Savva et al., 2010; Smagin, 2012). Physical soil properties are affected by human processes such as over-compaction, while chemical soil properties are affected by the processes of salinization and soil reaction variation (Vasenev et al., 2018).

3.5 Digital mapping of spatial variations of metal contaminants

The mapping of spatial variations of Pb contamination is shown in Fig. 4.4a. The higher Pb contamination amounts were found almost all over the area, while lower amounts of this contamination were observed in the southern and northern parts (Fig. 4.4a). The mapping of spatial variations of Cd contamination showed that higher amounts were observed in the central to western and northern parts of the study area (Fig. 4.4b).

According to Fig. 4.4, the Pb contamination mapping showed that the traffic and transportation systems in urban areas controlled the distribution of this element in the study area. The contamination of Pb could occur in urban areas due to atmospheric transportation of Pb-containing organic compound for example, lead tetraethyl and lead tetramethyl (Enjavinezhad & Kasraian, 2014). The Cd contamination mapping showed that traffic and transportation systems were common reasons for the distribution of this element over the area (Fig. 4.5). The Cd contamination is not as mobile as Pb heavy metal contamination by atmospheric factors and usually is found in engine batteries, alloys used in car carburetors, lubricating oils, and galvanized sheets (Enjavinezhad & Kasraian, 2014; Saeedi et al., 2009). Thus, its high concentrations were observed along the roads, particularly those experiencing heavy traffic and parking along the roadsides, such as ring belts or exit routes from the city, in addition

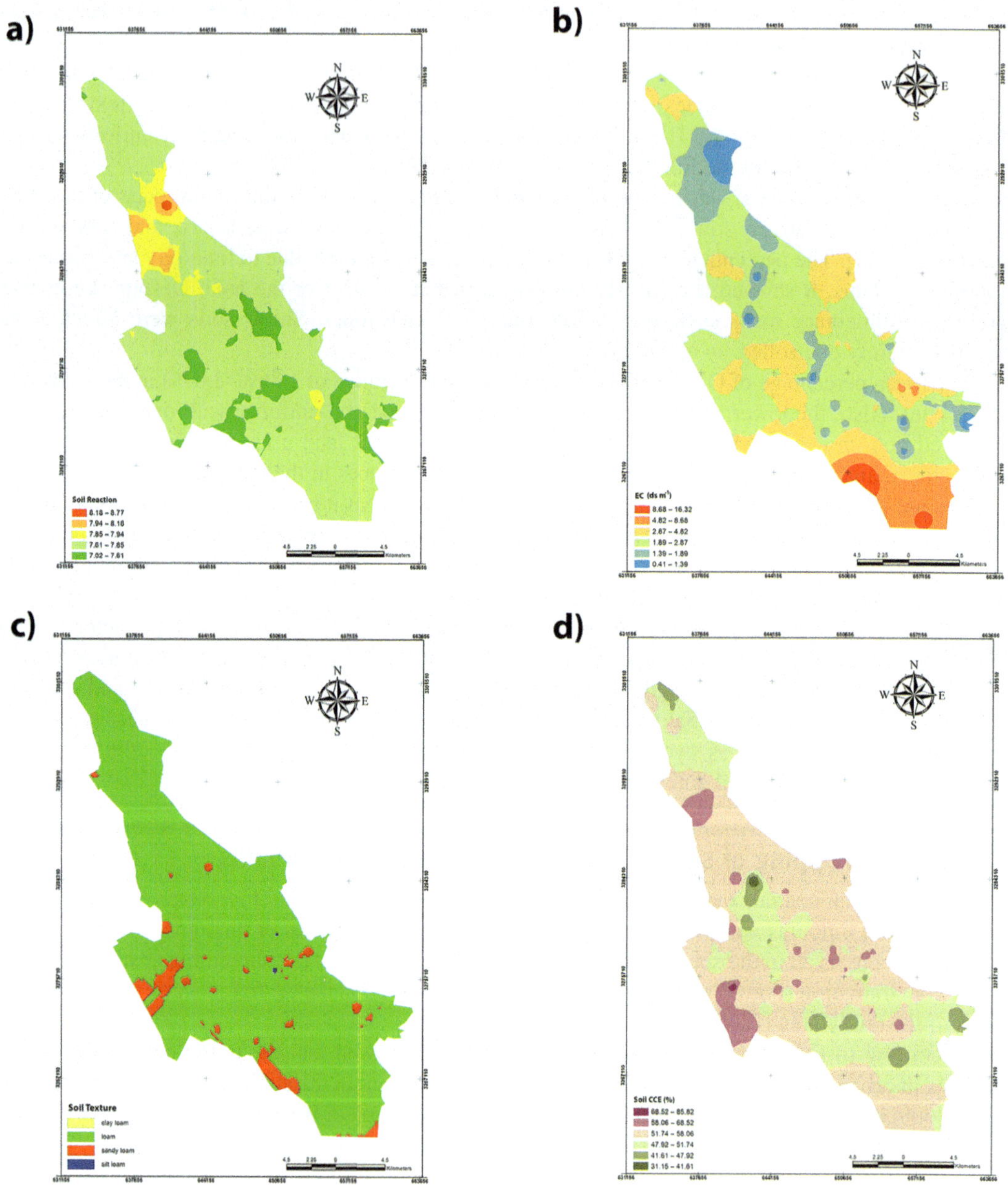

FIGURE 4.3

Mapping of spatial variations of soil properties. a) pH; b) electrical conductivity; c) soil texture; d) calcium carbonate equivalent; e) organic matter; f) cation exchange capacity; and g) iron, Fe. ***Map lines delineate study areas and do not necessarily depict accepted national boundaries***.

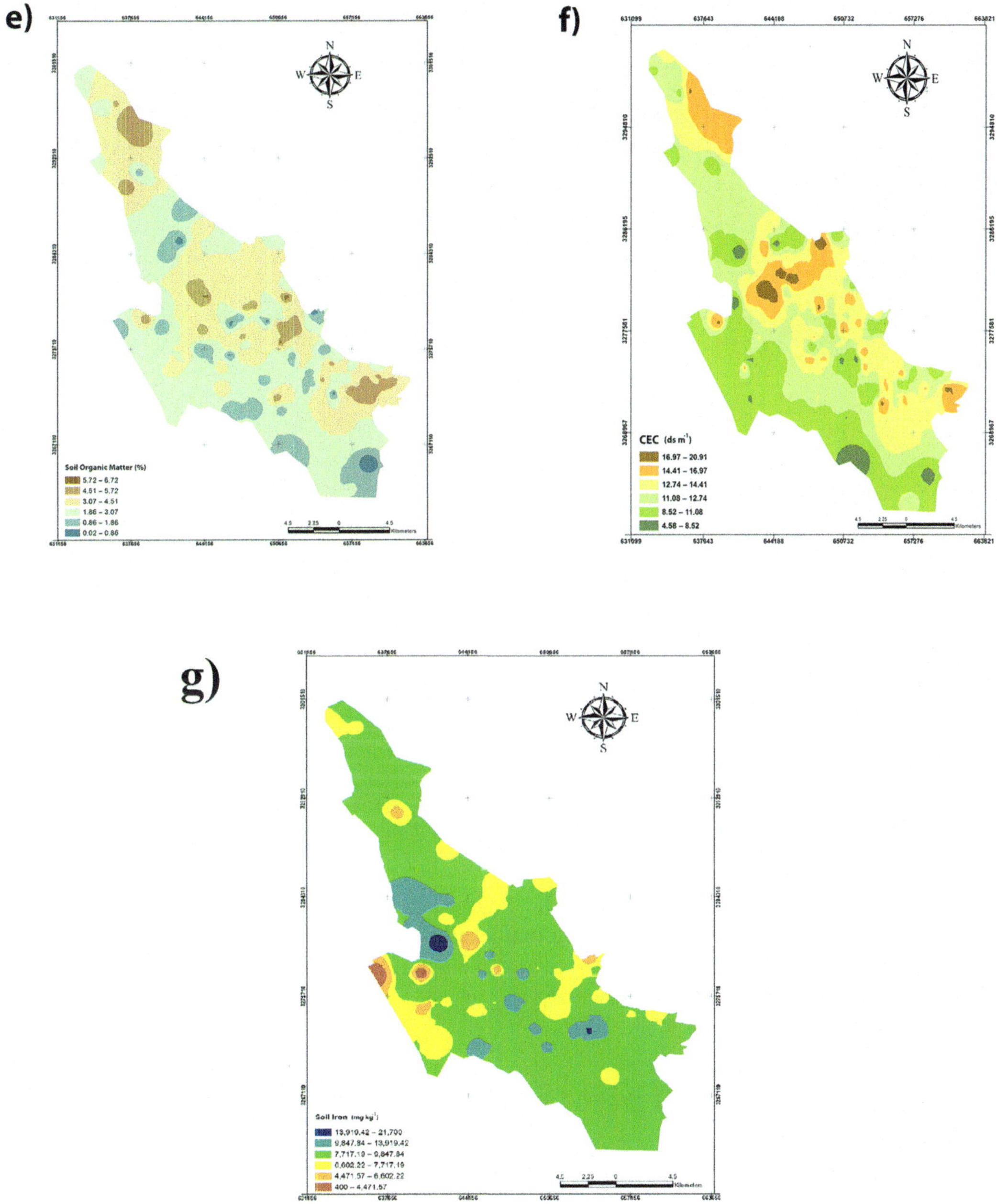

FIGURE 4.3 Cont'd

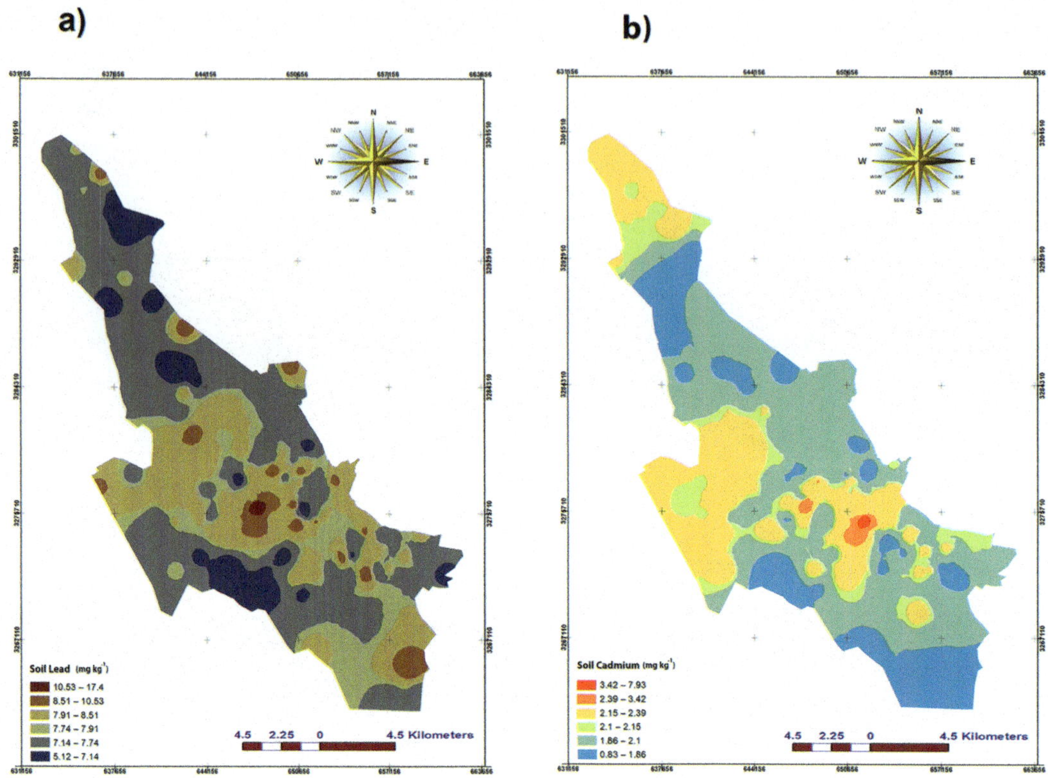

FIGURE 4.4

Mapping of heavy metal contamination in the study area: a) lead, Pb, and b) cadmium, Cd. ***Map lines delineate study areas and do not necessarily depict accepted national boundaries.***

to the central parts of the city with high traffic. This study showed that urban soils were subject to heavy metal contamination currently at very low levels and had the potential in the future to accumulate more under current land uses. Further accumulation of Pb and Zn in soils may endanger human health through further pollution of ground waters, suspended dust particles, and plant biomass produced by the roadside (Paltseva et al., 2020).

3.6 Land use effects on soil properties and metal contaminants

The descriptive analysis of soil properties and heavy metal contaminations concerning land use is shown in Table 4.5. The studied land uses were bare land, range land, irrigated agriculture, park, and garden. The soil pH trend within land uses was bare land < irrigated agriculture < range land - < garden < park, while EC had such a trend: irrigated agriculture < garden < park < bare land - < range land. The CCE trend was garden < park < irrigated agriculture < bare land < range land through different land use. The organic matter percentage trend through different land use was bare

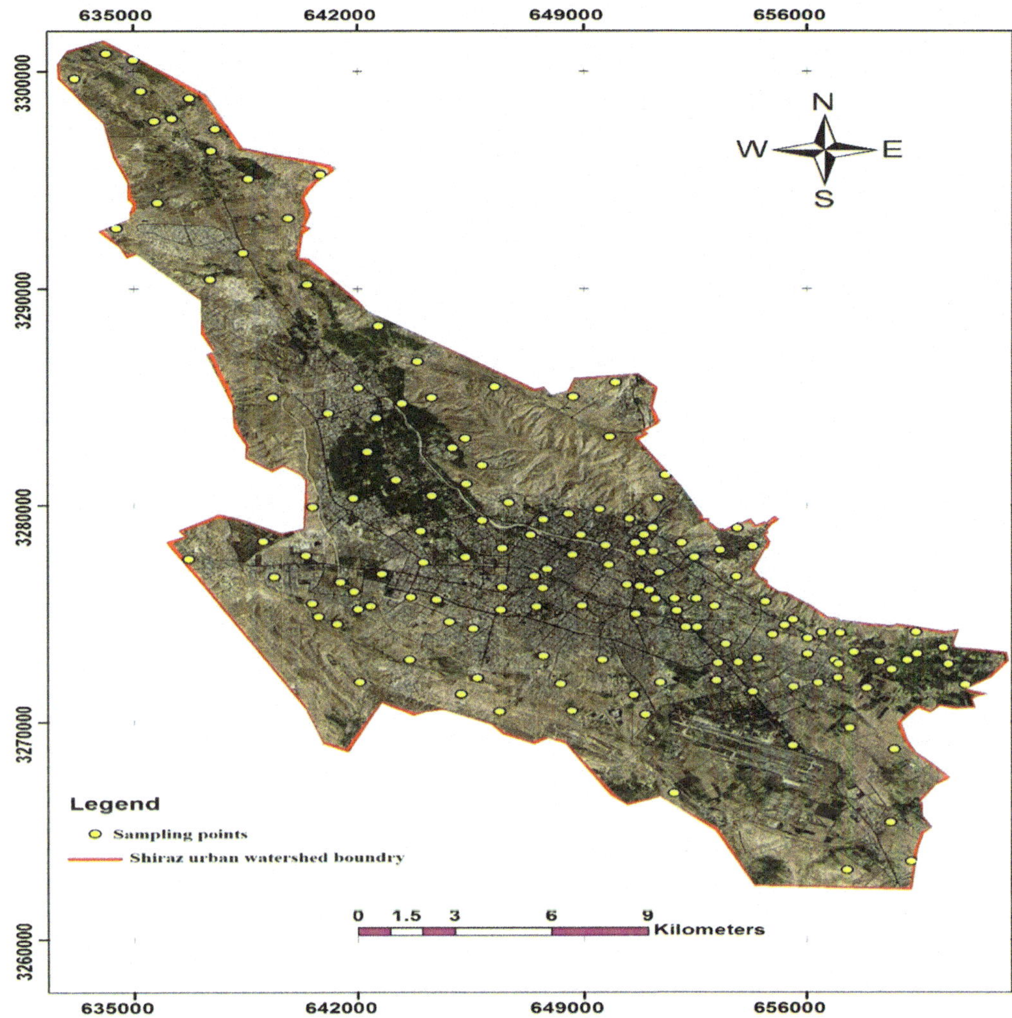

FIGURE 4.5

Estimation of distances from the road to the sampling locations. ***Map lines delineate study areas and do not necessarily depict accepted national boundaries.***

land < range land < irrigated agriculture < park < garden. The CEC amounts were bare land - < irrigated agriculture < range land < park < garden. According to Fig. 4.6, the mean values of pH, CCE, CEC, and OM were significantly different between the studied land uses ($P < .05$).

The distribution of sampling points among the different land uses in the study area is shown in Fig. 4.1. Although the sampling points were located either on calcareous soils or not, the differences in pH do not reflect that. The pH values might be more influenced by the application of irrigation water with unsuitable pH. Only topsoil was sampled and analyzed, which may be another reason for the

Table 4.5 Descriptive analysis of soil properties and metal concentrations in the studied land uses.

Soil properties/land use	Bare lands			Range lands			Irrigated agriculture			Park			Garden		
	Min	Max	Mean	Min	Max	Mean	Min	Max	Mean	Min	Max	Mean	Min	Max	Mean
pH of saturated paste	7.02	7.94	7.60	7.22	8.13	7.69	8.13	7.12	7.68	7.48	8.77	7.79	7.45	7.97	7.71
Electrical conductivity of saturated extract, EC (dS/m)	0.96	9.36	3.09	0.54	16.32	3.23	0.41	6.14	2.35	0.68	6.59	2.52	0.46	8.75	2.51
Sand (%)	23.60	72.60	44.09	22.80	68.30	44.10	29.40	63.70	47.09	25.90	65.10	42.39	21.90	59.40	43.86
Silt (%)	18.90	49.00	38.78	24.10	46.90	37.75	23.70	48.30	37.22	25.5	47.70	38.81	29.70	54.10	38.48
Clay (%)	8.50	29.10	17.17	7.60	33.50	18.15	9.40	25.40	15.69	9.40	29.00	18.70	10.90	28.70	17.66
Calcium carbonate equivalent, CCE (%)	39.25	65.57	52.85	38.15	85.82	55.72	33.17	68.81	52.41	42.08	65.57	52.16	31.15	65.57	50.25
Organic matter, OM (%)	0.05	5.76	1.33	0.07	6.72	2.79	0.10	6.38	3.53	0.03	5.86	3.64	0.02	6.50	4.21
Cation exchange capacity, CEC ($cmol_{(+)}$ kg^{-1})	5.14	18.53	10.08	4.58	20.27	12.30	4.90	19.05	11.91	6.32	20.91	13.57	7.40	20.04	13.72
Iron, Fe (mg/kg)	4700	12,700	8510	400	14,000	7968	6000	17,000	9362	3700	12,700	8243	1000	21,700	8127
Lead, Pb (mg/kg)	6.47	9.61	7.76	6.49	9.11	7.68	5.94	17.40	7.97	5.56	9.80	7.78	5.12	9.91	7.73
Cadmium, Cd (mg/kg)	0.83	7.93	2.20	0.92	2.68	1.95	0.87	3.10	2.12	0.94	2.44	1.99	0.89	3.20	2.08

Max, *Maximum;* Min, *Minimum.*

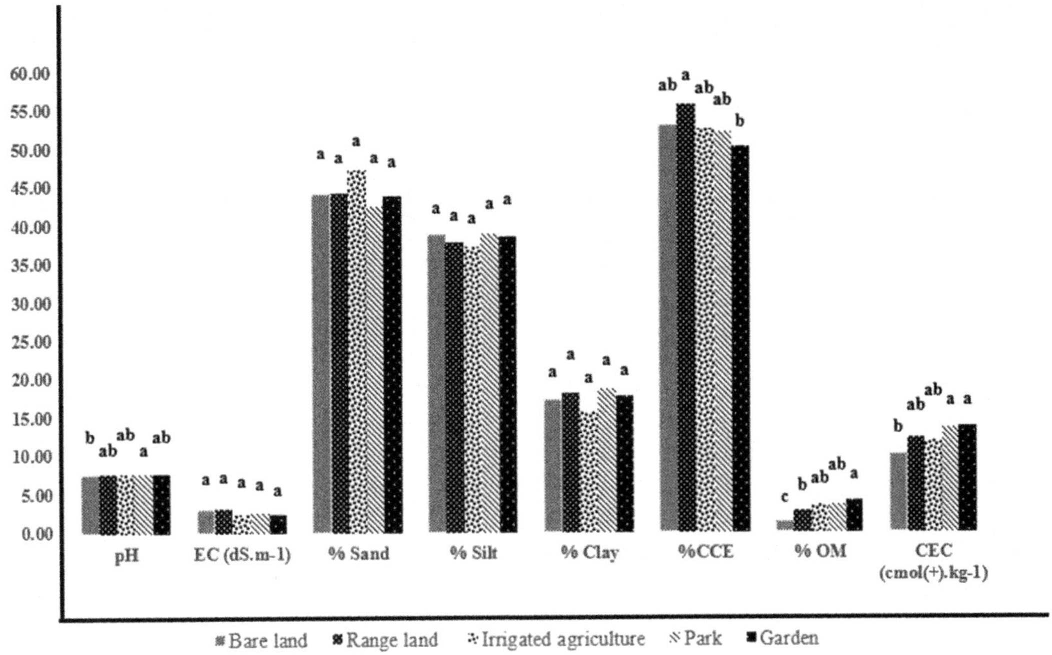

FIGURE 4.6

Comparing the studied soil properties among different land uses (means followed by different letters are significantly different at $P < .05$).

unexpected pH distribution. Other factors such as root zone secretions, fertilization, and the type of soluble salts in topsoil may be important. The EC amounts were low in the cultivated areas due to sensible land use planning, avoiding areas of high salinity. The CCE of cultivated land was also low for the same reason, where most land uses for cultivation mainly occurred on alluvial deposits rather than on highly calcareous geological formations. In addition, the soils with high $CaCO_3$ amounts might show micronutrient deficiency. Soil organic matter showed a logical variation with different land use in the study area, which indicates that as vegetation becomes denser, organic matter is also present in higher amounts. Two important factors controlling soil CEC variation are OM and clay content. The distribution of this property was consistent with these factors. Generally, in urban topsoil, the spatial variability of organic carbon is mainly controlled by standards and practices of land allocation and management, where soil with the highest organic matter content is designated for sustained green zones and residential areas (Shchepeleva et al., 2017; Weissert et al., 2016). Furthermore, some factors such as parent materials, land use type, and intensity of anthropogenic pressure control the texture and pH of urban soils (Vasenev et al., 2018; Yang & Zhang, 2015). The current study's findings were consistent with the mentioned processes and trends. As mentioned before, in urban areas, soil management practices and soil deposition from special places to improve the growing conditions of cultivated land had a special effect on soil OM and pH, so observing a significant difference ($P < .05$) between different land uses was logical. Since values of soil CEC were consistent with OM and clay content, and also CCE was controlled inherently by parent material, the observation of a significant difference in these parameters ($P < .05$) between different land uses was logical.

Regarding land use effects on metal contaminations, the trend of Fe contamination was range land < garden < park < bare land < irrigated agriculture. Through different land uses, Pb contamination had such a trend: range land < garden < bare land < park < irrigated agriculture. The observed trend for Cd contamination was range land < park < garden < irrigated agriculture < bare land. According to Fig. 4.7, the means of metal contamination were not significantly different across the studied land uses ($P < .05$). Range lands showed low Fe contents mainly due to the abundance of calcium carbonate in the studied soils. The higher contents of this element were found in irrigated agriculture, which might be due to Fe-rich irrigation waters. The abundance of $CaCO_3$ in the study area led to no observed significantly different Fe content ($P < .05$) in different land uses. The observed distribution of Pb and Cd heavy metal concentrations in soils showed that although traffic and transportation systems in urban areas may be the main reasons for controlling the distribution of these elements in the study area, no significant differences ($P < .05$) were found between different land uses. The slight differences in Cd contamination in different land uses may be related to the previous accumulation of Cd-containing wastes in open land and the presence of this element in P-fertilizers. This study's findings were partially consistent with the observations of previous studies (Alengebawy et al., 2021; Fan et al., 2018; Xiaobing et al., 2020).

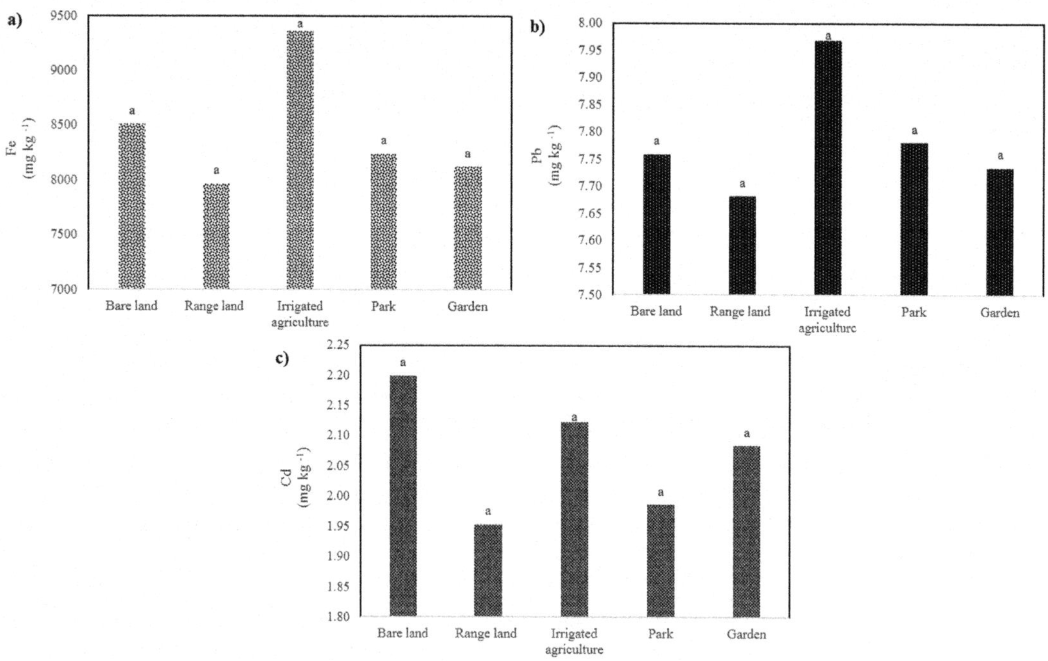

FIGURE 4.7

Iron (Fe), lead (Pb), and cadmium (Cd) concentration in relation to different land uses in the study area (means followed by different letters are significantly different at $P < .05$).

Table 4.6 Correlations between heavy metal concentrations and soil physicochemical properties.

	pH	EC (dS m^{-1})	CCE (%)	CEC ($cmol_{(+)}$ kg^{-1})	O.M. (%)	Clay (%)	Silt (%)	Sand (%)
Lead, Pb (mg/kg)	0.03	−0.05	0.02	−0.24**	−0.27**	−0.09	−0.05	0.08
Cadmium, Cd (mg/kg)	−0.05	−0.09	−0.08	0.11	−0.10	−0.02	0.02	0.00

CCE, *calcium carbonate equivalent;* CEC, *cation exchange capacity;* EC, *electric conductivity;* O.M, *organic matter.*
**: *show correlation coefficients are statistically significant at the probability levels of 0.01, respectively.*

3.7 Relationships between soil properties and metal contaminants

The correlations between soil properties and metal contaminants are shown in Table 4.6. None of the metal contaminants had significant correlations with sand, silt, and clay. The OM showed a significantly negative correlation with Pb ($P < .01$). Fe showed a significantly negative correlation with CCE ($P < .01$), while the other investigated metal contaminants had no significant correlations. Also, soil pH and EC had no significant correlations with the investigated metal contaminants. The CEC just had a significantly negative correlation with Pb ($P < .01$). The strong negative correlation ($P < .01$) between CCE and Fe concentrations may be related to the geological context and spatial distribution of the calcareous soils. This finding was consistent with previous studies (Taalab et al., 2019). The result of this study showed that Pb concentrations in soils of the study area negatively correlated with CEC and OM ($P < .01$). Although Cd did not have any significant correlation with soil properties, recent studies showed that soil pH and OM were the major factors affecting Cd accumulation in soils (Alengebawy et al., 2021). Eventually, high spatial heterogeneity could be named as a typical feature of urban soils, which was reported by many surveys (Paltseva et al., 2018), but the assessment of heavy metals contamination and the complex and non-linear effects of soil properties to achieve sustainable management should not be neglected.

4. Conclusions

The spatial distribution of soil properties and contaminants (Fe, Pb, and Cd) was well-described by spherical and exponential variogram models. The kriging model seems to be suitable for estimating and mapping soil physicochemical properties as well as heavy metal distribution in the study area. Although in some urban soils of Shiraz, only Cd concentrations among the studied metals were higher than the standards introduced by FAO and the UK, the concentrations of other metals should also be monitored regularly because urban environments change over a short period. Therefore, although among the studied physicochemical properties of the Shiraz city, CCE, OM, and CEC had special importance in controlling the trend of heavy metal pollution, the present study showed that proper land use management plays a key role in the management of urban soil characteristics as well as the management of metal pollution, which should be prioritized.

Conflict of interest

The authors declare no conflicts of interest.

References

Abd Elnabi, M. K., Elkaliny, N. E., Elyazied, M. M., Azab, S. H., Elkhalifa, S. A., Elmasry, S., Mouhamed, M. S., Shalamesh, E. M., Alhorieny, N. A., Abd Elaty, A. E., & Elgendy, I. M. (2023). Toxicity of heavy metals and recent advances in their removal: A review. *Toxics, 11*(7). https://doi.org/10.3390/toxics11070580. Article 580.

Alengebawy, A., Abdelkhalek, S. T., Qureshi, S. R., & Wang, M. Q. (2021). Heavy metals and pesticides toxicity in agricultural soil and plants: Ecological risks and human health implications. *Toxics, 9*(3). https://doi.org/10.3390/toxics9030042. Article 42.

Alghamdi, A. G., & Alasmary, Z. (2022). Fate and transport of lead and copper in calcareous soil. *Sustainability, 15*(1). https://doi.org/10.3390/su15010775. Article 775.

Appiah, S. K., Aidoo, E. N., Owusu, D. A., & Nuonabuor, M. W. (2018). Geostatistical analysis of contamination of soils in an urban area in Ghana. *International Journal of Computer Science and Mathematics, 12*(6), 139–147.

Asemave, K., & Anhwange, B. A. (2012). *Evaluation of heavy metals in waste dumpsites.* Lambert Academic Publishing.

Bouyoucos, G. J. (1962). Hydrometer method improved for making particle size analyses of soils 1. *Agronomy Journal, 54*(5), 464–465.

Chapman, H. D. (1965). Cation-exchange capacity. In C. A. Black (Ed.), *Methods of soil analysis: Part 2 chemical and microbiological properties, 9* (pp. 891–901). American Society of Agronomy.

Dehkordi, M. M., Nodeh, Z. P., Dehkordi, K. S., Khorjestan, R. R., & Ghaffarzadeh, M. (2024). Soil, air, and water pollution from mining and industrial activities: Sources of pollution, environmental impacts, and prevention and control methods. *Results in Engineering, 23*, Article 102729. https://doi.org/10.1016/j.rineng.2024.102729

Design, G. (2004). Geostatistics for the environmental science version 7 [computer software]. *Gamma Design.*

Enjavinezhad, S. M., & Kasraian, A. (2014). Comparison of the distribution of heavy elements of lead and cadmium around two highways in Shiraz. In *Proceedings of the 1st national conference on sustainable management of soil and environment resources (in Persian). Kerman, Iran.*

Enjavinezhad, S. M., Naghibi, S. J., Shirazi, M. P., Baghernejad, M., Fernández-Raga, M., & Rodrigo-Comino, J. (2025). From fields to cities: Innovating assessment of soil quality in Southern Iran's urban areas. *PLoS One, 20*(5), Article e0321312. https://doi.org/10.1371/journal.pone.0321312

Ersahin, S. (2003). Comparing ordinary kriging and cokriging to estimate infiltration rate. *Soil Science Society of America Journal, 67*(6), 1848–1855. https://doi.org/10.2136/sssaj2003.1848

Fan, Y., Li, Y., Li, H., & Cheng, F. (2018). Evaluating heavy metal accumulation and potential risks in soil-plant systems applied with magnesium slag-based fertilizer. *Chemosphere, 197*, 382–388. https://doi.org/10.1016/j.chemosphere.2018.01.073

Golia, E. E., Tsiropoulos, N. G., Dimirkou, A., & Mitsios, I. (2007). Distribution of heavy metals of agricultural soils of central Greece using the modified BCR sequential extraction method. *International Journal of Environmental Analytical Chemistry, 87*(13–14), 1053–1063. https://doi.org/10.1080/03067310701451012

Hong, A. H., Law, P. L., & Onni, S. S. (2014). Environmental burden of heavy metal contamination levels in soil from sewage irrigation area of Geriyo Catchment, Nigeria. *Civil and Environmental Engineering, 6*(10), 118–124.

Islam, S., Ahmed, K., & Al-Mamun, H. (2015). Distribution of trace elements in different soils and risk assessment: A case study for the urbanized area in Bangladesh. *Journal of Geochemical Exploration, 158*, 212–222. https://doi.org/10.1016/j.gexplo.2015.07.017

Islam, M. S., Ahmed, M. K., Al-Mamun, M. H., & Eaton, D. W. (2020). Human and ecological risks of metals in soils under different land-use types in an urban environment of Bangladesh. *Pedosphere, 30*(2), 201–213. https://doi.org/10.1016/S1002-0160(17)60305-5

Kishe, M. A., & Machiwa, J. F. (2003). Distribution of heavy metals in sediments of Mwanza Gulf of Lake Victoria, Tanzania. *Environment International, 28*(7), 619–625. https://doi.org/10.1016/S0160-4120(02)00102-1

Kuddus, M. A., Tynan, E., & McBryde, E. (2020). Urbanization: A problem for the rich and the poor? *Public Health Reviews, 41*. https://doi.org/10.1186/s40985-019-0116-0. Article 1.

Loeppert, R. H., & Suarez, D. L. (1996). Carbonate and gypsum. In D. L. Sparks (Ed.), *Methods of soil analysis: Part 3 Chemical methods* (pp. 437–474). Soil Science Society of America.

Lorenz, K., & Lal, R. (2015). Managing soil carbon stocks to enhance the resilience of urban ecosystems. *Carbon Management, 6*(1–2), 35–50. https://doi.org/10.1080/17583004.2015.1071182

Mihailović, A., Budinski-Petković, L., Popov, S., Ninkov, J., Vasin, J., Ralević, N. M., & Vasić, M. V. (2015). Spatial distribution of metals in urban soil of Novi Sad, Serbia: GIS based approach. *Journal of Geochemical Exploration, 150*, 104–114. https://doi.org/10.1016/j.gexplo.2014.12.017

Naghibi, S. J. (2023). *Genesis, classification and quality evaluation of shiraz urban watershed soils using integrated GIS, RS & GPS technology (3S)*. Ph.D. Thesis. Shiraz University (In Persian).

Nangia, S. B. (1991). *Soil pollution*. Ashish Publishing House.

Nelson, D. W., & Sommers, L. E. (1996). Total carbon, organic carbon, and organic matter. In D. L. Sparks (Ed.), *Methods of soil analysis: Part 3 Chemical methods, 5* (pp. 961–1010). Soil Science Society of America.

Page, A. L., Miller, R. H., & Keeney, D. R. (Eds.). (1982). *Methods of soil analysis: Part 2 Chemical and microbiological properties* (2nd ed.). American Society of Agronomy.

Paltseva, A., Cheng, Z., Deeb, M., Groffman, P. M., & Maddaloni, M. (2018). Variability of bioaccessible lead in urban garden soils. *Soil Science, 183*(4), 123–131. https://doi.org/10.1097/SS.0000000000000232

Paltseva, A. A., Cheng, Z., Egendorf, S. P., & Groffman, P. M. (2020). Remediation of an urban garden with elevated levels of soil contamination. *Science of the Total Environment, 722*, Article 137965. https://doi.org/10.1016/j.scitotenv.2020.137965

Ramirez-Rubio, O., Daher, C., Fanjul, G., Gascon, M., Mueller, N., Pajín, L., Plasencia, A., Rojas-Rueda, D., Thondoo, M., & Nieuwenhuijsen, M. J. (2019). Urban health: An example of a "health in all policies" approach in the context of SDGs implementation. *Global Health, 15*(1). https://doi.org/10.1186/s12992-019-0529-z. Article 87.

Reza, S. K., Baruah, U., Singh, S. K., & Das, T. H. (2015). Geostatistical and multivariate analysis of soil heavy metal contamination near coal mining area, Northeastern India. *Environmental Earth Sciences, 73*(9), 5425–5433. https://doi.org/10.1007/s12665-014-3797-1

Rezaee, L., Moosavi, A. A., Davatgar, N., & Sepaskhah, A. R. (2020). Soil quality indices of paddy soils in Guilan province of northern Iran: Spatial variability and their influential parameters. *Ecological Indicators, 117*, Article 106566. https://doi.org/10.1016/j.ecolind.2020.106566

Romzaykina, O. N., Vasenev, V. I., Paltseva, A., Kuzyakov, Y. V., Neaman, A., & Dovletyarova, E. A. (2021). Assessing and mapping urban soils as geochemical barriers for contamination by heavy metal(loid)s in Moscow megapolis. *Journal of Environmental Quality, 50*(1), 22–37. https://doi.org/10.1002/jeq2.20178

Saeedi, M., Hosseinzadeh, M., Jamshidi, A., & Pajooheshfar, S. P. (2009). Assessment of heavy metals contamination and leaching characteristics in highway side soils, Iran. *Environmental Monitoring and Assessment, 151*(1), 231–241. https://doi.org/10.1007/s10661-008-0264-z

Sager, M. (2020). Urban soils and road dust—Civilization effects and metal pollution—A review. *Environments, 7*(11). https://doi.org/10.3390/environments7110098. Article 98.

Savva, Y., Szlavecz, K., Pouyat, R. V., Groffman, P. M., & Heisler, G. (2010). Effects of land use and vegetation cover on soil temperature in an urban ecosystem. *Soil Science Society of America Journal, 74*(2), 469–480. https://doi.org/10.2136/sssaj2009.0107

Shchepeleva, A. S., Vasenev, V. I., Mazirov, I. M., Vasenev, I. I., Prokhorov, I. S., & Gosse, D. D. (2017). Changes of soil organic carbon stocks and CO2 emissions at the early stages of urban turf grasses' development. *Urban Ecosystems, 20*(2), 309–321. https://doi.org/10.1007/s11252-016-0594-1

Taalab, A. S., Ageeb, G. W., Siam, H. S., & Mahmoud, S. A. (2019). Some characteristics of calcareous soils: A review. *Middle East Journal of Agriculture Research, 8*(1), 96–105.

Vasenev, V. I., Van Oudenhoven, A. P. E., Romzaykina, O. N., & Hajiaghaeva, R. A. (2018). The ecological functions and ecosystem services of urban and technogenic soils: From theory to practice (a review). *Eurasian Soil Science, 51*(10), 1119–1132. https://doi.org/10.1134/S1064229318100130

Wan, G., Zhang, X., & Zhao, M. (2022). Urbanization can help reduce income inequality. *Npj Urban Sustainability, 2*. https://doi.org/10.1038/s42949-021-00040-y. Article 1.

Wang, Z., Hong, C., Xing, Y., Wang, K., Li, Y., Feng, L., & Ma, S. (2018). Spatial distribution and sources of heavy metals in natural pasture soil around copper-molybdenum mine in Northeast China. *Ecotoxicology and Environmental Safety, 154*, 329–336. https://doi.org/10.1016/j.ecoenv.2018.02.048

Wang, Y., Zhang, X., & Huang, C. (2009). Spatial variability of soil total nitrogen and soil total phosphorus under different land uses in a small watershed on the Loess Plateau, China. *Geoderma, 150*(1–2), 141–149. https://doi.org/10.1016/j.geoderma.2009.01.021

Weissert, L. F., Salmond, J. A., & Schwendenmann, L. (2016). Variability of soil organic carbon stocks and soil CO2 efflux across urban land use and soil cover types. *Geoderma, 271*, 80–90. https://doi.org/10.1016/j.geoderma.2016.02.012

Wu, J., Lu, J., Li, L., Min, X., & Luo, Y. (2018). Pollution, ecological-health risks, and sources of heavy metals in soil of the northeastern Qinghai-Tibet Plateau. *Chemosphere, 201*, 234–242. https://doi.org/10.1016/j.chemosphere.2018.02.122

Wu, S., Peng, S., Zhang, X., Wu, D., Luo, W., Zhang, T., & Wu, L. (2015). Levels and health risk assessments of heavy metals in urban soils in Dongguan, China. *Journal of Geochemical Exploration, 148*, 71–78. https://doi.org/10.1016/j.gexplo.2014.08.009

Xiaobing, W., Wuxing, L., Zhengao, L., Ying, T., Christie, P., & Yongming, L. (2020). Effects of long-term fertilizer applications on peanut yield and quality and plant and soil heavy metal accumulation. *Pedosphere, 30*(4), 555–562. https://doi.org/10.1016/S1002-0160(17)60341-9

Yang, J. L., & Zhang, G. L. (2015). Formation, characteristics and eco-environmental implications of urban soils–A review. *Soil Science & Plant Nutrition, 61*(Suppl. 1), 30–46. https://doi.org/10.1080/00380768.2015.1035622

Ye, L., Tan, W., Fang, L., Ji, L., & Deng, H. (2018). Spatial analysis of soil aggregate stability in a small catchment of the Loess Plateau, China: I. Spatial variability. *Soil and Tillage Research, 179*, 71–81. https://doi.org/10.1016/j.still.2018.01.012

Smagin, A. V. (2012). *Theory and practice of soil engineering*. MSU Press.

CHAPTER

5

Biofertilizers and organic amendments

Maryam Zahedifar

Department of Range and Watershed Management (Nature Engineering), Faculty of Agriculture, Fasa University, Fasa, Iran

1. Introduction

Climate change, global warming, and greenhouse gas emissions pose significant challenges to agricultural production. Meanwhile, the growing universal population has intensified the request for food and agricultural products, making this issue more critical than ever (Mthiyane et al., 2024). To enhance the productive capacity and promote the sustainable use of soils, it is vital to understand their properties and manage them accordingly. Implementing practices that improve these properties has a vital effect in increasing soil fertility. Traditional agriculture relies heavily on mineral and chemical fertilizers, along with various pesticides, to satisfy food demands. However, the overuse of inorganic fertilizers leads to significant waste, resulting in nutrients that are often unavailable to plants in various agricultural organizations. This overreliance on chemical fertilizers jeopardizes the sustainable health of soils and agricultural production. Consequences of this overuse include water pollution, nitrate leaching, soil acidification, distraction, and a deterioration in soil fertility and microbial populations. Given these issues, it is crucial to focus more on the use of organic compounds instead of chemical fertilizers to preserve soil fertility, supply essential nutrients, and enhance soil quality for sustainable agriculture (Fig. 5.1). According to Asghar et al. (2022), applying organic amendments is a vital strategy for improving and maintaining soil nutrients. Biofertilizers contain microorganisms that can be applied to seeds, plants, or soil. They form colonies around the roots or inside the plant, thereby increasing the plant's access to nutrients and promoting its growth (Vessey, 2003). Furthermore, biofertilizers offer an eco-friendly alternative for sustaining soil health and boosting crop production (Kour et al., 2020). Sustainable agricultural practices—particularly the use of biofertilizers and organic fertilizers—offer a promising way to boost crop yields while reducing the environmental impact of farming (Puglia et al., 2021). Biofertilizers increase soil health, soil quality, and nutrient cycling, which helps decrease the dependence on chemical fertilizers and greenhouse gas production (Bhardwaj et al., 2014; Matisic et al., 2022; Ngearnpat et al., 2023; Timmusk et al., 2017). This chapter aims to present different types of fertilizers derived from organic compounds, such as biochar, green manure, organic compost, and mulches, highlight their importance in maintaining soil and environmental health, and discuss their potential as alternatives to chemical fertilizers.

The Soil Nexus. https://doi.org/10.1016/B978-0-443-45286-4.00003-1

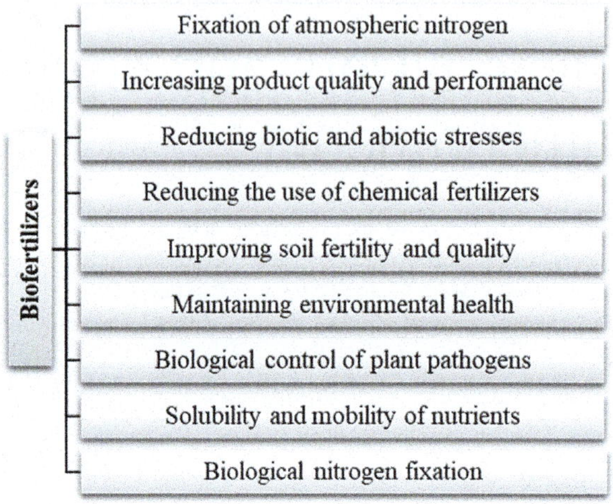

FIGURE 5.1

The effect of biofertilizers on the environment.

2. Types of biofertilizers

Biofertilizers are bioactive compounds produced through the fermentation of organic materials, containing either live or dormant microorganisms (such as bacteria, yeasts, bacilli, and algae) along with their metabolites and organo-mineral chelates (Xavier Júnior et al., 2021). These products are generated via fermentation, which can take place in either open or closed systems, and can be aerobic or anaerobic, based on the presence of air (Fig. 5.2).

2.1 Nitrogen-fixing biofertilizers

Nitrogen is a critical element found in amino acids, nucleic acids, adenosine triphosphate (ATP), and nicotinamide adenine dinucleotide in all living cells (Buren & Rubio, 2017). It is a key component of

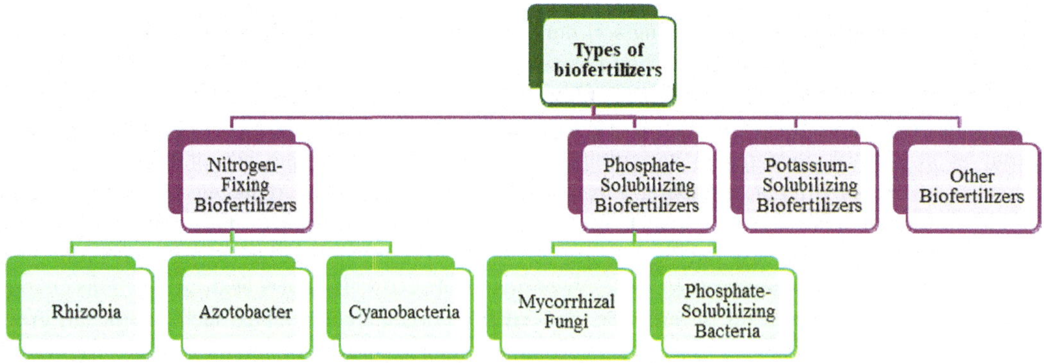

FIGURE 5.2

Types of biofertilizers.

chlorophyll production and photosynthesis, and is crucial for the growth of plants and productivity (Wagner, 2012). Dinitrogen (N_2) is the main gas of the atmosphere. But, only reactive or reduced N forms (NOx and NH_3) uptake by plants (Buren & Rubio, 2017). The production of nitrogen fertilizers, such as urea, through chemical synthesis consumes a significant amount of energy, leading to CO_2 emissions that can be tenfold or more (Zhang et al., 2013). Additionally, only 30%–40% of chemical fertilizers applied in agriculture are effectively used by plants (Prasad, 2009), while the remainder contributes to pollution and severe environmental issues. Alternatively, nitrogen can be supplied through biological methods. Biofertilizers are eco-friendly solutions that provide essential nutrients to plants and control plant pathogens. Biological nitrogen fixation is an enzymatic reaction within living organisms in which atmospheric dinitrogen gas is converted to ammonia. This conversion occurs more rapidly in the presence of the enzyme nitrogenase (Barney & Plunkett, 2022). Diazotrophs are a group of aerobic-anaerobic bacteria and cyanobacteria that produce the enzyme nitrogenase (Harding et al., 2022). According to Mukherjee &Sen (2021), there are a large number of free-living, symbiotic, and asymbiotic microbes that participate in biological nitrogen fixation.

2.1.1 Free-living nitrogen fixer

These bacteria possess a natural capability to fix nitrogen. They inhabit the soil, rhizosphere, and various plant-associated environments. In this category of bacteria, the fixation of atmospheric nitrogen in the soil occurs without direct interaction with plants. They derive the energy necessary for this process from the organic matter (OM) decomposition in the soil or through the oxidation of organic compounds by other microorganisms (Basu et al., 2021). Notably, some species, such as Azotobacter, can also obtain their energy from inorganic compounds (Kazemi et al., 2021).

2.1.2 Symbiotic nitrogen fixer

This form of N fixation, referred to as mutualism, involves a reciprocal interaction between plants and bacteria to fulfill their respective needs (Mus et al., 2016). Legumes are the main plants that utilize this method, allowing them to obtain the N they require. In exchange, legumes offer an ideal habitat for these N-fixing bacteria. These bacteria establish nodules within the plant tissues without any damage and reside there (Shamseldin, 2022). They acquire the energy required for N fixation from sugars produced by the plant through photosynthesis, and in return, they supply the plant with usable N (Doty, 2011). An example of these bacteria is rhizobia.

2.1.3 Associative symbiotic nitrogen fixer

In this type of symbiosis, nitrogen-fixing bacteria may attach to the roots of the host plant. Most of them reside on the roots, although some may penetrate into the plant. The host plant in this fixation is usually a cereal such as wheat, rice, corn, barley, or sorghum. This group of bacteria also uses atmospheric nitrogen and, after fixation, shares the product with the host plant. "Azospirillum" is an example of this group of fixers.

2.2 Phosphate-solubilizing biofertilizers

After nitrogen, phosphorus (P) is the second most abundant essential element for plants and plays a key role in influencing plant metabolism. Phosphate exists in the soil as soluble phosphate, insoluble organic phosphate, and insoluble inorganic phosphate. Although total phosphorus is abundant in the soil, its available form is very limited (Sharma et al., 2013). Given the vital work of this element in the functional roles of plants, adding fertilizers and amendments containing P is essential. Adding chemical fertilizers in acidic soils due to the formation of sediment with aluminum and iron cations, and in calcareous soils due to the formation of sediment with calcium ions, cannot be effective in providing the phosphorus required by the plant. Therefore, the use of alternative methods has attracted the attention of researchers. The dissolution of P in nature by phosphate-solubilizing microorganisms such as Pseudomonas and Bacillus can help provide phosphorus to plants (Kalayu, 2019). There are two categories of phosphate biofertilizers according to the use of "phosphate-solubilizing bacteria" and "phosphate-mobilizing microorganisms".

2.2.1 Phosphate-solubilizing microorganisms

This group includes strains of bacteria and fungi that dissolve insoluble inorganic phosphate compounds. The mechanism of action of bacteria and fungi in dissolving phosphate includes two theories: acid production and proton/enzyme production. In the acid-making system, phosphate-solubilizing microorganisms secrete organic acids such as citric, oxalic, malic, etc., which either directly dissolve phosphate rock by exchanging phosphate anions with acid anions or form chelates with calcium (Ca), aluminum (Al), and iron (Fe) ions to make phosphate soluble (Khan et al., 2009; Kishore et al., 2015). The secretion of organic acids by these microorganisms reduces soil pH and provides the basis for the dissolution of phosphates in the soil (Alori et al., 2017). The acid type and strength affect the released soluble phosphate. Mineral acids such as nitric and sulfuric acids, which are formed by some bacteria in the N oxidation or sulfur mineralization, can also help dissolve phosphate compounds in the soil (Fankem et al., 2006).

In the enzyme and proton theory, phosphate-solubilizing microorganisms help dissolve phosphate by producing the enzyme phosphatase. Sometimes, phosphate dissolution occurs without producing acid, as a result of the protons released along with respiration or ammonium absorption. The production of some chelating compounds, such as gibberellins and cytokinins, also leads to phosphate dissolution (Puri et al., 2020).

2.2.2 Phosphorus-mobilizing biofertilizers: "mycorrhiza"

This type of biofertilizer contains mycorrhizal fungi that settle in the roots of plants and establish a symbiotic life with most plants. These fungi establish a bridge-like connection between the roots and the soil and help the plant absorb nutrients. Ectomycorrhiza and endomycorrhiza are two important types of these fungi. The host plant of the fungus can access a larger volume of soil and obtain the nutrients it needs with the help of fungal hyphae. Mycorrhizae change the redox potential around the root by producing alkaline phosphatases and cause the conversion of insoluble phosphate into soluble forms. Hyphae of endomycorrhizal fungi help create stable soil aggregates in the soil by secreting sticky and sugary compounds called glomalin. Given the important and constructive role of mycorrhizae in achieving sustainable agricultural goals, attention to their use is increasing day by day.

2.3 Potassium-solubilizing biofertilizers

Potassium (K) is the third most abundant element required by plants and exists in enough amounts in the soil. This element plays a role in regulating osmotic pressure, opening and closing stomata, transporting compounds within the plant, activating enzymes, absorbing nutrients, making sugar and protein, photosynthesis, improving product quality, and creating resistance to environmental stresses

(Chaudhary et al., 2022). Soil K is observed in various forms, such as water-soluble, available, and unavailable, under the influence of physical and chemical properties (Basak et al., 2022). Immobile K is present in silicate minerals such as Illite, Muscovite, and Biotite. Varga et al. (2020) stated that potassium dissolution in the soil is carried out by bacteria and fungi. K-solubilizing bacteria are heterotrophic and provide their energy and carbon requirements from available OM in the soil. According to Zakaria (2009), these bacteria are effective in producing humus, increasing soil stability, and controlling root pathogens. K-solubilizing bacteria in the rhizosphere convert insoluble K into a soluble form by producing organic acids. Siderophores production, exchange reactions, and complex formation are other mechanisms by which these organisms solubilize soil K (Sattar et al., 2019). Research shows that inoculation of these bacteria can increase plant growth and nutrient concentrations in them (Basak & Biswas, 2010).

In addition to the earlier, there are other types of plant growth-promoting rhizobacteria. In general, it can be said that plant growth-promoting rhizobacteria are biological fertilizers that produce colonies in or on the root surface. Their effect on plant growth can be directed by absorbing nutrients and modulating growth by changing the level of plant hormones, or indirect by inhibiting various pathogens, which ultimately results in increased plant growth (Glick, 2012). The use of plant growth promoters should be a common practice in various agricultural systems (Basu et al., 2021). This is both environmentally friendly, safe for the consumer, and sustainable in the long term (Hyder et al., 2023).

3. Organic amendments

3.1 Definition and role in soil health

Due to the growing world population, the need for food production has increased. Providing these materials is not possible by increasing the cultivated area but by increasing the yield per unit area. In order to achieve this goal, double pressure has been put on the lands, and sometimes the excessive use of chemical fertilizers and various pesticides has caused inappropriate changes in the ecosystem and adverse environmental consequences. Maintaining soil health as an important component of the ecosystem should be considered. Soil health is the capacity of the soil to achieve optimal yield of agricultural products, maintain biological productivity, and support plant and animal health (Fausak et al., 2024). Today, the use of environmentally friendly organic compounds instead of chemical compounds has received great attention and interest in order to maintain soil health (Chojnacka et al., 2019). Currently, organic amendments are added as fertilizers or soil amendments. According to Kassam et al. (2009), organic amendments are materials that are added to improve various properties of the soil and increase plant growth. These materials usually result from organic sources such as animal manure, compost, and crop residues. These amendments improve the nutrient availability, soil structure, soil water, and biodiversity. In addition, they also have environmental benefits; chemical fertilizers increase the risk of heavy metals entering the ecosystem. Therefore, it can be concluded that the use of organic amendments is important to achieve sustainable agriculture since they promote the long-term health and productivity of agricultural soils.

3.2 Types of organic amendments

3.2.1 Animal manure

This organic compound consists of liquid and solid animal waste along with its bedding material, which has been used to improve soil fertility for many years. The amount of nutrients in animal manure depends on the type of animal, the length of storage, and the rate of decomposition. In addition to providing the nutrients needed by plants, animal manure increases soil OM and improves physical properties of the soil, such as its structure and stability. The use of fresh, and undecomposed animal manure may cause environmental problems, increased leaching and nitrogen loss, pollution of surface and groundwater, and an increase in dangerous pathogens (Omokaro et al., 2024). Drying and composting these fertilizers reduces the aforementioned risks, but it will increase global warming by releasing ammonia and carbon dioxide (Pell, 1997).

3.2.2 Compost

Composting is the decomposition of waste materials, shredded branches and leaves of plants, and plant waste under controlled conditions. It is a biological decomposition of organic materials that contain organic polymeric compounds such as lignin and cellulose. With this process, the mentioned polymeric compounds are decomposed into soluble organic carbon and converted into sugars that can be used by microorganisms. This process is a suitable alternative to landfill. Composting is an environmentally friendly waste treatment technology. In carrying out this process, the presence and function of microbes are important. These organisms include physical and chemical decomposers of waste. Bacteria, fungi, and actinomycetes are among the most important chemical decomposers of waste. Earthworms, ants, flies, and snails cause physical decomposition and crushing of waste. Composting organisms need four important factors for their activity: 1) Carbon (C) is required to provide energy, and C microbial oxidation to produce heat. 2) N is needed for the growth and reproduction of organisms. 3) Oxygen (O) is needed to oxidize carbon and the decomposition process. Aerobic bacteria that perform composting require oxygen levels of more than 5%. 4) Water is essential in adequate amounts. Living organisms can decompose OM if the four factors mentioned are present in sufficient quantities. The composting process includes the following stages (Hernandez-Rodriguez et al., 2017):

Mesophile Phase: Mesophilic organisms act at moderate temperatures.
Thermophilic Phase: Thermophilic bacteria act at high temperatures.
The cooling phase, which is called mesophile II.
Maturity Phase: At this stage, the temperature begins to decrease.

3.2.3 Green manure

Green manure is the enrichment of soil by tilling or mixing the soil with any green manure crop while it is green or shortly after the flowering stage begins. This fertilizer can be applied in situ, at the same planting site, or ex situ, by collecting plant residues at another site and then adding them to the soil a few days before planting (Meena et al., 2018). Application of green manure improves soil fertility by preserving soil nutrients, increasing nitrogen uptake and reducing leaching losses, increasing soil OM and microbial biomass, and helping to preserve the environment. Legumes as green manure release OM such as organic acids, amino acids, sugars, vitamins, and mucilage with the growth of the plant and the decomposition of plant residues into the soil (Shukla et al., 2011). These organic compounds increase soil resistance to erosion, increase soil porosity, water content, and infiltration in the soil. The use of cereal and legume mixtures as green manure, especially in rainfed almond orchards, reduces soil loss and runoff (Almagro et al., 2016), increases soil nutrient status, and carbon storage. Adding green manure to the soil creates a suitable environment for the activity and growth of beneficial soil organisms.

3.2.4 Biochar

To achieve higher production without harming the health of the environment and ecosystem, sustainable agriculture has been proposed. In sustainable agriculture, two limiting factors are known to be the lack of nutrients and the rapid mineralization of soil OM. In arid and semi-arid soils of the world, one of the most fundamental problems is the lack of OM. Soil nutrients can be supplied by adding compost and animal manure as a source of OM, but the decomposition of OM is recognized as the main source of carbon dioxide, which is one of the main greenhouse gas compounds and causes global warming (Zhang et al., 2010). One of the proposed compounds that has been introduced in recent years to improve the physical and chemical properties of soil is biochar. Biochar is a solid substance created through the thermochemical conversion of biomass in an oxygen-restricted environment.

This process is called pyrolysis. The physical and chemical properties of biochar affect its application. Biomass is biodegradable biological material originating from living or dead organisms. It also includes by-products, biodegradable municipal and industrial waste, plant and animal residues (Zahedifar & Najafian, 2017). Adding biochar to soil can reduce greenhouse gas releases to a great measure (Lucchini et al., 2014). The type of feedstock used and the pyrolysis techniques, fast or slow, used affect the properties of the resulting biochar (Oelbermann et al., 2020). Pyrolysis breaks down biomass into three main product categories (Fig. 5.3), with the amounts of "gas, bio-oil, and biochar" varying depending on the pyrolysis method applied (Tomczyk et al., 2020).

There are numerous technologies for biochar making (Oelbermann et al., 2020), but slow and fast pyrolysis are the most conventional methods for producing agricultural biochar (Brewer et al., 2012).

FIGURE 5.3

Biochar production process.

In fast pyrolysis, dry biomass is rapidly heated at rates exceeding 1000°C per second in an oxygen-free environment, resulting in a very short residence time of about 2 s. In contrast, slow pyrolysis, a more conventional method, involves the gradual heating of dry biomass at rates of 1–20°C per minute in the absence of oxygen, with residence times that can extend from hours to days (Dutta et al., 2012). The primary objective of fast pyrolysis is to maximize bio-oil yield, while slow pyrolysis aims to enhance biochar production.

3.2.5 Mulches

The process of covering the outer surface of the soil with a layer of external material is called mulching, and the material used for this purpose is called mulch. Mulching is especially suitable for crops of commercial value and gardens. The materials used to make mulch include two groups: organic and inorganic. Organic mulch includes any type of biodegradable organic material that can cover the soil. This mulch increases the soil humus and improves the soil quality, such as sugarcane bagasse, rice straw, dried grass leaves. Inorganic mulches can be made of polyethylene or a mixture of mineral compounds such as crushed stone, sand and gravel. Since this type of mulch is not biodegradable, it has a long life. Maintaining soil moisture, reducing the frequency of irrigation, reducing evaporation from the soil surface and regulating the temperature of plant root growth are some of the benefits of using mulch.

3.3 Benefits of organic amendments

Organic farming using organic amendments, compared to traditional agriculture that relies on the addition of chemical fertilizers and pesticides, increases crop productivity while maintaining ecosystem health. Here are some of the benefits of using organic amendments (Fig. 5.4).

3.3.1 Physical fertility

The use of organic amendments increases soil OM, which in turn improves "soil aggregate stability, water retention capacity, soil porosity, and ultimately soil quality". Karami et al. (2012) showed that the use of animal manure and wheat residues increased soil aggregate stability and reduced soil bulk density (BD). According to Zhao et al. (2009), organic amendments increased soil organic carbon and soil porosity, resulting in a reduction in soil BD.

3.3.2 Chemical and biological fertility

Following the addition of organic amendments, the organic carbon stock and cation exchange capacity of the soil are increased, which helps retain essential nutrient cations that can be made available to the plant over time. The solubility and therefore plant availability of P also increases in the presence of organic compounds. Some studies show that biochar is more effective than OM (Zheng et al., 2013). Compared to manure and compost, biochar is a more sustainable source of nutrients because it can slowly release and provide the nutrients needed by the plant, thus reducing the need for fertilization. In addition to the direct effect of biochar on nutrient supply, increasing soil quality by improving the efficiency of fertilizer use and retaining nutrients is an indirect effect of biochar (Lehmann et al., 2003; Zahedifar & Najafian, 2017). Based on its quality, biochar affects soil properties for example, cation exchange capacity (Uzoma et al., 2011), nutrient availability, efficiency of fertilizer use (Sohi et al., 2010), and heavy metal immobilization. The type of material used in biochar making, as well as the temperature of its manufacture, can affect the availability of metals. Research has shown that biochar

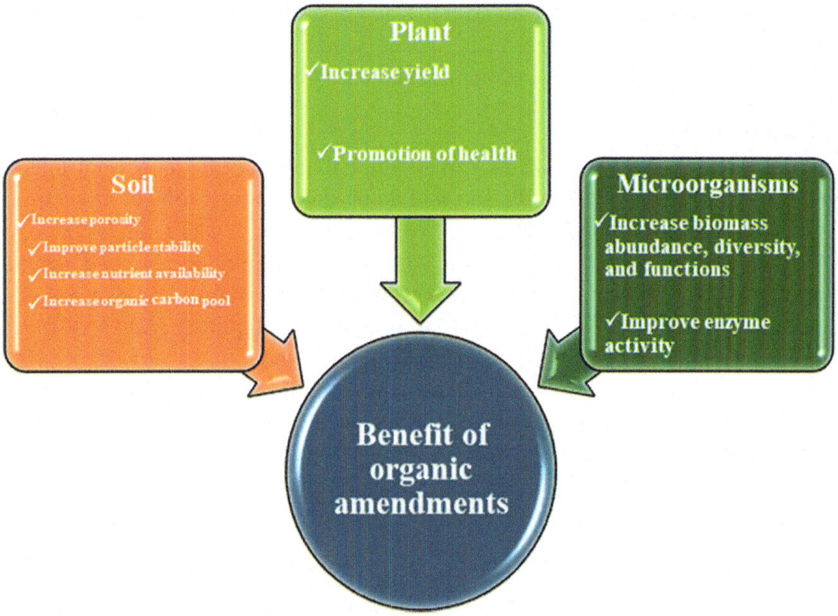

FIGURE 5.4

Benefit of organic amendments.

prepared from wheat has a greater effect on the availability of potassium compared to biochar prepared from corn, and if it is made at a higher temperature, it will have a greater effect on increasing the availability of phosphorus and potassium (Rasuli et al., 2021). Maintaining and increasing soil fertility is particularly important when facing environmental stresses. The use of organic amendments rich in amino acids and essential nutrients improved chlorophyll content and cellular metabolism, and subsequently the amount and value of tomato traits under water stress situations (Zahedifar et al., 2025b). Gavili et al. (2019) showed that under drought, the higher biochar applied, the higher grain yield of soybean gained.

Soil microorganisms affect the OM decomposition. The diversity of soil microorganism and mineralize of organic carbon are the key factors of fertility of soil and sustainability of agriculture. The addition of organic amendments to the soil increases the growth and microbial populations diversity. Soil biological properties are directly related to biomass, microbial activity and decomposition of organic compounds. The results show that enzymatic and biological activities in the soil increased after the addition of compost (Scotti et al., 2015). The positive effect of compost application on soil biological properties such as "soil respiration", "fluorescein diacetate hydrolases" and "phosphomonoester's activities" was stated by Iovieno et al. (2009) which indicates a change in microbial functions related to the C, N, and P cycles in the soil.

3.3.3 Plant disease suppression

Soil pathogens, for example, fungi, are the factors that reduce the yield of agricultural ecosystems and are frequently difficult to control using resistant varieties and synthetic fungicides. The use of organic amendments such as compost has been suggested as a approach for managing plant diseases resulted from soil-borne pathogens (Hossain et al., 2025). The suppressive effect of compost depends on the origin of the raw materials, the amount of application and the methods of its preparation.

3.3.4 Crop yield

Long-term use of chemical fertilizers and lack of attention to the addition of organic amendments have gradually reduced the organic input to the soil and its organic carbon storage, affecting soil fertility. The positive effect of organic amendments on the fertility of soil and plant yield is due to the provision of plant nutrients, improvement of soil structure, conservation and retention of soil water, and control of soil-borne pathogens. Foliar application of the organic manure improved fresh and dry weights of tomato significantly (Zahedifar et al., 2025a). It is assumed that liquid organic fertilizers can supply essential nutrient elements, and composites increase plant growth and yield. Moreover, they affect plant development positively and decrease the drought's contrary effects by regulating the plant's stomata (Zahedifar et al., 2025a). According to Bonanomi et al. (2020), continuing application of organic amendments improved crop yield, plant health, leaf quality, chemical and biological properties of the soil, while mineral fertilizers alone reduce microbial diversity and lead to soil acidity and salinization. Their study showed that while the addition of chemical fertilizers produced the highest crop yield after one year, it led to a significant decrease in yield over the long term. Nevertheless, the addition of organic amendments increased the activity and variety of microbial populations by providing a diverse food source. Their research showed that the application of organic materials can successfully reduce the negative effects of excessive use of chemical fertilizers in intensive farming systems.

4. Mechanisms of action

4.1 Nutrient availability

Nitrogen fixing bacteria for example, Rhizobium, change atmospheric N into a form that plants can readily absorb. Bacillus, Pseudomonas, and Aspergillus species increase the plant P absorption by dissolving insoluble forms of phosphate in the soil. These organisms reduce soil pH and dissolve phosphate. Pseudomonas species promote root growth and OM decomposition by producing hormones and antimetabolites, and increase nutrient availability. Potassium-solubilizing microorganisms, including Bacillus and Aspergillus niger species, decompose silicates in the soil by organic acids production and making the K ions in those minerals available to the plant. Thiobacillus also provides the plant with the sulfur it needs by oxidizing sulfur to sulfates that can be used by plants. Increased availability of iron and zinc in the rhizosphere is due to improved microbial activity. Increased uptake of these elements increases the amount of chlorophyll and consequently improves the efficiency of photosynthesis (Mthiyane et al., 2024).

4.2 Soil microbial activity enhancement

The type of soil management affects the amount of soil nutrients and the soil microenvironment characteristics, for example, the diversity, and microbial community structure (Liu et al., 2021). Du et al. (2022) stated that application of different fertilizers significantly changed the α-diversity, community composition and soil bacteria structure. The variation of fungal diversity is related to the soil properties for example, pH, OM, and nutrient availability. Organic fertilizers increase soil OM and control soil pH, so influencing soil microbial diversity and keystone taxa (Du et al., 2022).

4.3 Disease suppression mechanisms

Organic amendments affect the population, diversity, and activities of soil microbes by suppressing diseases and enhancing plant health. The effect of mulching on soil and rhizosphere microbiota has been constantly linked to soil suppression of several plant diseases. The introduction of organic compounds into the soil increases the soil's total microbial biomass and their activity and can suppress pathogens through competition for incomes or via other direct antagonism mechanisms (Weller et al., 2002). Physicochemical properties of the soil have been associated with disease destruction. Usually, an increase in microbial biomass makes a competitive environment that is detrimental to pathogens. The suppressive effect is likely related to a mixture of biotic and abiotic parameters, and not just a specific group of microorganisms is responsible for disease suppression. The production of antibiotic compounds by some Pseudomonas species is one of the methods of disease suppression. There is a great deal of variation in the efficiency of suppression, related to the nature of the organic compound, the type of crop, the pathogens, and environmental situations. Many of these relationships are still not well understood (Bonilla et al., 2012).

4.4 Improvement of soil physical properties

The effect of organic amendments on soil physical properties depends on the type of organic compound, duration of application, soil texture and tillage system. Research results show that adding organic amendments increases soil structural stability. This property depends on the type and amount of amendment. Organic amendments that contain more humic acid increase soil structural stability more than those that contain more fulvic acid. Humic substances with lower molecular weight and higher oxidation have a more effective role in stabilizing soil aggregates (Blanco-Canqui et al., 2024). According to Whalen et al. (2003), humic acids play a direct role in the formation of clay-organic complexes compared to fulvic acids, while fulvic acids are macromolecules with a lower polymerization index than humic acids, so they cannot form strong bonds with clays or mineral compounds, so the formed soil aggregates are not very stable. The presence of divalent cations such as calcium (Ca) and magnesium (Mg) affects positively on soil aggregate stability. The carbohydrate portion of organic matter performances as a glue to bind soil particles (Tejada & Gonzalez, 2007). Organic amendments may reduce soil BD and elasticity by increasing soil organic carbon concentration, thereby reducing soil compressibility. Adding organic carbon to soil in the long term reduces soil tensile and shear strength. It also increases the stability of wet aggregates and increases the saturated hydraulic conductivity of soil by increasing the porosity and organic carbon percentage of soil. As soil organic carbon and aggregate stability increase, water permeability in soil also improves. Another important result of adding organic compounds to soil is the increase in soil water retention capacity. This is mostly important in arid areas with water shortage problems. The use of organic amendments also affects the soil pores size distribution, which modulate water, gas, and heat flow and improving soil-root contact by increasing soil mesopores (Głąb et al., 2016).

5. Case studies of successful applications of biofertilizers

5.1 Impact of organic amendments on crop yields and quality

Biochar incorporation as an organic amendment, provide nutrients to plants. Zahedifar &Najafian (2017) investigated the effect of different levels of biochar (0%, 1.5% and 3% of cow manure source) and K nanochelate (0%, 0.03% and 0.06%) on growth characteristics and nutrient position of basil in a greenhouse experiment. Their results showed that the application of biochar significantly increased the shoot fresh weight. The shoot dry weight also increased significantly by 50% and 49%, respectively, with the addition of 1.5% and 3% biochar. Biochar can increase plant growth by providing nutrients and improving the physical, and biological properties of the soil. For example, biochar significantly increased the basil leaf area and height, as well as the concentrations and uptake of N, P, and K. The beneficial effects of biochar on growth and nutritional position of plant may be related to the nutrient content of the used biochar or to its positive impacts on the soil properties. Overall, the application of biochar not only enhances the quality of soils and helps mitigate greenhouse gas releases but also improves yield and the nutritional rank of basil plants. Additionally, biochar increased the leaves number and their surface areas, which are desirable traits in vegetables (Zahedifar & Najafian, 2017). The basil leaves are rich in essential oils with a strong aroma and high contents of phenolic composites, for example, rosmarinic and caffeic acid, which are known for their high antioxidant capacities (Bufalo et al., 2015; Grayer et al., 2004).

Zahedifar &Najafian (2023) studied the foliar spraying of different levels of humic (HA) and fulvic acids (FA) (0, 250, 500 and 1000 mg L^{-1}) on essential oil compounds and antioxidant activity of yarrow. Antioxidant compounds play an important role in controlling diseases caused by free radicals. “IC50” is an appropriate amount of oxidation development in oils, which is measured as a good index for the antioxidant efficiency evaluation. “Gallic acid” is used as a standard to show antioxidant activity. Stronger antioxidants result in an IC50 value that is closer to that of gallic acid. FA improved the repressive effects. So, the “IC50” reduced significantly with spraying different levels of “FA” compared to control (Table 5.1). Addition of greater applied level FA significantly increased the antioxidant action of “*A. millefolium* L.” extract. This positive effect was achieved by adding the greatest content of HA (Table 5.1).

Table 5.1 Effect of various contents of HA and FA on the antioxidant activity (IC50) of yarrow (Zahedifar & Najafian, 2023).

Factors	Content (mg L^{-1})	IC50 (mg g^{-1})
Control	0	454 b[a]
HA	250	878 a
	500	868 a
	1000	382 c
FA	250	409 BCE
	500	363 c
	1000	374 c

[a]*Means with the same lowercase letters are not significantly different according to Duncan’s Multiple Range Test* ($P \leq .05$).

Their findings are important and practical for better management of medicinal plant production and their products quality. Also, it should be noted that in sustainable agriculture, along with optimal crop production, environmental protection is of great importance, therefore, replacing these organic compounds instead of chemical fertilizers or together with chemical fertilizers will be of great interest. In another research, Zahedifar et al. (2025a) showed that water shortage declined fruit weights of tomato, while foliar spraying of organic fertilizer increased them significantly. Other growth traits for example, vitamin C content and citric acid improved after organic fertilizer application (Fig. 5.5a and b).

In addition, under water shortage, organic fertilizer improved the fruit Fe, sodium (Na) and K contents. The organic fertilizer addition significantly improved plant growth behaviors under water deficit. Thus, it can be stated incorporating organic fertilizer is an operative management plan to alleviate the negative effects of drought and enhance both the tomato fruit yield.

5.2 Soil quality (SQ) and organic amendments

The type and intensity of management and activities that prevent soil degradation or environmental contamination affect soil health. The critical limits of various soil properties are studied based on SQ indicators (Sanchez-Navarro et al., 2015). The ability of the soil to maintain the environment and the capacity of the ecosystem, along with increasing human, animal and plant health, are the main goals of SQ monitoring. By determining soil quality, in addition to obtaining information from the soil, it is possible to identify the correct methods of system sustainability (Obade & Gaya, 2021).

Zahedifar et al. (2025a) evaluated the combined effects of cattle manure-derived biochar (0 (B0), 25 (B25), 50 (B50), and 100 (B100) t ha^{-1}) and soil moisture levels (100%, 70%, and 55% of FC) on soybean yield and soil-plant nutrients status, quality indices in the postharvest soils and SQ grades. According to Fig. 5.6, with biochar application, soil BD showed a positive relationship with SQ indicators. Biochar increases soil water retention capacity by improving soil porosity and aggregate stability. Other results of this study included increased SQ, reduced adverse effects of BD in the soil, and increased plant growth and yield. As revealed by Fig. 5.6, adding 25 t ha^{-1} of biochar created a negative

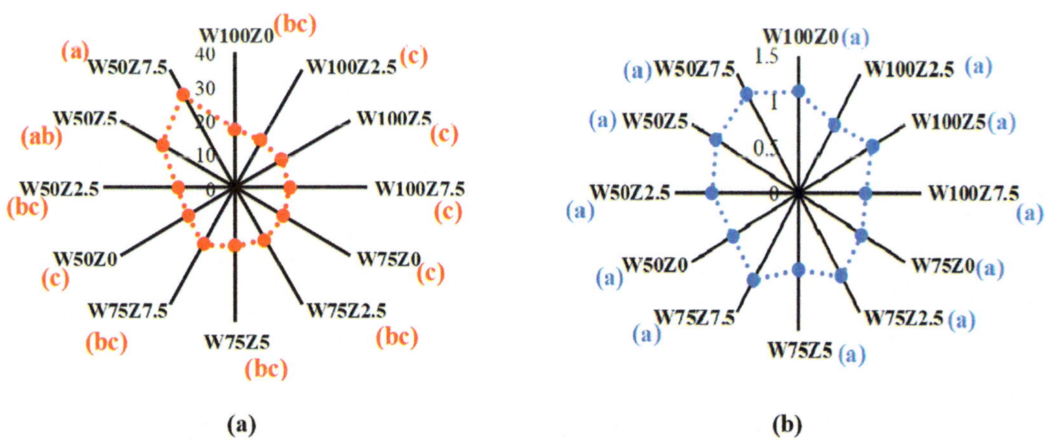

FIGURE 5.5

Vitamin C (a) and citric acid (b), based on mg 100 mL^{-1} and %, respectively as influenced by various levels of organic fertilizer (Z) and drought stress (W) (Zahedifar et al., 2025a).

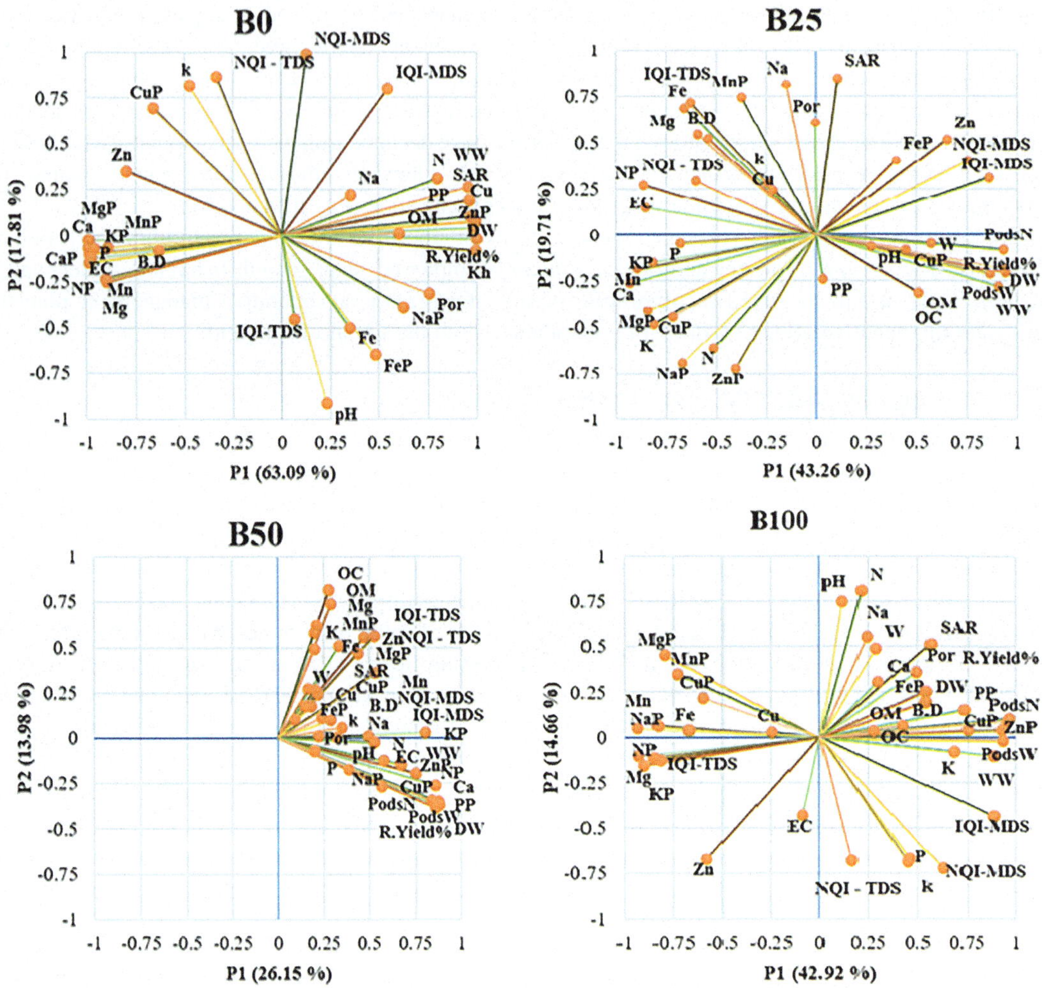

FIGURE 5.6

Pearson correlation coefficients for different soil traits as influenced by different biochar treatments (Zahedifar et al., 2025a).

relationship between organic matter and BD, while adding higher levels of this modifier created a positive correlation between the two characteristics mentioned earlier. The findings showed that the application of biochar reduced the adverse effects of salinity on SQ. The presence of a positive correlation between SQ indices and nutrient concentrations was one of the results of adding 25 t ha^{-1} of biochar.

5.3 Heavy metals (HMs) distribution in soil and organic amendments

Enhancing organic matter as a soil amendment can effectively improve the bioavailability of HMs. Recently, biochar, produced through the pyrolysis of various organic materials, has garnered significant attention in research due to its potential benefits and unique properties, including high surface area, organic matter content, porous structure, and cheapness (Dai et al., 2016). Minerals on the biochar surfaces and functional oxygen groups complex and precipitate with Zn (Dai et al., 2013). According to Raicevic et al. (2005), the formation of chemical mechanisms, including adsorption, precipitation, and complexation between modifiers and volatile organic compounds, reduces their mobility in the soil. Zahedifar (2017) evaluated the effect of three levels of biochar (0%, 1.5%, and 3% wt.) and two Zn levels (0 and 10 mg Zn kg^{-1} soil) on Zn fractionations of agricultural, rangeland, and forest soils. The results showed that the concentration of exchangeable Zn decreased due to the formation of complexes between the chelating groups of organic compounds and Zn. Furthermore, biochar increased soil alkalinity and high pH results in an increased negative charge and Zn^{+2} hydrolysis. The carbonate fraction of Zn was increased with 1.5% biochar addition. Biochar increased the organic Zn fraction significantly in rangeland soil. Bioavailability of Zn in agricultural and forest soils declined with biochar application, while it increased in rangeland soil. In all land uses, Zn bioavailability was improved with Zn addition. The evaluated mobility factor values indicated that the bioavailability of Zn was relatively low, suggesting that Zn in the studied soils is highly constant (see Fig. 5.7). It is important to note that the effect of biochar on Zn mobility varied with different land uses. This aspect should be taken into account when managing Zn-contaminated soils, as well as in remediation practices and soil fertility administration for agricultural soils.

6. Conclusion

Organic amendments and biofertilizers play a vital role in sustainable agriculture and soil health management. Their benefits can be summarized as follows:

1. **Soil Health Improvement**: Organic amendments improve soil structure, increase organic matter content, and water retention. This leads to improved soil ecosystems, which support diverse microbial communities.

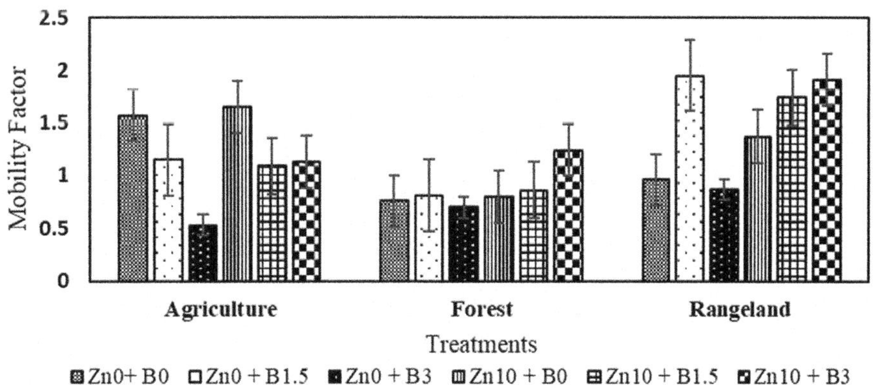

FIGURE 5.7

Zn mobility factor (MF) in the soils of different land use types as influenced by biochar and Zn application (Zahedifar, 2017).

2. **Nutrient Supply**: Biofertilizers, which include beneficial microorganisms, enhance nutrient availability, particularly nitrogen, phosphorus, and potassium. They help in the efficient uptake of these nutrients by plants, reducing the need for chemical fertilizers.
3. **Environmental Sustainability**: The use of organic amendments and biofertilizers reduces chemical inputs in agriculture, minimizing environmental pollution and promoting biodiversity. This is in accordance with the principles of organic farming and sustainable land management.
4. **Crop Productivity**: Research shows that integrating organic amendments and biofertilizers can lead to improved crop yields and quality. They not only support plant growth but also increase resilience against pests and diseases.
5. **Economic Viability**: By reducing reliance on synthetic fertilizers, farmers can lower input costs and improve profitability. Additionally, the long-term benefits of healthier soils contribute to sustainable agricultural practices.

In conclusion, the integration of organic amendments and biofertilizers into agricultural systems is essential for promoting sustainable practices, enhancing soil fertility, and ensuring food security while protecting the environment. Continued research and education in this area will further optimize their use and benefits in various agricultural contexts.

References

Almagro, M., de Vente, J., Boix-Fayos, C., García-Franco, N., Melgares de Aguilar, J., Gonzalez, D., Solé-Benet, A., & Martínez-Mena, M. (2016). . Sustainable land management practices as providers of several ecosystem services under rainfed mediterranean agroecosystems. *Mitigation and Adaptation Strategies for Global Change, 21*, 1029–1043. https://doi.org/10.1007/s11027-013-9535-2.21

Alori, E. T., Glick, B. R., & Babalola, O. O. (2017). Microbial phosphorus solubilization and its potential for use in sustainable agriculture. *Frontiers in Microbiology, 8*, 971. https://doi.org/10.3389/fmicb.2017.00971

Asghar, W., Akça, M. O., Akça, H., Jafari Tarf, O., Kataoka, R., & Turgay, O. C. (2022). Alternative strategies to synthetic chemical fertilizers: Revitalization of soil quality for sustainable agriculture using organic-based approaches. In H. B. Singh, & A. Vaishnav (Eds.), *New and future developments in microbial biotechnology and bioengineering* (pp. 1–30). Elsevier.

Barney, B. M., & Plunkett, M. H. (2022). Rnf1 is the primary electron source to nitrogenase in a high-ammonium-accumulating strain of azotobacter vinelandii. *Applied Microbiology and Biotechnology, 106*, 5051–5061.

Basak, B. B., & Biswas, D. R. (2010). Co-inoculation of potassium solubilizing and nitrogen fixing bacteria on solubilization of waste mica and their effect on growth promotion and nutrient acquisition by a forage crop. *Biology and Fertility of Soils, 46*, 641–648.

Basak, B. B., Maity, A., Ray, P., Biswas, D. R., & Roy, S. (2022). Potassium supply in agriculture through biological potassium fertilizer: A promising and sustainable option for developing countries. *Archives of Agronomy and Soil Science, 68*, 101–114. https://doi.org/10.1080/03650340.2020.1821191

Basu, A., Prasad, P., Das, S. N., Kalam, S., Sayyed, R. Z., Reddy, M. S., & El Enshasy, H. (2021). Plant growth promoting rhizobacteria (PGPR) as green bioinoculants: Recent developments, constraints, and prospects. *Sustainability, 13*(3), 1140. https://doi.org/10.3390/su13031140

Bhardwaj, D., Ansari, M. W., Sahoo, R. K., & Tuteja, N. (2014). Biofertilizers function as key players in sustainable agriculture by improving soil fertility, plant tolerance, and crop productivity. *Microbial Cell Factories, 13*, 66. https://doi.org/10.1186/1475-2859-13-66

Blanco-Canqui, H., Ruis, S. J., & Francis, C. A. (2024). Do organic farming practices improve soil physical properties? *Soil Use and Management, 40*, Article e12999.

Bonanomi, G., De Filippis, F., Zotti, M., Idbella, M., Cesarano, G., Al-Rowaily, S., & Abd-ElGawad, A. (2020). Repeated applications of organic amendments promote beneficial microbiota, improve soil fertility, and increase crop yield. *Applied Soil Ecology, 156*, Article 103714. https://doi.org/10.1016/j.apsoil.2020.103714

Bonilla, N., Gutiérrez-Barranquero, J. A., Vicente, A. D., & Cazorla, F. M. (2012). Enhancing soil quality and plant health through suppressive organic amendments. *Diversity, 4*(4), 475–491. https://doi.org/10.3390/d4040475

Brewer, C. E., Hu, Y. Y., Schmidt-Rohr, K., Loynachan, T. E., Laird, D. A., & Brown, R. C. (2012). Extent of pyrolysis impacts on fast pyrolysis biochar properties. *Journal of Environmental Quality, 41*, 1115–1122. https://doi.org/10.2134/jeq2011.0118

Bufalo, J., Cantrell, C. L., Astatkie, T., Zheljazkov, V. D., Gawde, A., & Boaro, C. S. F. (2015). Organic versus conventional fertilization effects on sweet basil (Ocimum basilicum L.) growth in a greenhouse system. *Industrial Crops and Products, 74*, 249–254.

Buren, S., & Rubio, L. M. (2017). State of the art in eukaryotic nitrogenase engineering. *FEMS Microbiology Letters, 365*, fnx274.

Chaudhary, P., Singh, S., Chaudhary, A., Sharma, A., & Kumar, G. (2022). Overview of biofertilizers in crop production and stress management for sustainable agriculture. *Frontiers in Plant Science, 13*, Article 930340. https://doi.org/10.3389/fpls.2022.930340

Chojnacka, K., Gorazda, K., Witek-Krowiak, A., & Moustakas, K. (2019). Recovery of fertilizer nutrients from materials-contradictions, mistakes and future trends. *Renewable and Sustainable Energy Reviews, 110*, 485–498.

Dai, Z., Meng, J., Muhammad, N., et al. (2013). The potential feasibility for soil improvement, based on the properties of biochars pyrolyzed from different feedstocks. *Journal of Soils and Sediments, 13*, 989–1000. https://doi.org/10.1007/s11368-013-0698-y

Dai, Z., Meng, J., Shi, Q., Xu, B., Lian, Z., Brookes, P. C., & Xu, J.-M. (2016). Effects of manure- and lignocellulose-derived biochars on adsorption and desorption of zinc by acidic types of soil with different properties. *European Journal of Soil Science, 67*.

Doty, S. L. (2011). Nitrogen-fixing endophytic bacteria for improved plant growth. In D. Maheshwari (Ed.), *Bacteria in agrobiology: Plant growth responses* (pp. 183–199). Berlin, Heidelberg: Springer.

Du, T. Y., He, H. Y., Zhang, Q., Lu, L., Mao, W. J., & Zhai, M. Z. (2022). Positive effects of organic fertilizers and biofertilizers on soil microbial community composition and walnut yield. *Applied Soil Ecology, 175*, Article 104457. https://doi.org/10.1016/j.apsoil.2022.104457

Dutta, B., Raghavan, G. S. V., & Ngadi, M. (2012). Surface characterization and classification of slow and fast pyrolyzed biochar using novel methods of pycnometry and hyperspectral imaging. *Journal of Wood Chemistry and Technology, 32*, 105–120. https://doi.org/10.1080/02773813.2011.607535

Fankem, H., Nwaga, D., Deubel, A., Dieng, L., Merbach, W., & Etoa, F. X. (2006). Occurrence and functioning of phosphate solubilizing microorganisms from oil palm tree *(Elaeis guineensis)* rhizosphere in Cameroon. *African Journal of Biotechnology, 5*(24).

Fausak, L. K., Bridson, N., Diaz-Osorio, F., Jassal, R. S., & Lavkulich, L. M. (2024). Soil health – a perspective. *Frontiers in Soil Science, 4*, Article 1462428.

Głąb, T., Pużyńska, K., Pużyński, S., Palmowska, J., & Kowalik, K. (2016). Effect of organic farming on a stagnic luvisol soil physical quality. *Geoderma, 282*, 16–25.

Gavili, E., Moosavi, A. A., & Zahedifar, M. (2019). Integrated effects of cattle manure-derived biochar and soil moisture conditions on soil chemical characteristics and soybean yield. *Archives of Agronomy and Soil Science, 65*(12), 1758–1774. https://doi.org/10.1080/03650340.2019.1576864

Glick, B. R. (2012). Plant growth-promoting bacteria: Mechanisms and applications. *Scientifica*, 1–15. https://doi.org/10.6064/2012/963401

Grayer, R. J., Vieira, R. F., Price, A. M., Kite, G. C., Simon, J. E., & Paton, A. J. (2004). Characterization of cultivars within species of ocimum by exudates flavonoid profiles. *Biochemical Systematics and Ecology, 32*, 1–10.

Harding, K. J., Turk-Kubo, K. A., Mak, E. W. K., Weber, P. K., Mayali, X., & Zehr, J. P. (2022). Cell-specific measurements show nitrogen fixation by particle-attached putative non-cyanobacterial diazotrophs in the North Pacific Subtropical Gyre. *Nature Communications, 13*(1), 6979.

Hernández-Rodríguez, A., Robles-Hernández, L., Ojeda-Barrios, D., Prieto-Luévano, J., González-Franco, A. C., & Guerrero-Prieto, V. (2017). Semicompost and vermicompost mixed with peat moss enhance seed germination and development of lettuce and tomato seedlings. *Interciencia, 42*(11), 774–779.

Hossain, M. M., Sultana, F., Mostafa, M., Rubayet, M. T., Mishu, N. J., Khan, I., & Mostofa, M. G. (2025). Biological management of soil-borne pathogens through tripartite rhizosphere interactions with plant growth promoting fungi. *Applied Microbiology, 5*, 123. https://doi.org/10.3390/applmicrobiol5040123

Hyder, S., Rizvi, Z. F., de los Santos-Villalobos, S., Santoyo, G., Gondal, A. S., Khalid, N., Fatima, S. N., Nadeem, M., Rafique, K., & Rani, A. (2023). Applications of plant growth-promoting rhizobacteria for increasing crop production and resilience. *Journal of Plant Nutrition, 46*, 2551–2580. https://doi.org/10.1080/01904167.2022.2160742

Iovieno, P., Morra, L., Leone, A., Pagano, L., & Alfani, A. (2009). Effect of organic and mineral fertilizers on soil respiration and enzyme activities of two mediterranean horticultural soils. *Biology and Fertility of Soils, 45*, 555–561.

Kalayu, G. (2019). Phosphate solubilizing microorganisms: Promising approach as biofertilizers. *International Journal of Agronomy, 2019*, Article 4917256. https://doi.org/10.1155/2019/4917256

Karami, A., Homaee, M., Afzalinia, S., Ruhipour, H., & Basirat, S. (2012). Organic resource management: Impacts on soil aggregate stability and other soil physico-chemical properties. *Agriculture, Ecosystems & Environment, 148*, 22–28. https://doi.org/10.1016/j.agee.2011.10.021

Kassam, A., Friedrich, T., Shaxson, F., & Pretty, J. (2009). The spread of conservation agriculture: Justification, sustainability and uptake. *International Journal of Agricultural Sustainability, 7*(4), 292–320.

Kazemi, E., Soofiyani, S. R., Ahangari, H., Eyvazi, S., Hejazi, M. S., & Tarhriz, V. (2021). Chemolithotroph bacteria: From biology to application in medical sciences. *Crescent Journal of Medical and Biological Sciences, 8*(2), 81–89.

Khan, A., Jilani, V., Akhtar, M. S., Naqvi, S. M. S., & Rasheed, M. (2009). Phosphorus solubilizing bacteria: Occurrence, mechanisms, and their role in crop production. *Journal of Agricultural and Biological Science, 1*, 48–58.

Kishore, N., Pindi, P. K., & Ram Reddy, S. (2015). Phosphate-solubilizing microorganisms: A critical review. In *Plant biology and biotechnology* (pp. 307–333). https://doi.org/10.1007/978-81-322-2286-6_12

Kour, D., Rana, K. L., Kaur, T., Sheikh, I., Yadav, A. N., Kumar, V., Dhaliwal, H. S., & Saxena, A. K. (2020). Microbe-mediated alleviation of drought stress and acquisition of phosphorus in great millet (Sorghum bicolour L.) by drought-adaptive and phosphorus-solubilizing microbes. *Biocatalysis and Agricultural Biotechnology, 23*, 1–9. https://doi.org/10.1016/j.bcab. 2020.101501

Lehmann, J., Pereira da Silva, J. J., Steiner, C., Nehls, T., Zec, W., & Glaser, B. (2003). Nutrient availability and leaching in an archaeological anthrosol and a ferralsol of the Central Amazon basin: Fertilizer, manure, and charcoal amendments. *Plant and Soil, 249*, 343–357.

Liu, J., Shu, A., Song, W., Shi, W., & Gao, Z. (2021). Long-term organic fertilizer substitution increases rice yield by improving soil properties and regulating soil bacteria. *Geoderma, 404*, Article 115287. https://doi.org/10.1016/j.geoderma.2021.115287

Lucchini, P., Quilliam, R., DeLuca, T., et al. (2014). Increased bioavailability of metals in two contrasting agricultural soils treated with waste wood-derived biochar and ash. *Environmental Science and Pollution Research, 21*, 3230–3240. https://doi.org/10.1007/s11356-013-2272-y

Matisic, M., Dugan, I., & Bogunovic, I. (2022). Challenges in sustainable agriculture—The role of organic amendments. *Agriculture, 14*, 643. https://doi.org/10.3390/agriculture14030643

Meena, R. S., Vijayakumar, V., Yadav, G. S., & Mitran, T. (2018). Response and interaction of bradyrhizobium japonicum and arbuscular mycorrhizal fungi in the soybean rhizosphere. *Plant Growth Regulation.* https://doi.org/10.1007/s10725-017-0334-8

Mthiyane, P., Aycan, M., & Mitsui, T. (2024). Integrating biofertilizers with organic fertilizers enhances photosynthetic efficiency and upregulates chlorophyll-related gene expression in rice. *Sustainability, 16*, 9297. https://doi.org/10.3390/su16219297

Mukherjee, R., & Sen, S. (2021). Role of biological nitrogen fixation (BNF) in sustainable agriculture: A review. *International Journal of Advanced Life Sciences Research, 4*(3), 1–7.

Mus, F., Crook, M. B., Garcia, K., Garcia Costas, A., Geddes, B. A., Kouri, E. D., Paramasivan, P., Ryu, M. H., Oldroyd, G. E. D., Poole, P. S., Udvardi, M. K., Voigt, C. A., Ané, J. M., & Peters, J. W. (2016). Symbiotic nitrogen fixation and the challenges to its extension to nonlegumes. *Applied and Environmental Microbiology, 82*(13), 3698–3710.

Ngearnpat, N., Chunhachart, O., Kotabin, N., & Issakul, K. (2023). Comparative assessment of gamma-polyglutamic acid and bacillus subtilis cells as biostimulants to improve rice growth and soil quality. *Journal of Ecology and Engineering, 24*, 46–59.

Obade, V. P., & Gaya, C. (2021). Digital technology dilemma: On unlocking the soil quality index conundrum. *Bioresources and Bioprocessing, 8*, 6.

Oelbermann, M., Berruti, F., & Lévesque, V. (2020). Biochar and its use in soil: Lessons from temperate agriculture. *World Journal of Agricultural and Soil Sciences, 5*, 1–9.

Omokaro, G. O., Osarhiemen, I. O., Idama, V., Airueghian, E. O., West, S. T., Igbigbi, F. E., Nnake, D. C., Obolokor, E., Ahmed, A., & Omoshie, V. O. (2024). The role of organic amendments and their impact on soil restoration: A review. *Asian Journal of Environment & Ecology, 23*(11), 41–52. https://doi.org/10.9734/ajee/2024/v23i11620

Pell, A. N. (1997). Manure and microbes: Public and animal health problem? *Journal of Dairy Science, 80*, 2673–2681. https://doi.org/10.3168/jds.S0022-0302(97)76227-1

Prasad, R. (2009). Efficient fertilizer use—The key to food security and better environment. *Journal of Tropical Agriculture, 47*, 1–17.

Puglia, D., Pezzolla, D., Gigliotti, G., Torre, L., Bartucca, M. L., & Buono, D. D. (2021). The opportunity of valorizing agricultural waste through its conversion into biostimulants, biofertilizers, and biopolymers. *Sustainability, 13*, 2710. https://doi.org/10.3390/su13052710

Puri, A., Padda, K. P., & Chanway, C. P. (2020). In vitro and in vivo analyses of plant-growth-promoting potential of bacteria naturally associated with spruce trees growing on nutrient-poor soils. *Applied Soil Ecology, 149*, Article 103538.

Raicevic, S., Kaludjerovic-Rdoicic, T., & Zouboulis, A. I. (2005). In situ stabilization of toxic metals in polluted soils using phosphates: Theoretical prediction and experimental verification. *Journal of Hazardous Materials, 117*, 41–53.

Rasuli, F., Owliaie, H., Najafi-Ghiri, M., & Adhami, E. (2021). Effect of biochar on potassium fractions and plant-available P, Fe, Zn, Mn and Cu concentrations of calcareous soils. *Arid Land Research and Management, 36* (1), 1–26. https://doi.org/10.1080/15324982.2021.1936689

Sanchez-Navarro, A., Gil-Vazquez, J. M., Delgado-Iniesta, M. J., Marin-Sanleandro, P., Blanco-Ber nardeau, A., & Ortiz-Silla, R. (2015). Establishing an index and identification of limiting parameters for characterizing soil quality in Mediterranean ecosystems. *Catena, 131*, 35–45.

Sattar, A., Naveed, M., Ali, M., Zahir, Z. A., Nadeem, S. M., Yaseen, M., et al. (2019). Perspectives of potassium solubilizing microbes in sustainable food production system: A review. *Agriculture, Ecosystems & Environment, 133*, 146–159. https://doi.org/10.1016/j.apsoil.2018.09.012

Scotti, R., D'Ascoli, R., Bonanomi, G., Caceres, M. G., Sultana, S., Cozzolino, L., Scelza, R., & Rao, M. A. (2015). Combined use of compost and wood scraps to increase carbon stock and improve soil quality in intensive farming systems. *European Journal of Soil Science, 66*, 463–475. https://doi.org/10.1111/ejss.12248

Shamseldin, A. (2022). Future outlook of transferring biological nitrogen fixation (BNF) to cereals and challenges to retard achieving this dream. *Current Microbiology, 79*(6), 171.

Sharma, S. B., Sayyed, R. Z., Trivedi, M. H., & Gobi, T. A. (2013). Phosphate solubilizing microbes: Sustainable approach for managing phosphorus deficiency in agricultural soils. *SpringerPlus, 2*, 587.

Shukla, K. P., Sharma, S., Singh, N. K., Singh, V., Tiwari, K., & Singh, S. (2011). Nature and role of root exudates: Efficacy in bioremediation. *African Journal of Biotechnology, 10*(48), 9717–9724.

Sohi, S. P., Krull, E., & Bol, R. (2010). A review of biochar and its use and function in soil. In *Advances in Agronomy* (1st ed.). Elsevier Inc. https://doi.org/10.1016/S0065-2113(10)05002-9

Tejada, M., & Gonzalez, J. L. (2007). Influence of organic amendments on soil structure and soil loss under simulated rain. *Soil and Tillage Research, 93*(1), 197–205. https://doi.org/10.1016/j.still.2006.04.002

Timmusk, S., Behers, L., Muthoni, J., Muraya, A., & Aronsson, A.-C. (2017). Perspectives and challenges of microbial application for crop improvement. *Frontiers in Plant Science, 8*, 49. https://doi.org/10.3389/fpls.2017.00049

Tomczyk, A., Sokołowska, Z., & Boguta, P. (2020). Biochar physicochemical properties: Pyrolysis temperature and feedstock kind effects. *Reviews in Environmental Science and Biotechnology, 19*, 191–215. https://doi.org/10.1007/s11157-020-09523-3

Uzoma, K. C., Inoue, M., Andry, H., Fijimaki, H., Zahoor, A., & Nishihara, E. (2011). Effect of cow manure biochar on maize productivity under sandy soil condition. *Soil Use and Management, 27*, 205–212.

Varga, T., Hixson, K. K., Ahkami, A. H., Sher, A. W., Barnes, M. E., Chu, R. K., et al. (2020). Endophyte-promoted phosphorus solubilization in populus. *Frontiers in Plant Science, 11*, Article 567918. https://doi.org/10.3389/fpls.2020.567918

Vessey, J. K. (2003). Plant growth promoting rhizobacteria as biofertilizers. *Plant and Soil, 255*(2), 571–586. https://doi.org/10.1023/A:1026037216893

Wagner, S. C. (2012). Biological nitrogen fixation. *Nature Education Knowledge, 3*, 15.

Weller, D. M., Raaijmakers, J. M., McSpadden, B. B., & Thomashow, L. S. (2002). Microbial populations responsible for specific soil suppressiveness to plant pathogens. *Annual Review of Phytopathology, 40*, 309–348.

Whalen, J. K., Hu, Q., & Liu, A. (2003). Compost application increases water-stable aggregates in conventional and no-tillage systems. *Soil Science Society of America Journal, 67*, 1842–184.

Xavier Júnior, O. S., Jardim, A. M., da, R. F., Souza, L. S. B. de, Souza, C. A. A. de, Cruz Neto, J. F. da, Araújo Júnior, G. do N., Leite, R. M. C., & Silva, T. G. F. da (2021). Utilização de biofertilizantes alternativos no cultivo de palma forrageira: Uma revisão. *Research, Society and Development, 10*(6), 1–15.

Zahedifar, M. (2017). Sequential extraction of zinc in the soils of different land use types as influenced by wheat straw derived biochar. *Journal of Geochemical Exploration, 182*, 22–31.

Zahedifar, M., Moosavi, A. A., & Gavili, E. (2025a). Monitoring soil quality indices and soybean yield as influenced by integrated biochar and drought stress. *Environment, Development and Sustainability, 27*, 1877–1903. https://doi.org/10.1007/s10668-023-03947-x

Zahedifar, M., Moosavi, A. A., Gavili, E., & Ershadi, A. (2025b). Tomato fruit quality and nutrient dynamics under water deficit conditions: The influence of an organic fertilizer. *PLoS ONE, 20*(1), Article e0310916. https://doi.org/10.1371/journal.pone.0310916

Zahedifar, M., & Najafian, S. H. (2017). Ocimum basilicum L. growth and nutrient status as influenced by biochar and potassium-nano chelate fertilizers. *Archives of Agronomy and Soil Science, 63*(5), 638–650. https://doi.org/10.1080/03650340.2016.1233323

Zahedifar, M., & Najafian, S. H. (2023). Evaluation of essential oil composition and antioxidant activity of yarrow as influenced by foliar application of humic substance-based products: Using multivariate exploratory method. *Scientia Horticulturae, 308*, Article 111557.

Zakaria, A. (2009). Growth optimization of potassium solubilizing bacteria isolated from biofertilizer. *Natural Resources Engineering University, 40*.

Zhang, A., Cui, L., Pan, G., Li, L., Hussain, Q., Zhang, X., Zheng, J., & Crowley, D. (2010). Effect of biochar amendment on yield and methane and nitrous oxide emissions from a rice paddy from Tai Lake plain, China. *Agriculture, Ecosystems & Environment, 139*, 469–475.

Zhang, W.-F., Dou, Z.-X., He, P., Ju, X.-T., Powlson, D., Chadwick, D., et al. (2013). New technologies reduce greenhouse gas emissions from nitrogenous fertilizer in China. *Proceedings of the National Academy of Sciences of the United States of America, 110*, 8375–8380.

Zhao, Y., Wang, P., Li, J., Chen, Y., Ying, X., & Liu, S. (2009). The effects of two organic manures on soil properties and crop yields on a temperate calcareous soil under a wheat-maize cropping system. *European Journal of Agronomy, 31*, 36–42.

Zheng, H., Wang, Z., Deng, X., Zhao, J., Luo, Y., Novak, J., Herbert, S., & Xing, B. (2013). Characteristics and nutrient values of biochars produced from giant reed at different temperatures. *Bioresource Technology, 130*, 463–471.

PART

Digital soil mapping

II

CHAPTER

Digital innovation and soil microbiome: Recent advances in ecosystem monitoring, management, and sustainable agriculture

6

Arun Karnwal[1], Neha Saklani[1], Abdel Rahman Mohammad Said Al-Tawaha[2], Gaurav Kumar[3], Deepak Kumar[4], Anupam Kumar[5], Natalia Nesterova[6] and Abdel Razzaq Al-Tawaha[7]

[1]*Department of Microbiology, Graphic Era (Deemed to be University), Dehradun, Uttarakhand, India;* [2]*Department of Biological Sciences, Al Hussein Bin Talal University, Ma'an, Jordan;* [3]*Department of Microbiology, Amity Institute of Microbial Technology, Amity University Rajasthan, Jaipur, Rajasthan, India;* [4]*Department of Chemistry, Manipal University Jaipur, Jaipur, Rajasthan, India;* [5]*Department of Biotechnology, Faculty of Applied, Pure and Health Sciences, Parul University, Quepem, Goa, India;* [6]*National University of Life and Environmental Sciences of Ukraine, Kyiv, Ukraine;* [7]*National Agricultural Research Center (NARC), Baq'a, Jordan*

1. Introduction

Various definitions identify the importance of soils, from the common perception of soil as a primary resource in agriculture production e.g., "that part of the upper layer of Earth's crust that supports plant growth"—to functioning as a complex ecosystem (Keesstra et al., 2016). Moreover, they provide habitat and support for fauna and flora and micro-organisms playing key roles in nutrient cycling (Bardgett & Van Der Putten, 2014). Soils fulfill many essential ecosystem functions contributing to the maintenance of biodiversity, regulation of climate, and physical characteristics of terrestrial ecosystems (Kibblewhite et al., 2008). Their healthy condition supports plant growth and other representative activities in the landscapes. It is increasingly appreciated that soils play a pivotal role in the functioning of global ecosystems. The habitat functions of soils are essential for soil biota, which includes a wide range of life—such as bacteria, fungi, micro-organisms, insects, worms, algae, and higher forms of life—and sustains the multi-functional soil services from natural regeneration of soil fertility to degradation and restoration of soil ecosystems (Lehmann et al., 2020).

Soils are part of a broader ecosystem, as illustrated in a simple life cycle of an organic material derived from photosynthesis. Oxidation of soil organic matter is a necessary process for coupled biogeochemical cycles such as C, N, and P, which occur hierarchically on the landscape (Schmidt et al., 2011). Water infiltration and drainage are key processes for the regeneration of terrestrial ecosystems. Soil structure and its associated oxic conditions maintain healthy cycling and

The Soil Nexus. https://doi.org/10.1016/B978-0-443-45286-4.00004-3

development of the soil biota (Bünemann et al., 2018). The plant–soil–microbe interaction forms the basis for ecosystems (van der Heijden et al., 2008). Human-made activities, particularly land-use change and agriculture, have modified these soil ecosystem functions and dynamics and created a new risk: degradation (FAO & ITPS, 2015). Various scenarios and indicators link degradation and restoration trajectories of soil ecosystems to these dynamics.

1.1 Importance of soils in global ecosystems

Soils are the cornerstone of terrestrial ecosystems, integrating functions that sustain the majority of global productivity and biodiversity. More than 99% of food production depends on productive terrestrial ecosystems and their underlying soils (Eswaran et al., 2019), which are essential for communities to flourish. Soils regulate chemical and physical processes that control water and nutrient availability and support habitat for life at different scales by providing substrates, nutrients, and microbial communities for plants and soil fauna. To highlight the importance of soils in global ecosystems, key functionalities and measurements are listed, including the international standards ("first why, later how"). Soils act as a base environmental factor, monitoring system, and regulator; they store, filter, and/or transform water; they maintain plant and animal diversity; are pools of nutrients and carbon; produce food, biomass, gas, energy, and mineral resources; and are repositories of archives and information, as a result of territory diversity and wide distribution of different terrestrial ecosystems led to the creation of five bigger soil categories or soil orders: histosols, aridosols, mollisols, andinazoems, and gelisols. The properties of these soils are encountered not only in these soils but also in the diverse other soil types. Fig. 6.1 highlights how plant roots, soil organisms, and decomposed organic matter interact within different soil layers to sustain nutrient cycling and soil health. These interconnected biological processes are essential for supporting plant growth and ecosystem function.

1.2 Soil as a dynamic and complex ecosystem

Soil must be regarded as a highly dynamic and complex ecosystem, a habitat teeming with life (Young & Crawford, 2004). Soil supports diverse multicellular organisms including fungi (multicellular forms), protozoa, micro-metazoa, earthworms, arthropods, vascular plants, and animals; these occupy interrelated micro-, meso-, and macro-scale habitats that adapt within a sensitive, dynamic biophysical environment (Bardgett & Wardle, 2010). Active biological, chemical, geomorphological, hydrological, meteorological, pedogenic, and physical systems/functions that combine to produce aboveground habitats linked in a complex and interdependent network with soils. These interfaces act as membranes at the contact point of the air, fresh water, ocean, and land surface environment. At the interface itself, respiration, inhalation/exhalation, and gaseous exchange occur in an invisible state. Soil respiration, comparable in importance to organismal respiration, drives key carbon and nutrient cycles through microbial, root, and physicochemical processes (Schimel, 2018).

Structure and soil ingredients determine many physical properties that, in conjunction with biotic activity over a range of scales, control fundamental functional processes such as organic matter degradation, nutrient cycling, and water movement and retention. Coarse-textured soils facilitate runoff concentration leading to gully erosion risks, while tunnel erosion often occurs in dispersive fine-textured subsoils; both depend on slope, vegetation, and management. The hydrology of soils controlled by the topography of the landscape directly affects the distribution of vegetation and their

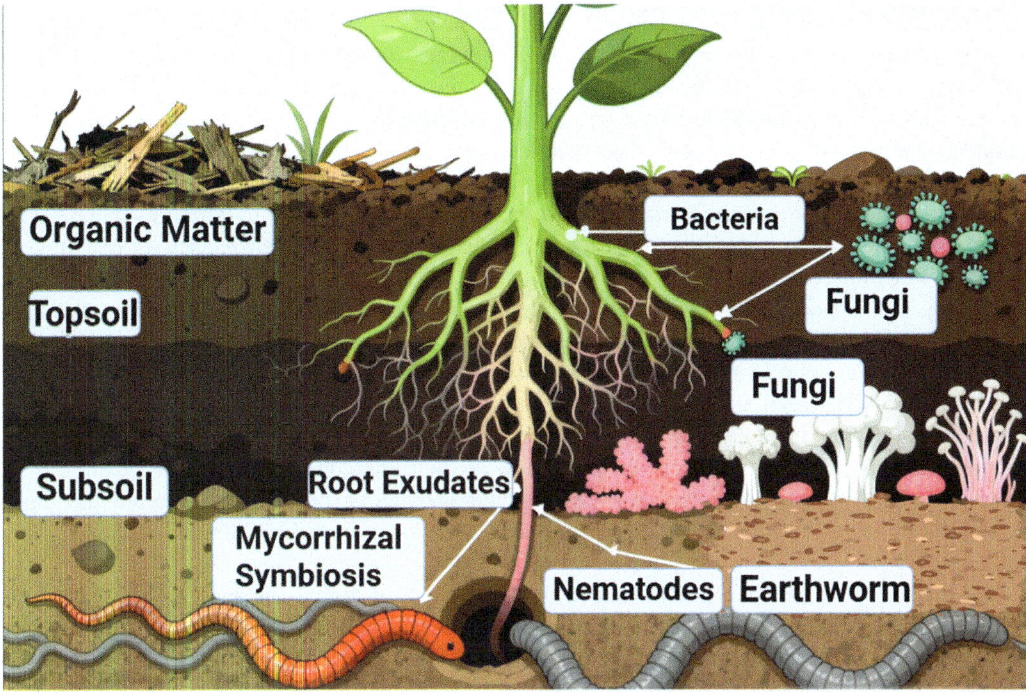

FIGURE 6.1

Schematic illustration of the soil micro-ecosystem depicting the interactions between organic matter, topsoil, subsoil, plant roots, bacteria, fungi, nematodes, earthworms, root exudates, and mycorrhizal symbiosis.

soil animals, thus directly influencing the composition of fauna and the sediments of river basins (Arrouays et al., 2017).

1.3 Historical perspective on soil study

Throughout its long history, soil study has reflected the norms, tools, and notions of environmental scholarship and has undergone several paradigm shifts. These culminated in a recognition of soil as a complex, living ecosystem, shaped by processes of biogeochemical alteration that define its properties. Such a holistic view can serve as the fulcrum for the integration of digital technologies into soil science. The first major shift took place in the late 19th and early 20th centuries, when pedology emerged as a distinct discipline concerned with soil spatial distribution and genesis (Dokuchaev, 1883). Early soil scientists such as Vasily Dokuchaev viewed soil from a dynamic, encompassing perspective: a system shaped by parent material and climate yet heavily modified by organisms. Building on this foundation, Hans Jenny (1941, 1944) proposed the universal soil-forming equation, describing how the interplay of environmental variables governs the development of soil properties. At the same time, work by Karl Russell and others revealed how soil's physical and chemical traits govern its biological life. Both perspectives collectively underscore the importance of considering soils as living systems responding to change in a complex, evolving manner.

1.4 The need for integrating digital technologies in soil science

With increasing demands and pressures on soil resources, integrating digital technologies in soil science offers avenues for knowledge generation and application (Arrouays et al., 2017). Advances in high-resolution remote sensing, together with in-situ sensors and big-data tools, open new possibilities for understanding soils and assessing their health (Mulder et al., 2011). Recent work has shown the potential of these tools for monitoring soil properties, mapping degradation trajectories, guiding land use decisions, and improving agricultural management. However, many opportunities remain unexploited, even though they address some of the grand challenges in soil science, such as assessing soil health and functions, detecting changes and degradation, accounting for variability, enabling precision agriculture, and supporting soil conservation. The fast development of many digital tools, platforms, and applications enables maintaining a holistic perspective on possible interconnections, functions, and synergies. Such a perspective is beneficial because the various innovations do not exist in isolation; rather, they are better seen as part of an ecosystem of rapidly developing technologies. A review of current digital technologies in environmental science provides a starting point to examine how digital innovations are transforming soil science and a view on their interconnectednes. Both qualitative and quantitative innovations in the research process and knowledge production underpin the digital transformation of environmental-data sciences and their integration into digital Earth ecosystem platforms.

Soil ecosystems have been overlooked in the current era of digital innovation despite the rapid advancement of digital soil mapping, monitoring, modeling, assessment, and decision-support systems (Arrouays et al., 2017). Accordingly, it is important to pinpoint gaps and explore opportunities for integrating digital technologies within more sections of soil science. The following research question guides this investigation: How can digital tools facilitate the understanding of soil as an ecosystem and promote its functioning as an ecosystem for reducing plant production risk while supporting resilience in ecosystems? Answering this question will contribute to defining a digitally enabled future perspective for soil science and offering avenues to put it into practice (Grunwald et al., 2011).

The analysis is based on a review of digital innovations that are transforming soil science, covering sensor networks for monitoring soil changing conditions; machine learning and big data for assessing soil health; digital soil maps and geospatial tools supporting the development of digital twins; mobile applications for farmers; and precision agriculture (Mulder et al., 2011). Several representative applications, including tailored management to increase plant production resilience, conservation agriculture to protect soil ecosystems, monitoring of soil ecosystem dynamics, and carbon management to mitigate climate change, illustrate the impact of digital innovations through real examples and in a variety of contexts, thus demonstrating the technology potential to foster a transformation in soil management and policy. Addressing technical challenges related to data standardization and integration, socioeconomic obstacles hindering digital technologies adoption, ethical issues surrounding data collection and sharing, environmental sustainability, and emerging research needs and technologies further support the full realization of this potential.

2. Soil: The living ecosystem

Soils are more than a substrate that supports plant life and delivers ecosystem services; they serve as a dynamic physicochemical matrix that hosts a vast diversity of biological life (Bardgett & Van Der

Putten, 2014). Far from being an inert substance, soil constitutes a complex ecosystem inhabited by millions of microorganisms and animals that drive the transformation of organic matter into bioavailable nutrients. Within this system, microbes, fungi, earthworms, beetles, and plants participate in intricate ecological interactions, with each functional group contributing distinct processes to overall ecosystem functioning (Lehmann et al., 2020). A clear understanding of the composition, functionality, and interactions of the soil ecosystem is essential to elucidate the role of soil in sustaining a healthy ecosystem (Wall et al., 2010). Soils are composed of minerals derived from rocks, organic matter derived from living organisms, water, and air. The interplay between climate, vegetation, animals, and microorganisms alters the relative proportions of these components (Jenny, 1941, 1944).

The soil ecosystem consists of a pool of microorganisms (bacteria, fungi, archaea), animals (microfauna, mesofauna, macrofauna), and plants (roots) interacting with each other and with the surrounding environment (the atmosphere, the hydrosphere, and the lithosphere). Soil microorganisms are the most numerous and abundant living organisms in any ecosystem; they outnumber soil animals by several orders of magnitude (Fierer, 2017). Healthy soils are also inhabited by a large diversity of animals such as protozoa, nematodes, tardigrades, rotifers, insect larvae, beetles, earthworms, and moles (Lavelle et al., 2006). Although these animals exhibit diverse life history strategies—ranging from the rapid reproductive cycles of microfauna to the longer lifespans of vertebrates—they collectively play essential roles in degrading organic matter, increasing soil fertility, and influencing the physical-chemical properties of soil (Bardgett & Wardle, 2010).

2.1 Soil formation processes and factors

Soil is a crucial resource for all life on Earth, and it, like any living organism, has a birth and an aging chronology. It is born from the weathering of rock. Its successive layers have evolved from the processes and factors prevalent at the site over a long period of geological time (Richter & Yaalon, 2012). The evolution of soil from the weathering of rock into proper soil takes thousands or even millions of years (Dokuchaev, 1883). New soil is continuously being formed by the weathering of rocks that are being uplifted due to tectonic activity. Soil is said to form when the products of weathering of bedrock—and soil-forming processes and factors—have matured sufficiently to support vegetation (Jenny, 1941, 1944).

Although climate has a dominating control, locally unusual parental material, relief, organisms and time can exert strong influence. Climate, parent material, topography, organisms, and time—especially combinations of climate and parent material—are considered the primary factors controlling the formation of soil as an ecosystem. Climate determines the degree, kind, and rate of weathering of rocks and contributes organic material, especially rainfall, snow, or groundwater. Parent material provides the minerals in the early stages of soil development. Topography determines the speed of run-off or drainage, the mass movement of mud and soil, and the rates of erosion. Organisms supply organic material and promote humus formation alongside mineral weathering; time allows horizonation and texture development to desired depths.

2.2 Physical and chemical properties of soil

Soil physical properties include texture (relative proportions of sand, silt, and clay), structure, permeability, density, porosity, and moisture retention. Chemical properties encompass soil acidity (pH), cation-exchange capacity, and electrical conductivity (Weil & Brady, 2017). While content and processes regulate soil characteristics, their distinct roles can be examined to clarify edaphic function and support efficient management of the soil's ecosystem.

Soil texture stems from the proportion of sand, silt, and clay in the material (Kravchenko & Guber, 2017). It may dictate the rate of many edaphic processes, including drainage, leaching, root penetration, and aeration. For example, sandy soils allow rapid drainage and leaching of dissolved nutrients but also rapid drying and limited moisture retention; clayey soils retain moisture but tend to be compacted, poorly drained, and difficult for roots to penetrate. Soils seldom consist entirely of particles of a single size, and the combination of particle dimensions may be more important than the relative proportions. Soil structure refers to the arrangement of particles into aggregates, or peds, of various sizes and shapes. Well-structured soils contain aggregates separated by spaces. These gaps, or pores, between peds are filled with air and water, facilitating root growth and supporting the activities of soil organisms.

2.3 Soil microbial communities: Bacteria, fungi, archaea

Microbial communities play a central role in every aspect of soil ecosystem health (Fierer, 2017). Constituents of the soil microbiome include Bacteria, Fungi, and Archaea (Delgado-Baquerizo et al., 2018). The soil-inhabiting archaea form a small but vital proportion of the overall community structure in terrestrial ecosystems (Bates et al., 2011). Highly diverse in composition, soil microbial communities carry out essential functional processes within both cultivated and natural ecosystems.

2.4 Functional roles of soil microbes in ecosystem health

Soils are critical to ecosystem health due to the support of plant life, regulation of atmospheric composition, and provision of ecosystem services (Bardgett & Van Der Putten, 2014). Microbes support these functions by driving nutrient cycling, maintaining soil structure, suppressing plant disease, and promoting plant growth (Lehmann et al., 2020). They mediate turnover and stabilization of soil organic matter (Schmidt et al., 2011). Soil microbial communities are the most diverse and abundant assemblages of living organisms on the planet (Delgado-Baquerizo et al., 2016). Their collective functionality underpins a large proportion of the known and yet-to-be-discovered biogeochemical processes occurring in soils (van der Heijden et al., 2008). Nutrient cycling in soils involves biogeochemical transformations of carbon (C), nitrogen (N), and phosphorus (P) (Krause et al., 2014). These cycles are fundamentally different, however, in their microbial mediation. The primary N-cycle components—nitrification, denitrification, and nitrogen fixation—are mediated by specialized bacteria and archaea (particularly AOA in nitrification) (Canfield et al., 2010) Specific fungi, principally Arbuscular Mycorrhizae and Ectomycorrhizae, integrate the cycling of P and C (Smith & Read, 2010). Humus plays a fundamental role in C storage under temperate ecosystems (Lehmann & Kleber, 2015), while the lack of such long-term stabilization mechanisms leads to rapid depletion of Soil Organic Carbon in tropical environments. Storage of soil water and structural support are key humus functions, alongside its vital role in nutrient retention, supply, and plant nutrition. Several aspects of soil properties and processes crucial for a better understanding of soil health and the roles of soil microbial communities are summarized next.

2.5 Soil fauna: Nematodes, earthworms, arthropods, and their ecological impact

The soil habitat supports a highly diversified and complex fauna, despite its apparent simplicity. Among the multitude of forms of soil life, nematodes, earthworms, and a wide range of arthropods play major roles. Nematodes and microscopic forms of soil life provide the greatest contribution to the total abundance and biodiversity of the biota, while soil-dwelling earthworms and arthropods are important for the function and resilience of the natural environment (Bardgett & Van Der Putten, 2014; Lavelle et al., 2006). The major impact of soil fauna lies in their considerable role in nutrient cycling and the improvement of soil functioning through their activities, especially by the processes that they perform during feeding, locomotion, and reproduction (Brussaard, 2012). While many species prefer to inhabit the soil or are only found in terrestrial cellulase-rich habitats during a small part of their life cycle, meso- and macrofauna (e.g., earthworms, arthropods) engineer soil structure at the macroscale, complementing microorganisms, mycorrhizal fungi, and nematodes—the prime actors in rapid organic matter turnover at the microscale (van der Heijden et al., 2008).

Nematodes are ubiquitous roundworms, commonly distributed in all ecosystems. Bacterial-feeding nematodes are the most abundant group, followed by fungal-feeding and plant-parasitic species (Yeates et al., 1993). The biogeographical patterns of nematode diversity largely reflect environmental conditions, and they respond quickly to disturbances. Changes in nematode community structure may be used as bioindicators of environmental stress or recovery (Ferris et al., 2001). Soil mites and collembolans constitute the main groups of soil microarthropods. Together with nematodes, they are the most abundant animals in most soils, and they play an important role in organic matter decomposition, nutrient cycling, and control of phytopathogenic fungi (Coleman et al., 2017; Hopkin, 1997). Important to the soil fauna community are earthworms that contribute significantly to the turnover of organic matter and the provision of a structure more favorable for plant growth (Edwards, 2004).

2.6 Symbiotic relationships: Mycorrhizae and nitrogen-fixing bacteria

The majority of terrestrial plant species enter symbiotic associations that improve nutrient availability and access. Mycorrhizal fungi form symbionts with plant roots that assist in the uptake of phosphorus and other nutrients (Smith & Read, 2010). In return, the host plant supplies energy-rich carbohydrates, amino acids, vitamins, and phenolic compounds that the fungus uses for growth. For the plant, mycorrhizae subsequently offer increased tolerance to low moisture, drought, salinity, and metal stress (Porcel et al., 2012). In addition, some ectomycorrhizal and arbuscular mycorrhizal fungi can protect host plants from root-pathogen damage by secreting antimicrobial compounds that promote defense mechanisms (Cameron et al., 2013).

The second major form of nutrient acquisition is through root nodules formed by nitrogen-fixing bacteria, where atmospheric nitrogen is retained and made available for other organisms and plants within the ecosystem (Oldroyd et al., 2011). These symbionts may also affect the quality and productivity of the ecosystem, as in the case of legumes that support shallow-rooted drainage of alkaline clay substrates. The interactions between mycorrhizae and plants are bacteria like *Bradyrhizobium* or *Rhizobium*; actinomycetes; *Frankia*; or the cyanobacteria *Nostoc*, *Anabaena*, or *Aulosira* present in the root nodules of leguminous plants play a significant role in determining the level and availability of nutrients to host plants in soil.

3. Ecosystem dynamics and soil functions

Summary of ecosystem functions and biogeochemical processes provided by soils, showing how physical–chemical transformations are regulated via microbial communities and presenting the impact of land use change and agriculture on process dynamics.

Piles of rocks remain from exposed mountain ranges, and of all the possible surfaces ocean waters are infested with billions of single-cell living beings together with huge animals. The soil stands among Earth's most dynamically rebuilt compartments, continuously recycled through bioturbation, microbial decomposition, and root turnover transforming it from perceived 'dead substrate' to living ecosystem matrix (Grunwald et al., 2024). These simplifications ignored the fact that soils are by far the most diverse ecosystems in the world, hosting all forms of life and the most diverse forms of matter: exhausted, recharged, hot, cold, acid, alkaline, rich, poor, and dry, and are the only terrestrial ecosystems capable of generating all other ecosystems—including deserts.

Soils deliver vital services to all terrestrial ecosystems through physical–chemical transformations driven by soil microbial communities (Fierer, 2017). Accumulation of organic matter through photosynthesis fuels the microbial communities that mainly govern organic matter degradation, nutrient cycling, and the balance between plant and water uptake. At the same time, plants grow roots and secrete organic exudates that regulate soil physical structure and drive water/nutrient flows within them. The health of terrestrial ecosystems depends on these processes, and the disturbance of ecological balance by land-use change and agriculture diminishes these ecosystem functions.

3.1 Biogeochemical cycles and nutrient transformations (carbon, nitrogen, phosphorus)

Soil sustains and drives terrestrial biogeochemical cycles of carbon, nitrogen, phosphorus, and other elements essential for all forms of life (Lehmann & Kleber, 2015). Biogeochemical cycling is itself intimately associated with soil formation processes, soil physical properties, biological activity, and pH. Although researchers have long recognized the significance of soils for elemental cycling, the importance of these constituent processes in driving soil ecosystem health is frequently neglected or underappreciated. Such cycling constitutes a critical set of provisioning, regulating, and supporting ecosystem services. Under many land-use scenarios, biogeochemical cycling also acts as a meter of soil ecosystem health and soil degradation risk. To fully delineate the complexity of soil ecosystems and their wide-reaching effects on ecosystem services and human well-being, it is crucial to consider the cycling of contaminants such as cadmium, chromium, mercury, nickel, and lead, in addition to the macronutrient cycles examined in this section (Ali et al., 2013). Soil also enables the biological transformation of hazardous chemicals, including toxins, endocrine-disruptors, and other micropollutants.

3.2 Soil organic matter and carbon sequestration

Both SOC quantity and quality strongly influence soil biodiversity and functionality. Their turnover is driven by microbial activity, and less stable pools, such as fresh organic matter or non-stabilized SOC, are actively consumed by soil microorganisms (Kögel-Knabner, 2002). At the same time, some intestinal products are also transformed into stable SOC, whose high turnover corroborates the important biochemical processes that take place in hotspots of SOC transformation activity involving non-living fractions (Lehmann et al., 2020). Therefore, even if the C storage function occurs over the long term and the C pools of different stabilities evolve in accordance with this characteristic of SOC, the requirement for constant replenishment of the less stable fractions remains essential for sustaining the biological activity responsible for C transformation and its stabilization.

Soils are the largest carbon sink in terrestrial ecosystems. They store more carbon than vegetation and the atmosphere combined (Batjes, 2016). Global SOC stock to a depth of 1 m soil is estimated to be 1500–1700 Pg, and approximately half of the C currently contained in soil is estimated to be potentially susceptible to degradation owing to land use and cover change. Ecosystem services associated with the SOC pool include CO_2 fixation from the atmosphere, food production, stabilization, structuration, and soil texture-conditioning processes, resistance to salinization and erosion, water filtration, and storage of nutrients. Increasing SOC storage would also facilitate adaptation to climate change and contribute to ameliorating extreme weather conditions. Given SOC heterogeneity, spatial and temporal distribution, the diverse factors that influence its dynamics, and its tight relations with other soil functions, SOC storage encompasses a wide range of management practices and land-use scenarios, through which SOC could be increased and contribute to climate-change mitigation.

3.3 Water dynamics: Infiltration, retention, and drainage

Water dynamics in soils encompass three fundamental aspects: infiltration, retention, and drainage. Infiltration refers to the vertical flow of water from the soil surface downward through the unsaturated zone toward the groundwater table (Jarvis, 2007). Its dynamic behavior depends on soil properties, rainfall characteristics, vegetation cover, land use, and initial conditions. Retention describes the mass of water stored in the soil after precipitation or irrigation, while drainage characterizes the horizontal outflow of water through the saturated zone toward surface water bodies, such as rivers, lakes, and oceans. Together, these processes constitute the most important controls on water movement and storage in terrestrial ecosystems (Hillel, 2003).

Physical soil properties exert the most significant control on infiltration capacity, retention, and drainage. Soil texture and structure determine porosity and pore-size distribution, while bulk density and organic matter content affect the elastic compliance of pore spaces. Organisms such as roots and earthworms can modify pore-size distribution and create preferential flow paths, enhancing hydraulic conductivity. Accordingly, dense surface crusts limit surface infiltration, impermeable layers within the soil profile restrict downward drainage, and compaction by machinery reduces soil porosity (Coleman et al., 2017).

3.4 Soil structure and its influence on plant growth

Soil structure has a pronounced impact on plant growth, both directly by modulating water and air retention and indirectly through its influence on the extent and distribution of rooting. A well-structured soil has a large number of macropores, which enables rapid infiltration, high pore drainage, and ensures fast gas exchange and diffusion of root signals (Lynch, 2019). In contrast, strong compaction can severely limit aeration and prolong periods of water saturation; planting-induced soil compaction beyond the natural bulk density is generally recognized as a common problem in conventionally managed croplands, especially in heavy-textured soils and in wet years (Hamza &

Anderson, 2005). Soil structural stability during growth is often underappreciated; dry sub-root zones create steep hydraulic gradients taxing plant water relations (Dexter, 2004). Beyond drainage/aeration, structure governs root proliferation and ecosystem function (Bronick & Lal, 2005). Aggregates protect OM while regulating microbial decomposition; organic matter and microbes prevent disaggregation, enabling growth in poorly structured soils.

3.5 Plant-soil-microbe interactions and feedback mechanisms

Plants, soils, and microorganisms interact closely to form a complex nexus in which organisms exchange resources, communicate, and regulate each other's capacity for growth and transport through signaling compounds (Bardgett & Wardle, 2010). Soil biota provide vital services such as nutrient cycling, pathogen suppression, seed predation, phytohormone production and nutrient solubilization (van der Heijden et al., 2008). These services vary widely across communities and depend on functional diversity, ultimately contributing to a healthy ecosystem (Philippot et al., 2013). Agriculture dramatically reshapes this system by changing soil structure, chemical composition, and hydrology, while selective cultivation of fast-growing plant species pressures soil and crop microbiota to maintain beneficial functions (Kuzyakov & Blagodatskaya, 2015). The pattern of interactions further dictates plant diversity, population dynamics, and primary production, with feedback mechanisms scaling from interspecific to intraspecific levels.

3.6 Impact of land use change and agriculture on soil ecosystem dynamics

Following the signature of the 1992 Convention on Biological Diversity, the Strategic Plan for Biodiversity 2011–20 proposed a target to reduce several drivers of biodiversity loss, including land use change. Land use change in general and the expansion of agriculture, especially the increase in monocultures, continue, however, to be identified as primary drivers of the loss of ecosystem services and ecosystem health. The adoption of sustainable land management practices can help reverse this trend and recover a part of the soil ecosystem health that has been lost. Because soils serve as a primary interface between terrestrial ecosystems and the atmosphere and provide services such as the management of water resources, regulation of local climates, and the storage of carbon, monitoring changes in soil ecosystem health can provide critical feedback on the overall health of terrestrial ecosystems.

Terrestrial ecosystems are in constant flux, changing with seasons, periods, and timescales that may involve decades or centuries: fluxes of matter and energy are transferred between ecosystems and the atmosphere over multiple pathways, mediating the transformation of nutrients and their assimilation, transportation, storage, succession, and ultimately decomposition and return to the atmosphere (Chapin et al., 2011). A multidimensional mapping framework to assess the dynamic changes in these connections and fluxes in relation to soil ecosystem health and their consequences for ecosystem services, and thus the overall health of terrestrial ecosystems, is presented. The relevance of this mapping for monitoring and enforcing the use of soil conservation measures and for supporting the UN 2030 Agenda for Sustainable Development is also highlighted (Shahmohamadloo et al., 2021).

4. Digital innovations transforming soil science

Digital innovation is generating an unprecedented flow of tools and data capable of reconceptualizing and revitalizing soil research. A comprehensive overview of these transformative advancements reveals the digital innovations currently influencing soil science, how they interact, and the resulting research and monitoring strategies (Arrouays et al., 2017). Such an integrated perspective aids those seeking to advance the digitalization of soil study by identifying key areas of specialization and integration, the generation and evolution of diverse types of data, and the establishment of new products.

Digital technologies have permeated nearly every aspect of environmental science, spawning previous catalogs of hardware, software applications, and analysis techniques specific to remote sensing, soil and vegetation sensors, and machine-learning approaches (Grunwald et al., 2024). A broader view of the new hardware, software, and analytical possibilities appears warranted to elucidate their complex interconnections and the intricate paths they create for knowledge generation and monitoring. From this standpoint, the innovations impacting soil science fall into four major categories: digital soil mapping, remotely sensed data, in situ soil sensors, and big-data strategies, supported by an ever-increasing collection of mobile applications designed to assist decision- These platforms and products are now facilitating the development of a variety of new approaches for assessing soil health, monitoring degradation, predicting ecosystem service provision, and evaluating environmental policy (Mulder et al., 2011). An increasing volume of information spanning both the digital soil-mapping and in situ-network categories—now provides the descriptive and semantic foundations for the construction of new maps, predictive models, and knowledge-enhancing sensing technologies.

Table 6.1 presents key digital technologies and their applications currently transforming soil science, highlighting their purposes, benefits, and challenges.

4.1 Overview of digital technologies in environmental science

Digital technologies are redefining capabilities in environmental science by enabling significant advances in understanding, assessing, and monitoring the environment. These technologies are being used to identify and characterize critical parameters in the environment and identify/quantify patterns hidden within vast amounts of heterogeneous data (Grunwald et al., 2024). New hardware, software, and analytic paradigms are being developed to enable data collection and analysis to be conducted at unprecedented scales, and information technology is being increasingly adopted by people from all walks of life and backgrounds (Arrouays et al., 2017). Applications enabled by new hardware lead to new sensing capabilities that support and automate human decision making, while enhanced software tools enable new data collection approaches and also support data analysis at hitherto unrealized scales.

The development of advanced and low-cost remote sensing platforms capable of collecting high-resolution spectral and spatial information continues at a rapid pace (Mulder et al., 2011). Satellites continue to provide high-resolution, temporally and spatially synchronous spectral information, and new environmental indices are being developed to facilitate joint soil-atmosphere-land cover modeling. However, advances to the automated algorithms that validate, calibrate, amalgamate, and link data products remain key to realising the full potential of satellite-based data products for soil surface monitoring (SSM).

4.2 Remote sensing and satellite imagery for soil analysis

Remote sensing technology, particularly satellite imaging, offers powerful quantitative capabilities. Various multispectral indices indirectly estimate SOC, pH, CEC, and texture with moderate accuracy;

Table 6.1 Digital technologies and applications in soil science.

Technology/ application	Purpose/Function	Key features/ Benefits	Challenges/ Limitations	References
Precision agriculture	Variable-rate input application	Real-time soil sensing, spatially explicit decisions	High-cost sensing and infrastructure are needed	Gebbers and Adamchuk (2010)
Remote sensing & satellite imagery	Large-scale soil property mapping	High-resolution spectral, multi-temporal datasets	Requires costly calibration and validation data	Mulder et al. (2011)
IoT soil sensors & networks	Continuous soil moisture, nutrient, and gas monitoring	Integration at multiple spatial scales, fine-tuning	Sensor maintenance and data quality assurance	Arrouays et al. (2017)
Machine learning & big data	Soil health prediction, feature selection	Automated data processing, uncertainty estimation	Need for large quality datasets, model reliability	
Digital soil mapping	Creation of high-resolution soil property maps	Integration of heterogeneous data, spatial interpolation	Standards and interoperability issues	Arrouays et al. (2017)
Soil conservation monitoring	Erosion risk assessment, restoration tracking	UAVs, spatial indicators, and modeling integration	Spatial heterogeneity limits broad applicability	Bronson and Knezevic (2016)
Mobile apps for farmers	Tailored soil management recommendations	User-friendly interfaces, real-time data feedback	Data privacy, user adoption barriers	Bronson and Knezevic (2016)
Soil carbon management	Soil carbon sequestration estimation & strategies	Regional carbon stock databases, mitigation planning	Site-specific effectiveness, water availability	Bronson and Knezevic (2016)
Digital soil health awareness	Integration of sensors, AI, and social platforms	Whole-of-society approach, stakeholder collaboration	Ethical data governance, capacity gaps	Baveye et al. (2016)
Interdisciplinary collaboration	Linking agronomy, ecology, and data science for soil health	Multidisciplinary knowledge sharing and policy design	Coordination complexity	Baveye et al. (2016)

many calibrated continentally despite prediction uncertainties. Earth-observing satellites operate at varying frequencies, resolutions, and costs. The latest generation uses ultra-high-resolution synthetic aperture radar (SAR) sensors, some of which have global coverage, to retrieve soil surface properties such as moisture content, roughness, and risk of crust formation. For other variables, especially SOC, optical images with finer spatial resolution (hyperspectral or multispectral) and digital surface models (e.g., from laser altimetry) prove most useful. Their products have limited coverage and uneven availability.

The spatially continuous characteristics of satellites support the development of soil and land-use-change models for different ecosites. The higher centers that analyze the images usually share and publish their results in a standardized archived format at various spatial, temporal, and spectral resolutions. However, despite the processing standardization, reconstructing local information with high accuracy requires ground data for calibration and validation. Local point samples are of limited use for spatial reconstruction, and therefore, many studies have become more regional and less local. Traditional machine-learning techniques require a large number of points even for regional reconstruction; recently, tropical ecosystems have emerged as a good training set for new data-driven approaches (Greinke & Lange, 2022; Hoek van Dijke et al., 2022; Pinheiro et al., 2025).

4.3 Soil sensors and internet of things devices for continuous monitoring

Data from soil sensors and Internet of Things (IoT) devices enable continuous monitoring of soil conditions and dynamics. Soil sensors measure key physicochemical parameters—water content, temperature, salinity, dissolved O_2, redox potential, nitrate, phosphorus—essential for understanding microbial habitat conditions. Networks of such sensors assist in integrating different spatial scales and fine-tuning model simulations. Sensor data can also be combined with photochemical measurements of soil gases.

Despite their potential advantages, maintaining sensor networks remains a major obstacle. Several crucial prerequisites must be fulfilled in order to ensure reliable data quality. First, the objectives of the measurements must be clearly defined, and the required precision of the measurements must be specified. Based on this information, a detailed monitoring design should be drawn up, which defines the number of measuring points, their positioning, spatial heterogeneity, the depth of measurements, and identifies possible sources of interference. For specific soil parameters, it is essential to define calibration procedures, carry out preliminary measurements in a test phase, and provide quality control. Finally, the regular expenditure related to maintenance, data retrieval, and data quality assurance should be evaluated in relation to the anticipated benefits of the sensor network.

4.4 Big data and machine learning approaches for soil health assessment

As the availability of Big Data increases for soil health assessment, appropriate Machine Learning models can be trained to extract useful information if the data features of any interest are informative enough and the right model is selected and validated for that purpose. Even when suitable compiled structured datasets are not available yet, but both Remote Sensing and IoT data from different sources fit the demanded predictive applications, it is possible to take advantage of the data and make assessment-based predictions. These are feature-selection processes to determine for any Big Data set which features of the validation of the considered prediction model can be used or excluded strictly. After these processes were applied, the prediction model can easily be evaluated both with susceptibility and uncertainty assessment.

The most demanding part is still the uncertainty estimation, since it cannot be avoided. A correct awareness of Data Quality and Data Availability should ideally be considered during the steps that lead to the final prediction evaluation. In particular, feature-selection methods are one of the key steps in predictive modelling, not yet adequately developed for soil health assessment based on either the cloud of data coming from the possibilities of Remote Sensing or those of Sensor-IoT networks. They also should consider multicollinearity and evaluate the final results of predictability of robust Neural Network models or of any other Machine Learning models, like Random Forest, for example, by means of susceptibility. Various types of models confirm the robustness of the prediction process by working in different approaches, yet they remain strictly connected together by the feature-selection aspect. In these terms, satisfactory test results are already available for soil salinity and Soil Organic Carbon content as first-order targets of Soil Health Assessment and appear for soil pH as a second-order target, while the uncertainty evaluations are currently in progress (Xiang et al., 2026).

4.5 Development and deployment of digital soil maps and geospatial tools

In the second half of the 1990s, the recommendation was raised to develop and use digital soil maps in the study of soil geography and hydrogeology to help in preliminary estimations of soil properties and to gauge their variation through arable land in the context of integrated groundwater, pollutants, farming, and non-farming studies. Inputting digital soil maps into models of surface and underground water and contaminant transport was also proposed. Hardware- and software-based Wall-to-Wall mappings have since made significant progress. These employed static models were progressively developed toward distributed, predictive, and dynamic models. Mass balance-based approaches have also made advances, yielding results subsequently tested against measured parameters. Digital soil maps and geospatial tools continue to develop, but many areas remain inadequately addressed, particularly the scale factor and interoperability issues mentioned above. A wide-ranging geospatial extension is needed.

A variety of methods have been employed, for example, as the basis of a soil-related quality index indicator, decision-support systems foresee the likely consequence of remedial measures on soil ecological status, rely on soil appropriateness mapping to provide spatially-referred guidance on resource-use options, and support the selection of sustainable soil management practices through integrated soil quality evaluations (de Paul Obade & Lal, 2016). Freely accessible web-based software facilitates the evaluation of spatially-referred predicted soil properties.

4.6 Mobile applications and platforms supporting farmer decision-making

Tools that facilitate farmers' adoption of tailored soil management practices need to be designed according to end-user demands to ensure they serve identified needs and functions, communicate reliable information, and encourage adoption uptakes (Bronson & Knezevic, 2016). The development and deployment of mobile applications are often motivated and guided by principles that support a farmer-centered focus (Wolfert et al., 2017). App interfaces and functionalities are co-designed with farmers and other stakeholders: testing and feedback provide formative evaluations that shape development stages and ensure identified functions are supported by user-friendly and intuitive interfaces. User accessibility to data is essential, generally taking the form of validation comments from farmers. The maintenance and regular update of the application have been supported through the establishment of open-access relational databases that contain the input variables required by the digital tools and by using data from local data networks that monitor such variables periodically.

The success of a user intervention strategy is a function of an understanding of farmers' behavioral patterns of smartphone use, including farmers' records of adopting similar mobile applications and technologies. The relevance of providing credible information is further emphasized when farmers' use of innovative technologies without significant results is explored: skeptical farmers are most influenced by data validation trends (Klerkx & Rose, 2020). These supporting user interaction tools demonstrate that the development of digital tools that require farmers' participation can only be achieved if a continuous farmer-developer interaction is maintained throughout all stages of tool-making.

5. Applications and case studies

Pragmatic uses of digital technologies for soils can be grouped as precision agriculture and tailored soil management, soil conservation, soil health degradation and restoration, soil organic carbon management, and climate change mitigation within a diverse set of geographic regions and ecosystems (Arrouays et al., 2017).

Precision agriculture harnesses data acquisition and analysis technologies; the integration of using sensors, remote sensing, and big data feeds decision-making into a variable-rate strategy for the spatial adjustment of inputs (such as fertilizer, pesticide, and water) in accordance with local conditions and resources (Gebbers & Adamchuk, 2010). Tailored soil management aims to adjust soil physical, chemical, and biological properties to the needs of the crops. Enabling technologies cover a variety of processes, from selecting suitable cultivation practices according to local conditions to assessing soil compressibility for selecting machinery with a suitable load for the defined activities.

Soil conservation exploits these same technologies to streamline monitoring, support erosion control actions, and guide soil restoration programs. Monitoring soil degradation and restoration trajectories is facilitated by improved data availability and analysis capabilities; approaches ranging from indicator and temporal relationship development to remote monitoring based on unmanned aerial vehicles (UAVs) are flourishing. The capacity of soils to act as carbon sinks and reduce the concentration of greenhouse gases in the atmosphere has been recognized, and has led to the formulation of guidelines and strategies to increase carbon sequestration. The results of spatial and temporal assessments at different scales of carbon stored and sequestered align with expectations and allow evaluations of the associated risks and opportunities.

The set of innovations, whenever supported by a diversity of actors of the agrifood system—farmers, researchers, policymakers, and private initiatives—also proves valuable. For example, in China, Mushrooms for Health links farmers and scientists to monitor and control the accumulation of heavy metals in crops. Currently, a stable coordination and communication network between all actors remains the crucial step for scaling up these and other experiences and increasing the impacts on soil health, biodiversity, and sustainability of the agrifood systems (Klerkx & Rose, 2020).

5.1 Precision agriculture and tailored soil management practices

Precision Agriculture enables continuous monitoring of soil physicochemical conditions using sensors and remote sensing satellites, supporting data-driven crop management decisions. Utilization of real-time data about soil chemical and physical properties, accompanied by crop response functions, enables targeted interventions, such as applying fertilizers to only the areas of the field that require such inputs. High spatial resolution data products, including soil texture, pH, cation exchange capacity, and organic matter, provide a basis on which fertilizers can be applied in a variable-rate manner, thereby optimizing crop productivity while reducing detrimental impacts of excess fertilizer application on the environment (Gebbers & Adamchuk, 2010).

Precision management also encompasses other soil amendments, including lime, gypsum, and organic waste. Sensor networks can be augmented with other IoT devices to closely monitor soil moisture status, while remote sensing data can be used to account for differences in disease or pest incidence across fields through site-specific management. Such high-resolution information can even be provided on a daily basis to support decisions for specific inputs like water, nutrients, and pest protection measures (Wolfert et al., 2017).

5.2 Soil conservation techniques powered by digital innovation

Precision agriculture relies on a combination of sensing, advanced data analytics, and machine learning to optimize and adapt soil management practices. Technology applications include methods to monitor soil properties and degradation processes, develop risk mitigation strategies, and control erosion. Impact estimates and spatial indicators provide useful support for adopting conservation practices and technologies (de Paul Obade & Lal, 2016). Digital innovations, integrated with ecological knowledge, empower farmers to develop regionally specific soil conservation solutions. Combined with local participatory processes and supportive policies, they overcome adoption barriers (costs, training, trust).

Integrated monitoring ensures proper implementation, assesses sustainability, and evaluates success. Users of decision-support applications engage in the evaluation process and are responsible for achieving conservation results. Potential climate benefits derive from reducing land use and productivity changes across Africa. Quantification requires downscaling models accounting for biological heterogeneity, careful ML feature selection, and validation with independent biological sources to reduce uncertainty.

5.3 Monitoring soil degradation and restoration efforts using technology

Soil degradation affects all aspects of the environment. It creates environmental challenges that undermine food security, biodiversity, and climate change mitigation (Rob et al., 2017). Countries and regions around the world have made various attempts to monitor degradation and restoration at local, national, and global scales. Decades- or centuries-long information is not always available. The need has arisen to assess soil degradation while concurrently monitoring restoration, using easy-to-apply indicators and clear trajectories, including pre-treatment baselines or trajectories before restoration commenced. Digital innovations integrate physicochemical erosion models (USLE, WEPP, RUSLE2) with AI, predicting sheet/rill erosion rates from precipitation, slope, erodibility, cover, and practices. Biological drivers like microbial aggregate stabilization require complementary assessment.

Capacity for carbon storage in soils, enhanced by cropland restoration, has been estimated. Stocks of organic carbon have been tracked and compared against pre-restoration baselines. Capture of retained historic samples through soil archives has enabled quantitative assessments of some restoration efforts.

5.4 Climate change mitigation through soil carbon management

Temperatures are rising as a result of climate change, which has catastrophic effects on the environment and society (Legg, 2021). GHG emissions must be reduced in order to meet climate targets established by international treaties on climate change and their implementation. It is vital to put into practice mitigation strategies in order to avoid emissions and increase carbon removal from the atmosphere. Increasing the amount of carbon stored in soils by soil carbon management (SCM) is one strategy that can decrease the amount of carbon dioxide in the atmosphere and possibly slow down climate change. SCM entails efficient cultivation strategies that support productivity while increasing the stocks of carbon in soils. Burn-out models have shown that SCM can take place within decades, and studies based on geological archives provide evidence that carbon can accumulate in soils on timescales of centuries to millennia.

The amount of carbon that can be sequestered in a specific area through SCM is being correctly approximated on a regional scale, which makes it possible to create databases that allow for the calculation of the opportunity to enhance soil carbon stock in any region. As many regions will contribute to climate change mitigation through SCM, a combination of the different areas' potential can then be evaluated for national and international targets. Experimental studies and robust meta-analyses reveal the effectiveness of management approaches based on the return of organic matter, the use of soil cover, and minimal tillage, all of which can enhance soil carbon stock and reduce carbon stock loss. However, at a global level, a huge proportion of agricultural areas in certain continents do not follow such practices. Scarcity of water resources makes it impossible to implement SCM in many areas, as does the necessity of maintaining the transport capacity of soil under crop for regions affected by flooding.

5.5 Examples from diverse geographic regions and ecosystems

Evidence from distinct spatial and ecological contexts highlights that digital innovations furnish the enabling framework for assessing soil health, tailoring management strategies, and verifying outcomes on the ground. In Jiangxi, China, the conventional model of passive fertilization and pest control is being transformed through sensors that provide real-time irrigation, nutrient, and disease feedback, enabling targeted in situ fertilization and precise pest management to enhance yield and revenue while lowering costs. In coastal cities, soil salinization is being monitored using remote sensing spectral indices, establishing a foundation for remediation planning (Angon et al., 2023; Prajapati et al., 2025). In Malaysia, rural farmers are harnessing mobile-based tools to monitor soil nutrition and health and tailor management practices.

In southwestern Australia and New Zealand, advanced soil information systems integrate big data sets, machine learning algorithms, and supported decision-making tools by farmers, agronomists, and soil scientists (Arrouays et al., 2017). In Kenya, dryland ecosystems are being monitored with soil-water sensors that drive a market-focused web application platform simplifying the adoption of moisture-regulated intelligent crop husbandry. In Canada, new soil management actions aimed at preserving drainage functionality, enhancing soil carbon stocks, supporting the revitalization of soil life, and diversifying the crop rotation are being experimentally evaluated. In the African Sahel, low-cost free-space optics and wireless sensor network technologies are monitoring soil nutrient status to improve crop yields and agricultural productivity while reducing fertilizer costs.

5.6 Multi-stakeholder involvement: Farmers, researchers, policymakers

Soils are a fundamental resource for agriculture and the maintenance of associated services; however, the extensive use of fertilizers, herbicides, and biocides has had negative consequences for both soil and water quality, leading to declines in consumer confidence and increasing regulatory pressure. A coordinated approach involving farmers, researchers, and policymakers across diverse geographical and environmental spheres is crucial to understanding and solving problems contributing to the decline in soil quality, while providing opportunities for large returns on investment—approaches that are well matched to the current technologies involving big data and machine learning applications. To support stakeholder engagement, tools and preventative measures—such as criteria indicators for soil degradation, maintenance, restoration, monitoring, and recovery metrics, as well as carbon-sequestration assessment—have been developed by soil managers, R&D organisations, and policy makers.

6. Challenges and future directions

Digital technologies hold immense promise for addressing soil-related challenges, but several barriers remain to fully realizing that potential. Standardization and integration issues hinder data synthesis and comprehensive interpretation, and a range of socioeconomic and policy barriers affect adoption across the diverse stakeholder communities engaged in soil management (Arrouays et al., 2017). Addressing these challenges during the design and development stages will ensure the tools deliver the expected benefits. The ethical implications of data ownership, privacy, and digital citizenship must also be considered (Wolfert et al., 2017).

A vision for a digitally enabled soil science ecosystem outlines an approach capable of creating, sharing, and implementing integrated knowledge across the full range of scales required to enable resiliency of soil resources and the ecological processes they support. The full potential of big data, IoT, and artificial intelligence will only be realized when standardization, interoperability, and accessibility challenges are overcome. Technology adoption requires consideration of local socioeconomic contexts and constraints and should be supported by incentives and capacity-building initiatives. Streamlining procedures for the use and sharing of data in specific areas can help reduce spatial disparities in access. Ethical and governance issues must also be considered, with the collection and use of sensitive information controlled according to clearly articulated rules (Bronson & Knezevic, 2016). Environmental sustainability remains paramount, and the resilience of soil ecosystems should therefore be assessed and monitored during the post-development phase (Adhikari & Hartemink, 2016).

Table 6.2 summarizes the major challenges facing digital soil science and outlines potential strategies and technological directions to overcome these barriers/.

6.1 Technical challenges: Data standardization, integration, and interpretation

Although promising, digital innovations in soil science face technical challenges that must be addressed for full realization. Data standardization, integration, and semantic interoperability remain unresolved problems (Arrouays et al., 2017). The lack of widely accepted protocols and standards hampers large-scale analysis of soil data and impedes the deployment of data-driven models (). Achieving soil digital twins requires converging the physical, biological, and chemical dimensions of soils and breaking down the artificial separation between soil properties and soil processes within these dimensions (Bronson, 2019).

Many soil characteristics are difficult and expensive to measure directly. Consequently, digital predictions based on soil properties often integrate multiple sources of information, usually through

Table 6.2 Challenges and future directions in digital soil science.

Challenge	Description	Impact on soil science	Potential solutions	References
Data standardization & integration	Lack of universal protocols limits data synthesis	Reduces model accuracy and data interoperability	Develop common standards, semantic interoperability	Wolfert et al. (2017)
Model uncertainty & validation	Soil properties are difficult to measure; prediction uncertainty	Limits trust in digital twins and management decisions	Advanced ML models with uncertainty quantification	Grunwald et al. (2024)
Socioeconomic barriers	Cost, knowledge gaps, and limited access to technologies	Unequal adoption slows digital transition	Training, funding, policy incentives	Wolfert et al. (2017)
Policy & governance challenges	Insufficient policies for data sharing and technology deployment	Hinders data accessibility and user confidence	Develop data governance frameworks, stakeholder engagement	Wolfert et al. (2017)
Ethical & privacy concerns	Unequal data ownership, risk of surveillance	Resistance among farmers and communities	Transparent data consent and anonymization policies	Bronson (2019)
Technical capacity gaps	Limited expertise in some regions	Slows adoption of innovations	Capacity building, interdisciplinary training	Wolfert et al. (2017)
Resilience monitoring	Need to track soil ecosystem stability under stress	Essential for long-term sustainability	Develop resilience metrics linked to climate risk	Bardgett and Wardle (2010)
Integration of diverse data	Integrating biological, physical, and chemical soil data	Complexity hampers holistic soil health assessment	Develop multi-domain digital twins	Arrouays et al. (2017)
Emerging technologies adoption	Low-cost hyperspectral sensors and AI under development	Tools not yet widely implemented	Investment in R&D and pilot studies	Rossel and Bouma (2016)
Engaging multi-stakeholders	Coordination between farmers, researchers, and policymakers	Critical for successful scaling and policy support	Facilitation platforms and participatory design	Baveye et al. (2016); Klerkx and Rose (2020)

regression models. These models predict previously unmeasured properties from known ones or predict properties that can only be estimated with uncertainty, such as soil moisture. Such complex prediction systems can be used to compute high-resolution maps of soil properties, but their value depends heavily on model quality. Recently, there has been growing interest in training machine learning algorithms to identify and weight the most relevant features automatically.

6.2 Socioeconomic and policy barriers to technology adoption in soil science

Understanding the barriers that inhibit the integration of digital innovations into the soil science domain is vital to the success of any future effort to create a digitally-enabled soil science ecosystem capable of responding to pressing technical needs and emerging global challenges. An analysis of obstacles identified by various stakeholders in Europe the Systemic Advisory Process reveals the existence of five major categories of impediment (policymaking, funding and financing, knowledge and understanding, adoption, and technologies), of which two policymaking and knowledge and understanding lie particularly within the scope of the present inquiry (Arrouays et al., 2017). Whether seeking to persuade policymakers to mandate training initiatives, to develop incentives designed to stimulate technology adoption, or to encourage the adoption of specialized policies that could facilitate access to soil-related information and resources, a fundamental understanding of the barriers encountered in adopting the currently available digital technologies is necessary. Clarity regarding technical, economic, or institutional trends is thus critical for the design of effective governance and business models, monitoring strategies, and international research agendas. Although the barriers identified differ from those prevalent in Europe, the numerous commonalities between the two sets point to extensive opportunities for the exchange of experience and good practices. Researchers classify impediments according to their effect on involvement in soil knowledge sharing and use, providing insight into relevant aspects of soil science governance for various types of agrarian politicians. The challenges of relevance, credibility, and accessibility represent major obstacles to the adoption of soil policies and practices addressing private, public, and broader concerns (Wolfert et al., 2017).

6.3 Ethical considerations in digital data use and privacy

The use of digital technologies in soil science raises ethical questions about data control, access, and costs. Remote sensing products, Internet of Things sensors, and soil health monitoring rely on digital information, and the digital economy generates valuable data, yet the benefits are often unevenly distributed, creating new inequalities (Bronson & Knezevic, 2016). For example, remote sensing products based on data from space agencies might be costly for developing countries; solutions might consist of international cooperation in data production or enabling low-cost or free access to tools such as Google Earth Engine for regional or local applications. In the context of Big Data, privacy is a major concern, and users might not be fully aware of the implications of sharing their location or activity data (Wolfert et al., 2017). Consent for data ownership, use, and sharing should therefore be governed by protocols defined according to the potential risks and benefits of the information flow.

Privacy and the use of data generated by farmers can also be a sensitive matter. Farmers uploading data on soil properties using mobile applications may want this information to be used only for the purpose of being sent tailored recommendations (Bronson, 2019). From a data protection point of view, it is usually necessary for the user to consent to the data access and use policy, which should define step-by-step how different actors will use the data transparently and clearly. Anonymization and aggregation mechanisms should be easily understandable, and data owners are more inclined to share information with a broader audience if data privacy considerations are made explicit (Wolfert et al., 2017).

6.4 Environmental sustainability and soil ecosystem resilience

Digital technology offers a unique opportunity to improve the management of soil ecosystems, both for productive and ecosystem-maintaining purposes (Adhikari & Hartemink, 2016). However, various technical, socioeconomic, and ethical challenges hinder its application. The technical issues relate primarily to the development of high-quality standardized data that can be used to analyze the adoption of these innovations and their effect on soil health. Socioeconomic challenges concern local and regional policy and the limited technical capacity of many users of digital innovation. Lastly, users of digital innovations are particularly sensitive to issues of privacy and the application of their data to purposes that they do not approve (Bronson, 2019).

The environmental dimension concerns the impact of digital innovation on the sustainability of the soil ecosystem, that is, its capacity to recover its functionality and its properties and maintain the delivery of ecosystem services after a perturbation (Keesstra et al., 2016). Within the soil acidification context, the natural capacity of the soil to recover after a decrease in pH depends on the maintenance of natural and healthy soil ecosystems that can efficiently provide nutrients, store moisture, sustain supporting services like plant growth, and regulate GHG flows. Hence, the assessment of resilience is of utmost importance for long-term ecosystem services' sustainability and is a major area of current research, exemplified by the large number of articles published in 2020 and 2021. A more environmentally oriented resilience analysis of soils focuses on understanding the stability of soil ecosystem services under specific scenarios of environmental changes predicted on global or local scales, like global warming, the influx of carbon dioxide or nitrogen in the atmosphere, land-use change, and the presence of invasive species.

6.5 Future research needs and emerging technologies

An outline of key research priorities points to areas that would significantly expand understanding and enhance soil management in a digital age. Many key aspects of soil remain poorly understood, hindering advances and posing obstacles for replication in other environments or ecosystems. Digital innovations provide an opportunity to fill some of these gaps and related technical challenges. The wide gap between current soil health and the full restoration potential and capacity to sustain long-term carbon sequestration illustrates the need for such advances (Arrouays et al., 2017).

- **Bridges.** Recommended investments in the development of "Bridges" aim to overcome current barriers in high-dimensional digital soil maps by creating digital analogue networks that facilitate new digital and non-digital prediction flows. Key drivers are reconstruction of soil GHG emissions and other ecological information using soil state as an indicator; provision of analogue locations for translation of soil health feature sets across regions; and intelligent design of soil restoration strategies. Emerging technologies will further shape the trajectory of soil studies. The digital soil health awareness fabric is envisioned as a whole-of-society approach to mobilizing data flows for assessing soil health and setting restoration trajectories. Potential priority areas for future investigation, development and integration include.
- **Soil ecology.** The need to push beyond top taxa and a few features is underscored. A high-dimensional description of soil state with incorporation of soil microfauna dynamics is identified as a priority for integration with long-term greenhouse gas emission data (Bardgett & Van Der Putten, 2014).

- **Artificial Intelligence and Natural Language Processing.** Artificial Intelligence has the potential to facilitate feature selection in broad data sets by linking independent variables with difficult-to-measure soil properties (). Natural Language Processing could also be applied to digital humid tropics soil maps and literature data. Development of other AI tools should proceed.
- **Digital soil health maps.** The state of the art in the prediction of soil health indicators is recognized with attention to standardization of processing and integration of environmental, cultural, and artificial intelligence set information from various chains.
- **Socioeconomic aspects.** Examining the socioeconomic impact of digital technologies in real estate and agribusiness models could support greater private-sector engagement in geography, such as soil carbon market risk analysis.
- **New measurement technologies.** Prototypes for low-cost hyperspectral sensors or proximal soil sensing systems could speed incorporation into future-sensing networks (Adamchuk et al., 2004).

Investment in soil ecology, artificial intelligence, natural language processing, digital soil health maps, or socioeconomic aspects backed by technical capacity should unlock the vast information contained in existing experimental data and models, accelerating the creation of cost-effective soil health awareness fabrics capable of satisfying all user segments at local, regional, and national levels.

6.6 Vision for a digitally-enabled soil science ecosystem

A digitally-enabled soil science ecosystem is required to facilitate integration and collaboration among diverse stakeholders and disciplines (Grunwald et al., 2011). The digitalization of soil science is creating an unprecedented volume of data with the promise of transformative change for Sustainable Development Goals. Society's ability to capitalize on this potential depends on three main aspects: (i) a sound technical infrastructure that enables coherent data flows across scales and domains; (ii) effective mechanisms to link multiple users with public and private data services; and (iii) future research that optimally exploits the new data streams (Eswaran et al., 2019).

An expansive vision encompasses standardization, interoperability and social innovation spanning techniques, tools and processes. Technical solutions focus on the harmonization of sensor data and remote-sensing products, improved data flows, and intelligent use of open-access data for predicting soil dynamics in time and space. Barriers to adoption are social and economic: offer stimulation, training and monitoring to enhance confidence in new techniques; promote financial support for coastal communities; and enable public–private initiatives that meet farmers' and commercial needs. The new use of digital information raises ethical and political issues, underscoring the need for a governance framework that defines data ownership and user rights. The integration of resilience considers soil resilience to climate change and other threats as an index to track progress toward climate-smart land management (FAO, 2022).

Emerging technologies accelerate prediction of soil health, digital soil mapping capacity, and the development of platforms that integrate advanced diagnostic services and farmer expertise (Arrouays et al., 2017). Implementing this vision will stimulate the digitalization of soil science and enhance both productivity and resilience across multiple scales.

7. Conclusion and call to action

Soils represent the ultimate biosphere resource for humankind. Their sustainability is a prerequisite for food production, water supply, biodiversity conservation, and climate stability. However, climate change, land use change, pollution, overexploitation, and other global threats are degrading soils everywhere (Smith & Read, 2010). Without proper management, the natural ecosystem service functions of soils are being compromised, leading to loss of resilience—the capacity to persist in the face of perturbation and change. Recent advances in digital technologies offer improved sensors, hardware, software, and big data analytics for real-time monitoring and assessment of soils at global, regional, and local scales. These digital innovations have great potential to expand understanding of the links between soil conditions, ecosystem dynamics, and environmental health.

Digital tools also facilitate the creation of tailored decision-support systems for soil conservation, restoration, and sustainable management. Nevertheless, digital technologies are not always beneficial, and several challenges must be addressed before widespread implementation. This work concludes with a call to action, emphasizing the need for interdisciplinary collaboration, the potential for global innovation, and the importance of continued research, education, and policy support. Achieving the interdependent ecosystem service intention will require mobilizing expertise from agriculture, ecology, data science, and sociology in cooperative frameworks at public, private, and non-profit levels.

7.1 Recap of key themes connecting ecosystem dynamics and digital tools

The interdependence of soil ecosystems with global ecosystem dynamics is proposed as a conceptual lens for examining emerging digital innovations influencing soil science research and enabling soil management interventions (Bünemann et al., 2018; Lehmann et al., 2020). Digital-enabling, or technology-driven, soil science tools, developed as stand-alone solutions, often overlook soil ecosystem dynamics. Their integration and application, at relevant scales, underpin their potential to address major soil ecosystem conservation and management challenges, such as vulnerability to degradation, resilience to climate change, and contribution to climate change mitigation via carbon sequestration. Examinations of soil ecosystem dynamics and the anticipated impact of digital-enabling tools can help identify key opportunities for technology-supported research and the development of novel soil management practices (Baveye et al., 2016).

Precision agriculture is the most commercially developed and widely adopted of the emerging technologies (Gebbers & Adamchuk, 2010). Other important digital-enabling solutions under development—monitoring of soil conservation practices, assessment of soil degradation and restoration trajectories, estimation of climate change mitigation via carbon sequestration, and guidance of technology-supported soil management by farmers—have yet to achieve widespread deployment and need further research. International developments in all applications should be carefully examined so that potential synergies can be identified and exploited. The involvement of multiple stakeholders in technology-supported soil management—from farmers to agronomists, soil scientists, remote sensing specialists, and policy-makers—is pivotal for successful implementation.

7.2 Emphasis on interdisciplinary collaboration

Three key disciplines must converge to develop a digitally enabled soil science ecosystem. First, agronomy is fundamental, as soils, their microorganisms, and plants are co-dependent (Baveye et al.,

2016). Novel understanding of dynamic soil ecosystems, driven by microbial interactions with one another and with plants, will guide the formulation of management techniques that restore and sustain soil health and ecosystem functions (Fierer, 2017). Second, ecology is crucial for assessing the state and health of global ecosystems through increasingly better-developed fingerprints for soil and ecosystem health and resilience. Restoration ecologists can identify changes in ecosystem state and function due to changing land use and land-use intensity and propose practices that can bring the system back to a healthy state. Finally, data science is pivotal, as a massive amount of research, commercial and public data, social networks, and boisterous communication are played out on digital platforms (Kamilaris et al., 2017). Data science can allow users to overlay data sources and produce knowledge in smarter, faster, and better ways.

Accomplishment of this multidisciplinary vision will require input from a range of stakeholders and disciplines, including agronomy, ecology, digital sciences, development, economics, and politics. Identification of enabling conditions, communicative and action agendas will consequently take place across a multitude of forums. A database of technology-piloted soil management practices will also enable the research community to understand the potential, barriers, impacts, and lessons learned from such practices.

Global digital innovations applied to soil management have the potential to achieve precision agriculture and precision soil management. As with all scientific advances, however, the development of such technologies must have an explicit ethical dimension: they ought to benefit societies as a whole (Rose & Chilvers, 2018). These technologies will assist farmers in planning their actions and enhance farmers' responses to science; evidence from several geographical regions confirms the practical gains obtained through digital technologies linked to soils.

7.3 Potential for innovation to transform soil management globally

The trajectory of soil science over the past decade underscores the importance of combining theory with technology and identifies digital solutions capable of enhancing global soil management. Afinra's "Digital Innovation to Transforming Soil Ecosystem Dynamics" initiative, supported by an exploratory grant program, compiled these insights and suggested a new research agenda in ecology. The potential for digital tools to revolutionize various aspects of soil cycle dynamics has been confirmed through ongoing research and their empirical application across diverse geographical areas and soil ecosystems.

To foster digital innovation in the sector globally, an analysis was undertaken of the technical, political, economic, and social aspects shaping or hindering the successful adoption of monitoring, prediction, and ecosystem restoration technologies by farmers and stakeholders (Wolfert et al., 2017). The conclusions point to innovative paths for overcoming these barriers namely, a coordinated multistakeholder effort involving agronomy, ecology, data science, and the political dimensions of environmental and resource economics (Basso & Antle, 2020).

7.4 Encouragement for continued research, education, and policy support

Innovative models and large-scale replications have the potential to transform soil management. Supporting the mobilization of developments and tools needs commitment from governments and private sector partners to create attractive environments for innovation and implementation of these new solutions (Bünemann et al., 2018). Continuous investment in research, education, and technology validation is vital to advance understanding of ecosystem functions and maintain links between external developments and practical uses on farms (Keesstra et al., 2016).

Addressing emerging trends, the current vision reinforces the necessity for integration of big digital datasets with omnipresent technologies capable of real-time continuous monitoring of soil and environmental variables (Rossel & Bouma, 2016). Recognition of the ethical implications of data ownership and concerns for the surveillance potential of technologies are part of the ongoing challenge of establishing citizen-centered digital democracies (Bronson, 2019). Tension between the development of subtle technologies that support a small number of farmers and the need for direct monitoring systems across wide ranges supports the idea that the latest sensors must be made affordable and widely available, keeping in mind the need for user-friendly operation.

Data Availability

Data available within the article.

Acknowledgments

We extend our gratitude to all the universities involved in this study for their support in facilitating my research.

References

Adamchuk, V. I., Hummel, J. W., Morgan, M. T., & Upadhyaya, S. K. (2004). On-the-go soil sensors for precision agriculture. *Computers and Electronics in Agriculture, 44*(1), 71–91.

Adhikari, K., & Hartemink, A. E. (2016). Linking soils to ecosystem services—A global review. *Geoderma, 262*, 101–111.

Ali, H., Khan, E., & Sajad, M. A. (2013). Phytoremediation of heavy metals—Concepts and applications. *Chemosphere, 91*(7), 869–881.

Angon, P. B., Mondal, S., Jahan, I., Datto, M., Antu, U. B., Ayshi, F. J., & Islam, M. S. (2023). Integrated pest management (IPM) in agriculture and its role in maintaining ecological balance and biodiversity. *Advances in Agriculture, 2023*(1), 5546373.

Arrouays, D., Leenaars, J. G., Richer-de-Forges, A. C., Adhikari, K., Ballabio, C., Greve, M., ... Rodriguez, D. (2017). Soil legacy data rescue via GlobalSoilMap and other international and national initiatives. *Geo Research Journal, 14*, 1–19.

Bardgett, R. D., & Van Der Putten, W. H. (2014). Belowground biodiversity and ecosystem functioning. *Nature, 515*(7528), 505–511.

Bardgett, R. D., & Wardle, D. A. (2010). *Aboveground-belowground linkages: Biotic interactions, ecosystem processes, and global change*. Oxford University Press.

Basso, B., & Antle, J. (2020). Digital agriculture to design sustainable agricultural systems. *Nature Sustainability, 3*(4), 254–256.

Bates, S. T., Berg-Lyons, D., Caporaso, J. G., Walters, W. A., Knight, R., & Fierer, N. (2011). Examining the global distribution of dominant archaeal populations in soil. *The ISME Journal, 5*(5), 908–917.

Batjes, N. H. (2016). Harmonized soil property values for broad-scale modelling (WISE30sec) with estimates of global soil carbon stocks. *Geoderma, 269*, 61–68.

Baveye, P. C., Baveye, J., & Gowdy, J. (2016). Soil "ecosystem" services and natural capital: Critical appraisal of research on uncertain ground. *Frontiers in Environmental Science, 4*, 41.

Bronick, C. J., & Lal, R. (2005). Soil structure and management: A review. *Geoderma, 124*(1–2), 3–22.

Bronson, K., & Knezevic, I. (2016). Big data in food and agriculture. *Big Data & Society, 3*(1), 2053951716648174.

Bronson, K. (2019). Looking through a responsible innovation lens at uneven engagements with digital farming. *NJAS - Wageningen Journal of Life Sciences, 90*, Article 100294.

Brussaard, L. (2012). Ecosystems services provided by the soil biota. In *Soil ecology and ecosystems services* (pp. 54–64). Oxford University Press.

Bünemann, E. K., Bongiorno, G., Bai, Z., Creamer, R. E., De Deyn, G., De Goede, R., ... Brussaard, L. (2018). Soil quality–A critical review. *Soil Biology and Biochemistry, 120*, 105–125.

Cameron, D. D., Neal, A. L., van Wees, S. C., & Ton, J. (2013). Mycorrhiza-induced resistance: More than the sum of its parts? *Trends in Plant Science, 18*(10), 539–545.

Canfield, D. E., Glazer, A. N., & Falkowski, P. G. (2010). The evolution and future of Earth's nitrogen cycle. *Science, 330*(6001), 192–196.

Chapin, F. S., III, Matson, P. A., & Vitousek, P. M. (2011). The ecosystem concept. In *Principles of terrestrial ecosystem ecology* (pp. 3–22). New York, NY: Springer New York.

Coleman, D. C., Callaham, M. A., & Crossley Jr, D. A. (2017). *Fundamentals of soil ecology.* Academic press.

de Paul Obade, V., & Lal, R. (2016). A standardized soil quality index for diverse field conditions. *Science of the Total Environment, 541*, 424–434.

Delgado-Baquerizo, M., Maestre, F. T., Reich, P. B., Jeffries, T. C., Gaitan, J. J., Encinar, D., ... Singh, B. K. (2016). Microbial diversity drives multifunctionality in terrestrial ecosystems. *Nature Communications, 7*(1), Article 10541.

Delgado-Baquerizo, M., Oliverio, A. M., Brewer, T. E., Benavent-González, A., Eldridge, D. J., Bardgett, R. D., ... Fierer, N. (2018). A global atlas of the dominant bacteria found in soil. *Science, 359*(6373), 320–325.

Dexter, A. R. (2004). Soil physical quality: Part I. Theory, effects of soil texture, density, and organic matter, and effects on root growth. *Geoderma, 120*(3–4), 201–214.

Dokuchaev, V. V. (1883). Russkiy chernozem (Russian black soil). St. Petersburg (in Russian).

Edwards, C. A. (2004). The importance of earthworms as key representatives of the soil fauna. In *Earthworm ecology* (pp. 3–11). Crc Press.

Eswaran, H., Lal, R., & Reich, P. F. (2019). Land degradation: An overview. *Response to Land Degradation*, 20–35.

FAO, I. T. P. S. (2015). Status of the world's soil resources (SWSR)–main report. In *Food and agriculture organization of the United Nations and intergovernmental technical panel on soils, Rome, Italy* (p. 650).

FAO. (2022). World food and agriculture—statistical yearbook 2022. Rome, Italy: FAO.

Ferris, H., Bongers, T., & de Goede, R. G. (2001). A framework for soil food web diagnostics: Extension of the nematode faunal analysis concept. *Applied Soil Ecology, 18*(1), 13–29.

Fierer, N. (2017). Embracing the unknown: Disentangling the complexities of the soil microbiome. *Nature Reviews Microbiology, 15*(10), 579–590.

Gebbers, R., & Adamchuk, V. I. (2010). Precision agriculture and food security. *Science, 327*(5967), 828–831.

Greinke, L., & Lange, L. (2022). Multi-locality in rural areas–an underestimated phenomenon. *Regional Studies, Regional Science, 9*(1), 67–81.

Grunwald, S., Murad, M. O. F., Farrington, S., Wallace, W., & Rooney, D. (2024). Multi-sensor soil probe and machine learning modeling for predicting soil properties. *Sensors, 24*(21), 6855.

Grunwald, S., Thompson, J. A., & Boettinger, J. L. (2011). Digital soil mapping and modeling at continental scales: Finding solutions for global issues. *Soil Science Society of America Journal, 75*(4), 1201–1213.

Hamza, M. A., & Anderson, W. K. (2005). Soil compaction in cropping systems: A review of the nature, causes and possible solutions. *Soil and Tillage Research, 82*(2), 121–145.

Hillel, D. (2003). *Introduction to environmental soil physics.* Elsevier.

Hoek van Dijke, A. J., Herold, M., Mallick, K., Benedict, I., Machwitz, M., Schlerf, M., ... Teuling, A. J. (2022). Shifts in regional water availability due to global tree restoration. *Nature Geoscience, 15*(5), 363–368.

Hopkin, S. P. (1997). *Biology of the springtails:(Insecta: Collembola).* OUP Oxford.

Jarvis, N. J. (2007). A review of non-equilibrium water flow and solute transport in soil macropores: Principles, controlling factors and consequences for water quality. *European Journal of Soil Science, 58*(3), 523–546.

Jenny, H. (1941). *Factors of soil formation, 28.* New York: McGraw-Hill.

Jenny, H. (1994). *Factors of soil formation: a system of quantitative pedology. Courier Corporation.*

Kamilaris, A., Kartakoullis, A., & Prenafeta-Boldú, F. X. (2017). A review on the practice of big data analysis in agriculture. *Computers and Electronics in Agriculture, 143*, 23–37.

Keesstra, S. D., Bouma, J., Wallinga, J., Tittonell, P., Smith, P., Cerdà, A., ... Fresco, L. O. (2016). Forum paper: The significance of soils and soil science towards realization of the UN sustainable development goals (SDGS). *Soil Discussions, 2016*, 1–28.

Kibblewhite, M. G., Ritz, K., & Swift, M. J. (2008). Soil health in agricultural systems. *Philosophical Transactions of the Royal Society B: Biological Sciences, 363*(1492), 685–701.

Klerkx, L., & Rose, D. (2020). Dealing with the game-changing technologies of agriculture 4.0: How do we manage diversity and responsibility in food system transition pathways? *Global Food Security, 24*, Article 100347.

Kögel-Knabner, I. (2002). The macromolecular organic composition of plant and microbial residues as inputs to soil organic matter. *Soil Biology and Biochemistry, 34*(2), 139–162.

Krause, S., Le Roux, X., Niklaus, P. A., Van Bodegom, P. M., Lennon, J. T., Bertilsson, S., ... Bodelier, P. L. (2014). Trait-based approaches for understanding microbial biodiversity and ecosystem functioning. *Frontiers in Microbiology, 5*, 251.

Kravchenko, A. N., & Guber, A. K. (2017). Soil pores and their contributions to soil carbon processes. *Geoderma, 287*, 31–39.

Kuzyakov, Y., & Blagodatskaya, E. (2015). Microbial hotspots and hot moments in soil: Concept & review. *Soil Biology and Biochemistry, 83*, 184–199.

Lavelle, P., Decaëns, T., Aubert, M., Barot, S., Blouin, M., Bureau, F., ... Rossi, J. P. (2006). Soil invertebrates and ecosystem services. *European Journal of Soil Biology, 42*, S3–S15.

Legg, S. (2021). IPCC, 2021: Climate change 2021-the physical science basis. *Interaction, 49*(4), 44–45.

Lehmann, J., & Kleber, M. (2015). The contentious nature of soil organic matter. *Nature, 528*(7580), 60–68.

Lehmann, J., Bossio, D. A., Kögel-Knabner, I., & Rillig, M. C. (2020). The concept and future prospects of soil health. *Nature Reviews Earth & Environment, 1*(10), 544–553.

Lynch, J. P. (2019). Root phenotypes for improved nutrient capture: An underexploited opportunity for global agriculture. *New Phytologist, 223*(2), 548–564.

Mulder, V. L., De Bruin, S., Schaepman, M. E., & Mayr, T. R. (2011). The use of remote sensing in soil and terrain mapping—A review. *Geoderma, 162*(1–2), 1–19.

Oldroyd, G. E., Murray, J. D., Poole, P. S., & Downie, J. A. (2011). The rules of engagement in the legume-rhizobial symbiosis. *Annual Review of Genetics, 45*(1), 119–144.

Philippot, L., Raaijmakers, J. M., Lemanceau, P., & Van Der Putten, W. H. (2013). Going back to the roots: The microbial ecology of the rhizosphere. *Nature Reviews Microbiology, 11*(11), 789–799.

Pinheiro, F. L., Balland, P. A., Boschma, R., & Hartmann, D. (2025). The dark side of the geography of innovation: relatedness, complexity and regional inequality in Europe. *Regional Studies, 59*(1), 2106362.

Porcel, R., Aroca, R., & Ruiz-Lozano, J. M. (2012). Salinity stress alleviation using arbuscular mycorrhizal fungi. A review. *Agronomy for Sustainable Development, 32*(1), 181–200.

Prajapati, C. S., Priya, N. K., Bishnoi, S., Vishwakarma, S. K., Buvaneswari, K., Shastri, S., ... Jadhav, A. (2025). The role of participatory approaches in modern agricultural extension: Bridging knowledge gaps for sustainable farming practices. *Journal of Experimental Agriculture International, 47*(2), 204–222.

Richter, D. D., & Yaalon, D. H. (2012). "The changing model of soil" revisited. *Soil Science Society of America Journal, 76*(3), 766–778.

Rob, V., Giovanni, B. L., Kostas, S., Boyd, H., Aysen, T., Martin, P., ... Mulu, K. (2017). *The future of food and agriculture (Doctoral dissertation, Food and Agriculture Organization of the United Nations (FAO)).*

Rose, D. C., & Chilvers, J. (2018). Agriculture 4.0: Broadening responsible innovation in an era of smart farming. *Frontiers in Sustainable Food Systems, 2*, Article 387545.

Rossel, R. A. V., & Bouma, J. (2016). Soil sensing: A new paradigm for agriculture. *Agricultural Systems, 148*, 71–74.

Schimel, J. P. (2018). Life in dry soils: Effects of drought on soil microbial communities and processes. *Annual Review of Ecology, Evolution, and Systematics, 49*(1), 409–432.

Schmidt, M. W., Torn, M. S., Abiven, S., Dittmar, T., Guggenberger, G., Janssens, I. A., ... Trumbore, S. E. (2011). Persistence of soil organic matter as an ecosystem property. *Nature, 478*(7367), 49–56.

Shahmohamadloo, R. S., Febria, C. M., Fraser, E. D., & Sibley, P. K. (2021). The sustainable agriculture imperative: A perspective on the need for an agrosystem approach to meet the United Nations Sustainable Development Goals by 2030. *Integrated Environmental Assessment and Management, 18*(5), 1199–1205.

Smith, S. E., & Read, D. J. (2010). *Mycorrhizal symbiosis*. Academic press.

Van Der Heijden, M. G., Bardgett, R. D., & Van Straalen, N. M. (2008). The unseen majority: Soil microbes as drivers of plant diversity and productivity in terrestrial ecosystems. *Ecology Letters, 11*(3), 296–310.

Wall, D. H., Bardgett, R. D., & Kelly, E. (2010). Biodiversity in the dark. *Nature Geoscience, 3*(5), 297–298.

Weil, R. R., & Brady, N. C. (2017). *The nature and properties of soils* (15th ed.). New York: Pearson.

Wolfert, S., Ge, L., Verdouw, C., & Bogaardt, M. J. (2017). Big data in smart farming–a review. *Agricultural Systems, 153*, 69–80.

Xiang, K., Chen, H., Chen, L., Che, X., Hanzhe, C., Ren, H., & Zhou, J. (2026). Development and validation of an explainable machine learning model for prediction of biochar impacts on soil properties and crop yield across diverse regions. *Soil Use and Management, 42*(2), e70238.

Yeates, G. W., Bongers, T., De Goede, R. G., Freckman, D. W., & Georgieva, S. (1993). Feeding habits in soil nematode families and genera—An outline for soil ecologists. *Journal of Nematology, 25*(3), 315.

Young, I. M., & Crawford, J. W. (2004). Interactions and self-organization in the soil-microbe complex. *Science, 304*(5677), 1634–1637.

CHAPTER

7

A novel AI-driven framework for farmer segmentation: A digital innovation case study linking socioeconomic profiles to perceived soil and agronomic challenges

Moein Tosan[1], Afshin Shayeghi[2], Mahsa Mardani[3], Amin Sabbagh[4] and Morteza Yaqubi[5]

[1]*Department of Water Science and Engineering, Ferdowsi University of Mashhad, Mashhad, Iran;* [2]*Department of Geography & Environmental Sustainability, University of Oklahoma, Norman, OK, United States;* [3]*Department of Water Resources Engineering and Management, University of Birjand, Birjand, Iran;* [4]*Department of Civil Engineering, Faculty of Engineering, Shahid Bahonar University of Kerman, Kerman, Iran;* [5]*Saffron Institute, University of Torbat Heydarieh, Torbat-e Heydarieh, Iran*

1. Introduction

Agriculture continues to function as the economic backbone and primary livelihood source for rural populations in many developing nations (Dalir et al., 2024; Gholian-Jouybari et al., 2023). Within this context, saffron (*Crocus sativus* L.), frequently referred to as red gold, maintains a unique strategic importance (Acharya et al., 2025). Given its status as a high-value crop with low water requirements, saffron represents a sustainable option for arid regions confronting water scarcity (Moudi et al., 2025; Razmavaran et al., 2024). With regard to edaphic requirements, optimal growth is observed in deep, well-drained clay-calcareous soils that contain sufficient organic matter (Acharya et al., 2025; Ghorbani & Koocheki, 2017). Beyond its physical needs, the crop frequently exhibits superior economic and ecological efficiency relative to staple alternatives (Mohammadzadeh et al., 2018), thereby creating a critical opportunity for income diversification in marginal rural territories (Cicco et al., 2025; Giupponi et al., 2023).

Currently, Iran commands the global market, accounting for over 90% of total world production (Kafi & Showket, 2007; Moghaddam et al., 2017). The undisputed center of this industry is the Khorasan province, specifically the Torbat Heydarieh region, which is the focus of this research (Yaqubi et al., 2024). Here, saffron is not merely an agricultural product but a vital socioeconomic engine that supports employment and economic resilience for thousands of rural households (Dalir et al., 2024; Hosseini et al., 2010). Nevertheless, the industry is impeded by persistent challenges

The Soil Nexus. https://doi.org/10.1016/B978-0-443-45286-4.00013-4

throughout the value chain (Acharya et al., 2025; Nehvi et al., 2007). Producers struggle with significant agronomic and soil-related limitations, including water stress driven by climate change (Aziz & Sadok, 2015; Moudi et al., 2025), nutrient management difficulties, and declining soil fertility (Koocheki & Seyyedi, 2020). Yield trends have also been negatively impacted by biotic stressors, particularly soil-borne pathogens and corm rot diseases like Fusarium oxysporum (Caligiore-Gei et al., 2023; Gupta et al., 2021). These agronomic challenges are further compounded by an extensive reliance on expensive manual labor (Ghorbani & Koocheki, 2017; Hayati & Marzban, 2024), resulting in a noticeable decline in productivity over time (Kafi & Showket, 2007).

Post-harvest limitations present significant hurdles as well; specifically, the scarcity of modern processing infrastructure and inadequate technical expertise frequently lead to quality deterioration (Nehvi et al., 2007). Significant barriers also exist at the market level, including supply chain fragmentation, weak institutional support (Pitoska & Gatzioufa, 2010), and extensive product adulteration (Dai et al., 2023). Consequently, despite its dominance in production volumes, Iran fails to secure the primary share of value-added profits in the international arena (Yaqubi et al., 2024).

Previous research in the region has effectively identified and ranked these multifaceted challenges (Shahroudi et al., 2007; Tosan, Khashei Siuki, et al., 2024). However, these studies, often relying on factor analysis, inherently treat the farmer population as a single, homogenous group. This one-size-fits-all approach masks the critical human element of the soil nexus: *why* do certain farmers perceive specific challenges as more severe than others? Studies confirm that socioeconomic characteristics like education and age are dominant variables in farmer decision-making and technology adoption (Lambarraa-Lehnhardt et al., 2022; Tohidi et al., 2023). The information-seeking behavior of elite farmers, for instance, differs significantly from that of ordinary farmers in the same region (Radad et al., 2017).

This chapter addresses a critical gap by applying a novel analytical framework. While many digital innovations currently focus on monitoring the physical aspects of the soil nexus using hardware (e.g., IoT sensors, UAVs), this chapter utilizes machine learning (ML) as an analytical tool to model the human perception of this system. The primary objective is to demonstrate, as a case study, how ML can be used to: 1) segment (or typologize) saffron farmers based on their perceived challenges regarding soil/production, domestic markets, foreign markets, and processing; 2) identify the key socioeconomic features (e.g., education, farm size) that predict these perceptions; and 3) provide a data-driven blueprint for targeted extension programs and future Decision Support Systems (DSS) that effectively connect soil ecosystem dynamics to human-centric decision-making.

2. Materials and methods

2.1 Study area and sampling strategy

This investigation adopted a quantitative, cross-sectional field survey design. The fieldwork was carried out in the principal saffron-cultivating areas of the Torbat-Heydarieh and Zaveh counties, which are situated within Khorasan Razavi province, Iran. This geographical zone holds recognized international importance, often cited as the world's primary center for saffron production (Rezvani Moghaddam et al., 2017; Tosan, Khashei-Siuki, et al., 2024). The selected research area is geographically positioned at roughly 34° 17 ′ N latitude and 59° 12 ′ E longitude, maintaining an elevation of 1451 m above mean sea level. This region exhibits a cold-arid climate pattern. Long-term meteorological data indicate that the area receives an average annual rainfall of approximately 113.8 mm. Furthermore, the mean minimum and maximum temperatures recorded for the region are 7.6 and 21.4°C, respectively. Importantly, these two specific counties are responsible for the annual production of over 70 tons of saffron, making them the most significant production cluster within the entire province (Feizi & Tosan, 2017) (Fig. 7.1).

The target population for this investigation comprised saffron growers situated within the primary agricultural districts of Torbat-Heydarieh (specifically, the Central district, Jolgeh Rokh, Kadkan, and Bayg) and Zaveh (Central district and Feyzabad). The demographic characteristics and agricultural scale of these specific districts are summarized in the accompanying Table 7.1. A convenience sampling method was employed for data collection, which was carried out during the 2023–24 (1402) cultivation season (Yaqubi et al., 2024). Following the collection phase, the raw data underwent a rigorous cleaning process to systematically identify and remove incomplete or otherwise invalid entries. This detailed process resulted in a final, robust dataset consisting of 89 valid questionnaires retained for the subsequent comprehensive analysis.

2.2 Data collection instrument

The primary instrument selected for data acquisition was a structured questionnaire. This crucial tool was subjected to meticulous development and rigorous validation by specialized experts affiliated with

FIGURE 7.1

Location of the study area in Khorasan Razavi province, Iran, highlighting the surveyed districts.

Table 7.1 Demographic profile of the saffron production districts in the study area.

County	District	No. of village districts	No. of villages	Households (n)	Population (n)
Torbat-Heydarieh	Central district	3	104	43,029	140,019
	Jolgeh Rokh	4	141	11,588	37,706
	Kadkan	2	87	3752	12,225
	Bayg	1	46	1270	3392
Zaveh	Central district	4	83	20,063	67,695
Total surveyed area	-	14	461	80,702	261,037

the Saffron Institute at the University of Torbat Heydarieh (Yaqubi et al., 2024). To ensure adequate and comprehensive coverage of the defined target population, the survey employed a strategic dual administration methodology. This approach effectively integrated traditional in-person interviews with the use of contemporary online platforms. The complete document was systematically arranged into four principal sections designed to capture diverse aspects of saffron cultivation. The initial two segments were dedicated to establishing the foundational farmer profile, focusing on Sociodemographic features (including location, age, education, and experience) and farm and economic features (synthesizing data on farm size, net profit, gross income, yield (kg), and production costs). The third segment, recognized as the core analytical component, was titled Perception of challenges and utilized 24 individual items across four specialized domains (production, domestic market, foreign market, and processing). Finally, the last section defined the response scale, recording all challenge items on a consistent 7-point Likert scale, which ranged from the lowest anchor of 1 ('very low'/'strongly disagree') to the highest anchor of 7 ('very high'/'strongly agree').

2.3 Analytical framework and statistical modeling

This study intentionally diverges from traditional analytical approaches such as Exploratory Factor Analysis or nonparametric procedures like the Friedman test by implementing a multi-stage ML framework. This innovative methodology is meticulously designed for robust farmer segmentation and predictive analysis, offering a more nuanced, multidimensional, and practical assessment of farmer behavior, which is essential for modern information and communication technology (ICT) applications. The entirety of the analytical workflow was executed using the Python programming language (v. 3.x), leveraging the computational capabilities of the Pandas, Scikit-learn, Seaborn, and Matplotlib libraries.

The analytical workflow was intentionally commenced with an emphasis on rigorous data preprocessing and subsequent feature engineering. The raw information acquired directly from the field survey underwent a comprehensive sequence of steps for cleaning and standardization. The dataset was verified to be complete for all primary sociodemographic predictors (e.g., age, education, farm size), requiring no imputation. As a necessary component of this vital preparatory effort, cost ranges that were originally presented in a descriptive textual format (such as 'between 1-2 million Toman') were systematically transformed into precise numerical representations. This specialized conversion method relied upon leveraging the median value of the stated range to provide the final numerical input (e.g., 1.5). In a parallel and systematic procedure, the textual responses collected from the administered Likert scales were accurately mapped to their designated ordinal numerical values (covering the full span from 1 through 7). A pivotal action that concluded this initial phase involved the creation of four new dependent variables (DVs). These specific DVs were constructed by calculating the mean score derived from all individual survey items that contributed to each of the four principal challenge categories previously identified in the study.

Subsequent to this thorough data preparation, the research moved toward establishing an intuitive, visual understanding of the cohort's characteristics by conducting a visual descriptive analysis using the Seaborn library. This initial descriptive effort encompassed the generation of count plots for discrete categorical features (such as education level) and box plots. to facilitate the visual comparison of the distribution of key economic outcomes (like 'net profit') across distinct 'locations'. To significantly enhance this initial descriptive overview and allow for the quantification of statistically significant linear associations, two established classical statistical methods were subsequently implemented: Pearson correlation and analysis of variance (ANOVA). A correlation matrix was computed to accurately measure the direction and strength of linear relationships existing between the core economic/production metrics and the four challenge DVs. These patterns were clearly accentuated using a heatmap. Complementing this, a one-way ANOVA test was conducted to specifically ascertain whether the mean perception scores for each challenge group varied significantly (at a predetermined threshold of $P < 05$) across differing farmer sociodemographic categories (e.g., 'education' and 'farm size').

The next critical phase focused on predictive modeling and assessing feature importance. To accurately uncover the most influential nonlinear predictors of challenge perception, four distinct random forest (RF) classifier models were constructed. This ensemble methodology was purposefully selected owing to its inherent high predictive accuracy and its robust capability for calculating Gini Importance (Tohidi et al., 2023). For this stage, categorical features (e.g.,'Education') were first transformed into binary dummy variables utilizing One-Hot Encoding. Prior to model training, a collinearity diagnostic was performed using the variance inflation factor (VIF). All predictor variables exhibited VIF values below 5.0, confirming the absence of significant multicollinearity and justifying the inclusion of all selected socio-demographic features in the final models. To facilitate the classification task, the continuous challenge scores were binned into two distinct classes ('High_Challenge' vs. 'Low_Challenge') based on the median value of each domain. Due to the discrete nature of Likert-scale responses and ties at the median, this resulted in class imbalances for some domains (e.g., 25.8% High Challenge for the Domestic Market vs. 47.2% for Processing). To mitigate potential bias arising from these distributions, stratified sampling was employed for all train-test splits, ensuring that the class proportions in the training (75%) and testing (25%) subsets remained identical to the original population. Four separate RF models were then trained one for predicting each of the four challenge types and the resulting Gini Importance scores were visualized using Bar Charts to clearly identify the dominant predictive features. A sensitivity analysis was also conducted by retraining the models using the mean score as an alternative threshold. The results confirmed the stability of the feature importance rankings, particularly for the 'Processing' challenge, where education and location consistently emerged as the top predictors regardless of the threshold used. To optimize model performance and prevent overfitting, hyperparameters were rigorously tuned using GridSearchCV with 5-fold cross-validation. For the RF ensemble, the optimal configuration was identified as n_estimators = 300, max_depth = 4, and min_samples_split = 2. For the single interpretable Decision Tree, the hyper-parameters were set to max_depth = 2, criterion = 'entropy', min_samples_leaf = 4, and class_-weight = 'balanced' to maximize both predictive accuracy and model transparency.

The concluding phase of the analytical methodology concentrated on addressing the inherent black box nature of the high-performing RF ensemble, thereby ensuring comprehensive analytical transparency. To accomplish this, an interpretability and validation study was executed, with its primary focus placed on the predictive model developed for the 'Processing Challenge'. The comprehensive dataset was initially segregated into dedicated subsets: a training set (75%) and an independent, unseen testing set (25%). A single DT model was then trained exclusively on the training data, purposefully limited to a maximum depth of three (with max_depth = 3) to preserve maximum clarity and ease of interpretation. This highly interpretable model was subsequently evaluated utilizing the reserved independent test data. The resulting performance metrics were rigorously assessed using a Confusion Matrix alongside standard classification metrics namely accuracy, precision, and recall in order to robustly validate the fundamental predictive power initially observed in the RF ensemble approach.

2.4 Construct reliability and validity

The reliability of the challenge constructs was assessed using Cronbach's alpha. The 'Processing' challenge demonstrated high internal consistency (alpha > 0.95), confirming that farmers perceive post-harvest technological and capital constraints as a unified, systemic barrier. Conversely, the 'Production', 'Domestic Market', and 'Foreign Market' domains exhibited lower alpha values. This is attributed to their nature as formative indices, where individual indicators (e.g., climate change vs. labor costs) represent distinct, independent causes of the composite challenge rather than correlated manifestations of a single latent trait (Diamantopoulos & Winklhofer, 2001). Consequently, the aggregation of these items into mean scores remains a valid approach for capturing the cumulative pressure of these heterogeneous constraints.

3. Results and discussion

This chapter presents the analytical findings of the study. It begins with a comprehensive descriptive analysis of the farmer cohort and their economic performance, moves to a correlational analysis to identify underlying relationships, and culminates in a series of predictive ML models to segment farmers based on their perceived challenges.

3.1 Descriptive profile of the saffron growers

The study sample consists of 89 farmers from Torbat Heydarieh, a primary hub of global saffron production. The demographic and professional profile of the respondents (Figs. 7.2 and 7.3) reveals a mature, experienced, and established cohort.

The majority of farmers possess a 'diploma' (41.6%, n = 37) or 'primary' education (20.2%, n = 18), with only a small fraction (13.4% combined) holding an 'Associate' or 'Bachelor' degree (see Table 7.2). This profile contrasts sharply with emerging saffron cultivation regions, such as Italy,

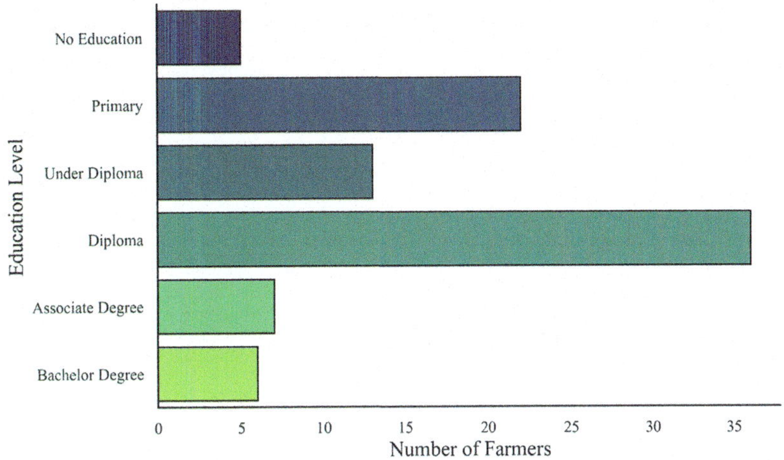

FIGURE 7.2

Distribution of farmer education levels (n = 89).

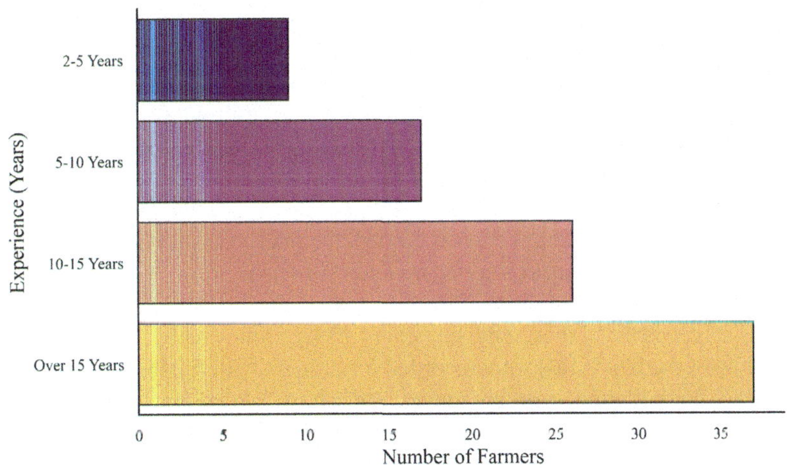

FIGURE 7.3

Distribution of farmer work experience (n = 89).

where farmers are often young, new entrants with a high level of literacy, including bachelor's or master's degrees (Giupponi et al., 2023).

Furthermore, the cohort is highly experienced, with 47.2% (n = 42) having over 15 years of experience in saffron cultivation. This suggests that the production systems in the region are deeply rooted in traditional and indigenous knowledge (Ghorbani & Koocheki, 2017), rather than being driven by new entrants.

Table 7.2 Frequency distribution of farmer socio-demographic characteristics (n = 89).

Category	Variable	Frequency	Percentage (%)
Education	Diploma	37	41.6
	Primary	18	20.2
	Under diploma	14	15.7
	No education	8	9.0
	Associate degree	6	6.7
	Bachelor degree	6	6.7
Experience	Over 15 years	42	47.2
	2–5 years	17	19.1
	10–15 years	16	18.0
	5–10 years	14	15.7

3.2 Economic performance and socioeconomic linkages

The economic viability of saffron is evident from the descriptive statistics of farm performance (Table 7.3). The mean net profit was 219.06 million Toman (approx. 3650 USD at 60,000:1 exchange rate), with a mean dry stigma yield of 4.54 kg. This yield is consistent with onfarm studies in Southern Khorasan, which report frequent mean yields between 4 and 6 kg/ha (Jami-alahmadi et al., 2010).

However, the distribution of net profit is highly right-skewed, as seen in Fig. 7.4. The median profit (170 million Toman) is significantly lower than the mean (219 million Toman), indicating that a few highly successful farms pull the average up, while the majority of farmers earn a more modest income. This economic disparity highlights the varying levels of success and efficiency among farmers in the same region (Mahdian Moghadam & Tosan, 2015). To investigate the source of this performance variance, we explored the relationship between farmer characteristics and farm outputs. A key finding emerged when linking farmer education to farm yield (Fig. 7.5).

As shown in Fig. 7.5 and detailed in Table 7.4, a clear positive trend is visible. Farmers with a 'bachelor degree' achieved the highest median yield (4.25 kg), followed by 'associate degree' (3.5 kg). Conversely, the lowest median yields were recorded for farmers with 'under diploma' (1.3 kg) and 'no education' (3.0 kg). This finding strongly suggests that higher education levels are correlated with improved agronomic performance and yield. This aligns with studies highlighting that low productivity is often linked to a lack of awareness and that technology transfer programs and training (which are more accessible to educated farmers) are critical for improving economic gains (Hakeem et al., 2018; Nagoo et al., 2018; Radad et al., 2017).

Table 7.3 Descriptive statistics for key economic and production variables (n = 89).

Variable	Mean	Median	Std. dev.	Min	Max
Net profit (M toman)	219.06	170.0	241.03	0.0	1700.0
Yield (dry kg)	4.54	3.75	3.21	0.15	13.0

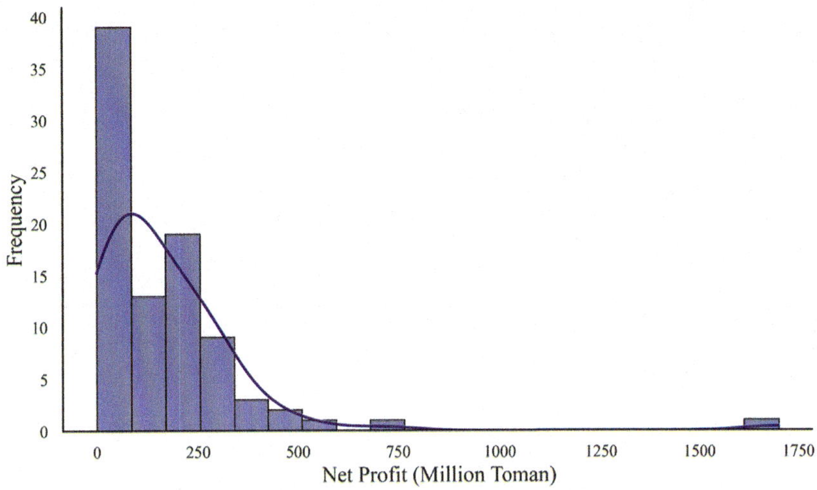

FIGURE 7.4

Distribution of net profit among farmers, showing a right-skewed pattern.

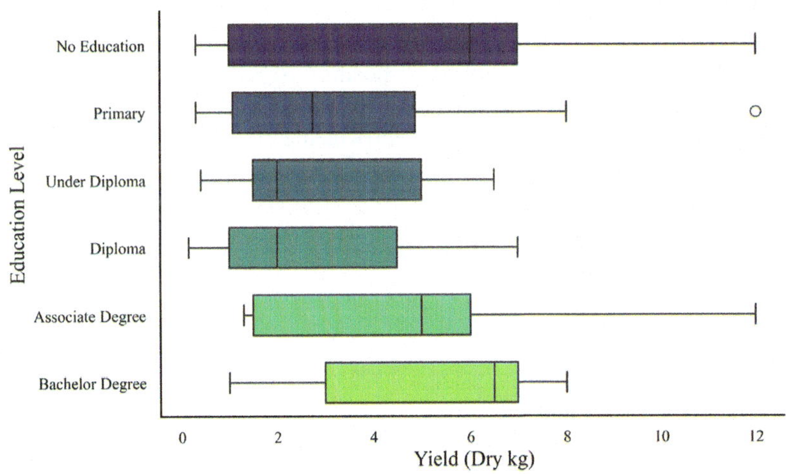

FIGURE 7.5

Boxplot of saffron yield (dry kg) by farmer education level.

Furthermore, farm performance is significantly linked to geography (Fig. 7.6). The boxplot analysis reveals substantial variability in net profit based on the farm's location. Central district shows the highest mean (296.8 M Toman) and median (220 M Toman) profit, while also displaying the widest variance. 'Jolgeh Zaveh' shows a lower mean profit (191.7 M Toman) but is far more consistent. This geographical disparity may be linked to differing microclimates, soil types (Naim et al., 2023), or variations in market access and the influence of intermediaries in different subregions (Lambarraa-

Table 7.4 Descriptive statistics for saffron yield (kg) by farmer education Level.

Education level	Mean	Median (50%)	Count
Bachelor degree	4.33	4.25	6
Associate degree	4.41	3.50	6
Diploma	5.37	4.00	37
Primary	3.63	3.00	18
Under diploma	3.09	1.30	14
No education	4.02	3.00	8

Lehnhardt et al., 2022). This finding strongly supports the inclusion of 'location' as a key predictive feature in the subsequent ML models.

3.3 Correlational analysis and statistical linkages

While the descriptive statistics profile the average farmer, they do not explain the heterogeneity in performance and perception. To move from description to analysis, we first employed a Pearson correlation analysis to map the linear relationships between key economic/production variables and the farmers' perceived challenge levels.

The correlation matrix (Fig. 7.7 and Table 7.5) reveals several critical underlying relationships that are not immediately obvious. First, the strongest positive correlation ($r = 0.79$) exists between the perception of processing challenges and foreign market challenges. This is a highly significant finding, suggesting that these two concepts are deeply intertwined in the farmers' minds. Farmers who are

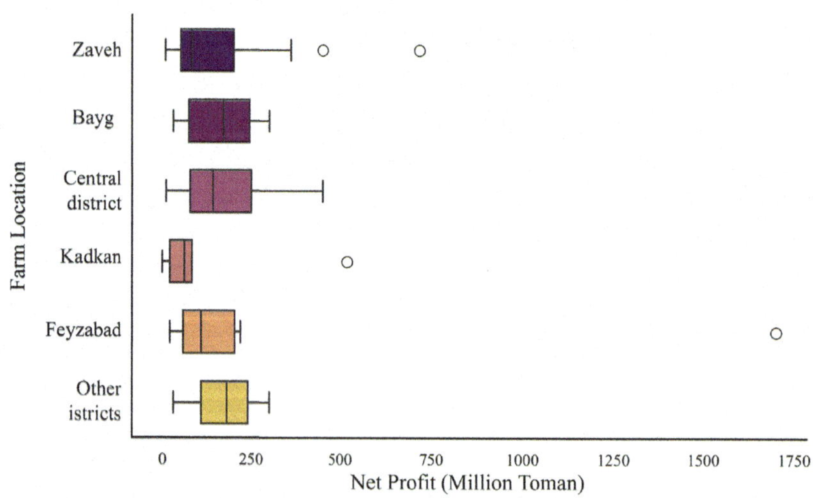

FIGURE 7.6

Boxplot of net profit (million toman) by farm location.

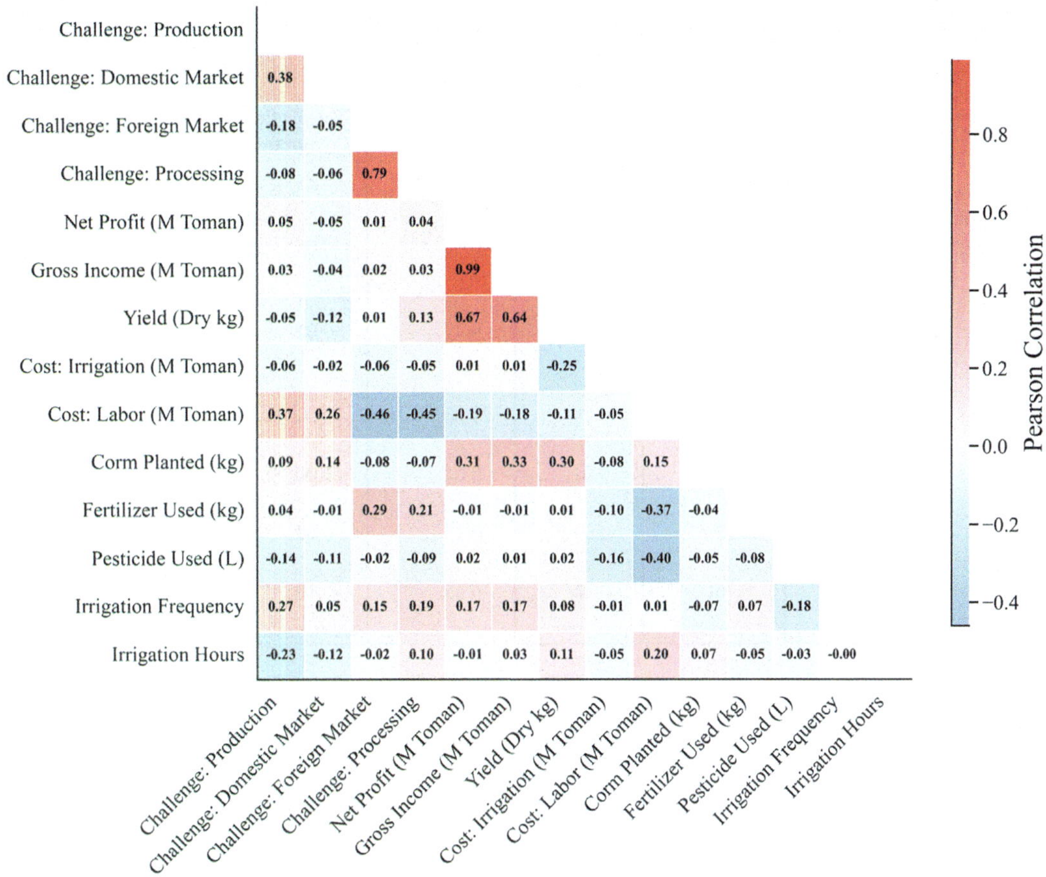

FIGURE 7.7

Correlation heatmap of key economic, production, and challenge perception variables.

concerned with export-level barriers (e.g., global standards, adulteration, trade logistics) are the *same* farmers who are most concerned with domestic value-chain limitations (e.g., lack of modern equipment, poor post-harvest handling). This links the technical act of processing directly to the economic goal of exportation and value addition (Acharya et al., 2025; Giupponi et al., 2023).

Second, labor cost (cost: Labor (M Toman)) acts as a powerful proxy for traditional versus modern farming mentalities. It shows a moderate positive correlation with the perception of production challenges ($r = 0.37$) and domestic market challenges ($r = 0.26$). This is logical, as high labor cost is a primary component of traditional, labor-intensive cultivation (Ghorbani & Koocheki, 2017; Hayati & Marzban, 2024). Conversely, labor cost has a strong negative correlation with foreign market challenges ($r = -0.46$) and processing challenges ($r = -0.45$). This suggests a clear typology: farmers

Table 7.5 Pearson correlation matrix of key study variables.

Variable	Challenge: Production	Challenge: Domestic Mkt	Challenge: Foreign Mkt	Challenge: Processing	Net profit (Mtoman)	Yield (dry kg)	Cost: Labor (Mtoman)
Challenge: Production	1.00	0.38	−0.18	−0.08	0.05	−0.05	0.37
Challenge: Domestic Mkt	0.38	1.00	−0.05	−0.06	−0.05	−0.12	0.26
Challenge: Foreign Mkt	−0.18	−0.05	1.00	0.79	0.01	0.01	−0.46
Challenge: Processing	−0.08	−0.06	0.79	1.00	0.04	0.13	−0.45
Net profit (Mtoman)	0.05	−0.05	0.01	0.04	1.00	0.67	−0.19
Yield (dry kg)	−0.05	−0.12	0.01	0.13	0.67	1.00	−0.11
Cost: Labor (Mtoman)	0.37	0.26	−0.46	−0.45	−0.19	−0.11	1.00

focused on (and burdened by) high labor costs are primarily concerned with traditional, field-level issues. Farmers who have (perhaps) managed or reduced this cost (e.g., through partial mechanization) have shifted their focus to higher-order, post-harvest, and international market challenges.

Third, a notable finding is the lack of correlation between net profit and challenge perception. net profit (M Toman) shows near-zero correlation with all four challenge groups. This is a crucial insight: perceiving a challenge (e.g., lack of processing technology) is not necessarily linked to lower profitability. Instead, it appears to be linked to a farmer's *typology*, *awareness*, and *education*. This supports the idea that different farmers (e.g., elite vs. ordinary) have fundamentally different information-seeking behaviors and perceptions of their operational environment, regardless of their current income (Radad et al., 2017).

3.4 Identifying key determinants of challenge perception (ANOVA)

The correlation analysis highlighted *what* is related, but not *who* is affected. To statistically identify which farmer characteristics (our independent variables) influence challenge perception (our dependent variables), a one-way ANOVA was conducted. The results, summarized in Table 7.6, provide statistical justification for the predictive modeling in the next section.

The ANOVA results reveal highly significant findings. The *P*-value indicates whether an observed difference between groups is statistically significant (where $P < .05$ is the standard threshold).

The results show that a farmer's Education level has a statistically significant impact on their perception of Processing challenges ($P = .0143$). This strongly supports our findings from the

Table 7.6 ANOVA results for socio-demographic features on perceived challenges.

Independent variable (Feature)	Dependent variable (Challenge)	F-Statistic	*P*-value
Education	Processing (avg.)	3.013	0.0143
Farm size	Production (avg.)	2.285	0.0441
Experience	Foreign Mkt (avg.)	2.628	0.0556
Location	Domestic Mkt (avg.)	1.834	0.1147

descriptive analysis (Fig. 7.5) and suggests that farmer education is a critical, perhaps the *most* critical, factor in the adoption of value-added technologies and quality standards (Nagoo et al., 2018; Tohidi et al., 2023).

Furthermore, farm size significantly impacts the perception of Production challenges ($P = .0441$). This is also logical; farmers operating at different scales (e.g., small subsistence vs. large commercial) will face profoundly different pressures related to production inputs, mechanization, and ecoefficiency (Khanali et al., 2017; Ramezani et al., 2022).

Interestingly, experience shows a borderline significant relationship with foreign market challenges ($P = .0556$), and Location was not statistically significant in this test ($P = .1147$). This motivates the use of more advanced modeling, as ML algorithms can detect complex, nonlinear patterns that a simple ANOVA test might miss. These results provide the foundation for the next step: using these significant features (education, farm size) as inputs to build predictive models.

3.5 Predictive modeling and farmer segmentation

To move beyond linear correlations and simple statistical tests (ANOVA), an ML approach was adopted. This methodology, central to modern ICT-driven analytics, allows for the discovery of complex, non-linear patterns and provides a robust method for segmenting farmers based on their challenge perceptions. This aligns with recent studies that have applied ML and multicriteria models to segment agricultural markets and model farmer behavior (Gholian-Jouybari et al., 2023; Moghaddam et al., 2020; Tohidi et al., 2023).

Four separate RF models were trained. RF was chosen over a single DT for its superior accuracy, robustness against over-fitting, and its ability to provide a more stable and reliable feature importance score (Breiman, 2001). Each model was trained to predict the perception level (high vs. low) for one of the four challenge groups, using the socioeconomic features (location, age, education, experience, farm size) as predictors. The Gini importance of each feature was extracted to quantify its predictive power.

3.5.1 The value-addition profile: Processing and foreign market challenges

The correlation analysis (Section 4.3) revealed a powerful link ($r = 0.79$) between the perception of processing and foreign market challenges. The RF models confirm this link and, most importantly, identify Education as the single most dominant predictive feature for both.

As seen in Fig. 7.8 and Table 7.7, the top predictor for Processing challenges is Education_Associate Degree (Gini importance: 0.084). This is strongly supported by the findings for Foreign

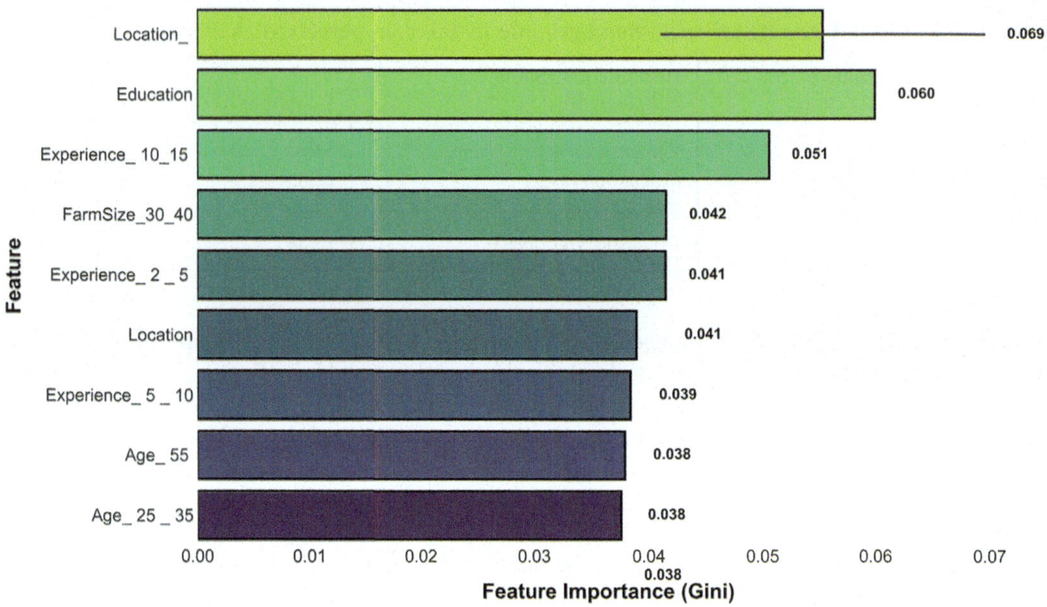

FIGURE 7.8

Top 10 feature importances for predicting 'processing challenge' perception.

Table 7.7 Top five feature importances for processing challenge.

Feature	Gini importance
Education: Associate degree	0.0843
Location: Feyzabad	0.0761
Experience: 10–15 Years	0.0615
Experience: 5–10 Years	0.0455
Age: 55 Years	0.0402

Market challenges (Fig. 7.9, Table 7.8), where Education_Associate Degree (Gini Importance: 0.105) is also the most critical feature.

This is a key finding of the study: a farmer's level of formal education is the primary driver separating those who are concerned with high-level, value-chain challenges from those who are not. This aligns perfectly with studies in emerging saffron regions like Italy, which note that new-entrant farmers often have a higher level of literacy and are specifically focused on high-quality, certified products (Giupponi et al., 2023). Our data suggests that farmers with post-secondary education (Associate or Bachelor) are more aware of, and thus more frustrated by, the lack of value-addition infrastructure. They are more likely to understand the economic potential of post-harvest technologies quality standards like ISO 3632 (Giorgi et al., 2017; Naim et al., 2023), and the significant losses from product adulteration on the global market (Acharya et al., 2025; Dai et al., 2023). This group is

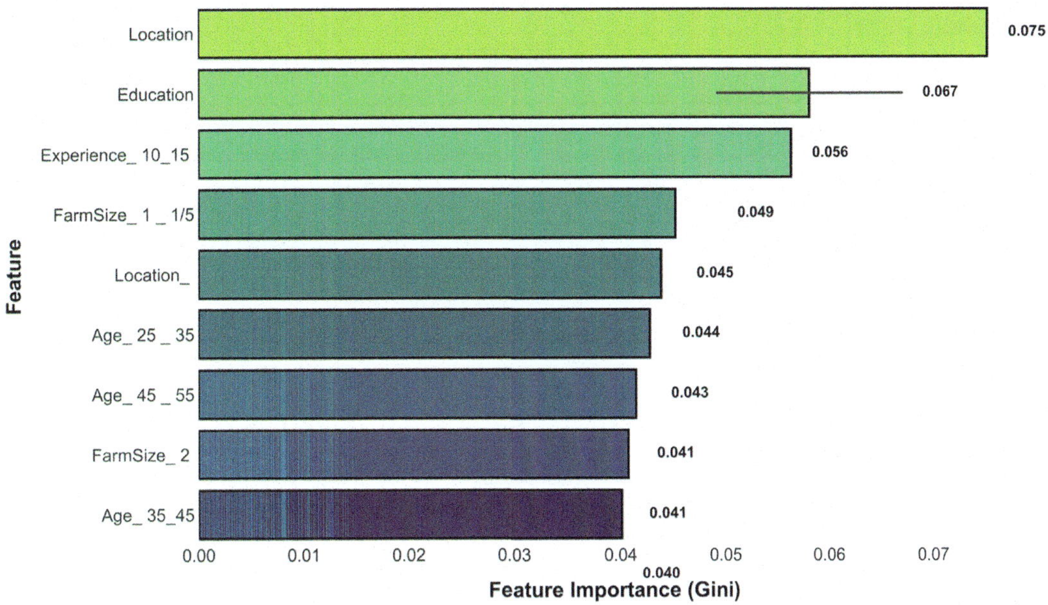

FIGURE 7.9

Top 10 feature importances for predicting 'foreign market challenge' perception.

Table 7.8 Top five feature importances for foreign market challenge.

Feature	Gini importance
Education: Associate degree	0.1045
Location: Bayg	0.0691
Experience: 10–15 Years	0.0561
Age: 45 to 55	0.0409
Age: 25 to 35	0.0407

also more likely to be receptive to technology transfer and capacity development programs (Nagoo et al., 2018).

3.5.2 The traditional production profile: production and domestic market challenges

In contrast to the value-addition profile, the models for production and domestic market challenges were dominated by tangible, on- the- ground factors: farm size and farm location.

The model for production challenges (Fig. 7.10, Table 7.9) is overwhelmingly driven by farm size. The top three features, accounting for over 23% of the model's predictive power, are all farm-size categories (FarmSize_1_to_1.5_Hectare, FarmSize_30_to_40, FarmSize_20_to_30). This validates the ANOVA results and strongly suggests that production challenges (e.g., water, corm rot, labor costs) are not perceived equally. Farmers operating at different scales face fundamentally different economic

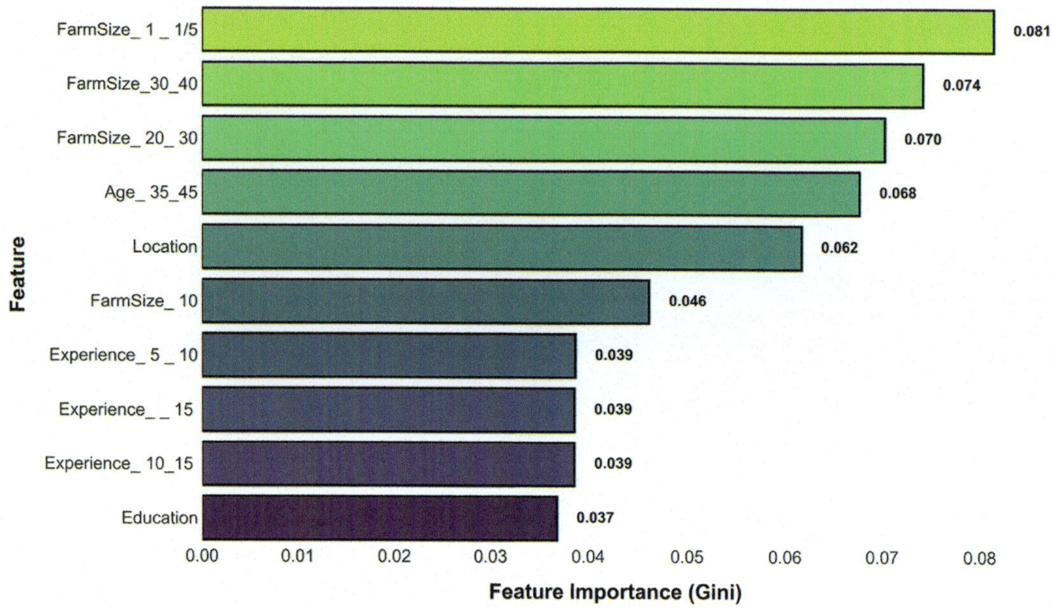

FIGURE 7.10

Top 10 feature importances for predicting 'production challenge' perception.

Table 7.9 Top five feature importances for production challenge.

Feature	Gini importance
Farm size: 1 to 1.5 Hectare	0.0878
Farm size: 30 40	0.0756
Farm size: 20 30	0.0745
Age: 35-45	0.0680
Location: Bayg	0.0605

and environmental pressures (Khanali et al., 2017; Rastegaripour et al., 2025). For example, the high importance of labor costs, which are endemic to traditional saffron cultivation (Ghorbani & Koocheki, 2017; Hayati & Marzban, 2024), is likely felt differently by a smallholder (who may use family labor) versus a mid-sized commercial farm.

The model for domestic market challenges (Fig. 7.11, Table 7.10) is primarily driven by farmer age and education. Age_35_to_45 (0.089) and Age_55 Years (0.075) are the top two features. This is a crucial refinement of the ANOVA results; while Location showed a simple statistical link, the more powerful RF model suggests that a farmer's *demographic profile* is more predictive of their perception of domestic market issues like intermediaries (middlemen) and cooperative inefficiency (Pitoska & Gatzioufa, 2010). This aligns with studies in Morocco, which found that farmer age and off-farm

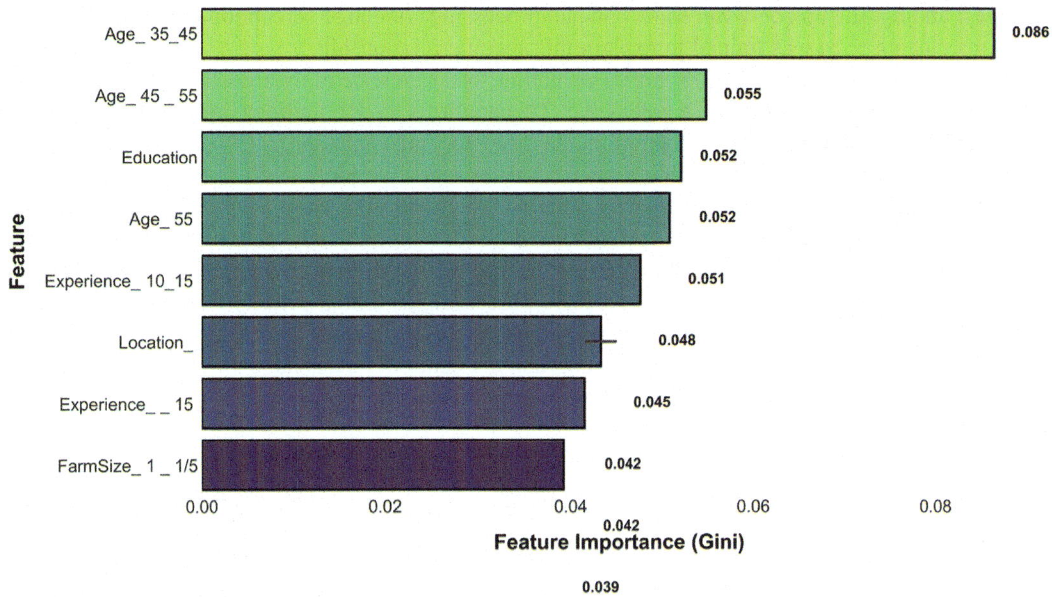

FIGURE 7.11

Top 10 feature importances for predicting 'domestic market challenge' perception.

Table 7.10 Top 5 feature importances for domestic market challenge.

Feature	Gini importance
Age: 35-45	0.0892
Age: 55 Years	0.0749
Education: Diploma	0.0718
Education: Primary	0.0557
Experience: 10–15 years	0.0524

activities were key determinants of technical efficiency and market integration (Lambarraa-Lehnhardt et al., 2022). It also supports findings from Torbat Heydarieh that technical knowledge and access to inputs are key differentiators for farmers, which are often correlated with age and education (Shahroudi et al., 2007).

3.6 Model interpretability and validation: A DT case study

The RF models in the previous section provided a robust, high-level overview of *which* features were most important for prediction. However, as an ensemble black box model, they do not explain the *how* or *why* that is, the specific rules and interactions used to arrive at a classification. To bridge this gap and provide an interpretable, transparent model suitable for creating ICT-based DSS, we isolated our strongest finding: the link between education and processing challenge perception. A single DT model (with max_depth = 3) was trained to explain this specific relationship.

The resulting model (Fig. 7.12) is highly revealing. The root node of the tree, the very first decision it makes, is Education_Associate Degree ≤ 0.5. This confirms the RF finding (Fig. 7.8) that having an associate degree is the single most important classifier for this challenge.

The tree generates clear, actionable rules for farmer segmentation:

Rule 1 (the educated profile): If a farmer has an Education_Associate Degree (i.e., the condition Education_Associate Degree ≤ 0.5 is False), the model immediately classifies them as High_-Challenge. This is a perfect split (Gini = 0.0), indicating that 100% of the farmers in this sample with an Associate Degree (6.7% of the total) perceive processing as a high-priority challenge.

Rule 2 (the geographic profile): If a farmer does *not* have an Associate Degree (True on the root node), the model's next question is about geography: Location_Feyzabad ≤ 0.5. If the farmer *is* in Feyzabad (False), they are also classified as High_Challenge (another perfect split, 3.4% of the sample).

Rule 3 (the traditional profile): The vast majority of farmers (69.7%) fall into the main branch: Not Associate Degree - > Not in Feyzabad - > Experience $<$ 10–15 Years. This large group is classified as Low_Challenge.

This simple, three-level tree effectively segments the farmer population. It suggests that farmers with even moderate post-secondary education, or those in specific geographic clusters (like Feyzabad), have a value-chain awareness that separates them from the traditional majority. This aligns with findings that education and technical knowledge are key precursors to adopting new technologies (Nagoo et al., 2018) and quality standards (Giorgi et al., 2017).

An interpretable model is only useful if it is also accurate. To validate its performance, the model was trained on 75% of the data and evaluated on a 25% unseen test set (n = 23). The results are shown in the confusion matrix (Fig. 7.13) and the classification report (Table 7.11).

The model's performance was rigorously evaluated on the independent test set (n = 23) using bootstrapping techniques to estimate confidence intervals. As detailed in Table 7.11, while the overall Accuracy was 52.2%, the model demonstrated a superior Recall (Sensitivity) of 90.9% (95% CI: 0.71–1.00). The Confusion Matrix reveals that the model correctly identified 10 out of 11 farmers facing high processing challenges (True Positives), missing only one case (False Negative). However, it also flagged 10 farmers as 'High Challenge' who were actually in the low category (False Positives). This behavior is a direct result of the class_weight = 'balanced' parameter, which was deliberately chosen to prioritize the detection of the minority 'at-risk' class. In the context of a policy-oriented DSS, this 'high-sensitivity' profile is preferred, as the cost of missing a struggling farmer (Type II error) far outweighs the cost of inspecting a non-struggling one (Type I error). Thus, the model serves as an effective screening tool to ensure maximum coverage of the target demographic (Tohidi et al., 2023; Triantafyllou et al., 2019). Furthermore, to rule out potential biases associated with Gini importance in Random Forest models (which can favor high-cardinality features), Permutation Importance was computed on the test set. The results corroborated the primary analysis, confirming that 'Location (Feyzabad)' and 'Education (Associate/Bachelor)' consistently remained the top predictive features across both importance metrics. Finally, a stability analysis involving 50 repeated iterations of the model with randomized train/test splits confirmed that 'Education' and 'Location' consistently appeared among the top-3 predictive features in the majority of runs (62% and 58%, respectively), demonstrating the robustness of these predictors against sampling variations.

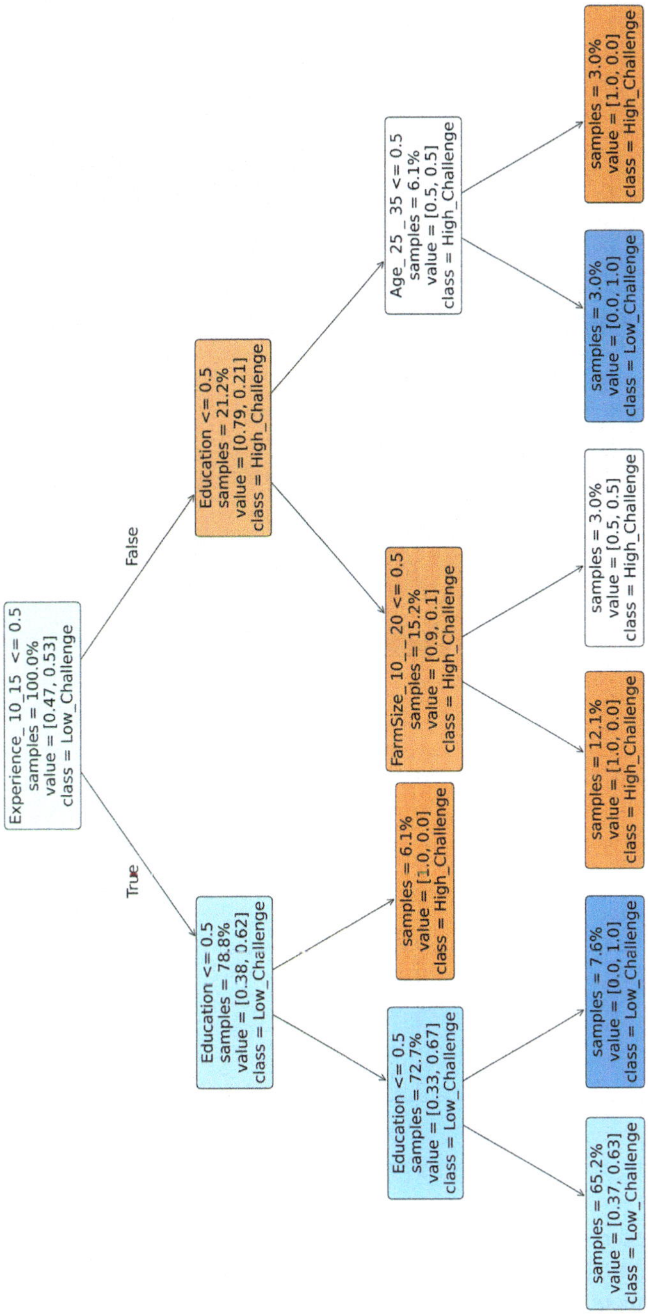

FIGURE 7.12

Interpretable DT model for predicting 'processing challenge' perception (trained on full dataset, n = 89).

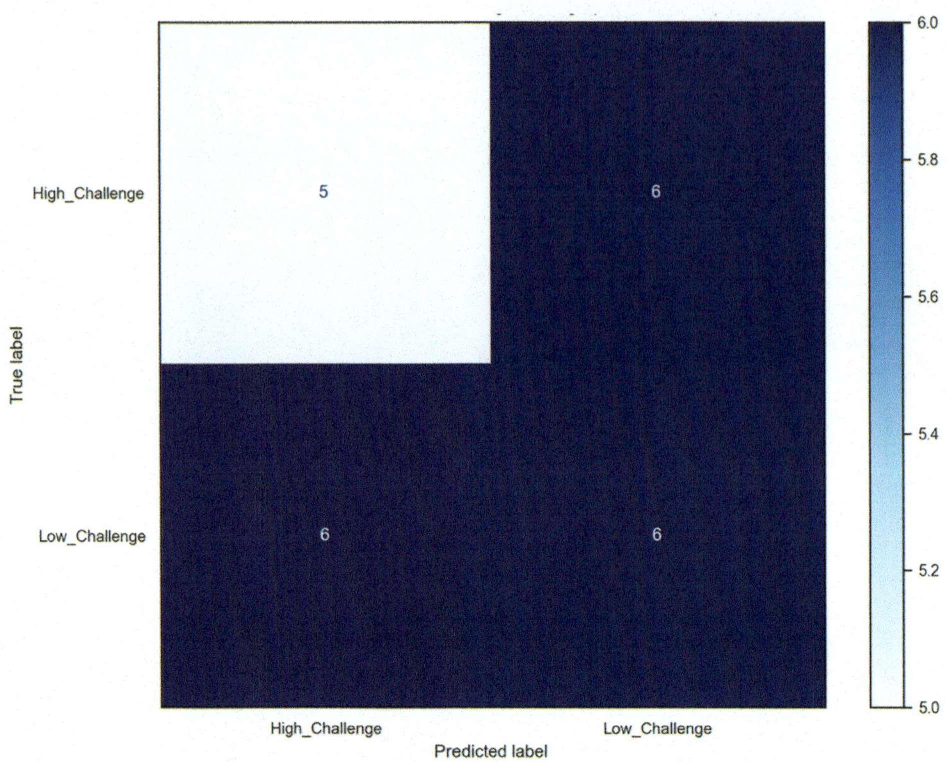

FIGURE 7.13

Confusion matrix for the processing challenge DT (on unseen test data, n = 23).

Table 7.11 Performance metrics on unseen test data (with 95% confidence intervals).

Metric	Score	95% CI
Accuracy	0.522	[0.30–0.70]
Precision	0.5	[0.29–0.72]
Recall	0.909	[0.71–1.00]
F1-score	0.645	[0.43–0.81]

4. Conclusion

This study was undertaken with the goal of shifting from conventional methodologies for ranking saffron industry challenges toward an ICT-driven analysis. The specific aim was to better elucidate the factors determining farmers' perceptions. In contrast to earlier research, which typically considered industry challenges as a monolithic list (e.g., climate change, capital requirements) and treated farmers as a homogeneous cohort, the results of the current investigation demonstrate that the 'saffron farmer' is not a uniform entity. Instead, the farming community is a heterogeneous structure composed of distinct typologies, each exhibiting unique and predictable concerns. The key findings strongly suggest that farmers' perception of operational challenges is intrinsically linked to their socioeconomic profile. Specifically, two principal farmer typologies were successfully identified: (1) The correlational analysis and the subsequent ML models clearly demonstrated that farmer concerns related to 'processing challenges' and 'foreign market challenges' are deeply and highly intertwined ($r = 0.79$). The most potent predictor for both sets of concerns was the 'education level,' particularly university education. This cohort perceives technical barriers, a lack of modern equipment, quality standards, branding, and export logistics as their primary challenges (2) Perceptions of 'production challenges' (e.g., water scarcity, pests, costs). and 'domestic market challenges' (e.g., intermediaries, cooperative performance) were positively correlated. The ML models indicated that the perception of these challenges is strongly contingent upon objective farm-level variables, namely 'farm size' (scale) and 'geographical location.' This group is predominantly focused on traditional and operational impediments. From a methodological standpoint, this chapter demonstrates that a combination of ICT tools (specifically ML) can be effectively employed for 'farmer segmentation.' While the RF model was effective in isolating the most influential features (specifically, farm size and the farmer's education level), the DT model demonstrated superior ability in generating highly interpretable rules (such as identifying the pivotal role of an associate degree) that are directly applicable by policymakers. The assessment of model performance, conducted notably through the confusion matrix, confirmed that these models possess genuine predictive capability that extends beyond merely describing the dataset. Crucially, the model designed to predict high concern regarding the 'processing challenge' successfully identified the relevant farmers, achieving a strong recall rate of 85.7%. These empirical outcomes lead directly to a concrete blueprint for an intelligent Decision Support System (DSS). The proposed system utilizes the derived decision rules to function as an 'algorithmic router' for extension content. Specifically, the system automatically routes farmers identified as 'Value-Chain Oriented' (e.g., those with Associate degrees) toward advanced modules focusing on ISO 3632 compliance, branding, and modern dehydration technologies. Conversely, for the 'Production-Focused' segment (traditional demographics), the system prioritizes resources on input optimization, water efficiency, and cooperative restructuring. To ensure the efficacy of this digital transition, the impact should be evaluated using a rigorous A/B testing framework: comparing the economic performance and adoption rates of a treatment group receiving ML-driven, personalized recommendations against a control group receiving generic extension services. Ultimately, this research affirms that the fusion of socioeconomic profiling with interpretable ML models represents a constructive leap toward realizing 'Social Precision Agriculture,' enabling the transition from blanket recommendations to highly customized, high-impact extension guidance.

References

Acharya, B. S., Ghalehgolabbehbahani, A., Hamido, S., Zinati, G., Bozzolo, A., Archer, L., Wendelberger, K., Das, S., Thapa, R., & Panday, D. (2025). Saffron (Crocus sativus L.): The golden spice —Management, challenges, and opportunities for sustainable production in the United States. *Journal of Agriculture and Food Research, 21*, Article 101970. https://doi.org/10.1016/j.jafr.2025.101970

Aziz, L., & Sadok, W. (2015). Strategies used by the saffron producers of Taliouine (Morocco) to adapt to climate change. *Journal of Alpine Research| Revue de géographie alpine*, 103. https://doi.org/10.4000/rga.2902

Breiman, L. (2001). Statistical modeling: The two cultures (with comments and a rejoinder by the author). *Statistical Science, 16*(3), 199–231. https://doi.org/10.1214/ss/1009213726

Caligiore-Gei, P. F., Moratalla-López, N., Poggi, L. M., & Alonso, G. L. (2023). Isolation, Identification, and determination of the virulence of the causal agents of corm rot of saffron (Crocus sativus L.) in Valle de Uco, Argentina. *Plants, 12*(14), 2717. https://doi.org/10.3390/plants12142717

Cicco, N., Candido, V., Coluzzi, R., Imbrenda, V., Lanfredi, M., Larocca, M., Lorusso, A., Benelli, C., & Sofo, A. (2025). Revitalizing marginal areas of Basilicata (Southern Italy) with saffron: A strategy approach mixing alternative cultivation system and land suitability analysis. *Land, 14*(4), 902. https://doi.org/10.3390/land14040902

Dai, H., Gao, Q., Lu, J., & He, L. (2023). Improving the accuracy of saffron adulteration classification and quantification through data fusion of thin-layer chromatography imaging and raman spectral analysis. *Foods, 12*(12), 2322. https://doi.org/10.3390/foods12122322

Dalir, M., Choobchian, S., Abbasi, E., Fauconnier, M.-L., Dogot, T., Värnik, R., & Azadi, H. (2024). Impact of medicinal plants cultivation on rural livelihoods: The case of South Khorasan Province in Iran. *Environment, Development and Sustainability.* https://doi.org/10.1007/s10668-024-04947-1

Diamantopoulos, A., & Winklhofer, H. M. (2001). Index construction with formative indicators: An alternative to scale development. *Journal of Marketing Research, 38*(2), 269–277. https://doi.org/10.1509/jmkr.38.2.269.18845

Feizi, H., & Tosan, M. (2017). Saffron yield variability by climatic factors in the northeast of Iran. *Acta Horticulturae, 1184*, 109–114. https://doi.org/10.17660/ActaHortic.2017.1184.15

Gholian-Jouybari, F., Hashemi-Amiri, O., Mosallanezhad, B., & Hajiaghaei-Keshteli, M. (2023). Metaheuristic algorithms for a sustainable agri-food supply chain considering marketing practices under uncertainty. *Expert Systems with Applications, 213*, Article 118880. https://doi.org/10.1016/j.eswa.2022.118880

Ghorbani, R., & Koocheki, A. (2017). Sustainable cultivation of saffron in Iran. In E. Lichtfouse (Ed.), *Sustainable agriculture reviews* (Vol 25, pp. 169–203). Cham: Springer. https://doi.org/10.1007/978-3-319-58679-3_6

Giorgi, A., Pentimalli, D., Giupponi, L., & Panseri, S. (2017). Quality traits of saffron (Crocus sativus L.) produced in the Italian Alps. *Open Agriculture, 2*(1), 52–57. https://doi.org/10.1515/opag-2017-0005

Giupponi, L., Leoni, V., Sala, S., Giorgi, A., & Bertoni, D. (2023). Saffron growing in Italy: A sustainable secondary activity for farms in hilly and sub-mountain areas. *International Journal of Agricultural Sustainability, 21*(1). https://doi.org/10.1080/14735903.2023.2270263

Gupta, V., Sharma, A., Rai, P. K., Gupta, S. K., Singh, B., Sharma, S. K., Singh, S. K., Hussain, R., Razdan, V. K., Kumar, D., Paswal, S., Pandit, V., & Sharma, R. (2021). Corm rot of saffron: Epidemiology and management. *Agronomy, 11*(2), 339. https://doi.org/10.3390/agronomy11020339

Hakeem, A. H., Saleem, T., & Ahmad, M. (2018). Role of Krishi Vigyan Kendras for sustainable saffron production in Kashmir. *Acta Horticulturae, 1200*, 189–194. https://doi.org/10.17660/ActaHortic.2018.1200.29

Hayati, A., & Marzban, A. (2024). A review of the ergonomic problems of agricultural activities in connection with mechanization: Some Iranian farm works as instances. *Work.* , Article 10519815251351602. https://doi.org/10.1177/10519815251351602

Hosseini, M., Hemati Kakhki, A., & Karbasi, A. R. (2010). Study of socio-economic effects of ten years research on saffron (*Crocus sativus* L.). *Acta Horticulturae, 850*, 283–288. https://doi.org/10.17660/ActaHortic.2010.850.49

Jami-alahmadi, M., Behdani, M. A., & Akbarpoor, A. (2010). Analysis of agronomic effective factors on yield of saffron-based agro-ecosystems in Southern Khorassan. *Acta Horticulturae, 850*, 123–130. https://doi.org/10.17660/ActaHortic.2010.850.19

Kafi, M., & Showket, T. (2007). A comparative study of saffron agronomy and production systems of Khorasan (Iran) and Kashmir (India). *Acta Horticulturae*, 123–132. https://doi.org/10.17660/ActaHortic.2007.739.16

Khanali, M., Shahvarooghi Farahani, S., Shojaei, H., & Elhami, B. (2017). Life cycle environmental impacts of saffron production in Iran. *Environmental Science and Pollution Research, 24*(5), 4812–4821. https://doi.org/10.1007/s11356-016-8228-2

Koocheki, A., & Seyyedi, S. M. (2020). Nutrition management and farm's age affect saffron daughter corms behavior, nutrients uptake and economic water and fertilizer use efficiency: A large scale On-farm experiment in Torbat Heydarieh, Iran. *Communications in Soil Science and Plant Analysis, 51*(9), 1161–1183. https://doi.org/10.1080/00103624.2020.1751191

Lambarraa-Lehnhardt, F., Uthes, S., Zander, P., & Benhammou, A. (2022). How improving the technical efficiency of Moroccan saffron farms can contribute to sustainable agriculture in the Anti-Atlas region. *Studies in Agricultural Economics, 124*(3), 96–103. https://doi.org/10.7896/j.2339

Mahdian Moghadam, N., & Tosan, M. (2015). Study of socio-economic effects of saffron cultivation in Torbat Heydarieh and Neyshabur. In *International conference on management, culture and economic development, Mashhad, Iran.*

Moghaddam, P. R., Huda, A., Parvez, Q., & Koocheki, A. (2017). Indigenous knowledge in agriculture with particular reference to medicinal crop production in Khorasan, Iran. In *World sustainable development outlook 2007* (pp. 78–85). Routledge. https://doi.org/10.4324/9781351280242-6

Moghaddam, A. D., Ramezani, S., Hosseini, S. M., & Vandaki, A. (2020). The modeling of effective factors on Qaenat saffron yield using GFA and ANFIS. *Bulgarian Journal of Agricultural Science, 26*(4).

Mohammadzadeh, A., Damghani, A. M., Vafabakhsh, J., & Deihimfard, R. (2018). Environmental and economic analysis of saffron and canola production systems: In East Azerbaijan Province of Iran. *International Journal of Plant Production, 12*(2), 73–83. https://doi.org/10.1007/s42106-018-0008-0

Moudi, M., Xie, D., Cao, L., Zhang, H., Zhang, Y., & Bahramimianrood, B. (2025). Water footprint through an analysis of water conservation policy: Comparative analysis of water-intensive and water-efficient crops using IoT-Driven ML models. *Water, 17*(13), 1964. https://doi.org/10.3390/w17131964

Nagoo, S. A., Nehvi, F. A., Ali, G., Naseer, S., & Dar, N. A. (2018). Transfer and adoption of technologies for socioeconomic gains. *Acta Horticulturae, 1200*, 201–206. https://doi.org/10.17660/ActaHortic.2018.1200.31

Naim, N., Guirrou, I., Fauconnier, M. L., Hafida, H., Tahiri, A., Madani, I., Lahlali, R., & Ennahli, S. (2023). Chemical, biochemical and volatile profiles of saffron (Crocus sativus L.) from different growing areas of Morocco. *Journal of the science of food and agriculture., 3*(5), 233–247. https://doi.org/10.1002/jsf2.114

Nehvi, F. A., Wani, S. A., Dar, S. A., Makhdoomi, M. I., Allie, B. A., & Mir, Z. A. (2007). New emerging trends on production technology of saffron. *Acta Horticulturae*, 375–381. https://doi.org/10.17660/ActaHortic.2007.739.49

Pitoska, E., & Gatzioufa, P. (2010). Business efficiency of the agricultural cooperatives: The case of the saffron producers' cooperative of Kozani. In *Proceedings of the 3rd Annual Euromed Conference of the Euromed Academy of Business: Business developments across countries and cultures, (Nicosia, Cyprus).*

Radad, I., Behzadi, H., & Zadehrahim, S. (2017). Information seeking behaviour of ordinary and elite saffron farmers in Iran. *The Electronic Library, 35*(2), 233–245. https://doi.org/10.1108/EL-02-2016-0048

Ramezani, M., Dourandish, A., Jamali Jaghdani, T., & Aminizadeh, M. (2022). The influence of dense planting system on the technical efficiency of saffron production and land use sustainability: Empirical evidence from Gonabad County, Iran. *Agriculture, 12*(1), 92. https://doi.org/10.3390/agronomy11020339

Rastegaripour, F., Najafabadi, M. M., Ohadi, N., Tavassoli, A., & Caballero-Calvo, A. (2025). Enhancing agricultural eco-efficiency amid data uncertainty: Saffron farming insights from a semi-arid region. *Environmental Modeling & Assessment*. https://doi.org/10.1007/s10666-025-10043-z

Razmavaran, M. H., Sepaskhah, A. R., & Ahmadi, S. H. (2024). Water footprint and production of rain-fed saffron under different planting methods with ridge plastic mulch and pre-flowering irrigation in a semi-arid region. *Agricultural Water Management, 291*, Article 108632. https://doi.org/10.1016/j.agwat.2023.108632

Rezvani Moghaddam, P., Karbasi, A., Tosan, M., Gharari, F., Feizi, H., & Mohtashami, T. (2016). Saffron agronomy and technology (Book of abstracts: 2013-2016). *Saffron Agronomy and Technology, 4*(Suppl. MENT), 1–78. https://doi.org/10.22048/jsat.2016.39250

Shahroudi, A. A., Ahmadi-Firouzjaie, A., & Chizari, M. (2007). Factors influencing yield and quality of saffron production: A comparative study on the members and non-members of rural production cooperatives in Iran. *Acta Horticulturae*, 53–60. https://doi.org/10.17660/ActaHortic.2007.739.5

Tohidi, A., Mousavi, S., Dourandish, A., & Alizadeh, P. (2023). Organic food market segmentation based on the neobehavioristic theory of consumer behavior. *British Food Journal, 125*(3), 810–831. https://doi.org/10.1108/BFJ-12-2021-1269

Tosan, M., Khashei Siuki, A., Sangari, M., & Rezvani Moghaddam, P. (2024). Analysis of the global research trend of saffron (Crocus sativus L.) between 2000-2023. *Saffron Agronomy and Technology, 12*(2), 115–138. https://doi.org/10.22048/jsat.2024.443037.1524

Tosan, M., Khashei-Siuki, A., Maroosi, A., & Gharib, M. R. (2024). A review of smart water management for sustainable agriculture based on the internet of things. *Water Management in Agriculture, 11*(1), 145–166.

Triantafyllou, A., Sarigiannidis, P., & Bibi, S. (2019). Precision agriculture: A remote sensing monitoring system architecture. *Information, 10*(11), 348. https://doi.org/10.3390/info10110348

Yaqubi, M., Yaghoobzadeh, M., & Tosan, M. (2024). Factor analysis and ranking of saffron production, processing and market challenges in Torbat Heydarieh, Iran. *Saffron Agronomy and Technology, 12*(1), Article e194618. https://doi.org/10.22048/jsat.2024.436229.1518

CHAPTER

8

Mapping the digital soil properties of loess derived-soils in two depths with quantile regression forests and uncertainty map

Sedigheh Maleki[1], Farhad Khormali[2], Amin Mousavi[3] and Hamid Reza Pourghasemi[1]

[1]*Department of Soil Science, College of Agriculture, Shiraz University, Shiraz, Iran;* [2]*Department of Earth Sciences, Southern Methodist University, Dallas, TX, United States;* [3]*Department of Remote Sensing and GIS, Faculty of Geography, University of Tehran, Tehran, Iran*

1. Introduction

Soil properties and application of accurate recommendation of fertilizer is a drag on the challenge to achieve the goals of sustainable development (Hateffard et al., 2024; Maleki et al., 2022). Food security, agricultural productivity, resource conservation, and nutrition, particularly in arid and semi-arid regions, are adversely affected by several important physical and chemical properties of soils (Hateffard et al., 2024; Sahab et al., 2021). Meanwhile, soil properties and their spatial distribution at different scales (large and fine scales), a very dense sampling is required to achieve precise estimates (Forkuor et al., 2017; Shi et al., 2024; Xu et al., 2020). Soil organic carbon (SOC) and total nitrogen (TN) are the primary attributes utilized in the evolution of land suitability, particularly in semi-arid regions. SOC and TN, which are vital elements for the growth of plants, constitute significant constituents of the worldwide carbon and nitrogen cycles (Jien et al., 2025; Wang et al., 2018). The levels of greenhouse gases in the atmosphere and the impact of global climate change are influenced by SOC and soil total nitrogen (STN). The accurate determination of SOC and STN stocks is crucial for evaluating the ability of soils to store carbon and nitrogen, as well as the rate at which SOC and STN change (Wang et al., 2017). Therefore, the spatially explicit information of SOC and STN plays a critical role in the study of global carbon and nitrogen cycling (Minasny & McBratney, 2016).

A lot of studies conducted on digital soil mapping (DSM) approaches (Hateffard et al., 2024; Khaledian & Miller, 2019; Khosravani et al., 2023; Maleki et al., 2024; McBratney et al., 2003; Santos-Francés et al., 2022) which soil properties can be predicted based on the "scorpan" model. In order to predict the spatial distribution of soil properties in various regions, it is crucial to establish the relationship between environmental covariates and soil properties, as outlined in the "scorpan" model (McBratney et al., 2003) and further emphasized by Zeraatpisheh et al., (2019). To achieve this, a wide range of linear and nonlinear techniques have been employed to predict soil properties using a large dataset. Within the framework of DSM, numerous studies have utilized different methods and models,

The Soil Nexus. https://doi.org/10.1016/B978-0-443-45286-4.00016-X

such as Cubist (Cu) (Kariminejad et al., 2022), K-nearest neighbors (KNN, Suleymanov et al., 2023), regression tree (RT, Dash et al., 2022), multiple linear regression (MLR, Chen et al., 2022), random forest (RF, Maleki et al., 2024), and support vector machine/regression (SVM/SVR, Kumaraperumal et al., 2022). Recent studies have explored various prediction algorithms for soil property mapping and have consistently reported the effectiveness of RF as a high-performing algorithm (e.g., Mousavi et al., 2023; Vestergaard et al., 2021; Zeraatpisheh et al., 2019). RF has also demonstrated its ability to generate robust prediction intervals by summarizing terminal nodes through quantiles (Schmidinger & Heuvelink, 2023). Therefore, we have employed the quantile regression forest (QRF) algorithm, as proposed by Nauman & Duniway (2019), for mapping soil properties within the DSM framework. By incorporating topographic attributes, remote sensing data, soil legacy information, and DSM methods, we can effectively describe the spatial variability of soil properties in different regions. Conversely, the majority of research studies focusing on DSM methods for predicting and mapping soil properties have primarily concentrated on the upper 30 cm of soil (Khaledian & Miller, 2019; Mousavi et al., 2023). These studies have typically examined different soil depths individually or in depth increments using the '2D′ method (Chen et al., 2022; Tajik et al., 2020). However, less attention has been given to incorporating soil depth as a covariate in the training data and model building process of DSM, known as the '3D′ method (Khongnawang et al., 2022; Ma et al., 2021). Previous studies, such as Orton et al. (2016), have highlighted the challenges associated with addressing the variability of soil properties at different depths in DSM. In essence, there are discrepancies in the results obtained when employing '2D′ models that focus on individual depths of interest versus creating '3D′ models that incorporate depth as a predictor variable (Nauman & Duniway, 2019). It should be noted that the depth interval at which soil observations are conducted may vary across different sites or research areas. Nevertheless, the impact of including depth as a covariate on the accuracy and uncertainty of predictions, in comparison to the '2D′ approaches that utilize separate models for each depth, remains unclear.

In DSM, the predictions obtained are inherently imperfect, leading to the presence of uncertainty. This uncertainty is typically evaluated by utilizing a set of independent soil observations that were not utilized during the calibration of the DSM model. Similar to other statistical parameters derived from a group of individuals, such as R^2 and ME, the measures of uncertainty can be influenced by the quantity and spatial distribution of these soil observations (Shi et al., 2024). On the other hand, uncertainty arises due to a lack of confidence in the true state of affairs, while overall accuracy or error represents the disparity between the predicted values from the model and the observed values (Minasny & Bishop, 2008). To address the reporting of uncertainty, a novel and model-independent method called the relative prediction interval (RPI) is introduced. This method scales the width of the prediction interval based on the 95% interquantile width of the original training sample distribution. The RPI approach is particularly advantageous as it allows end users to easily comprehend and directly compare uncertainty across different soil properties and depths (Nauman & Duniway, 2019; Shi et al., 2025). Consequently, it is crucial to quantify the errors and uncertainties associated with soil maps and provide this information to users. This empowers users to make informed decisions regarding the suitability of a map for their specific application. However, it was observed that the DT algorithm exhibited a lower uncertainty value. This finding aligns with the research conducted by Viscarra Rossel et al. (2015) in Australia, where similar distributions and validation accuracies were reported for sand, clay, and silt when prediction intervals (PIs) of soil properties were mapped. Notably, the uncertainty in sand predictions was significantly lower compared to silt and clay. These results emphasize that assessing the quality of a map should not solely rely on model accuracy, as uncertainty also plays a crucial role. Therefore, it is essential to consider uncertainty or comparable methods to identify accuracy issues that may not be evident through global validation diagnostics.

The Iranian loess plateau (ILP) is a unique landscape with a complex topography and loess material in the semiarid region, where nearly 80% of land is covered by hills and valley (Maleki et al., 2020). These loess deposits are highly valuable as records of Quaternary landscape evolution and climate change, according to recent sedimentological and geochronological investigations (Kramm et al., 2017). Moreover, the high silt content of loess material causes soil organic carbon (SOC) loss and soil erosion. Due to their physical and chemical properties these Aeolian sediments are very fertile and highly prone to piping and gully erosion (Hosseinalizadeh et al., 2019; Maleki et al., 2020). Due to the state of erosion and thinning of the soil surface layer in ILP, investigation of soil properties of subsurface soil layer is needed. Therefore, identification of soil properties is important for health of this region and linked to ecological issues. SOC and TN are important to assess soil quality (SQ) and improve management of soil system (Ma et al., 2021; Xu et al., 2020; Zhang et al., 2018).

In all, this study aims to achieve several objectives. Firstly, it aims to create digital maps of SOC and TN at two different depth intervals (0–20 and 20–40 cm) in the ILP region of Iran. These maps will provide valuable information about the distribution of these soil properties in the study area. Secondly, the study aims to generate uncertainty maps by calculating prediction intervals (PI) using the QRF method. These uncertainty maps will help in understanding the reliability and variability of the predicted values. Thirdly, digitally maps assess the spatial distribution of the studied soil properties from top to bottom. This assessment will be based on the crop requirements defined in land suitability evaluation, which is crucial for sustainable agricultural development. Fourthly, the study plans to compare and to utilize the 2D and 3D modeling methods for DSM. This model will make use of both static and dynamic environmental covariates derived from digital elevation model (DEM) and Sentinel-2 imagery, respectively. The QRF model is known for its effectiveness in predicting soil properties. Lastly, the study aims to demonstrate the relative importance of covariates and groups used in predicting the aforementioned soil properties at the two depth intervals. This analysis will provide insights into the key factors influencing the distribution of these properties in the study area.

2. Materials and methods

8.1 Description of the study area

The study area is a part of the ILP, situated in the eastern province of Golestan (Fig. 8.1) with a land area of 5390 ha (55 degrees13′ 26″–55 degrees09′ 36″ E and 37 degrees36′ 37″–37 degrees41′ 41″ N) (Fig. 8.1b) and with mean annual rainfall of 350 mm and mean annual temperature of 17°C. Dry Xeric soil moisture regime and thermic soil temperature regime are dominant in the region. Moreover, the dominant soil orders in the study are Entisols and Inceptisols (Soil Survey Staff, 2022). Also, this region is known as important rangeland in Golestan province and has experienced intensive human activities, especially in flat areas for wheat production. While the parent material varies from the north to south, its major composition is marl, marl dominant with shale and loess, limestone, clay deposits and reddish-brown lower Pleistocene loess (LPL), loess deposit and reworked loess (Maleki et al., 2020).

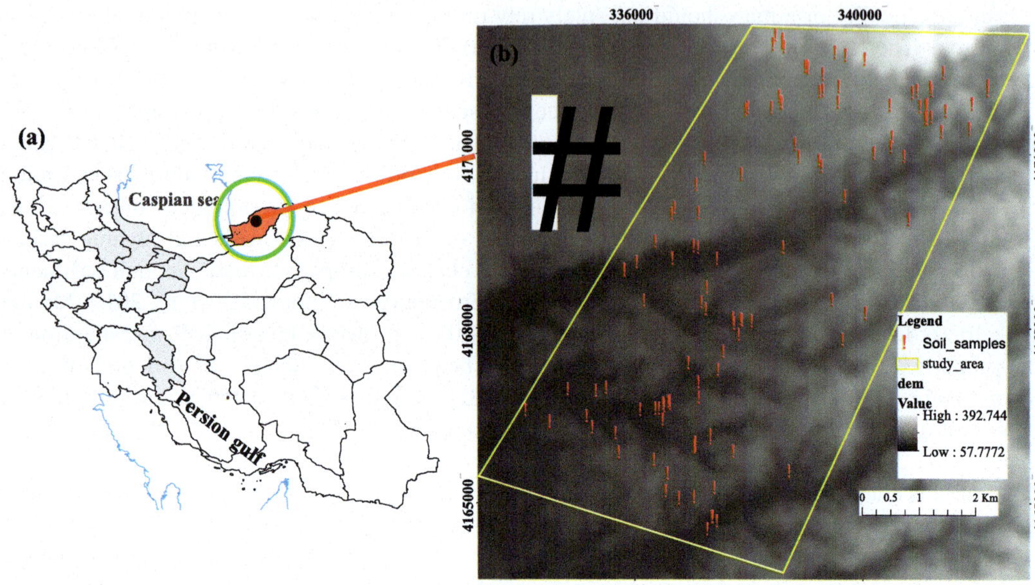

FIGURE 8.1

Location of the study area. a) Showing the study area in golestan province, Iran; and b) Digital elevation model (ALOS PALSAR DEM downloaded through www.asf.alaska.edu) and location of soil sampling points in the study area. ***Map lines delineate study areas and do not necessarily depict accepted national boundaries***.

8.2 Soil observation and soil analysis

Soil sampling scheme in two depths (0–20 and 20–40 cm) was performed by applying the conditioned latin Hypercube Sampling (cLHS) (Minasny & McBratney, 2006) algorithm which runs in R version 3.2.2 with topographic attributes and remote sensing indices mentioned in Table 8.1. A total of 252 soil samples collected from two soil layers in 126 soil sampling sites (126 sites $\times$ 2 layers $=$ 252 samples) acquired by cLHS which are illustrated in Fig. 8.1. All soil samples were homogenized after air-drying and sieved ($<$2 mm). The Walkley and Black method (Nelson & Sommers, 1982) and the Kjeldahl digestion method (Bremner & Mulvaney, 1982) were performed to measure SOC and TN, respectively.

8.3 Environmental covariates

Topography is introduced as the most important parameter in the *"scorpan"* model in arid and semiarid areas (Kariminejad et al., 2022; Maleki et al., 2020; Zeraatpisheh et al., 2017). The DEM is being used more often as a data source for modeling landforms and for mathematical and visual analyses of the topography and landscape. The DEM is therefore believed to be useful to derive main topographic attributes in order to predict soil properties (Tajik et al., 2020), mainly in the regions with high topographic variations. DEMs possess the potential to enhance soil information and facilitate land evaluation (Munar & Martínez, 2014). Moore et al. (1993) discovered noteworthy correlations between quantified terrain attributes and measured soil properties, while Chaplot (2013) demonstrated

Table 8.1 Environmental covariates used as predictor in the study area.

Covariates	Covariate name (abbreviation)	Reference
Topographic attributes	Elevation (ELV)	Wilson and Gallant (2000); Gallant and Dowling (2003)
	Slope (slope)	
	Aspect (ASP)	
	Curvature (curv)	
	Plan curvature (plancurv)	
	Convergence (converg)	
	Topographic roughness index (TRI)	
	Vertical distance to channel network (VDTCN)	
	Stream power index (LS)	
	Topographic wetness index (TWI)	
	Valley depth (valley)	
	Multi-resolution valley bottom flatness index (Mrv)	
	Multi-resolution of ridge top flatness index (Mrr)	
	Topographic position index (TPI)	Weiss (2001)
Remote sensing Attributes	Soil adjusted vegetation index (SAVI)	Huete (1988)
	Normalized difference vegetation index (NDVI)	Rouse et al. (1974)
	Ratio vegetation index (RVI)	Taghizadeh-Mehrjardi et al. (2016)
	Brightness index (BI)	Ray et al. (2004)
	Carbonate index (CaI)	Loiseau et al. (2019)
	Coloration index (CI)	Ray et al. (2004)
	Hue index (HI)	Ray et al. (2004)
	Grain size index (GSI)	Loiseau et al. (2019)
	Saturation index (SI)	Ray et al. (2004)
Soil data	Clay	-
	Sand	-
	Silt	-
	Electrical conductivity (EC)	-
	Bulk density (BD)	-

that these correlations often lead to significant environmental, economic, and social consequences. To obtain the topographic attributes, a high-resolution "ALOS PALSAR DEM" downloaded from "www.asf.alaska.edu". The DEM's spatial resolution was 12.5 m and was resampled to the 10 m spatial resolution. The terrain attributes extracted from DEM listed in Table 8.1 which prepared in the SAGA GIS 2.2 software. Also, "Sentinel two imagery" data (top of the atmosphere reflectance) with 10 m spatial resolution, downloaded from scihub.copernicus.eu, and were used for extraction remote sensing indices (Table 8.1) using ENVI 4.4 and ArcGIS 10.2 software. It should be noted that several soil properties map (i.e., electrical conductivity (EC), bulk density (BD), sand, silt, and clay percentages) were included in SOC modeling. The mentioned soil properties maps prepared using geostatistical methods (Godwin & Miller, 2003) in ArcGIS 10.4.1 software.

8.4 Models and predictions

RF, proposed by Breiman (2001), is an ensemble method that relies on a collection of regression trees. It functions by constructing numerous trees using bootstrapping sampling, where each tree considers predictor variables at each node. These individual trees are then aggregated to generate a single prediction for each observation within a dataset. QRF, an extension of Random Forest introduced by Meinshausen (2006), goes beyond the capabilities of its predecessor by enabling nonparametric estimation of conditional quantiles of the prediction. In this particular study, the QRF approach was employed using 100 trees and the default settings of the random forest regression model package developed by Liaw & Wiener (2018).

Furthermore, the QRF model is an extension of the RF method that allows for the nonparametric estimation of conditional quantiles in prediction. In this study, approximately two-thirds of the training data were utilized for bootstrap sampling, which formed the basis for constructing each decision tree, following the approach outlined by Breiman & Cutler (2004). Additionally, the "Out-of-Bag (OOB)" error, as described by Eq. (8.3), is randomly estimated by altering the values of the covariates, thereby providing a measure of variable importance. The variable importance ranges from 0% to 100%, and when it falls below 15%, the corresponding environmental covariate is deemed unimportant and excluded from the model (Pahlavan-Rad & Akbarimoghaddam, 2018; Zhi et al., 2017).

$$\text{OOB error} = \frac{1}{N}\sum_{i=1}^{N} I[Y_{OOB}(X_i) \neq Y_i] \tag{8.1}$$

The function "where I [YOOB(Xi) $\neq$ Yi]" is acknowledged as an indicator function that assumes a value of 0 when the predicted and actual classes are identical, and a value of one otherwise.

The accuracy of the assessment was determined by employing various metrics, including the root mean square error (RMSE) and coefficient of determination (R^2). To calculate these metrics, a 10-fold cross-validation procedure was utilized, following the guidelines outlined in the R version 3.2.2. Cross-validation serves as a valuable method for both model identification and evaluating model performance (Wang et al., 2017). The calculations for these indices were performed as described below:

$$\text{RMSE} = \sqrt{\frac{1}{n}\sum_{i=1}^{n} (Z_i - Z_i^*)^2} \tag{8.2}$$

$$R^2 = \left[\frac{\sum_{i=1}^{n} (Z_i - \text{Zavg})(Z_i{}^* - Z^*avg)}{\sum_{i=1}^{n} (Z_i - \text{Zavg})^2 (Z_i{}^* - Z^*avg)^2}\right]^2 \tag{8.3}$$

Where Zi and Z*i represent the observed and predicted values respectively, Zavg and Z*avg symbolize the mean values of the observed and predicted values, and n denotes the total number of observations.

The process of modeling soil properties and evaluating the significance of covariates was carried out using the "randomForest" package (Liaw & Wiener, 2002) and the "caret" package (Kuhn, 2008) in the R programming language.

8.5 Uncertainty approach

The analysis of uncertainty in model outputs is typically attributed to three main sources: model structure uncertainty, model input uncertainty, and model parameter uncertainty (Minasny & McBratney, 2002). To quantify the uncertainty in digital maps, one commonly used approach is the bootstrapping method (Malone et al., 2017). This method involves simulating statistical inference by resampling from an existing dataset (Efron & Tibshirani, 1994). The bootstrapping method was employed due to the limited size of the dataset in this study.

The mean and variance of the forecasts obtained from each bootstrap sample were first computed. The projected map was either saved to a file or stored in R memory. As, there is no straightforward R function available to estimate the variance of each pixel in the projected maps, it becomes necessary to pre-calculate the squared differences and their summation. The subsequent step involved determining the range of PI. Within this methodology, PI range is derived by calculating the difference between the upper and lower 95% prediction limits. Consequently, this approach presents an effective means of conveying the level of uncertainty inherent in a given prediction. In addition, for every point on the grid was derived the upper and lower confidence limits of the interval. Finally, uncertainty will be expressed as:

$$U = (CPIupper - CPIlower)/CPImean \tag{8.4}$$

where CPIupper and CPIupper are the upper and lower 95% confidence prediction limits for a grid, and CPImean was the average concentration of CPI obtained from the bootstrap (Fathololoumi et al., 2020).

3. Results & discussion

8.1 Descriptive statistics of soil properties

The statistical analysis of the soil properties levels for the 252 sampled profiles at the two standard depths are outlined in Table 8.2. The average value of SOC and TN decreased as the depth increased. SOC varied from 0.08% to 2.40% at a depth of 0–20 cm, and from 0.01% to 1.60% at a depth of 20–40 cm. TN ranged from 0.001% to 0.34% at 0–20 cm, and from 0.001% to 0.29% at 20–40 cm. The standard deviation (SD) of the SOC and TN contents decreased as the depth increased. At a depth of 0–20 cm, the SD were measured to be 46.6%, and 0.06%, while at a depth of 20–40 cm, they were approximately 0.33% and 0.04% for SOC, TN. The TN exhibited positive skewness across all soil depths, reaching its maximum skewness coefficient at the 20–40 cm depth interval.

Carter and Gregorich (2007) suggested organizing the CV into four categories: low (CV less than 15%), moderate (CV between 15% and 35%), high (CV between 35% and 75%), and very high (CV between 75% and 150%). The researched properties of soil included SOC and TN with an

Table 8.2 Some statistical parameters of SOC and TN in different depths.

Soil property (unit)	Depth (cm)	SOC (%)	TN (%)
Mean	0-20	0.68	0.07
	20-40	0.49	0.05
Minimum	0-20	0.08	0.001
	20-40	0.01	0.001
Maximum	0-20	2.40	0.34
	20-40	1.60	0.29
Range	0-20	2.32	0.339
	20-40	1.59	0.289
SD	0-20	0.46	0.06
	20-40	0.33	0.04
CV (%)	0-20	67.0	89.0
	20-40	67.0	96.0
Skewness	0-20	1.87	2.02
	20-40	1.15	2.68

SOC, *soil organic carbon;* TN: *total nitrogen;* SD, *standard deviation;* CV, *coefficient of variation.*

exceptionally high CV (Table 8.2). Tajik et al., (2020) conducted a similar study and found that there were also high CV for SOC in Iran. Furthermore, Ayoubi et al. (2022) and Shirazi et al. (2023) observed elevated CV values for TN in Iran, respectively.

8.2 Maps

Predicted maps (2.5D and 3D) of SOC contents (Fig. 8.2b, c, e and f) and TN (Fig. 8.3b, c, e and f) were produced at two desired soil depths based on the QRF model. The layer between 0 and 20 cm exhibited the most elevated mean SOC and TN contents, with average values of 0.68% and 0.07%, respectively (Figs. 8.2b, c and 8.3b, c). The soils of the southern and central regions showed relatively high levels of SOC and TN compared to the remaining parts of the study area. Across the northeastern regions, soil SOC and TN values were lower (Figs. 8.2 and 8.3). The primary factor contributing to the diminished SOC content in the southern region of the area can be attributed to the concomitant scarcity of vegetation covers. This scarcity subsequently results in insufficient nutrient inputs to the soil in the absence of appropriate management.

Similar patterns of spatial distribution were also evident in the SOC and TN distribution maps (Figs. 8.2 and 8.3). The topographic variables, particularly slope and Converg, showed a strong correlation with the spatial distribution pattern of SOC and TN (Figs. 8.2 and 8.3). As predicted, the concentrations of SOC and TN were primarily concentrated in the topsoil (0–20 cm) and decreased as one descended. Previous studies (Podwojewski et al., 2011; Tian et al., 2007; Wang et al., 2010) have shown that topographic variables including elevation and slope have an impact on SOC and TN. Tsui et al. (2013) found that as elevation increased, the SOC content rose noticeably. Variations in elevation

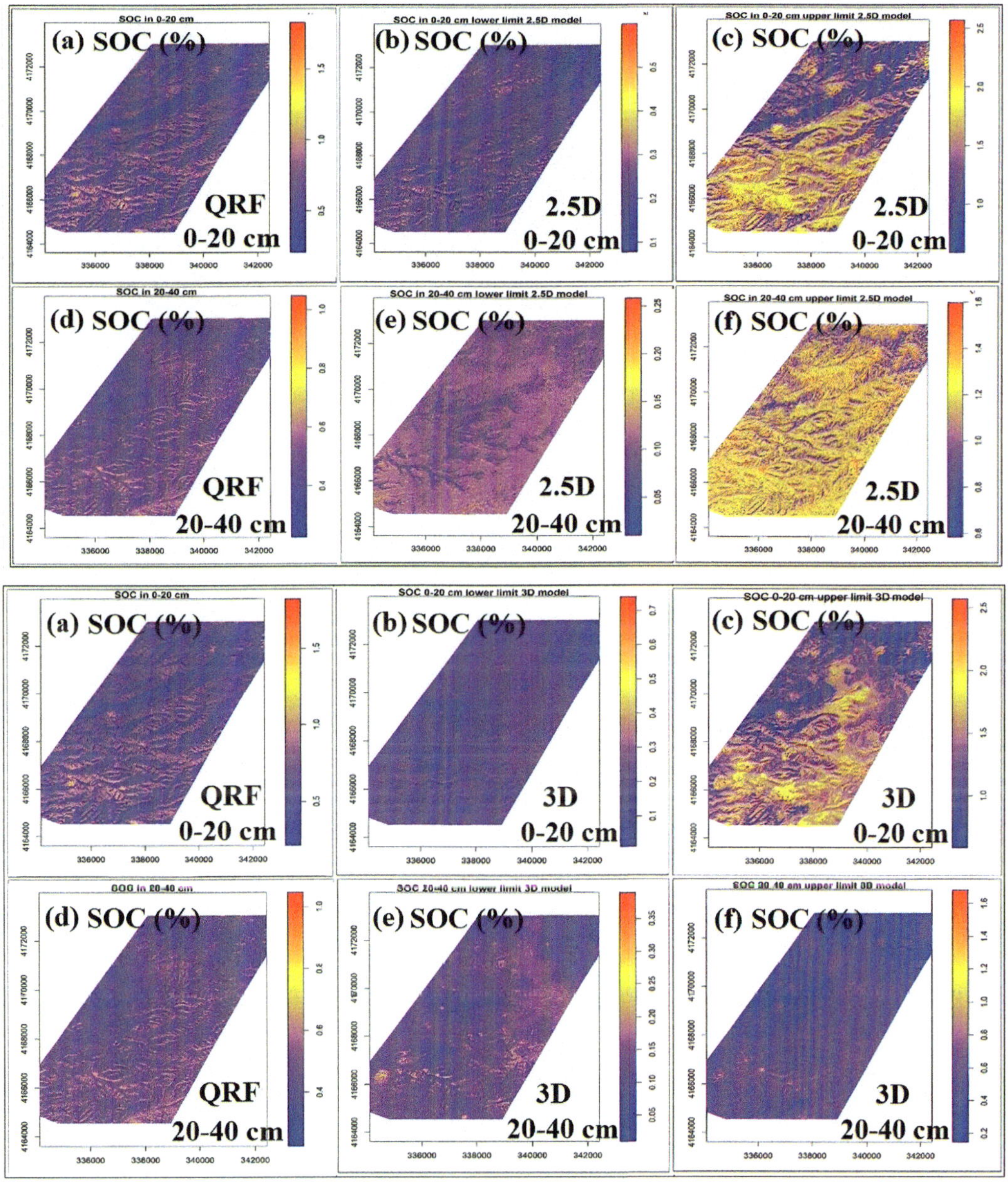

FIGURE 8.2

The spatial distribution of SOC at two different depths is predicted using the QRF model. Subsections (a), (b), and (c) pertain to the SOC content at a depth of 0–20 cm (2.5D), while subsections (d), (e), and (f) correspond to the SOC content at a depth of 20–40 cm (3D).

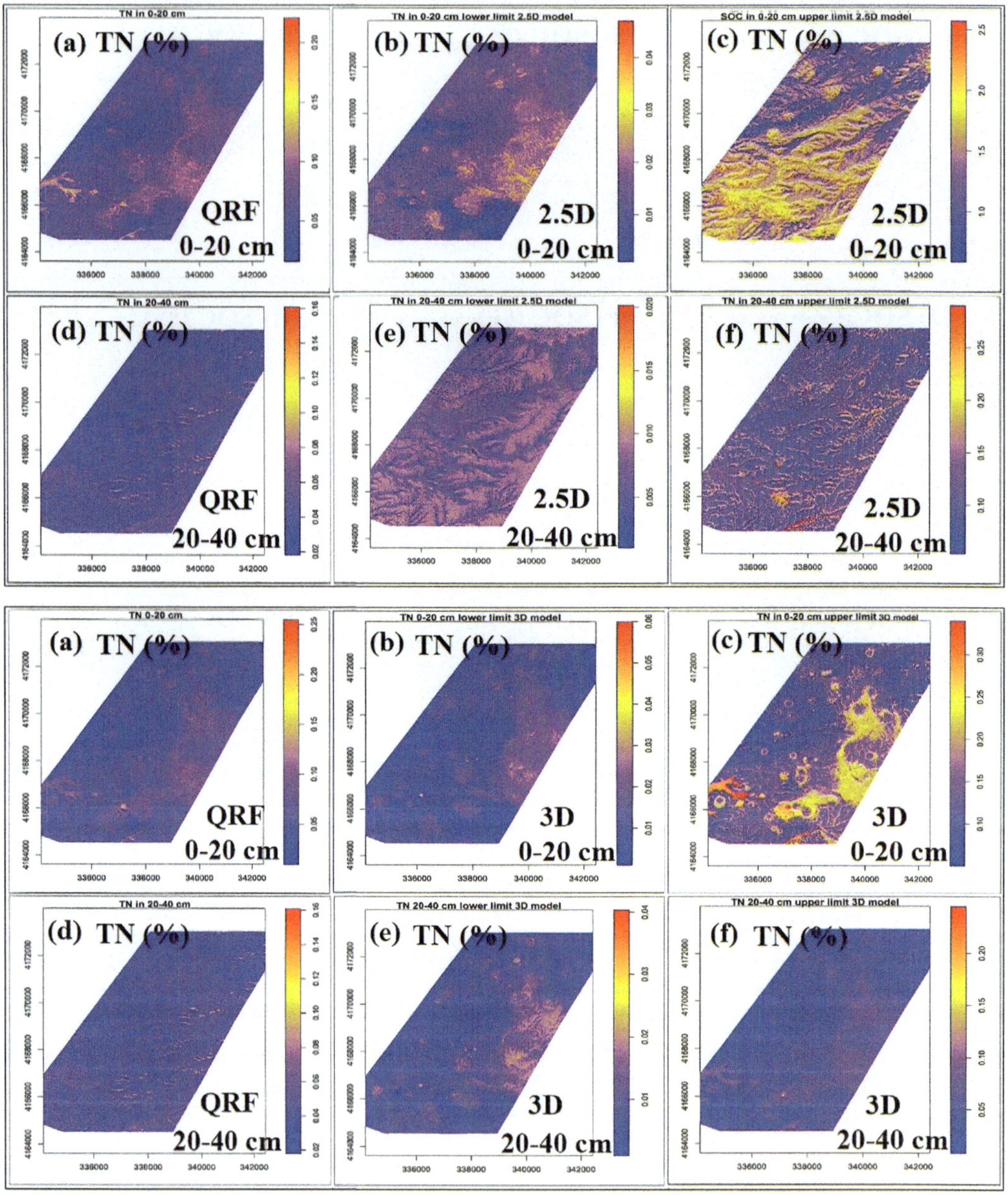

FIGURE 8.3

The spatial distribution of TN at two different depths is predicted using the QRF model. Subsections (a), (b), and (c) pertain to the SOC content at a depth of 0–20 cm (2.5D), while subsections (d), (e), and (f) correspond to the TN content at a depth of 20–40 cm (3D).

gradients primarily impacted SOC and TN and loss through indirect controls like temperature and precipitation (Garten Jr & Hanson, 2006; Wang et al., 2010).

8.3 Model performance of quantile regression forest

Table 8.3 presents a summary of the model's effectiveness evaluation, as measured by R^2 and RMSE.

This evaluation was conducted through 100 iterations of the model and by employing a 10-fold cross-validation technique (Table 8.3). The findings demonstrated that the QRF model exhibited remarkable prognostic potential, with consistently elevated R^2 and diminished RMSE values in the prediction of SOC and TN. The model parameters that yielded the best results differed significantly for SOC and TN, as presented in Table 8.3. Each model employed distinct split rules, with the SOC model requiring a higher optimal mtry compared to the TN model. Moreover, the R^2 value for the SOC model (0.38) vastly exceeded that of the TN model (0.22), while the SOC model exhibited an RMSE of 0.36%, more than six times higher than the TN model's RMSE of 0.054%. In addition, the levels of precision exhibited a considerable degree of similarity across regions characterized by high elevation and regions characterized by a high-water table. The levels of precision exhibited comparable patterns for SOC. In addition, the accuracy levels showed a similar pattern with R^2 equal to 0.16 for SOC at a depth of 20–40 cm. The R^2 value was found to be the lowest in TN (0.066) when compared to TN (0.16). However, it is important to highlight that the R^2 demonstrated the highest values in the uppermost layer of soil (0.60) and subsequently decreased as the depth increased. Accuracy for the studied features was generally different from the results of previous soil mapping studies. Adhikari et al. (2013) as well as Adhikari et al. (2014) documented R^2 values ranging from 0.40 to 0.55 in relation to texture fractions and log-transformed organic carbon stocks in the uppermost layer of soil. Furthermore, our minimum R^2 for TN (0.066), specifically within the depth range of 20–40 cm, also aligns with the R^2 range of 0.066–0.32 reported for the same depth interval in these aforementioned studies.

The spatial cross-validation strategy may be accountable for the generally low to moderate accuracies. An example of this can be found in the cautionary words of Wadoux et al. (2021), who warned that spatial cross-validation can yield an excessively pessimistic estimation of predictive accuracy. However, the diminished accuracy of the TN model is also likely attributed to the pronounced positive skewness of the variable, as evidenced by its skewness values ranging from 2.02 to 2.68 (Table 8.3). Moreover, the R^2 value for TN was higher compared to the alternative (0.22 vs. 0.066).

Table 8.3 Some statistical parameters of studied soil properties in different depths.

			Calibration		Validation	
Soil properties	**Depth**	**Mtry**	**RMSE**	**R^2**	**RMSE**	**R^2**
SOC	0-20	29	0.15	0.93	0.36	0.38
	20-40	15	0.13	0.94	0.30	0.16
TN	0-20	15	0.023	0.94	0.054	0.22
	20-40	2	0.02	0.93	0.41	0.066

8.4 Covariate importance and influence

Fig. 8.4 illustrates the significance of auxiliary variables in determining SOC and TN through the RF model, considering two standard depths of 0–20 cm and 20–40 cm.

Based on Fig. 8.4a, it is important to emphasize that the slope direction and EC had the greatest effect on SOC model prediction at 0–20 cm depth and 20–40 cm depth. Furthermore, SOC was influenced by multiple distinct collections of factors. These factors encompassed covariates associated with parent materials (carbonate index (CI); grain size index (GSI); coloration index (CI) and saturation index (SI) and topography (slope and aspect). The impact of the topographic roughness index (TRI) on SOC was minimal in comparison to the other covariates, as it ranked as the 28th most significant covariate. The finding presented in this location is not unexpected and strongly implies that the accumulation of SOC in the ILP is primarily limited by the aspect and slope, either directly or indirectly, within the upper layer of the soil. This finding is consistent with the extensive knowledge of experts in the field of soil sciences within this specific region, as demonstrated in previous research conducted by Kehl & Khormali (2014), Yang et al. (2021) and Xie et al. (2021). The significant significance of elevation within the SOC model is primarily attributable to the diminished SOC contents within these comparatively recent, semi formed deposits. This also accounts for the notable significance of the geological classification, which frequently aligns with landscapes. Furthermore, the predicted SOC is particularly elevated at elevated topographic positions in soils that are well-drained, and in regions that experience substantial precipitation. These impacts are generally challenging to separate, since higher elevations tend to receive greater amounts of precipitation and typically possess favorable drainage conditions (Møller et al., 2017). The predominant influence of elevation on SOC levels may be primarily associated with its impact on vegetation, slope, hydrological processes, and the transportation of nutrients, thereby constituting the principal source of spatial variability in SOC. Furthermore, the predicted SOC exhibited a predominantly elevated value within the range of 0–40 cm in depth (as depicted in Fig. 8.2). This observation was particularly notable at greater elevation and within soils that experience well drainage. The covariates associated with soil texture (silt, clay, and GSI) and the relative slope position generally indicate elevated SOC contents in wetland regions. For instance, soils containing more than 30% silt, which provides more surface for maintaining SOC (Konen et al., 2003) typically exhibit high levels of SOC. The origin of the initial substance is attributed to parent materials, which act as the principal suppliers of the original material. These parent materials demonstrate a clear connection with diverse soil attributes, comprising soil texture, pH, and clay mineralogy. As a result, these parent materials subsequently exert a substantial impact on other vital soil characteristics, including the amount of SOC and the soil's ability to sequester carbon (Ingram & Fernandes, 2001).

According to the data presented in Fig. 8.4b, it can be observed that LS, ELV, ASP are the three covariates that hold the highest significance in explaining 80% of the TN content within the layer spanning from 20 to 40 cm. Similarly, in the layer ranging from 0 to 20 cm, the covariates ASP exhibit the most influential impact, accounting for 34% of the overall TN content, as depicted in Fig. 8.4b.

It is worth noting that the initial one covariate is associated with remote sensing variable, whereas the latter two covariates are linked to topography. Hence, it is possible to state that the terrain attributes, particularly the aspect, hold greater significance in relation to all soil properties through both soil depths. Despite the absence of consensus regarding the factors that contribute to the variability of soil properties at the surface and subsurface levels, as well as the interactions between soil properties

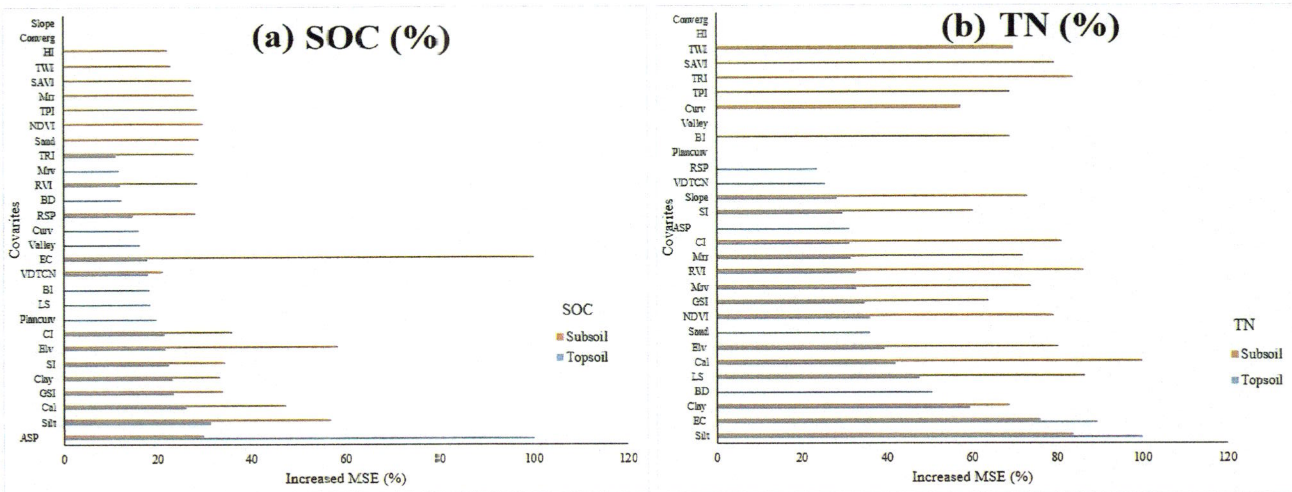

FIGURE 8.4

Illustrates the variability of importance in QRF models for SOC (a) and TN (b) in both the topsoil (0–20 cm) and subsoil (20–40 cm). The variable importance below 10% was not observed for all the covariates at both depths.

and these factors, several researchers have validated the influence of topographic attributes on the variability of soil properties (Dai et al., 2022; Nikodem et al., 2021). Topographic variables were extensively utilized in the estimation of SOC density, particularly in regions characterized by significant alterations in terrain (Grimm et al., 2008; Minasny & McBratney, 2016; Wang et al., 2020). These factors govern the movement of water and nutrients across the terrain, thereby accounting for the majority of the spatial distribution of SOC and clay content in sediment transfer and erosion. Consequently, the relative importance of the slope aspect becomes evident in subsequent degrees.

In addition, Chagas et al. (2018) have identified a significant spatial relationship between CEC and remote sensing indices. The association between soil properties and vegetation indices may be indicative of the contribution of vegetation cover to the fluctuations in SOC, as observed in this particular study. Correspondingly, the findings of the present investigation align with those of Bhunia et al. (2019) and Nie et al. (2021). The variations in SOC were predominantly influenced by vegetation indices and soil covariates in similar climatic conditions. Moreover, NDVI was identified as the primary determinant of SOC variability. In relation to the ELV factor and SAVI, Zeraatpisheh et al. (2019) have presented compelling evidence of noteworthy associations between SOC and these factors in surface soils of a semiarid locale within the Chaharmahal-Va Bakhtiari Province, situated in central Iran. Qi et al. (2019) reached the conclusion that SAVI and NDVI are essential remote sensing variables for the prediction of topsoil SOC density in the coastal forest area of northeast China. Shi et al., (2025) said that SOC exhibits distinct distribution patterns depending on the type of vegetation and clay. As, local QRF using MIR data captures chemical information, specifically, clay minerals related to carbon storage.

The findings presented in this study demonstrate the influence of microclimate patterns in relation to the aspect on the density of vegetation and subsequent accumulation of SOC and TN in hillslopes. It is observed that there may be variations in these parameters from the northern to the southern aspect. The significant correlation between topographic factors and soil properties underscores the vulnerability of soil properties to degradation in areas with elevated, south-facing slopes, where vegetation density is diminished and there is an increase in erosional surfaces due to over-grazing.

Thus, it can be asserted that terrain attributes, particularly aspect, play a significant role in the characteristics of soil at various depths. While there is a lack of consensus regarding the factors that contribute to the variability of soil properties on the surface and subsurface, as well as the interactions between these properties and said factors, a multitude of scholars have affirmed the influence of topographic attributes on the variability of soil properties (Dai et al., 2022; Nikodem et al., 2021). Topographic attributes, including TWI and ELV, can exert an influence on the stability of soil aggregation (SAS). This influence can occur through direct effects or indirect effects mediated by their impact on various soil properties. These properties include the fixation and decomposition of SOC, hydrological conditions, and the spatial distribution of vegetation cover and land use. The relationship between soil properties and vegetation indices can also reflect the role of vegetation cover in the variations of certain soil properties, such as SOC and MWD, as observed in this study. In this regard, the findings of the present study align with previous studies conducted by Bhunia et al. (2019), Tajik et al. (2020), Nie et al. (2021). Previous research has established that there exists a connection between the presence of vegetation and the quantity of SOC in the uppermost layer of soil. In a study carried out by Kim & Grunwald (2016), an RF model was formulated to prognosticate SOC density, utilizing field observations, supplementary environmental data, and spectral data extracted from remote sensing images.

The researchers arrived at the conclusion that vegetation indices have a substantial impact on SOC models. The spectral reflectance of remote sensing data and the resultant vegetation index furnish insights into the coverage extent of vegetation, which is intimately linked to the density of SOC. Zeraatpisheh et al. (2019) have demonstrated significant associations between SOC and the ELV factor and SAVI in the surface soils of a semiarid region in Chaharmahal-Va-Bakhtiari Province, central Iran. Qi et al. (2019) have concluded that SAVI and NDVI are essential variables in remote sensing for predicting the density of topsoil SOC in the coastal forest area of northeast China. Additionally, NDVI can also act as an indicator of vegetation productivity and biomass, which has been extensively used in previous spatial predictions of SOC density.

According to the findings, the role of aspect in this investigation is confirmed at two different soil depths, which aligns closely with the expertise of a knowledgeable individual in the field of soil science who is familiar with this specific area. This alignment is also supported by previous studies conducted by Kehl & Khormali (2014), Maleki et al. (2018), and Yu et al. (2020). It is essential to compare the opinions of experts with the outcomes of the modeling process, particularly due to the growing usage of machine learning (ML) techniques in digital soil mapping. This is because the valuable insights provided by experts may be overshadowed by the extensive number of predictors used for prediction purposes, leading to the replacement of expert knowledge by the blind application of ML approaches. In contrast, these findings illustrate the influence of microclimate patterns resulting from the aspect (Maleki et al., 2018) that governs vegetation density, subsequently affecting SOC and TN accumulation in hillslopes, which may vary from the north to the south aspect. The strong reliance of soil properties on topographic factors emphasizes the susceptibility of soil properties to degradation in areas with an elevated south aspect, caused by the reduction in vegetation density, TPI, and the increase in erosional surfaces due to over-grazing (Bouguerra et al., 2023).

These factors collectively dictate the movement of water and nutrients within the landscape, thus accounting for the majority of the spatial distribution of SOC and clay content in relation to sediment transfer and erosion. As a result, the significance of slope aspect becomes apparent in subsequent degrees of importance. Previous studies have demonstrated a noteworthy correlation between elevation and the spatial arrangement of SOC density, particularly in regions characterized by mountainous topography.

The relationship between elevation and microclimate was attributed to explaining the distribution of SOC density. In regions that experience significant changes in terrain, topographic variables were extensively used to estimate SOC density. These variables were considered particularly important in studies conducted by Minasny & McBratney (2016), and Wang et al. (2020). The results of this research are not only a scientific document for understanding the paleo-environmental changes of the Quaternary deposits of the region, but also a practical and essential tool for future planning in the field of sustainable watershed management, combating desertification, and preserving the natural capital of Golestan Province. It is recommended to continue this research by focusing on the role of clay minerals, organic matter and TN in the interaction with prevention of desertification.

4. Conclusions

The anticipated accuracy was generally high for SOC and TN. In addition, the predictions were generally reasonably accurate in the topsoil (0–20 cm) and inaccurate in the subsoil (20–40 cm). Nonetheless, the fusion of QRF generally produced reliable predictions for SOC, and TN. Particularly in the topsoil (0–20 cm), predictions for SOC demonstrated significantly elevated levels, primarily due to the substantially fluctuating quantities of silt. The modeling and prediction processes were executed utilizing machine learning algorithm (QRF), thereby enabling the evaluation of their individual performances. The prediction performance of the QRF algorithm, as evaluated by the R^2 metric, exhibits a decrease with increasing soil depth. The QRF algorithm is consistently identified as the top-performing ML method for all depths and scenarios. Notably, a two-fold enhancement in performance is achieved when utilizing S1 instead of S2, underscoring the crucial role of soil variables in accurately mapping SOC content. From a comparison of the relative importance of the environmental covariates and soil variables at different depths within the soil, it is evident that soil variables including LS, ELV, ASP, SAVI, Converg and TPI play a dominant role below the soil surface, while the significance of the environmental covariates is comparatively lower. This contrasts with the topsoil layer of the soil (0–20 cm), where the SAVI and NDVI were identified as the most influential covariates. At the same time, the utilization of pprediction accuracy allowed us to investigate the impact of each covariate on the SOC content, producing results that closely align with those reported in existing literature. In the realm of SOC and TN mapping, it can be inferred that the utilization of both environmental covariates. This finding bears significant implications for the meticulous planning of sampling campaigns. Since SOC measurements necessitate collection for the training and validation of ML algorithm, it is deemed worthwhile to allocate additional efforts and resources toward the simultaneous measurement of other crucial soil properties.

Acknowledgments

The author gratefully acknowledges the financial support provided by the Iran's National Elites Foundation (INEF) for this research.

References

Adhikari, K., Hartemink, A. E., Minasny, B., Kheir, R. B., Greve, M. B., & Greve, M. H. (2014). Digital mapping of soil organic carbon contents and stocks in Denmark. *PLoS One, 9*(8), Article e105519. https://doi.org/10.1371/journal.pone.0105519

Adhikari, K., Kheir, R. B., Greve, M. B., Bøcher, P. K., Malone, B. P., Minasny, B., McBratney, A. B., & Greve, M. H. (2013). High-resolution 3-D mapping of soil texture in Denmark. *Soil Science Society of America Journal, 77*(3), 860–876. https://doi.org/10.2136/sssaj2012.0275

Ayoubi, S., Mohammadi, A., Abdi, M. R., Abbaszadeh Afshar, F., Wang, L., & Zeraatpisheh, M. (2022). Assessment of soil redistribution following land rehabilitation with an apple orchard in hilly regions of central Iran. *Agronomy, 12*(2), 451. https://doi.org/10.3390/agronomy12020451

Bhunia, G. S., Shit, P. K., & Pourghasemi, H. R. (2019). Soil organic carbon mapping using remote sensing techniques and multivariate regression model. *Geocarto International, 34*(2), 215–226. https://doi.org/10.1080/10106049.2017.1381179

Bouguerra, H., Tachi, S. E., Bouchehed, H., Gilja, G., Aloui, N., Hasnaoui, Y., … Navarro-Pedreño, J. (2023). Integration of high-accuracy geospatial data and machine learning approaches for soil erosion susceptibility mapping in the mediterranean region: A case study of the macta basin, Algeria. *Sustainability, 15*(13), Article 10388. https://doi.org/10.3390/su151310388

Breiman, L. (2001). Random forests. *Machine Learning, 45*(1), 5–32. https://doi.org/10.1023/A:1010933404324

Breiman, L., & Cutler, A. (2004). *Random forests*. http://www.stat.berkeley.edu/users/breiman.RandomForests/cc_papers.htm.

Bremner, J. M., & Mulvaney, C. S. (1982). Methods of soil analysis. In A. L. Page, R. H. Millar, & D. R. Keneey (Eds.), *nitrogen-total., part 2, chemical and microbiological*.

Carter, M. R., & Gregorich, E. G. (Eds.). (2007). *Soil sampling and methods of analysis*. CRC press.

Chagas, C. D. S., Carvalho Júnior, W. D., Pinheiro, H. S. K., Xavier, P. A. M., Bhering, S. B., Pereira, N. R., & Calderano Filho, B. (2018). Mapping soil cation exchange capacity in a semiarid region through predictive models and covariates from remote sensing data. *Revista Brasileira de Ciência do Solo, 42*. https://doi.org/10.1590/18069657rbcs20170183

Chaplot, V. (2013). Impact of terrain attributes, parent material and soil types on gully erosion. *Geomorphology, 186*, 1–11. https://doi.org/10.1016/j.geomorph.2012.10.031

Chen, S., Arrouays, D., Mulder, V. L., Poggio, L., Minasny, B., Roudier, P., … Walter, C. (2022). Digital mapping of globalsoilmap soil properties at a broad scale: A review. *Geoderma, 409*, Article 115567. https://doi.org/10.1016/j.geoderma.2021.115567

Dai, L., Ge, J., Wang, L., Zhang, Q., Liang, T., Bolan, N., … Rinklebe, J. (2022). Influence of soil properties, topography, and land cover on soil organic carbon and total nitrogen concentration: A case study in qinghai-tibet plateau based on random forest regression and structural equation modeling. *Science of the Total Environment, 821*, Article 153440. https://doi.org/10.1016/j.scitotenv.2022.153440

Dash, P. K., Panigrahi, N., & Mishra, A. (2022). Identifying opportunities to improve digital soil mapping in India: A systematic review. *Geoderma Regional, 28*, Article e00478. https://doi.org/10.1016/j.geodrs.2021.e00478

Efron, B., & Tibshirani, R. J. (1994). *An introduction to the bootstrap*. CRC press.

Fathololoumi, S., Vaezi, A. R., Alavipanah, S. K., Ghorbani, A., & Biswas, A. (2020). Comparison of spectral and spatial-based approaches for mapping the local variation of soil moisture in a semi-arid mountainous area. *Science of the Total Environment, 724*, Article 138319. https://doi.org/10.1016/j.scitotenv.2020.138319

Forkuor, G., Hounkpatin, O. K., Welp, G., & Thiel, M. (2017). High resolution mapping of soil properties using remote sensing variables in South-Western Burkina Faso: A comparison of machine learning and multiple linear regression models. *PLoS One, 12*, Article e0170478.

Gallant, J. C., & Dowling, T. I. (2003). A multiresolution index of valley bottom flatness for mapping depositional areas. *Water resources research, 39*(12).

Garten Jr, C. T., & Hanson, P. J. (2006). Measured forest soil C stocks and estimated turnover times along an elevation gradient. *Geoderma, 136*(1–2), 342–352. https://doi.org/10.1016/j.geoderma.2006.03.049

Godwin, R. J., & Miller, P. C. H. (2003). A review of the technologies for mapping within-field variability. *Biosystems Engineering, 84*(4), 393–407. https://doi.org/10.1016/S1537-5110(02)00283-0

Grimm, R., Behrens, T., Marker, M., & Elsenbeer, H. (2008). Soil organic carbon concentrations and stocks on barro Colorado Island: Digital soil mapping using random forests analysis. *Geoderma, 146*, 102–113. https://doi.org/10.1016/j. geoderma.2008.05.008

Hateffard, F., Steinbuch, L., Gerard, B. M., & Heuvelink, G. (2024). Evaluating the extrapolation potential of random forest digital soil mapping. *Geoderma, 441*, Article 116740. https://doi.org/10.1016/j.geoderma.2023.116740

Hosseinalizadeh, M., Kariminejad, N., Chen, W., Pourghasemi, H. R., Alinejad, M., Mohammadian Behbahani, A., & P. Tiefenbacher, J. (2019). Gully headcut susceptibility modeling using functional trees, naïve bayes tree, and random forest models. *Geoderma, 342*, 1–11. https://doi.org/10.1016/j.geoderma.2019.01.050

Huete, A. R. (1988). A soil-adjusted vegetation index (SAVI). *Remote sensing of environment, 25*(3), 295–309. https://doi.org/10.1016/0034-4257(88)90106-X

Ingram, J. S. I., & Fernandes, E. C. M. (2001). Managing carbon sequestration in soils: Concepts and terminology. *Agriculture, Ecosystems & Environment, 87*(1), 111–117. https://doi.org/10.1016/S0167-8809(01)00145-1

Jien, S.-H., Minasny, B., Yang, B., Zhang, Y., Yen, C., Liu, Y., Yen, C., Ocba, M. A., Zhang, Y. T., & Syu, C. (2025). Enhancing soil carbon storage: Developing high-resolution maps of topsoil organic carbon sequestration potential in Taiwan. *Geoderma, 459*, Article 117369.

Kariminejad, N., Pourghasemi, H. R., Maleki, S., & Hosseinalizadeh, M. (2022). Digital soil mapping and modeling in loess-derived soils of Iranian Loess Plateau. *Geocarto International, 37*(26), 11633–11651. https://doi.org/10.1080/10106049.2022.2060317

Kehl, M., & Khormali, F. (2014). *Excursion book of international symposium on loess, soils & climate change in Southern Eurasia. 15-19 October.* Gorgan, Iran: Gorgan University of Agricultural Sciences and Natural Resources.

Khaledian, Y., & Miller, B. A. (2019). Selecting appropriate machine learning methods for digital soil mapping. *Applied Mathematical Modelling.* https://doi.org/10.1016/j.apm.2019.12.016

Khongnawang, T., Zare, E., Srihabun, P., Khunthong, I., & Triantafilis, J. (2022). Digital soil mapping of soil salinity using EM38 and quasi-3d modelling software (EM4Soil). *Soil Use and Management, 38*(1), 277–291. https://doi.org/10.1111/sum.12778

Khosravani, P., Baghernejad, M., Moosavi, A. A., & FallahShamsi, S. R. (2023). Digital mapping to extrapolate the selected soil fertility attributes in calcareous soils of a semiarid region in Iran. *Journal of Soils and Sediments*, 1–23. https://doi.org/10.1007/s11368-023-03548-1

Kim, J., & Grunwald, S. (2016). Assessment of carbon stocks in the topsoil using random forest and remote sensing images. *Journal of Environmental Quality, 45*(6), 1910–1918. https://doi.org/10.2134/jeq2016.03.0076

Konen, M. E., Burras, C. L., & Sandor, J. A. (2003). Organic carbon, texture, and quantitative color measurement relationships for cultivated soils in north central Iowa. *Soil Science Society of America Journal, 67*(6), 1823–1830. https://doi.org/10.2136/sssaj2003.1823

Kramm, T., Hoffmeister, D., Curdt, C., Maleki, S., Khormali, F., & Kehl, M. (2017). Accuracy assessment of landform classification approaches on different spatial scales for the Iranian Loess Plateau. *ISPRS International Journal of Geo-Information, 6*(366), 1–22.

Kuhn, M. (2008). Building predictive models in R using the caret package. *Journal of Statistical Software, 28*, 1–26. https://doi.org/10.18637/jss.v028.i05

Kumaraperumal, R., Pazhanivelan, S., Geethalakshmi, V., Nivas Raj, M., Muthumanickam, D., Kaliaperumal, R., … Tarun Kshatriya, T. V. (2022). Comparison of machine learning-based prediction of qualitative and quantitative digital soil-mapping approaches for Eastern districts of Tamil Nadu, India. *Land, 11*(12), 2279. https://doi.org/10.3390/land11122279

Liaw, A., & Wiener, M. (2002). Classification and regression by randomforest. *R News, 2*(3), 18–22.

Liaw, A., & Wiener, M. (2018). *Package: Randomforest. title: Breiman and cutler's random forests for classification and regression. (R package version: 4.6–14).*

Loiseau, T., Chen, S., Mulder, V. L., Roman Dobarco, M., Richer-de-Forges, A. C., Lehmann, S., Bourennane, H., Saby, N. P. A., Martin, M. P., Vaudour, E., Gomez, C., Lagacherie, P., & Arrouays, D. (2019). Satellite data integration for soil clay content modelling at a national scale. *International Journal of Applied Earth Observation and Geoinformation, 82*, Article 101905. https://doi.org/10.1016/j.jag.2019.101905

Møller, A. B., Beucher, A., Iversen, B. V., & Greve, M. H. (2017). Prediction of soil drainage classes in Denmark by means of decision tree classification. *Geoderma, 352*, 314–329. https://doi.org/10.1016/j.geoderma.2017.10.015

Ma, G., Ding, J., Han, L., Zhang, Z., & Ran, S. (2021). Digital mapping of soil salinization based on sentinel-1 and sentinel-2 data combined with machine learning algorithms. *Regional Sustainability, 2*(2), 177–188.

Maleki, S., Fathizad, H., Karimi, A., Taghizadeh-Mehrjardi, R., & Pourghasemi, H. R. (2022). Monitoring of spatiotemporal changes of soil salinity and alkalinity in eastern and central parts of Iran. In *Computers in Earth and environmental sciences* (pp. 547–561). Elsevier Inc. https://doi.org/10.1016/B978-0-323-89861-4.00011-7

Maleki, S., Khormali, F., Bagheri Bodaghabadi, M., Mohammadi, J., Hoffmeister, D., & Kehl, M. (2018). Role of geomorphic surface on the above-ground biomass and soil organic carbon storage in a semi-arid region of Iranian Loess Plateau. *Quaternary International, 552*, 111–121. https://doi.org/10.1016/j.quaint.2018.11.001

Maleki, S., Khormali, F., Mohammadi, J., Bogaert, P., & Bagheri Bodaghabadi, M. (2020). Effect of the accuracy of topographic data on improving digital soil mapping predictions with limited soil data: An application to the Iranian Loess Plateau. *Catena, 195*, Article 104810. https://doi.org/10.1016/j.catena.2020.104810

Maleki, S., Soheila Pouyan, S., Mohammadi Khoshoui, M., Tiefenbacher, P., & Pourghasemi, J. (2024). Detecting soil salinization, sodicity, and alkalization hazards within cultivated lands using digital soil mapping approaches. In *Advanced tools for studying soil erosion processes* (pp. 435–460). Elsevier Inc. https://doi.org/10.1016/B978-0-443-22262-7.00019-9

Malone, B. P., Minasny, B., McBratney, A. B., Malone, B. P., Minasny, B., & McBratney, A. B. (2017). Continuous soil attribute modeling and mapping. *Using R for Digital Soil Mapping*, 117–149. https://doi.org/10.1007/978-3-319-44327-0_5

McBratney, A. B., Mendoça Santos, M. L., & Minasny, B. (2003). On digital soil mapping. *Geoderma, 117*, 3–52. https://doi.org/10.1016/S0016-7061(03)00223-4

Meinshausen, N. (2006). Quantile regression forests. *Journal of Machine Learning Research, 7*, 983–999.

Minasny, B., & Bishop, T. F. A. (2008). *Analysing uncertainty. guidelines for surveying soil and land resources*. Melbourne: CSIRO Publishing.

Minasny, B., & McBratney, A. B. (2002). The neuro-m method for fitting neural network parametric pedotransfer functions. *Soil Science Society of America Journal, 66*(2), 352–361. https://doi.org/10.2136/sssaj2002.0352

Minasny, B., & McBratney, A. B. (2006). A conditioned Latin hypercube method for sampling in the presence of ancillary information. *Computers & Geosciences., 32*(9), 1378–1388. https://doi.org/10.1016/j.cageo.2005.12.009

Minasny, B., & McBratney, A. B. (2016). Digital soil mapping: A brief history and some lessons. *Geoderma, 264*, 301–311. https://doi.org/10.1016/j. geoderma.2015.07.017

Moore, I. D., Gessler, P. E., Nielsen, G. A. E., & Peterson, G. A. (1993). Soil attribute prediction using terrain analysis. *Soil science society of america journal, 57*(2), 443–452. https://doi.org/10.2136/sssaj1993.03615995005700020026x

Mousavi, A., Karimi, A., Maleki, S., Safari, T., & Taghizadeh-Mehrjardi, R. (2023). Digital mapping of selected soil properties using machine learning and geostatistical techniques in mashhad plain, northeastern Iran. *Environmental Earth Sciences, 82*(9), 234. https://doi.org/10.21203/rs.3.rs-1728200/v1

Munar, O. J., & Martínez, L. J. (2014). Relief parameters and fuzzy logic for land evaluations of mango crops (Mangifera indica L.). *Agronomía Colombiana*, 246–254. https://doi.org/10.15446/agron.colomb.v32n2.43988

Nauman, T. W., & Duniway, M. C. (2019). Relative prediction intervals reveal larger uncertainty in 3D approaches to predictive digital soil mapping of soil properties with legacy data. *Geoderma, 347*, 170–184. https://doi.org/10.1016/j.geoderma.2019.03.037

Nelson, D. W., & Sommers, L. E. (1982). Total carbon, organic carbon, and organic matter. In A. L. Page, & et al. (Eds.) (2nd ed.,*Agronomy Monograph 9Methods of soil analysis. Part 2* (pp. 539–577). Madison, WI: ASA and SSSA.

Nie, T., Dong, G., Jiang, X., & Lei, Y. (2021). Spatio-temporal changes and driving forces of vegetation coverage on the Loess Plateau of Northern Shaanxi. *Remote Sensing, 13*(4), 613. https://doi.org/10.3390/rs13040613

Nikodem, A., Kodešová, R., Fér, M., & Klement, A. (2021). Using scaling factors for characterizing spatial and temporal variability of soil hydraulic properties of topsoils in areas heavily affected by soil erosion. *Journal of Hydrology, 593*, Article 125897. https://doi.org/10.1016/j.jhydrol.2020.125897

Orton, T. G., Pringle, M. J., & Bishop, T. F. A. (2016). A one-step approach for modelling and mapping soil properties based on profile data sampled over varying depth intervals. *Geoderma, 262*, 174–186. https://doi.org/10.1016/j.geoderma.2015.08.013

Pahlavan-Rad, M. R., & Akbarimoghaddam, A. (2018). Spatial variability of soil texture fractions and pH in a flood plain (case study from eastern Iran). *Catena, 160*, 275–281. https://doi.org/10.1016/j.catena.2017.10.002

Podwojewski, P., Poulenard, J., Nguyet, M. L., De Rouw, A., Pham, Q. H., & Tran, D. T. (2011). Climate and vegetation determine soil organic matter status in an alpine inner-tropical soil catena in the Fan Si Pan Mountain, Vietnam. *Catena, 87*(2), 226–239. https://doi.org/10.1016/j.catena.2011.06.002

Qi, L., Wang, S., Zhuang, Q., Yang, Z., Bai, S., Jin, X., & Lei, G. (2019). Spatial-temporal changes in soil organic carbon and pH in the Liaoning Province of China: A modeling analysis based on observational data. *Sustainability, 11*(13), 3569. https://doi.org/10.3390/su11133569

Ray, S. S., Singh, J. P., Das, G., & Panigrahy, S. (2004). Use of high resolution remote sensing data for generating site-specific soil management plan. *International Archives of Photogrammetry, Remote Sensing and Spatial Information Sciences, B35*(7), 127–131.

Rouse, J. W., Hass, R. H., Schell, J. A., & Deering, D. W. (1974). Monitoring vegetation systems in the great plaints with ERTS. *Proceedings 3rd Earth Resource Technology Satellite (ERTS) Symposium, 1*, 48–62.

Sahab, S., Suhani, I., Srivastava, V., Chauhan, P. S., Singh, R. P., & Prasad, V. (2021). Potential risk assessment of soil salinity to agroecosystem sustainability: Current status and management strategies. *Science of the Total Environment, 764*, Article 144164. https://doi.org/10.1016/j.scitotenv.2020.144164

Santos-Francés, F., Martínez-Graña, A., Ávila-Zarza, C., Criado, M., & Sánchez-Sánchez, Y. (2022). Soil quality and evaluation of spatial variability in a semi-arid ecosystem in a region of the Southeastern Iberian Peninsula (Spain). *Land, 11*(1), 5. https://doi.org/10.3390/land11010005

Schmidinger, J., & Heuvelink, G. B. (2023). Validation of uncertainty predictions in digital soil mapping. *Geoderma, 437*, Article 116585. https://doi.org/10.1016/j.geoderma.2023.116585

Shi, L., Daly, K., & O'Rourke, S. (2025). Estimating mineral-associated organic carbon saturation and sequestration potential using MIR spectral based local quantile regression Longnan. *Geoderma, 454*, Article 117181.

Shi, G., Shangguan, W., Zhang, Y., Li, Q., Wang, C., & Li, L. (2024). Reducing location error of legacy soil profiles leads to improvement in digital soil mapping Gaosong. *Geoderma, 447*, Article 116912.

Shirazi, R. A., Shahbazi, F., Rezaei, H., & Biswas, A. (2023). Digital assessments of soil organic carbon storage using digital maps provided by static and dynamic environmental covariates. *Soil Use & Management, 39*, 948–974. https://doi.org/10.1111/sum.12900

Soil Survey Staff. (2022). *Keys to soil taxonomy, 13th edition. USDA natural resources conservation service.*

Suleymanov, A., Tuktarova, I., Belan, L., Suleymanov, R., Gabbasova, I., & Araslanova, L. (2023). Spatial prediction of soil properties using random forest, k-nearest neighbors and cubist approaches in the foothills of the Ural Mountains, Russia. *Modeling Earth Systems and Environment*, 1–11. https://doi.org/10.1007/s40808-023-01723-4

Taghizadeh-Mehrjardi, R., Nabiollahi, K., & Kerry, R. (2016). Digital mapping of soil organic carbon at multiple depths using different data mining techniques in Baneh region, Iran. *Geoderma, 266*, 98–110. https://doi.org/10.1016/j.geoderma.2015.12.003

Tajik, S., Ayoubi, S., & Zeraatpisheh, M. (2020). Digital mapping of soil organic carbon using ensemble learning model in mollisols of hyrcanian forests, Northern Iran. *Geoderma Regional, 20*, Article e00256.

Tian, G., Justicia, R., Coleman, D. C., & Carroll, C. R. (2007). Assessment of soil and plant carbon levels in two ecosystems (woody bamboo and pasture) in montane Ecuador. *Soil Science, 172*(6), 459–468. https://doi.org/10.1097/ss.0b013e31804fa21a

Tsui, C. C., Tsai, C. C., & Chen, Z. S. (2013). Soil organic carbon stocks in relation to elevation gradients in volcanic ash soils of Taiwan. *Geoderma, 209*, 119–127. https://doi.org/10.1016/j.geoderma.2013.06.013

Vestergaard, R. J., Vasava, H. B., Aspinall, D., Chen, S., Gillespie, A., Adamchuk, V., & Biswas, A. (2021). Evaluation of optimized preprocessing and modeling algorithms for prediction of soil properties using vis-nir spectroscopy. *Sensors, 21*(20), 6745. https://doi.org/10.3390/s21206745

Viscarra Rossel, R. A. V., Chen, C., Grundy, M. J., Searle, R., Clifford, D., & Campbell, P. H. (2015). The Australian three dimensional soil grid: Australia's contribution to the globalsoilmap project. *Australian Journal of Soil Research, 53*(8), 845–864. https://doi.org/10.1071/SR14366

Wadoux, A. M. J. C., Heuvelink, G. B. M., de Bruin, S., & Brus, D. J. (2021). Spatial cross-validation is not the right way to evaluate map accuracy. *Ecological Modelling, 457*. https://doi.org/10.1016/j.ecolmodel.2021.109692

Wang, S., Gao, J., Zhuang, Q., Lu, Y., Gu, H., & Jin, X. (2020). Multispectral remote sensing data are effective and robust in mapping regional forest soil organic carbon stocks in a northeast forest region in China. *Remote Sensing, 12*(3), 393. https://doi.org/10.3390/rs12030393

Wang, H., Liu, S. R., Mo, J. M., Wang, J. X., Makeschin, F., & Wolff, M. (2010). Soil organic carbon stock and chemical composition in four plantations of indigenous tree species in subtropical China. *Ecological Research, 25*, 1071–1079. https://doi.org/10.1007/s11284-010-0730-2

Wang, B., Waters, C., Orgill, S., Cowie, A., Clark, A., Li Liu, D., Simpson, M., McGowen, I., & Sides, T. (2018). Estimating soil organic carbon stocks using different modelling techniques in the semi-arid rangelands of Eastern Australia. *Ecological Indicators, 88*, 425–438. https://doi.org/10.1016/j.ecolind.2018.01.049

Wang, S., Zhuang, Q., Wang, Q., Jin, X., & Han, C. (2017). Mapping stocks of soil organic carbon and soil total nitrogen in Liaoning Province of China. *Geoderma, 305*, 250–263. https://doi.org/10.1016/j.geoderma.2017.05.048

Weiss, A. (2001). Topographic position and landforms analysis. In , *Vol 200. Poster presentation, ESRI user conference, San Diego, CA.*

Wilson, J. P., & Gallant, J. C. (2000). Digital terrain analysis. *Terrain analysis: Principles and applications, 6*(12), 1–27.

Xie, E., Zhang, Y., Huang, B., Zhao, Y., Shi, X., Hu, W., & Qu, M. (2021). Spatiotemporal variations in soil organic carbon and their drivers in Southeastern China during 1981-2011. *Soil and Tillage Research, 205*, Article 104763. https://doi.org/10.1016/j.still.2020.104763

Xu, Y., Wang, X., Bai, J., Wang, D., Wang, W., & Guan, Y. (2020). Estimating the spatial distribution of soil total nitrogen and available potassium in coastal wetland soils in the yellow river delta by incorporating multi-source data. *Ecological Indicators, 111*, Article 106002. https://doi.org/10.1016/j.ecolind.2019.106002

Yang, C., Zhang, X., Ni, H., Gai, X., Huang, Z., Du, X., & Zhong, Z. (2021). Soil carbon and associated bacterial community shifts driven by fine root traits along a chronosequence of moso bamboo (Phyllostachys edulis) plantations in subtropical China. *Science of the Total Environment, 752*, Article 142333. https://doi.org/10.1016/j.scitotenv.2020.142333

Yu, H., Zha, T., Zhang, X., Nie, L., Ma, L., & Pan, Y. (2020). Spatial distribution of soil organic carbon may be predominantly regulated by topography in a small revegetated watershed. *Catena, 188*, Article 104459. https://doi.org/10.1016/j.catena.2020.104459

Zeraatpisheh, M., Ayoubi, S., Jafari, A., & Finke, P. (2017). Comparing the efficiency of digital and conventional soil mapping to predict soil types in a semi-arid region in Iran. *Geomorphology, 285*, 186–204. https://doi.org/10.1016/j. geomorph.2017.02.015

Zeraatpisheh, M., Ayoubi, S., Jafari, A., Tajik, S., & Finke, P. (2019). Digital mapping of soil properties using multiple machine learning in a semi- arid region, Central Iran. *Geoderma, 338*, 445–452. https://doi.org/10.1016/j.geoderma.2018.09.006

Zhang, X., Zhang, F., Wang, D., Fan, J., Hu, Y., Kang, H., Chang, M., Pang, Y., Yang, Y., & Feng, Y. (2018). Effects of vegetation, terrain and soil layer depth on eight soil chemical properties and soil fertility based on hybrid methods at urban forest scale in a typical loess hilly region of China. *PLoS One, 13*(10), Article e0205661. https://doi.org/10.1371/journal.pone.0205661

Zhi, J., Zhang, G., Yang, F., Yang, R., Liu, F., Song, X., & Li, D. (2017). Predicting mattic epipedons in the northeastern Qinghai-Tibetan Plateau using random forest. *Geoderma Regional, 10*, 1–10. https://doi.org/10.1016/j.geodrs.2017.02.001

CHAPTER

9 Monitoring soil degradation warning zones using remote sensing data

Narges Kariminejad[1], Abdolhossein Boali[2], Michele Delchiaro[3] and Mohammad Moumeni[4]

[1]*Department of Natural Resources and Environmental Engineering, College of Agriculture, Shiraz University, Shiraz, Iran;* [2]*Department of Arid Zone Management, Gorgan University of Agricultural Sciences and Natural Resources, Gorgan, Iran;* [3]*Department of Earth Sciences, Sapienza University of Rome, Rome, Italy;* [4]*Department of Science, University of Roma Tre, Rome, Italy*

1. Introduction

Soil degradation refers to the process of land degradation in which fertile and cultivable areas are transformed into arid and desert landscapes. Land degradation is recognized as a major global environmental challenges that require continuous attention and monitoring (Boali et al., 2019). This phenomenon is typically caused by human activities such as unsustainable agricultural practices, deforestation, and improper use of water resources, while natural factors, such as climate change, also play a significant role. The consequences of soil degradation include reduced soil fertility, a decline in vegetation cover, increased frequency of dust storms, and a deterioration in the quality of life for rural communities. Monitoring soil degradation trends using classical methods demands substantial time, financial, and human resources, and such approaches typically offer limited spatial and temporal coverage (Birhanu et al., 2019). Moreover, both environmental and human-induced changes occur rapidly, and outdated data lead to reduced effectiveness of mitigation and prevention programs (Lee et al., 2019). Furthermore, the lack of a continuous and effective monitoring system further complicates the accurate assessment of soil degradation drivers and impacts, consequently delaying the implementation of scientific solutions (Ontel et al., 2023).

In recent years, the application of remote sensing technologies and satellite data has fundamentally transformed the monitoring of environmental phenomena, including soil degradation or desertification (Ebrahimi-khusfi et al., 2021). By providing diverse and repetitive datasets that cover vast areas, these technologies offer powerful tools for monitoring trends, identifying hazard zones, and analyzing land surface dynamics (Wei et al., 2020). The integration of satellite imagery with GIS further enhances these capabilities by enabling the production of up-to-date and reliable maps, facilitating trend analysis and scenario modeling, and supporting decision-making processes (Boali et al., 2025). Consequently, remote sensing has become a powerful tool for generating spatial information at regular intervals and assessing complex environmental processes (Akbari et al., 2020). In soil degradation/desertification

The Soil Nexus. https://doi.org/10.1016/B978-0-443-45286-4.00008-0

monitoring, various indices are extracted from satellite images that provide a comprehensive overview of desertification status across different regions (Guo et al., 2020).

In remote sensing indices, warning ranges refer to a span of values indicating unfavorable or critical conditions within a region (Alibakhshi, 2023). In the context of desertification, these ranges are established by defining specific thresholds to identify areas at high risk of desertification (Akbari et al., 2020). Accurate determination of these thresholds and warning zones enables more precise monitoring and timely intervention to prevent further desertification, thereby helping managers and policymakers optimize resource management. For instance, Akbari et al. (2020) reported a 78% increase in desertification warning zones in arid regions of northeastern Iran in 2015 compared to the beginning of the study period, based on 15 years of data and thresholds defined for key land degradation indices. Moreover, Lamchin et al. (2016) employed Landsat satellite imagery and indices like NDVI, TGSI, and albedo to assess land cover changes and desertification trends in the Hovokhan Protected Area of Mongolia. Their findings demonstrated the effectiveness of these indices in diagnosing land condition and monitoring desertification dynamics.

Sabzevar County is recognized as a region prone to soil degradation in northeastern Iran, due to factors such as pressure on groundwater resources, recurrent droughts, degradation of native vegetation cover, and prolonged dry cycles. The future of agriculture and the sustainability of the ecosystems in this area largely depend on proper management and timely identification of critical zones. The present study investigates soil degradation warning zones in Sabzevar County by integrating remote sensing data, temporal monitoring of changes, and spatial pattern analysis. This research aims to identify the trend and intensity of soil degradation from 2005 to 2024 using multiple remote sensing indices and to precisely delineate areas at greater risk. The goal of this approach is to provide a remote sensing-based platform to support management and decision-making and promoting sustainable natural resource management in Sabzevar County.

2. Study area

Sabzevar County, covering an area of over 12,420 km^2, is located in northeastern Iran, and mostly lies within the Central Desert Basin. The main drainage direction in the area is from east to west, dominated by the Kalshur River. Two important towns in this county are Davarzan and Rudab. The elevation ranges vary from 796 m in the plains to 2924 m in the Joghatai Mountains. Average annual precipitation is approximately 180 mm, increasing to 400 mm in the highlands and decreasing to below 150 mm in desert areas. The mean annual temperature is 17.2°C, and the annual evaporation ranges between 2.200 and 3.000 mm (Fig. 9.1).

3. Materials and methods

3.1 Data collection

The first step in conducting this research was data collection. To evaluate the physical and chemical properties of the soil, 30 soil samples were collected from the study area in July 2023. Vegetation cover was assessed using one-square-meter plots systematically established along 100 m transects (three transects per land use type). Within each plot, the percentage of vegetation cover was estimated by expert assessment and subsequently generalized to represent the corresponding land use category.

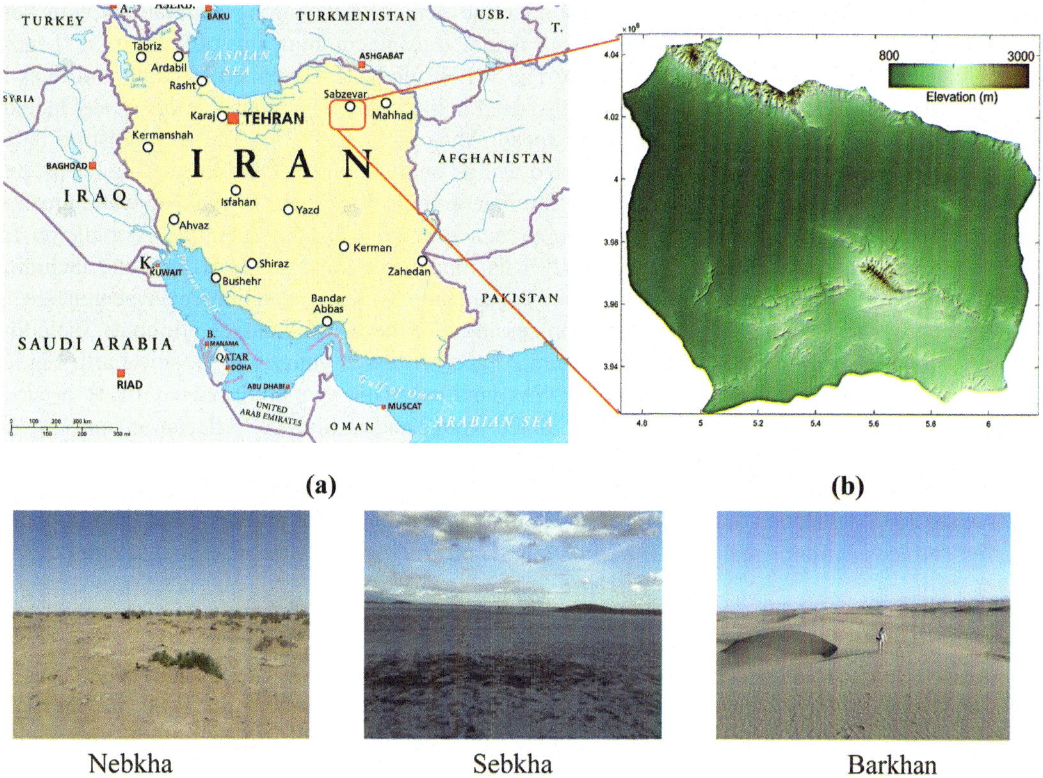

FIGURE 9.1

Location of the study area (A: Map of Iran and B: Digital elevation model).

To assess soil degradation warning zones, three remote sensing indices, NDVI, NDSI, and TGSI, and two remote sensing-based models, WEHI and ICONA, were employed. Satellite imagery from the Landsat 5, 7, and 8 series, preprocessed for atmospheric and geometric corrections and cross-calibration between sensors, served as the primary data source (Dwyer et al., 2018). The indices were extracted annually for the period 2005 – 2024, and field data were used to validate the remote sensing results. All image processing and data analyses were performed using R statistical software (Gorelick et al., 2017).

Annual Landsat 5, 7, and 8 Surface Reflectance imagery was used to derive NDVI, NDSI, and TGSI, WEHI and ICONA over the period 2005–2024. Image preprocessing, cloud and shadow masking, and index calculation were conducted using the Google Earth Engine (GEE) platform, applying consistent quality assurance (QA) criteria across the entire time series. The Landsat 7 SLC-off issue after 2003 was addressed by compositing multiple images within each annual window using a median-based approach. Validation was performed in 2024 using field data.

3.2 Calculation of remote sensing indices

Following data collection, the next step involved selecting appropriate indices capable of accurately reflecting environmental changes. The chosen indicators were selected based on the climatic and geomorphological characteristics of the study area and demonstrated strong correlations with field data. The TGSI, NDSI, and NDVI indices were calculated using combinations of Landsat 5, 7, and 8 spectral bands (Table 9.1) (Boali et al., 2024). The WEHI model, first applied by Yang and Lees in 2014 to assess wind erosion in Australia, was also used in this study. The inputs for this model include soil moisture, bare soil percentage, and wind speed (Yang & Leys, 2014). The soil moisture index (NDMI) and bare soil index (BSI) were extracted using combinations of Landsat satellite bands, and wind speed data, which were obtained from local meteorological stations. To evaluate water erosion, the ICONA model, a remote sensing–based approach developed by the Scientific Association for Nature Conservation in Spain, was employed. This model integrates multiple datasets, including remote sensing indices, geological information, slope, land use, and vegetation cover percentage. A key component of the study was spatial data processing. For the ICONA model, all inputs, including geology, slope, land use, and NDVI were generated and processed entirely in Google Earth Engine (GEE). Slope was derived from a 30 m SRTM DEM, land-use maps were produced in GEE at the same resolution, and annual NDVI values were calculated from Landsat Surface Reflectance imagery. All layers were normalized, weighted according to ICONA guidelines, and combined using a weighted overlay approach. The ICONA and WEHI models, being remote sensing-based, are well suited for monitoring soil degradation and erosion over large areas and long time periods. Their integration of satellite-derived environmental data allows consistent and efficient assessment across the study region.

3.3 Determination of warning thresholds

To identify areas exposed to potential risks, a warning threshold was set for each selected index mentioned earlier. These thresholds were determined based on a review of scientific literature and similar studies addressing environmental change, allowing the identification of high-risk zones.

Table 9.1 Selection of soil degradation indicators based on the review of sources and field studies.

		Classes			
Indicators	**Formula**	**Low**	**Medium**	**Severe**	**References**
TGSI	(R − B/R + B + G)	(−0.50) − (−0.20)	(−0.20) − (−0.10)	(−0.10) − (0.20)	(Lamchin et al., 2016; Meng et al., 2021)
NDSI	(R − NIR/R + NIR)	(−1) − 0	0–0.5	0.5–1	(Lamchin et al., 2016)
NDVI	(NIR + R/NIR − R)	>0.40	0.40–0.25	0.25≥	(Jiang & Lin, 2018; Lamchin et al., 2016)
WEHI[a]	(NDMI/(wind speed[a] MBI))	0–0.25	0.26–0.5	0.5 <	(Yang and Leys, 2014)
ICONA[b]	(Geology[a] slope)[a] (Land use[a] NDVI)	Class 1 and 2	Class3	Class4	(Reis et al., 2017)

[a]*Wind erosion hazard index.*
[b]*Institute conservation of nature.*

Defining precise warning levels helps environmental managers to implement preventive actions before critical degradation occurs.

3.4 Trend analysis of time series changes

To examine temporal changes in environmental indices and quantify the intensity and direction of these changes, time series trend analysis was employed. The non-parametric Kendall's Tau statistic was chosen as the main trend analysis index (Elzopy et al., 2021). This statistic allows the determination of both the direction (increasing or decreasing) and strength of trends independently of data distribution, making it suitable for environmental datasets. Higher Kendall's Tau values indicate stronger trends, and analysis of these trends can reveal patterns of environmental change, such as desertification, drought, rising temperature, or declining rainfall in the Sabzevar region. The results of these analyses lead to a deeper understanding of current and future climatic and environmental consequences.

3.5 Detecting abrupt changes in data trends using Pettitt's test

Environmental time series may not always exhibit gradual changes. Sudden shifts or abrupt changes in trends can occur, which are critical for risk assessment and management. To detect such abrupt changes, Pettitt's Test was utilized. This non-parametric test is particularly effective for identifying change points in environmental and climatic time series of satellite data (Pettitt, 1979). The results allow researchers to determine the exact timing of major changes in index trends and to explore their potential relationships with environmental or human-induced phenomena.

4. Results and discussion

4.1 Evaluation of remote sensing indicators

Quantitative validation was performed by correlating remote sensing indices and model outputs with field measurements. Specifically, NDSI was compared with soil EC ($R^2 = 0.65$), NDVI with vegetation cover plots ($R^2 = 0.72$), TGSI with soil particle size distribution ($R^2 = 0.58$), and both WEHI and ICONA with independent field-based erosion evidence ($R^2 = 0.50$–0.78). These results indicate satisfactory agreement between satellite-derived indices and ground data. Fig. 9.2 illustrates the key remote sensing indices used to assess the ecological status of the study area. The NDVI index reveals the spatial distribution of vegetation density, indicating higher values in the northern and northeastern parts of Sabzevar County, while the central and southern regions show lower values, reflecting degradation or reduced ecosystem functionality. Conversely, the NDSI and TGSI indices represent soil salinity and surface soil texture, respectively. Higher NDSI values in the northern and central parts indicate saline conditions and reduced soil fertility, both of which can seriously hinder plant growth and increase the risk of land degradation. The TGSI results highlight a predominance of fine soil particles in the northern and eastern parts, making these areas more susceptible to wind and water erosion. The spatial overlap of NDVI, NDSI, and TGSI patterns suggests that the combined effects of weak vegetation cover, high soil salinity, and fine-textured soils create fragile environmental conditions in specific parts of the region (Wei et al., 2018).

Findings derived from the WEHI index and the ICONA model further support this trend. The concentration of high WEHI values in the central and northern parts indicates the sensitivity of these

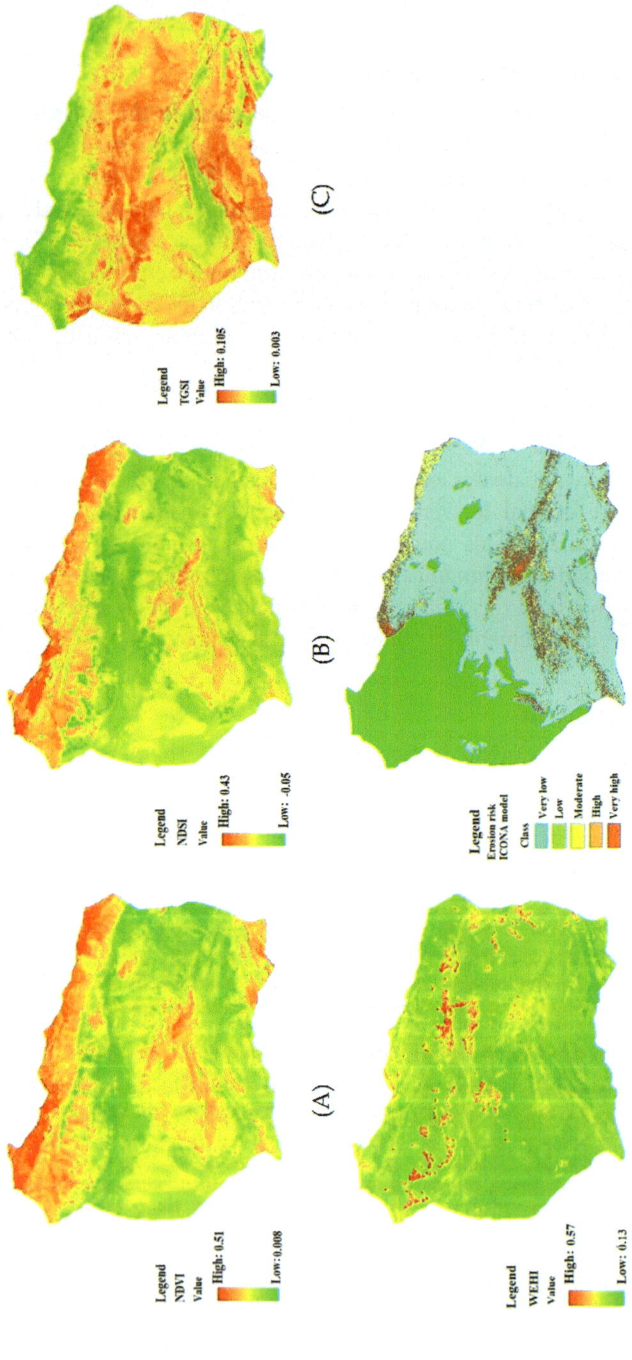

FIGURE 9.2

Distribution of remote sensing indicators in the study area (A: NDVI, B: NDSI, C: TGSI, D: WEHI, and E: ICONA).

areas to wind erosion, especially where reduced vegetation cover and dominance of fine-textured soils are observed. In contrast, results from the ICONA model show that, although large parts of the region are at low risk of water erosion, the northern and eastern areas face moderate to high susceptibility. These patterns reveal the synergistic effects of multiple degradation factors and emphasize the necessity of integrated management strategies to reduce environmental stress. Overall, the results indicate that soil salinity, weak vegetation cover, and fine-textured soils are the principal drivers intensifying erosion processes in the region (Reis et al., 2017). Therefore, particular attention to protective and restorative measures in these areas is essential.

4.2 Determining the alarm threshold for remote sensing indicators

For each remote sensing index, a warning threshold was set to identify high-risk areas. NDVI values below 0.3 were classified as indicative of weak vegetation cover, whereas NDSI values greater than 0.2 and TGSI values more than −0.1 were interpreted as markers of saline and fine-textured soils, respectively. Additionally, WEHI values higher than 0.3 were used to identify areas susceptible to wind erosion, and ICONA class ≥3 was designated to represent regions with moderate to high water erosion risk (Table 9.2). Application of these thresholds indicates that the combined presence of weak vegetation cover, high soil salinity, and fine-textured soils creates critical hotspots for wind erosion. Conversely, in areas characterized by steep slopes and sensitive lithology, the combination of decreased NDVI and inappropriate land use practices highlights regions of high risk for water erosion (ICONA ≥3). These results emphasize the importance of simultaneously managing vegetation cover, remediating saline and fine-textured soils, and implementing soil and water conservation measures in high-risk zones (Rodríguez-Caballero et al., 2019).

4.3 Monitoring soil degradation warning areas

Fig. 9.3 shows the temporal changes in warning zones for each remote sensing index across the study area between 2005 and 2024. All indices exhibit a pronounced increasing trend in the extent of warning zones (Table 9.3). The ICONA index shows the greatest increase in warning zone area among all indices, with an expansion of 29,937 ha (24.09%), indicating a significant intensification of water-induced soil erosion in the region. The NDVI ranks second, with an increase of 21,962 ha (17.67%),

Table 9.2 Determining the warning threshold of remote sensing indicators in the studied area.

Indicators	Threshold	Ground surface data	References
TGSI	More than −0.1	Soil surface profile	(Xiao et al., 2006)
NDSI	More than 0.2		(Dehni & Lounis, 2012)
NDVI	Less than 0.3	Vegetation percentage in plots	(Lamchin et al., 2016)
WEHI	More than 0.3	IRIFR model map based on area data	(Yang and Leys, 2014)
ICONA	Class 3	EPM model map based on area data	(Reis et al., 2017)

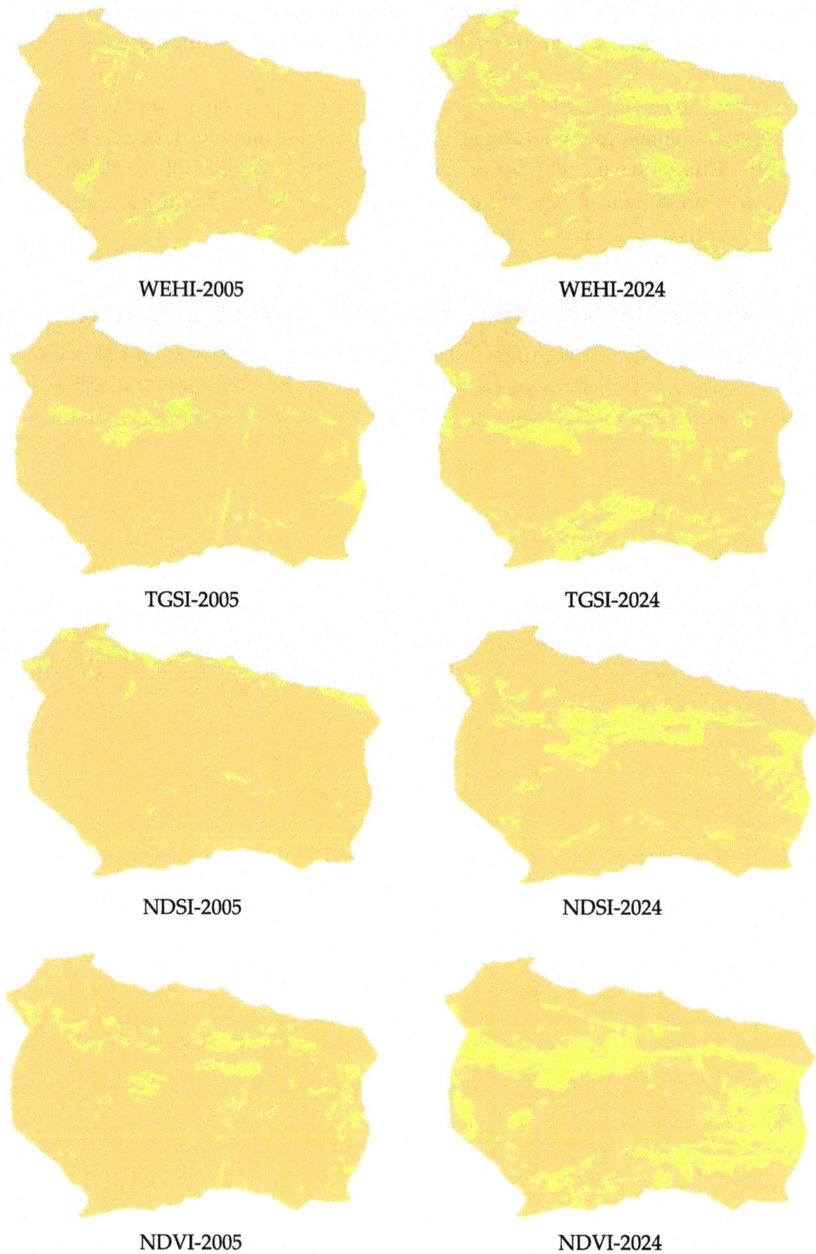

FIGURE 9.3

Investigating the trend of changes in the area of warning zones for each of the indicators under study.

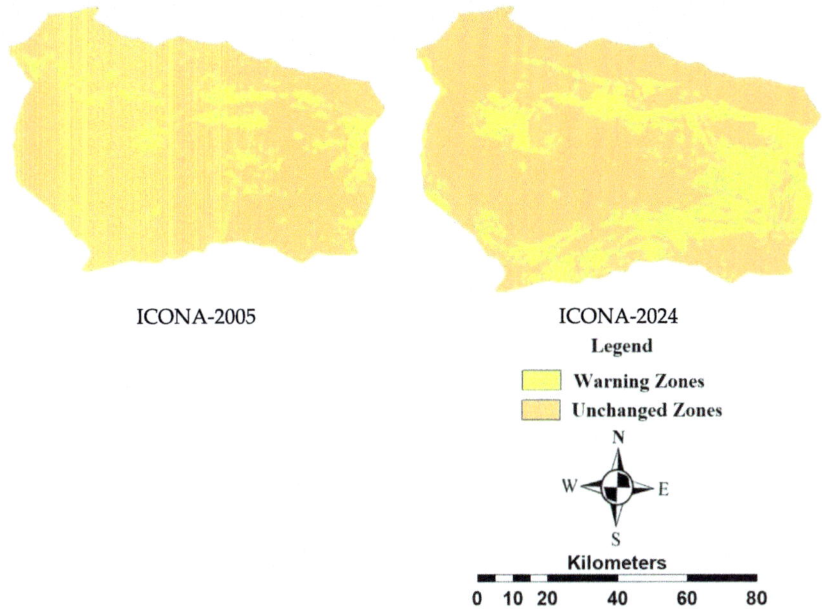

FIGURE 9.3 Cont'd

Table 9.3 Trend of changes in warning ranges of remote sensing indicators.

Indicators	2005 (Hectares)	2024 (Hectares)	Changes (2005–2024)	
			Hectares	Percent (%)
TGSI	4676	13,417	8741	7.03
NDSI	3488	14,497	11,009	8.86
NDVI	5724	27,687	21,962	17.67
WEHI	4696	16,331	11,634	9.36
ICONA	10,446	40,384	29,937	24.09

reflecting the expansion of areas characterized by weak and stressed vegetation cover. The NDSI and wind erosion (WEHI) indices also show concerning trends, with increases of 8.86% and 9.36%, respectively. Although, the soil surface texture index (TGSI) demonstrates the smallest relative growth (7.03%), it still shows a considerable expansion by adding 87,405 ha to warning areas.

The ICONA is likely associated with increased severe rainfall events, loss of protective vegetation cover, and land use changes. Simultaneous increase in the NDVI, which signifies declining plant health, is both a cause and consequence of these degradation processes. A predictable spatial overlap among warning zones identified by WEHI, TGSI, and NDSI indices with areas of weak NDVI forms a vicious cycle of degradation (Fig. 9.3). Reduced vegetation cover leads to increased wind erosion and salinity, which, in turn, further inhibits vegetation recovery. Overall, these data reveal a critical environmental situation that requires urgent intervention and planning for sustainable land management.

4.4 Analysis of Kendall Tau and Petit tests

By examining the trend graphs and statistical test results, it is clearly observed that all indices (except TGSI) exhibit a strong and statistically significant increasing trend in the area of warning zones from 2005 to 2024 (Fig. 9.4). The Kendall's Tau test, which evaluates the magnitude and direction of the trend, produced high positive tau values for NDSI (0.811), WEHI (0.791), and NDVI (0.751), indicating a strong and consistent increasing trend over the 20-year period. Pettitt's test, which detects abrupt change points in trends, confirms the significance of these changes for NDSI, WEHI, and NDVI with a *P*-value $<.05$ (Table 9.4). Specifically, the wind erosion index (WEHI) shows the strongest abrupt change point with a *P*-value of 0.002.

In contrast, although the ICONA and TGSI indices also show increasing trends in the graphs, their statistical significance and consistency are comparatively weaker. Pettitt's test identified 2010 as the change-point year for NDSI, WEHI, and NDVI, indicating a significant shift in the trend of warning zones ($P < .05$). This change corresponds with documented drought events and land-use shifts in Sabzevar County. The results highlight a marked intensification of desertification processes, particularly in soil salinity, vegetation cover degradation, and wind erosion.

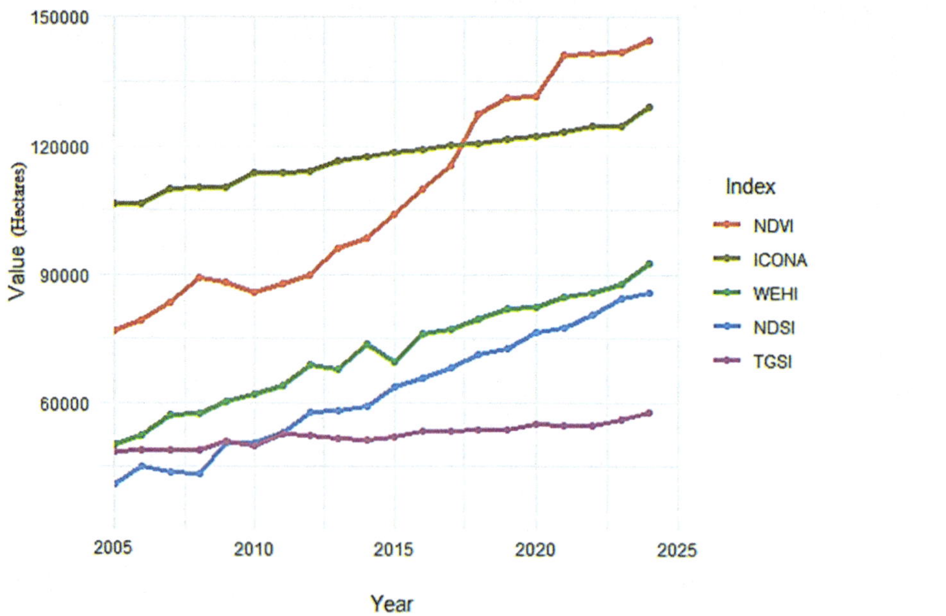

FIGURE 9.4

Time series review of areas under remote sensing indicator warnings (2005–2024).

Table 9.4 Investigating statistical tests of trends and jumps in the warning areas.

Remote sensing indicators	NDVI	NDSI	TGSI	WEHI	ICONA
Kendall's tau	0.751	0.811	0.232	0.791	0.312
Pettitt (*P*-value)	0.039	0.031	0.153	0.002	0.422

5. Conclusion

The results of this study clearly indicate an alarming and intensifying trend of soil degradation in the study area over the past 2 decades. Spatiotemporal analyses based on remote sensing revealed overlapping risk patterns, showing that reductions in vegetation cover, increased soil salinity, and dominance of fine-textured soils have exacerbated a vicious cycle of degradation. Strong and statistically significant upward trends in the area of warning zones, especially in indicators of water erosion (24.09% increase), vegetation degradation (17.67% increase), and wind erosion (9.36% increase), along with the detection of abrupt change points by Pettitt's test, confirm the current critical state.

This study demonstrates that a complex interplay of natural and human factors has driven a significant acceleration in degradation processes. The spatial overlap of areas with weak vegetation, saline soils, and erosion susceptibility emphasizes the need for an integrated management approach. The findings highlight the necessity of continuous monitoring using remote sensing technologies, combined with the implementation of sustainable land management strategies. These results underscore the urgent need for immediate intervention and integrated management measures, including vegetation restoration, saline soil amelioration, and erosion control, to combat desertification and move toward long-term ecological sustainability.

Acknowledgments

The authors would like to thank DeepSeek for assistance with the grammatical editing of the manuscript.

References

Akbari, M., Shalamzari, M. J., Memarian, H., & Gholami, A. (2020). Monitoring desertification processes using ecological indicators and providing management programs in arid regions of Iran. *Ecological Indicators, 111* (December 2019), Article 106011. https://doi.org/10.1016/j.ecolind.2019.106011

Alibakhshi, S. (2023). A robust approach and analytical tool for identifying early warning signals of forest mortality. *Ecological Indicators, 155*(June), Article 110983. https://doi.org/10.1016/j.ecolind.2023.110983

Birhanu, A., Masih, I., Zaag, P., Van Der Nyssen, J., & Cai, X. (2019). Impacts of land use and land cover changes on hydrology of the Gumara catchment , Ethiopia. *Physics and Chemistry of the Earth, March 2018*, 1–10. https://doi.org/10.1016/j.pce.2019.01.006

Boali, A., Bashari, H., & Jafari, R. (2019). Evaluating the potential of Bayesian networks for desertification assessment in arid areas of Iran. *Land Degradation and Development, 30*(4), 371–390. https://doi.org/10.1002/ldr.3224

Boali, A., Hosseinalizadeh, M., Kariminejad, N., Asgari, H. R., Behbahani, A. M., Naimi, B., … Rad, M. M. (2025). Evaluation of early warning signals for soil erosion using remote sensing indices in northeastern Iran. *Scientific Reports, 15*(9742), 1–13.

Boali, A., Asgari, H.R., Mohammadian, A., Salmanmahiny, A., & Naimi, B. (2024). Remote sensing applications : Society and environment remotely sensed desertification modeling using ensemble of machine learning algorithms. *Remote Sensing Applications: Society and Environment, 34*(January), 1-16.

Dehni, A., & Lounis, M. (2012). Remote sensing techniques for salt affected soil mapping: Application to the Oran region of Algeria. *Procedia Engineering, 33*, 188–198. https://doi.org/10.1016/j.proeng.2012.01.1193

Dwyer, J. L., Roy, D. P., Sauer, B., Jenkerson, C. B., Zhang, H. K., & Lymburner, L. (2018). Analysis ready data: Enabling analysis of the landsat archive. *Remote Sensing, 10*(9). https://doi.org/10.3390/rs10091363

Ebrahimi-Khusfi, Z., Taghizadeh-Mehrjardi, R., & Mirakbari, M. (2021). Evaluation of machine learning models for predicting the temporal variations of dust storm index in arid regions of Iran. *Atmospheric Pollution Research, 12*(1), 134–147. https://doi.org/10.1016/j.apr.2020.08.029

Elzopy, K. A., Chaturvedi, A. K., Chandran, K. M., Gopinath, G., Naveena, K., & Surendran, U. (2021). Trend analysis of long-term rainfall and temperature data for Ethiopia. *South African Geographical Journal, 103*(3), 381–394. https://doi.org/10.1080/03736245.2020.1835699

Gorelick, N., Hancher, M., Dixon, M., Ilyushchenko, S., Thau, D., & Moore, R. (2017). Remote sensing of environment Google Earth Engine : Planetary-scale geospatial analysis for everyone. *Remote Sensing of Environment, 202*(2016), 18–27. https://doi.org/10.1016/j.rse.2017.06.031

Guo, B., Zang, W., Han, B., Yang, F., Luo, W., He, T., … Chen, S. (2020). Dynamic monitoring of desertification in Naiman Banner based on feature space models with typical surface parameters derived from LANDSAT images. *Land Degradation and Development, 31*(12), 1573–1592. https://doi.org/10.1002/ldr.3533

Jiang, M., & Lin, Y. (2018). Desertification in the South Junggar Basin, 2000–2009: Part I. Spatial analysis and indicator retrieval. *Advances in Space Research, 62*(1), 1–15. https://doi.org/10.1016/j.asr.2017.11.038

Lamchin, M., Lee, J. Y., Lee, W. K., Lee, E. J., Kim, M., Lim, C. H., … Kim, S. R. (2016). Assessment of land cover change and desertification using remote sensing technology in a local region of Mongolia. *Advances in Space Research, 57*(1), 64–77. https://doi.org/10.1016/j.asr.2015.10.006

Lee, E. J., Piao, D., Song, C., Kim, J., Lim, C. H., Kim, E., … Lee, W. K. (2019). Assessing environmentally sensitive land to desertification using MEDALUS method in Mongolia. *Forest Science and Technology, 15*(4), 210–220. https://doi.org/10.1080/21580103.2019.1667880

Meng, X., Gao, X., Li, S., Li, S., & Lei, J. (2021). Monitoring desertification in Mongolia based on Landsat images and Google Earth Engine from 1990 to 2020. *Ecological Indicators, 129*. https://doi.org/10.1016/j.ecolind.2021.107908

Ontel, I., Cheval, S., Irimescu, A., Boldeanu, G., Amihaesei, V.-A., Mihailescu, D., … Craciunescu, V. (2023). Assessing the recent trends of land degradation and desertification in Romania using remote sensing indicators. *Remote Sensing Article, 4842*(15), 1–20.

Pettitt, A. N. (1979). A Non-Parametric Approach to the Change-Point Problem. *Applied Statistics, 28*(2), 126–135.

Reis, M., Dutal, H., Bolat, N., & Savaci, G. (2017). Soil erosion risk assessment using GIS and ICONA: A case study in Kahramanmaras, Turkey Mahmut. *Journal of Agricultural Faculty of Gaziosmanpasa University, 34* (2017–1), 64–75. https://doi.org/10.13002/jafag4208

Rodríguez-caballero, E., Raúl, J., Fernández, R., & Chamizo, S. (2019). Biocrust landscape-scale spatial distribution is strongly controlled by terrain attributes : Topographic thresholds for colonization in a semiarid Badland system. *Earth Surface Processes and Landforms, 44*(14), 2771–2779. https://doi.org/10.1002/esp.4706

Wei, H., Wang, J., Cheng, K., Li, G., Ochir, A., Davaasuren, D., & Chonokhuu, S. (2018). Desertification information extraction based on feature space combinations on the Mongolian plateau. *Remote Sensing, 10*(10). https://doi.org/10.3390/rs10101614

Wei, H., Wang, J., & Han, B. (2020). Desertification information extraction along the China-Mongolia railway supported by multisource feature space and geographical zoning modeling. *IEEE Journal of Selected Topics in Applied Earth Observations and Remote Sensing, 13*, 392–402. https://doi.org/10.1109/JSTARS.2019.2962830

Xiao, J., Shen, Y., Tateishi, R., & Bayaer, W. (2006). Development of topsoil grain size index for monitoring desertification in arid land using remote sensing. *International Journal of Remote Sensing, 27*(12), 2411–2422. https://doi.org/10.1080/01431160600554363

Yang, X., & Leys, J. (2014). Mapping wind erosion hazard in Australia using MODIS-derived ground cover, soil moisture and climate data. *IOP Conference Series: Earth and Environmental Science, 17*. https://doi.org/10.1088/1755-1315/17/1/012275

CHAPTER

10

Digital soil moisture mapping using vegetation indices and land surface temperature in Mashhad Plain, Northeastern Iran

Amin Mousavi[1], Sedigheh Maleki[2], Alireza Karimi[3], Seyed Kazem Alavipanah[1] and Tayebeh Safari[3]

[1]*Department of Remote Sensing and GIS, Faculty of Geography, University of Tehran, Tehran, Iran;* [2]*Department of Soil Science, College of Agriculture, Shiraz University, Shiraz, Iran;* [3]*Department of Soil Science, College of Agriculture, Ferdowsi University of Mashhad, Mashhad, Iran*

1. Introduction

Soil moisture refers to the water contained between soil particles and plays an important role in the regeneration of vegetation cover. It controls the spatial and temporal patterns of water infiltration, evaporation, and distribution of vegetation cover (Chaney et al., 2015; Penna et al., 2013). Surface soil moisture is a key variable in hydrological processes that affect water and energy exchanges at the land surface and the atmosphere. However, soil moisture varies greatly at different spatial and temporal scales (Xu et al., 2021) and is strongly influenced by topography, temperature, land use, vegetation cover type, water demand, and soil properties (Kassaye et al., 2021). Accurate estimation of temporal and spatial characteristics of soil moisture is essential for various environmental studies and recent advances in remote sensing (RS) techniques using satellite observations have shown that surface soil moisture can be measured using RS methods (Wang & Qu, 2009). Soil moisture is recognized as a key variable in many environmental, meteorological, hydrological, agricultural and climate change studies (Adekanmbi et al., 2023; Entekhabi et al., 2023). Accurate estimation of soil moisture is essential in studies including agricultural drought monitoring (Imtiaz et al., 2024), vegetation growth (Nawaz et al., 2023) and water resources management (Singh et al., 2023). The ongoing trend of climate change has caused weather events and increased the risk of floods and droughts in the agricultural sector (Adekanmbi et al., 2023; Samaniego et al., 2018). Therefore, accurate monitoring of soil moisture content (SMC) is necessary to estimate crop drought stress and the amount of water required by plants, timely warn farmers to prevent a decrease in agricultural production.

RS is used to monitor large-scale landscapes across a variety of land uses and covers (Corwin, 2021). RS data provide valuable information on various environmental parameters, including vegetation cover, soil composition, and moisture content, which are related to salt content (Rahmanian et al., 2023). Analysis and interpretation of SMC and soil salinity data allow researchers and practitioners to produce accurate spatial maps and models of soil salinity and moisture distribution. These

The Soil Nexus. https://doi.org/10.1016/B978-0-443-45286-4.00010-9

maps and models can provide managers with useful information about soil water status and can be used in management decisions and the development of effective strategies to reduce drought and soil salinity risks (Silakhori et al., 2024; Singh, 2022). Over the past few decades, RS has undergone significant advances, enabling the collection of high-resolution data at various scales (Taghadosi et al., 2019). With the advancement of research, various tools such as satellites, image processing technology, and different spectral bands have emerged (Kasim et al., 2020). Among these tools are multispectral imaging sensors that capture information at different wavelengths and provide more accurate results with higher spatial resolution (Azabdaftari & Sunar, 2016). Soil moisture is an important component of the hydrological cycle that regulates runoff, vegetation cover production, and transpiration (Kim et al., 2016). Soil moisture is a key soil indicator for defining and identifying drought (Keesstra et al., 2016). Soil moisture estimation is useful for detecting early-stage water deficits and tracking evolving drought events. It plays a vital role in addressing crop yield uncertainty, ensuring food security, supporting agricultural insurance, informing policy and decision-making, and facilitating crop planning especially in arid and semi-arid regions (Keesstra et al., 2018; Nawaz et al., 2023; Imtiaz et al., 2024). In drought conditions, water scarcity affects water resources and this can lead to reduced agricultural yields, reduced farmer incomes, and increased unemployment. The lack of rainfall may bring about social and economic issues that cause increased social pressures, potentially resulting in social and political conflicts in developing countries (Kelley et al., 2015). Consequently, soil moisture modeling and measurement is a growing interest. Monitoring the spatial and temporal changes of soil moisture is fundamental to the impact and adaptation to climate change for the sustainability of cropping systems and development of precision agriculture and food security (Zareie et al., 2016). Surface soil moisture is typically defined as the amount of water contained in the upper 5–15 cm of the soil layer (Wang & Qu, 2009). Although water in this thin surface soil layer represents only a small portion of the global water reserves, it plays a critical role in key hydrological, biochemical, physiological, agricultural, and geological processes. Accurate assessment of the spatial and temporal dynamics of surface soil moisture is essential for effective groundwater policy and management (Arshad et al., 2025). Soil moisture at the field level can be estimated using a wide range of methods (Rasheed et al., 2022). Among measurement methods, the point method is the most accurate method for measuring soil moisture (Zhang et al., 2014), but their installation and maintenance are very time-consuming and costly (Arshad et al., 2025; Benninga et al., 2018; Rasheed et al., 2022). One of the advantages of RS data over other indirect methods is that it allows for the creation of a large archive of data with high resolution (Abdulraheem et al., 2023). Different sensors are capable of estimating soil moisture at varying spatial resolutions. Numerous studies have been conducted for this purpose. For example, Zhao et al., (2017) demonstrated that incorporating the Modified Normalized Difference Water Index (MNDWI) can improve the performance of the triangle-based empirical model for estimating surface soil moisture using Land Surface Temperature (LST), the Normalized Difference Vegetation Index (NDVI), and surface albedo. However, the method still faces high daily variability and considerable uncertainty. Li et al., (2011) also used two active and two passive sensors and measurements of Brightness Temperature (BT) to estimate soil moisture in China which They concluded data from microwave sensors had a strong potential for soil moisture estimation. In agricultural areas of Shushtar, soil moisture estimation based on satellite imagery and RS indices indicated that surface soil moisture had a positive relationship with NDVI, SAVI, NDMI, and Soil Moisture Shortwave Infrared (SM_{SWIR}) indices, but an inverse relationship with the LST index. The SM_{SWIR} index provided a relatively accurate representation of soil moisture, suggesting that the SM_{SWIR} index was a significant contributor to representing soil surface moisture values (Abiyat et al., 2022). Additionally, Bagheri et al. (2019) conducted ground sampling from 128 points and Landsat-8 and Sentinel-2 sensors to estimate soil moisture in southern Tehran. The results showed that these sensors can play an effective role in estimating surface soil moisture. Asghari et al., (2022) examined surface soil moisture dynamics based on satellite imagery from Landsat-8 and Sentinel-1. While the TOTRAM algorithm was applied to the Landsat data, Support Vector Regression (SVR) was used to evaluate the spatial distribution patterns and statistical relationships of the Sentinel-1 data. They found that soil moisture distribution is affected strongly by environmental conditions and spatial autocorrelation or the observed clustered structure indicating that areas located closely to each other have similar soil moisture values and there were clearly defined soil moisture clusters that could be important for precision agriculture and environmental studies.

The literature reviewed show that satellite-based indices as LST/NDVI and other moisture-based indices, SMI, NDMI, and NDWI, are very important metrics for assessing soil moisture conditions and crop performance. The interaction between these parameters enables farmers to observe crop conditions, modify irrigation practices, and improve agricultural productivity. The Mashhad Plain is one of the largest and most important plains in Khorasan Razavi Province, Iran. This plain, in addition to supplying the water needed for agriculture, it also supplies drinking water to the cities of Mashhad, Chenaran, Shandiz and Torqebe in eastern Iran. Hence, this research intends to estimate soil moisture using Landsat-8 images over areas within the Mashhad Plain and to monitor soil moisture changes while assessing the effects of soil moisture and salinity status in agricultural land and areas with sparse vegetation through RS indices. This study also evaluates correlations between vegetation indices and environmental factors derived from Landsat-8 data and the relationships between proposed indices and surface soil moisture.

2. Materials and methods

2.1 Study area

The study area, Mashhad plain, is located in the northeastern part of Iran, between the Binalood and Hezarmasjid mountain ranges, and forms part of the Kashaf Roud River basin in the northern section of Khorasan Razavi Province. Geographically, it extends from 58 degrees20′ to 60 degrees08′ E longitude and 35 degrees40′ to 36 degrees03′ N latitude, with a central point near 36 degrees05′28″ N and 60 degrees44′27″ E. The Kashaf Roud River flows from the northwest to the southeast across the plain. Fig. 10.1 shows the location of the study area (Lashkaripour et al., 2014). The Mashhad Plain has an average annual precipitation of 260 mm and a mean temperature of 13.7°C (Islamic Republic of Iran Meteorological Organization, 2017). According to the DeMartonne aridity index, the region is classified as semi-arid. The Mashhad Plain is in the Kopeh Dagh and Binalood geological structural zones, and features as an open syncline bounded on both sides by longitudinal faults. The groundwater aquifers of the Mashhad Plain are in excess of 3000 km^2 in area and progress southeastward toward the Binalood and Hezarmasjid areas (Gharechelou et al., 2021). The aquifer at the southern slope of the plain has been found to be thicker and more permeable. Different locations throughout grounds of the Mashhad Plain have shown greatly varying thickness of aquifers, ranging from 2 m where the aquifer outlets (the outlet zone) to 150 m thick at the southern foothills of the plain (Regional Water Authority of Khorasan Razavi, 2015).

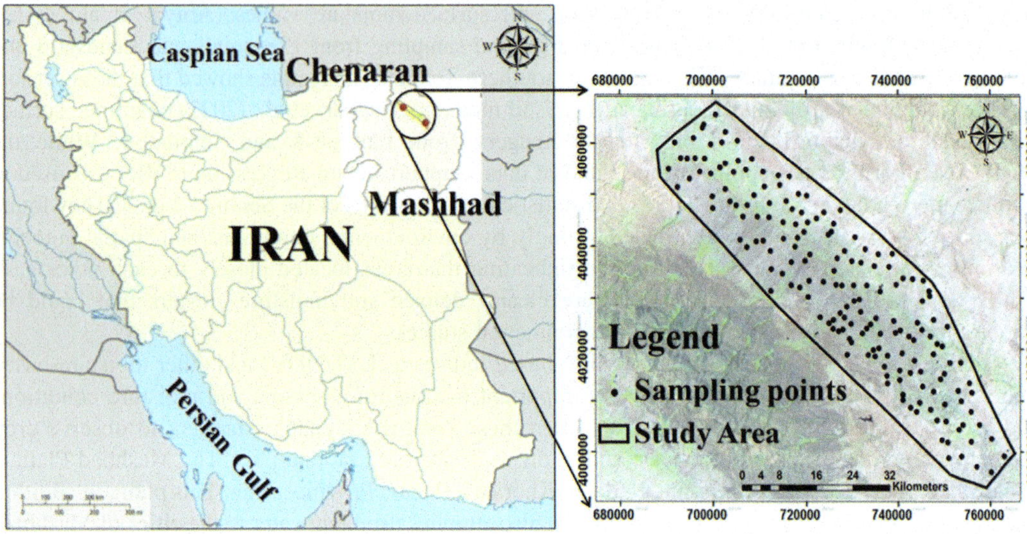

FIGURE 10.1

Location of the study area in Iran with showing of soil sampling points. ***Map lines delineate study areas and do not necessarily depict accepted national boundaries.***

2.2 Datasets collection of satellite imagery

The imagery for this study was gathered in June 2019 with Landsat eight satellites. Landsat eight satellites are equipped with two sensors, the Operational Land Imager (OLI) and the Thermal Infrared Sensor (TIRS). The OLI gathers information in nine shortwave spectral bands (Table 10.1) and also the TIRS gathers information in two thermal infrared bands. The images received from these sensors have a spatial resolution of up to 30 m and are extremely useful for spatial analysis. In this study, these sensors images underwent atmospheric correction, geometric refinement, alongside with thorough radiometric calibration, and pruning of nonessential components. All these steps are critical for achieving better trustworthiness and precision for the dataset (Table 10.2).

2.3 Research implementation process and stages

The rate of evaporation and transpiration, as well as wind and atmospheric currents, have a big effect on how much moisture is in the soil surface. The Google Earth Engine (GEE) environment was used for all steps of the calculation and image processing. GEE is an open-source platform that was chosen because it is very fast at doing calculations and has a large library of spatial datasets (Silakhori et al., 2024). The platform addresses one of the major challenges in RS and Earth system science, namely the management and processing of petabyte-scale satellite imagery datasets. GEE includes a rich and organized multi-petabyte data catalog that includes important public archives such as Landsat,

Table 10.1 Spectral characteristics of the Landsat 8 sensor.

Bands	Wavelength (μm)	Spatial resolution (m)
B1- Coastal/Aerosol	0.43–0.45	30
B2- Blue (B)	0.45–0.51	30
B3- Green (G)	0.53–0.59	30
B4- Red (R)	0.64–0.67	30
B5- Near-infrared (NIR)	0.85–0.88	30
B6- Short wave infrared (SWIR 1)	1.57–1.65	30
B7- Short wave infrared (SWIR 2)	2.11–2.29	30
B8- Panchromatic	0.50–0.68	15
B9- Cirrus	1.36–1.38	30
B10- Thermal infrared (TIRS) 1	10.60–11.19	100
B11- Thermal infrared (TIRS) 2	11.50–12.51	100

Table 10.2 Selection and preprocessing parameters for Landsat-8 images.

Bands	Description
Landsat 8 (OLI/TIRS)	Satellite
LANDSAT/LC08/C02/T1_L2	Dataset
Collection 2, Level-2 surface reflectance	Processing level
LaSRC	Atmospheric correction
June 2019	Acquisition period
<15%	Cloud cover threshold
B7- Short wave infrared (SWIR 2)	Cloud masking method
B8- Panchromatic	Processing platform

Sentinel, and MODIS imagery in a continuous and uptodate manner. Using the relationships suggested by different researchers, each of the vegetation indices in Table 10.3 was examined at and estimated. Satellite images were used to see how they were related to soil moisture. Since the study's sampling happened in June 2019, we tried to choose images with as little cloud cover as possible. Based on the criteria, the images used had less than 15% cloud cover.

NDVI, Soil Adjusted Vegetation Index (SAVI), Normalized Difference Moisture Index (NDMI), Normalized Difference Tillage Index (NDTI), and the (SM_{SWIR}) (Table 10.3) were all used to figure out how wet the soil surface was. NDVI indicates the vitality and health of the plant which is directly linked to the changes in soil moisture. SAVI is an index that corrects NDVI for areas with deep soils and high vegetative cover. It is very important in estimating soil moisture in regions with diverse high vegetative cover. The NDMI was specifically used to estimate soil moisture at a level that is directly related to water. NDTI is responsive to changes in soil moisture after planting. Information captured in

Table 10.3 Indicators used in soil moisture prediction and mapping

Index	Formula	Reference
Normalized Difference Vegetation Index (NDVI)	$NDVI = \frac{(NIR - R)}{(NIR+R)}$	Rouse Jr et al. (1973)
Soil-Adjusted Vegetation Index (SAVI)	$SAVI = \frac{(NIR - R)}{(NIR+R+L)(1+L)}$	Pettorelli et al. (2005)
Normalized Difference Moisture Index (NDMI)	$NDMI = \frac{(NIR - SWIR1)}{(NIR+SWIR2)}$	Van de Griend and Engman (1985)
Normalized Difference Tillage Index (NDTI)	$NDTI = \frac{(SWIR1 - SWIR2)}{(SWIR1+SWIR2)}$	Lobell and Asner (2002)
Soil Moisture shortwave infrared (SM_{SWIR})[a]	$SM_{SWIR} = (SWIR1/RED)^a$	Fabre et al. (2015)
Land surface temperature (LST)[b]	$LY = ML \times Q_{cal} + AL,$ $T = \frac{(K2)}{Ln\left(\frac{K1}{L\lambda} + 1\right)}$	Chavez Jr (1989)

[a] *0.5–1.5.*

[b] *where LY is the spectral radiance (watts per square meter per micrometer), ML and AL are the pixel-to-reflectance conversion factors, which have different values for each band, Qcal is the digital number (DN) of the corresponding band, T is the Earth's surface temperature, K1 and K2 are the thermal constants for each band, whose units are watts per square meter per micrometer and degrees Kelvin (°K), respectively.*

the short wave and near infrared bands of the spectrum is helpful in assessing soil moisture. This has resulted in the design of an index, SM_{SWIR}, which is capable of estimating changes in soil moisture. To assess temperature at the surface level, the LST Index serves as a suitable physical indicator. Temperature varies may reflect changes in soil moisture, as soil moisture directly affects ground surface temperature. The study area included a considerable part of the sampling conducted in the Mashhad Plain, which had sparse vegetation cover, and some parts of the sampling included lands covered by the irrigation network with appropriate vegetation density. It must be mentioned that in the Landsat-8 sensor, besides the above indicators, the LST index was also investigated.

2.4 Field data (Ground data)

The Mashhad Plain, a modern agricultural center in Iran, can help agriculture get better and grow by keeping an eye on and managing salinity and moisture in a smart way. This will also stop saline lands from spreading. In this research, 180 surface soil samples were extracted and measured from a depth of 0–10 cm in a regular grid sampling arrangement (Fig. 10.1). It should be noted that the selection of soil depth is related to the maximum reflectance of electromagnetic waves from the soil surface to increase the accuracy of mapping and prediction by the model and selected auxiliary variables. At each sampling point, a square with a side length of 30 m was considered and a composite sample was taken after mixing the five subsamples from the center and four from the vertices. Most of the samples were found in alluvial sediments originating from watersheds with different geological compositions.

It is imperative to acknowledge that the determination of soil depth is intrinsically linked to the optimal reflectance of electromagnetic waves emanating from the soil surface, thereby enhancing the precision of spatial mapping and predictive modeling in conjunction with the chosen auxiliary variables. At each designated sampling location, a square area with a side length of 30 m was delineated, and a composite sample was obtained subsequent to the amalgamation of five subsamples collected from the center and four additional subsamples acquired from the vertices. The majority of the samples were identified within alluvial sediments that were derived from watersheds exhibiting diverse geological compositions. A total of soil samples was air-dried, passed through a 2-mm sieve and analyzed for soil moisture by direct method and weight moisture percentage (Soil Survey Staff, 1996). Soil moisture was asse ssed as gravimetric moisture content (%) through weight reduction following oven drying. Soil moisture measurements in the field were gathered within ±2 days of the Landsat overpass, with the majority of samples collected either on the same day or the next day. Sampling was deliberately timed to steer clear of intervals right after rainfall or irrigation occurrences. Records of precipitation from local meteorological stations and field observations confirmed prior dry conditions, and locations influenced by recent irrigation were removed from the analysis.

2.5 Accuracy assessment

Following the extraction of desired indicators from satellite images and checking the normality of the resultant data, to check the accuracy and validity of the findings, the correlation relation of each indicator with field data was checked. Next, through linear regression equations, a surface soil moisture model was developed based on the investigated indicators. For validation and assessment purposes, 80% of the data was utilized for training the model and 20% of the data for validation and assessment purposes. Due to the restricted quantity of field samples, a completely independent spatial cross-validation was not implemented, which must be taken into account when analyzing the provided accuracy metrics.

In this research, several statistical indicators were employed, including root mean square error (RMSE), normalized coefficient of determination (R^2), and percentage bias (PBIAS), which are presented in Eqs. (10.1) to (10.3). In addition, NRMSE was determined by normalizing the RMSE against the average of the observed soil moisture measurements.

$$R^2 = \left\{ \frac{\sum_{i=1}^{n} (O_i - \overline{O})(P_i - \overline{P})}{\left[\sum_{i=1}^{n} (O_i - \overline{O})^2\right]^{0.5} \left[\sum_{i=1}^{n} (P_i - \overline{P})^2\right]^{0.5}} \right\}^2 \tag{10.1}$$

$$\text{RMSE} = \sqrt{\frac{\sum_{i=1}^{n} (O_i - \overline{P})^2}{n - 1}} \tag{10.2}$$

$$\text{PBIAS} = 100 \frac{\sum_{i=1}^{n} (O_i - P_i)}{\sum_{i=1}^{n} O_i} \tag{10.3}$$

O_i is the observed soil moisture, $\overline{O}$ is the average of the soil moisture, P_i is the model prediction at the sampling point, and $\overline{P}$ is the average of the relevant values. This is the level of correlation between the two data sets. R^2 values near 1 show that the model accounts for the variation of the observations

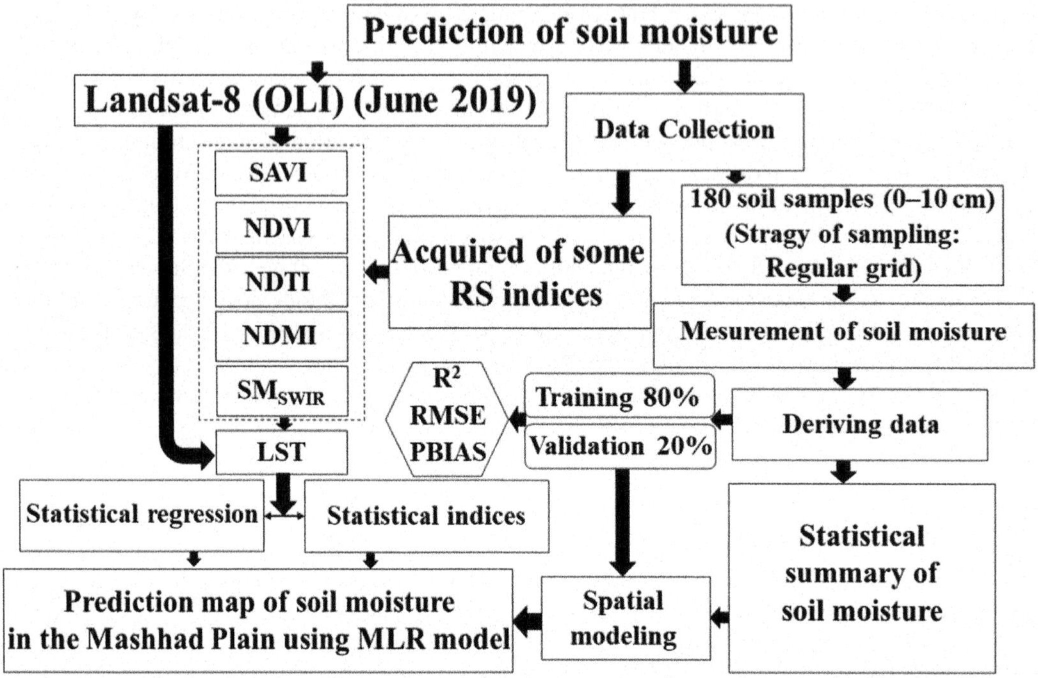

FIGURE 10.2

Overview of the research workflow.

very well. In the same way, and values near zero express good model accuracy. Fig. 10.2 shows the workflow used in this research.

3. Results and discussion

After calculating the RS indices, the Kolmogorov–Smirnov test was used in SPSS software to examine the normal distribution of the data. An examination of the outputs from the Kolmogorov–Smirnov test shows that the data measured and calculated from the Landsat-8 sensor follow a normal distribution (Table 10.4). The Pearson correlation test was used to examine the relationship between soil moisture values and the indicators under study (Table 10.5).

To assess the relationship between field-measured parameters and the indices derived from each sensor, an analysis was performed to investigate the correlation between each index and soil moisture. The results obtained using the Landsat-8 sensor indicated that, among the evaluated indices, NDVI showed strong correlation at 0.01% level with soil moisture, with a coefficient of 0.83. Soil moisture and NDVI are inherently related to plant growth and biological productivity; as soil moisture increases, plants can more effectively use water and nutrients, leading to increased vegetation cover and, consequently, higher NDVI values. Statistical analysis further showed that NDVI had a significant correlation with soil moisture, air temperature, and LST ($P < .001$). After NDVI, the NDTI index had

Table 10.4 Results of normality test for the used indices in soil moisture estimation model

	NDVI	SAVI	NDTI	NDMI	SM_{SWIR}	LST (°C)	Soil moisture (%)
Mean	0.23	0.17	0.08	0.08	0.43	21.44	15.76
Std. Deviation	0.07	0.14	0.11	0.05	0.14	1.85	4.08
Extreme Absolute	0.38	0.36	0.49	0.22	0.51	0.18	0.12
Positive	0.35	0.35	0.49	0.24	0.37	0.15	0.11
Negative	−0.29	−0.31	0.34	−0.15	−0.53	−0.18	−0.08
Kolmogorov-Smirnov Z	0.67	0.58	0.49	0.47	0.53	0.70	0.55
Asymp. Sig. (2-tailed)	0.74	0.84	0.36	0.65	0.68	0.58	0.43

Table 10.5 Correlation between soil moisture values and estimated indices derived from landsat eight significance level = 0.001.

NDVI	NDVI							
SAVI	0.63	SAVI						
NDTI	0.66	0.78	NDTI					
NDMI	0.54	0.53	0.54	NDMI				
SM_{SWIR}	0.60	0.71	0.48	0.75	SM_{SWIR}			
LST	−0.73	−0.87	−0.70	−0.68	−0.67	LST		
Soil moisture	0.83	0.65	0.77	0.56	0.58	−0.60	Soil moisture	
Temperature	0.73	0.69	0.66	0.53	0.68	−0.81	0.74	Temperature

the second lowest correlation with soil moisture, with a coefficient of 0.77 using the Landsat-8 data. Finally, the relationship between LST and soil moisture was evaluated.

LST is an important parameter in soil moisture monitoring and is a useful indicator for predicting soil moisture and analyzing its changes over time. LST can have a significant impact on soil moisture because as temperature increases, evapotranspiration intensifies from the ground surface, resulting in reduced soil moisture. As soil moisture decreases, plants cannot absorb enough water from the soil, resulting in reduced growth and development. Therefore, LST showed high negative correlation (−0.73) with soil moisture.

For this purpose, the study of the correlation between LST and vegetation cover indices showed that all the indices studied, including SAVI, SM_{SWIR}, NDMI, NDTI, and NDVI, had a high correlation, and SAVI and NDVI indices had the lowest negative correlation with values of −0.87 and −0.73, respectively. One of the reasons for the negative correlation between SAVI and LST is the effect of heat on plant health; because with increasing LST, plants are exposed to thermal effects, which can reduce plant growth and health. As a result, vegetation cover in the region decreases and causes a decrease in the SAVI index (Jagadish et al., 2021). Similarly, with increasing LST, water evaporation from the soil and plant surfaces also increases, and this can reduce NDVI and, as a result, the amount of vegetation cover. Another factor affecting the negative correlation between vegetation cover indices and LST is plant phenological changes, because with increasing temperature, plant gas exchange decreases, which causes a decrease in plant photosynthetic activity, growth, and plant health (Ruiz-Vera et al., 2015). Besides analyzing the relationship between indices, regression models were created to forecast soil moisture. Thus, regression model (2) was developed utilizing two main variables, NDTI and LST, that demonstrated significant correlations with soil moisture. Merging these two indices improved the model's performance, allowing for a more precise estimation of soil moisture with Landsat-8 imagery. Fig. 10.3 presents the scatter map of the estimated indices by the Landsat-8 sensor. The study of the spatial NDVI changes of the index showed that this index in Landsat-8 identified only a limited part of the upstream lands of salinity as areas with relatively appropriate vegetation density. These observations are further corroborated by the spatial distribution of the indices, as depicted in Fig. 10.3. Examination of NDVI variations indicates that only a negligible portion of the upstream saline regions exhibits relatively sufficient vegetation cover. This pattern can be explained by the strong constraints that soil salinity imposes on vegetation growth, resulting in reduced NDVI values (Stavi et al., 2021; Shi et al., 2023). Likewise, the SAVI index seems underestimated, likely due to the reduced sensitivity of Landsat-8's spectral bands in regions characterized by sparse vegetation or salt-affected soils. The SMSWIR index, using short-wave infrared bands six and seven, shows high sensitivity to soil moisture, improving the accuracy of soil moisture estimation. Overall, while the term "upstream lands of salinity" highlights areas with the most pronounced salinity effects, the majority of the upstream catchment exhibits varying degrees of salinity, consistent with prior studies in similar semi-arid environments. In addition, in the study of the changes and scatter of the SAVI index, it was also observed that this index in Landsat-8 was underestimated. The reason for this could be due to the low sensitivity of the red and near-infrared bands; therefore, it may be one of the reasons for the low estimate of Landsat-8. Also, in addition to the above, in SAVI, the parameter L (as the earth constant) was used to determine the intensity of infrared light, which should be adjusted based on different environmental conditions. So, in areas with less vegetation and limited soil, the value of L should be smaller in order to reduce the differences in light reflectance in the near-infrared and blue bands. In Landsat-8, due to possible dispersion in the images and the presence of noise, it is difficult to determine the optimal value for L, which reduces the accuracy of estimating vegetation cover with SAVI. An examination of the changes in the SM_{SWIR} index in Landsat-8 showed that the infrared bands in the short-wave range include bands six and seven. Accordingly, the presence of infrared bands in this range with high sensitivity increases the accuracy of estimating soil moisture. The NDMI map indicated differences in canopy and soil moisture, with decreased NDMI values linked to drier and saltier areas, aligning with diminished water content in vegetation and soil experiencing salinity stress (Ceccato et al., 2002). The NDTI indicated bare soil and disrupted areas, typically linked to thin vegetation and soils affected by salinity. The LST maps indicated elevated temperatures in saline and sparsely vegetated zones because of reduced evapotranspiration and heightened soil albedo, further confirming the identification of areas vulnerable to stress (Wang et al., 2020). The SM-SWIR, employing Landsat-8 bands six and seven, demonstrated high sensitivity to changes in surface soil moisture, detecting minor variations influenced by soil texture and salinity. Elevated SM-SWIR values typically aligned with increased moisture availability, while lower values signified dry and possibly saline regions (Dong & Na, 2021). The integration of NDVI, NDMI, NDTI, LST, SAVI, and SM-SWIR indices allows for a thorough evaluation of vegetation stress, soil moisture shortages, and

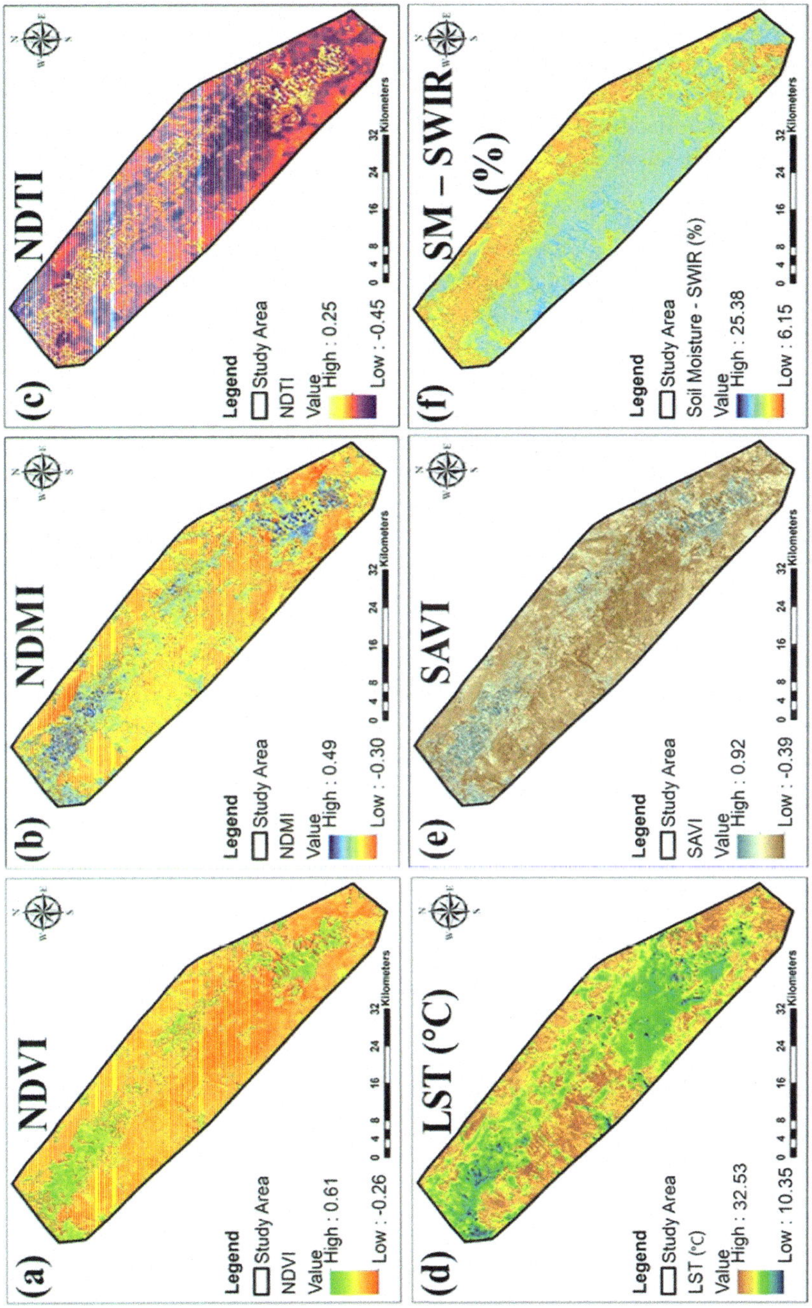

FIGURE 10.3
Scatter map of indices which estimated using Landsat-8 images.

areas susceptible to salinity, providing a more dependable spatial mapping of upstream lands influenced by salinity. Overall, while the term "upstream lands of salinity" emphasizes regions with the most pronounced salinity effects, field observations, historical soil surveys, and previous studies indicate that a substantial portion of the upstream catchment exhibits varying levels of salinity, consistent with semi-arid environment patterns (Lugassi et al., 2017; Wen et al., 2022). The multi-index RS approach thus effectively highlights the combined impacts of salinity, moisture, and vegetation stress across the study area.

3.1 Soil moisture models

Table 10.6 shows the results of the multiple regression analysis between the mean measured soil moisture values with the predicted soil moisture data using regression model. Model one illustrated the predicted model for the region, which is based solely on the SAVI index (Table 10.6). Saline soils lower the osmotic potential of the soil, restricting water absorption by plants, which subsequently affects vegetation indices like SAVI (Allbed & Kumar, 2013; Munns & Tester, 2008). As a result, areas with increased salinity often show reduced plant cover and modified SAVI values, essential for precisely estimating soil moisture in semi-arid and saline-affected regions (Silakhori et al., 2024). Thus, including soil salinity in the model improves its predictive ability and accounts for the observed differences in vegetation indices. Soil electrical conductivity (EC) data measured in the field were utilized to aid in understanding areas impacted by salinity; nonetheless, EC was not directly incorporated as an independent variable in the regression models since the main emphasis of the research was on estimating soil moisture through remote sensing.

In general, the statistical indices indicate that model 1, with an R^2 of 0.61 and an RMSE of 5.26, exhibits a lower correlation with observed values than model 2, while still providing acceptable accuracy. Examination of the PBIAS index as one of the statistical indices shows how much is the average difference between the predicted values and the actual observations.

The PBIAS value in model 1 is 0.31 and in model 2 is 0.25, indicating that both models were able to provide predicted values with acceptable accuracy. The analysis of the normalized root mean square error (NRMSE = 0.28) shows that Model 2 provides acceptable accuracy for estimating soil moisture. Overall, when Landsat-8 data are used for soil moisture estimation, Model 2 demonstrates substantially higher accuracy than Model 1, as reflected by the NRMSE values. The error magnitude achieved is suitable for regional-scale soil moisture evaluation and comparative analysis of moisture variation, but not for point-scale irrigation planning. The error magnitude achieved is suitable for regional-scale

Table 10.6 Results of the regression model obtained from the landsat 8 OLI sensor.

Model		R^2	RMSE	NRMSE	PBIAS
1	Soil moisture = 42.81–120.8 SAVI – 133.5 $SAVI^2$	0.61	5.26	0.33	0.31
2	Soil moisture = 27.28–31.42 LST – 81.85 NDTI	0.69	4.27	0.28	0.25

soil moisture evaluation and comparative analysis of moisture variation, but not for point-scale irrigation planning.

To provide additional context for the performance of the proposed models, a quantitative comparison was performed using the NDVI–LST–based Soil Moisture Index (SMI) as a reference method. The SMI, based on the trapezoidal NDVI–LST feature space, exhibited reduced predictive accuracy relative to the suggested regression Model 2, especially in regions marked by limited vegetation and salinity stress. This outcome indicates that combining land surface temperature with soil disturbance data (NDTI) enhances soil moisture estimation more than relying solely on vegetation-based methods.

Fig. 10.4 also shows the scatterplot of the predicted moisture values using the two fitted models and the measured moisture values. Therefore, considering that regression model 1 predicted moisture based only on the SAVI index, it may have ignored some factors affecting soil moisture. On the other hand, regression Model 2 simultaneously incorporates the factors affecting soil moisture using the two indices, NDTI and LST, which may result in higher accuracy when estimating soil moisture from Landsat-8 images. In a study, it was found that the use of the NDTI and LST indices in soil moisture analysis was shown to be more accurate when combined with other factors (Sánchez et al., 2015). Similarly, Wang et al. (2020) showed that combining SAR data with multi-source optical data, including NDTI index and LST, can improve the accuracy of soil moisture retrieval (Figs. 10.5 and 10.6).

In general, the Landsat 8 OLI sensor has adequate accuracy in estimating soil moisture. The accuracy of the OLI sensor in estimating soil moisture has been studied extensively. Given that part of the study area is located within saline areas with sparse vegetation cover, the ground surface temperature increases due to water shortage and soil salinity and is an important factor in reducing soil moisture. In fact, ground surface temperature can be considered as an important indicator for studying soil moisture in dry and saline areas. As ground surface temperatures rise, the rates of evaporation and transpiration from both soil and plant surfaces experience an increase, leading to a reduction in soil moisture; conversely, the decline in soil moisture coupled with elevated temperatures results in heightened soil salinity, which can adversely affect plant growth and agricultural yield. Consequently, regression

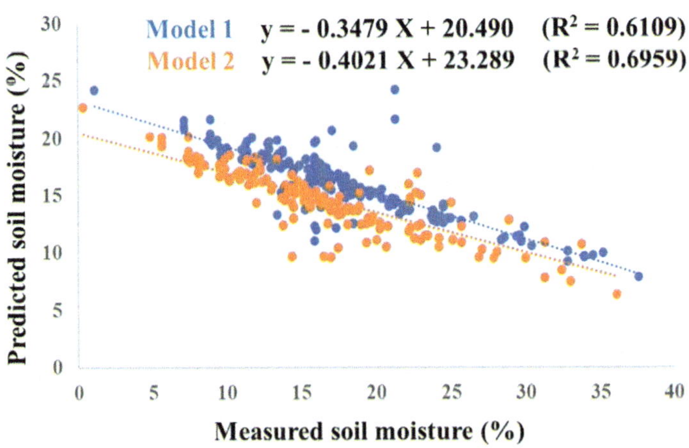

FIGURE 10.4

Scatterplot of predicted soil moisture values and measured soil moisture for model 1 and model 2 with fitted linear regression lines and corresponding slopes.

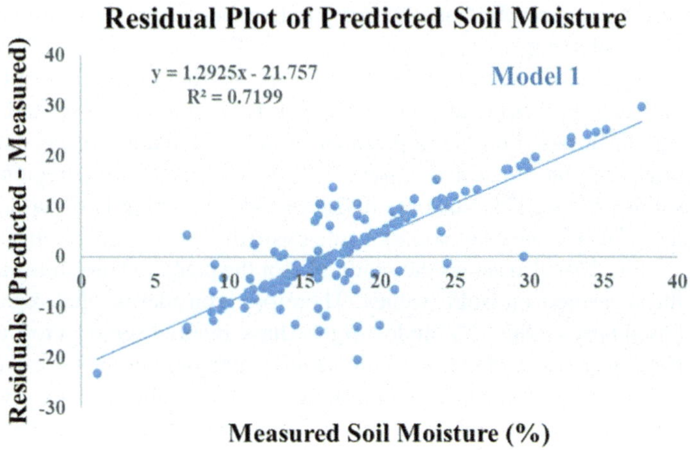

FIGURE 10.5

Residual plots of predicted soil moisture values and measured soil moisture for model 1 for evaluate model bias and dispersion.

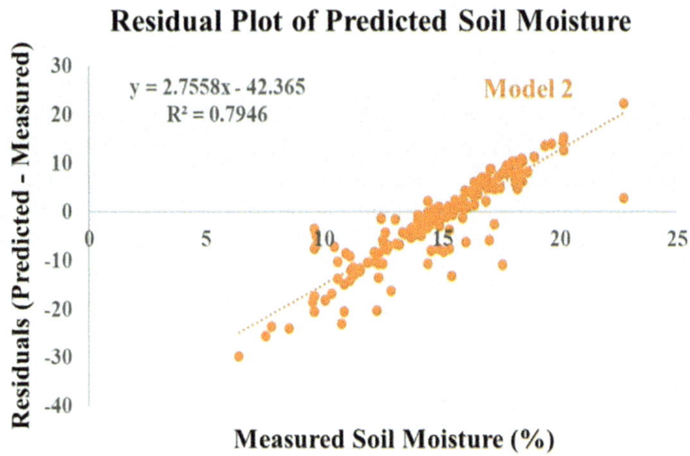

FIGURE 10.6

Residual plots of predicted soil moisture values and measured soil moisture for model 2 for evaluate model bias and dispersion.

model 1 recorded an R^2 value of 0.61 for the designated study area, whereas model 2 exhibited enhanced precision with an R^2 of 0.69.

Overall, the methods employed to examine the spatiotemporal fluctuations of soil moisture in farming areas will facilitate subsequent investigations and aid in efficiently administering agricultural techniques. Findings also show that satellite images and the indices extracted from them can be used to monitor soil moisture throughout the year and achieve more accurate management in the field. In addition to examining the correlation of indices, regression models were also prepared to predict soil moisture. The regression model (2) was prepared by simultaneously using two indices, NDTI and LST, as factors affecting soil moisture. This model can provide higher accuracy in assessing soil moisture using Landsat-8 images.

4. Conclusions

Surface soil moisture plays a very important role in the global water and energy balance by affecting hydrological, ecological and meteorological processes. The use of satellite images and RS indices as reliable and accurate tools provides the possibility of estimating and zoning surface soil moisture in hot and dry regions. In this study, to understand the relationship between soil moisture and its related indicators, the Landsat-8 OLI sensor has been used to estimate the indices. Therefore, this high-precision regression model was introduced in the region. The results of this study showed that the use of RS data as an effective tool in estimating soil moisture and soil salinity in agricultural and salt-affected areas of Mashhad Plain can be useful. In addition, the results showed that the use of RS data continuously and in short time intervals is very useful as a tool for accurate and continuous monitoring of soil moisture. Considering that soil moisture is one of the most important factors in achieving food security and producing effective agricultural products, the use of RS data can be used as an effective tool in soil management and sustainable agriculture.

References

Abdulraheem, M. I., Zhang, W., Li, S., Moshayedi, A. J., Farooque, A. A., & Hu, J. (2023). Advancement of remote sensing for soil measurements and applications: A comprehensive review. *Sustainability, 15*(21), Article 15444.

Abiyat, M., Abiyat, M., & Abiyat, M. (2022). Estimation of soil surface moisture in agricultural lands using satellite images and remote sensing indicators (case study: Shushtar County). *Iranian Journal of Soil and Water Research, 53*(5), 957–970. https://doi.org/10.22059/ijswr.2022.341981.669251

Adekanmbi, T., Wang, X., Basheer, S., Nawaz, R. A., Pang, T., Hu, Y., & Liu, S. (2023). Assessing future climate change impacts on potato yields—A case study for Prince Edward Island, Canada. *Foods, 12*(6), 1176. https://doi.org/10.3390/foods12061176

Allbed, A., & Kumar, L. (2013). Soil salinity mapping and monitoring in arid and semi-arid regions using remote sensing technology: A review. *Advances in Remote Sensing, 2*(4), 373–385. https://doi.org/10.4236/ars.2013.24040

Arshad, M. Z., Iqbal, Z., Saeed, M., Saleem, M., Mansha, R., Abdullah-Al-Wadud, M., & Tariq, A. (2025). Evaluating the effect of land use land cover changes on soil moisture index and vegetation indices using multi-spectral satellite data: MZ Arshad et al. *Journal of Earth System Science, 134*(2), 126. https://doi.org/10.1007/s12040-025-02595-1

Asghari, S. S., Esfandayari Darabad, F., Mollanouri, E., & Safary, S. (2022). Investigation of soil surface moisture in Ardabil city using Landsat 8 and Sentile 1 satellite data. *Journal of Water and Soil Resources Conservation, 11*(4), 89–103. https://doi.org/10.30495/WSRCJ.2022.20006

Azabdaftari, A., & Sunar, F. (2016). Soil salinity mapping using multitemporal Landsat data. *The International Archives of the Photogrammetry, Remote Sensing and Spatial Information Sciences, 41*, 3–9. https://doi.org/10.5194/isprs-archives-XLI-B7-3-2016

Bagheri, K., Bagheri, M., Hoseinzadeh, A., & Parvin, M. (2019). Estimation of soil moisture using optical, thermal and radar Remote Sensing Case Study: South of Tehran. *Iranian Journal of Watershed Management Science and Engineering, 13*(47), 63–74.

Benninga, H. J. F., Carranza, C. D., Pezij, M., van Santen, P., van der Ploeg, M. J., Augustijn, D. C., & van der Velde, R. (2018). The Raam regional soil moisture monitoring network in the Netherlands. *Earth System Science Data, 10*(1), 61–79. https://doi.org/10.5194/essd-10-61-2018

Ceccato, P., Flasse, S., & Grégoire, J. M. (2002). Designing a spectral index to estimate vegetation water content from remote sensing data. *Remote Sensing of Environment, 82*(2–3), 198–207. https://doi.org/10.1016/S0034-4257(02)00036-6

Chaney, N. W., Roundy, J. K., Herrera-Estrada, J. E., & Wood, E. F. (2015). High-resolution modeling of the spatial heterogeneity of soil moisture: Applications in network design. *Water resources research, 51*(1), 619–638. https://doi.org/10.1002/2013WR014964

Chavez, P. S. (1989). Radiometric calibration of Landsat Thematic Mapper multispectral images. *Photogrammetric Engineering and Remote Sensing, 55*(9), 1285–1294.

Corwin, D. L. (2021). Climate change impacts on soil salinity in agricultural areas. *European Journal of Soil Science, 72*(2), 842–862. https://doi.org/10.1111/ejss.13010

Dong, R., & Na, X. (2021). Quantitative retrieval of soil salinity using Landsat 8 OLI imagery. *Applied Sciences, 11*(23), Article 11145. https://doi.org/10.3390/app112311145

Entekhabi, D., Nakamura, H., & Njoku, E. G. (2023). Retrieval of soil moisture profile by combined remote sensing and modeling. In *Passive microwave remote sensing of Land–Atmosphere interactions* (pp. 485–498). CRC Press.

Fabre, S., Briottet, X., & Lesaignoux, A. (2015). Estimation of soil moisture content from the spectral reflectance of bare soils in the 0.4–2.5 μm domain. *Sensors, 15*(2), 3262–3281. https://doi.org/10.3390/s150203262

Gharechelou, S., Akbari Ghoochani, H., Golian, S., & Ganji, K. (2021). Evaluation of land subsidence relationship with groundwater depletion using Sentinel-1 and ALOS-1 radar data (case study: Mashhad plain). *Journal of RS and GIS for Natural Resources, 12*(3), 11–14.

Imtiaz, F., Farooque, A. A., Randhawa, G. S., Wang, X., Esau, T. J., Acharya, B., & Garmdareh, S. E. H. (2024). An inclusive approach to crop soil moisture estimation: Leveraging satellite thermal infrared bands and vegetation indices on Google Earth engine. *Agricultural Water Management, 306*, 109172. https://doi.org/10.1016/j.agwat.2024.109172

Islamic Republic of Iran Meteorological Organization (IRIMO) (2017). (Tehran, Iran).

Jagadish, S. V., Way, D. A., & Sharkey, T. D. (2021). Scaling plant responses to high temperature from cell to ecosystem. *Plant, Cell and Environment, 44*(7), 1987–1991. https://doi.org/10.1111/pce.14082

Kasim, N., Maihemuti, B., Sawut, R., Abliz, A., Dong, C., & Abdumutallip, M. (2020). Quantitative estimation of soil salinization in an arid region of the Keriya Oasis based on multidimensional modeling. *Water, 12*(3), 880. https://doi.org/10.3390/w12030880

Kassaye, K. T., Boulange, J., Saito, H., & Watanabe, H. (2021). Soil water content and soil temperature modeling in a vadose zone of Andosol under temperate monsoon climate. *Geoderma, 384*, Article 114797. https://doi.org/10.1016/j.geoderma.2020.114797

Keesstra, S. D., Bouma, J., Wallinga, J., Tittonell, P., Smith, P., Cerdà, A., … Fresco, L. O. (2016). Forum paper: The significance of soils and soil science towards realization of the UN sustainable development goals (SDGS). *Soil Discussions*, 1–28. https://doi.org/10.5194/soil-2-111-2016

Keesstra, S., Mol, G., De Leeuw, J., Okx, J., De Cleen, M., & Visser, S. (2018). Soil-related sustainable development goals: Four concepts to make land degradation neutrality and restoration work. *Land, 7*(4), 133. https://doi.org/10.3390/land7040133

Kelley, C. P., Mohtadi, S., Cane, M. A., Seager, R., & Kushnir, Y. (2015). Climate change in the fertile crescent and implications of the recent Syrian drought. *Proceedings of the national Academy of Sciences, 112*(11), 3241–3246. https://doi.org/10.1073/pnas.1421533112

Kim, S., Liu, Y., Johnson, F., & Sharma, A. (2016). A temporal correlation based approach for spatial disaggregation of remotely sensed soil moisture. *AGU Fall Meeting Abstracts, 2016*, H51H–1606.

Lashkaripour, G. R., Ghafoori, M., & Maddah, S. M. (2014). An investigation on the mechanism of land subsidence in the northwest of Mashhad city, NE Iran. *Journal of Biodiversity and Environmental Sciences, 5*(3), 321–327.

Li, Q., Zhong, R., Huang, J., & Gong, H. (2011). Comparison of two retrieval methods with combined passive and active microwave remote sensing observations for soil moisture. *Mathematical and Computer Modelling, 54* (3–4), 1181–1193. https://doi.org/10.1016/j.mcm.2010.11.052

Lobell, D. B., & Asner, G. P. (2002). Moisture effects on soil reflectance. *Soil Science Society of America Journal, 66*(3), 722–727. https://doi.org/10.2136/sssaj2002.7220

Lugassi, R., Goldshleger, N., & Chudnovsky, A. (2017). Studying vegetation salinity: From the field view to a satellite-based perspective. *Remote Sensing, 9*(2), 122. https://doi.org/10.3390/rs9020122

Nawaz, R. A., Wang, X., Basheer, S., Sonier, K., Pang, T., & Adekanmbi, T. (2023). Spatiotemporal trends in temperature and precipitation for Prince Edward Island over 1971–2020. *Canadian Journal of Civil Engineering, 50*(11), 892–910. https://doi.org/10.1139/cjce- 2023-0186

Penna, D., Brocca, L., Borga, M., & Dalla Fontana, G. (2013). Soil moisture temporal stability at different depths on two alpine hillslopes during wet and dry periods. *Journal of Hydrology, 477*, 55–71. https://doi.org/10.1016/j.jhydrol.2012.10.052

Pettorelli, N., Vik, J. O., Mysterud, A., Gaillard, J. M., Tucker, C. J., & Stenseth, N. C. (2005). Using the satellite-derived NDVI to assess ecological responses to environmental change. *Trends in Ecology and Evolution, 20* (9), 503–510. https://doi.org/10.1016/j.tree.2005.05.011

Rahmanian, S., Nasiri, V., Amindin, A., Karami, S., Maleki, S., Pouyan, S., & Borz, S. A. (2023). Prediction of plant diver sity using multi-seasonal remotely sensed and geodiversity data in a mountainous area. *Remote Sensing, 15*, 387. https://doi.org/10.3390/rs15020387

Munns, M., & Tester, M. (2008). Mechanisms of salinity tolerance. *Annual Review of Plant Biology, 59*(1), 651–681. https://doi.org/10.1146/annurev.arplant.59.032607.092911

Rasheed, M. W., Tang, J., Sarwar, A., Shah, S., Saddique, N., Khan, M. U., Imran Khan, M., Nawaz, S., Shamshiri, R. R., Aziz, M., et al. (2022). Soil moisture measuring techniques and factors affecting the moisture dynamics: A comprehensive review. *Sustainability, 14*, Article 11538. https://doi.org/10.3390/su141811538

Regional Water Authority of Khorasan Razavi (RWAKR). (2015). *Kashfarud rescue plan with the participation of the people, the deputy for planning and management improvement* (Vol. 2, p. 72).

Rouse, J. W., Jr., Haas, R. H., Schell, J. A., & Deering, D. W. (December 1973). Paper a 20. *Third earth resources technology satellite-1 symposium: The proceedings of a symposium held by Goddard space flight center at Washington, DC on, 351*, 309.

Ruiz-Vera, U. M., Siebers, M. H., Drag, D. W., Ort, D. R., & Bernacchi, C. J. (2015). Canopy warming caused photosynthetic acclimation and reduced seed yield in maize grown at ambient and elevated. *Global Change Biology, 21*(11), 4237–4249. https://doi.org/10.1111/gcb.13013

Samaniego, L., Thober, S., Kumar, R., Wanders, N., Rakovec, O., Pan, M., … Marx, A. (2018). Anthropogenic warming exacerbates European soil moisture droughts. *Nature Climate Change, 8*(5), 421–426. https://doi.org/10.1038/s41558-018-0138-5

Sánchez, N., Alonso-Arroyo, A., Martínez-Fernández, J., Piles, M., González-Zamora, Á., Camps, A., & Vall-Llosera, M. (2015). On the synergy of airborne GNSS-R and landsat 8 for soil moisture estimation. *Remote Sensing, 7*(8), 9954–9974. https://doi.org/10.3390/rs70809954

Shi, X., Song, J., Wang, H., Lv, X., Tian, T., Wang, J., … Jiang, M. (2023). Improving the monitoring of root zone soil salinity under vegetation cover conditions by combining canopy spectral information and crop growth parameters. *Frontiers in Plant Science, 14*, Article 1171594. https://doi.org/10.3389/fpls.2023.1171594

Silakhori, E., Maleki, S., & Pourghasemi, H. R. (2024). Spatiotemporal variations in land use of Mahvelat plain in Iran using google Earth engine from 2011 to 2030. In *Advanced tools for studying soil erosion processes* (pp. 413–433). Elsevier Inc. https://doi.org/10.1016/B978-0-443-22262-7.00032-

Singh, A. (2022). Soil salinity: A global threat to sustainable development. *Soil Use and Management, 38*(1), 39–67. https://doi.org/10.1111/sum.12772

Singh, A., Gaurav, K., Sonkar, G. K., & Lee, C.-C. (2023). Strategies to measure soil moisture using traditional methods, automated sensors, remote sensing, and machine learning techniques: Review, bibliometric analysis, applications, research findings, and future directions. *IEEE Access*. https://doi.org/10.1109/ACCESS.2023.3243635

Soil Survey Staff. (1996). Report No. *Soil survey laboratory methods manual* (Vol 42). USDA.

Stavi, I., Thevs, N., & Priori, S. (2021). Soil salinity and sodicity in drylands: A review of causes, effects, monitoring, and restoration measures. *Frontiers in Environmental Science, 9*, Article 712831. https://doi.org/10.3389/fenvs.2021.712831

Taghadosi, M. M., Hasanlou, M., & Eftekhari, K. (2019). Soil salinity mapping using dual-polarized SAR Sentinel-1 imagery. *International Journal of Remote Sensing, 40*(1), 237–252. https://doi.org/10.1080/01431161.2018.1512767

Van de Griend, A. A., & Engman, E. T. (1985). Partial area hydrology and remote sensing. *Journal of Hydrology, 81*(3–4), 211–251. https://doi.org/10.1016/0022-1694(85)90033-2

Wang, Q., Li, J., Jin, T., Chang, X., Zhu, Y., Li, Y., … Li, D. (2020). Comparative analysis of Landsat-8, Sentinel-2, and GF-1 data for retrieving soil moisture over wheat farmlands. *Remote Sensing, 12*(17), 2708. https://doi.org/10.3390/rs12172708

Wang, L., & Qu, J. J. (2009). Satellite remote sensing applications for surface soil moisture monitoring: A review. *Frontiers of Earth Science in China, 3*(2), 237–247. https://doi.org/10.1007/s11707-009-0023-7

Wen, W., Timmermans, J., Chen, Q., & van Bodegom, P. M. (2022). Monitoring the combined effects of drought and salinity stress on crops using remote sensing. *Hydrology and Earth System Sciences Discussions, 2022*, 1–15. https://doi.org/10.5194/hess-26-4537-2022

Xu, Y., Zhu, G., Wan, Q., Yong, L., Ma, H., Sun, Z., … Qiu, D. (2021). Effect of terrace construction on soil moisture in rain-fed farming area of Loess Plateau. *Journal of Hydrology: Regional Studies, 37*, Article 100889. https://doi.org/10.1016/j.ejrh.2021.100889

Zareie, A., Amin, M. S. R., & Amador-Jiménez, L. E. (2016). Thornthwaite moisture index modeling to estimate the implication of climate change on pavement deterioration. *Journal of Transportation Engineering, 142*(4), Article 04016007. https://doi.org/10.1061/(ASCE)TE.1943-5436.0000840

Zhang, D., Tang, R., Zhao, W., Tang, B., Wu, H., Shao, K., & Li, Z. L. (2014). Surface soil water content estimation from thermal remote sensing based on the temporal variation of land surface temperature. *Remote Sensing, 6*(4), 3170–3187. https://doi.org/10.3390/rs6043170

Zhao, W., Li, A., & Zhao, T. (2017). Potential of estimating surface soil moisture with the triangle-based empirical relationship model. *IEEE Transactions on Geoscience and Remote Sensing, 55*(11), 6494–6504. https://doi.org/10.1109/TGRS.2017.2728815

PART

III

Economic valuation of soil and policy frameworks for soil management and sustainable development

CHAPTER

11

Spatial modeling of soil organic carbon using satellite imagery and machine learning for sustainable land management

Abdolhossein Boali[1], Narges Kariminejad[2] and Morteza Akbari[1]

[1]*Department of Arid and Desert Region Management, Faculty of Natural Resources and Environment, Ferdowsi University of Mashhad, Mashhad, Iran;* [2]*Department of Natural Resources and Environment Engeenering, College of Agriculture, Shiraz University, Shiraz, Iran*

1. Introduction

Soil, as one of the most vital yet most vulnerable natural resources on the Earth, plays a fundamental role in supporting ecosystems, ensuring food security, and regulating biogeochemical cycles (Padarian et al., 2020). At the heart of this complexity, SOC holds a dual importance: not only as a primary indicator of soil health and fertility but also as the planet's second-largest carbon reservoir (Ben Ghorbal et al., 2025). The intelligent management of this reservoir is a key strategy for enhancing agricultural sustainability and simultaneously mitigating the adverse effects of climate change through the sequestration and permanent storage of atmospheric carbon (Nauman et al., 2023). However, effectively harnessing this potential requires a precise understanding of the spatial distribution and dynamics of SOC at large scales, which in itself poses a significant challenge in soil science. Traditional methods of SOC monitoring, based on discrete sampling and costly laboratory analyses, despite their high accuracy at sampling points, lack the capability to produce continuous and practical maps at regional and national scales (Li et al., 2025). This limitation necessitates the development of spatial models. However, modeling SOC is fraught with numerous inherent challenges. For example, the high spatial variability of soil, influenced by a complex set of non-linear factors such as climate, topography, parent material, vegetation, and land use, is a significant challenge. Furthermore, the subsurface nature of organic carbon is another issue that must be addressed. Satellite sensors primarily capture surface information, and the direct, strong relationship between surface spectral reflectance and carbon concentration in deeper soil layers is weak, especially in areas with dense vegetation (Han et al., 2010). In response to these challenges, Digital Soil Mapping has opened new horizons by leveraging the combined capabilities of remote sensing and machine learning (Santra et al., 2017; Xia et al., 2022). Acknowledging these limitations, this study presents an innovative and robust modeling framework.

In recent years, the integration of remote sensing data and machine learning algorithms has become a standard approach in the field of Digital Soil Mapping for estimating SOC. Various researchers have

The Soil Nexus. https://doi.org/10.1016/B978-0-443-45286-4.00005-5

shown that machine learning models, particularly models such as Random Forest, Gradient Boosting, SVM, perform significantly better than classical statistical methods due to their ability to model complex non-linear relationships (Li et al., 2025; Padarian et al., 2020). Bispo et al. (2023) emphasized that machine learning models are now considered the forefront of predicting soil properties.

The core idea of this research is the development and evaluation of a machine learning-based modeling framework that exclusively uses remote sensing indices to estimate and predict the spatial distribution of SOC. The innovative approach of this research lies in this specific focus; instead of combining complex, multi-source data (such as topographic, climatic, and pedological data), we hypothesize that spectral indices extracted from satellite images (like NDVI, EVI, and soil-sensitive indices) alone contain sufficient information about the key factors influencing soil carbon, including vegetation cover, surface moisture, and overall ecosystem health. The machine learning model is tasked with uncovering and learning the non-linear and hidden relationships between these surface proxies and the actual carbon values measured in the field (Boali et al., 2024a, 2024b; Yin et al., 2024). The ultimate goal is to prove the hypothesis that reliable estimates of soil carbon can be achieved with a simplified, low-cost, and purely satellite-based model, making large-scale monitoring for sustainable land management significantly more practical and accessible.

This research is conducted with a focus on the Maharloo watershed as a case study; an area that has faced serious environmental crises in recent years due to consecutive droughts, land use changes, and improper resource management, including a decline in lake levels, salinity, and soil degradation. These conditions have transformed the precise assessment of soil health and its carbon storage potential into an undeniable necessity for any management planning. Therefore, by producing a spatial map of soil organic carbon in this watershed, this research provides a key scientific tool for identifying critical areas and laying the groundwork for effective measures toward the sustainable restoration of this sensitive ecosystem.

2. Materials and methods

2.1 Study area

The Maharloo watershed in Fars Province, with an average elevation of 1211 m above sea level, is located in the southwestern region of Iran, within the Zagros Mountain range. The study area is situated between latitudes 29 degrees 1′ to 29 degrees 58′ N and longitudes 53 degrees 12′ to 53 degrees 28′ E (Fig. 11.1). The Maharloo watershed covers an area of 4274 square kilometers, of which 2056 square kilometers is plain and the remainder consists of highlands.

2.2 Research data

In this study, to validate the remotely sensed indices, a comprehensive field survey was conducted in July 2025. A total of 80 soil samples were collected across different land use types, with sampling performed at a depth of 0–10 cm. Samples from each land use were composited to obtain representative laboratory subsamples for analysis of Soil Organic Carbon (SOC) as well as key soil properties including salinity, moisture, and texture. SOC was measured using standard laboratory procedures (e.g., dry combustion or Walkley-Black method, depending on the protocol), ensuring high analytical accuracy. The geolocation of each sample was recorded using GPS devices with sub-meter accuracy.

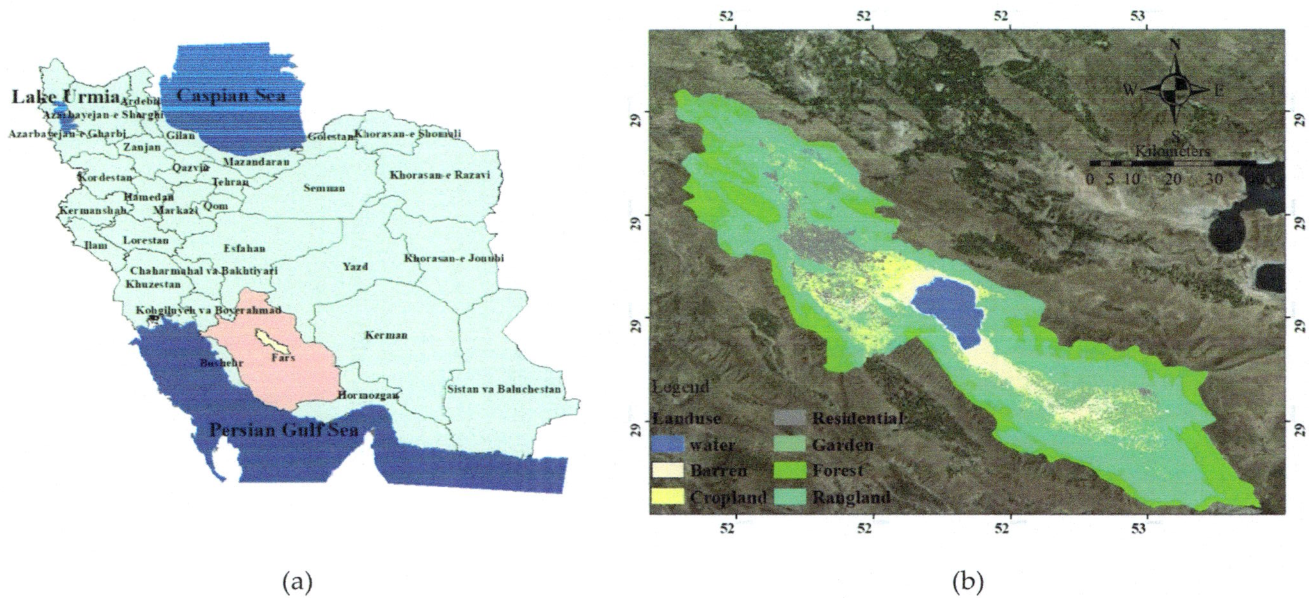

FIGURE 11.1

Location of the study area in Iran (a) and google earth image and land use of the maharloo watershed (b). ***Map lines delineate study areas and do not necessarily depict accepted national boundaries.***

To align point-based SOC measurements with the 30 m resolution of Landsat-derived indices, a circular buffer of 15 m radius was applied around each sampling point, and the mean value of all pixels within the buffer was extracted. This approach minimized spatial mismatch between point observations and raster data, allowing reliable validation. Vegetation cover and bare soil percentage were also quantified in situ using one-square-meter plots along two 100-meter transects per land use type. Additionally, long-term climatic variables (rainfall and temperature; 2010–24) from five meteorological stations were incorporated to contextualize the environmental conditions of the study area.

Landsat 8 and 9 Tier 1 Surface Reflectance collections were used, with images selected to match the July 2025 field survey period. All imagery underwent atmospheric correction and cross-sensor calibration to ensure consistency across sensors. Cloud and shadow masking was performed using the CFMASK algorithm in Google Earth Engine, providing reliable input data for SOC modeling. The land use map was generated from Landsat imagery processed in the Google Earth Engine platform using a supervised classification approach. The map corresponds to July 2025, consistent with the SOC field sampling and remote sensing data, and achieved an overall classification accuracy exceeding 85%.

2.3 Remote sensing indices

In this study, to model the spatial distribution of SOC, a set of auxiliary predictor variables representing the key factors influencing the processes of carbon formation and stabilization was selected. These variables included spectral indices and land use (Table 11.1). In this context, the Normalized Difference Vegetation Index (NDVI), as the primary source of carbon input to the soil through the decomposition of organic matter, was employed to assess vegetation density and health (Alibakhshi et al., 2017). Conversely, the Bare Soil Index (BSI) was calculated to quantify surfaces where organic matter input is negligible and the potential for carbon release through mineralization is high (Gorelick et al., 2017). The physicochemical properties of the soil, which play a decisive role in carbon stabilization, were accounted for using the Normalized Difference Moisture Index (NDMI) to estimate moisture (a regulator of microbial activity), the Topsoil Grain Size Index (TGSI) to estimate soil texture (a primary factor in the physical protection of carbon), and the Normalized Difference Salinity Index (NDSI) to measure salinity stress (which limits plant growth and, consequently, organic matter production) (Boali

Table 11.1 Remote sensing indices for soil quality modeling.

Indicators	Formula	References
NDVI	((NIR-R))/((NIR+R))	Guo et al. (2020)
NDSI	R–NIR/R + NIR	Dehni and Lounis (2012)
TGSI	R–B/R + B + G	Wei et al. (2018)
NDMI	(NIR + SWIR)/(NIR−SWIR)	Serrano et al. (2019)
Albedo	$0.3\times R_{Red}+0.3\times R_{NIR}+0.4\times R_{SWIR}$	Wei et al. (2018)
BSI	(NIR-Red)/(NIR + Red)	Mohammed et al. (2020)
Land use	–	Desta and Fetene (2020)

et al., 2024a, 2024b). Furthermore, Albedo was included in the model as an indicator of soil color, which is directly related to the amount of organic carbon (darker soils have more organic carbon) (Wen et al., 2020). Finally, the land use variable was also added to the set of variables as a comprehensive factor that encompasses the effects of land management on soil organic carbon dynamics (Xiao et al., 2006).

2.4 Correlation of ground and satellite data

To assess the accuracy and validate the results derived from satellite indices for estimating SOC, the values estimated by these indices will be compared with field data (as reference and actual values). For this purpose, two statistical metrics will be used: the coefficient of determination (R^2) and the Root Mean Square Error (RMSE). The R^2 is used to measure the degree of correlation and the explanatory power of the model) (Boali et al., 2025). This coefficient indicates the percentage of variation in soil organic carbon that is explained by the given satellite index; a value of R^2 closer to one signifies a stronger correlation. The RMSE will be used to evaluate prediction accuracy and measure the model's average error. RMSE quantitatively expresses the difference between the values predicted by the model and the actual values measured during the field survey; lower values for this metric indicate higher accuracy and less model error. The combination of these two metrics provides a comprehensive evaluation of the indices' performance, whereby R^2 shows the strength of the relationship and RMSE shows its absolute accuracy.

2.5 Machine learning models

In this study, to estimate and map the spatial distribution of soil organic carbon, three machine learning algorithms with different approaches will be used: SVM, RF, and BRT. SVM, by creating a separating hyperplane in the feature space and using kernel functions, is capable of modeling complex and non-linear relationships between predictor variables and soil organic carbon (Padarian et al., 2020). RF, as an ensemble-based algorithm, builds a large number of decision trees and, by averaging their results, increases model stability and prevents overfitting; furthermore, this method allows for the assessment of the importance of each input variable (Breiman, 2001). Finally, BRT also operates on the basis of decision trees, but sequentially; in a way that each new tree is built to correct the errors of the previous ones, and this boosting process leads to increased model prediction accuracy (Boali et al., 2024a, 2024b). The model parameters were set based on optimized configurations determined during the modeling process. In the SVM model, the RBF kernel was employed due to its high capability in capturing nonlinear relationships, making it suitable for environmental data. For the RF model, 500 trees were used, which provided the best performance based on testing various values. In the BRT model, a learning rate of 0.01 was applied to prevent overfitting and ensure a stable learning process. The concurrent use of these three approaches allows for a comprehensive comparison of their performance and the selection of the most accurate model for mapping soil organic carbon in the study area.

3. Results

2.1 Variations of remote sensing indices in the study area

The integrated spatial analysis of the seven key indices presented in Fig. 11.2 depicts a multidimensional and comprehensive picture of the ecological conditions of the study area. This dataset reveals a clear spatial dichotomy, in which patches with dense vegetation and high moisture are contrasted with vast areas of dry and barren land. This pattern is the product of a complex interaction between natural factors and human activities.

High TGS I values signify the dominance of coarser soil particles (such as sand and gravel) on the soil surface. This phenomenon is a direct result of the process of selective erosion, wherein fine, fertile soil particles (silt and clay) are displaced by wind and water, leaving behind coarser, less valuable particles. Therefore, the TGS I map is a qualitative representation of the extent of soil degradation progression in the region. This degradation process is clearly traceable through its spatial correlation with other indicators. Areas with high TGS I precisely overlap with areas where the NDVI and NDMI are at their lowest. This pattern reveals the mechanism of the degradation cycle: the loss of vegetation (low NDVI) renders the soil surface vulnerable to erosive agents, leading to a high Bare Soil Index (BI). The resulting erosion from this exposure removes fine soil particles, causing an increase in TGS I.

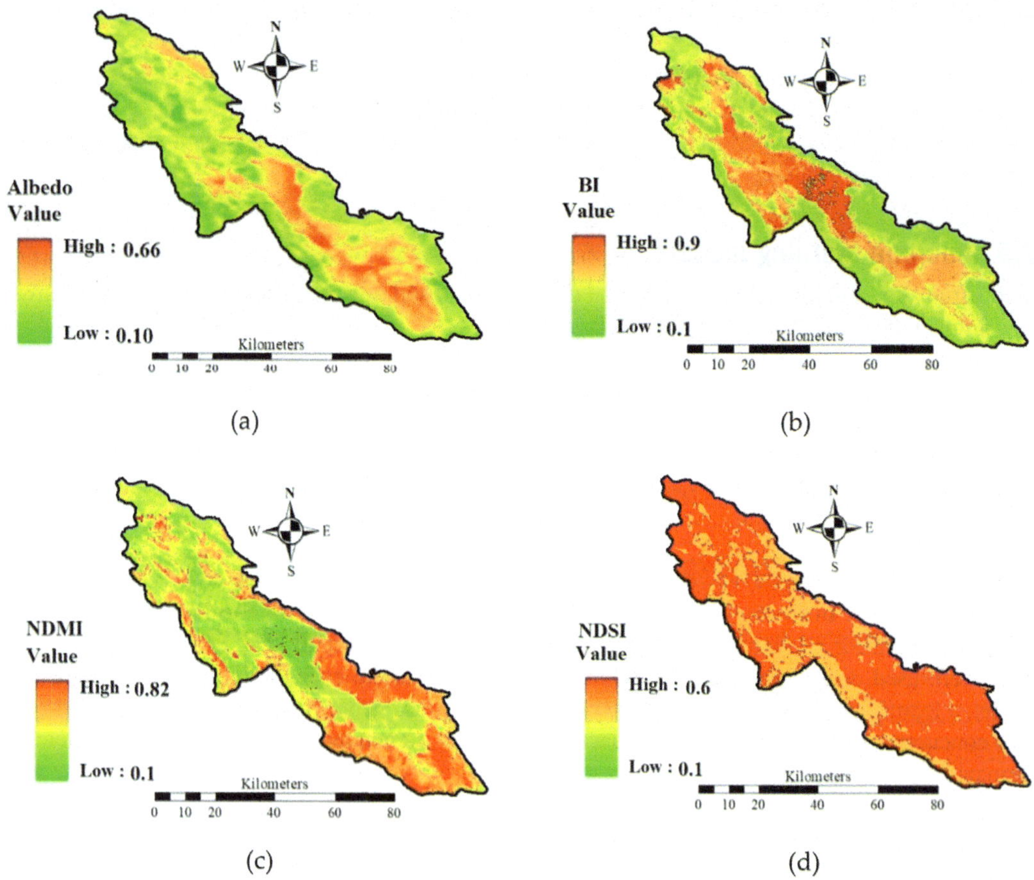

FIGURE 11.2

Map of remote sensing indicators in the study area (albedo (a), BI (b), NDMI (c), NDSI (d), NDVI (e), TGSI (f), and landuse (g). ***Map lines delineate study areas and do not necessarily depict accepted national boundaries.***

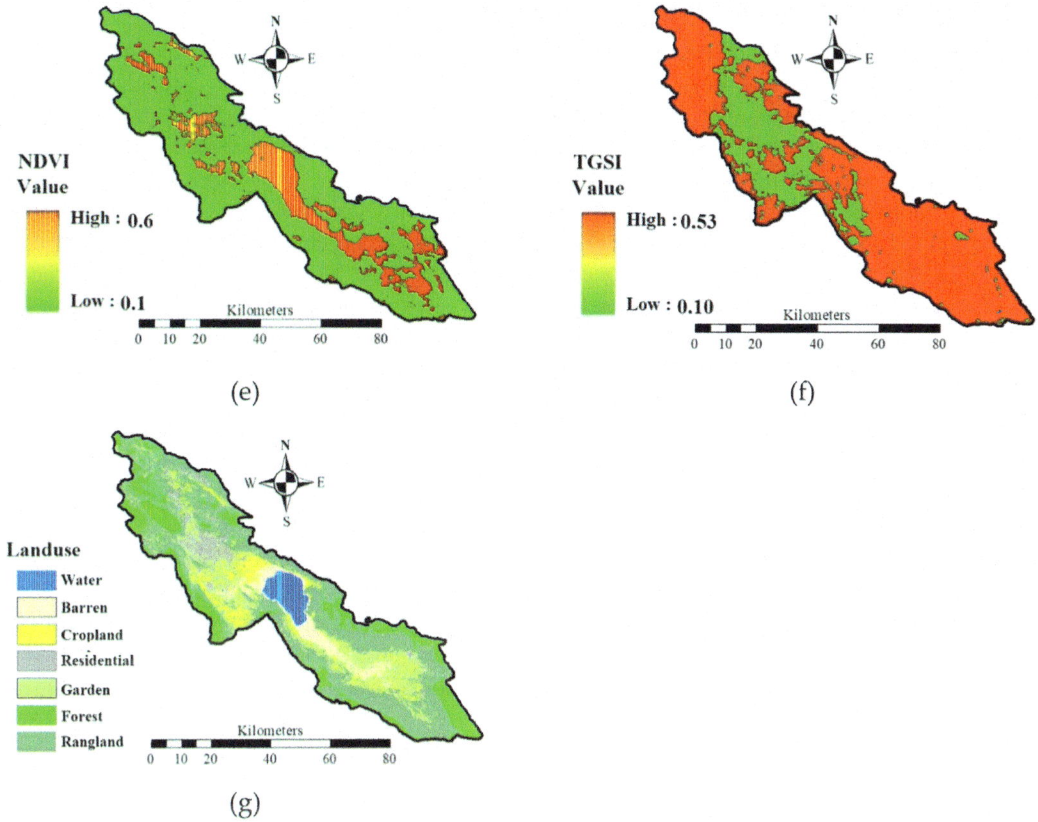

FIGURE 11.2 Cont'd

The degraded, coarse soil has a very low water-holding capacity, which leads to low NDMI, and exhibits a high surface albedo, contributing to the warming of the local microclimate. This cycle constitutes a positive feedback loop of degradation, where each stage intensifies the next, making a return to the original state increasingly difficult.

In contrast to these degrading areas, there exist zones with low TGS I, which are indicative of healthy and stable soil. These areas, identified as "Cropland" on the land use map, exhibit a positive, yet stabilizing, feedback loop. Dense vegetation (high NDVI) protects the soil surface, and its roots stabilize soil particles. This vegetation also adds organic matter to the soil and improves its structure, which prevents erosion (low BI). Consequently, fine soil particles are preserved, and TGS I remains at a low level. This healthy soil retains moisture more effectively (high NDMI) and, due to its darker surface, has lower reflectance (low Albedo). Meanwhile, the NDSI map distinctly highlights salinity hotspots within the study area. Regions showing high values on this map precisely overlap with zones where the BI and the TGS I peak. This spatial correlation indicates that in areas lacking vegetation and exposed to erosion, the process of salinization is also underway.

2.2 Results of the machine learning models

Spatial cross-validation (block-based) was applied to evaluate model performance under spatially independent conditions. The results showed a slight reduction in accuracy metrics compared to random data splitting, reflecting the effect of spatial autocorrelation. Nevertheless, the Random Forest model consistently outperformed BRT and SVM across all folds. Importantly, the spatial patterns of SOC distribution and the ranking of predictor importance remained stable, with Albedo identified as the most influential variable. The results of modeling the spatial distribution of SOC using three algorithms BRT, RF, SVM are presented in Fig. 11.3. A consistent and pronounced spatial pattern is clearly observed across all three models: higher concentrations of organic carbon are predominantly located in the central and eastern parts of the region, which were previously identified as agricultural lands. This spatial overlap corroborates the direct relationship between dense vegetation (high NDVI), high NDMI, and the accumulation of organic carbon in the soil. Conversely, the lowest SOC values have been consistently predicted in the northern and western areas, which are characterized by barren land use and high degradation indices (BI, TGS I, and NDSI). This strong agreement among the models lends high credibility to the analysis, confirming the hypothesis that land degradation status and land use are the primary determinants of organic carbon distribution in the region.

However, a more detailed comparison of the models reveals key differences in their performance. The BRT and SVM models exhibit a wider dynamic range, predicting organic carbon values from a minimum of approximately 0.45%–0.52% to a maximum of 3.44%. This broader range indicates the higher sensitivity of these two algorithms to input variables (such as environmental indices) and their ability to capture extreme values (both maximum and minimum). In other words, these models were better able to capture subtle and severe fluctuations in carbon levels, particularly in the most fertile and most degraded points. In contrast, the RF model provides a more limited prediction range (from 0.86% to 2.92%), which indicates a "smoothing effect" in its results. By averaging the predictions of multiple decision trees, this algorithm naturally reduces variance and tends to underestimate extreme values. While this feature can contribute to the model's stability and generalizability, it may lead to overlooking hotspots with very high carbon or severely carbon-poor areas. Visually, the BRT and SVM maps also show more precise spatial heterogeneities within the agricultural and barren land zones, whereas the RF map depicts a more uniform and homogeneous surface.

2.3 Evaluation of machine learning model performance

The RF model clearly emerges as the most accurate model in this study. In the plot, the data points exhibit a high density around the 1:1 agreement line, indicating a strong correlation and low prediction error. This convergence signifies that the RF model has successfully learned the complex, non-linear relationships between the predictor variables and soil organic carbon. However, a subtle yet significant nuance lies alongside this high accuracy: at very high organic carbon values (above 3%, representing ecological hotspots), the data points show a notable tendency to fall below the agreement line. This phenomenon, known as under-prediction at extremes, is a reflection of the smoothing nature of this algorithm; by averaging predictions from multiple trees, RF tends to pull outlier values toward the dataset's mean, thereby underestimating the true intensity of these critical points.

In contrast, the BRT model exhibits a competitive performance, albeit with distinct characteristics. Although the overall scatter of points in its plot is slightly greater than that of RF, the BRT model

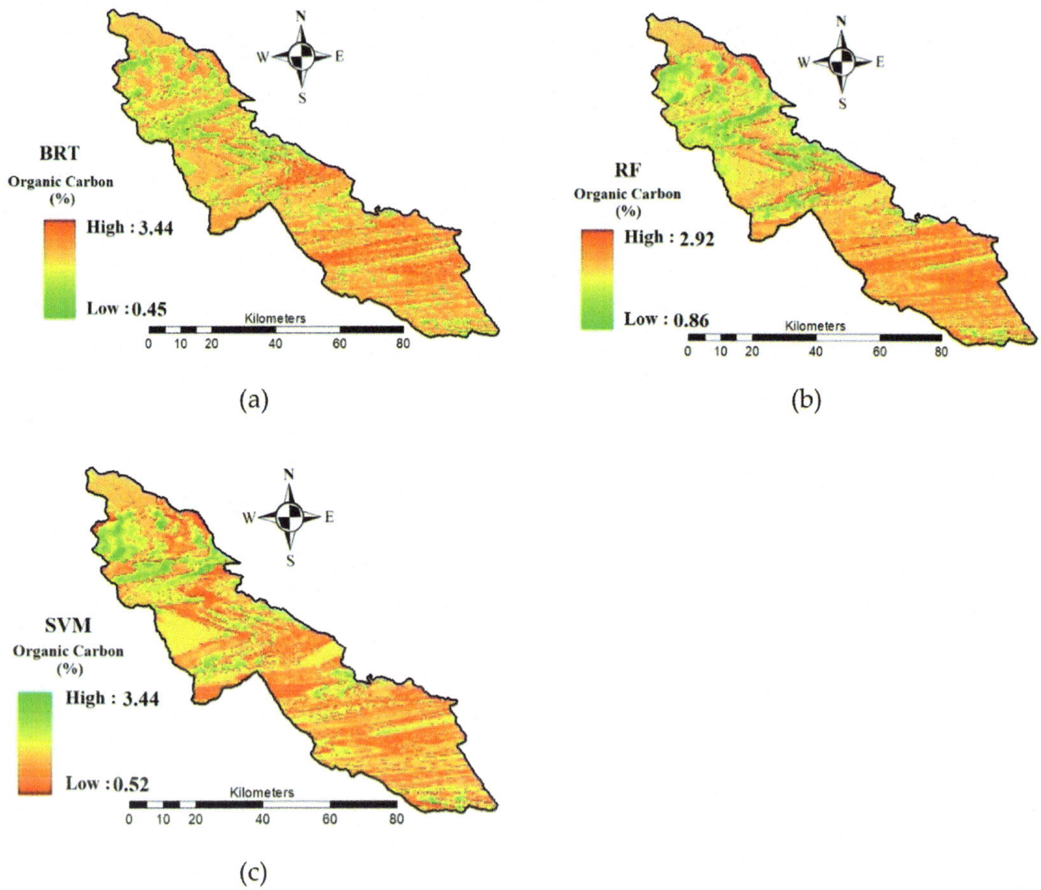

FIGURE 11.3

Machine learning models of BRT (a), RF (b), and SVM (c). ***Map lines delineate study areas and do not necessarily depict accepted national boundaries.***

demonstrates a relative advantage in a key area: predicting extreme values. Unlike RF, this model is less prone to smoothing, and the data points at the highest and lowest organic carbon values show a more balanced distribution around the 1:1 line. This characteristic makes the BRT model a valuable tool for identifying specific areas with high or very low carbon potential, even if it comes at the cost of lower overall accuracy. The SVM model, in this comparison, demonstrates acceptable but weaker performance. The more significant scatter of data points around the agreement line indicates higher uncertainty in this model's predictions. Particularly in the mid-range of organic carbon values, SVM shows less ability to capture subtle patterns compared to the other two models, and its predictions are associated with more noise Fig. 11.4.

Model performance was evaluated using the coefficient of determination (R^2) and root mean square error (RMSE). These metrics were calculated for Random Forest (RF), Boosted Regression Trees

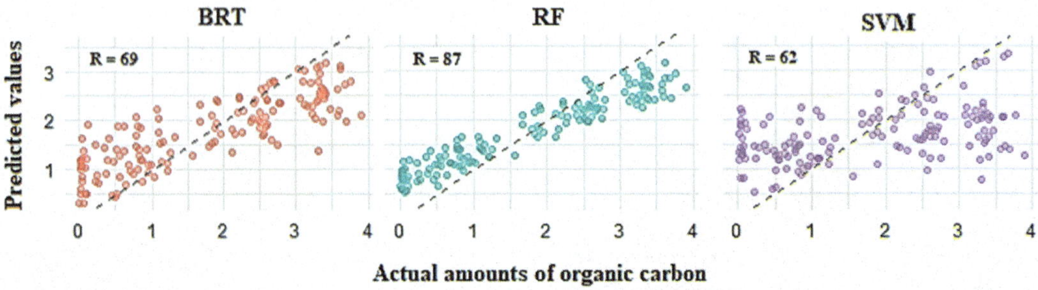

FIGURE 11.4

Correlating real data with machine learning model predictions.

(BRT), and Support Vector Machine (SVM) models to enable a quantitative comparison of predictive accuracy. As shown in Table 11.2, the RF model achieved the highest R^2 and the lowest RMSE, confirming its superior performance in modeling the spatial distribution of SOC.

2.4 Variable importance in the study area based on the superior model

The analysis of predictor variable importance, based on the superior model (RF), reveals the key drivers controlling the spatial variations of soil organic carbon (Fig. 11.5). The preeminence of the Albedo index, with a value of 0.45, as the most important variable, provides a deep insight into the region's dynamics; this index alone is a powerful proxy for the primary dichotomy between stabilized (vegetated and dark) ecosystems and degraded (barren and bright) areas, indicating that the surface energy balance is a determining factor in carbon distribution. This key role is reinforced by a cohesive group of indicators including NDMI (0.40), BI (0.37), and land use (0.35), which together delineate the biophysical and anthropogenic boundaries governing the ecosystem; moisture, soil exposure, and the type of human activity form a triad that, alongside Albedo, constitutes the main predictive framework. In contrast, the lesser importance of NDVI (0.31), NDSI (0.25), and TGSI (0.22) suggests that although these indices describe specific processes (such as greenness, salinity, and soil textural degradation), their impact at the regional scale is less pronounced compared to the aforementioned dominant factors. Ultimately, the RF model indicates that soil organic carbon in the study area primarily revolves around a main axis: the contrast between moisture and cover versus dryness and exposure, all of which are directly influenced by the type of land use.

Table 11.2 Performance comparison of machine learning models for SOC prediction.

Model	R^2	RMSE
RF	87	3.7
BRT	69	4.6
SVM	62	5.1

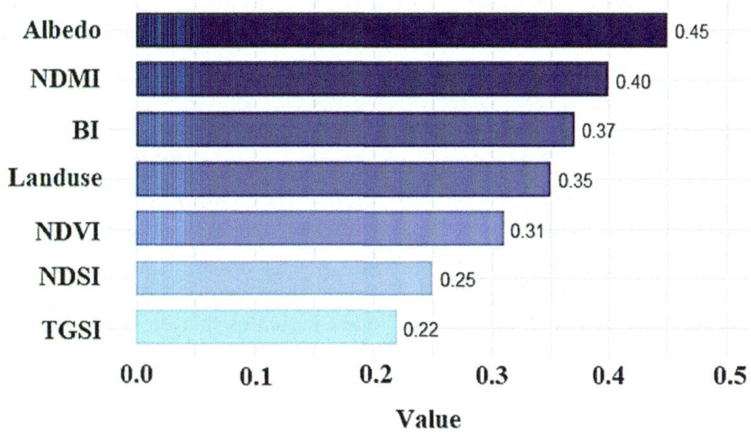

FIGURE 11.5

The importance of variables affecting soil organic carbon based on the model RF.

4. Discussion

The findings of the present research introduce the spatial distribution of SOC as a powerful diagnostic indicator for assessing ecosystem health in the study area. The results reveal a clear dichotomy in organic carbon storage capacity: patches with high potential for SOC accumulation versus extensive areas that have undergone severe carbon depletion. This spatial pattern is not merely a difference in surface cover but a direct reflection of biogeochemical cycles that have been activated or deactivated by land use type (agriculture vs. bare soil). The overlap of regions with degradation indicators (TGS I, BI, Albedo) and low SOC values, and conversely, the concentration of SOC in vegetated and high-moisture lands, confirms the hypothesis that land management is the determining factor in controlling the soil carbon budget in this region (Ben Ghorbal et al., 2025).

A deeper analysis of feedback cycles reveals the driving mechanisms behind the depletion and accumulation of SOC. In degraded areas, a "vicious cycle of carbon depletion" is at play: the loss of vegetation cover (low NDVI) cuts off the input of organic matter (the primary source of SOC). Simultaneously, exposed soil (high BI) leads to erosion and the loss of carbon-rich particles, a process reflected in an increase in TGS I. The degraded soil loses its moisture retention capacity (low NDMI), and with increased reflectivity (high Albedo), it contributes to the warming of the microclimate and accelerates the microbial decomposition of the remaining SOC (Emadi et al., 2020).

Modeling the spatial distribution of SOC using machine learning algorithms successfully translated these ecological complexities into comprehensible and actionable maps (Li et al., 2025). The strong agreement among the three models in identifying the main hotspots of SOC accumulation (center and east) and depletion (north and west) strengthens the credibility of the analysis in confirming the hypothesis that land degradation status and land use are the primary determining factors of organic carbon distribution in the region (Nabiollahi et al., 2021). Beyond a mere visual display, these maps represent the "functional potential" of the soil for carbon storage, offering a key tool for assessing the current status and planning for the future.

The comparison of model performance provides crucial insights for selecting the appropriate analytical tool tailored to SOC management objectives. The superiority of the RF model in terms of overall accuracy makes it a reliable tool for regional-scale estimates and large-scale SOC budgeting (Li et al., 2025). However, this model's tendency to "smooth" extreme values renders it less suitable for the precise identification of "carbon hotspots" with very high potential for sequestration requiring immediate intervention. In contrast, the superior performance of the ERT model in predicting extreme values positions it as a valuable tool for site-specific management and carbon offset projects, as this model is capable of accurately identifying areas with the highest economic or ecological return on investment in the SOC sector (Emadi et al., 2020).

The importance of variables was analyzed based on the superior model (RF) for the spatial distribution of SOC. The superiority of the Albedo index indicates that the surface energy balance is a determining factor in the fate of SOC. High albedo (bare soil) accelerates the microbial decomposition of SOC by heating the surface, while low albedo (vegetation cover) contributes to carbon stability by cooling the environment. The high importance of NDMI, BI, and land use also outlines this "vital quartet" for SOC management (Nabiollahi et al., 2021). Soil moisture (which affects decomposition and plant growth), soil exposure (the main pathway for erosion and carbon loss), land use type (which determines carbon input), and energy balance (which controls the rate of carbon output) are the four main pillars that shape the soil carbon budget in this region.

Ultimately, this research provides an evidence-based framework for the sustainable management of soil organic carbon. The findings indicate that remote sensing indicators can quantitatively and spatially elucidate the mechanisms of SOC depletion and accumulation, while machine learning models leverage these complexities to accurately predict SOC distribution. The importance of key variables defines management objectives; actions that lead to reducing soil exposure, increasing vegetation cover, and improving the region's energy and water balance will directly contribute to enhancing SOC storage capacity and ecosystem resilience to climate change. Therefore, this study offers a scientific roadmap for transitioning from purely descriptive approaches to effective, site-specific management strategies aimed at the restoration and sustainability of soil carbon resources as a vital natural asset.

5. Conclusion

This study demonstrates that modeling SOC by integrating remote sensing indicators and machine learning algorithms is a powerful tool for understanding and predicting the ecological dynamics of the region. Among the models used, the RF was identified as the most accurate, simulating the spatial distribution of SOC with the highest precision by utilizing key indicators: Albedo, NDMI, BI, NDVI, NDSI, TGSI, and land use. The primary outcome of this modeling was the revelation of a clear spatial pattern, wherein the highest organic carbon values are concentrated in agricultural lands (center and east of the region) and the lowest values are found in bare and degraded areas (north and west). This finding indicates that the distribution of soil organic carbon in the region primarily revolves around the dichotomy between moisture and vegetation versus dryness and exposed soil. Therefore, this study provides a scientific roadmap for site-specific management, demonstrating that actions to improve vegetation cover and moisture are the most critical strategies for increasing carbon storage and restoring soil health in the study area.

Acknowledgments

The authors would like to thank DeepSeek for assistance with the grammatical editing of the manuscript.

References

Alibakhshi, S., Groen, T. A., Rautiainen, M., & Naimi, B. (2017). Remotely-sensed early warning signals of a critical transition in a wetland ecosystem. *Remote Sensing, 9*(4). https://doi.org/10.3390/rs9040352

Ben Ghorbal, A., Grine, A., Eid, M. M., & El-kenawy, E. S. M. (2025). Sustainable soil organic carbon prediction using machine learning and the ninja optimization algorithm. *Frontiers in Environmental Science, 13*. https://doi.org/10.3389/fenvs.2025.1630762

Bispo, A., et al. (2023). Machine learning for soil organic carbon stock prediction: A review. *Geoderma, 430*, Article 116312.

Boali, A., Asgari, H. R., Mohammadian Behbahani, A., Salmanmahiny, A., & Naimi, B. (2024a). Remotely sensed desertification modeling using ensemble of machine learning algorithms. *Remote Sensing Applications: Society and Environment, 34*. https://doi.org/10.1016/j.rsase.2024.101149

Boali, A. K. N., Hosseinalizadeh, & Mohsen. (2024b). Enhancing wind erosion risk assessment through remote sensing techniques: Modeling, monitoring, and forecasting. *PLoS One, 10*(19), 1–19.

Boali, A., Hosseinalizadeh, M., Kariminejad, N., Asgari, H. R., Mohamadian Behbahani, A., Naimi, B., Shafaie, V., & Movahedi Rad, M. (2025). Evaluation of early warning signals for soil erosion using remote sensing indices in northeastern Iran. *Scientific Reports, 15*(1). https://doi.org/10.1038/s41598-025-94926-x

Breiman, L. (2001). Random forests. *Machine Learning, 45*(1), 5–32. https://doi.org/10.1109/ICCECE51280.2021.9342376

Dehni, A., & Lounis, M. (2012). Remote sensing techniques for salt affected soil mapping: Application to the Oran region of Algeria. *Procedia Engineering, 33*, 188–198. https://doi.org/10.1016/j.proeng.2012.01.1193

Desta, H., & Fetene, A. (2020). Land-use and land-cover change in Lake Ziway watershed of the Ethiopian Central Rift Valley Region and its environmental impacts. *Land Use Policy, 96*(April), Article 104682. https://doi.org/10.1016/j.landusepol.2020.104682

Emadi, M., Taghizadeh-Mehrjardi, R., Cherati, A., Danesh, M., Mosavi, A., & Scholten, T. (2020). Predicting and mapping of soil organic carbon using machine learning algorithms in Northern Iran. *Remote Sensing, 12*(14). https://doi.org/10.3390/rs12142234

Gorelick, N., Hancher, M., Dixon, M., Ilyushchenko, S., Thau, D., & Moore, R. (2017). Remote Sensing of Environment Google Earth engine: Planetary-scale geospatial analysis for everyone. *Remote Sensing of Environment, 202*, 18–27. https://doi.org/10.1016/j.rse.2017.06.031

Guo, B., Zang, W., Han, B., Yang, F., Luo, W., He, T., & Chen, S. (2020). Dynamic monitoring of desertification in Naiman Banner based on feature space models with typical surface parameters derived from LANDSAT images. *Land Degradation & Development, 31*(12), 1573–1592.

Han, F., Hu, W., Zheng, J., Du, F., & Zhang, X. (2010). Spatial variability of soil organic carbon in a catchment of the Loess Plateau. *Acta Agriculturae Scandinavica Section B Soil and Plant Science, 60*(2), 136–143. https://doi.org/10.1080/09064710902721354

Li, Y., Li, J., Tan, J., Ma, T., Yan, X., Chen, Z., & Li, K. (2025). Fine resolution mapping of forest soil organic carbon based on feature selection and machine learning Algorithm. *Remote Sensing, 17*(12). https://doi.org/10.3390/rs17122000

Mohammed, S., Al-Ebraheem, A., Holb, I. J., Alsafadi, K., Dikkeh, M., Pham, Q. B., … Szabo, S. (2020). Soil management effects on soil water erosion and runoff in central Syria—A comparative evaluation of general linear model and random forest regression. *Water, 12*(9), 2529.

Nabiollahi, K., Shahlaee, S., Zahedi, S., Taghizadeh-Mehrjardi, R., Kerry, R., & Scholten, T. (2021). Land use and soil organic carbon stocks—change detection over time using digital soil assessment: A case study from kamyaran region, Iran (1988–2018). *Agronomy, 11*(3). https://doi.org/10.3390/agronomy11030597

Nauman, T. W., Munson, S. M., Dhital, S., Webb, N. P., & Duniway, M. C. (2023). Synergistic soil, land use, and climate influences on wind erosion on the Colorado Plateau: Implications for management. *Science of the Total Environment, 893*(December 2022), Article 164605. https://doi.org/10.1016/j.scitotenv.2023.164605

Padarian, J., Minasny, B., & McBratney, A. B. (2020). Machine learning and soil sciences: A review aided by machine learning tools. *SOIL, 6*(1), 35–52. https://doi.org/10.5194/soil-6-35-2020

Santra, P., Kumar, M., & Panwar, N. (2017). Digital soil mapping of sand content in arid western India through geostatistical approaches. *Geoderma Regional, 9*, 56–72. https://doi.org/10.1016/j.geodrs.2017.03.003

Serrano, J., Shahidian, S., & Marques da Silva, J. (2019). Evaluation of normalized difference water index as a tool for monitoring pasture seasonal and inter-annual variability in a Mediterranean agro-silvo-pastoral system. *Water, 11*(1), 62.

Wei, H., Wang, J., Cheng, K., Li, G., Ochir, A., Davaasuren, D., & Chonokhuu, S. (2018). Desertification information extraction based on feature space combinations on the Mongolian plateau. *Remote Sensing, 10*(10). https://doi.org/10.3390/rs10101614

Wen, Y., Guo, B., Zang, W., Ge, D., Luo, W., & Zhao, H. (2020). Desertification detection model in Naiman Banner based on the albedo-modified soil adjusted vegetation index feature space using the Landsat8 OLI images. *Geomatics, Natural Hazards and Risk, 11*(1), 544–558. https://doi.org/10.1080/19475705.2020.1734100

Xia, Y., McSweeney, K., & Wander, M. M. (2022). Digital mapping of agricultural soil organic carbon using soil forming factors: A review of current efforts at the regional and national scales. *Frontiers in Soil Science, 2*. https://doi.org/10.3389/fsoil.2022.890437

Xiao, J., Shen, Y., Tateishi, R., & Bayaer, W. (2006). Development of topsoil grain size index for monitoring desertification in arid land using remote sensing. *International Journal of Remote Sensing, 27*(12), 2411–2422. https://doi.org/10.1080/01431160600554363

Yin, W., Hu, Q., Liu, J., He, P., Zhu, D., & Boali, A. (2024). Assessing climate and land-use change scenarios on future desertification in Northeast Iran: A data mining and Google Earth engine-based approach. *Land, 13*(11). https://doi.org/10.3390/land13111802

CHAPTER

12

Costs of soil degradation and economic loss calculations with implementation challenges in soil conservation laws

Elham Sirjani

Department of Soil Science, Ferdowsi University of Mashhad, Mashhad, Iran

1. Introduction

Soil degradation has emerged as one of the most important environmental challenges of the early 21st century. This problem has significant implications on food production, ecological integrity, and economic growth. Healthy soils are essential for terrestrial ecosystems and agricultural production. This is because they support plant growth, control water and nutrient cycles, and sustain biodiversity. However, global assessments indicate that soil resources are increasingly getting degraded. Approximately 33% of the world's land is reported to be moderately to severely degraded (FAO & ITPS, 2015b) and that can increase to 90% by 2050 if unsustainable activities like intensive farming, overgrazing, deforestation, urban expansion, and inadequate land management continue (Robinson et al., 2024). Human activities mainly affect agricultural areas. Although croplands cover only about 13% of the global land surfaces, they account for nearly 29% of all degraded lands. Furthermore, about one-third of rainfed croplands and nearly half of irrigated agricultural fields show signs of degradation, resulting from anthropogenic pressures (FAO, 2021).

Soil is a vital natural resource that sustains agricultural production, ecosystem functioning, and human health. It serves as a foundation for all terrestrial ecosystems by regulating water and nutrient cycles, storing carbon, and also supporting biodiversity. The development of even 1 cm of soil takes hundreds to thousands of years; therefore, soil is basically a non-renewable source for humans (Pimentel et al., 1995). Nevertheless, soil is lost in large amounts each year worldwide as a result of various processes of degradation. Processes such as water and wind erosion, salinization, nutrient depletion, compaction, and organic matter decline affect vast areas of land worldwide, particularly in arid and semi-arid regions, where soils are inherently vulnerable. Despite growing scientific consensus on the severity of soil degradation and its role in exacerbating food insecurity and environmental instability, the problem continues to intensify due to population growth, land-use pressure, unsustainable agricultural practices, and climate change.

Soil degradation represents a critical global challenge with far-reaching consequences for agricultural productivity, ecosystem functioning, water security, and climate regulation. When soil

The Soil Nexus. https://doi.org/10.1016/B978-0-443-45286-4.00017-1

degradation occurs, soil loses its ability to carry out critical ecological functions within the ecosystem; ultimately resulting in a decline of ecosystem services provided to society. FAO estimates that 12 million hectares of productive land are lost every year as a result of land degradation events. The associated financial costs are significant. It is estimated that soil degradation costs several hundred billion dollars each year due to decreased productivity in agriculture, cost of repairing or rehabilitating degraded soils, and cost of mitigating environmental damage. Developing countries are particularly vulnerable to land degradation; thus, they experience a much higher rate of soil degradation than developed nations primarily because of their lack of resources available to implement sound land management. Thus, it is of great importance to minimize or completely prevent soil degradation in order to preserve the services provided by soils, as this approach is considerably more cost-effective than restoring soils after degradation has occurred (FAO & ITPS, 2015b).

While the biophysical mechanisms and environmental impacts of soil degradation have been extensively studied, its economic dimension remains insufficiently integrated into decision-making processes. Soil degradation generates both direct economic losses, such as reduced crop yields, increased input requirements, and land rehabilitation costs, and indirect losses, including off-site sedimentation, water quality deterioration, infrastructure damage, and the loss of ecosystem services. These economic impacts often extend beyond farm boundaries and accumulate over time, creating long-term societal costs that are rarely captured by conventional market indicators. As a result, the true magnitude of soil degradation losses is frequently underestimated, leading to underinvestment in soil conservation and restoration measures.

Estimating the economic costs of soil degradation is therefore essential for translating environmental damage into policy-relevant and comparable metrics. Economic loss calculations provide a critical bridge between scientific assessments of land degradation and the formulation of effective soil conservation policies. By expressing degradation impacts in monetary terms, these assessments enable policymakers to evaluate trade-offs, prioritize interventions, and compare the costs of prevention and restoration with the costs of inaction. However, the valuation of soil degradation is inherently complex due to spatial heterogeneity, long time horizons, data limitations, uncertainty, and the difficulty of monetizing non-market ecosystem services.

In addition to methodological challenges, the implementation of soil conservation laws faces significant institutional, legal, and economic constraints. In many regions, weak enforcement mechanisms, fragmented governance structures, limited financial incentives, and the absence of reliable economic evidence hinder the effective application of soil protection policies. Without robust and transparent economic assessments, soil conservation laws often lack the empirical foundation needed to justify regulatory measures and public investment.

Accordingly, this chapter aims to systematically examine the economic dimensions of soil degradation by integrating biophysical understanding with economic valuation and policy analysis. It reviews the major soil degradation processes and presents and compares the principal methods for estimating direct and indirect economic costs. Through the use of illustrative case studies, including an original empirical assessment of wind erosion–induced productivity losses in arid and semi-arid regions, the chapter demonstrates how economic loss quantification can support more effective soil conservation policies. It further explores implementation barriers within soil conservation laws and offers recommendations for improving policy design and practice. By integrating economic valuation with environmental science and legal considerations, this chapter aims to contribute to a more effective and evidence-based framework for soil conservation at both national and international levels.

2. Types and processes of soil degradation

Soil degradation include a wide range of physical, chemical, and biological processes that diminish soil's ability to function as a vital natural resource. These processes can be classified into several categories, each of which associated with distinct causes and consequences.

2.1 Physical degradation

Physical soil degradation refers to the deterioration of soil structure and texture, often resulting in reduced infiltration, compaction, surface sealing, and crusting. It is mostly related to the breakdown of soil aggregates and changes of the volume and size distribution of pores (Singer & Scalenghe, 2006), which limits root penetration and water movement, thereby reducing plant growth. This form of human-induced soil degradation represents about 4% of the total degraded area worldwide. There are several factors affecting this type of degradation, among which improper agricultural practices and excessive grazing are the primary contributors to human-induced physical degradation of soils (Oldeman, 1994).

2.2 Soil erosion

Regarding various land degradation processes, soil erosion is one of the most important contributors to land degradation worldwide, with significant impacts on soil quality and ecosystem services (Borrelli et al., 2017). Generally, soil erosion occurs through the action of various erosive agents such as water, wind, gravity, and glaciers (Bayata, 2024); however, it mostly occurs in two forms, including:

- Water erosion (such as sheet, rill, and gully erosion)
- Wind erosion (especially in arid and semi-arid regions)

Water- and wind-induced soil erosion is among the major factors contributing to global soil degradation. It affects nearly 84% of degraded lands globally, of which 56% is attributed to water erosion and 28% to wind erosion (Gholami et al., 2024; Oldeman, 1994). Both forms have a detrimental effect on crop productivity, either directly through removal of topsoil/A-horizon, organic matter, and plant-available nutrients or indirectly through reduced water retention and increased yield variability. Field studies consistently show substantial yield declines following significant topsoil loss. Small amounts of soil removals are associated with little or no yield response, whereas yield losses become increasingly severe beyond certain thresholds of topsoil depletions. Recent empirical studies suggest that soil erosion can lead to important but often non-linear yield reductions. A hierarchical meta-analysis of 290 observations of topsoil-removal studies found no statistically significant yield losses up to an erosion depth of 5 cm. However, productivity becomes difficult to sustain beyond approximately 25 cm of the A-horizon loss, and yields are often not recoverable after losses of around 20 cm of the A-horizon (Zhang et al., 2021). The widely-cited method-comparison by Bakker et al. (2004) synthesized experimental studies and showed that different experimental approaches produce systematically different estimates of crop productivity responses to topsoil loss. For example, crop yields have been reported to decline by approximately 4.3% per 10 cm of topsoil loss in comparative-plot studies, about 10.9% in transect studies, and up to 26.6% in desurfacing or topsoil-removal experiments. Crop-specific and regional evidence supports similar magnitudes of yield response.

Gao et al. (2015) estimated that soybean yield decreased by approximately 6% per 10 cm of topsoil loss. Likewise, a spatial agroecological study in the US Corn Belt reported yield reductions of around $6 \pm 2\%$ associated with cumulative A-horizon loss (Thaler et al., 2021). At larger scales, integrated assessments that combine effects of erosion and compaction suggest yield declines due to soil loss in the range of 10%–20%, with the strongest impacts observed in low-input agricultural systems (Sonderegger & Pfister, 2021). This range of effects should be viewed as process-informed benchmarks for productivity reduction rather than exact values. Ultimately, local soil depth, crop sensitivity, climate, and management practices are the main predictors of yield response.

On the other hand, as the off-site consequences, the sediment being produced during soil erosion processes can mostly lead to severe environmental damages. In fact, sediment particles and their associated pollutants like different types of elements, heavy metals, pesticides, and even microplastics can make water resources polluted and thereby endanger the lives of so many living things (Sirjani et al., 2022).

Recently, numerous studies have shown that both water (Rehm et al., 2021) and wind erosion (Rezaei et al., 2019) are as main pathways through which microplastics can be transported and dispersed in the environment. This has already raised concerns in terms of human health especially due to the ingestion and inhalation of dust particles containing microplastics, especially in arid and semi-arid regions (Rezaei et al., 2019). Wind erosion is mostly visible during the occurrence of dust storms in which wind-blown sediments are transported by suspension. Once soil erosion occurs, various sized particles are transported through different mechanisms of suspension, saltation, and bed load (BL) (Sirjani et al., 2022). Due to small size and low density, fine particles can be transported in long distances by suspension and may contaminate even the remote regions. This emphasizes that soil aggregate size distribution plays a significant role in soil susceptibility to wind erosion and its associated impacts (Sirjani et al., 2024).

Global estimates of soil loss by wind and water erosion indicate that soil erosion is a significant and ongoing threat to soil resources and productivity in agriculture. For instance, Table 12.1 represents the estimated values of soil erosion by water provided by the Global Soil Erosion Modeling platform (GloSEM) developed by Borrelli et al. (2022). The GloSEM model applies the Revised Universal Soil

Table 12.1 Mean annual water erosion rates by continent for cropland areas, as modeled by GloSEM (RUSLE-based global assessment). Date from Borrelli et al. (2022).

Continent	Soil erosion rate (t ha^{-1} y^{-1})
Africa	17.1
Asia	12.8
Europe	2.6
North America	11.0
Oceania	2.7
South America	22.0
Total	12.3

Borrelli, P., Ballabio, C., Yang, J. E., Robinson, D. A., & Panagos, P. (2022). GloSEM: High-resolution global estimates of present and future soil displacement in croplands by water erosion. Scientific Data, 9(1), 406.

Loss Equation (RUSLE) framework to estimate long-term average soil loss using high-resolution datasets on rainfall erosivity, soil erodibility, topography, and land cover. The analysis uses 2019 as the baseline year. Results are expressed as mean annual soil loss (t ha^{-1} yr^{-1}) and are limited to croplands, which are the land use category most affected by soil erosion globally. There is considerable variation in the intensity of erosion among continents due to differences in topography, rainfall erosivity, vegetation cover, and management practices. South America and Africa exhibit the highest average soil loss rates for cropland (≈17–22 t ha^{-1} yr^{-1}), driven by steep slopes, high rainfall erosivity, and limited implementation of soil conservation practices in some regions. Europe and Oceania show the lowest average rates (≈2–3 t ha^{-1} yr^{-1}), reflecting relatively stable landscapes and more widespread adoption of soil conservation-oriented management. Asia and North America display intermediate values, reflecting strong contrasts between highly intensive agricultural areas and more stable landscapes.

Figure 12.1 illustrates the spatial distribution of mean annual soil loss by water across the globe as modeled by GloSEM framework for 2012. According to the map, the largest increases and highest risks of soil erosion are concentrated in Sub-Saharan Africa, South America (particularly parts of Brazil and other Amazon-frontier areas) and Southeast Asia, with additional intense hot-spots occurring in China, India, the south-eastern United States, and parts of Central and South America.

Figure 12.1 also indicates that very high erosion rates (>20 Mg ha^{-1} yr^{-1}) are spatially clustered particularly in China, Brazil, equatorial African territories, India, south-eastern USA, Central-Eastern

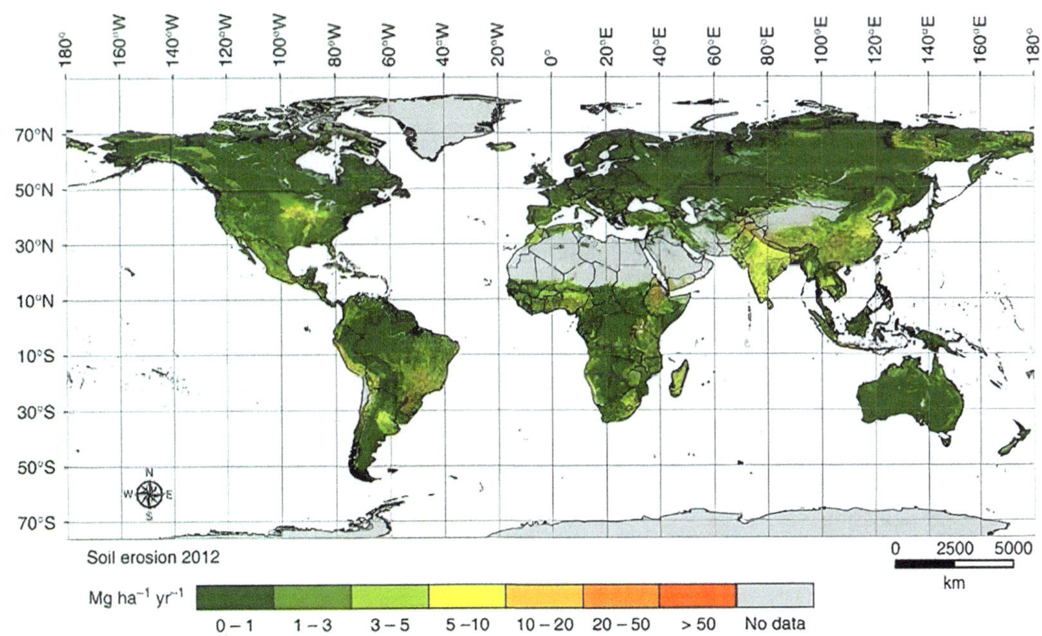

FIGURE 12.1

Global distribution of the estimated soil erosion by water for the year 2012.

-Reprinted from Borrelli et al. (2017), under a CC BY license.

Ethiopia, Mexico, Indonesia, Peru and parts of the Mediterranean region. These patterns are largely driven by cropland expansion and land-use change, with developing economies bearing a disproportionate share of the erosion burden.

It can be concluded that areas where intense rainfall interacts with steep slopes and poor ground cover are associated with the highest risk of soil erosion. This pattern reflects the spatial inequality of soil degradation, and highlights that developing countries experience disproportionately high losses in agricultural productivity due to soil erosion.

Recently, large-scale assessments have provided valuable insights into the spatial and temporal dynamics of soil loss by wind, as well. For example, Sun et al. (2024) estimated a total modeled wind-driven soil erosion of about 312.5 Pg yr^{-1} for 1982–2019. In model-based global assessments like that conducted by Borrelli et al. (2021), the global median rate of wind erosion is estimated to be approximately 6.3 t ha^{-1} yr^{-1}. However, local rates vary considerably, ranging from near-zero in well-protected lands to tens to hundreds of t ha^{-1} yr^{-1} in severely eroding drylands.

According to Oldeman (1994), wind erosion potentially threatens a total of 549 million ha of land. Moreover, as pointed out by Sun et al. (2024), the spatial variation of wind erosion around the world shows maximum values in major drylands, including the Sahara, Arabian Peninsula, Central Asia, northern China, and Australia. Temperate agricultural areas have moderate to low erosion rates, while the tropicals and humid areas show no or negligible values. These spatial patterns demonstrate the substantial influence of climate drivers and land use on global wind erosion distributions.

In conclusion, although water and wind erosion exhibit distinct global patterns, they can be complementary. Water erosion, as shown by Borrelli et al. (2017), mainly occurs in humid and sub-humid regions characterized by heavy rainfall and sloping cultivated lands. This usually happens in regions such as tropical South America, sub-Saharan Africa, and South and Southeast Asia. However, wind erosion as reported by Sun et al. (2024), mostly occurs in arid and semi-arid areas such as the Sahara, Central Asia, northern China and Australia, where poor vegetation and strong winds promote soil loss. Since wind and water erosion usually do not spatially coincide, transition lands such as the Mediterranean basin and parts of southern Africa, may be affected by both processes. Overall, these pattrens highlight the need for regionally adapted, process-specific soil conservation strategies to reduce global soil degradation and agricultural productivity losses.

2.3 Chemical degradation

Chemical degradation involves alterations in soil chemical properties that negatively impact soil productivity. This type of degradation affects nearly 240 million hectares across the globe, representing around 12% of all areas impacted by human-caused soil degradation (Oldeman, 1994). Common processes of this kind of degradation include:

- Nutrient depletion
- Salinization and sodification
- Acidification
- Contamination

These processes impair nutrient availability and may lead to soil toxicity.

2.3.1 Nutrient depletion (deficiency and mining of soil nutrients)

Soil nutrient depletion refers to the extensive and long-term net loss of plant available essential elements principally nitrogen (N), phosphorus (P), and potassium (K) as well as secondary and micro elements from the root zone of plants. This loss reduces nutrient concentrations to the levels that affect crop productivity and ecosystem functioning. It is important to recognize that nutrient depletion and temporary nutrient deficiencies are different problems. In fact, depletion describes a situation where nutrient inputs (e.g., fertilizer and biological fixations) do not sustain the plant nutrient uptake needs or are not enough to balance nutrient losses occurring over several consecutive years through removal with harvest, leaching, erosion, gaseous emissions, etc., (Smith et al., 2024).

Soil nutrient depletion is attributed to a variety of factors which include:

- removing crops but not adding nutrients back to appropriate levels,
- erosion that carries away nutrient-laden topsoil particles,
- monoculture crop growth in intensified soils,
- discarding organic residues rather than recycling, and
- socioeconomic factors such as limited access to fertilizers and effective nutrient management.

Climate-induced changes (e.g., temperature and precipitation extremes) contribute to nutrient loss by altering decomposition and leaching processes. Many global studies report nitrogen and phosphorus cycles are in an imbalanced state and agricultural systems are regularly witnessing potassium and micronutrient depletion (Bhardwaj et al., 2024).

Soil nutrient deficiencies cause various negative effects which decrease agricultural output, degrade crop nutritional value and weaken plant defense systems against environmental challenges. The process of nutrient mining at a landscape scale leads to the development of marginal areas which in turn perpetuates the ongoing cycle of land deterioration (Smith et al., 2024).

To manage this detrimental phenomenon, sustainable solutions consist of balanced fertilizer application based on soil tests and crop needs (precision nutrient management), using slow-released or control-released fertilizers, recycling organic resources (manure, compost, crop residues), integrated nutrient management combining mineral and biological inputs (e.g., legumes for N fixation), and erosion control to retain topsoil. Policy measures that improve farmer access to inputs and knowledge, and incentives for soil-restoring rotations, are crucial (Khambalkar et al., 2025).

2.3.2 Salinization and sodification of soils

Salinization is a type of chemical degradation of the soil that significantly reduces soil productivity. Salinization occurs when there is an accumulation of water-soluble salts (mainly Na^+, Ca^{2+}, Mg^{2+}, and K^+) in the root zone that reaches thresholds which limit the amount of available water for plants and in certain situations can produce ion-specific toxicity (Rengasamy, 2006). Some potential sources of soluble salts in soils, other than irrigation water, include mineral weathering, fertilizers, atmospheric depositions, and the lateral movement of groundwater from saline regions (Rengasamy, 2006; Shrivastava & Kumar, 2015). Estimates by Jamil et al. (2011) indicate that more than 50% of all agricultural land will be affected by salinization by the year 2050.

Sodification (sodicity) specifically refers to the elevated amount of exchangeable sodium on the cation-exchange sites in soil when compared to calcium and magnesium. Sodicity is harmful to soil structure since sodium lower the aggregation of clay particles, which finally limits infiltration and aeration into the soil (Qadir & Schubert, 2002).

Salt-affected soils are normally designated as saline, sodic, or saline - sodic depending on the most dominant condition (FAO, 2021). The measurable indicators used to identify soil types consist of:

- Electrical conductivity (EC) of the saturation paste extract (or EC of the soil-water extract): This criterion signifies the concentration of soluble salts. A soil is classified as saline when EC of its saturation extract exceeds 4 dS m^{-1} in the root zone (Shrivastava & Kumar, 2015).
- The exchangeable sodium percentage (ESP) and sodium adsorption ratio (SAR) are used to assess sodicity risk: Generally, a soil is considered as sodic when ESP >15% and SAR >13 $(meq/L^{)-0.5}$.
- Soil pH: Sodic soils are usually alkaline and show a pH > 8.5.

Natural factors which contribute to soil salinization include arid to semi-arid climate (where evaporation is higher than precipitation), geological deposits of salts, marine transgression, and the upward movement of saline groundwater. Anthropogenic factors which are often more rapid and widespread than natural factors, include improper irrigation management (application of saline irrigation water and insufficient leaching and drainage), over-extraction of groundwater (resulting in elevated saline water tables), and poor land use decisions that interfere with hydrology (Tarolli et al., 2024). In sodic soils, continued input of sodium (or leaching of Ca^{2+} and or Mg^{2+}) raises the ESP/SAR of the soil, resulting in dispersion of clays (Qadir & Schubert, 2002).

Salt-affected soils pose a significant production risk. According to the UN/FAO, there are on the order of hundreds of million to over one billion hectares of formerly productive land affected or at risk due to these soils around the world. Thus, salinity must be considered a significant issue for food security. Crop yield reductions will vary by crop and severity; in the case of severe salinity the loss can be in excess of 50%–70% in some extreme local cases; however, moderate salinity can develop into chronic yield penalties and changes to more tolerable crops or abandonment. The structural damage of sodicity can last a significant period of time due to requiring divalent cations and time to re-aggregate the soil back to a more stable structure (FAO, 2021).

There are some common approaches to remediate saline and sodic soils:

- For saline soils initially make sure that drainage is adequate and then implement leaching which is applying sufficient good quality water to move salts below the root-zone without causing waterlogging. Improve irrigation scheduling and, where feasible, use good quality water. Grow salt tolerant cultivars or adjust cropping systems until salinity has decreased (Singh, 2021).
- For sodic soils different apply calcium amendments (typically gypsum) to displace exchangeable Na^{+} with Ca^{2+}. It should be noted that this will subsequently require leaching to remove the sodium ions. In this process, addition of organic matter such as farm yard manure and compost would improve aggregation as well as the efficacy of gypsum. In fact, the combination of organic amendments and gypsum would often perform better than gypsum alone (Stavi et al., 2021).
- Integrated and ecosystem approaches: the methods like saline agriculture, halophyte-based remediation, better water governance, and farmer training are necessary for long-term sustainability and to develop solutions that match local socio-economic realities (Batlle-Sales, 2023).

2.3.3 Soil contamination

Soil contamination occurs when toxic substances or unusual amounts of natural elements appear in the soil at levels which damage soil organisms and plants and also threaten water systems and human health. The contaminants possess varying levels of movement through the environment and different persistence rates and toxic effects which require specific monitoring and remediation techniques (Das et al., 2023). There are several possible sources and processes of soil pollution, some of which include as following:

Heavy metals: Heavy metals enter the environment through mining activities, smelting operations, metallurgical processes, coal burning, sewage sludge and fertilizer use, and atmospheric deposition. The soil particles create strong bonds with metals but these bonds can break down when the soil experiences changes in redox potential and pH levels (Angon et al., 2024).

Agrochemical residues (pesticides, some fertilizer-derived contaminants): Repeated pesticide applications, improper timing and dosage, and persistent metabolites lead to accumulation of pollutants. Moreover, some certain practices also mobilize residues to groundwater (Pathak et al., 2022).

Microplastics (MPs): MPs in soil originate from plastic mulches, sludge and biosolid application, wastewater irrigation, atmospheric deposition and fragmentation of larger plastics. Actually, soils are now recognized as a major sink for MPs (Gundogdu et al., 2025).

Soil pollution can have serious consequences in many aspects. In other words, soil contaminants can disrupt microbial community composition, lower enzymatic functions, and disrupt nutrient cycling. Microplastics also change soil physical properties and shift microbial groups (Wang et al., 2025). Heavy metals and some pesticides will bioaccumulate into edible plant parts presenting food chain risks. Additionally, metal-contaminated soils have well documented detriments to crop yield and quality (Das et al., 2023). Leaching and runoff transfer contaminants to groundwater and surface waters. Ultimately, inhalation or ingestion of contaminated dust or crops is a primary human exposure route (Acharya et al., 2025).

To remediate polluted soils, several different ways can be applied depending on the classes of contaminants. Some of the most important methods include phytoremediation and phytoextraction for certain metals, soil washing or stabilization/immobilization (i.e., with amendments), bioremediation for many organics, and hotspot removal or containment (Das et al., 2023). For microplastics, research that attempts to eliminate them is in its early stages; thus, prevention (i.e., source reduction and improved waste and wastewater management) is the best path forward for now.

2.3.4 Soil acidification

Soil acidification is the slow decline of soil pH (or increased activity of hydrogen ions) due to both natural processes and human activities that exceed the soil's buffering capacity. It is an important form of chemical degradation of soil, especially in humid areas and agricultural lands under intensive management. Soil acidification is driven by: (i) natural pedogenic processes such as the leaching of basic cations (Ca^{2+}, Mg^{2+}, K^{+}, Na^{+}) and the accumulation of acidic entities (Al^{3+}, H^{+}), (ii) anthropogenic impacts especially when acids are applied in fertilizer, and (iii) acid rains (FAO & ITPS, 2015a).

Major contributors to soil acidification are:

- Excessive use of nitrogen fertilizers especially ammonium-based fertilizers (such as ammonium sulfate) that release H^{+} ions during nitrification (Guo et al., 2010; Vitousek et al., 1997).
- Removal of base cations (Ca^{2+}, Mg^{2+}, K^{+}, Na^{+}) by crop harvesting and leaching and the accumulation of acidic entities (Al^{3+}, H^{+}): intensive cropping with little or no liming depletes soil base cations to the point of acidification (Fageria et al., 2010).

- Acidic compounds, that is, SO_2, NO_x that are deposited from the atmosphere during acid rains and subsequently form sulfuric and nitric acids (Ding et al., 2024).
- Acid parent materials and high rainfall with promote leaching of base cations in natural ecosystems (Regasa et al., 2025).

Acidification poses serious biogeochemical and biological impacts, including:

- *Toxicity:* The solubility of Al^{3+} and Mn^{2+} ions, increases at low pH levels which can harm root systems and restrict growth (Von Uexküll & Mutert, 1995).
- *Nutrient imbalances:* Soil acidification occurs due to leaching and reduced availability of basic and essential nutrients (Fageria & Baligar, 2008).
- *Microbial inhibition:* Reduction of soil microbial diversity and activity can also occur, particularly for nitrifiers and decomposers, leading to a limitation in the rate of organic matter turnover (Wang et al., 2022).
- *Structural decline:* The loss of organic matter and cations can contribute to structural decline and reduced water-holding capacity (Lal, 2015).
- *Agronomic impacts:* Depending on crop and severity, different yield losses have been reported in acidified soils. A recent global meta-analysis has found that soil acidification reduces average crop yield by about 13.7% (Du et al., 2024).

To remediate acidified soils, liming is the most common method, in which alkaline materials (e.g., calcium carbonate, dolomitic lime) are added to neutralize soil acidity and replenish base cations to the soil (Goulding, 2016). In addition, it is recommended to periodically apply organic amendments (e.g., compost, biochar) to help improve buffering capacity and cation exchange capacity (Naramabuye & Haynes, 2006).

2.4 Biological degradation

Biological soil degradation refers to the deterioration in the abundance, diversity, and functional capacity of soil biota as well as the associated loss of biological processes necessary to uphold soil fertility and ecosystem processes. This type of degradation will mean long-term losses of soil organic matter and microbial biomass, reduced soil enzymatic activity, and the establishment of dysfunctional processes involved in decomposition, nutrient cycling, and disease suppression. Large assessments (FAO & ITPS, 2015b; IPBES, 2018) identify the loss of soil biodiversity, loss of soil organic carbon and loss of soil fertility as critical threats to soil health on a global scale.

Biological degradation occurs through a combination of factors that will manifest simultaneously:

Intensive land use and cultivation practices: Repeated tillage and removal of crop residues, monoculture, and low input of organic materials result in a loss in habitat complexity for soil microbes and fauna and in food resources, promoting losses of soil organic carbon and microbial biomass. Tilling and continuous disturbances will also lessen the connectivity of fungal networks and removed beneficial symbionts (i.e., mycorrhizae) from the soil food web (Derpsch et al., 2024; Lori et al., 2025).

Soil compaction and physical disturbance: Heavy machinery and overgrazing (or overstocking) by grazers will compact soil pores, will reduce aeration and infiltration of water, and will disrupt burrowing fauna and aerobic microbes. These actions will lead to reduced biological activity and litter decay rates (Nassir et al., 2024).

***Chemical stressors (pesticides, fertilizers, pollutants)*:** Frequent or excessive application of insecticides, fungicides, certain herbicides, and fertilizers not only will alter the composition of the soil microbial community, but also will inhibit beneficial species and can decrease important enzymatic activities associated with nutrient cycling (e.g., cellulose activity). Like other stressors, heavy metals and industrial pollutants can kill soil organisms or alter a community to one that is resistant to the stressor with reduced functional capabilities (Swaine et al., 2025).

Emerging contaminants — microplastics and nanoplastics: Generally, microplastics (<5 mm) are classified as large MPs (1–5 mm) and small MPs (0.3–1 mm), and nanoplastics are plastic particles <100 nm. Microplastics and nanoplastics have entered the soil matrix via various pathways, including landfill disposal, plastic mulching, atmosphere, and the land application or recycling of wastewater sludge (Rillig, 2012). These particles can influence physical habitat and interacts with soil microbial communities. Experimental and field experiments have documented changes in microbial biomass, inhibition of growth or reproduction of soil fauna, and changes in enzymatic rates (Sajjad et al., 2022).

Climate change: The variation of temperature and precipitation periods will affect decomposition rates, soil moisture regimes and disturbance frequencies. This leads to changes in biological communities and their resilience. Drought and extreme wetting events alter soil biodiversity and the ratios of functional groups (Kopittke et al., 2025).

Losses in soil biological integrity cause a series of functional declines, including:

- Slower decomposition of litter.
- Reduced nutrient mineralization and retention.
- Loss of symbiotic services (N fixation, mycorrhizal P uptake).
- Weakened natural suppression of soil pathogens.
- Weaker soil structure due to reduced aggregate formation and bioturbation.

These processes contribute to reduced crop yield, increased reliance on external inputs (e.g., fertilizers, pesticides), reduction of resilience to stress (drought, pests), and declines in soil quality that once initiated, they can be expensive and time-consuming to remedy. Global assessments estimated that a large share of productivity decline is attributed to losses in soil organic matter and biodiversity linked to unsustainable management (Derpsch et al., 2024; Smith et al., 2024).

Biological health in soil can be restored using a combined approach to restore not only habitat and food resources but also alleviate chemical and physical stresses:

- *Increase organic matter and maintain the minimum necessary removal of residues.* Soil organic carbon (SOC) restoration is also possible by adding crop residues, composts, manures, and cover crops that are able to rebuild diversified soils with microbial and faunal communities.
- *Conservation agriculture (or reduced tillage).* There are applications of reduced soil disturbance to retain pre-existing fungal networks and soil structure and increase fungal and microbial biomass.
- *A variety of crop rotations and agroecological practices.* The diversity of crop species, such as legumes, agroforestry, and intercropping, increase nutritional complexity and functional redundancy.
- *Reduce pesticide use and integrated pest management (IPM).* Reducing reliance on broad spectrum pesticides and using IPM eliminates some chemical workload from soil organisms, including microbes and earthworms, and their functions.

- *Avoid and remediate soil compaction.* Controlled traffic, reduced axle loads, and deep loosening where necessary restore pore networks for fauna and roots.
- *Tackle contaminants and microplastics at source.* Improved waste-sorting, better regulation of biosolids, alternatives to plastic mulches, and circular-economy measures reduce long-term pollutant loads to soil.

Experimental studies and meta-analyses show that various combinations of these strategies can increase SOC, microbial biomass, and other functional indicators from years to decades, while providing significant co-benefits for productivity and resilience (Lori et al., 2025; Robinson et al., 2024).

2.5 Comparative overview of soil degradation processes

Soil degradation occurs through multiple and often interacting pathways that vary significantly in spatial scale and economic importance. Among these processes, soil erosion (primarily caused by water, although wind-erosion hotspots play an important role) is the main driver of measurable soil loss and productivity decline and is consistently reported as the greatest threat in global assessments. Many studies have demonstrated that soil erosion, particularly water erosion, accounts for the majority of mapped soil degradation worldwide, causing the most extensive on-site loss of fertile topsoil and significant off-site sedimentation costs (Borrelli et al., 2021; Panagos et al., 2015, 2024). Wind erosion, which is particularly severe in arid and semi-arid regions, can result in substantial local soil loss and significant dust-related impacts. Therefore, it is often considered as one of the most important soil erosion processes in many dryland assessments (FAO & ITPS, 2015b). Salinization and sodification contribute to chemical degradation with a highly concentrated but highly significant impact on irrigated and coastal lowlands. A recent FAO's global assessment (FAO, 2024) indicates that 1.380 billion ha ($\approx$10.7% of land) are already salt affected and about 10% of irrigated and rainfed cropland may show salinity impacts, making salinization one of the greatest factors reducing production if it occurs. Declines in soil organic matter and compaction produce broad, landscape-scale reductions in resilience, water-holding capacity, and nutrient supply. These effects per hectare can be less severe, but are significant overall as they exacerbate vulnerability to erosion, drought and yield losses across large agricultural areas (FAO & ITPS, 2015a). Lastly, chemical contamination (i.e., heavy metals, microplastics, persistent organic pollutants, agrochemical residues) is generally much more spatially heterogeneous, and also much more localized in concentration (e.g., point sources, hotspots). But when present, it imposes a disproportionate cost in much stronger health, remediation, and market-access (FAO & UNEP, 2021).

As a result, many global assessments have ranked water erosion first, wind erosion (in drylands) second, salinity (a very evident issue in irrigated/coastal systems) third, and finally organic-matter loss, soil compaction, and more localized contamination last. This broad ranking reflects both the extent of land affected and the economic impacts commonly reported in the literature.

3. Methods for calculating soil degradation costs

Evaluating the costs associated with soil degradation provides a basis for understanding its impacts on agricultural productivity, ecosystem services, and public infrastructure. However, assessing the economic value of soil degradation is a challenging and complex process, due to the multifaceted nature of soil functions, the variability of degradation processes, and the long-term, often indirect nature of their effects. Many impacts, such as the loss of soil biodiversity, reductions in carbon sequestration, and downstream sedimentation, are difficult to isolate, quantify, and monetize. In addition, uncertainties in valuation methods, data limitation, spatial and temporal variability, and the need to account for both market and non-market values further complicate economic assessments. Despite these challenges, a range of frameworks and methodologies have been developed and improved to evaluate the diverse economic dimensions of soil degradation, from estimating productivity losses to assessing broader environmental and societal costs (Borrelli et al., 2021; ELD Initiative, 2015; FAO & ITPS, 2015b; Panagos et al., 2015; Panagos et al., 2024). The following section reviews the main methods used to assess the costs of soil degradation, with particular emphasis on their theoretical foundations, practical implementation, and limitations.

Economic assessments of soil degradation are commonly organized into two broad groups:

- Direct costs: costs that can be directly attributed to the loss of goods and services resulting from soil degradation; and
- Indirect costs: off-site, economy-wide, or long-term impacts that occur as secondary consequences of soil degradation.

3.1 Direct economic losses

Direct economic costs arise from soil degradation via reduced soil productivity, the most immediate effect of soil loss on agricultural products and income. When soil fertility is reduced due to soil erosion, nutrient depletion, and salinization, agricultural yields also decline. Once yields reduces, farmer incomes decrease, and the cost of inputs increases (e.g., fertilizers, irrigation). The economic costs related to these losses can often be conservatively evaluated by monitoring current or historical productivity to productivity that can be attributed to "healthy" soil. According to numerous studies, soil erosion has the capacity to decrease agricultural productivity by 50% in some of the hardest-hit places. This indicates that soil erosion causes a major economic loss in these areas. Furthermore, soil restoration (e.g., through reforestation, soil amendment or terracing) is associated with financial costs that must also be considered at an economic level for the government and landowners. In addition to these financial costs to rehabilitate degraded soils; direct costs associated with reduced productivity are typically the first costs associated with soil degradation. The only way to provide a global estimate of these costs is to rely on average productivity by establishing productivity levels locally. For these reasons, assessing the economic loss from soil degradation can be very challenging, as their very specific measurement of total losses requires a high level of understanding of soil health, productivity, and management practices. When it comes to developing countries, data for accurately estimating these losses are often lacking or inconsistent.

Within the framework of direct costs, a useful functional classification distinguishes between market-based, non-market based, and hybrid or integrated approaches. In fact, researchers and practitioners often need purely market-based indicators, expressed preference assessments, or a combination of both to capture the full value of soil functions.

3.1.1 Market-based approaches

Market-based methods value the observable, traded outputs or replacement goods lost because of soil degradation (e.g., reductions in crop output, additional fertilizer required, market costs of dredging).

These are often the first line of assessment because they rely on observable prices and quantities (Panagos et al., 2024). In fact, these methods rely on actual market prices to estimate the costs associated with productivity loss, increased inputs, or restoration expenses. Market-based approaches are called “pricing” (Ghafari et al., 2025). Common techniques of this approach include the following.

(a) Productivity reduction method

This approach values on-site soil degradation through reduced agricultural production. It is assumed that the differences in yield are directly attributable to soil degradation processes such as erosion, nutrient loss, and salinity. In fact, this method calculates the reduction in crop yield caused by degradation and multiplies it by market prices to estimate the economic losses.

The formula used in this method is:

$$C_{pd} = \left(Y_p - Y_d\right) \times P_c$$

where,

- C_{pd}: Economic cost of productivity loss (USD per hectare or region)
- Y_p: Potential yield under non-degraded soil conditions (ton ha^{-1})
- Y_d: Actual yield under degraded soil conditions (ton ha^{-1})
- P_c: Market price of the crop (USD ton^{-1})

Key limitations of this approach include:

- Attribution of yield loss to a single degradation process is difficult.
- Sensitive to crop price variability and market fluctuations.
- Ignoring off-site and non-market impacts.

Example:

To illustrate the practical application of the productivity reduction method, the economic costs of wind-induced soil degradation are estimated for three representative regions selected from a previous wind tunnel study conducted by Sirjani et al. (2019). While the original study focused on measuring wind-induced soil loss, the economic cost assessment presented here is performed for the first time to demonstrate the practical implementation of the proposed methodology.

The selected regions are located in the northern, eastern, and southern drylands of Fars Province, southern Iran. Wind-induced soil loss (g m^{-2} s^{-1}) in the selected regions was measured using a portable wind tunnel under standardized experimental conditions. Figure 12.2 illustrates the portable wind tunnel used in the field measurements. Moreover, the relevant soil physicochemical properties required for the economic calculations were determined using standard laboratory methods. A detailed description of the wind tunnel design, experimental procedures, soil sampling, and analytical methods is provided by Sirjani et al. (2019). Therefore, only the data required for the present economic calculations are used in this section.

Due to the limited availability of long-term hourly wind speed data for each study site, the duration of erosive winds was assumed to be 20 h yr^{-1}. This assumption represents moderate wind activity typical of the upland plains of northern Fars Province and provides a consistent basis for comparing soil loss and the associated economic costs among the three study regions. Accordingly, Table 12.2

FIGURE 12.2

Field views of the portable wind tunnel setup used in the study.

Table 12.2 The annual soil loss and lost soil depth measured using bulk density and wind data at each study site.

Site	Sediment mass flux ($g\ m^{-2}\ s^{-1}$)	Soil loss by wind ($ton\ ha^{-1}\ h^{-1}$)	Annual soil loss by wind ($ton\ ha^{-1}\ yr^{-1}$)	Bulk density ($Kg\ m^{-3}$)	Depth of soil loss (mm)
1	0.3	10.68	213.53	1173.3	18.20
2	0.86	30.78	615.60	1430	43.05
3	1.00	35.91	718.20	1543.3	46.54

presents the estimated annual soil loss rate and soil loss depth for each study region, calculated using sediment flux and soil bulk density.

Calculation Steps:

The economic cost estimation follows the below sequential steps:

Step 1: Converting the measured wind tunnel data to annual soil loss using the following equation:

$$\text{Annual soil loss } (\text{ton ha}^{-1}\ \text{yr}^{-1}) = \text{Q} \times 3600 \times \text{H} \times 0.01.$$

where, Q is sediment mass flux from wind tunnel experiments ($g\ m^{-2}\ s^{-1}$), H is annual erosive wind hours (hypothesized 20 h yr^{-1}), and 3600 and 0.01 are conversion factors to ton ha^{-1}.

Step 2: Converting soil mass loss to soil depth loss via the below equation:

$$\text{Depth loss } (mm) = \frac{\text{Annual soil loss} \times 100}{\text{Bulk density}}$$

where, annual soil loss is in ton ha^{-1}, bulk density is in $Kg\ m^{-3}$, and 100 is the conversion factor to mm.

Step 3: Estimating Yield Reduction.

A simple and widely used approach expresses yield loss as a percentage of the baseline yield per unit of soil depth removed (Pimentel et al., 1995). The relationship is typically linear for small depth losses (<5 cm). Therefore, yield loss is commonly approximated as:

$$\text{Yield reduction (\%)} = \text{Soil depth loss (mm)} \times \text{k}$$

where, k is the yield reduction coefficient (% per mm soil loss), derived from literature. According to Pimentel et al. (1995) and Lal (1998), typical yield reduction values range from 0.2% to 0.4% yield loss per 1 mm of soil removed (depending on the crop and region). So, based on these empirical estimates, we conservatively applied a coefficient 0.3% yield reduction per mm of soil depth loss in this study.

Step 4: Calculating Annual Yield Loss.

In the next stage, the annual yield loss was calculated via the following equation:

$$\text{Yield loss}\left(\text{ton ha}^{-1}\text{yr}^{-1}\right) = \text{Yield reduction (\%)} \times \text{baseline yield}\left(\text{ton ha}^{-1}\text{yr}^{-1}\right)$$

Step 5: Economic Cost of Soil Loss.

Finally, the cost of soil loss was obtained through the below equation:

$$\text{Economic cost}\left(\text{USD ha}^{-1}\text{yr}^{-1}\right) = \text{Yield loss}\left(\text{ton ha}^{-1}\text{yr}^{-1}\right) \times \text{wheat price (USD / ton)}$$

Results of the cost estimations: The application of the productivity reduction method combined with portable wind tunnel measurements revealed clear spatial differences in the economic cost of wind-induced soil loss across the three studied regions (Table 12.3). Region 3 with the highest wind-induced soil loss (Table 12.2) showed the greatest estimated cost (157.1 USD ha^{-1} yr^{-1}). This region probably has the most erodible soils which can be easily detached and transported by wind (Sirjani et al., 2024). Region 2 also indicated a high soil loss and economic impact (145.3 USD ha^{-1} yr^{-1}); however, the lowest soil loss and estimated cost (61.4 USD ha^{-1} yr^{-1}) were recorded for region 1. This site likely has a more stable and less erodible soil system and greater resistance to the wind induced detachment. These spatial differences emphasize that economic costs derived using yield-loss functions will not be uniform across regions (Santra et al., 2017). While conditions were standardized and constant across sites (wind speed, exposure duration, erosive-wind hours, and crop yield), this resulted in varying values of economic losses. Thus, it seems the main factor of differences in costs was soil erodibility and variability in surface stability. The results demonstrate how quantitative estimation of

Table 12.3 Summary of calculated yield reduction (%), baseline wheat yield, corresponding yield losses, wheat price, and resulting economic costs for each study region.

Site	Yield reduction (%)	Mean wheat yield (ton ha^{-1})	Yield loss (ton ha^{-1})	Market price of wheat (USD/ton)	Cost of soil loss (USD ha^{-1} yr^{-1})
1	5.46	4.5	0.25	250	61.4
2	12.91	4.5	0.58	250	145.3
3	13.96	4.5	0.63	250	157.1

the economic costs of wind-induced soil degradation can support more effective soil conservation policies and evidence-based decision-making. Moreover, the integration of wind tunnel measurements with economic valuation provides a practical and reliable framework for conservation planning, investment prioritization, and sustainable soil management.

(b) Replacement cost method

This approach estimates the cost of returning degraded land to its former productivity through some inputs. In fact, this method calculates how much it would cost to replace lost nutrients, organic matter, or retention capacity by artificial inputs such as fertilizers, compost, and soil amendments.

The general formula used in this method is:

$$C_{RC} = \sum_{j=1}^{m} Q_j \times C_j$$

where,

- C_{RC}: total replacement cost (USD ha^{-1} yr^{-1}).
- Q_j: amount of nutrient/organic matter lost (kg $ha^{-1}yr^{-1}$)
- C_j: unit price of the replacement input (USD $unit^{-1}$)

Key limitations of this method are:

- The possibility that replacement may not restore other soil functions such as structure and biota.
- The omission of off-site costs.
- Assumes full substitutability between natural soil functions and artificial inputs.
- Ignores ecosystem complexity and irreversibility of soil degradation.
- May underestimate long-term and non-market damages.

However, as a strength, this method is useful for budgeting and estimating immediate remediation costs.

Example:

Using the Replacement Cost Method, Lemma et al. (2017) determined the Economic Value of Soil Nutrient Losses via the estimated cost of unnaturally replacing eroded Nitrogen (N) and Phosphorus (P) via the application of commercial mineral fertilizers. This research was conducted in a semi arid watershed in Northern Ethiopia, where intensive agriculture occurs and rill erosion is pronounced during the rainy seasons. In this study, rill-induced eroded sediments were analyzed for N and P content. Then, nutrient losses were converted to economic costs using current market prices of mineral fertilizers (urea for N and diammonium phosphate for P).

Their results showed that:

- Average soil loss by rill erosion was 3.17 ton ha^{-1} yr^{-1}.
- Nutrient losses were ~2.4 kg N ha^{-1} yr^{-1} and 0.02 kg P ha^{-1} yr^{-1}.
- Replacement cost (for N + P) was USD 1341 per year for the study area.

Thus, Replacement Cost Method allows translation of physical nutrient losses into monetary terms and provide a tangible measure of economic damage caused by soil erosion.

(c) Restoration or rehabilitation cost method

Soil Restoration Cost (SRC) is an approach to evaluate the economics of soil degradation by quantifying the cost of restoring degraded soils back to their original (functional) condition using engineering or biological interventions (Barbier, 2007). Typical restoration activities include topsoil replacement, adding organic amendments, re-vegetation and reforestation, erosion control structures, terracing and soil fertility improvement programs.

The formula used in this method is:

$$\text{Restoration Cost} = \sum_{i=1}^{n} C_i \times A$$

where,

- C_i: cost per unit area for each restoration activity (i) (e.g., re-vegetation, terracing, organic amendments).
- A: total degraded area requiring restoration.
- n: number of restoration interventions

This approach is especially used when direct market valuation is not possible, ecosystem services are not traded in markets, or when rapid policy decisions require cost-based estimates.

Key limitations of this method include:

- Restoration may not fully recover original soil functions.
- If full restoration is impossible, losses may be underestimated.
- The effectiveness of restoration and outcome are uncertain.
- Costs are highly site-specific and time-dependent.
- Long time horizons complicate discounting and evaluation.

Example:

Korneeva (2021) used the Restoration (Rehabilitation) Cost Method to assess the financial impact of restoring degraded soils utilizing conservation afforestation systems. The research area was located in the southern Russian Plain, where soils are subject to extreme wind erosion, depletion of soil organic matter, and degradation due to the impacts of continuous cultivation and excessive deflation-prone conditions. The purpose of evaluating protective afforestation as a means of rehabilitation was twofold: i) to improve soil quality and reduce the risk of soil erosion, and ii) to achieve the restoration of agricultural productivity.

The substantial economic costs associated with restoring degraded soils included:

- Annual cost of soil fertility restoration was EUR 4333–17,176 per hectare.
- Annual avoided-damage cost (including climate/runoff variability) was EUR 2167–8588 per hectare per year.

This study clearly demonstrates how the Restoration/Rehabilitation Cost Method can be used to estimate the actual financial investment required to restore degraded land, rather than merely replacing lost nutrients According to this study, restoration is substantially more expensive than nutrient replacement approaches, highlighting the value of preventive soil conservation. The integration of ecological measures (afforestation) and economic valuation provides a strong justification for policy-level investment in land rehabilitation.

Difference Between Replacement Cost Method and Restoration or Rehabilitation Cost Method: Although Replacement Cost Method and Restoration Cost Method are conceptually related and often appear similar in practical applications, they should not be used interchangeably. Each method captures a distinct dimension of soil degradation costs. The former focuses on valuing the nutrients or soil components lost to degradation by estimating the cost of substituting them with commercial inputs. However, the latter quantifies the investment which is needed to restore the land to a functional or pre-degradation state through ecological or engineering interventions.

(d) Avoided-damage/cost–benefit analysis (CBA) method

The Avoided-Damage method, often operationalized through Cost–Benefit Analysis (CBA), is a widely used economic approach for valuing soil degradation and assessing the economic efficiency of soil conservation measures. Instead of directly measuring productivity losses or market impacts, this method evaluates the costs that would occur in the absence of conservation practices and compares them with the costs of implementing interventions. Soil degradation is thus valued based on the damage avoided when conservation practices are adopted (Barbier, 1996; Boardman et al., 2018).

The use of this methodology has successfully enabled the evaluation of large programs and interventions in numerous fields, including but not limited to watersheds, erosion prevention, and policies aimed at reducing sedimentation, nutrient loss or yield losses. Besides this, it is widely used in environmental economics, the policy development of natural resources, and research related to land degradation.

The formula used in this method is:

$$\mathrm{NPV} = \sum_{t=0}^{T} \frac{(B_t - C_t)}{(1+r)^t}$$

where,

- *NPV*: net present value (benefit of soil conservation) (USD)
- B_t: benefits or avoided damage costs in year t
- C_t: total costs of conservation in year t
- r: discount rate
- T: time horizon (years)
- So, if NPV >0 → conservation measure is economically efficient.

In fact, CBA evaluates whether soil conservation investments (e.g., contour plowing, terraces) are economically justified over time by comparing implementation costs with the avoided productivity and sediment losses.

Key limitation of this method include:

- Requires reliable estimation of baseline damage, which is often uncertain in soil degradation studies.
- Need for long-term data and assumptions (uncertainty).
- Sensitive to discount rate selection and time horizon assumptions.

- Non-market damages (e.g., biodiversity) often underestimated.

However, since it integrates both on-site and off-site effects, it is considered as one of the most comprehensive economic assessment tools.

Example:

Balana et al. (2012) conducted an analysis to examine the economic viability of building exclosures on degraded lands in the northern region of Ethiopia. Exclosure is a type of area where grazing is restricted in order to promote natural vegetation recovering. The authors used a Cost-Benefit Analysis (CBA) to determine the monetary value of the costs that can be avoided through reducing soil erosion and sedimentation from developing exclosures (avoided cost of soil erosion and sediment removal) as well as the ecological benefits associated with the exclosures, including biomass regeneration, increased soil fertility, and other ecosystem services.

The findings indicated that constructing exclosures on degraded and marginal land has a positive net present value (ETB[1] 5620 per hectare) and a benefit-to-cost ratio greater than 1.00. The net present value of the ecological benefits derived from constructing and maintaining the exclosures outweigh the cost of construction and maintenance. In contrast, converting productive cropland to exclosures does not generate net present value and incurs opportunity costs.

In summary, this study has demonstrated the potential of cost-benefit analysis to identify whether land restoration investments are economically viable based on the avoided costs of soil erosion and the long-term ecological benefits associated with those investments.

3.1.2 Non-market-based approaches

Non-market-based methods aim to value the loss of ecosystem services that are not traded in formal markets, such as biodiversity, carbon storage, or water regulation. In fact, these approaches are called "valuation" (Ghafari et al., 2025). Key techniques of this method include.

(a) Contingent valuation method (CVM)

This method uses surveys asking individuals about their willingness to pay (WTP) to prevent degradation or their willingness to accept (WTA) compensation for land loss.

The general formula used in this method is:

$$V_{NM} = \overline{w} \times N$$

where,

- $\overline{w}$: WTP per household/person (currency yr^{-1}),
- N: Affected or relevant population (households or individuals).

Key limitations of this method include:

- Exposure to hypothetical and strategic bias, as responses may differ from actual behavior.
- Results may be influenced by strategic bias and income constraints.
- Sensitive to survey design, framing effects, and respondents' understanding.
- High cost for large samples.

[1]The official currency of Ethiopia

Example:

Areal (2024) estimated the general public's WTP for a soil conservation and improvement program targeting agricultural lands in Spain and the UK, using a dual-restricted two-choice CVM. To do this, a cross-sectional online survey (via Qualtrics) was conducted in March–April 2021 with 1000 respondents in Spain and 984 in the UK. Respondents were asked whether they would pay additional monthly tax (over the next 10 years) to fund soil quality improvement through sustainable agricultural practices. Offer levels ranged from £/€5 to £/€45, and follow-up offers being made depending on the first response. The survey introduced a treatment group that received extra information about the uncertainty of policy's success (number of farmers who actually adopt the practices and level of environmental benefit). This allowed authors to test how perceived uncertainty affects WTP.

Key Findings:

There was significant public support for soil-quality protection and improvement in both countries. The average annual WTP was estimated to be GBP 223 for the UK and EUR 293 for Spain in the control group. Under uncertainty, WTP changed slightly but stayed strong. The reasons for WTP differed in each country. In Spain, trust in government and pro-social behavior were more important. However, in the UK, risk and time preferences played a greater role when uncertainty was introduced.

The author argues that these WTP estimates can help design public payment schemes or agri-environmental policies to encourage farmers to improve soil quality.

(b) Hedonic pricing

This method assesses how soil quality (or degradation) affects land prices, particularly in agricultural or real estate markets.

Key limitations of this method include:

- This method is only applicable when behavior reveals a clear relationship to soil attributes.
- It requires large, high-quality market data.
- Reliable estimation is impossible when housing or land markets are thin or data are inconsistent.
- Difficult to isolate soil effects from other correlated property characteristics.

Example:

In their study, Palmquist & Danielson (1989) used the Hedonic Pricing Method (HPM) to determine the economic value of soil degradation. In measuring the impact of soil erosion on capitalizing land value, the authors analyzed a comprehensive set of detailed plot-level data from the state of North Carolina. They found that erosion level, erosion control costs, and drainage characteristics significantly impacted the value of agricultural land. Fields with severe erosion sold for 8%–12% less than similar fields with minimal erosion, holding all other characteristics constant. Land with moderate erosion experienced a price discount of 3%–5%. Investments in erosion-control practices (contour farming, terracing) increased the value of farmland by approximately 6%–9%, showing that buyers view expected future productivity gains as an investment.

The study indicated that market prices internalize at least some of the long-term productivity losses caused by soil degradation, making hedonic pricing a suitable tool for estimating implicit economic costs.

(c) Cost of ecosystem service loss

This approach estimates the loss in regulating/supporting services like flood control, water purification, and climate regulation.

The general formula used in this method is (generic willingness-to-pay model):

$$V_{ES} = \frac{1}{n} \sum_{i=1}^{n} WTP_i$$

where,

- V_{ES} is the average value of the soil-related ecosystem service (USD household^{-1} yr^{-1}),
- WTP_i is willingness to pay of respondent i for maintaining soil quality.

This category captures non-market services (carbon storage, biodiversity, flood regulation) using contingent valuation (CV), choice experiments (CE), or benefit transfer.

Key limitations of this method include:

- Complexity of ecological processes: Ecosystem services involve multiple interacting functions (e.g., nutrient cycling, biodiversity, water regulation), making it difficult to isolate and value individual services accurately.
- High data requirements: Economic valuation of ecosystem service loss requires extensive biophysical and socioeconomic data, which are often unavailable, especially in data-scarce regions.
- Methodological inconsistency: Many studies rely on cost-based proxies (e.g., replacement cost) that are easy to apply but not fully consistent with economic theory, leading to imprecise valuations.
- Market price distortion: When using market prices as proxies for ecosystem service values, distortions such as subsidies, taxes, or imperfect markets can bias results.
- Partial representation of services: Valuation studies often focus on a limited set of ecosystem services (e.g., crop productivity, carbon storage) while ignoring others (e.g., cultural or supporting services), resulting in underestimates.
- Non-use values are hard to quantify: Services that have non-market values (e.g., existence, cultural significance) are difficult to monetize reliably, especially without robust preference data.
- Uncertainty and risk: The inherently uncertain nature of ecosystem responses and future environmental change introduces risk and variability into cost estimates.
- Context specificity: Values of ecosystem services depend heavily on local ecological, social, and economic conditions, limiting the transferability of estimates across regions.

Example:

Romanazzi et al. (2024) quantified the economic value of several ecosystem services lost as a result of soil erosion in a Mediterranean river basin. The services considered included: carbon storage (regulation), nutrient retention (supporting), habitat quality (supporting), crop production (provisioning), and the land's suitability for tourism (cultural).

They measured soil loss rates over a 3-year period (2017–19) across 25 plots. Laboratory soil analyses determined the content of potassium (K), phosphorus (P), and organic matter, which allowed the calculation of mineral-element loss per hectare. For each service, they assigned a monetary value (for example via current market values for nutrients, social cost of carbon for carbon losses, estimated values of habitat loss, agricultural yield losses, etc.).

The valuation was carried out on each plot and then aggregated to estimate the total annual cost of lost ecosystem services across the basin. Based on their results, the estimated economic loss per hectare varied by service. For instance:

- Losses of mineral elements (P and K) ranged up to €1641 ha^{-1}.
- Losses of habitat quality (supporting services) reached up to ≈ €5000 ha^{-1} in some plots.
- Crop-production losses (providing service) per hectare reached €35,320.50 ha^{-1} (for severely damaged plots).
- Carbon-storage losses (regulating service) were also monetized, with values up to €348.94 per ton of CO_2 per hectare.

The results provide a strong economic argument for implementing soil conservation and sustainable land-management practices, as the avoided losses across diverse services are substantial.

3.1.3 Hybrid or integrated approaches

Hybrid approaches combine market and non-market techniques to capture both traded outputs and non-market ecosystem services (e.g., reduced productivity + carbon sequestration + WTP for landscape protection). These approaches are considered the gold standard for regional and national assessments, as direct single-method estimates (market-only or survey-only) of societal cost often do not account adequately for large on-site or off-site components of the "real" societal cost.

(a) Biophysical models plus economic modules

Since no single method can capture the full range of impacts, integrated models are often used, combining biophysical models with economic assessment frameworks. Some of the most important models in this method include:

- RUSLE (an empirical model): this model produces pixel-level or field estimates of long-term annual soil loss (ton ha^{-1} yr^{-1}). It is widely used for continental and regional mapping (e.g., EU assessments) and is often coupled with simple damage functions or productivity modules.
- WEPP (a process-based model): this model simulates runoff, detachment, transport, and deposition at plot to watershed scales with temporal resolution (daily to event). This model is useful for identifying critical source areas, timing of sediment pulses, and the effectiveness of specific BMPs (Best Management Practices).
- SWAT (a semi-distributed watershed model): this model simulates hydrology, erosion, and sediment delivery in large watersheds and relates land use scenarios (simulations) to sediment flow into water bodies or reservoirs, often used in cases where reservoir and impoundment impacts are assessed.

Key limitations of this method are:

- High data requirements: It require detailed and high-quality biophysical and economic inputs that are often unavailable, especially in data-scarce regions.
- Model uncertainty propagation: Uncertainty in biophysical model outputs is transferred and amplified in economic valuation results.

- Strong dependence on assumptions: Results are highly sensitive to parameterization, scenario design, and damage functions.
- Technical complexity: High level of expertise is required, limiting transparency and usability for policymakers.

Example:

Panagos et al. (2018) applied the Hybrid 'Biophysical Models and Economic Module' approach to determine the economic impact of soil erosion on agricultural productivity in the European Union. They estimated soil erosion rates using the RUSLE2015 biophysical model, and derived corresponding yield losses of major crops (wheat, maize, and barley) from these soil erosion rates. The resulting biophysical output data were then incorporated into a computable general equilibrium (CGE) model of the EU to capture the associated impacts on market prices, production, trade flows and overall economic welfare caused by soil erosion.

According to their analysis, the annual direct cost of yield losses due to soil erosion across the EU is estimated to be approximately €1.25–1.40 billion. However, through the CGE model the total economic burden of yield losses attributed to soil erosion is significantly increased as a result of accounting for all of the indirect impacts caused by soil erosion at the macroeconomic level which extends beyond the farm gate. The analysis indicates that southern and eastern EU member states incur the majority of these economic losses due to higher rates of soil erosion and the greater vulnerability of their agricultural systems to soil erosion.

This study showed that combining the use of biophysical erosion models and macroeconomic models allows for a more complete assessment of soil degradation costs and gives more accurate information to policymakers about the economic impact of erosion on the economy. This combination will help to provide a more holistic picture of the economic implications of soil degradation.

(b) Spatially explicit GIS-based approaches

Spatially explicit GIS-based methods integrate biophysical models, economic valuation techniques, and geospatial analysis to assess the economic costs of soil degradation across landscapes. These approaches allow researchers and policymakers to quantify how the spatial variability of soil properties, land use, climate, and erosion processes influences economic losses at local, regional, or national scales. The approach is being used in the EU, China, and India for mapping regional costs of erosion.

Key limitations of this method include:

- Sensitivity to spatial data quality: Errors or coarse resolution in soil, land-use, or erosion maps directly affect cost estimates.
- Scale and aggregation issues: Spatial aggregation may mask local heterogeneity and site-specific degradation processes.
- Reliance on modeled or secondary data: Often depends on indirect estimates rather than field measurements, increasing uncertainty.
- Limited representation of socio-economic dynamics: GIS-based outputs may inadequately capture behavioral responses and institutional factors.

Example:

Panagos et al. (2024) compiled spatially explicit data on sediment accumulation in reservoirs across the EU by combining estimates of soil loss, sediment delivery to waterways, and reservoir capacity loss (i.e., a GIS-based mapping of the depositional site). Then, they monetized the cost of that sediment accumulation using available published costs for sediment removal (dredging, disposal, reservoir maintenance).

The authors estimated that approximately 135 million m^3 per year of sediment caused by water erosion is delivered to reservoirs across the EU + UK. The cost associated with removing this sediment (dredging, disposal, and maintenance) is estimated to be more than €2.3 billion annually. Additionally, the authors believe that the maximum cost could be around €5-8 billion per year if all sediment accumulation and removal is included in the total figure.

Therefore, the key strength of a spatially explicit GIS method is that it allows for the combination of information about biophysical erosion and sediment-delivery modes with the economic data available at the reservoir-scale, allowing for an estimate of off-site costs resulting from soil erosion to be made with geographic specificity and in a way that is relevant to policy.

3.2 Indirect economic costs

Indirect economic costs refer to the off-site and secondary economic impacts of soil degradation that occur outside the boundaries of the degraded land itself. Unlike direct costs which focus on yield loss, nutrient depletion, or rehabilitation expenses on the affected farm, indirect costs capture the broader societal consequences generated by sediment export, hydrological disruption, environmental pollution, and damage to downstream infrastructure. For instance, if erosion occurs and the soil washes into a reservoir or river and causes sedimentation, this sediment may be responsible for clogging waterways, and creating more costs as the water is treated. Sedimentation may also decrease aquatic habitat and threaten fisheries and tourism. These secondary impacts are valuable to consider in economically assessing soil degradation.

3.2.1 Off-site sedimentation cost method

This method captures indirect losses such as reservoir siltation, water quality deterioration, or road damage from sediment movement. Off-site costs are widely recognized in soil economics as "externalities"; costs borne by society rather than by the individual landholder. This includes the cost of dredging accumulated sediments, maintenance of hydraulic structures, water treatment, reduced reservoir capacity, and other public expenditures related to sediment deposition.

The general formula used in this method is (Telles et al., 2013):

$$C_{\text{off-site}} = \sum_{i=1}^{n} S_i \times C_i$$

where,

- $C_{\text{off-site}}$: total off-site sedimentation cost (USD yr^{-1}),
- S_i: quantity of sediment delivered to the downstream receptor i (e.g., $m^3\ yr^{-1}$),
- C_i: unit cost associated with sediment handling at receptor i (e.g., dredging cost per m^3 or USD m^{-3}),

- n: number of affected receptors (reservoirs, waterways, irrigation channels, etc.).

Some of the key limitations of this method include:

- Data-intensive: This method requires detailed sediment and hydrological data.
- Source uncertainty: It is hard to distinguish sediment origin.
- Partial valuation: Many ecological and water-quality impacts are ignored.
- Long-term processes: Sedimentation occurs over years, not annually.
- Complex sediment transport: It is difficult to accurately model deposition.

Example:
Panagos et al. (2024) can also be cited as a case study for the Off-site Sedimentation Cost method, since the study explicitly monetizes the economic impacts of sediment delivery and accumulation in reservoirs across the EU.

3.3 Overview and comparison of soil degradation cost assessment methods

To enhance clarity and facilitate a systematic comparison of the economic approaches discussed in this chapter, Table 12.4 provides a comparative synthesis of the main economic methods used to estimate the costs of soil degradation, highlighting their respective advantages, limitations, and typical applications. The comparison illustrates that no single method can comprehensively capture the full economic impacts of soil degradation, as each approach emphasizes different dimensions, spatial scales, and types of costs. Market-based and productivity-oriented methods are generally more suitable for estimating direct, on-site losses, while indirect and ecosystem-based approaches are better suited for capturing off-site impacts and non-market values. Hybrid and spatially explicit methods offer more comprehensive assessments by integrating biophysical processes with economic valuation, but they require greater data availability and technical capacity. Overall, Table 12.4 underscores the importance of selecting valuation methods based on study objectives, data constraints, and policy relevance, and highlights the growing need for integrated approaches to support effective soil conservation planning and implementation.

Building on the comparative assessment of economic valuation methods presented in this part, the following section applies the *Productivity Reduction Method* in a detailed case study focusing on soil degradation caused by wind erosion in arid and semi-arid regions. Given the strong link between wind-driven soil loss, nutrient depletion, and crop yield reduction, this method is particularly appropriate for quantifying direct on-site economic losses. The case study integrates controlled wind tunnel experiments with field-based soil and crop data to estimate erosion-induced productivity decline and associated economic costs. By combining empirical erosion measurements with economic valuation, the next section demonstrates how experimental data can be translated into policy-relevant cost estimates, thereby operationalizing the methodological framework discussed in this chapter within a real-world dryland context.

4. Frameworks for economic loss estimation

For effective land management and policymaking purposes, it is critical to quantify the economic losses caused by soil degradation. While theoretical models provide a foundation, real-world data and

Table 12.4 Comparative overview of economic methods for estimating soil degradation costs.

Method	Typical application	Main advantages	Main limitations
Productivity reduction method	Estimating yield loss due to erosion, nutrient depletion, or salinization at farm or regional scale.	Direct link between soil degradation and yield loss; Simple and transparent calculations; Applicable under limited soil data availability; Uses observed market prices; Easily understood by policymakers.	Captures only on-site impacts; Ignores off-site and non-market effects; Sensitive to yield and price variability.
Replacement cost method	Estimating soil degradation cost based on the expense required to replace lost soil functions or inputs with market substitutes.	Simple and transparent; Relies on observable market prices; Relatively easy to apply with limited data.	Assumes perfect substitutability between natural soil functions and artificial inputs; May underestimate ecological complexity.
Restoration/ rehabilitation cost method	Estimating cost based on expenses needed to restore or rehabilitate degraded land to a functional or acceptable condition.	Reflects actual field interventions; Captures site-specific conditions; Suitable for engineering and ecological interventions.	Restoration may not fully recover original soil functions; Costs are context-specific and time-dependent.
Avoided-damage/ cost–benefit analysis (CBA)	Evaluation of soil conservation and erosion control projects; Assessment of economic benefits of preventive measures.	Explicitly links conservation actions to avoided losses; strong policy relevance; allows comparison of alternative management scenarios.	Uncertain baseline damage estimates; Sensitive to discount rate and time horizon; Non-market benefits may be underestimate.
Contingent valuation method (CVM)	Valuing public willingness to pay for soil protection, land restoration, and ecosystem services.	Captures non-market and intangible values; Directly reflects individual preferences and willingness to pay (WTP) for soil conservation; Useful when market data are absent.	Subject to survey bias; hypothetical responses; costly to implement.
Hedonic pricing	Impact of soil quality or erosion risk on land or property values.	Based on real market behavior; Avoids hypothetical bias; Useful for estimating the implicit value of soil quality attributes.	Requires active land or property markets; Difficult to isolate soil effects; High data requirements.

Continued

Table 12.4 Comparative overview of economic methods for estimating soil degradation costs.—cont'd

Method	Typical application	Main advantages	Main limitations
Cost of ecosystem service loss	Values soil degradation based on the monetary losses of multiple ecosystem services that are reduced or lost when soil quality deteriorates.	Captures a broad range of ecosystem functions beyond direct on-site productivity; reflects social and environmental values that are not traded in markets; supports holistic policy decisions.	Requires extensive interdisciplinary data; Risk of double counting; higher methodological complexity and uncertainty in non-market valuation.
Biophysical models + economic modules	Estimating productivity losses from erosion and nutrient depletion; assessing long-term economic impacts of land degradation; Evaluating policy scenarios at regional to national scales.	Integrates physical soil processes with economic valuation in a consistent framework; Captures dynamic and long-term impacts of soil degradation; Suitable for scenario analysis and policy simulations.	High data and parameter requirements; Results are sensitive to model assumptions and calibration; Technically complex and less transparent for non-experts.
Spatially explicit GIS-based approaches	Mapping spatial distribution of soil erosion costs; Regional and national land degradation assessments; targeting conservation and restoration interventions.	Explicitly represents spatial heterogeneity of soil degradation; Effective for identifying hotspots and priority areas; Strong visualization and communication tool for policymakers.	Highly dependent on spatial data quality and resolution; scale and aggregation effects may mask local processes; often relies on secondary or modeled data.
Off-site sedimentation cost method	Reservoir dredging and irrigation maintenance costs; Damage assessment to hydropower and waterways; Indirect cost estimation of soil erosion.	Captures downstream economic damages; Reflects real infrastructure and maintenance costs Suitable for watershed-scale analysis.	Difficult source attribution; High data and modeling requirements; Partial valuation of ecological impacts.

case-specific frameworks are necessary to reflect the spatial and temporal variability of degradation processes. This section outlines commonly used estimation frameworks.

Several frameworks have been developed to estimate economic losses from soil degradation. The most commonly used frameworks include the following.

(a) Total economic value (TEV) framework

Total Economic Value (TEV) is a broad conceptual framework for organizing the complete array of values that society extracts from an environmental asset (here: soil). TEV is not a singular valuation method, but rather a classification that suggests which particular valuation methods should be employed to capture the composite social value of soil functions (production, water regulation, carbon storage, biodiversity, cultural/landscape values, etc.). TEV enables policymakers to assess the full societal cost of soil degradation, beyond market transactions. This comprehensive approach includes:

- Direct use values: agricultural productivity, biomass, fuel
- Indirect use values: nutrient cycling, water filtration, carbon storage
- Option values: potential future uses (e.g., medicinal plants)
- Non-use values (existence and bequest): cultural, spiritual, or biodiversity-related significance

Therefore, there is no single mathematical formula for TEV because it is an aggregation framework (Pearce, 1992):

$$\text{TEV} = \text{Direct use value} + \text{Indirect use value} + \text{Option value} + \text{Existence value}$$

TEV is mostly used when the goal is to communicate the comprehensive societal value of soil (policy briefs, national assessments, ELD-style country studies). TEV helps ensure that non-market and future values are not neglected (Somda & Awaïss, 2013). However, TEV summaries can give a false impression of precision if component valuations are uncertain. It must be ensured that methods, assumptions, and uncertainty boundaries are transparent, double counting is avoided and private (farm level) values are clearly separated from social values (Pearce, 1992).

(b) Cost-benefit analysis (CBA)

Cost-benefit analysis is a tool that compares soil conservation investment costs (e.g., terracing, cover crops) with the costs of avoided soil degradation. This approach is often used in development and aid-funded projects to determine return on investment.

CBA is an ex-ante or ex-post decision-support framework that compares the present value of benefits (including avoided damages) with the present value of costs for a proposed action (e.g., a soil-conservation intervention). In soil contexts, CBA is commonly used to appraise investments in sustainable land management (SLM), terracing, reforestation, or policy instruments. CBA is also the natural operational companion to TEV when the policy question is: Is the intervention economically justified?

CBA is used to prioritize investments, allocate scarce public funds, or compare alternative SLM interventions. It gives explicit decision metrics (NPV, BCR) and supports spatial targeting (identify pixels/areas with positive NPV). Nevertheless, its results are sensitive to the discount rate, time horizon, and the set of benefits included. Attribution of avoided damages to interventions often relies on counterfactual assumptions and models of effectiveness. Non-market benefits can be uncertain and can dominate outcomes if included without robust estimates. Transparency and sensitivity analysis are essential (ELD Initiative, 2015).

(c) Damage function models

These models estimate the relationship between soil degradation (e.g., topsoil loss in ton/ha) and economic impacts (e.g., yield reduction, water treatment costs). The damage-function approach develops empirically based (or modeled) functions that relate a physical measure of degradation (e.g., tons of soil lost per hectare per year, salinity level, depth of removed topsoil) to an economic damage metric (e.g., percent yield loss, remediation cost, or reduced land value). Damage functions are attractive because they make the link between biophysical models (RUSLE, sediment delivery, salinity maps) and economic outcomes explicit and reproducible. Walker's classical formulation for erosion damage is an early and widely cited example (Walker, 1982).

Damage-function models are used when there are: (a) measurable biophysical outputs from erosion/salinity models (e.g., RUSLE soil loss, salinity indices), and (b) empirical data linking those physical measures to economic endpoints (yields, remediation cost, land value). Damage functions transparently convert modeled environmental change into monetary terms and are well suited for spatially explicit mapping of costs (pixel-level damage estimates). They are used extensively in watershed and regional studies (Panagos et al., 2018).

However, there are some limitations. In fact, damage functions require empirical calibration and mis-specification can bias cost estimates. They may not capture complex, lagged, or nonlinear ecological responses (e.g., loss of soil organic matter that affects yields years later) unless the functional form and time horizon explicitly include such dynamics. Care must be taken with aggregation (pixel → watershed → national) as non-linear damages do not aggregate simply (Kefi & Yoshino, 2010).

5. Key challenges in cost estimation

Accurate evaluation of economic costs of soil degradation is needed to support policy making, to prioritize conservation investments, and inform land-use decisions. Despite significant advances in methodology, estimating such costs remains inherently complicated. This is because soil degradation involves complex multidimensional, nonlinear ecological processes and is embedded into dynamic economic systems. These ecological processes are often spatially variable and limited data may increase the likelihood of inaccuracies when making connections between degradation and specific driving forces. This complexity generates significant inconsistencies across studies or regional assessments. These matters require a common framework as well as appropriate assessment methods to provide estimates that can be interpreted and compared reliably across different studies oriented toward policymakers.

5.1 Data scarcity and spatial heterogeneity

A central challenge is obtaining long-term, high-resolution data which is needed to quantify degradation processes and their impacts on economic valuation. Soil properties, degradation intensity, crop response, and off-site effects vary temporally and spatially, complicating efforts to scale from select field data to national or regional levels. Additionally, spatial issues hinder the valuation of ecosystem services, as the same degradation type - e.g., salinization - can have distinct economic valuations based on local climate, soil texture, farming systems, and market conditions (e.g., payment for ecosystem services) (Gibbs & Salmon, 2015; Nkonya et al., 2016).

5.2 Attribution of degradation causes

It remains a significant challenge to disentangle the effects of soil degradation from those of other biophysical and socioeconomic factors. For example, reductions in crop yields may be due to climate variation, pests, and input use, as well as economic factors, in addition to soil degradation. It is difficult to distinguish causation, a scenario that leads to over- or under-estimation of agricultural economic losses associated with degradation (Berry et al., 2003). Advanced statistical or process-based models can address some of these challenges, but these models use extensive data that may be hard to gather for developing countries.

5.3 Valuing non-market ecosystem services

Several impacts of soil degradation, including lower biodiversity, reduced carbon storage capacity, loss of water filtration, and diminished cultural ecosystem services, are services for which armchair economists cannot assign market values. This market valuation of services will employ economic techniques such as contingent valuation, replacement cost, or benefit transfer, all of which will carry some degree of methodological uncertainty and bias (Russi et al., 2013). Thus, economic assessments of soil degradation will primarily undervalue the true societal costs of soil degradation.

5.4 Off-site costs are difficult to measure

Degraded soils often produce externalities like sedimentation, dust storms, eutrophication, or water quality impairments. Measuring off-site costs of degradation would be much more difficult than measuring on-site costs due to the need to integrate hydrology, atmospheric, and economic models. Off-site costs are often diffuse, affecting multiple sectors (agriculture, infrastructure, public health), and could even be geographically distant from the point of degradation, making cost allocation and assessment more complex (Pimentel & Burgess, 2013).

5.5 Differences in methodological approaches

Studies of economic losses related to soil degradation employ various approaches to estimate the costs of soil degradation, including but not limited to replacement cost method, production function method, or integrated assessment models. The approaches differ in their underlying assumptions, data requirements, and the types of ecosystem services assessed (including factors such as the individual vs. aggregate value of ecosystem services). As a result of these variations in approaches, the global estimates of soil degradation costs exhibit significant variations, which in turn limit the comparability of results between studies and the options for policymakers or practitioners (Nkonya et al., 2016).

5.6 Time-scale challenges and discounting

Economic assessments must consider the timescales associated with soil degradation processes. While the impact of soil erosion, nutrient depletion, salinization, or soil pollution occurs slowly over long spans of time, contributing to the substantial uncertainty in estimates, the benefits of soil conservation practices may not be realized for years or decades following the initial implementation. The choice of discount rates can have a significant influence on present value estimates of costs. Higher discount rates may even reduce the costs of future losses and benefits which may lead to diminish the value of making investments in soil conservation practices (Boardman et al., 2018). This continues to be a source of significant argument in environmental economics.

5.7 Lack of standardized frameworks

Even with the extensive efforts, to date, there is no a universally accepted approach to costing soil degradation. Lack of consistency in the definitions of soil degradation, indicators of degradation, categories of ecosystem services, selection of valuation methods & public data reporting formats, among other factors, hinders the opportunity to compare across studies. Therefore, the lack of consistency among frameworks, limits the ability of governments or international agencies to compare data or create similar social policies.

6. Implementation challenges in soil conservation laws

Even though many countries have passed legislation and policies for soil protection, the gap between legislation and practices is still significant. This section discusses the main institutional, technical, social, economic, and political constraints to enforce action for soil protection, particularly in developing and in transition countries.

6.1 Legal and regulatory design problems

Frequently, legislations which are not framed cohesively or are scattered become an obstacle. Laws that are either too broadly framed and idealistic without operational rules and benchmarks, or too narrowly focused on limited types of land use or degradation, diminish the ability to enforce or relevant policies. In some jurisdictions, soil protection legal measures sit in a variety of legal instruments (i.e., environment, agriculture, water, planning), leading to gaps and overlaps. In comparative reviews, stronger definitions (i.e., soil health), measurable targets, and alignment of legal mandates governing soil (or some are just related to land) across different sectors are suggested to improve local soil stakeholder engagement (Altvater et al., 2015).

6.2 Institutional fragmentation and weak governance

A major challenge for soil conservation is institutional fragmentation, which is essential for effective implementation, although institutions need to be clear about their mandates and how activities need to be coordinated, and also need adequate staffing and decision-making power across sectors. Many countries have no single empowered agency to coordinate soil policy and the roles of soil governance and management are divided between ministries of agriculture, environment, water, and local authorities. This fragmentation acts as an impediment to the monitoring, enforcement, and financing of soil actions. Studies of national regimes show that governance gaps are identified for soil and systemic recommendations for integrated institutional structures, or coordination mechanisms, to implement soil protection objectives are made repeatedly (Burt et al., 2022).

6.3 Insufficient financing and perverse incentives

Soil conservation typically provides diffuse, long-term public benefits with associated short-term costs to land managers, such as labor, lost yield, or equipment. Adoption rates of conservation practices are usually low without direct financial incentives such as targeted subsidies, payments for ecosystem services, or market rewards. Analysis of experience (for instance, subsidy programs or other incentive instruments) suggests that well-constructed financial incentives can increase the adoption of conservation practices, but in many areas the appropriate financial instruments are under-resourced or inconsistently applied. The lack of sustained funding undermines both initial adoption and long-term maintenance of measures (Smith, 1998).

6.4 Monitoring, data and indicator shortages

Laws dictate measurable indicators and monitoring systems to measure compliance and progress. However, there are generally inconsistent and low-quality data for soils (soil organic carbon, erosion rates, salinity, contamination) and national monitoring systems are incomplete or occur infrequently. As a result, hotspots may not be identified and targeted for intervention and there is a lack of assurance that legal obligations are being addressed. The proposed EU Soil Health Law and other policy initiatives all stress the need to establish harmonized monitoring systems and observatories specifically to address this issue (Pieper et al., 2023).

6.5 Enforcement capacity and political economy constraints

Even when clear rules exist, their enforcement requires significant expenditures of time, resources, and political sensitivity. Enforcing conservation law generally needs a wide range of activities, including site inspections, detection via satellite and proximal sensing, and discipline against offenders through administrative sanctions or legal action; all of which demand human and technical capacity. Political resistance from powerful stakeholders, weak sanctioning powers, or local governance constraints can blunt enforcement. Cross-country surveys identify "poor enforcement of conservation law" as one of the most frequent barriers experienced by practitioners (Burt et al., 2022).

6.6 Farmer decision-making, knowledge and short-term losses

The acceptance of conservation practices depends on how farmers perceive risk, their tolerance for risk, their labor availability, and any anticipated short run income associated with possibilities to enhance how soils are protected (e.g., terracing, contour bunds, cover crops, conservative tillage practices). Some soil practices provide long-term benefits to soil health, but short-run income can be reduced and new skillsets developed to be effective. Social studies also suggest that without extension services, demonstration projects, and upfront (partial) cost-sharing of practices that result in soil protection, adoption will be limited. Therefore, behavioral constraints associated with tradition and distrust in authorities and opportunity costs will be some of the main implementation constraints specific to social learning (Savari et al., 2022).

6.7 Spatial targeting and mismatch of scale

Soil issues and solutions are spatially heterogenous - national regulations need to be applied to local contexts quickly. If national targets or similar national rules are mandated without considering local agro-ecological conditions, land tenure systems or the spatial heterogeneity of degradation drivers, failure to implement action is likely. Programs that successfully manage soil issues generally combine national laws with local planning and stakeholder participation and adaptive management (Nugroho et al., 2022).

6.8 Integrating soil law with other policy domains (water, climate, land use)

Soil health is affected by and affects other policy areas (water quality, land-use planning, climate mitigation). Isolated policymaking leads to contradictory incentives, for example water subsidies that encourage excessive irrigation and salinization, or planning rules that permit sealing of fertile soils. Better policy coherence, for instance coupling soil law with water and climate instruments and using economic instruments to align incentives, improves implementation prospects. Recent policy analyses of the EU and national reforms stress this need for cross-sectoral integration (Pieper et al., 2023; Reinholz & Rodi, 2025).

6.9 Legal transposition and administrative burden

International or transnational directives, such as the proposed EU Soil Health Law, that need to be made applicable in different countries can encounter problems regarding the time and the manner of implementation due to the differences in administrative capacities and legal traditions. Administrative complexity might also make it difficult for local implementers to take decisions if the processes of permitting, reporting, and compliance involve high costs or are lengthy. Consequently, reducing the number of administrative procedures while keeping the accountability is a major issue in successful implementation (Pieper et al., 2023).

6.10 Addressing legacy contamination and complex technical problems

Some soil challenges such as industrial contamination, advanced salinization, or compaction require complex remediation (that may or may not be costly) and could require long-term commitments. Laws that need remediation without the associated funds, a technical framework, or liability for practitioners, leave remediation efforts to languish when landowners cannot pay or organize the logistics of the cleanup. National case studies of contaminated land regimes indicate that practical remediation pathways like risk-based standards, phased clean-ups, public–private financing are needed for effective remediation (Rodrigues et al., 2009).

Therefore, the successful implementation of soil conservation laws depends not only on the content of the law, but also on cohesive institutional frameworks, sufficient funding, proper enforcement, public participation, and scientific data for making laws. These capacities will be necessary to ensure that soil conservation laws develop beyond symbolic rules resulting in value added to the environmental and economic resources.

7. Conclusion

This chapter demonstrated that a wide range of valuation frameworks and methodologies, including market-based assessments, avoided-cost approaches, production-function models, and ecosystem service valuation, provide valuable tools for estimating the economic costs of soil degradation. However, no single method can fully capture the complex, multidimensional nature of soil degradation, particularly given the challenges associated with data availability, spatial variability, non-market valuation, and uncertainties in linking biophysical processes with economic outcomes.

Beyond methodological limitations, the chapter highlighted that the effectiveness of soil conservation policies is strongly influenced by institutional and governance factors. Weak enforcement, limited financial resources, fragmented land tenure, inadequate monitoring systems, and insufficient stakeholder participation continue to hinder the successful implementation of soil conservation measures. The illustrative application of the Productivity Reduction Method further demonstrated how integrating biophysical measurements with economic valuation can generate site-specific estimates of soil degradation costs, providing practical evidence to support conservation planning and informed decision-making.

Overall, effective soil conservation requires the integration of sound scientific evidence, robust economic valuation, and supportive governance frameworks. Strengthening national soil information systems and incorporating soil ecosystem services into decision-making and natural capital accounting are essential for improving the effectiveness of soil conservation policies. Equally important are promoting integrated policy instruments that combine regulatory measures with economic incentives, enhancing cross-sectoral coordination, and investing in long-term monitoring, capacity building, and sustainable land management practices.

Ultimately, recognizing soil as a finite and economically valuable natural asset is fundamental to developing resilient agricultural systems and achieving long-term environmental sustainability and food security.

References

Acharya, L. K., Paramaguru, P. K., Tripathi, K., Bhoi, T. K., Seth, P., & Birah, A. (2025). Pesticide contamination in groundwater: Processes, risks, and mitigation strategies. *Discover Agriculture, 3*(1), 1–25.

Altvater, S., Dooley, E., & Roberts, E. (2015). Legal instruments to implement the objective "land degradation neutral world" in International Law. German Environment Agency (UBA), Dessau-Roßlau, Germany.

Angon, P. B., Islam, M. S., Das, A., Anjum, N., Poudel, A., & Suchi, S. A. (2024). Sources, effects and present perspectives of heavy metals contamination: Soil, plants and human food chain. *Heliyon, 10*(7).

Areal, F. J. (2024). Willingness to pay for agricultural soil quality protection and improvement. *Land, 13*(8), 1118.

Bakker, M. M., Govers, G., & Rounsevell, M. D. A. (2004). The crop productivity–erosion relationship: An analysis based on experimental work. *Catena, 57*, 55–76.

Balana, B. B., Muys, B., Haregeweyn, N., Descheemaeker, K., Deckers, J., Poesen, J., ... Mathijs, E. (2012). Cost-benefit analysis of soil and water conservation measure: The case of exclosures in Northern Ethiopia. *Forest Policy and Economics, 15*, 27–36.

Barbier, E. (1996). Economics of soil erosion: Theory, methodology and examples. Economy and Environment Program for Southeast Asia (EEPSEA), Special and Technical Paper No. SP199601T2.

Barbier, E. B. (2007). Valuing ecosystem services as productive inputs. *Economic Policy, 22*(49), 178–229.

Batlle-Sales, J. (2023). Salt-affected soils: A sustainability challenge in a changing world. *Italian Journal of Agronomy, 18*(2), 2188.
Bayata, A. (2024). Soil degradation: Contributing factors and extensive impacts on agricultural practices and ecological systems-systematic review. *Journal of Agriculture and Environmental Sciences, 13*, 16–34.
Berry, L., Olson, J., & Campbell, D. (2003). *Assessing the extent, cost and impact of land degradation at the national level: Findings and lessons learned from seven pilots*. World Bank.
Bhardwaj, R. L., Parashar, A., Parewa, H. P., & Vyas, L. (2024). An alarming decline in the nutritional quality of foods: The biggest challenge for future generations' health. *Foods, 13*(6), 877.
Boardman, A. E., Greenberg, D. H., Vining, A. R., & Weimer, D. L. (2018). *Cost–benefit analysis: Concepts and practice* (5th edn). Cambridge University Press.
Borrelli, P., Alewell, C., Alvarez, P., Anache, J. A. A., Baartman, J., Ballabio, C., ... Panagos, P. (2021). Soil erosion modelling: A global review and statistical analysis. *Science of the Total Environment, 780*, Article 146494.
Borrelli, P., Ballabio, C., Yang, J. E., Robinson, D. A., & Panagos, P. (2022). GloSEM: High-resolution global estimates of present and future soil displacement in croplands by water erosion. *Scientific Data, 9*(1), 406.
Borrelli, P., Robinson, D. A., Fleischer, L. R., Lugato, E., Ballabio, C., Alewell, C., ... Panagos, P. (2017). An assessment of the global impact of 21st century land use change on soil erosion. *Nature Communications, 8*(1), 2013.
Burt, A., Nuno, A., Turnbull, L., Fleischer-Dogley, F., & Bunbury, N. (2022). An international assessment of the barriers influencing the effectiveness of island ecosystem management. *People and Nature, 5*(1).
Chepil, W. S. (1950). Properties of soil which influence wind erosion: II. dry aggregate structure as an index of erodibility. *Soil Science, 69*(5), 403–414.
Derpsch, R., Kassam, A., Reicosky, D., Friedrich, T., Calegari, A., Basch, G., ... dos Santos, D. R. (2024). Nature's laws of declining soil productivity and conservation agriculture. *Soil Security, 14*, Article 100127.
Ding, Y., Sun, L., Li, C., Chen, M., Zhou, Y., Meng, M., ... Liu, X. (2024). Effects of short-term simulated acid rain and nitrogen deposition on soil nutrients and enzyme activities in cunninghamia lanceolata plantation. *Frontiers in Ecology and Evolution, 12*, Article 1365954.
Du, L., Zhang, Z., Chen, Y., Wang, Y., Zhou, C., Yang, H., & Zhang, W. (2024). Heterogeneous impact of soil acidification on crop yield reduction and its regulatory variables: A global meta-analysis. *Field Crops Research, 319*, Article 109643.
ELD Initiative. (2015). The value of land: Prosperous lands and positive rewards through sustainable land management. Available from: www.eld-initiative.org.
Fageria, N. K., & Baligar, V. C. (2008). Ameliorating soil acidity of tropical oxisols by liming for sustainable crop production. *Advances in Agronomy, 99*, 345–399.
Fageria, N. K., Baligar, V. C., & Jones, C. A. (2010). *Growth and mineral nutrition of field crops*. CRC press.
FAO & ITPS. (2015a). Status of the World's soil resources (SWSR): Technical Summary. FAO, Rome, Italy. ISBN 978-92-5-108960-6.
FAO & ITPS. (2015b). Status of the World's soil resources (SWSR): Main report. Food and Agriculture Organization of the United Nations. Food and Agriculture Organization of the United Nations (FAO) and Intergovernmental Technical Panel on Soils (ITPS), Rome, Italy.
FAO. (2021). *Global map of salt-affected soils (GSASmap)*. FAO Global Soil Partnership.
FAO. (2024). *Assessment of salt-affected soils in 50 years*.
FAO & UNEP. (2021). *Global assessment of soil pollution: Report*.
Gao, X., Xie, Y., Liu, G., Liu, B., & Duan, X. (2015). Effects of soil erosion on soybean yield as estimated by simulating gradually eroded soil profiles. *Soil and Tillage Research, 145*, 126–134.
Ghafari, H., Parvizi, Y., & Arabkhedri, M. (2025). An overview of methods for economic valuation of soil erosion damages: Fundamentals, approaches, and operational methodology. *Land Management Journal, 12*(2), 187–208.

Gholami, H., Mohammadifar, A., Song, Y., Li, Y., Rahmani, P., Kaskaoutis, D. G., ... Borrelli, P. (2024). An assessment of global land susceptibility to wind erosion based on deep-active learning modelling and interpretation techniques. *Scientific Reports, 14*(1), Article 18951.

Gibbs, H. K., & Salmon, J. M. (2015). Mapping the world's degraded lands. *Applied Geography, 57*, 12–21.

Goulding, K. W. T. (2016). Soil acidification and the importance of liming agricultural soils with particular reference to the United Kingdom. *Soil Use and Management, 32*(3), 390–399.

Gundogdu, S., Akca, M. O., Gürsoy, M., Yılmaz, M., Zhang, X., Rodríguez-Seijo, A., ... Velimirovic, M. (2025). Microplastics in soil: A comprehensive review of analytical techniques. *Frontiers in Soil Science, 5*, Article 1614075.

Guo, J. H., Liu, X. J., Zhang, Y., Shen, J. L., Han, W. X., Zhang, W. F., ... Zhang, F. S. (2010). Significant acidification in major Chinese croplands. *Science, 327*(5968), 1008–1010.

IPBES. (2018). *Report Land regradation & restoration assessment. The assessment report on land degradation and restoration (summary and findings).*

Jamil, A., Riaz, S., Ashraf, M., & Foolad, M. R. (2011). Gene expression profiling of plants under salt stress. *Critical Reviews in Plant Sciences, 30*(5), 435–458.

Kefi, M., & Yoshino, K. (2010). Evaluation of the economic effects of soil erosion risk on agricultural productivity using remote sensing: Case of watershed in Tunisia. International archives of the photogrammetry. *Remote Sensing and Spatial Information Science, 38*(Part 8), 930.

Khambalkar, P. A., Agrawal, S., Dhaliwal, S. S., Yadav, S. S., Sadawarti, M. J., Singh, A., ... Afreen, N. (2025). Sustainable nutrient management balancing soil health and food security for future generations. *Applied Food Research*, Article 101087.

Kopittke, P. M., Harper, S. M., Asio, L. G., Asio, V. B., Batalon, J. T., Batuigas, A. M. T., ... Sanchez, P. B. (2025). Soil degradation: An integrated model of the causes and drivers. *International Soil and Water Conservation Research.*

Korneeva, E. A. (2021). Economic evaluation of ecological restoration of degraded lands through protective afforestation in the south of the Russian plain. *Forests, 12*(10), 1317.

Lal, R. (1998). Soil erosion impact on agronomic productivity and environment quality. *Critical Reviews in Plant Sciences, 17*(4), 319–464.

Lal, R. (2015). Restoring soil quality to mitigate soil degradation. *Sustainability, 7*(5), 5875–5895.

Lemma, B., Kebede, F., Mesfin, S., Fitiwy, I., Abraha, Z., & Norgrove, L. (2017). Quantifying annual soil and nutrient lost by rill erosion in continuously used semiarid farmlands, North Ethiopia. *Environmental Earth Sciences, 76*(5), 190.

Lori, M., Leitao, R., David, F., Imbert, C., Corti, A., Cunha, L., ... Vazquez, C. (2025). Response of soil biota to agricultural management practices: A systematic quantitative meta-data-analysis and method selection framework. *Soil Biology and Biochemistry, 207*, Article 109815.

Naramabuye, F. X., & Haynes, R. J. (2006). Effect of organic amendments on soil pH and Al solubility and use of laboratory indices to predict their liming effect. *Soil Science, 171*(10), 754–763.

Nassir, A., Mishall, A., Sabah, Q., Al-mothefer, A., Awad, H., & Alwan, A. (2024). Soil compaction induced by different tillage systems and its impact on growth and yield of maize (zea mays L.): Soil compaction induced by different tillage systems and its impact on growth and yield of maize (zea mays L.): A review. *University of Thi-Qar Journal of agricultural research, 13*(1), 185–200.

Nkonya, E., Mirzabaev, A., & Von Braun, J. (2016). *Economics of land degradation and improvement–a global assessment for sustainable development.* Springer Nature.

Nugroho, H. Y. S. H., Basuki, T. M., Pramono, I. B., Savitri, E., Purwanto, Indrawati, D. R., ... Nandini, R. (2022). Forty years of soil and water conservation policy, implementation, research and development in Indonesia: A review. *Sustainability, 14*(5), 2972.

Oldeman, L. R. (1994). The global extent of soil degradation. *Soil Resilience and Sustainable Land Use, 9.*

Palmquist, R. B., & Danielson, L. E. (1989). A hedonic study of the effects of erosion control and drainage on farmland values. *American Journal of Agricultural Economics, 71*(1), 55–62.
Panagos, P., Borrelli, P., Poesen, J., Ballabio, C., Lugato, E., Meusburger, K., … Alewell, C. (2015). The new assessment of soil loss by water erosion in Europe. *Environmental Science & Policy, 54*, 438–447.
Panagos, P., Matthews, F., Patault, E., De Michele, C., Quaranta, E., Bezak, N., … Borrelli, P. (2024). Understanding the cost of soil erosion: An assessment of the sediment removal costs from the reservoirs of the European Union. *Journal of Cleaner Production, 434*, Article 140183.
Panagos, P., Standardi, G., Borrelli, P., Lugato, E., Montanarella, L., & Bosello, F. (2018). Cost of agricultural productivity loss due to soil erosion in the European Union: From direct cost evaluation approaches to the use of macroeconomic models. *Land Degradation & Development, 29*(3), 471–484.
Pathak, V. M., Verma, V. K., Rawat, B. S., Kaur, B., Babu, N., Sharma, A., … Cunill, J. M. (2022). Current status of pesticide effects on environment, human health and it's eco-friendly management as bioremediation: A comprehensive review. *Frontiers in Microbiology, 13*, Article 962619.
Pearce, D. W. (1992). *Economic valuation and the natural world* (Vol 988). World Bank Publications.
Pieper, S., Fraunstein, J., Ginzky, H., Glante, F., Grimski, D., Kotschik, P., … Biegel-Engler, A. (2023). *The upcoming European soil health law—Chances and challenges for an effective soil protection.* German Environment Agency, Dessau-Roßlau.
Pimentel, D., & Burgess, M. (2013). Soil erosion threatens food production. *Agriculture, 3*(3), 443–463.
Pimentel, D., Harvey, C., Resosudarmo, P., Sinclair, K., Kurz, D., McNair, M., … Blair, R. (1995). Environmental and economic costs of soil erosion and conservation benefits. *Science, 267*(5201), 1117–1123.
Qadir, M., & Schubert, S. J. L. D. (2002). Degradation processes and nutrient constraints in sodic soils. *Land Degradation & Development, 13*(4), 275–294.
Regasa, A., Haile, W., & Abera, G. (2025). Variation in soil acidity across different land uses, soil types and altitudinal gradients in Western Oromia, Ethiopia. *Scientific Reports, 15*(1), Article 31565.
Rehm, R., Zeyer, T., Schmidt, A., & Fiener, P. (2021). Soil erosion as transport pathway of microplastic from agriculture soils to aquatic ecosystems. *Science of the Total Environment, 795*, Article 148774.
Reinholz, T., & Rodi, M. (2025). *Soil protection and land use as challenges for climate law and policy– Germany and the European Union*. The Institute for Climate Protection. Energy and Mobility (IKEM).
Rengasamy, P. (2006). World salinization with emphasis on Australia. *Journal of Experimental Botany, 57*(5), 1017–1023.
Rezaei, M., Riksen, M. J., Sirjani, E., Sameni, A., & Geissen, V. (2019). Wind erosion as a driver for transport of light density microplastics. *Science of the Total Environment, 669*, 273–281.
Rillig, M. C. (2012). *Microplastic in terrestrial ecosystems and the soil.* ACS Publications.
Robinson, J. M., Liddicoat, C., Muñoz-Rojas, M., & Breed, M. F. (2024). Restoring soil biodiversity. *Current Biology, 34*(9), R393–R398.
Rodrigues, S. M., Pereira, M. E., da Silva, E. F., Hursthouse, A. S., & Duarte, A. C. (2009). A review of regulatory decisions for environmental protection: Part I—Challenges in the implementation of national soil policies. *Environment International, 35*(1), 202–213.
Romanazzi, G. R., Ottomano Palmisano, G., Cioffi, M., Leronni, V., Toromani, E., Koto, R., … Roma, R. (2024). A cost–benefit analysis for the economic evaluation of ecosystem services lost due to erosion in a Mediterranean River Basin. *Land, 13*(9), 1512.
Russi, D., ten Brink, P., Farmer, A., Badura, T., Coates, D., Förster, J., … Davidson, N. (2013). The economics of ecosystems and biodiversity for water and wetlands. *IEEP, London and Brussels, 78*, Article 118p.
Sajjad, M., Huang, Q., Khan, S., Khan, M. A., Liu, Y., Wang, J., … Guo, G. (2022). Microplastics in the soil environment: A critical review. *Environmental Technology & Innovation, 27*, Article 102408.
Santra, P., Moharana, P. C., Kumar, M., Soni, M. L., Pandey, C. B., Chaudhari, S. K., & Sikka, A. K. (2017). Crop production and economic loss due to wind erosion in hot arid ecosystem of India. *Aeolian Research, 28*, 71–82.

Savari, M., Yazdanpanah, M., & Rouzaneh, D. (2022). Factors affecting the implementation of soil conservation practices among Iranian farmers. *Scientific Reports, 12*(1), 8396.

Shao, Y. (Ed.). (2008). *Physics and modelling of wind erosion. Dordrecht*. Netherlands: Springer.

Shrivastava, P., & Kumar, R. (2015). Soil salinity: A serious environmental issue and plant growth promoting bacteria as one of the tools for its alleviation. *Saudi Journal of Biological Sciences, 22*(2), 123–131.

Singer, M. J., & Scalenghe, R. (2006). Physical degradation of soils. In G. Certini (Ed.), *Soils: Basic concepts and future challenges* (pp. 223–234). Cambridge: Cambridge University Press.

Singh, A. (2021). Soil salinization management for sustainable development: A review. *Journal of environmental management, 277*, 111383.

Sirjani, E., Mahmoodabadi, M., & Cerdà, A. (2022). Sediment transport mechanisms and selective removal of soil particles under unsteady-state conditions in a sheet erosion system. *International Journal of Sediment Research, 37*(2), 151–161.

Sirjani, E., Sameni, A., Mahmoodabadi, M., Moosavi, A. A., & Laurent, B. (2024). In-situ wind tunnel experiments to investigate soil erodibility, soil fractionation and wind-blown sediment of semi-arid and arid calcareous soils. *Catena, 241*, Article 108011.

Sirjani, E., Sameni, A., Moosavi, A. A., Mahmoodabadi, M., & Laurent, B. (2019). Portable wind tunnel experiments to study soil erosion by wind and its link to soil properties in the fars province, Iran. *Geoderma, 333*, 69–80.

Smith, P. (1998). *The use of subsidies for soil and water conservation: A case study from Western India. Agricultural Research and Extension Network (AgREN), Network Paper No*. London, UK: Overseas Development Institute (ODI.

Smith, P., Poch, R. M., Lobb, D. A., Bhattacharyya, R., Alloush, G., Eudoxie, G. D., … Hallett, P. (2024). Status of the world's soils. *Annual Review of Environment and Resources, 49*(2024), 73–104.

Somda, J., & Awaïss, A. (2013). Economic valuation of ecological functions and services of natural ecosystems: Guide on the use of simple methods. In *Ouagadougou, Burkina Faso: IUCN*, Article 32pp.

Sonderegger, T., & Pfister, S. (2021). Global assessment of agricultural productivity losses from soil compaction and water erosion. *Environmental Science & Technology, 55*, 12162–12171 (Global integrated assessment; potential long-term productivity losses ~10–20%).

Stavi, I., Thevs, N., & Priori, S. (2021). Soil salinity and sodicity in drylands: A review of causes, effects, monitoring, and restoration measures. *Frontiers in Environmental Science, 9*, Article 712831.

Sun, R., He, H., Jing, Y., Leng, S., Yang, G., Lü, Y., … Fu, B. (2024). Global wind erosion reduction driven by changing climate and land use. *Earth's Future, 12*(10), Article e2024EF004930.

Swaine, M., Bergna, A., Oyserman, B., Vasileiadis, S., Karas, P. A., Screpanti, C., & Karpouzas, D. G. (2025). Impact of pesticides on soil health: Identification of key soil microbial indicators for ecotoxicological assessment strategies through meta-analysis. *FEMS Microbiology Ecology, 101*(6), Article fiaf052.

Tarolli, P., Luo, J., Park, E., Barcaccia, G., & Masin, R. (2024). Soil salinization in agriculture: Mitigation and adaptation strategies combining nature-based solutions and bioengineering. *iScience, 27*(2).

Telles, T. S., Dechen, S. C. F., Souza, L. G. A. D., & Guimarães, M. D. F. (2013). Valuation and assessment of soil erosion costs. *Scientia Agricola, 70*, 209–216.

Thaler, E. A., Larsen, I. J., & Yu, Q. (2021). The extent of soil loss across the US Corn Belt. *Proceedings of the National Academy of Sciences, 118*(8), Article e1922375118.

Vitousek, P. M., Aber, J. D., Howarth, R. W., Likens, G. E., Matson, P. A., Schindler, D. W., … Tilman, D. G. (1997). Human alteration of the global nitrogen cycle: Sources and consequences. *Ecological Applications, 7* (3), 737–750.

Von Uexküll, H. R., & Mutert, E. (1995). Global extent, development and economic impact of acid soils. *Plant and Soil, 171*(1), 1–15.

Walker, D. J. (1982). A damage function to evaluate erosion control economics. *American Journal of Agricultural Economics, 64*(4), 690–698.

Wang, T., Cao, X., Chen, M., Lou, Y., Wang, H., Yang, Q., … Zhuge, Y. (2022). Effects of soil acidification on bacterial and fungal communities in the Jiaodong Peninsula, Northern China. *Agronomy, 12*(4), 927.

Wang, K., Wang, F., Yu, Y., Yang, S., Han, Y., & Yao, H. (2025). Microplastics and soil microbiomes. *BMC Biology, 23*(1), 273.

Zhang, L., Huang, Y., Rong, L., Duan, X., Zhang, R., Li, Y., & Guan, J. (2021). Effect of soil erosion depth on crop yield based on topsoil removal method: A meta-analysis. *Agronomy for Sustainable Development, 41*, 63. https://doi.org/10.1007/s13593-021-00718-8

CHAPTER 13

The role of NGOs and civil society in soil protection: A review

Foroogh Golkar

Department of Water Engineering & Oceanic and Atmospheric Research Centre, College of Agriculture, Shiraz University, Shiraz, Iran

1. Introduction

Soil functions as a crucial life-support system for the Earth, underpinning food production, livelihoods, climate regulation, water quality, biodiversity, and long-term planetary stability (Fig. 13.1). Approximately 95%–99% of human caloric consumption derives from soil-based agriculture (Kopittke et al., 2021, 2022), and soil health, defined by its physical, chemical, and biological properties, directly affects crop yields, resilience to environmental stressors, and the sustainability of food systems (Hou et al., 2020; Lehmann et al., 2020; Tahat et al., 2020). Healthy soil is fundamentally linked to the United Nations Sustainable Development Goals (SDGs), including the eradication of hunger and the alleviation of poverty, owing to its role in enhancing rural economies and reducing poverty (Bach et al., 2020; Hou et al., 2020).

Soils have a crucial role in climate regulation, sequestering more carbon than both the atmosphere and terrestrial vegetation combined, so acting as a major global carbon sink (Kopittke et al., 2022; Lal et al., 2021; Telo da Gama, 2023). Degraded or improperly managed soils release carbon dioxide and other greenhouse gases, significantly exacerbating climate change and other Earth-system impacts, such as ocean acidification and ozone depletion (Kopittke et al., 2021, 2022; Lal et al., 2021). Moreover, soils regulate hydrological processes by filtering and purifying water, controlling runoff and flooding, and influencing regional water cycles, so ensuring a clean water supply and reducing ca tastrophe risk (Hou et al., 2020; Lal et al., 2021; Plaas et al., 2019; Telo da Gama, 2023).

Soil constitutes the most biologically diverse ecosystem on Earth, harboring approximately one-fourth of global biodiversity and supporting critical functions such as nutrient cycling, decomposition, and disease regulation, thereby connecting soil health to the comprehensive "One Health" framework that includes humans, animals, and ecosystems (Bach et al., 2020; Delgado Baquerizo et al., 2020, 2025). Soil affects a minimum of five planetary boundaries through its linked processes, including biogeochemical cycles, land-system change, and climate control, and is increasingly acknowledged as a "master variable" of Earth-system stability (Kopittke et al., 2021, 2022). Soil, a predominantly non-renewable natural resource essential for environmental resilience and human welfare (Telo da Gama, 2023), relies significantly on management practices for its ability to provide ecosystem services. While high-input agricultural systems may enhance short-term yields, they

The Soil Nexus. https://doi.org/10.1016/B978-0-443-45286-4.00014-6

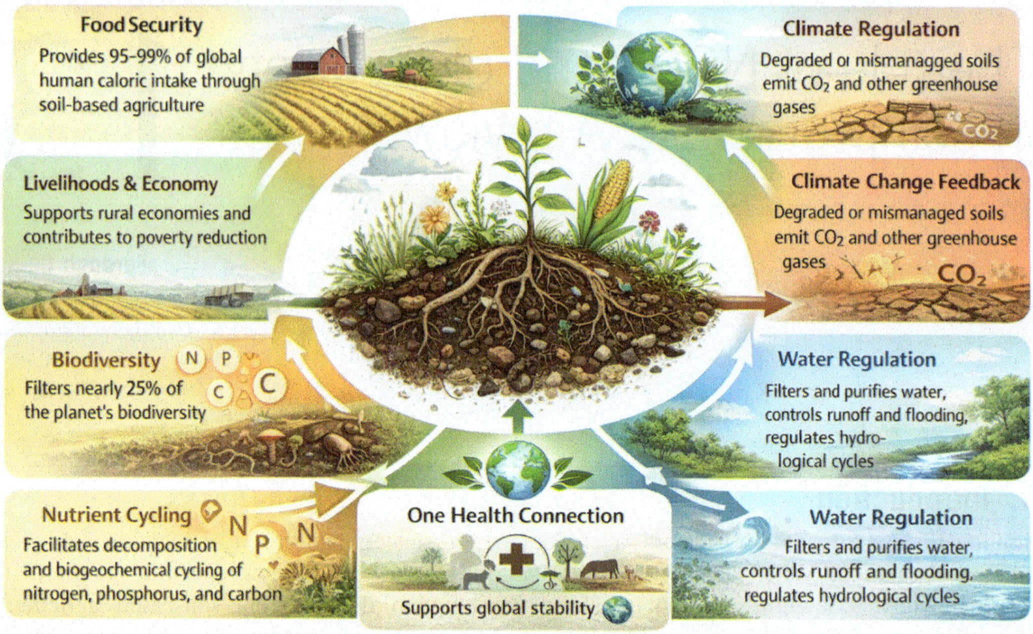

FIGURE 13.1

Soil as a master variable of earth-system stability.

(created by Notebook LM).

frequently result in soil degradation through erosion, compaction, salinization, and loss of organic matter (Cárceles Rodríguez et al., 2022; Kopittke et al., 2019; Powlson et al., 2011). In contrast, conservation and integrated strategies—such as reduced tillage, residue retention, improved nutrient and pest management, crop rotation, cover cropping, and agroforestry—improve soil fertility, resilience, and agro-environmental performance, often achieving yields that match or surpass those of conventional methods while ensuring the provision of food, fodder, and fuelwood (Kihara et al., 2020; Kumar et al., 2024; Powlson et al., 2011; Țopa et al., 2025).

The management of soil organic carbon is crucial to these benefits, as it governs soil structure, infiltration, erosion resistance, and carbon storage, hence influencing climate mitigation and water management (Pereira et al., 2018; Powlson et al., 2011; Telo da Gama, 2023). Sustainable practices, such as cover crops, organic amendments, residue retention, and reduced tillage, improve carbon sequestration, reduce runoff and erosion—often by over 30%–40%—and lower nutrient losses compared to conventional systems (Altobelli et al., 2020; Kihara et al., 2020; Xiao et al., 2021). Inadequate soil management practices, including deforestation, excessive tillage, and imbalanced fertilization, intensify emissions of carbon dioxide and nitrous oxide, facilitate nutrient leaching, and lead to water pollution, thereby undermining climate and water quality services on a broader scale (Pandian et al., 2024; Pereira et al., 2018; Powlson et al., 2011). Soil-friendly management enhances the sustainability and cultural ecosystem services by increasing soil biodiversity and microbial activity, which are essential for nutrient cycling, disease suppression, and drought resilience (Kihara et al., 2020; Omer et al., 2024; Tahat et al., 2020). Nature-based and regenerative practices, such as conservation agriculture, adaptive grazing, and organic farming, restore soil cover, improve habitat quality, and simultaneously increase diverse ecosystem services, including carbon sequestration, water regulation, and biodiversity (Cárceles Rodríguez et al., 2022; Keesstra et al., 2018). This collection of evidence illustrates that conservation-oriented, integrated, and soil-carbon-centric management practices are vital for preserving ecosystem services and achieving lasting global sustainability goals (Fontaine et al., 2023; Kopittke et al., 2019; Pereira et al., 2018; Powlson et al., 2011; Telo da Gama, 2023).

Soil degradation entails interrelated physical, chemical, and biological impairments caused by human activities that surpass the soil's ability to regenerate, hence diminishing ecosystem benefits. Physical degradation is primarily caused by water and wind erosion, which poses a significant global threat to fertile topsoil (Borrelli et al., 2017; Jie et al., 2002; Karlen & Rice, 2015). Additionally, compaction from heavy machinery and overgrazing impedes infiltration and root development (Baumhardt et al., 2015; Lal, 2024), while soil sealing permanently removes land from agricultural use (Karlen & Rice, 2015). Chemical degradation arises from nutrient imbalances and the depletion of organic matter due to intensive agriculture (Gupta, 2019; Karlen & Rice, 2015; Lal, 2015), as well as salinization, acidification, and contamination resulting from inadequate irrigation, fertilization, heavy metals, effluents, and pesticides, which adversely affect soil biota, water quality, and food safety (Baumhardt et al., 2015; Ferreira et al., 2022; Gupta, 2019; Karlen & Rice, 2015). Biological degradation entails the depletion of soil organic carbon and biodiversity resulting from deforestation, residue extraction, monoculture practices, and overgrazing, which heightens erosion risk and exacerbates desertification when land-use pressures surpass recovery potential (Bhattacharyya et al., 2023; Gupta, 2019; Karlen & Rice, 2015; Lal, 2015, 2024; Mesele et al., 2025). These processes are intricately associated with deforestation, inappropriate agricultural practices, overgrazing, and urban, industrial, and mining activities (Baumhardt et al., 2015; Bhattacharyya et al., 2023, 2015; Chen et al., 2002; Gupta, 2019; Karlen & Rice, 2015; Kumar, 2023; Lal, 2015), collectively compromising the soil's ability to support food production, biodiversity, and vital ecosystem services.

Global research suggests that soil degradation is escalating worldwide, directly linked to the expansion and intensification of high-input agricultural systems (Gomiero, 2016). Multiple assessments associate intensive agriculture with significant land degradation, including increased erosion, decreased fertility, diminished biodiversity, and rising greenhouse gas emissions, largely due to intensive tillage, monocropping, and excessive agrochemical application (Francaviglia et al., 2023; Kopittke et al., 2019; Trivedi et al., 2016). The conversion of natural ecosystems to agricultural land and subsequent intensification generally reduces soil organic carbon stocks by 40%–60%, undermining soil structure, resilience, and long-term productivity (Trivedi et al., 2016). Quantitative assessments indicate that soil compaction and water erosion associated with global crop production could lead to lasting productivity reductions of 10%–20%, particularly in regions experiencing rapid automation and intensification (Sonderegger & Pfister, 2021). Global erosion models suggest that the principal factors driving the increase in soil erosion this century are the expansion of farmland and heightened land use, with the most substantial anticipated increases occurring in Sub-Saharan Africa, South America, and Southeast Asia (Borrelli et al., 2017; Prăvălie, 2021). Currently, at least 24% of global agricultural land is damaged due to overexploitation and unsustainable intensive practices, with this figure expected to rise (Bedolla-Rivera et al., 2023). At the field scale, intensive agricultural systems exacerbate deterioration by repeated machinery traffic, which compacts soil, reduces porosity and infiltration, and can potentially decrease yields by up to 50% in extreme cases (Shah et al., 2017; Shaheb et al., 2021). Moreover, monocultures and insufficient residue return further diminish organic matter and microbial diversity, intensifying erosion and nutrient depletion (Bedolla-Rivera et al., 2023; Kopittke et al., 2019; Trivedi et al., 2016). This body of evidence demonstrates that intensive agriculture, marked by heavy tillage, monocropping, and extensive use of machinery and inputs, significantly accelerates global soil degradation, threatening long-term productivity and ecosystem services unless production is decoupled from ongoing soil deterioration through sustainable management practices.

Soil protection prioritizes the reduction of erosion, the maintenance of soil organic matter, and the preservation of biodiversity to ensure sustainable productivity and resilience. Conservation and regenerative agriculture are essential techniques, as low soil disturbance by no-till or reduced tillage reduces erosion and preserves soil structure and organic carbon (Cárceles Rodríguez et al., 2022; Francaviglia et al., 2023; Țopa et al., 2025). Furthermore, the implementation of permanent soil cover through crop residues, mulches, or cover crops protects against the effects of wind and raindrop impact, thereby diminishing erosion and runoff by around 70%–80% (Du et al., 2022; Rajbanshi et al., 2023; Xiong et al., 2018). Crop rotation and diversification enhance nutrient cycling, disrupt pest and disease cycles, and strengthen soil biota (Cárceles Rodríguez et al., 2022; Francaviglia et al., 2023; Țopa et al., 2025). Controlled or rotational grazing alleviates compaction and overgrazing, facilitating the repair of soil organic carbon and improving ecosystem resilience (Lal, 2015; Rathi et al., 2024). Preserving and recovering soil organic matter is a primary goal of conservation, as sustaining soil organic carbon levels above critical thresholds (about 10–15 g/kg) improves soil structure, fertility, and resilience to degradation (Bhaduri et al., 2022; Lal, 2015; Rathi et al., 2024). Organic supplements such as manure, compost, green manures, and biochar enhance soil aggregation, water retention, and microbial activity (Bhaduri et al., 2022; Hou et al., 2020; Kimmell et al., 2023). Integrated nutrient management, which combines balanced mineral and organic inputs, sustains yields and alleviates acidification, salinization, and pollution (Lal, 2015; Narayanan & Ma, 2024; Rathi et al., 2024). Physical and vegetative erosion-control methods, such as terraces, contour bunds, hedgerows, strips, and trenches, can reduce soil loss by roughly 60%–85% across diverse contexts (Djillo et al., 2024; Rajbanshi et al., 2023; Xiong et al., 2018; Ebabu et al., 2022; Silva et al., 2023). Biological methods such afforestation, shelterbelts, and permanent grass cover are especially effective on slopes and degraded terrains (Rajbanshi et al., 2023; Xiong et al., 2018; Du et al., 2022). Sustainable land management and soil preservation policies aim to integrate these strategies, evaluate soil health, and encourage widespread implementation through integrated frameworks and alliances (Francaviglia et al., 2023; Hou et al., 2020). These methods demonstrate that effective soil preservation depends on the integration of minimal disturbance, continuous cover, improvement of organic matter, judicious fertilizer treatment, and both physical and vegetative management, supported by policies that promote long-term soil health.

Soil protection is supported by a coalition of international organizations, global agreements, and national laws that promote sustainable soil management and regulate land use and pollution. The FAO-led Global Soil Partnership (GSP) functions as a UN-hosted voluntary coalition that orchestrates global initiatives in sustainable soil management and has developed the Voluntary Guidelines for Sustainable Soil Management (VGSSM), in addition to regional platforms such as the European Soil Partnership (Baritz et al., 2017; Erdoğan et al., 2021; Montanarella, 2015). The European Soil Partnership and its sub-regional groups facilitate information exchange and support the implementation of soil-related SDGs across Europe and Central Asia (Erdoğan et al., 2021). Scientific organizations such as the European Society for Soil Conservation (ESSC) conduct research, education, and policy advisory activities concerning soil degradation and conservation (Dazzi et al., 2019), while the International Union of Soil Sciences (IUSS) explicitly links soil science to the UN Sustainable Development Goals, emphasizing soil health as crucial to SDGs 2, 6, 13, and 15 (Lal et al., 2021).

International legal and policy frameworks enhance soil conservation. The UN Convention to Combat Desertification (UNCCD) is a crucial treaty aimed at addressing land and soil degradation and advancing land degradation neutrality goals, while the Convention on Biological Diversity (CBD) and protected-area legislation indirectly safeguard soils by conserving natural habitats. The UN Sustainable Development Goals, although not legally enforceable, profoundly influence national soil policies by setting objectives concerning food security, water resources, climate, and terrestrial ecosystems (Lal et al., 2021). Additionally, the VGSSM offers a globally recognized, albeit non-binding, framework for directing sustainable soil management practices and policies (Baritz et al., 2017).

Soil conservation is integrated into environmental and agricultural policy at both regional and national tiers. Soil-related initiatives in the European Union are integrated into the Common Agricultural Policy (CAP), rural development programs, the EU Green Deal, and proposed frameworks such as the EU Soil Monitoring and Resilience Directive (Costantini et al., 2025; Futa et al., 2024; Löbmann et al., 2022; Martinho et al., 2024). Germany's Federal Soil Protection Act, fertilizer regulations, and building codes regulate soil conservation and contamination (Ronchi et al., 2019), while the Netherlands implements soil protection through its Soil Protection Act, soil quality decree, and erosion regulations (Li et al., 2019; Ronchi et al., 2019). The Soil Pollution Prevention and Control Law of China (2018) established a risk-based soil management framework (Li et al., 2019), whereas in the United States, soil conservation has been historically shaped by the Soil Conservation Act (1935) and continues through the efforts of the USDA–NRCS and the Soil Health Institute (Honeycutt et al., 2020; Peake & Robb, 2021).

Non- Governmental Organizations (NGOs) dedicated to soil protection are private, non-profit, voluntary entities that operate independently of governmental and commercial influences to preserve and sustainably manage soil resources. NGOs represent a component of civil society and are sometimes classified as part of the "third sector," distinct from the state (first sector) and the private market (second sector) (Abiddin et al., 2022; Lewis, 2023). They are participants from civil society engaged in public-interest issues, encompassing environmental conservation, development, and human rights (Lewis, 2023). NGOs often do not distribute profits to proprietors, function independently, are formally recognized, and pursue communal or public welfare instead of commercial gain (Abiddin et al., 2022).

In the domain of soil and natural resource management, NGOs typically engage through direct implementation and community participation. They implement soil and water conservation measures—such as terraces, bunds, afforestation, and gully rehabilitation—in collaboration with farmers and local communities (Dangiso & Wolka, 2023; Klingner, 2021), and provide financial, technical, and logistical support for land restoration initiatives (Klingner, 2021; Peterson et al., 2018). In various situations, NGOs are more involved in policy implementation than in formal policy creation, particularly when legislative frameworks restrict advocacy roles (Ariti et al., 2018; Broeckhoven et al., 2020; Hammadi, 2025). They also augment awareness, cultivate local skills, and strengthen farmers' capacity to maintain conservation practices and address soil erosion and land degradation challenges

(Ariti et al., 2018; Dangiso & Wolka, 2023). Fig. 13.2 illustrates a summary of governance structures and the responsibilities of NGOs in soil protection.

This review paper intends to thoroughly analyze the function of NGOs and civil society in soil conservation and sustainable land management. It also evaluates their participation in policy advocacy and the incorporation of soil health into climate, biodiversity, and sustainable development initiatives. NGOs provide a vital intermediary role by connecting local people, scientific expertise, and governance frameworks, promoting the implementation of sustainable practices, and enhancing institutional and social capacity for enduring soil management. This study consolidates disparate evidence to elucidate the contributions, benefits, and constraints of NGOs and civil society in promoting resilient and sustainable soil systems.

2. Methodology

This review followed established standards for systematic and evidence-based literature reviews, emphasizing transparency, replicability, and comprehensive database coverage. A systematic search strategy was developed and executed across three primary multidisciplinary databases: Web of Science Core Collection, Scopus, and Google Scholar. Web of Science and Scopus were selected for their

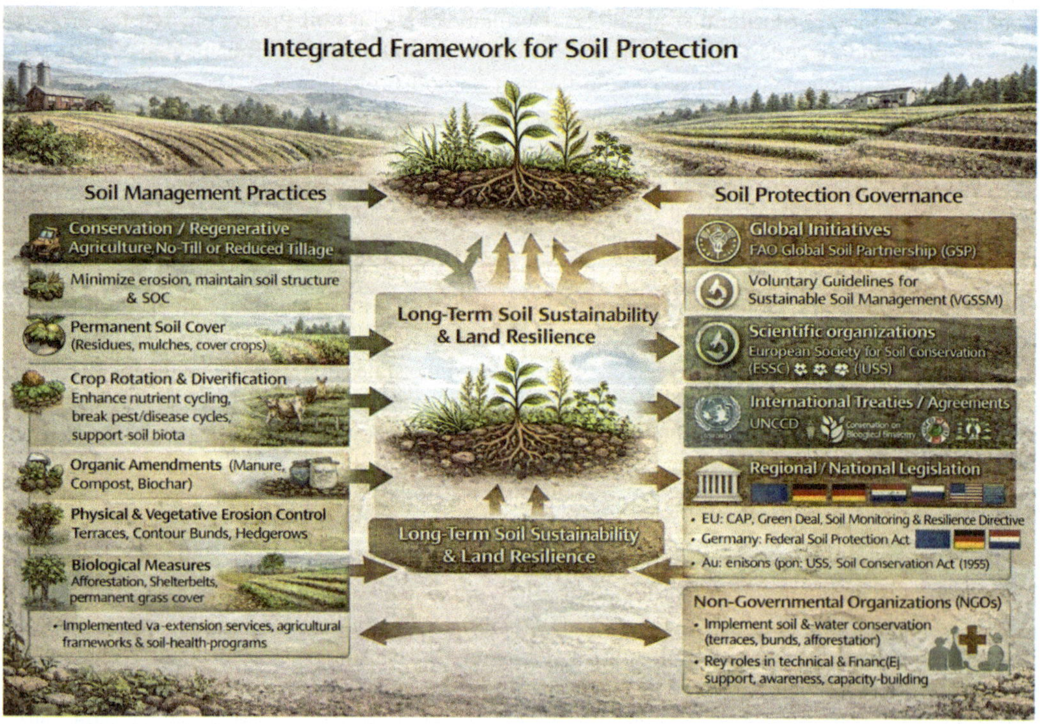

FIGURE 13.2

Governance frameworks and NGO roles in soil protection.

Created by AI tools; Notebook LM and ChatGPT.

comprehensive indexing of high-quality, peer-reviewed research in environmental and social sciences, and employing at least two primary databases is recommended to enhance coverage and reduce selection bias. Google Scholar was utilized to identify recent publications and relevant materials that may not yet be indexed in traditional databases or are more difficult to reach using standard indexing methods.

The analysis included 121 peer-reviewed studies published between 2000 and 2025. Imposing a time limit is considered appropriate and lawful when adequately justified and recorded. The year 2000 was selected as the starting point because of the significant advancement in research and policy formulation concerning NGOs and environmental or soil governance that occurred after the late 1990s, while earlier literature is less likely to reflect current institutional, legal, and sustainability frameworks. All database searches were last updated on January 2025.

Explicit inclusion and exclusion criteria were devised and applied throughout two screening phases: an initial evaluation of titles and abstracts, followed by an extensive full-text assessment, in accordance with recognized recommendations for systematic and quick reviews. Studies were eligible if they were peer-reviewed journal articles, books, or book chapters; focused on NGOs or civil society organizations and their roles in soil, land, or environmental protection and management; and provided empirical evidence, methodological analysis, or conceptual discourse relevant to governance, interventions, or advocacy. Publications were restricted to English-language sources, a common and typically low-risk limitation when sufficiently justified. Table 13.1 offers a succinct summary of NGO operations across different nations, derived from the publications examined in this study, serving as an overview preceding the Results section (Sec. 3).

3. Results

The examination of the 121 assessed papers indicates that NGOs and civil society groups fulfill multifaceted and context-dependent functions in soil conservation at local, national, and global levels. The research illustrates a transition from predominantly project-based interventions in the early 2000s to more integrated governance, advocacy, and knowledge-brokering roles in subsequent years. NGO operations are concentrated in three interconnected domains: (1) practical methods for soil conservation and restoration; (2) governance, networking, and institutional facilitation; and (3) structural and operational restrictions affecting efficacy. The subsequent sections consolidate these findings by analyzing (3.1) the operational role of NGOs in soil conservation, (3.2) their governance roles and geographical disparities, and (3.3) the primary obstacles and constraints influencing their efficacy.

3.1 The role of NGOs in soil protection

An examination of the 121 analyzed papers demonstrates a constant trend in the contributions of NGOs and civil society to soil conservation. The literature reveals consistent mechanisms through which NGOs exert influence, including advocacy, governance involvement, community action, knowledge creation, and project execution, by analyzing empirical case studies, policy evaluations, and participatory initiatives across various regions. These studies jointly demonstrate the extensive range of NGO operations and the interrelations of policy, practice, and capacity-building initiatives. Based on these insights, the subsequent sections delineate the primary functions of NGOs in soil conservation into five interconnected domains: (a) advocacy for policy and governance, (b)

Table 13.1 Examples of NGOs, civil society organizations, and global partnerships engaged in soil protection and sustainable land management worldwide.

NGO/Partnership/ Organization	Country/Region	Activities in soil protection	Year	Author(s)
Coalition of action 4 soil health (CA4SH)	Global	Multistakeholder engagement to scale soil health globally; implementing practices for carbon sequestration.	2025	Winowiecki et al.
Civil society organizations (CSOs)	Oman	Environmental preservation and community-led sustainability initiatives under new legal trajectories.	2025	Hammadi
Gully rehabilitation trusts	Kenya	Community participation to rehabilitate eroded gullies and manage pastoral land through local action.	2018	Peterson et al.
European soil partnership (ESP)	Europe & central Asia	Implementing sustainable soil management (SSM) and aligning soil conservation with UN SDGs.	2021	Erdoğan et al.
PREPSOIL project (citizen science)	Europe	Engaging citizens in monitoring soil health and mapping degradation for the European soil mission.	2025	Charvát et al.
Civil society organizations (CSOs)	Ethiopia	Policy advocacy for natural resource management and bridging the gap between state and community.	2020	Broeckhoven et al.
European society for soil conservation (ESSC)	Europe	Promoting scientific research, soil education, and sustainability policies across the continent.	2019	Dazzi et al.
Global soil partnership (GSP)	Global/Italy	Coordinating the "Italian soil partnership" and implementing global voluntary soil guidelines.	2020	Altobelli et al.

Table 13.1 Examples of NGOs, civil society organizations, and global partnerships engaged in soil protection and sustainable land management worldwide.—cont'd

NGO/Partnership/ Organization	Country/Region	Activities in soil protection	Year	Author(s)
Savingscapes project	Italy & Spain	Mobilizing civil society to protect landscape soil and improve ecosystem services.	2023	Delponte et al.
Living labs	Europe (Denmark, Germany, etc.)	Co-creating soil-improving management practices between researchers, NGOs, and farmers.	2025	Taskin et al.
International Union of soil sciences (IUSS)	Global	Scientific advocacy for soil's role in meeting 21st-century planetary boundaries and SDGs.	2021	Lal et al.
NGOs & local Cooperatives	Mexico	Knowledge governance and promoting the adoption of conservation practices among local farmers.	2019	Cotler & Cuevas
Civil society initiatives (climate-Smart)	Global/South Asia	Implementing climate-smart agriculture to restore soil functions and improve resilience.	2024	Davila et al.
Joint research-civil society approach	Romania	Monitoring and protecting soil in high nature value (HNV) agricultural areas.	2023	Dumitrascu et al.
NGOs & CSOs (land rights)	Argentina	Collaborating with the state to protect soil resources from unsustainable land grabbing.	2019	Busscher et al.
NGOs	Malaysia	Capacity building for sustainable community development and local soil stewardship.	2022	Abiddin et al.
Participatory science frameworks	Global (urban focus)	Organizing public engagement for urban soil research and remediation of contaminated soils.	2025	Paltseva
NGOs & human rights groups	Global/Egypt	Preserving the environment and human rights related to land access and soil health.	2022	Henry et al.

community-based soil conservation, (c) capacity building and education, (d) research, monitoring, and knowledge dissemination, and (e) project execution and funding mobilization. This framework illustrates how the examined literature reveals NGOs converting global and national soil initiatives into tangible, context-specific actions.

a. Advocacy for policy and governance

NGOs and civil society platforms are increasingly promoting the incorporation of soil health into national and international policy frameworks. Global and regional coalitions, such as the Coalition of Action 4 Soil Health (CA4SH) and the European Soil Partnership, involve governments, research institutions, agricultural entities, and NGOs to advocate for enforceable soil protection regulations, promote sustainable soil management goals, and integrate soil considerations into policies regarding climate, biodiversity, and food systems (Winowiecki et al., 2025; Erdoğan et al., 2021; Lal et al., 2021; Yin & Pereira, 2021). These alliances have impacted soil-related outcomes in UNFCCC and UNCCD deliberations and have promoted the recognition of soil health as an essential element of ecosystem restoration and the implementation of the SDGs (Winowiecki et al., 2025; Lal et al., 2021; Nolan et al., 2021; Yin & Pereira, 2021).

The participation of civil society concurrently highlights persistent government shortcomings. Existing soil and biodiversity policies are often fragmented and lack enforceable goals for soil ecosystem functionality and biodiversity, underscoring the need for improved governance mechanisms and monitoring systems specifically targeting soils (Erdoğan et al., 2021; Zeiss et al., 2022). Through advocacy, agenda-setting, and participation in multistakeholder dialogs, NGOs facilitate the redefinition of soils from a simple agricultural input to a public good that requires coordinated "soil health care" systems, akin to public health models that emphasize prevention, monitoring, and remediation (Lal et al., 2021).

b. Community-based soil conservation

At the local level, NGOs and grassroots organizations promote soil conservation using context-specific, community-focused solutions designed to address specific degradation challenges. Data from Kenya indicates that local NGOs have enabled the formation of Gully Rehabilitation Trusts, allowing farmers to collaboratively devise and execute strategies to mitigate gully erosion while concurrently connecting communities to external organizations and higher-level institutions (Peterson et al., 2018). Comparable participatory methodologies recorded in Mexico and southern Italy demonstrate that engaging farmers and social organizations in the co-design and governance of conservation practices enhances adoption rates and promotes the development of local institutional frameworks conducive to sustainable land management (Salvia et al., 2021a, 2021b). Fig. 13.3 presents images of NGO activities worldwide.

In the European context, the research conducted by Salvia et al. (2021a), (2021b) indicates that Environmental NGOs enhance soil conservation and climate mitigation chiefly by impacting Local Climate Plans (LCPs) in 327 cities. Nevertheless, these entities—namely Climate Alliance, the Global Covenant of Mayors for Climate and Energy, C40, and the Carbon Neutral Cities Alliance—primarily function as transnational urban climate networks. Their function is thus primarily focused on coordination, agenda-setting, and knowledge sharing among municipalities rather than direct project execution (Reckien et al., 2018; Salvia et al., 2021a, 2021b, 2023).

Similar dynamics are evident in Iran, where environmental NGOs (ENGOs) assume an increasingly significant albeit institutionally limited role in soil conservation (Amiraslani & Caiserman, 2018). National entities include Cenesta, Earth Watchers Center, Green Front of Iran, and the Iranian Society of Environmentalists (IRSEN), together with several NGOs focused on forestry, rangeland, and anti-desertification efforts, play a role in mitigating land degradation (Amiraslani & Caiserman, 2018; Moosavi et al., 2015). Local efforts, such as the Women's Society Against Environmental Pollution in the Zayandeh Rud basin, participate in catchment-level activities (Tabrizi, 2020). Their initiatives include anti-desertification programs, erosion-sensitive plantations, promotion of drought-resistant crops, rural environmental education, and women's empowerment in provinces like Mazandaran and Khorasan Razavi (Amiraslani & Caiserman, 2018; Amiraslani & Dragovich, 2010; Bouzarjomehri & Javani, 2020; Tabrizi, 2020). Despite constitutional acknowledgment of environmental preservation, inadequate coordination with governmental entities and legal limitations—especially concerning litigation—constrain its wider policy impact.

In these contexts, community-based initiatives often transcend top-down, incentive-driven models by aligning conservation efforts with farmers' priorities—such as enhancing food quality, diversifying production, and increasing social recognition—thereby reinforcing social capital and promoting enduring commitment to soil stewardship (Salvia et al., 2021a, 2021b). When properly acknowledged and assisted, community-based organizations can serve as principal agents in erosion management, land rehabilitation, and the improvement of soil-related ecosystem services in accordance with various SDGs (Peterson et al., 2018; Yin & Pereira, 2021).

c. Capacity building and education

Capacity building and awareness-raising are essential activities through which NGOs and civil society advocate for soil protection. Various programs emphasize "soil literacy," clarifying soil's crucial significance in food security, water management, climate mitigation, and biodiversity, as well as the consequences of degradation (Hou et al., 2020; Lal et al., 2021; Nolan et al., 2021). Assessments of sustainable soil management reveal that changes in land user behavior depend on information management and social learning processes, in which scientists, policymakers, and practitioners together generate and share knowledge (Bouma, 2019; Hou et al., 2020).

NGOs frequently operate as intermediaries in collaborative learning processes, enabling farmer-to-farmer exchanges, field demonstrations, and participatory training that translate complex soil–water–plant–climate interactions into practical recommendations (Bouma, 2019; Salvia et al., 2021a, 2021b). Educational activities include both educational institutions and the broader community, promoting a cultural shift that recognizes soils as dynamic systems and vital infrastructure for achieving the SDGs. Forming a social constituency is crucial for sustained political and financial support for soil conservation (Bouma, 2019; Hou et al., 2020; Lal et al., 2021; Nolan et al., 2021).

d. Research, monitoring, and knowledge sharing

NGOs and civil society organizations are progressively involved in applied research, soil monitoring, and knowledge creation in both rural and urban settings. In High Nature Value (HNV) regions, Romanian soil scientists collaborated with the ADEPT Foundation at the Târnava Mare Natura 2000 site to assess farmers' perceptions, collect soil samples, evaluate fertility, and create the inaugural national HNV soil database, along with a rapid soil-monitoring system utilizing locally pertinent indicators; this initiative illustrates how NGOs collaboratively select indicators such as erosion evidence, crop yield, and land management practices to inform agri-environmental strategies (Dumitrascu et al., 2023). Comparable partnerships in HNV landscapes throughout Europe have established foundational soil databases in under-explored areas, integrating scientific evaluation with stakeholder contributions to enable accelerated monitoring systems (Dumitrascu et al., 2023).

Throughout Europe, citizen science activities coordinated by NGOs have broadened the spatial and temporal scope of soil data. The PREPSOIL effort aggregated 96 volunteer-led studies assessing soil organic carbon, nutrients, contaminants, biodiversity, soil structure, and vegetation cover utilizing economical kits and mobile instruments (Charvát et al., 2025). A evaluation of 24 citizen-science projects in agriculture revealed that the majority produced data on soil biodiversity, vegetation cover, and soil organic carbon, frequently adhering to norms established by NGOs (Mason et al., 2024). Earthwatch Europe instructed volunteers to assess soil organic carbon based on soil color in parks across the UK and France, yielding results equivalent to laboratory analyses (Jorge et al., 2021). Evaluations of these initiatives indicate that participatory and citizen science frameworks augment evidence-based policymaking, collaboratively generate knowledge with farmers and citizens, and enhance data reliability through continuous engagement and training (Charvát et al., 2025; Head et al., 2020; Mason et al., 2024).

Urban soil initiatives exemplify the contributions of NGOs to research and public involvement. Our Soil in Troy, USA, collaborated with community organizations to map lead contamination, integrating laboratory analysis with local expertise (Price et al., 2024). The Solenville program in Strasbourg engaged people, educational institutions, NGOs, and artists to assess urban soil biodiversity, evaluate perceptions, and incorporate findings into municipal planning via 750 pitfall traps (Glatron et al., 2025). These urban and agricultural participatory frameworks underscore the significance of co-creation, continuous communication, and knowledge-sharing networks in fostering trust, maintaining engagement, and incorporating local insights into institutional soil information systems (Bouma, 2019; Charvát et al., 2025; Erdoğan et al., 2021; Hou et al., 2020; Paltseva, 2025; Winowiecki et al., 2025).

Through various rural and urban activities, NGOs offer essential technical support while promoting social learning, behavioral transformation, and institutional progress. By connecting local stakeholders, scientists, and policymakers, NGOs establish sustainable knowledge networks that enhance local soil conservation methods and broader evidence-based governance structures, demonstrating their crucial role in modern soil protection initiatives.

e. Project implementation and funding mobilization

NGOs and civil society frequently act as implementers and mediators for financing soil conservation initiatives. Comprehensive coalitions such as CA4SH aim to overcome economic and institutional barriers to adoption by promoting public and private investment in soil health, featuring dedicated working groups that concentrate on finance and implementation (Winowiecki et al., 2025). At the project level, NGOs design and manage restoration and sustainable soil management programs funded by governments, development agencies, or philanthropic organizations, providing technical assistance while incorporating social safeguards and considerations for gender and social inclusion (Abiddin et al., 2022; Nolan et al., 2021; Peterson et al., 2018; Zikargae et al., 2022).

Findings from community environmental projects in rural Ethiopia and comparable developing areas suggest that NGO-led initiatives grounded in genuine participation—defined by consultation,

(a)

(b)

(c)

FIGURE 13.3

Images depicting NGO activities in soil conservation from FAO website and google search; (a) Forest expansion in Turkey, (b) Environmental preservation in Kenya, (c) Soil day and NGO efforts against climate change in India.

collaboration, and partnership—are more likely to cultivate trust, enhance environmental security, and maintain livelihood benefits (Abiddin et al., 2022; Peterson et al., 2018). In Ethiopia, the Organization for Rehabilitation and Development in Amhara (ORDA) exemplifies such approaches by executing Environment and Forest Development Programs that incorporate community greening, seedling production, fuel-efficient stoves, biogas, watershed establishment, and informal environmental education to empower marginalized groups and strengthen local institutions (Zikargae et al., 2022). These activities are consistent with national environmental policies and assist in addressing deficiencies in governmental competence and coordination (Zikargae et al., 2022).

The collaboration between the government and NGOs is exemplified in Ethiopia's Productive Safety Net Program (PSNP), a public food security initiative that employs terracing, exclosures, and soil and water conservation public works to rehabilitate degraded lands and improve resilience (Rapiya et al., 2024; Woolf et al., 2018). NGOs and international entities, notably United Nations agencies, provide co-financing, technical assistance, and community mobilization (Woolf et al., 2018). In addition to these initiatives, the FAO supplies soil data, maps, and methodological guidelines for assessing erosion and degradation, and facilitates integrated watershed and sustainable land management planning in collaboration with ministries, local NGOs, farmers, and extension services (Muche et al., 2024; Workie & Teku, 2025). Conceptual analyses of restoration finance highlight that prioritizing soil in funding and planning frameworks improves long-term resilience and multifunctional outcomes, thereby reinforcing the argument for continued investment in soil-focused interventions (Hou et al., 2020; Nolan et al., 2021; Yin & Pereira, 2021).

By executing and financing initiatives, NGOs and civil society translate global soil and Sustainable Development Goal objectives into tangible actions, demonstrating the ecological, economic, and social advantages of investing in soil health. The previously listed themes are encapsulated inTable 13.2.

3.2 Functions, governance, and regional perspectives

Empirical case studies from various regions reveal that, despite differences in strategies influenced by socio-ecological and governance contexts, several common patterns emerge: substantial participation, multi-level coordination, knowledge co-production, and ongoing engagement are essential for attaining sustainable soil protection outcomes.

At both global and regional levels, NGOs and civil society organizations play a vital role in emphasizing soil health in policy agendas. Multi-stakeholder initiatives such as the Coalition of Action 4 Soil Health (CA4SH) and the European Soil Partnership involve governments, research institutions, agricultural organizations, and NGOs to promote improved soil protection frameworks, further sustainable soil management goals, and integrate soil considerations into climate, biodiversity, and food system policies (Erdoğan et al., 2021; Lal et al., 2021; Winowiecki et al., 2025; Yin & Pereira, 2021). These coalitions have elevated the recognition of soil issues within the UNFCCC and UNCCD frameworks and have reinforced the definition of soil health as a critical component of ecosystem restoration and the achievement of the SDGs (Erdoğan et al., 2021; Lal et al., 2021; Yin & Pereira, 2021).

Civil society actors often emphasize that soil administration is disjointed and frequently devoid of enforceable objectives for soil ecosystem functionality and biodiversity (Bhattacharyya et al., 2023; Davila et al., 2024). Through agenda-setting, advocacy, and participation in multistakeholder dialogs,

Table 13.2 A review of the functions of NGOs in soil protection and their effects.

Role/Function	Key actions	Impact
Policy advocacy	Promote soil inclusion in national & global policies, engage alliances (e.g., CA4SH, European soil partnership)	Stronger soil governance, SDG integration, ecosystem restoration
Community conservation	Support local soil projects, co-design practices with farmers	Increased adoption of sustainable practices, erosion control, local institution building
Capacity building & education	Raise soil literacy, facilitate training, participatory learning	Behavioral change, social awareness, long-term support for soil health
Research & knowledge sharing	Conduct applied research, soil monitoring, citizen science programs	Improved soil data, evidence-based policies, co-production of knowledge
Project implementation & funding	Execute soil projects, mobilize public/private funding, integrate social safeguards	Concrete soil restoration, livelihoods support, sustainable investment in soil health

NGOs facilitate the redefinition of soils from a mere technical agricultural resource to a public good that requires coordinated "soil health care" systems—analogous to public health models that emphasize prevention, monitoring, and remediation (Romero-Brito et al., 2016; Lal et al., 2021).

Regional experiences highlight the potential and constraints of advocacy. In Bangladesh, environmental civil society organizations are actively involved in grassroots mobilization and media outreach regarding salinity intrusion, drought, and riverbank erosion; however, their influence on national environmental policymaking remains comparatively constrained, underscoring a persistent governance gap between local activism and central decision-making (Bhattacharyya et al., 2023). Comparative research from Italy and Spain (Genoa and Girona) illustrates that local environmental movements can impact planning dialogs and territorial identities, despite erratic formal policy alterations, via sustained civic participation and divergent viewpoints on land and landscape governance (Delponte & co-authors, 2023).

At the local level, NGOs and grassroots organizations advocate for community-focused soil conservation that tackles specific degradation issues and social conditions. In western Kenya, the NGO NYADEC created "Gully Rehabilitation Trusts" that engage farmers to combat erosion-affected gullies by combining technical measures with the establishment of new community organizations for land restoration (Peterson et al., 2018). This participatory paradigm contrasts with top-down initiatives by emphasizing community decision-making, local leadership, and horizontal collaboration within villages. Nonetheless, enduring success depends on linking grassroots initiatives with higher-level organizations for financial support, technical aid, and policy alignment (Peterson et al., 2018).

In northern Tanzania, the joint co-design of erosion-control strategies with Maasai pastoralists addressed biophysical challenges alongside cultural identities, insufficient governance structures, and resource constraints (Blake et al., 2018; Kelly et al., 2020). By integrating scientific diagnoses with indigenous knowledge and priorities, NGOs enhanced community preparedness to alter grazing practices and created a reproducible model for erosion hotspots in other areas of the Global South (Blake et al., 2018; Kelly et al., 2020).

Comparable participative processes are evident in Southern Europe, Latin America, and Iran. In Mexico and southern Italy, the involvement of farmers and social organizations in the co-design of conservation strategies has enhanced adoption rates and facilitated the establishment of new local institutions for sustainable land management (Salvia et al., 2021a, 2021b). In the Santiago del Estero region of Argentina, NGOs partnered with provincial authorities to enhance land tenure for rural communities confronting industrial agricultural expansion, while also fostering conditions favorable to sustainable soil and land management (Busscher et al., 2019). In Iran, co-management frameworks in the Galazchai watershed, empowerment initiatives in Kerman, and community-based irrigation management in central regions have fostered trust, social networks, and local institutions, thus facilitating the adoption of soil conservation practices (Adhami et al., 2019; Avazpour et al., 2024). These examples illustrate that soil conservation is frequently linked to overarching concerns of land governance, rights, and institutional capacity, and that the efficacy of community-based conservation relies on fostering trust, enhancing social capital, and bolstering locally rooted institutions that sustain and evolve conservation practices over time (Blake et al., 2018; Kelly et al., 2020; Salvia et al., 2021a, 2021b).

Capacity building and awareness-raising are essential functions that link local initiatives to broader governing structures. NGOs promote "soil literacy" by clarifying the importance of soil in food security, water management, climate mitigation, and biodiversity, while also outlining the consequences of degradation (Lal et al., 2021; Yin & Pereira, 2021). Studies on sustainable soil management indicate that behavioral change depends on social learning processes in which scientists, policymakers, and practitioners together create knowledge (Blake et al., 2018; Lal et al., 2021).

NGOs frequently operate as intermediaries in collaborative learning frameworks, enabling farmer-to-farmer exchanges, interactive field demonstrations, and training programs that translate complex soil-water-plant-climate relationships into practical guidance (Salvia et al., 2021a, 2021b). In Europe, the collaboration between science and civil society in Romania's High Nature Value areas demonstrates how joint soil assessments and stakeholder involvement can bolster support for agri-environmental projects and national soil legislation (Dumitrascu et al., 2023). These approaches augment technical proficiency and normative dedication, fostering a culture shift that recognizes soils as living systems and vital infrastructure for achieving the SDGs (Lal et al., 2021; Yin & Pereira, 2021).

Furthermore, NGOs and civil society organizations are increasingly engaged in applied research and soil monitoring systems. Partnerships between soil scientists and environmental NGOs have yielded innovative soil fertility data, recognized relevant monitoring indicators, and established baseline soil databases in underexplored regions (Dumitrascu et al., 2023). Citizen science has emerged as a significantly important innovation. Volunteer-led soil monitoring initiatives across Europe—predominantly documented under the PREPSOIL project—collect data on soil organic carbon, nutrients, pollutants, structure, biodiversity, and vegetation cover (Caggiano et al., 2025; Charvát et al., 2025). These initiatives improve spatial and temporal data coverage while including communities and farmers in information generation (Charvát et al., 2025). Participatory research frameworks emphasize the significance of co-creation, established protocols, sustained communication, and transparent governance of citizen-generated data to ensure data quality, trust, and continuous engagement (Kazanskaia, 2025).

NGOs, via monitoring and research initiatives, furnish empirical insights and institutional expertise, so augmenting the knowledge foundation for soil policy at both regional and national tiers (Davila et al., 2024; Lal et al., 2021).

Besides campaigning and information enhancement, NGOs frequently act as implementers and financial intermediaries for soil protection. Coalitions such as CA4SH seek to enhance public and private investment in soil health by addressing economic and institutional barriers to adoption through dedicated finance and implementation working groups (Winowiecki et al., 2025). At the project level, NGOs design and administer restoration and sustainable soil management initiatives funded by governments, development agencies, or philanthropic organizations, often integrating female participation and social safeguards (Henry et al., 2022; Yin & Pereira, 2021).

Data from rural Ethiopia and comparable developing areas indicate that NGO-led projects grounded on genuine engagement and collaboration are more effective in cultivating trust, improving environmental security, and guaranteeing the durability of livelihood benefits (Henry et al., 2022). Studies on restoration financing demonstrate that prioritizing soil within financial frameworks enhances long-term resilience and multifunctional ecosystem outcomes (Yin & Pereira, 2021; Lal et al., 2021).

However, scalability relies on the establishment of legal and institutional structures. In such situations, NGOs are confined to implementation roles, exerting diminished influence on national policy development, which may impede the broader institutionalization of effective community models (Henry et al., 2022).

3.3 Challenges and limitations in NGO activity in soil protection

The reviewed literature consistently indicates that although NGOs and civil society play a crucial role in soil conservation and governance, their effectiveness is affected—and often constrained—by institutional, political, economic, and socio-cultural barriers. These constraints influence the longevity of grassroots initiatives and the capacity of NGOs to induce formal policy and institutional change.

Evidence from multiple sources suggests that the application of soil and water conservation (SWC) measures does not guarantee long-term sustainability. In an Ethiopian watershed, 65% of farmers initially implemented soil and water conservation (SWC) structures funded by NGOs or the government; however, 79% later failed to sustain them, resulting in considerable degradation of the installations (Dangiso & Wolka, 2023). Identified causes included personnel shortages, insufficient technical skills, limited perceived economic benefits, and the perception that conservation measures were externally imposed through campaigns or incentive-based programs (Dangiso & Wolka, 2023). Similar findings in Europe demonstrate that even well-supported conservation projects face considerable economic and technological challenges, particularly when requiring fundamental changes in production systems, technology, or agricultural management practices (Henry et al., 2022). The findings suggest that NGO-supported programs may be technically feasible but misaligned with local social or economic circumstances, hence limiting their long-term efficacy.

The research underscores considerable political and legal limitations on NGO involvement in soil governance. In Ethiopia, a 2009 statute curtailed NGO participation in agenda-setting and policy formulation for natural resource management. Most NGOs reported encountering legal and administrative obstacles, restricted access to information, inadequate collaboration with the government, and apprehensions regarding state oversight, thereby relegating them to implementation roles, such as soil and water conservation, instead of facilitating extensive governance reforms (Ariti et al., 2018). The lack of binding international soil treaties and fragmented soil policies restricts NGOs' capacity to promote enforceable frameworks. Within multistakeholder platforms, limited legal authority and inadequate institutional integration frequently diminish the influence of NGO involvement on official policy results (Erdoğan et al., 2021).

The lack of binding global soil treaties and the dispersion of soil policy at the international level weaken civil society's ability to advocate for cohesive and enforceable frameworks (Erdoğan et al., 2021). Despite NGOs participating in multistakeholder forums, limited legal authority and insufficient institutional integration may undermine their influence on enforceable policy outcomes.

Stakeholder consultations throughout Europe demonstrate that weak science–policy–practice interfaces, inefficient soil monitoring systems, and inadequate knowledge dissemination are important hurdles to sustainable soil management (Pulido-Moncada et al., 2025; Vanino et al., 2023). NGOs often function within these disjointed governance frameworks and encounter resource constraints, ephemeral project funding cycles, and limited access to high-quality soil data, all of which hinder their capacity to facilitate sustainable, evidence-based soil management (Taskin et al., 2025; Vanino et al., 2023).

In African contexts, extensive structural impediments—such as insufficient institutional capacity, restricted education and awareness, feeble extension systems, and inadequate economic incentives—diminish the scalability and sustainability of NGO-supported soil management initiatives (Haregeweyn et al., 2015; Tetteh, 2021; Tetteh et al., 2025). These statistics imply that the efficacy of NGOs relies on both organizational proficiency and supporting institutional and economic settings.

Socio-cultural and socio-technical elements additionally hinder the implementation of soil conservation methods. In northern Tanzania, awareness of erosion difficulties, coupled with robust cattle-centric identities, weak governance frameworks, and resource limits, limited collective action and the readiness to execute suggested improvements in grazing ways (Blake et al., 2018). In Europe, socio-technical barriers—including inadequate machinery, perceived financial risks, entrenched agricultural traditions, and market pressures—impede the adoption of conservation agriculture and other soil-enhancing practices (Heller et al., 2024; Pulido-Moncada et al., 2025). The findings imply that NGO-led awareness campaigns or participatory approaches alone may be inadequate to remove established production systems and cultural norms without accompanying economic incentives and structural reforms.

The studied evidence reveals that the contributions of NGOs and civil society to soil protection are considerable yet contingent upon context. Long-term efficacy needs not only technical proficiency and participatory design but also supportive legal frameworks, consistent funding, institutional coordination, and conformity with socio-economic reality (Taskin et al., 2025; Tetteh et al., 2025).

4. Challenges and future directions

The reviewed literature demonstrates a steady transition from disconnected, project-oriented NGO initiatives to more integrated, multi-tiered governance systems. In many scenarios and policy evaluations, NGOs indicate best efficacy when integrated into coordinated governance frameworks that have stated mandates, trustworthy funding ways, and established roles in knowledge co-production and policy formation. Sustainable effect seems to require on the incorporation of NGO activities into regional, national, and global soil governance frameworks, rather than largely functioning through short-term, externally financed projects (Erdoğan et al., 2021; Montanarella et al., 2012; Ronchi et al., 2019).

Data from regional soil partnerships shows the expanded benefits of umbrella networks that unite governments, scientists, farmers, and civil society organizations. These platforms encourage knowledge transfer, promote policy alignment, and advocate for more enforceable and coordinated frameworks for sustainable soil management (Altobelli et al., 2020; Erdoğan et al., 2021; Montanarella et al., 2012; Ronchi et al., 2019). The results reveal that NGOs wield higher impact while participating in coordinated, multi-actor coalitions rather than working as standalone implementers.

In the European context, mission-oriented governance tools—such as "Living Labs" and "Lighthouses" related with soil missions—are acknowledged as creative platforms for experimentation and extension. These programs promote collaboration among farmers, NGOs, researchers, and public authorities to co-develop and assess soil-health practices before their adoption into advisory systems and policy instruments (Bouma & Veerman, 2022; Pulido-Moncada et al., 2025). The literature underlines that these frameworks allocate specific roles to NGOs in innovation facilitation, stakeholder coordination, and sharing of best practices, therefore enhancing their contributions beyond local pilot programs.

Citizen science efforts, such as those reported in the PREPSOIL project, indicate the capacity of NGOs to expand soil data coverage by engaging volunteers and local populations (Charvát et al., 2025). The efficacy of these approaches depends on defined protocols, uniform training, persistent funding, and specific governance mechanisms addressing data quality and ownership (Charvát et al., 2025; Evans et al., 2021). These findings underline that NGO contributions to monitoring and knowledge systems are more sustainable when explicitly integrated into institutional frameworks rather than depending on temporary activities.

Multiple strategic directions arise across several policy sectors. Initially, NGOs function as intermediates by connecting local soil projects to the Sustainable Development Goals—specifically Zero Hunger, Climate Action, and Life on Land—while also supporting stronger global soil governance frameworks (Lal et al., 2021). The literature advocates for the integration of binding soil legislation with volunteer initiatives and communication campaigns, wherein NGOs act as essential partners for implementation and outreach (Altobelli et al., 2020; Montanarella & Panagos, 2020; Panagiotakis & Dermatas, 2022; Ronchi et al., 2019). Third, within the EU policy framework, NGOs are poised to play a role in the formulation and execution of eco-schemes, soil monitoring systems, and participatory processes as outlined in the Soil Strategy, the Green Deal, and CAP strategic plans (Bouma & Veerman, 2022; Erdoğan et al., 2021; Francaviglia et al., 2023; Montanarella & Panagos, 2020; Powlson et al., 2011; Pulido-Moncada et al., 2025). Ultimately, upcoming frameworks prioritize integrated soil monitoring, open data platforms, and collaborative procedures, with NGOs serving as conveners, communicators, and custodians of data (Altobelli et al., 2020; Charvát et al., 2025; Evans et al., 2021; Hou et al., 2020; Zeiss et al., 2022).

The findings imply that (i) enhancing NGO efficacy in soil protection relies more on integrating civil society into cohesive, multi-tiered governance structures than on increasing the quantity of individual conservation efforts. NGOs display optimal efficacy when they participate in multi-tiered collaborations that are coherent with regional and global plans (Altobelli et al., 2020; Erdoğan et al., 2021; Montanarella et al., 2012; Ronchi et al., 2019).

(ii) integrate advocacy with structured implementation roles through both binding and voluntary policy instruments (Altobelli et al., 2020; Montanarella & Panagos, 2020; Panagiotakis & Dermatas, 2022; Ronchi et al., 2019); (iii) engage in co-designed innovation platforms that promote scalability (Bouma & Veerman, 2022; Pulido-Moncada et al., 2025); and (iv) support institutionalized monitoring and open data systems that facilitate evidence-based governance (Altobelli et al., 2020; Charvát et al., 2025; Evans et al., 2021; Hou et al., 2020; Zeiss et al., 2022).

5. Conclusion

This review indicates that NGOs and civil society actors execute multiple, interrelated roles in soil conservation at numerous scales. They function as proponents, boosting soil health within climate, biodiversity, and food system projects while exposing governance shortcomings in disconnected policy structures. They function as community facilitators, organizing locally customized soil and water conservation programs and strengthening social capital through participatory techniques. As capacity builders and educators, NGOs enhance soil literacy, social learning, and behavioral transformation among farmers and the broader people. They further engage in research, monitoring, and citizen science, expanding soil data coverage and boosting knowledge co-production. Ultimately, NGOs work as executors and financial mediators, translating global restoration and sustainability goals into tangible actions at the field level.

The impacts of these functions are substantial yet contingent upon circumstance. When real participation, institutional backing, and economic incentives coincide, NGO interventions increase erosion control, recover soil organic carbon, conserve biodiversity, and bolster livelihood resilience. Nonetheless, sustained efficacy is typically limited by restricted finance periods, poor integration of science and policy, prohibitive regulatory frameworks, and socio-economic hurdles to implementation. The study consistently reveals that technical solutions alone are inadequate; successful soil conservation relies on institutional integration, trust development, and alignment with local contexts.

NGOs and civil society greatly contribute to soil protection by linking local land users to international policy frameworks and integrating scientific knowledge with community activities. They aid in redefining soil from a limited agricultural resource to a public asset vital for ecosystem functioning, climate management, food security, and sustainable development. NGOs promote the social and institutional frameworks necessary for sustainable land management through advocacy, partnership creation, participatory implementation, and knowledge exchange.

The study underlines that NGO efficacy is enhanced when integrated into cohesive, multi-tiered governance frameworks that contain enforceable legislation, voluntary activities, stable funding, and coordinated innovation platforms. NGOs are most effective when incorporated into long-term soil policies, monitoring systems, and policy processes, rather than serving as solo project implementers.

Soil protection is not only a technical or governmental responsibility; it is a societal undertaking needing cooperation across all sectors and levels. NGOs and civil society are crucial actors in this process. They engage communities, apply scientific concepts, develop and share knowledge, and support enhanced governance frameworks. However, their efficacy relies on supportive institutional structures, continual investment, and real participatory techniques.

The basic thesis of this study is that strengthening soil conservation involves overcoming disconnected, short-term methods in favor of coherent, multi-tiered governance that incorporates civil society as a crucial component. With unified policies, long-term finance, and inclusive collaborations, NGOs may significantly contribute to the protection of soils as a crucial resource for food security, climate stability, biodiversity, and sustainable development.

Declaration of generative AI and AI-assisted technologies in the manuscript preparation process

During the preparation of this work, the authors used ChatGPT to write and edit the original manuscript in English used, and then used Quilbot to improve the clarity and quality of the English text by paraphrasing. After using these tools, the authors reviewed and edited the content as needed and take full responsibility for the content of the published article.

References

Abiddin, N. Z., Ibrahim, I., & Abdul Aziz, S. A. (2022). Non-governmental organisations (NGOs) and their part towards sustainable community development. *Sustainability, 14*(8), 4386. https://doi.org/10.3390/su14084386

Adhami, M., Sadeghi, S., Duttmann, R., & Sheikhmohammady, M. (2019). Best soil comanagement practices for two watersheds in Germany and Iran using game theory-based approaches. *The Science of the Total Environment, 698*, Article 134265. https://doi.org/10.1016/j.scitotenv.2019.134265

Altobelli, F., Vargas, R., Corti, G., Dazzi, C., Montanarella, L., Monteleone, A., Caon, L., Piazza, M. G., Calzolari, C., Munafò, M., & Benedetti, A. (2020). Improving soil and water conservation and ecosystem services by sustainable soil management practices: From a global to an Italian soil partnership. *Italian Journal of Agronomy, 15*(4), 1765. https://doi.org/10.4081/ija.2020.1765

Amiraslani, F., & Caiserman, A. (2018). Multi-stakeholder and multi-level interventions to tackle climate change and land degradation: The case of Iran. *Sustainability, 10*(6), 2000. https://doi.org/10.3390/su10062000

Amiraslani, F., & Dragovich, D. (2010). Combating desertification in Iran over the last 50 years: An overview of changing approaches. *Journal of Environmental Management, 92*(1), 1–13. https://doi.org/10.1016/j.jenvman.2010.08.012

Ariti, A. T., Van Vliet, J., & Verburg, P. H. (2018). What restrains Ethiopian NGOs to participate in the development of policies for natural resource management? *Environmental Science and Policy, 89*, 292–299. https://doi.org/10.1016/j.envsci.2018.08.008

Avazpour, L., Ghorbani, M., Naderi, A., Izadi, N., Azadi, H., & Yazdanparast, M. (2024). Dryland co-management in Kerman province, Iran: A dynamic analysis of social networks. *Environment, Development and Sustainability, 27*, 12259–12279. https://doi.org/10.1007/s10668-023-04346-y

Bach, E. M., Ramirez, K. S., Fraser, T. D., & Wall, D. H. (2020). Soil biodiversity integrates solutions for a sustainable future. *Sustainability, 12*(7), 2662. https://doi.org/10.3390/su12072662

Baritz, R., Hiederer, R., & Tóth, G. (2017). Voluntary guidelines for sustainable soil management: Global action for healthy soils. In *Proceedings of the global symposium on soil organic carbon (GSOC17)*. Food and Agriculture Organization of the United Nations (FAO). https://doi.org/10.1007/978-3-319-68885-5_3

Baumhardt, R. L., Stewart, B. A., & Sainju, U. M. (2015). North American soil degradation: Processes, practices, and mitigating strategies. *Sustainability, 7*(3), 2936–2960. https://doi.org/10.3390/su7032936

Bedolla-Rivera, H. I., Ponce-Mendoza, A., Mancilla-Villa, Ó. R., & Velasco-Velasco, J. (2023). Analyzing the impact of intensive agriculture on soil quality: A systematic review and global meta-analysis of quality indexes. *Agronomy, 13*(8), 2166. https://doi.org/10.3390/agronomy13082166

Bhaduri, D., Ghosh, D., & Meena, V. S. (2022). A review on effective soil health bio-indicators for ecosystem restoration and sustainability. *Frontiers in Microbiology, 13*, Article 865232. https://doi.org/10.3389/fmicb.2022.938481

Bhattacharyya, R., Bhatia, A., Ghosh, B. N., Das, T. K., Agarwal, P., Maity, A., Bisai, A., & Rai, R. K. (2023). Soil degradation and mitigation in agricultural lands in the Indian Anthropocene. *European Journal of Soil Science, 74*(3), Article e13396. https://doi.org/10.1111/ejss.13388

Bhattacharyya, R., Ghosh, B. N., Mishra, P. K., Mandal, B., Rao, C. S., Sarkar, D., Das, K., Anil, K. S., Subba Rao, A., Sen, H. S., & Lalitha, M. (2015). Soil degradation in India: Challenges and potential solutions. *Sustainability, 7*(4), 3528–3570. https://doi.org/10.3390/su7043528

Blake, W. H., Rabinovich, A., Wynants, M., Kelly, C., Nasseri, M., Ngondya, I., Patrick, A., Mtei, K., Munishi, L., Boeckx, P., Navas, A., Smith, H. G., Gilvear, D., Wilson, G., Roberts, N., & Ndakidemi, P. (2018). Soil erosion in East Africa: An interdisciplinary approach to realising pastoral land management change. *Environmental Research Letters, 13*(12), Article 124014. https://doi.org/10.1088/1748-9326/aaea8b

Borrelli, P., Robinson, D. A., Fleischer, L. R., Lugato, E., Ballabio, C., Alewell, C., Meusburger, K., Modugno, S., Schütt, B., Ferro, V., Bagarello, V., Van Oost, K., Montanarella, L., & Panagos, P. (2017). An assessment of the global impact of 21st century land use change on soil erosion. *Nature Communications, 8*, 2013. https://doi.org/10.1038/s41467-017-02142-7

Bouma, J. (2019). How to communicate soil expertise more effectively in the information age when aiming at the UN SDGs. *Soil Use and Management, 35*(1), 32–38. https://doi.org/10.1111/sum.12415

Bouma, J., & Veerman, C. P. (2022). Developing management practices in: "Living Labs" that result in healthy soils for the future, contributing to sustainable development. *Land, 11*(12), 2178. https://doi.org/10.3390/land11122178

Bouzarjomehri, K., & Javani, K. (2020). The role of Non-Governmental Organizations (NGOs) in women's empowerment and environmental protection in the rural areas (case study: rural areas of county Roshtkhar). *Journal of Research and Rural Planning, 9*, 69–82. https://doi.org/10.22067/jrrp.v9i2.82728

Broeckhoven, N., Desta, G., Gidey, M., Reda, K., Townsend, D., & Verschuuren, J. (2020). CSOs in sustainable development in Ethiopia: Past practices and new trajectories. *African Journal of Legal Studies*. https://doi.org/10.1163/17087384-12340063

Busscher, N., Vanclay, F., & Parra, C. (2019). Reflections on how state–civil society collaborations play out in the context of land grabbing in Argentina. *Land, 8*(8), 116. https://doi.org/10.3390/land8080116

Cárceles Rodríguez, B., Durán-Zuazo, V. H., Soriano Rodríguez, M., Carvajal Sánchez, R., & Francia Martínez, J. R. (2022). Conservation agriculture as a sustainable system for soil health: A review. *Soil Systems, 6*(4), 87. https://doi.org/10.3390/soilsystems6040087

Caggiano, M., Zingg, C., Abiven, S., Bampa, F., Bünemann, E. K., Debeljak, M., ... Van den Brink, A. (2025). *Training and citizen engagement to tackle soil health challenges*. Open Access Government.

Charvát, K., Šmejkal, J., Horák, P., Kollerová, M., Horáková, S., & Renault, P. (2025). Citizen science for soil monitoring and protection in Europe: Insights from the PREPSOIL project under the European soil mission. *Sustainability, 17*(11), 5042. https://doi.org/10.3390/su17115042

Chen, J., Chen, J.-Z., Tan, M.-Z., & Gong, Z.-T. (2002). Soil degradation: A global problem endangering sustainable development. *Journal of Geographical Sciences, 12*, 243–252. https://doi.org/10.1007/BF02837480

Costantini, E. A., Baritz, R., & Montanarella, L. (2025). Recent developments in global soil health protection policies. *Soil Security, 14*, Article 100322. https://doi.org/10.1016/j.soisec.2025.100191

Dangiso, Y., & Wolka, K. (2023). Assessing sustainable management practices of expert-introduced physical soil and water conservation measures. *Environmental Challenges, 13*. https://doi.org/10.1016/j.envc.2023.100780. Article 100780.

Davila, F., Jacobs, B., Nadeem, F., Kelly, R., & Kurimoto, N. (2024). Finding climate smart agriculture in civil-society initiatives. *Mitigation and Adaptation Strategies for Global Change, 29*(2), 58. https://doi.org/10.1007/s11027-024-10108-6

Dazzi, C., Cornelis, W., Constantini, A. C., Dumitru, M., Fullen, A., Gabriels, D., Kasparinskis, R., Kertész, A., Lo Papa, G., Pérès, G., Rickson, J., Rubio, J. L., Sholten, T., Theocharopoulos, S., & Vasenev, I. (2019). The contribution of the European Society for Soil Conservation (ESSC) to scientific knowledge, education and sustainability. *International Soil and Water Conservation Research, 7*(1), 102–107. https://doi.org/10.1016/j.iswcr.2018.11.003

Delgado-Baquerizo, M., Eldridge, D. J., Liu, Y.-R., Liu, Z.-W., Coleine, C., & Trivedi, P. (2025). Soil biodiversity and function under global change. *PLoS Biology, 23*(3), Article e3003093. https://doi.org/10.1371/journal.pbio.3003093

Delgado-Baquerizo, M., Reich, P. B., Trivedi, C., Eldridge, D. J., Abades, S., Alfaro, F. D., ... Singh, B. K. (2020). Multiple elements of soil biodiversity drive ecosystem functions across biomes. *Nature Ecology and Evolution, 4*(2), 210–220. https://doi.org/10.1038/s41559-019-1084-y

Delponte, I., & co-authors. (2023). Civil society mobilizations shaping landscape in Genoa and Girona areas: Results and lessons learnt from the savingscapes project. *Land, 12*(4), 922. https://doi.org/10.3390/land12040922

Djillo, S. C., Ayalew, D. A., & Zeleke, T. T. (2024). Assessing soil erosion and farmers' decision of reducing erosion for sustainable soil and water conservation in Burji Woreda, southern Ethiopia. *Scientific Reports, 14*, 11234. https://doi.org/10.1038/s41598-024-59076-6

Du, X., Jian, J., Du, C., & Stewart, R. D. (2022). Conservation management decreases surface runoff and soil erosion. *International Soil and Water Conservation Research, 10*(2), 188–196. https://doi.org/10.1016/j.iswcr.2021.08.001

Dumitrascu, M., Ştefănescu, S. L., Vidican, R., & Untesu, A. A. (2023). A research–civil society joint approach for soil conservation in high natural value areas. *Bulletin of University of Agricultural Sciences and Veterinary Medicine Cluj-Napoca - Agriculture, 80*(2). https://doi.org/10.15835/buasvmcn-ag:2023.0020

Ebabu, K., Verheijen, F. G. A., Haregeweyn, N., Poesen, J., Zemadim, B., Nyssen, J., Adimassu, Z., Taye, G., Vanmaercke, M., Demissie, B., Tolla, M., Nigussie, A., Kassa, A. K., Tilahun, S. A., & Meshesha, D. T. (2022). Global analysis of cover management and support practice factors that control soil erosion and conservation. *International Soil and Water Conservation Research, 10*(2), 161–176. https://doi.org/10.1016/j.iswcr.2021.12.002

Erdoğan, H., Havlicek, E., Dazzi, C., Montanarella, L., Van Liedekerke, M., Vrščaj, B., Krasilnikov, P., Khasankhanova, G., & Vargas, R. (2021). Soil conservation and sustainable development goals (SDGs) achievement in Europe and Central Asia: Which role for the European soil partnership? *International Soil and Water Conservation Research, 9*(3), 95–106. https://doi.org/10.1016/j.iswcr.2021.02.003

Evans, D. L., Janes-Bassett, V., Borrelli, P., Chenu, C., Ferreira, C. S. S., Griffiths, R. I., Kalantari, Z., Keesstra, S., Lal, R., Panagos, P., Robinson, D. A., Seifollahi-Aghmiuni, S., Smith, P., Steenhuis, T. S., Thomas, A. D., & Visser, S. (2021). Sustainable futures over the next decade are rooted in soil science. *European Journal of Soil Science, 72*(4), 1401–1427. https://doi.org/10.1111/ejss.13145

Ferreira, C. S. S., Seifollahi-Aghmiuni, S., Destouni, G., Ghajarnia, N., & Kalantari, Z. (2022). Soil degradation in the European mediterranean region: Processes, status and consequences. *Science of the Total Environment, 805*, Article 150106. https://doi.org/10.1016/j.scitotenv.2021.150106

Fontaine, S., Angers, D. A., Barré, P., Bloor, J. M. G., Freschet, G. T., Guenet, B., Majdi, H., Mary, B., Reichstein, M., & Wardle, D. A. (2023). Plant–soil synchrony in nutrient cycles: Learning from ecosystems to design sustainable agrosystems. *Global Change Biology, 29*(3), 541–559. https://doi.org/10.1111/gcb.17034

Francaviglia, R., Almagro, M., & Vicente-Vicente, J. L. (2023). Conservation agriculture and soil organic carbon: Principles, processes, practices and policy options. *Soil Systems, 7*(1), 17. https://doi.org/10.3390/soilsystems7010017

Futa, B., Kowalska, A., & Nowak, P. (2024). Innovative soil management strategies for sustainable agriculture. *Sustainability, 16*(21), 9481. https://doi.org/10.3390/su16219481

Glatron, S., Auclerc, A., Descarpentries, C., Ertlen, D., Imfeld, G., Franck-Neumann, F., Philippot, V., Vincent, Q., & Ponton, R. (2025). *Working together to observe, understand, and improve living urban soils: Solenville, an interdisciplinary participatory research program in Strasbourg. ARPHA conference abstracts*. https://doi.org/10.3897/aca.8.e154673

Gomiero, T. (2016). Soil degradation, land scarcity and food security: Reviewing a complex challenge. *Sustainability, 8*(3), 281. https://doi.org/10.3390/su8030281

Gupta, G. (2019). Land degradation and challenges of food security. *Review of European Studies, 11*(3), 52–61. https://doi.org/10.5539/res.v11n1p63

Hammadi, B. (2025). Non-Governmental Organizations (NGOs) and civil society in Oman. *Civil Szemle*. https://doi.org/10.62560/csz.2025.04.3

Haregeweyn, N., Tsunekawa, A., Nyssen, J., Poesen, J., Tsubo, M., Tsegaye Meshesha, D., Schütt, B., Adgo, E., & Tegegne, F. (2015). Soil erosion and conservation in Ethiopia: A review. *Progress in Physical Geography, 39* (6), 750–774. https://doi.org/10.1177/0309133315598725

Head, J. S., Crockatt, M. E., Didarali, Z., Woodward, M.-J., & Emmett, B. A. (2020). The role of citizen science in meeting SDG targets around soil health. *Sustainability, 12*(24), Article 10254. https://doi.org/10.3390/su122410254

Heller, O., Bene, C. D., Nino, P., Huyghebaert, B., Arlauskienė, A., Castanheira, N. L., Higgins, S., Horel, A., Kir, A., Kizeková, M., Lacoste, M., Munkholm, L. J., O'Sullivan, L., Radzikowski, P., Rodríguez-Cruz, M. S., Sandén, T., Šarūnaitė, L., Seidel, F., Spiegel, H., … Vanwindekens, F. (2024). Towards enhanced adoption of soil-improving management practices in Europe. *European Journal of Soil Science, 75*(2), Article e13483. https://doi.org/10.1111/ejss.13483

Henry, G., et al. (2022). Role of civil society organizations in preserving environment and human rights. *Sinai Journal of Applied Sciences*. https://doi.org/10.21608/sinjas.2022.148306.1123

Honeycutt, C. W., Morgan, C. L. S., Elias, P., Doane, M., Mesko, J., Myers, R., Odom, L., Moebius-Clune, B., & Nichols, R. (2020). Soil health: Model programs in the USA. *Frontiers of Agricultural Science and Engineering, 7*(3), 356–361. https://doi.org/10.15302/J-FASE-2020340

Hou, D., Bolan, N. S., Tsang, D. C. W., Kirkham, M. B., & O'Connor, D. (2020). Sustainable soil use and management: An interdisciplinary and systematic approach. *Science of the Total Environment, 729*, Article 138961. https://doi.org/10.1016/j.scitotenv.2020.138961

Jie, C., Chen, J., Tan, M., & Gong, Z. (2002). Soil degradation: A global problem endangering sustainable development. *Journal of Geographical Sciences, 12*(2), 243–252. https://doi.org/10.1007/BF02837480

Jorge, N., Clark, J., Cárdenas, M., Geoghegan, H., & Shannon, V. (2021). Measuring soil colour to estimate soil organic carbon using a large-scale citizen science-based approach. *Sustainability, 13*(19), Article 11029. https://doi.org/10.3390/su131911029

Karlen, D. L., & Rice, C. W. (2015). Soil degradation: Will humankind ever learn? *Sustainability, 7*(9), 12490–12501. https://doi.org/10.3390/su70912490

Kazanskaia, A. (2025). Case studies of collaborative strategies in the non-profit sector: Global, regional, and local perspectives. *NEYA Global Journal of Non-Profit and Social Impact*. https://doi.org/10.64357/neya-gjnps-ntallstrscimp-07

Keesstra, S. D., Nunes, J. P., Novara, A., Finger, D., Avelar, D., Kalantari, Z., & Cerdà, A. (2018). The superior effect of nature-based solutions in land management for enhancing ecosystem services. *Science of the Total Environment, 610–611*, 997–1009. https://doi.org/10.1016/j.scitotenv.2017.08.077

Kelly, C., Wynants, M., Munishi, L. K., Nasseri, M., Patrick, A., Mtei, K. M., Mkilema, F., Rabinovich, A., Gilvear, D., Wilson, G., Blake, W., & Ndaki, P. A. (2020). Mind the gap': Reconnecting local actions and multi-level policies to bridge the governance gap. An example of soil erosion action from East Africa. *Land, 9* (10), 352. https://doi.org/10.3390/land9100352

Kihara, J., Bolo, P., Kinyua, M., Nyawira, S. S., & Sommer, R. (2020). Soil health and ecosystem services: Lessons from Sub-Sahara Africa (SSA). *Geoderma, 370*, Article 114342. https://doi.org/10.1016/j.geoderma.2020.114342

Kimmell, L., Fagan, J., & Havrilla, C. (2023). Soil restoration increases soil health across global drylands: A meta-analysis. *Journal of Applied Ecology*. https://doi.org/10.1111/1365-2664.14459

Klingner, D. (2021). Non-governmental organizations. *Public Integrity, 23*(4), 345–348. https://doi.org/10.1080/10999922.2021.1911151

Kopittke, P. M., Berhe, A. A., Carrillo, Y., Cavagnaro, T. R., Chen, D., Chen, Q. L., … Minasny, B. (2022). Ensuring planetary survival: The centrality of organic carbon in balancing the multifunctional nature of soils. *Critical Reviews in Environmental Science and Technology, 52*(23), 4308–4324. https://doi.org/10.1080/1080/10643389.2021.2024484

Kopittke, P. M., Menzies, N. W., Wang, P., McKenna, B. A., & Lombi, E. (2019). Soil and the intensification of agriculture for global food security. *Environment International, 132*, Article 105078. https://doi.org/10.1016/j.envint.2019.105078

Kopittke, P. M., Menzies, N. W., Wang, P., McKenna, B. A., & Lombi, E. (2021). The role of soil in defining planetary boundaries and the safe operating space for humanity. *Environment International, 146*, Article 106245. https://doi.org/10.1016/j.envint.2020.106245

Kumar, N. (2023). Soil degradation and its causes. *International Journal for Multidisciplinary Research, 5*(1), 45–52. https://doi.org/10.36948/ijfmr.2023.v05i01.1443

Kumar, P., Singh, R., & Yadav, A. (2024). Influence of soil on crop performance and ecosystem services: A review. *Journal of Experimental Agriculture International, 46*(8), 640–650. https://doi.org/10.9734/jeai/2024/v46i82746

Löbmann, M., Fiener, P., Zarei, M., & Schneider, K. (2022). Systems knowledge for sustainable soil and land management. *Science of the Total Environment, 825*, Article 153891. https://doi.org/10.1016/j.scitotenv.2022.153389

Lal, R. (2015). Restoring soil quality to mitigate soil degradation. *Sustainability, 7*(5), 5875–5895. https://doi.org/10.3390/su7055875

Lal, R. (2024). Soil degradation effects on human malnutrition and under-nutrition. *Medical Research Archives, 12*(1), 1–20. https://doi.org/10.18103/mra.v12i10.5753

Lal, R., Bouma, J., Brevik, E., Dawson, L., Field, D., Glaser, B., Herrmann, R., Lyu, H., Machmuller, M. B., Mol, S. T., Pelter, A. S., Rumpel, C., Song, X., Steffens, M., Tanveer, M., Wall, D. H., & Zhang, R. (2021). Soils and sustainable development goals of the United Nations: An international union of soil sciences perspective. *Geoderma Regional, 25*, Article e00398. https://doi.org/10.1016/j.geodrs.2021.e00398

Lehmann, J., Bossio, D. A., Kögel-Knabner, I., & Rillig, M. C. (2020). The concept and future prospects of soil health. *Nature Reviews Earth and Environment, 1*, 544–553. https://doi.org/10.1038/s43017-020-0080-8

Lewis, D. (2023). Non-governmental organizations, definitions, and history. In R. A. List, H. K. Anheier, & S. Toepler (Eds.), *International encyclopedia of civil society*. Cham: Springer. https://doi.org/10.1007/978-3-319-99675-2_3-1

Li, T., Liu, Y., Lin, S., Liu, Y., & Xie, Y. (2019). Soil pollution management in China: A brief introduction. *Sustainability, 11*(3), 556. https://doi.org/10.3390/su11030556

Martinho, V. J. P. D., Ferreira, A. J. D., Cunha, C., Pereira, J. L. da S., Sánchez-Carreira, M. del C., Castanheira, N. L., & Brito Ramos, T. (2024). Soil legislation and policies: Bibliometric analysis, systematic review and quantitative approaches with an emphasis on the specific cases of the European Union and Portugal. *Heliyon, 10*(3), Article e25431. https://doi.org/10.1016/j.heliyon.2024.e25431

Mason, E., Gascuel-Odoux, C., Aldrian, U., Sun, H., Miloczki, J., Götzinger, S., Burton, V. J., Rienks, F., Di Lonardo, S., & Sandén, T. (2024). Participatory soil citizen science: An unexploited resource for European soil research. *European Journal of Soil Science, 75*(2), Article e13470. https://doi.org/10.1111/ejss.13470

Mesele, S. A., Mechri, M., Okon, M. A., Isimikalu, T. O., Wassif, O. M., Asamoah, E., Al-Shammari, D., Bashir, S., & Khurshid, C. (2025). Current problems leading to soil degradation in Africa: Raising awareness and finding potential solutions. *European Journal of Soil Science, 76*(1), Article e70069. https://doi.org/10.1111/ejss.70069

Montanarella, L. (2015). The global soil partnership. *IOP Conference Series: Earth and Environmental Science, 25*(1), Article 012001. https://doi.org/10.1088/1755-1315/25/1/012001

Montanarella, L., & Panagos, P. (2020). The relevance of sustainable soil management within the European Green Deal. *Land Use Policy, 100*, Article 104950. https://doi.org/10.1016/j.landusepol.2020.104950

Montanarella, L., Pennock, D. J., McKenzie, N., Badraoui, M., Chude, V., Baptista, I., Mamo, T., Yemefack, M., Singh Aulakh, M., & Zhang, G.-L. (2012). Global governance of soil resources as a necessary condition for sustainable development. *Current Opinion in Environmental Sustainability, 4*(5), 559–564. https://doi.org/10.1016/j.cosust.2012.06.007

Moosavi, M., Koosha, J., Abbas, P., & Bavand, D. (2015). Non-Governmental actors and the evolution of international environmental law (an overview of the NGOs in Iran). *Mediterranean Journal of Social Sciences, 6*, 46. https://doi.org/10.5901/mjss.2015.v6n6s6p46

Muche, A., Ketsela, Y., & Ali, B. (2024). Assessing the effectiveness of integrated watershed management practices and suggesting innovative strategies in southern Ethiopia. *Heliyon, 10*. https://doi.org/10.1016/j.heliyon.2024.e38619

Narayanan, M., & Ma, Y. (2024). Recent progress on conservation and restoration of soil fertility for horticulture. *Chemosphere, 356*, Article 142599. https://doi.org/10.1016/j.chemosphere.2024.142599

Nolan, M., Stanton, K. J., Evans, K., Pym, L., Kaufman, B., & Duley, E. (2021). From the ground up: Prioritizing soil at the forefront of ecological restoration. *Restoration Ecology, 29*(S2), Article e13453. https://doi.org/10.1111/rec.13453

Omer, E., Szlatenyi, D., Csenki, S., Alrwashdeh, J., Czakó, I., & Láng, V. (2024). Farming practice variability and its implications for soil health in agriculture: A review. *Agriculture, 14*(12), 2114. https://doi.org/10.3390/agriculture14122114

Paltseva, A. A. (2025). Participatory science in urban soil research: A framework for overcoming challenges and expanding public engagement. *iScience, 28*, Article 112361. https://doi.org/10.1016/j.isci.2025.112361

Panagiotakis, I., & Dermatas, D. (2022). New European Union soil strategy: A potential worldwide tool for sustainable waste management and circular economy. *Waste Management and Research: The Journal for a Sustainable Circular Economy, 40*(3), 245–247. https://doi.org/10.1177/0734242X221079114

Pandian, K., Firnass, M. M. R. A., Govindaraj, M., Arunachalam, P., Abishek, R., Manimaran, G., Rahman, R. A., & Varanasi, S. T. (2024). Synergistic conservation approaches for nurturing soil, food security and human health towards sustainable development goals. *Journal of Hazardous Materials Advances.* , Article 100479. https://doi.org/10.1016/j.hazadv.2024.100479

Peake, L., & Robb, C. (2021). Saving the ground beneath our feet: Establishing priorities and criteria for governing soil use and protection. *Royal Society Open Science, 8*(11), Article 201994. https://doi.org/10.1098/rsos.201994

Pereira, P., Bogunović, I., Muñoz-Rojas, M., & Brevik, E. C. (2018). Soil ecosystem services, sustainability, valuation and management. *Current Opinion in Environmental Science & Health, 5*, 7–13. https://doi.org/10.1016/j.coesh.2017.12.003

Peterson, R. B., Kapiyo, R., Campbell, E., & Nyabua, P. O. (2018). Gully rehabilitation trusts: Fighting soil erosion through community participation in Western Kenya. *Journal of Rural Studies, 58*, 122–134. https://doi.org/10.1016/j.jrurstud.2017.12.028

Plaas, E., Meyer-Wolfarth, F., Banse, M., Bengtsson, J., Bergmann, H., Faber, J., Potthoff, M., Runge, T., Schrader, S., & Taylor, A. (2019). Towards valuation of biodiversity in agricultural soils: A case for earthworms. *Ecological Economics, 159*, 291–300. https://doi.org/10.1016/j.ecolecon.2019.02.003

Powlson, D., Gregory, P. J., Whalley, W. R., Quinton, J., Hopkins, D., Whitmore, A., Hirsch, P., & Goulding, K. (2011). Soil management in relation to sustainable agriculture and ecosystem services. *Food Policy, 36*(Suppl. 1), S72–S87. https://doi.org/10.1016/j.foodpol.2010.11.025

Prăvălie, R. (2021). Exploring the multiple land degradation pathways across the planet. *Earth-Science Reviews, 220*, 103689. https://doi.org/10.1016/j.earscirev.2021.103689

Price, H., Mauro, S., High, K., Ramírez-Andreotta, M., Ureta, S., Walls, D., Xu, C., & Kinchy, A. (2024). Soil publics: Regenerating relations with urban soils through citizen science. *Local Environment, 29*, 480–494. https://doi.org/10.1080/13549839.2023.2300977

Pulido-Moncada, M., Thorsøe, M., Miranda Vélez, J. F., Graversgaard, M., & Munkholm, L. (2025). Soil health and challenges to sustainable soil management in Denmark: Stakeholder perceptions. *European Journal of Soil Science, 76*. https://doi.org/10.1111/ejss.70038

Rajbanshi, J., Das, S., & Paul, R. (2023). Quantification of the effects of conservation practices on surface runoff and soil erosion in croplands and their trade-off: A meta-analysis. *Science of the Total Environment, 864*, Article 161015. https://doi.org/10.1016/j.scitotenv.2022.161015

Rapiya, M., Truter, W., & Ramoelo, A. (2024). The integration of land restoration and biodiversity conservation practices in sustainable food systems of Africa: A systematic review. *Sustainability, 16*(20), 8951. https://doi.org/10.3390/su16208951

Rathi, A., Kumar, P., Nangla, S., Sharma, S., & Sharma, S. (2024). Soil restoration strategies for sustaining soil productivity: A review. *Asian Research Journal of Agriculture, 17*(1), 33–48. https://doi.org/10.9734/arja/2024/v17i1408

Reckien, D., Salvia, M., Heidrich, O., Church, J., Pietrapertosa, F., De Gregorio-Hurtado, S., D'Alonzo, V., Foley, A., Simoes, S., Lorencová, E., Orru, H., Orru, K., Wejs, A., Flacke, J., Olazabal, M., Geneletti, D., Feliu, E., Vasilie, S., Nador, C., … Dawson, R. (2018). How are cities planning to respond to climate change? Assessment of local climate plans from 885 cities in the EU-28. *Journal of Cleaner Production*. https://doi.org/10.1016/j.jclepro.2018.03.220

Romero-Brito, T. P., Buckley, R. C., & Byrne, J. (2016). NGO partnerships in using ecotourism for conservation: Systematic review and meta-analysis. *PLoS ONE, 11*(11). https://doi.org/10.1371/journal.pone.0166919. Article e0166919.

Ronchi, S., Salata, S., Arcidiacono, A., Piroli, E., & Montanarella, L. (2019). Policy instruments for soil protection among the EU member states: A comparative analysis. *Land Use Policy, 82*, 763–780. https://doi.org/10.1016/j.landusepol.2019.01.017

Salvia, M., Pietrapertosa, F., D'Alonzo, V., Maestosi, P., Simoes, S., & Reckien, D. (2023). Key dimensions of cities' engagement in the transition to climate neutrality. *Journal of Environmental Management, 344*, Article 118519. https://doi.org/10.1016/j.jenvman.2023.118519

Salvia, M., Reckien, D., Pietrapertosa, F., Eckersley, P., Spyridaki, N.-A., Krook-Riekkola, A., Olazabal, M., De Gregorio Hurtado, S., Simões, S. G., & Geneletti, D. (2021a). Will climate mitigation ambitions lead to carbon neutrality? An analysis of the local-level plans of 327 cities in the EU. *Renewable and Sustainable Energy Reviews, 135*, Article 110253. https://doi.org/10.1016/j.rser.2020.110253

Salvia, M., Reckien, D., Pietrapertosa, F., Eckersley, P., Spyridaki, N., Krook-Riekkola, A., Olazabal, M., Hurtado, S., Simoes, S., Geneletti, D., Viguié, V., Fokaides, P., Ioannou, B., Flamos, A., Csete, M., Buzási, A., Orru, H., Boer, C., Rižnar, K., … Heidrich, O. (2021b). Data for: Climate mitigation ambition towards carbon neutrality? An analysis of local-level plans of 327 cities in the EU [data set]. *Mendeley Data*, (1)https://doi.org/10.17632/65h7t7sdd7

Shah, A. N., Tanveer, M., Shahzad, B., Yang, G., Abbas, G., Ashraf, U., Wang, W., Sikandar, S., Maqsood, S., & Abbas, S. (2017). Soil compaction effects on soil health and crop productivity: An overview. *Environmental Science and Pollution Research, 24*, 10056–10067. https://doi.org/10.1007/s11356-017-8421-y

Shaheb, M. R., Venkatesh, R., & Shearer, S. A. (2021). A review on the effect of soil compaction and its management for sustainable crop production. *Journal of Biosystems Engineering, 46*, 417–439. https://doi.org/10.1007/s42853-021-00117-7

Silva, T. P., Keesstra, S., Ribeiro, B. T., et al. (2023). Best management practices to reduce soil erosion and change water balance components in watersheds under grain and dairy production. *International Soil and Water Conservation Research, 11*(3), 373–388. https://doi.org/10.1016/j.iswcr.2023.06.003

Sonderegger, T., & Pfister, S. (2021). Global assessment of agricultural productivity losses from soil compaction and water erosion. *Environmental Science and Technology, 55*(18), 12162–12171. https://doi.org/10.1021/acs.est.1c03774

Țopa, D.-C., Căpșună, S., Calistru, A.-E., & Ailincăi, C. (2025). Sustainable practices for enhancing soil health and crop quality in modern agriculture: A review. *Agriculture, 15*(9), 998. https://doi.org/10.3390/agriculture15090998

Tabrizi, K. (2020). Environmental capacity building program for the residents of the Zayandeh Rud Basin. In *Sustainable management of water resources in the Zayandeh Rud Basin*. Springer. https://doi.org/10.1007/978-3-030-50684-1_13

Tahat, M. M., Alananbeh, K. M., Othman, Y. A., & Leskovar, D. I. (2020). Soil health and sustainable agriculture. *Sustainability, 12*(12), 4859. https://doi.org/10.3390/su12124859

Taskin, E., Rastorgueva, N., Foley, L., Noto, R., Borruso, L., Cesco, S., & Mimmo, T. (2025). Living labs for sustainable soil management: A systematic review of global practices and perspectives. *Journal of Soils and Sediments, 25*, 3342–3359. https://doi.org/10.1007/s11368-025-04100-z

Telo da Gama, J. (2023). The role of soils in sustainability, climate change, and ecosystem services: Challenges and opportunities. *Ecologies, 4*(3), 552–567. https://doi.org/10.3390/ecologies4030036

Tetteh, F. M. (2021). Challenges to soil protection and sustainable management in Africa. In H. Ginzky, et al. (Eds.), *International Yearbook of Soil Law and Policy 2019* (Vol 2019, pp. 1–15). Cham: Springer. https://doi.org/10.1007/978-3-030-52317-6_1

Tetteh, S., Amoabeng-Nimako, S., Leeuwis, C., Freeman, C., Maat, H., Macnaghten, P., Doukkali, R., Bindraban, P., Ingenbleek, P., Dittoh, S., & Gouzaye, A. (2025). Nourishing the nexus: System analysis of the enablers and barriers in the Ghanaian soil health system. *Nutrient Cycling in Agroecosystems, 131*, 651–672. https://doi.org/10.1007/s10705-025-10439-z

Trivedi, P., Delgado-Baquerizo, M., Anderson, I. C., & Singh, B. K. (2016). Response of soil properties and microbial communities to agriculture: Implications for primary productivity and soil health indicators. *Frontiers in Plant Science, 7*, 990. https://doi.org/10.3389/fpls.2016.00990

Vanino, S., Pirelli, T., Di Bene, C., Bøe, F., Castanheira, N., Chenu, C., Cornu, S., Feiza, V., Fornara, D., Heller, O., Kasparinskis, R., Keesstra, S., Lasorella, M. V., Madenoglu, S., Meurer, K., O'Sullivan, L., Peter, N., Piccini, C., Siebielec, G., & Farina, R. (2023). Barriers and opportunities of soil knowledge to address soil challenges: Stakeholders' perspectives across Europe. *Journal of Environmental Management, 325*, Article 116581. https://doi.org/10.1016/j.jenvman.2022.116581

Winowiecki, L., Linden, H., Alexander, S., Bargués Tobella, A., Campari, J., Christensen, C., Choudhury, D., Duijn, H., Fleckenstein, M., Harz-Pitre, Y., Hussain, Z., Kakade, B., Kamau, D., Luu, P., Magaju, C., Makiyi, V., Mbabazi, P., Morgan, C., Snapp, S., & Lal, R. (2025). Multistakeholder engagement to scale soil health globally: The coalition of action 4 soil health. *European Journal of Soil Science, 76*. https://doi.org/10.1111/ejss.70128

Woolf, D., Solomon, D., & Lehmann, J. (2018). Land restoration in food security programmes: Synergies with climate change mitigation. *Climate Policy, 18*, 1260–1270. https://doi.org/10.1080/14693062.2018.1427537

Workie, M., & Teku, D. (2025). Assessing soil erosion hotspots and land degradation extent in Beshilo watershed, Northeastern Ethiopia. *Journal of Agricultural Sciences*.

Xiao, L., Kuhn, N. J., Zhao, R., & Cao, L. (2021). Net effects of conservation agriculture principles on sustainable land use: A synthesis. *Global Change Biology, 27*(24), 6321–6330. https://doi.org/10.1111/gcb.15906

Xiong, M., Sun, R., & Chen, L. (2018). Effects of soil conservation techniques on water erosion control: A global analysis. *Science of the Total Environment, 645*, 753–760. https://doi.org/10.1016/j.scitotenv.2018.07.124

Yin, C., & Pereira, P. (2021). Soil conservation service underpins sustainable development goals. *Global Ecology and Conservation, 33*, Article e01974. https://doi.org/10.1016/j.gecco.2021.e01974

Zeiss, R., Eisenhauer, N., Orgiazzi, A., Rillig, M., Buscot, F., Jones, A., Lehmann, A., Reitz, T., Smith, L., & Guerra, C. (2022). Challenges of and opportunities for protecting European soil biodiversity. *Conservation Biology, 36*(5), Article e13930. https://doi.org/10.1111/cobi.13930

Zikargae, M., Woldearegay, A., & Skjerdal, T. (2022). Empowering rural society through non-formal environmental education: An empirical study of environment and forest development community projects in Ethiopia. *Heliyon, 8*(3), Article e09127. https://doi.org/10.1016/j.heliyon.2022.e09127

CHAPTER

14

Groundwater governance gaps as a driver of land degradation in arid and semiarid agricultural systems of Iran

Raziyeh Shamshirgaran[1], Parviz Rezvani Moghaddam[2], Moein Tosan[3], Afshin Shayeghi[4] and Ali Nasirian[1]

[1]*Department of Water Resources Engineering and Management, University of Birjand, Birjand, Iran;* [2]*Faculty of Agriculture, Ferdowsi University of Mashhad, Mashhad, Iran;* [3]*Department of Water Science and Engineering, Ferdowsi University of Mashhad, Mashhad, Iran;* [4]*Department of Geography & Environmental Sustainability, University of Oklahoma, Norman, OK, United States*

1. Introduction

For arid and semiarid zones, groundwater stands as the critical basis for water security, supporting a full range of essential activities from potable consumption to agricultural and industrial operations (Akbarpour et al., 2024). The ecological status of soil in these water-scarce areas is intrinsically tied to the sustainability of groundwater management (Acevedo et al., 2022; Quandt et al., 2023). Nevertheless, intensive and uncontrolled pumping has led to a steep decline in water table elevations (Shamshirgaran et al., 2025), precipitating severe environmental harm. This includes pervasive soil salinization (Feng et al., 2025), significant land subsidence (Li et al., 2023; Sajjad et al., 2023), and a measurable reduction in microbial biodiversity all definitive markers of degraded soil health. Despite extensive technical endeavors, such as aquifer balancing plans, aimed at sustainable resource control (Scanlon et al., 2023), empirical data and several studies demonstrate that these initiatives have achieved only minimal success (Farzaneh et al., 2019; Hashemi et al., 2023). The persistent nature of this issue does not stem from a deficiency in technology or capital, but rather from a substantial perception gap among key stakeholder groups (Bonì et al., 2022; Yadav et al., 2025). The ongoing challenge is rooted in a fundamental perceptual divide, which is articulated through three separate and distinct viewpoints. For instance, the academic community invariably frames the crisis using the lens of ecological models and long-term sustainability (Xue et al., 2024).

In sharp divergence from other perspectives, government experts and state institutions, guided by a managerial and technical rationale, primarily focus on executing control-oriented solutions. These governmental measures incorporate technical advances, such as calibrating irrigation schedules using soil moisture sensors, alongside mandates for infrastructure, including the deployment of smart water meters (Gholamrezai et al., 2023; Nourani et al., 2025) and the mandatory decommissioning of illegal

The Soil Nexus. https://doi.org/10.1016/B978-0-443-45286-4.00018-3

wells (Haji et al., 2024). Conversely, the farmers, who are the central end users, assess the water scarcity issue predominantly through the lens of livelihoods and immediate economic viability. Their focus centers on operational challenges, such as fine-tuning irrigation levels to secure crop yields (Rezvani Moghaddam et al., 2022). When management policies neglect this reality, the typical outcomes are resistance, apathy, and the eventual failure of technically sound projects. This conflict transcends a mere difference of opinion, exposing a more profound, structural "governance gap" (Mahdavi, 2024; Nouri et al., 2023). Fundamentally, the core issue is a system that fails to successfully integrate the distinct logics: the scientific logic of academics, the managerial logic of experts, and the practical logic of farmers. While prior research has emphasized weak regulation or inadequate participation (Farzaneh et al., 2019; Samani, 2020), the governance gap framework utilized herein reinterprets these failures as secondary symptoms of a broader fragmentation (du Plessis, 2023; Rodzynek & Houston, 2021). It's a system where academic knowledge stays academic, policies are handed down from above (Kamal et al., 2021), and local wisdom is overlooked (Hidayati et al., 2025; Shunglu et al., 2022). The path forward, then, isn't more technology or tighter controls. It's a fundamental redesign of governance to connect people and ideas through genuine dialog and shared decision-making (Koebele et al., 2024; Nagata et al., 2022). The urgency of this is clear in places like South Khorasan Province, Iran. With 31 of its 44 basins in critical condition and an average water level drop of 0.21 m per year (Havangi & Hhozeymehnezhad, 2023), the province has turned to aquifer balancing projects to fight illegal pumping, improve irrigation, and recharge its groundwater. Despite the efforts of the South Khorasan Regional Water Company, which earned a top national ranking in 2021 for its implementation efforts, a comprehensive evaluation of the success of these projects from the perspective of diverse stakeholders especially farmers has been largely neglected.

This chapter aims to elucidate how the divergent attitudes of stakeholders in groundwater management cultivate a significant governance gap. Subsequently, this study reveals the intrinsic link between this governance gap and the ineffective implementation of technical solutions. It is argued that this institutional disconnect is a key driver of land degradation in Iran's arid agricultural systems.

2. Materials and methods

2.1 Study area

The Boshrouyeh study area, spanning an expanse of 7900 km^2, is located in South Khorasan Province. It is bordered by Bardaskan and Bajestan Counties to the north, Tabas County to the west and south, and Ferdows County to the east. The region is situated within Iran's central desert, between longitudes 57 degrees3′E and 58 degrees12′E and latitudes 33 degrees23′N and 34 degrees18′N (Fig. 14.1 and Table 14.1). The heart of the study area is the main Boshrouyeh aquifer, which covers about 620 km^2. Here, the water table depth varies dramatically, ranging from as shallow as 13 m in the south to 100 m in the north, with an average depth estimated at 48 m (Havangi & Hhozeymehnezhad, 2023; Feizi & Tosan, 2016). For a visual overview of the region, Fig. 14.1 maps Boshrouyeh County's location, the aquifer's boundaries, and local climatic conditions. It is important to note that due to the high aridity index and potential evaporation rates far exceeding precipitation (less than 120 mm/year), natural recharge from rainfall is negligible compared to the volume of groundwater abstraction. Consequently, the groundwater dynamics in this basin are driven primarily by anthropogenic pumping rather than short-term climatic variability.

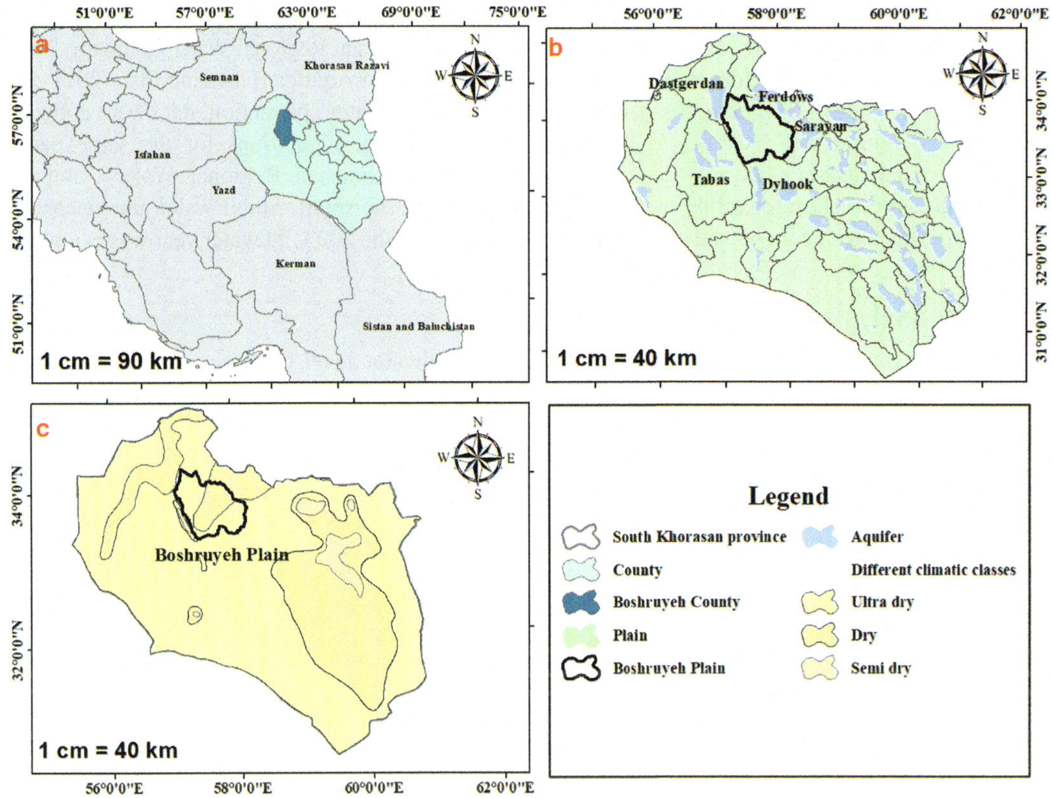

FIGURE 14.1

a) Location of Boshruyeh county in South Khorasan province, Iran, b) The Boshruyeh plain (study area) and regional aquifers, c) Climatic classification of the Boshruyeh plain based on the De Martonne method. ***Map lines delineate study areas and do not necessarily depict accepted national boundaries.***

Table 14.1 Geographical and hydrogeological characteristics of the Boshruyeh plain aquifer.

Types of geological formations	Average rainfall (mm)	Geographical location (Grade)	The state of exploitation of the plain	Aquifer area (Km^2)	Total area (Km^2)	Range name
Quaternary, Tertiary, and Miocene	119.1	53.7–58.12 33.23–34.18	Prohibited area	620	7900	4720

2.2 Data used

Two types of data were utilized in this research:

Groundwater level data: To analyze changes in groundwater levels within the study area, data from piezometric (observation) wells were obtained from the South Khorasan Regional Water Company. These data cover the period from before to after the implementation of the Aquifer Balancing Plan (2011–24).

Groundwater abstraction data: To assess the status of groundwater abstraction, data were collected from 10 agricultural wells subject to the abstraction adjustment regulations for the same period (2011–22). This dataset was also acquired from the South Khorasan Regional Water Company. Additionally, information regarding the three meter types (Volume smart, Smart water and electricity, and Volume) installed on these wells and their discharge rates for the 2023–24 water year was examined.

2.3 Data analysis methods

To assess the impact of the Water Balancing Plan, groundwater level data were analyzed for two distinct periods: the pre-plan era (2011) and the post-plan era (2022). Piezometric maps (depth to groundwater) were generated for these 2 years using Kriging interpolation (Spherical model) with

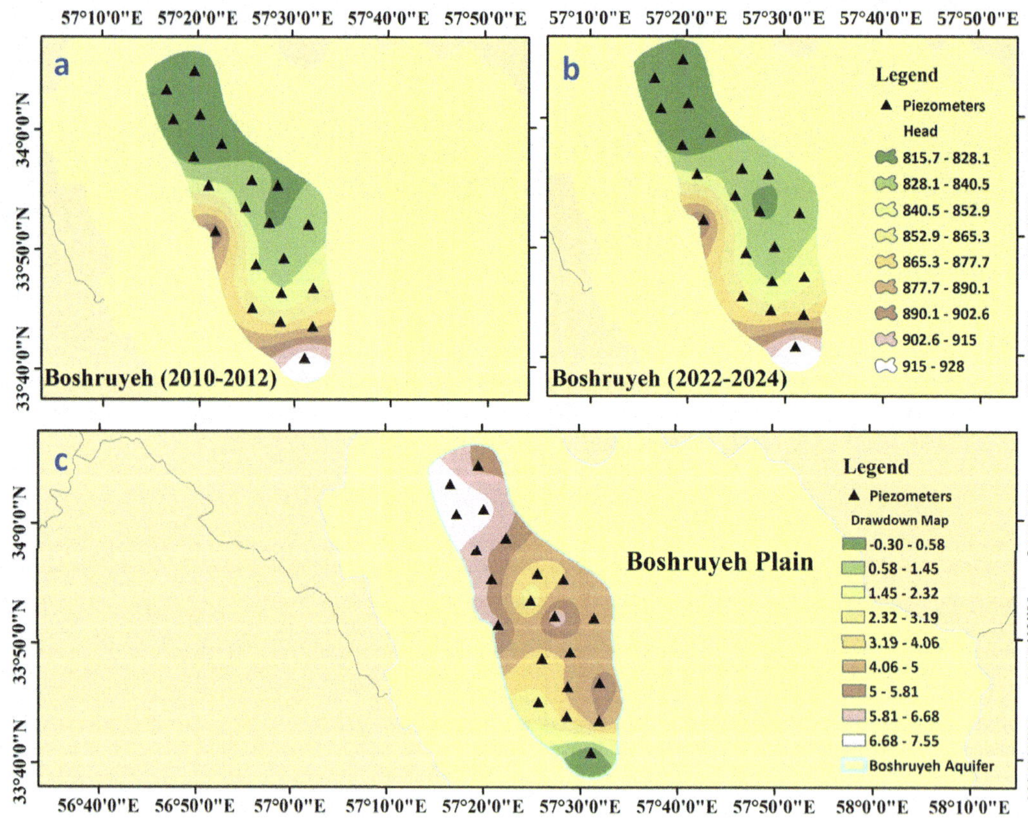

FIGURE 14.2

Hydrogeological baseline and change in the Boshruyeh plain: (a) Groundwater head map before water balancing plan (2010–12); (b) Groundwater head map after water balancing plan (2022–24); (c) Groundwater drawdown map showing net decline in water table elevationn (2010–24). ***Map lines delineate study areas and do not necessarily depict accepted national boundaries.***

cross-validation on data collected from 18 representative monitoring wells to visualize the spatial distribution of the water table (Fig. 14.2a and b). Crucially, a Groundwater Drawdown Map was subsequently generated by subtracting the piezometric level of 2022 from that of 2011 to quantify the net spatial decline in the aquifer (Fig. 14.2c). Additionally, the operational status of all 199 active wells was examined for the 2023–24 water year. In accordance with Table 14.2, 10 agricultural wells were selected from the 199 active wells using a stratified random sampling approach. This selection ensured representation across critical and semicritical zones, varying well depths, and different meter types to provide a reliable snapshot of compliance behavior. Data regarding the installed meters (Volume smart, Smart water and electricity, and Volume) and recorded discharge rates for these 10 wells during the 2023–24 water year were compiled and compared against the licensed permits to evaluate the effectiveness of the metering program in controlling abstraction (Fig. 14.3).

Table 14.2 Characteristics of 10 agricultural wells in Boshruyeh plain.

Agricultural well number	The village	Utm X	Utm Y
1	Newbie	3,742,458	526,520
2	Motokianizadeh	3,745,328	547,479
3	Ersk	3,730,643	545,632
4	Raqqa	3,741,848	525,390
5	Krand	3,755,616	514,534
6	Rig Boshruyeh	3,746,503	546,479
7	Rig Boshruyeh	3,754,241	543,124
8	Rig Boshruyeh	3,747,679	547,307
9	Siyah Rig Boshruyeh	3,729,470	545,850
10	Rig Boshruyeh	3,749,671	548,870

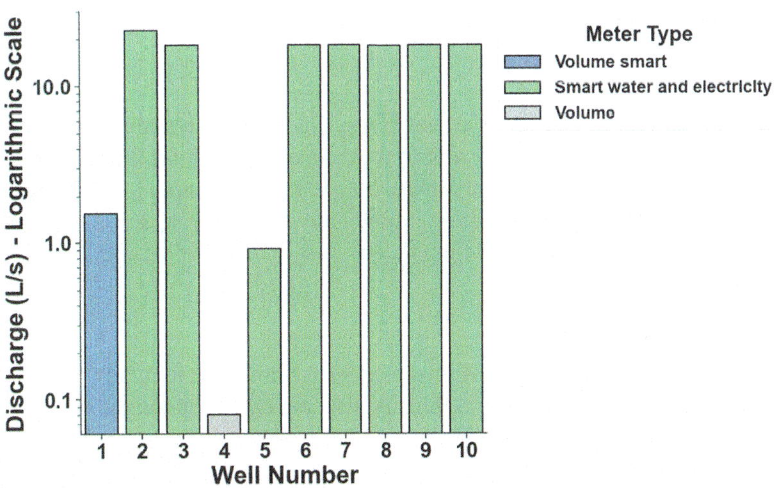

FIGURE 14.3

The status of metering and discharge of agricultural wells included in the adjustment plan in the Boshruyeh plain aquifer (2023–24).

2.4 Data collection instruments

The primary data collection instrument for this research was a questionnaire. Separate questionnaires were designed for the three stakeholder groups: farmers, experts from relevant government agencies, and university academics.

2.4.1 Questionnaire design

The questionnaire was designed specifically to measure the perception gap between the groups. Its questions aimed to capture how each respondent viewed the water crisis, the Balancing Plan, and its overall success. To ensure the tool's validity, it was reviewed by experts for both content and clarity.

Farmer questionnaire: This questionnaire comprised three main sections:

Section one: Demographic information and agricultural characteristics of respondents (e.g., age, education, crop type, cultivated area, and farming experience).

Section two: Questions regarding the perceived severity of the water crisis, its contributing factors, and the effectiveness of the Balancing Plan.

Section Three: An open-ended question for suggestions on improving the Balancing Plan.

Expert and academic questionnaire: This questionnaire consisted of four main sections:

Section one: Demographic and professional information (e.g., education, specialization, work experience, and institutional affiliation).

Section two: Questions on the perceived severity of the water crisis and its contributing factors.

Section three: Questions regarding the effectiveness of the Aquifer Revival and Balancing Plan.

Section four: An open-ended question for suggestions on improving the Balancing Plan.

Questions in the second and third sections of the expert and academic questionnaires were designed using a five-point Likert scale (ranging from "Very low" to "Very high," with an option for "No opinion") (Yaqubi et al., 2024). Table 14.3 outlines the content of the designed questionnaires.

2.4.2 Questionnaire validity and reliability

To establish content validity, the questionnaires were reviewed by specialists and university professors in the fields of water engineering, civil engineering, agriculture, management, sociology, and hydrology. Necessary revisions were made based on their feedback. Furthermore, to ensure face validity, the questionnaires were pilot-tested with a small number of individuals from the target populations (employees, academics, and farmers) to gather feedback on appearance, appeal, clarity, and ease of comprehension.

The reliability of the questionnaires was assessed using Cronbach's alpha coefficient (Pakzad & Alaeddini, 2017). The calculated values were 0.75 for the experts, 0.59 for the academics, and 0.56 for the farmers. While the alpha for farmers is slightly below the conventional 0.7 threshold, this reflects the high heterogeneity within the rural farming community and diverse agricultural practices rather than a flaw in the instrument. To mitigate this, face-to-face interviews were conducted to ensure accurate comprehension.

2.4.3 Data collection procedure

Data were collected from 2023 to 2024. A multistage cluster sampling method (Movallali et al., 2017) was employed to gather data from the geographically dispersed farmer population. First, the villages of Shahrak-e Emam, Ghaniabad, Dasht-e Ali Jamal, and Dasht-e Sharestan in Boshrouyeh were selected as sample clusters. Subsequently, active farmers within these villages were chosen through simple

Table 14.3 The content of the designed questionnaires.

Questionnaire of farmers	Questionnaire of employees of departments related to balancing plan.	Questionnaire of university professors and experts
Screening people		
Product type Education Age Agricultural work experience Cultivated area The number of farmers' share of water	Department of service Education/ specializationDegree Work history Activity unit	Type of university of service Education Scientific rank Work history Faculty of service
The importance of the water crisis and its solutions		
Factors contributing to the Intensification of the water crisis Crop pattern in agriculture Purchase prices of agricultural products Aquifer balancing plan Approved by the ministry of energy Irrigation methods/ traditional irrigation practices Access to government subsidies Number of training programs Related to the water crisis Rainfall over the past five years The link between migration and water scarcity Alternative livelihoods to agriculture	Perspectives on the water crisis in the region Population growth and its impact on the water crisis Mismanagement and development thirst in the water crisis Regional agricultural practices and the water crisis The link between migration and water scarcity Scoring and evaluating solutions for the regional water crisis.	Perspectives on the water crisis in the region Population growth and its impact on the water crisis Mismanagement and development thirst in the water crisis Regional agricultural practices and the water crisis The link between migration and water scarcity Scoring and evaluating solutions for the regional water crisis.
The necessity of implementing the aquifer balancing plan		
–	Reaction to the effectiveness of the Aquifer balancing plan Impact on farmers' income Effectiveness in improving the status of groundwater resources	Reaction to the effectiveness of the aquifer balancing plan Impact on farmers' income Effectiveness in improving the status of groundwater resources in the region Role of the plan in reducing and eliminating illegal water abstraction Effectiveness of smart monitoring (telemetry)

Continued

Table 14.3 The content of the designed questionnaires.—cont'd

Questionnaire of farmers	Questionnaire of employees of departments related to balancing plan.	Questionnaire of university professors and experts
	in the region Role of the plan in reducing and eliminating illegal water abstraction Effectiveness of smart monitoring (telemetry) Impact on water quality Effectiveness of the Qanat Restoration and rehabilitation project Factors and organizations contributing to the unsuccessful Implementation of the aquifer balancing plan Impact of artificial recharge projects on balancing Alternative plans to the aquifer balancing project Evaluation of the zoning and risk assessment of land subsidence Opinions on local stakeholder education and collaboration Effectiveness of the balancing plan in reducing groundwater salinity Evaluation of the revision of drilled well permits Assessment of the program for reducing licenses for well and Qanat construction Opinions on the plan for completing the piezometric well network Evaluation of the implementation of smart and volumetric flow meter installation	Impact on water quality Effectiveness of the Qanat restoration and rehabilitation project Factors and organizations contributing to the unsuccessful implementation of the aquifer balancing plan Impact of artificial recharge projects on balancing Alternative plans to the aquifer balancing project Evaluation of the zoning and risk assessment of land subsidence Opinions on local stakeholder Education and collaboration Effectiveness of the balancing plan in reducing groundwater salinity Evaluation of the revision of drilled well permits Assessment of the program for reducing licenses for well and Qanat construction Opinions on the plan for completing the piezometric well network Evaluation of the implementation of smart and volumetric flow meter installation

random sampling. For experts and academics, purposive sampling was used (Babak et al., 2024), given their limited numbers and the requirement for relevant expertise. Our sample of experts was drawn from key organizations like the Regional Water Company, the Water and Wastewater Company, and Agriculture Jihad, along with university academics from the region. Data from farmers were collected through face-to-face interviews, while experts and academics completed the questionnaires independently. The sample size for each group was determined using different methods Cochran's formula for farmers and expertise criteria for academics and agency experts and is presented in Table 14.4.

The collected data were analyzed using SPSS Statistics version 27. Descriptive statistics, including mean, standard deviation, and frequency, were used to summarize the data. To compare the mean scores of the different groups, a one-way analysis of variance (ANOVA) was conducted. The significance level for all tests was set at $P < .05$.

3. Results

3.1 Hydrogeological evidence of overexploitation

Crucial hydrogeological evidence confirming the overexploitation of the Boshruyeh plain aquifer is derived from the temporal analysis of piezometric level data and direct abstraction records. This section presents quantitative evidence that the water crisis in this region is an anthropogenic physical phenomenon driven by unsustainable anthropogenic pressures.

The collected groundwater hydrographs from observation wells throughout the plain explicitly demonstrate a distinct and critical alteration in the fundamental condition of the aquifer. Specifically, the data clearly reveals an inflection point that precisely aligns with the official activation of the "Aquifer Balancing Plan" in 2014 (Fig. 14.4). During the initial stage (pre-2014), the measured piezometric levels consistently exhibited a steep, negative slope, conclusively showing a pronounced and protracted declining trend within the aquifer. This consistent resource depletion serves as a definitive indicator of severe hydraulic disequilibrium, a condition established when the volume of groundwater abstraction significantly exceeds the aquifer system's inherent natural recharge capacity. Consequently, this period is characterized by intensive and unregulated pumping, resulting in a quantifiable reduction in groundwater storage and unequivocally confirming a state of critical overdraft.

Table 14.4 Sample size of each group.

General category	Specific group name	Sample size (n)
Department experts	Regional water department	24
Department experts	Water and sewage department	15
Department experts	Agricultural jihad department	10
Independent experts	University professors	13
Stakeholders	Farmers of the region	16
–	**Sum**	**78**

Following the execution of the Balancing Plan and the mandatory installation of smart groundwater meters after 2014, the hydrographs (Fig. 14.4) register a discernible turning point. The most conclusive proof of the plan's efficacy lies in the reversal of the sustained water level decline trend across the majority of observation wells. Instead of ongoing depletion, the water levels in most areas either began to rise (indicating recovery) or, at minimum, the rate of decline markedly decelerated or stabilized. This palpable enhancement in the Piezometric Head post-2014, confirmed by the positive change in the trend lines' slope, furnishes compelling hydrogeological evidence of the plan's success in mitigating aquifer depletion and fostering initial recovery. A formal trend analysis confirmed this shift: the average groundwater drawdown rate significantly decreased from 0.53 m/year (pre-2014) to 0.33 m/year (post-2014). A paired t-test verified that this reduction in depletion rate is statistically significant ($t = -7.52$, $P < .001$), with 17 out of 18 observation wells recording a positive shift in trend.

Pumping records from 10 representative agricultural wells were rigorously examined across the pre- and postintervention phases (Figs. 14.5 and 14.6) to directly establish the causal link between abstraction volumes and the documented fall in piezometric head. Fig. 14.5 (Heatmap) provides visual confirmation of the consistently high abstraction rates that characterized the period before 2014. Crucially, the regulatory enforcement actions implemented subsequent to 2014 successfully induced a discernible reduction in the pumped water volumes throughout most of the sampled wells. This reduction is clearly visualized in the Heatmap (Fig. 14.5) by the shift toward lighter shades.

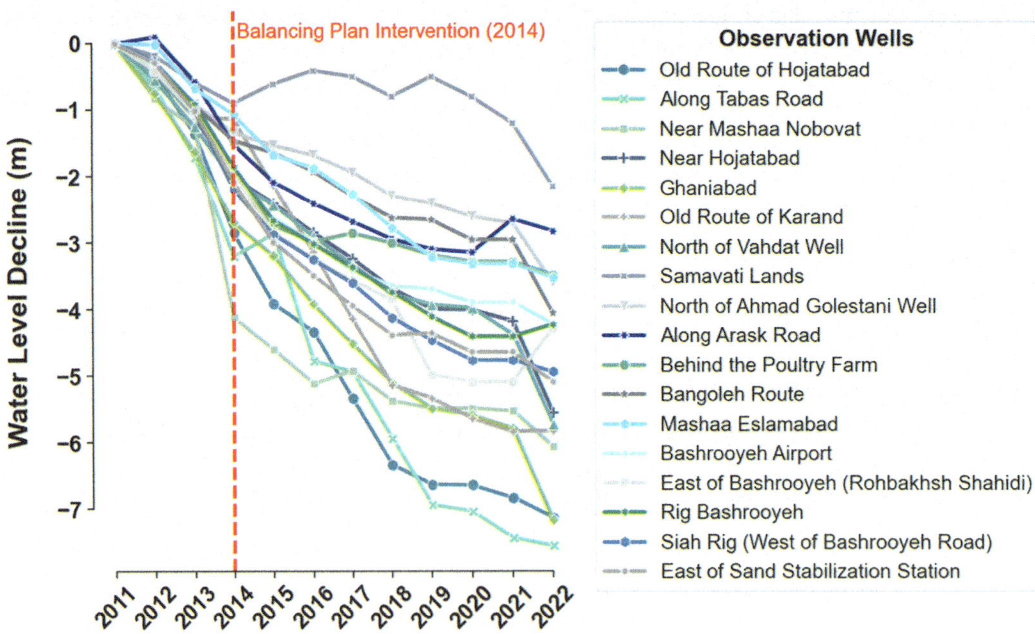

FIGURE 14.4

Cumulative water level decline (drawdown) in the Boshruyeh plain aquifer (2011–22), showing the impact of the aquifer balancing plan.

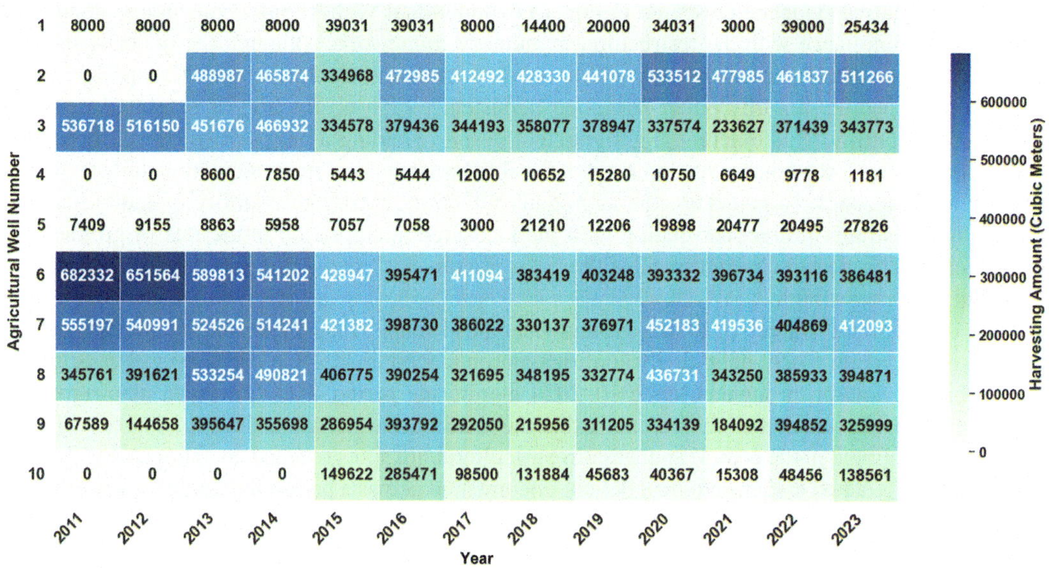

FIGURE 14.5

Groundwater abstraction volumes from the Boshruyeh plain aquifer (2011–23): A heatmap analysis of 10 representative wells subject to regulation.

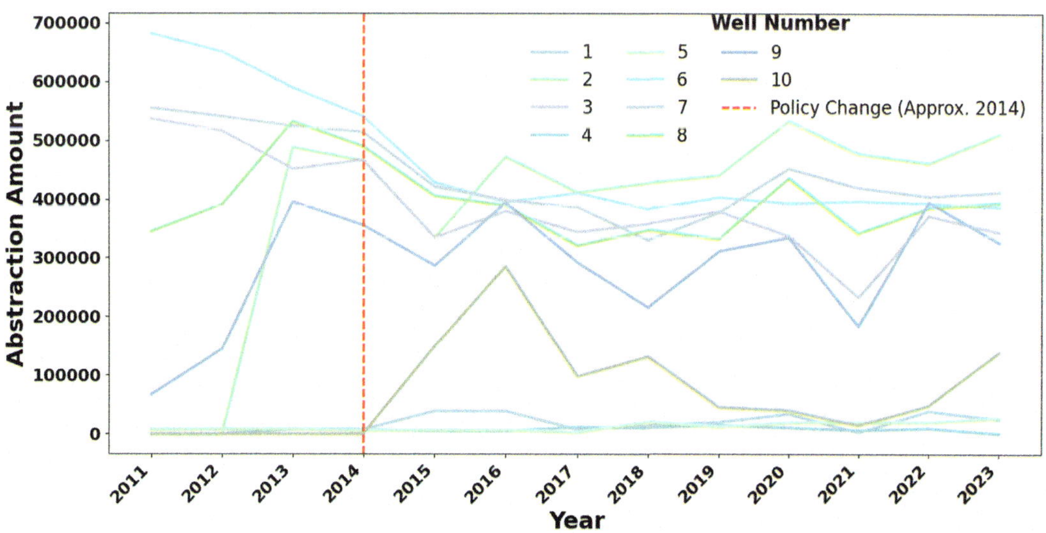

FIGURE 14.6

Quantitative trends in groundwater abstraction for 10 representative wells: Pre- and post-2014 policy analysis.

The quantitative evidence presented in Fig. 14.6 (Individual Well Trends) nonetheless documents nonuniform compliance with the restructured regulatory framework. Although the majority of wells recorded a clear and persistent reduction in annual water withdrawal after the 2014 policy implementation, a specific minority either sustained persistently high abstraction or exhibited increased withdrawal volatility. This outcome highlights that despite yielding a net positive effect, the regulation's efficacy was not consistent, indicating possible resistance or implementation failures within specific aquifer zones. Nonetheless, the aggregate reduction across the majority of individual well records demonstrates a direct correlation with the measured amelioration in the rate of groundwater level decline. This decisively confirms that the crisis is a physical reality that measurably responded to the regulatory intervention.

The comparative evaluation of the groundwater head maps (Fig. 14.2a and b) and the derived drawdown map (Fig. 14.2c) definitively confirms a pronounced net decline in the overall condition of the Boshruyeh aquifer throughout the 2010–2024 monitoring period. The drawdown map rigorously quantifies this observed depletion, indicating that the maximum water table elevation losses which are notably concentrated in the central and western agricultural hot spots reached an approximate range of 5–6 m. This protracted hydrogeological deficit, documented across a period that includes the initiation of the water balancing plan, confirms that the ongoing rate of overextraction remains a critical concern that has not yet been effectively constrained. Consequently, these findings unequivocally highlight the paramount policy challenge inherent in simultaneously satisfying the sustained requirements of agricultural water demand while concurrently guaranteeing the long-term hydrological sustainability of the aquifer system.

3.2 Quantifying the governance gap

3.2.1 Sociodemographic characteristics of respondents

The findings related to the sociodemographic characteristics of the respondents in the three groups farmers, experts, and academics are as follows:

Experts/agency employees: Among the experts, 59% held degrees related to water engineering, and 56% had obtained a Master's degree. A significant portion, 43%, possessed more than 15 years of professional experience. This composition suggests that the respondents possess a substantial level of knowledge and experience in water resource management, enabling them to offer expert and precise perspectives.

Academics and university experts: Of the academic respondents, 93% were affiliated with national universities, 64% had educational backgrounds in water engineering, and 43% held the academic rank of Associate Professor. This demographic highlights the presence of highly knowledgeable specialists and scholars among the respondents, whose viewpoints can cover the scientific and research dimensions of the subject.

Farmers: The farmer group was composed of experienced agriculturalists. Their average age was 49, with most participants between 40 and 50 years old. All were male, and just over half (56%) had a primary or middle school education. This demographic profile points to a community of seasoned, older farmers, whose local knowledge and hands-on experience in farming and water management represent a valuable resource.

3.2.2 Assessment of the water crisis

The results concerning the assessment of the water crisis severity from the perspective of the three respondent groups are presented in Table 14.5.

The results of the ANOVA test indicated a statistically significant difference in the mean assessment of the water crisis severity among the three groups ($F = 10.108$, $P < .001$). University academics had the highest mean score for perceiving the crisis's importance, whereas farmers showed the lowest. This disparity can be attributed to differences in understanding the water crisis based on educational level, profession, and the degree of direct dependence on water resources. Academics, with a more scientific and comprehensive viewpoint, assign greater importance to the water crisis. In contrast, farmers, who directly confront the consequences of the crisis, have a more practical understanding but may not perceive its broader dimensions to the same extent as academics. This finding is consistent with the results of studies by Farzaneh et al. (2019) concerning the low social acceptance of orchard desiccation. This divergence in perspectives underscores the necessity for targeted awareness-raising programs for farmers and other stakeholders.

3.2.3 Factors contributing to the worsening water crisis

As illustrated in Fig. 14.7, a significant gap exists in the perspectives of different stakeholders regarding the primary drivers of the water crisis. From the farmers' viewpoint, "outdated agricultural practices" (31.2%), "inappropriate cropping patterns" (25%), and "low irrigation efficiency" (18.8%) were identified as the most critical factors. This indicates that farmers perceive operational challenges as the principal cause of the crisis.

In contrast, experts and university academics identified "inefficient agriculture" (93.9% and 84.6%, respectively) and "mismanagement"" (91.8% and 100%, respectively) as the primary causal factors, with overwhelmingly high percentages. This difference in perspective reveals that experts and academics tend to focus on macro-level management and policy issues, whereas farmers concentrate on daily operational problems. Furthermore, both the expert and academic groups identified "population growth" as another influential factor with similar attribution rates (approximately 46%).

3.2.4 Solutions for the water crisis

By integrating strategies derived from scientific literature and expert opinions, a list of proposed solutions for addressing the water crisis in the Boshrouyeh plain was compiled (Hannah et al., 2022; He

Table 14.5 Assessing the importance of the water crisis from the point of view of farmers, employees and university professors.

The importance of the water crisis	Percentage of agreement/ confirmation (%)	Average	Standard deviation	Anova	
				Sig	F
Organ experts	62.82	4.45	0.867	10.108	>0.001
Academic professors and experts	66.16	4.92	0.885	10.108	>0.001
Farmers	20.51	3.63	0.277	10.108	>0.001

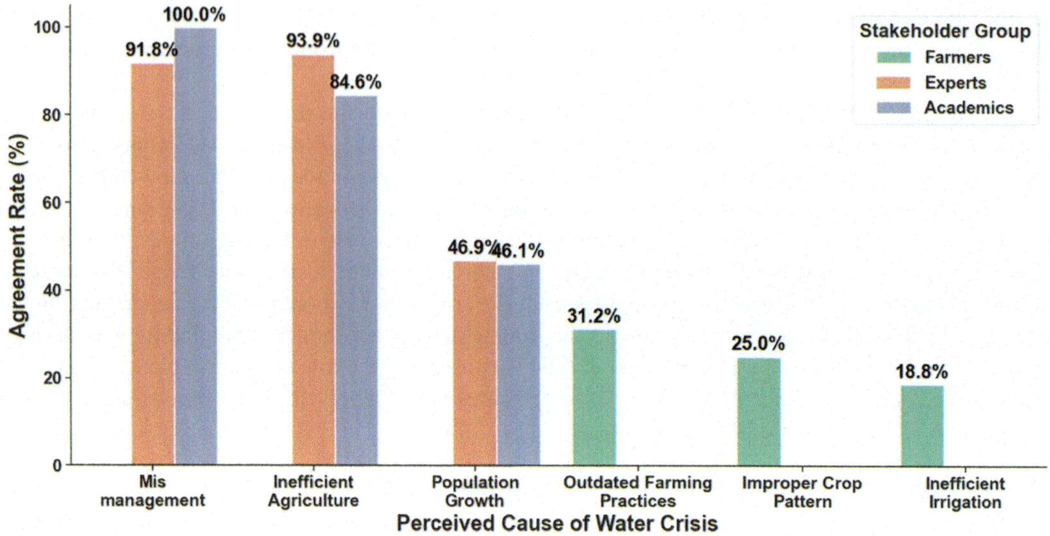

FIGURE 14.7

The perception gap on the primary drivers of the water crisis.

et al., 2021). The results presented in Table 14.6 are based on participants' scoring of these solutions, followed by the calculation of mean scores and their subsequent ranking.

The consensus between agency employees and academics regarding the importance of "reforming the management system" and "improving infrastructure" indicates that these two factors play a pivotal role in managing the water crisis. Efficient management of water resources and the enhancement of distribution and irrigation infrastructure can help reduce water loss and increase its efficient use (Tosan et al., 2025). A divergence of opinion was noted regarding "establishing agricultural management cooperatives," which may stem from different perceptions of the functionality and effectiveness of such organizations. The employees, who are active in the executive sector and may have more experience with the advantages and challenges of these cooperatives, ranked this solution second in importance.

3.2.5 Effectiveness of the aquifer balancing plan

The results concerning the evaluation of the overall effectiveness of the aquifer balancing plan from the perspective of the three surveyed groups are presented in Table 14.7. Further details from the perspectives of the employees and academics are provided in Table 14.8.

Table 14.7 presents the results from the evaluation of the overall effectiveness of the Boshrouyeh plain's Groundwater Aquifer Balancing Plan from the perspective of the three stakeholder groups. Although effectiveness ratings varied, a One-Way ANOVA on the 'Plan Necessity' scores revealed no statistically significant difference between groups ($F = 0.855$, $P = .429$), with Farmers (Mean = 3.33), Experts (Mean = 3.27), and Academics (Mean = 3.61) all agreeing on the general necessity of the intervention. The sharp split observed in effectiveness ratings (Table 14.7), therefore, reflects a disagreement on *implementation outcomes* rather than the *principle* of the policy. An examination of

Table 14.6 Scoring solutions to solve the water crisis from the point of view of employees and professors.

Solutions to solve the water crisis	Average score	Rank (professors)	Average score	Rank (Employees)
Reviewing population growth policies	3	10	3.65	10
Management of macro ecosystems	3.56	9	4.65	9
Revision of cultivation patterns and irrigation methods	4.44	8	6.53	5
Increase in the price of energy and water carriers	7.33	5	5.41	6
Establishing agricultural management cooperative organizations in each region	6	7	7.59	2
Creating a water market	6.67	6	5	8
Inventing new water saving technologies	8.44	3	7.11	4
Changing the management system from passive management to efficient management	10	1	8.29	1
Improving distribution infrastructure	9.33	2	7.41	3
Education for consumption and lifestyle change	8	4	5.47	7

Table 14.7 The effectiveness of the balancing plan from the point of view of farmers, employees and professors.

Balancing plan of ministry of energy	Frequency	Average	Percentage of agreement/ confirmation (%)	Rank earned
Department experts	46	0.93	93.9	1
Academic professors and experts	12	0.92	92.3	2
Farmers	48	0.75	75	3

Table 14.8 Evaluating the effectiveness of the balancing plan from the point of view of employees and professors.

Effective items in the success of the balancing plan	Average		Standard deviation		Percentage of agreement/ confirmation (%)	
	Emp.	Prof.	Emp.	Prof.	Emp.	Prof.
Increasing the income of farmers	24.78	32.77	40.37	45.96	14.29	46.15
Improving the status of groundwater resources	23.27	18.38	39.27	35.78	34.69	76.92
Reduction and elimination of unauthorized withdrawals	16.06	11.92	31.69	26.17	75.51	84.62

the table reveals that the plan, implemented by the Ministry of Energy, was perceived as most effective by agency experts, with a 93.9% positive response rate, earning it the first rank. This signifies that the overwhelming majority of experts assessed the plan as effective in restoring and balancing groundwater resources.

Academics and university experts ranked second, a result which indicates that their perspective on the plan's effectiveness is also positive and largely aligns with that of the agency experts. However, the slightly lower response percentage suggests that minor disagreements regarding the degree of the plan's effectiveness may exist within this group. Finally, farmers, with a 75% positive response rate and a rank of three, attributed the lowest level of effectiveness to the Ministry of Energy's plan. This significant discrepancy between the views of farmers and those of experts and academics warrants closer examination. The more positive perception of the plan's effectiveness among experts and academics compared to farmers can be attributed to several factors, including:

Variations in plan implementation across different areas: The execution of the plan may have differed in the regions where farmers operate, resulting in less tangible benefits for them (Mirabolghasemi, 2022).

The extent of farmers' access to water resources: The plan's implementation may have directly impacted farmers' access to water, with some facing new limitations (Yousefi et al., 2022).

The economic impacts of the plan on their livelihoods: The plan may have had varied economic consequences for farmers' livelihoods, leading to dissatisfaction among some.

Their level of awareness and understanding of the plan's objectives and details: Farmers may lack sufficient information about the plan's goals and specifics, thus preventing a complete understanding of its effectiveness (Zare et al., 2017).

Table 14.8, which details the evaluation of the Balancing Plan's effectiveness from the viewpoints of employees and university academics, reveals nuanced differences in their perceptions of the plan's impact. While academics (mean = 32.77, SD = 45.96) are more optimistic than employees (mean = 24.78, SD = 40.37) regarding the plan's effect on increasing farmer income, employees (mean = 23.27) perceive its impact on improving the condition of groundwater resources more positively than academics do (mean = 18.38). Nonetheless, both groups show considerable agreement on the importance of reducing unauthorized abstractions. This aspect was deemed critical by 75.51% of employees and 48.62% of academics, although the mean response from employees was slightly higher than that of academics (16.06 > 11.92).

Table 14.9 examines the perspectives of two groups agency experts (representing the executive sector) and university academics (representing the scientific and research sector) on various restoration and balancing plans for the Boshrouyeh plain aquifer. This evaluation was conducted based on four criteria: "Frequency," "Mean," "Percentage of Importance/Effectiveness," and "Rank Obtained." The data analysis is as follows:

Smart monitoring (telemetry) and installation of volumetric and smart meters: These two plans received the highest percentage of importance/effectiveness from both groups. This indicates that the use of modern technologies in water resource management for aquifer balancing is considered highly significant by both sectors (Tosan et al., 2024). The "Installation of Volumetric and Smart Meters" had the highest frequency among all plans for employees (frequency = 35). Meanwhile, "Smart monitoring (telemetry)" showed a higher mean score from employees (mean = 26.02), which could suggest the greater perceived success of this particular plan (Zare et al., 2017).

Table 14.9 Evaluation of different plans in line with the aquifer restoration and balancing plan.

Various plans for the restoration and balancing of the aquifer	Frequency		Average		Percentage of importance/ effectiveness (%)		Rank earned	
	Emp.	Prof.	Emp.	Prof.	Emp.	Prof.	Emp.	Prof.
Smart monitoring (telemetry)	22	10	26.02	11.54	44.90	76.92	2	1
Restoration and rehabilitation of qanats	20	7	17.35	11.08	40.82	53.84	3	3
Artificial recharge projects	10	6	19	10.62	21.41	46.15	5	4
Zoning and risk assessment of land subsidence	18	8	23.81	26	36.73	61.54	4	2
Evaluation and quality monitoring of groundwater resources	8	4	14.75	10.69	16.32	30.77	7	5
Reduction and revision of drilled well permits	9	7	15.19	25.77	18.37	53.84	6	3
Completion of the piezometric well network plan	8	2	23.85	32.62	16.32	15.38	7	6
Installation of smart and volumetric flow meters	35	10	8.06	11.54	71.42	76.92	1	1

Restoration and rehabilitation of Qanats: This plan was also assessed as effective by both groups, with an importance/effectiveness rating of 40.8% from employees and 53.8% from academics. Furthermore, the identical rank obtained (3 for employees and three for academics) indicates that both academics and employees assign a similar priority to the success of this plan.

Artificial recharge projects: This plan received a moderate rating in terms of "Percentage of Importance/Effectiveness" compared to other plans. This may reflect uncertainty or skepticism regarding the efficacy or sustainability of this method under the specific conditions of the study area:

1. With an importance/effectiveness rating of 36.73% from employees and 61.54% from academics, and obtaining ranks of four and 2, respectively, this plan was deemed relatively important by both groups. This consensus is likely due to the growing significance of land subsidence and its impact on groundwater resources, which makes the implementation of this plan critical (Asadi et al., 2022).
2. These plans were rated at a moderate level of importance/effectiveness, with considerable differences in opinion observed between employees and academics. For instance, in the plan for "Reducing permits and amending licenses for existing wells," ranks of 3 and 6 were assigned by academics and employees, respectively. This suggests that academics priorities on the effectiveness of this plan for restoring the Boshrouyeh plain aquifer.

In conclusion, the evaluation of the Boshrouyeh aquifer restoration and balancing plans from the perspectives of agency experts and university academics reveals a consensus on the importance of implementing projects such as smart monitoring and volumetric meters, as well as on the relative significance of the plan for zoning land subsidence risks (Mirabolghasemi, 2022). However, disagreements exist regarding the effectiveness of plans such as the restoration and rehabilitation of qanats and the reduction of permits and amendments to well licenses. These findings emphasize the importance of considering the perspectives of both groups in the planning and implementation of aquifer restoration and balancing projects, underscoring the need for further research to investigate the reasons behind these differences and to propose suitable solutions (Yousefi et al., 2022).

4. Discussion

4.1 The disconnect between top-down policymaking and bottom-up realities

The findings from this investigation furnish convincing evidence of a profound governance deficit within the Boshrouyeh plain's water resource management framework. The governance deficit in the region primarily emerges from a clear discrepancy between the centralized, top-down perspective embraced by policymakers and experts and the concrete, bottom-up realities faced by the local farming community. While the hydrogeological data (detailed in Section 3.1) clearly confirm the physical effectiveness of the regulatory action—the "Aquifer Balancing Plan"—in stabilizing the groundwater table, a parallel assessment of stakeholder perspectives simultaneously shows an inability to foster the crucial social legitimacy and alignment needed for sustained, long-term operational viability.

4.2 Divergence in perceiving the crisis and identifying its causes

This fundamental manifestation of the deficit is clearly observable in the differing perspectives on the water crisis itself. Drawing upon macro-level data and a systemic outlook, university academics and technical experts typically classify the crisis as maximally severe, assigning its root causes predominantly to mismanagement and inefficient agricultural methods. This distinct strategy embodies a structural and managerial analysis, which fundamentally locates the problem's origins in macro-level organizational frameworks and existing large-scale policies.

In contrast, farmers, who are directly affected by the limitations of water resources, perceive the crisis less severely than the other two groups and identify its causes in tangible, operational issues such as outdated agricultural practices and inappropriate cropping patterns (Fig. 14.5). This divergence shows that policymakers view the crisis through the lens of a management problem, for which the solution lies in policy reforms and technical oversight (such as meter installation). Farmers, however, view the crisis through the prism of their livelihood challenges and daily operational constraints. This difference in problem framing means that any top-down policy is likely to be met with resistance and nonacceptance from the local community, as each group is essentially trying to solve a different problem.

4.3 Contradictory assessments of the aquifer balancing plan's effectiveness

The climax of the governance gap appears in the conflicting evaluations of the "Aquifer Balancing Plan's" effectiveness. Experts responsible for implementing the plan deem it highly effective, with 93.9% approval. This positive assessment from experts and academics is likely tied to physical data, like the slowing rate of water level decline confirmed in the hydrogeological analysis. They also affirm the plan's effectiveness from a scientific standpoint. Farmers, however, saw things differently. With only 75% approval, their lower rating of the plan (Table 14.7) is not a result of being uninformed but a reflection of their lived experience. From their perspective, a plan that restricts access to water, potentially cuts into their income, and adds regulatory burdens is hard to call "effective," regardless of improvements in hydrogeological data. This demonstrates that the hydrological success of a plan does not necessarily translate to its socioeconomic success. Policymakers define success as balancing the aquifer, while farmers define it as the sustainability of their livelihoods; within the current framework, these two definitions are in conflict.

4.4 Differing priorities in solution selection

The disparity becomes highly apparent in the contrasting prioritization of solutions. A pronounced consensus exists among technical experts and academic researchers, one that strongly favors technological and regulatory interventions. These measures routinely include the deployment of smart monitoring systems and the mandatory installation of volumetric meters. This distinct preference clearly underscores the pervasive influence of a technocratic management paradigm. Conversely, the broad disagreement surrounding alternative remedies, such as the establishment of agricultural management cooperatives or the necessary reduction and modification of existing permits, highlights the considerable social and logistical complexities often disregarded by top-down decision-making processes. To illustrate, while academics may regard permit reduction as a direct and effective policy instrument, both executive experts and farmers possess a much deeper appreciation for the resulting practical challenges and potential social consequences.

In conclusion, the water crisis in the Boshrouyeh plain is fundamentally a governance crisis, extending far beyond a simple scarcity of resources. This clear perceptual gap between high-level policymakers and local farmers is not merely an academic observation; it entails tangible physical consequences for the regional ecosystem. When the farming community lacks confidence in, or alignment with, the rationale behind policies like the installation of smart meters, they tend to actively resist the measures or look for alternative approaches to meet their water demands. This specific behavior directly impedes the success of water conservation initiatives, thereby ensuring the continued depletion of groundwater reserves. In turn, this process exacerbates critical issues such as land subsidence and soil salinization, which ultimately endangers the very basis of agricultural production and livelihoods in the area.

These findings are well aligned with key principles established in the global literature concerning water governance and the effective management of common-pool resources (CPRs). This body of knowledge emphasizes that the sustainable management of resources like groundwater is nearly impossible without strong local institutions, collectively agreed-upon rules, and the active participation of users. The current top-down, control-oriented approach in the Boshrouyeh plain overlooks these principles, and as a result, has failed to create a sustainable, community-based solution, despite its intermittent technical successes. Therefore, bridging this gap and transitioning toward a participatory and adaptive governance model one that integrates scientific knowledge with the experience and needs of the local community is not a choice but an undeniable necessity for saving the ecological and economic future of the Boshrouyeh plain.

Furthermore, the connection to land degradation is evidenced by the increasing salinity trends observed in agricultural wells and farmers' reports of declining soil fertility. These indicators serve as proxies for degradation, confirming that governance gaps are directly translating into ecological deterioration.

5. Conclusion

The conclusions derived from this research unambiguously affirm that the ongoing water crisis in the Boshrouyeh plain is primarily rooted in a governance failure, thereby representing an issue that significantly exceeds the typical constraints of resource scarcity alone. This inherent crisis is clearly manifested through a profound divergence of viewpoints: specifically, the separation that exists between the macro-level perspectives employed by policymakers and technical experts, and the practical, on-the-ground realities experienced by the agricultural community. This resultant perceptual disconnect is further evidenced by significant variations in how stakeholders apprehend the crisis's severity, attribute its underlying root causes, and assess the efficacy of proposed solutions. Although a top-down management style may secure certain intermittent technical achievements, these successes are highly improbable to be sustained over the long term, owing to a critical deficit in both social acceptance and operational legitimacy.

To overcome this challenge, the focus must shift from purely technical solutions to bridging the governance gap. In this regard, the following recommendations are proposed as practical steps:

1. Instead of one-way training, joint dialog workshops should be organized among farmers, experts, and government managers. The goal of these workshops is to foster mutual understanding of each group's perspectives and constraints. More importantly, within these platforms, technical knowledge must be translated into language that is tangible to farmers. This means the direct and

documented link between groundwater decline and phenomena such as land subsidence and soil salinization on their own farms must be demonstrated to them objectively. This will help create a shared understanding of the crisis's urgency.

2. A shift from merely "consulting" farmers to truly "involving" them in decision-making is needed. This requires creating mechanisms that give farmer representatives a real voice and vote in meetings that decide policies on crop patterns, water allocation, and management. Giving the local community a sense of ownership and responsibility in this way will produce more realistic policies that align with the region's socioeconomic landscape.
3. A foundational shift in policy is required, wherein support packages are redesigned from compensatory handouts to strategic incentives that promote sustainable agriculture. A more effective strategy would be to transform financial support from a passive subsidy into an active tool for change. In practice, this approach necessitates directly tying benefits—such as low-interest loans or subsidized equipment—to on-the-ground performance metrics. Farmers would gain access to these crucial resources by undertaking specific, measurable actions, including the installation of modern irrigation technologies, transitioning to agricultural crops that demand less water, or demonstrably adhering to established water quotas. This methodology effectively establishes a direct and powerful linkage between the fundamental economic self-interest of the farmer and the overall environmental health and sustainability of the broader region.

Ultimately, the primary obstacle to the sustainable management of water and soil in the plains of South Khorasan is not a technological deficit but a deep and persistent governance gap. Bridging this gap is a fundamental prerequisite for simultaneously protecting groundwater resources and the agricultural lands whose very existence depends on them.

References

Acevedo, S. E., Waterhouse, H., Barrios-Masias, F., Dierks, J., Renwick, L. L., & Bowles, T. M. (2022). How does building healthy soils impact sustainable use of water resources in irrigated agriculture? *Elementa: Science of the Anthropocene, 10*(1), Article 00043.

Akbarpour, A., Dastourani, M., Tosan, M., & Gharib, M. R. (2024). Performance analysis of finite element method in groundwater studies based on web of science using R Biblioshiny. *Journal of Aquifer and Qanat, 4*(2), 131 148.

Asadi, E., Sadeghzade, P., Majnouni Heris, A., & Samadian, M. (2022). Evaluation of the effectiveness of groundwater reclamation and balancing plan based on fuzzy-TOPSIS method (case study: Urmia plain). *Water and Irrigation Management, 12*(3), 483–496.

Babak, N., Hosseinian, S., Younesi, J., & MardaniRad, M. (2024). The structural model of predicting the professional development of consultants based on causal conditions with the mediation of individual and collaborative strategies with the grounded theory method. *Medical Journal of Mashhad University of Medical Sciences, 66*(6).

Bonì, R., Teatini, P., Zoccarato, C., Guardiola-Albert, C., Ezquerro, P., Bru, G., Tomás, R., Valdes-Abellan, J., Pla, C., & Navarro-Hernández, M. I. (2022). Stakeholders' perspective on groundwater management in four water-stressed Mediterranean areas: Priorities and challenges. *Land, 11*(5), 738.

du Plessis, A. (2023). Fragmented water governance, institutional problems and questionable decisions. In *South Africa's water predicament: Freshwater's unceasing decline* (pp. 67–87). Springer.

Farzaneh, M. R., Bagheri, A., & Momeni, F. (2019). A criticism to framework of groundwater resources reclamation and suggesting alternative method to the implement in Rafsanjan region around. *Journal of Water and Soil Conservation, 26*(1), 169–185.

Feizi, H., & Tosan, M. (November 2016). Saffron yield variability by climatic factors in the northeast of Iran. In *V international symposium on saffron biology and technology: Advances in biology, technologies, uses and market, 1184* (pp. 109–114).

Feng, Z., Dai, L., Miao, Q., Gonçalves, J. M., Shi, H., Li, Y., & Feng, W. (2025). Effects of groundwater depth on soil water and salinity dynamics in the Hetao irrigation district: Insights from laboratory experiments and HYDRUS-1D simulations. *Agronomy, 15*(9), 2025.

Gholamrezai, S., Azadi, H., Karamian, F., Khosravi, E., Moghaddam, S. M., Goli, I., & Scheffran, J. (2023). Smart control of agricultural water wells in western Iran: Application of the Q-methodology. *Journal of Soil and Water Conservation, 78*(1), 58–69.

Haji, L., Momenpour, Y., & Choobchian, S. (2024). Moving toward salvaging Iran's groundwater: A psychological analysis of blocking unauthorized agricultural wells. *Agricultural Water Management, 303*, Article 109035.

Hannah, D. M., Abbott, B. W., Khamis, K., Kelleher, C., Lynch, I., Krause, S., & Ward, A. S. (2022). Illuminating the 'invisible water crisis' to address global water pollution challenges. *Hydrological Processes, 36*(3).

Hashemi, M., Zadeh, H. M., Zarghami, M., Demeke, B. W., & Delgarm, R. T. (2023). An analysis of why rehabilitation and balancing programs for aquifers do not meet water organizations' targets (a case study of the Qazvin aquifer in Iran). *Agricultural Water Management, 281*, Article 108258.

Havangi, M., & Hhozeymehnezhad, H. (2023). Evaluation of the implementation of the restoration and resilience plan of groundwater resources with an emphasis on the installation of smart meters (case study: Boshroyeh Plain, South Khorasan). *Journal of Water and Soil Conservation, 30*(2), 161–171.

He, C., Liu, Z., Wu, J., Pan, X., Fang, Z., Li, J., & Bryan, B. A. (2021). Future global urban water scarcity and potential solutions. *Nature Communications, 12*(1), 4667.

Hidayati, E., Septalinda, F., Fardian, M. I., & Zidniyati, Z. (2025). Integrating community-based AI with local knowledge for sustainable water governance: A case from the osing people. *Jurnal Penelitian IPTEKS, 10*(2), 79–99.

Kamal, M. M., Amiri, H., Moghadam, V., & Rahimi, D. (2021). Institutional analysis of top-down regulatory: Evidence from Iran local governance. *Water Policy, 23*(4), 930–945.

Koebele, E. A., Méndez-Barrientos, L. E., Nadeau, N., & Gerlak, A. K. (2024). Beyond engagement: Enhancing equity in collaborative water governance. *Wiley Interdisciplinary Reviews: Water, 11*(2), Article e1687.

Li, P., Wu, J., Zhou, W., & LaMoreaux, J. W. (2023). Land subsidence and groundwater seepage. In *Hazard hydrogeology* (pp. 45–74). Springer.

Mahdavi, T. (2024). Assessing water governance gaps with a four-layer governance model and OECD principles. *Water Abstraction Research, 7*(2), 314–333.

Mirabolghasemi, H. (2022). Analytical note: Restoring and transpooning the subterranean water supply system from another look. *Journal of Water and Sustainable Development, 9*(3), 155–157.

Movallali, G., Nesayan, A., & Asadi Gandomani, R. (2017). Psychometric properties of Dunn's sensory profile school companion. *Archives of Rehabilitation, 18*(3), 194–201.

Nagata, K., Shoji, I., Arima, T., Otsuka, T., Kato, K., Matsubayashi, M., & Omura, M. (2022). Practicality of integrated water resources management (IWRM) in different contexts. *International Journal of Water Resources Development, 38*(5), 897–919.

Nourani, V., Tosan, M., Huang, J. J., Gebremichael, M., Kantoush, S. A., & Dastourani, M. (2025). Advances in multi-source data fusion for precipitation estimation: Remote sensing and machine learning perspectives. *Earth-Science Reviews*, Article 105253.

Nouri, M., Homaee, M., Pereira, L. S., & Bybordi, M. (2023). Water management dilemma in the agricultural sector of Iran: A review focusing on water governance. *Agricultural Water Management, 288*, Article 108480.

Pakzad, R., & Alaeddini, F. (2017). Misuse and misconception of Cronbach's alpha coefficient as an index of internal consistency of measuring tools. *Iranian Journal of Epidemiology, 12*(4), 64–71.
Quandt, A., Larsen, A. E., Bartel, G., Okamura, K., & Sousa, D. (2023). Sustainable groundwater management and its implications for agricultural land repurposing. *Regional Environmental Change, 23*(4), 120.
Rezvani Moghaddam, P., Khorramdel, S., & Moalem Benhangi, F. (2022). Optimization of manure and irrigation levels on flower and corm yields of saffron by using a central composite design. *Journal of Saffron Research, 10*(1), 45–63.
Rodzynek, A., & Houston, K. B. (2021). *The fallacy of water governance: A comparative analysis of predominant approaches.*
Sajjad, M. M., Wang, J., Afzal, Z., Hussain, S., Siddique, A., Khan, R., Ali, M., & Iqbal, J. (2023). Assessing the impacts of groundwater depletion and aquifer degradation on land subsidence in Lahore, Pakistan: A PS-InSAR approach for sustainable urban development. *Remote Sensing, 15*(22), 5418.
Samani, S. (2020). Providing sustainable global groundwater resources management models to improve the sustainability plan in Iran. *Iran-Water Resources Research, 16*(2), 271–291.
Scanlon, B. R., Fakhreddine, S., Rateb, A., de Graaf, I., Famiglietti, J., Gleeson, T., Grafton, R. Q., Jobbagy, E., Kebede, S., & Kolusu, S. R. (2023). Global water resources and the role of groundwater in a resilient water future. *Nature Reviews Earth & Environment, 4*(2), 87–101.
Shamshirgaran, R., Tosan, M., & Nasirian, A. (2025). Investigating the functional problems of ground water studies in dry areas: A case of Boshruyeh Plain, South Khorasan, Iran. *Journal of Aquifer and Qanat, 5*(2), 99–120.
Shunglu, R., Köpke, S., Kanoi, L., Nissanka, T. S., Withanachchi, C. R., Gamage, D. U., Dissanayake, H. R., Kibaroglu, A., Ünver, O., & Withanachchi, S. S. (2022). Barriers in participative water governance: A critical analysis of community development approaches. *Water, 14*(5), 762.
Tosan, M., Khashei-Siuki, A., Maroosi, A., & Gharib, M. R. (2024). A review of smart water management for sustainable agriculture based on the internet of things. *Water Management in Agriculture, 11*(1), 145–166.
Tosan, M., Nourani, V., Kisi, O., & Dastourani, M. (2025). Evolution of ensemble machine learning approaches in water resources management: A review. *Earth Science Informatics, 18*(2), 416.
Xue, Y., Niu, Z., Zhang, R., Jia, L., & Guo, S. (2024). Current status and future research of groundwater under climate change: A bibliometric analysis. *Water, 16*(23), 3438.
Yadav, M., Vashisht, B., Jalota, S., Kaur, S., & Yadav, S. (2025). Stakeholders participation and groundwater management: Raising the awareness. In *Water sustainability and hydrological extremes* (pp. 293–314). Elsevier.
Yaqubi, M., Yaghoobzadeh, M., & Tosan, M. (2024). Factor analysis and ranking of saffron production, processing and market challenges in torbat heydarieh, Iran. *Saffron Agronomy and Technology, 12*(1), Article e194618.
Yousefi, A., Sarai, T. M., Porhemmat, J., & Babazadeh, H. (2022). *Studying groundwater resources balancing plan under climate change conditions (case study: Hashtgerd study area).*
Zare, S., Mohammadi, H., & Sabouhi, M. (2017). Simulation of developing modern irrigation systems on groundwater resources balance of Khorasan Razavi. *Journal of Agricultural Economics & Development, 31*(2), 179–195.

PART

Soil and emerging technologies IV

CHAPTER

Technologies for soil pollution reduction and restoration

15

Arun Karnwal[1], Neha Saklani[1], Abdel Rahman Mohammad Said Al-Tawaha[2], Abdel Razzaq Al-Tawaha[3], Natalia Nesterova[4], Gaurav Kumar[5], Anupam Kumar[6] and Deepak Kumar[7]

[1]*Department of Microbiology, Graphic Era (Deemed to be University), Dehradun, Uttarakhand, India;* [2]*Department of Biological Sciences, Al Hussein Bin Talal University, Ma'an, Jordan;* [3]*National Agricultural Research Center (NARC), Baq'a, Jordan;* [4]*National University of Life and Environmental Sciences of Ukraine, Kyiv, Ukraine;* [5]*Amity Institute of Microbial Technology, Amity University Rajasthan, Jaipur, Rajasthan, India;* [6]*Department of Biotechnology, Faculty of Applied, Pure and Health Sciences, Parul University, Quepem, Goa, India;* [7]*Department of Chemistry, Manipal University Jaipur, Jaipur, Rajasthan, India*

1. Introduction to soil pollution

Soil is a complex, heterogeneous system comprising minerals, organic matter, gases, liquids, and a myriad of organisms (Fagnano, 2018). Air, water, and soil together comprise the trio of essential environmental media for all terrestrial life forms. At this interphase, soil degradation is an extremely widespread phenomenon capable of altering the composition and ecotoxicological properties of the other two media. Soil pollution occurs when soil contains contaminants at concentrations exceeding natural levels due to human activities (Rodriguez-Eugenio et al., 2018). The problem deserves attention because, compared with air and water pollution, soil pollution is often persistent, frequently irreversible, and capable of general climatic balances (Cachada et al., 2018). In fact, since the soil represents an environmental matrix that supports and regulates water and air ecosystems, its degradation has large-scale environmental consequences. Soil contaminants originate from the uncontrolled disposal of solid/liquid waste, as well as anthropogenic activities such as intensive agriculture, intensive breeding, land use changes, heavy metal extraction from ores, the use of fertilizers, and the application of toxic heavy metals in industrial activities. Various contaminants, including polychlorinated biphenyls (PCBs), pharmaceuticals, and veterinary drugs, result from these activities.

2. Types of soil pollutants

Soil pollution, degradation, and contamination have been linked to the discharge of waste and the use of hazardous chemicals from industrial processes and agricultural activities. The major contaminants fall into four main categories: heavy metals, pesticides, industrial chemicals, and microbial pathogens. Heavy metals include arsenic (As), cadmium (Cd), chromium (Cr), copper (Cu), mercury (Hg), lead

The Soil Nexus. https://doi.org/10.1016/B978-0-443-45286-4.00006-7

(Pb), and zinc (Zn) originating primarily from industrial manufacturing and waste discharge (Roy et al., 2022). Their environmental persistence results in contamination of surface and groundwater, leading to accumulation in the food chain and posing serious health risks. Pesticides, including herbicides, bactericides, fungicides, and rodenticides, are widely used to protect crops but can also poison non-target species, present in the environment, and contaminate drinking water (Ali et al., 2019). Industrial chemicals encompass polycyclic aromatic hydrocarbons (PAHs), organic halogens, polychlorinated biphenyls (PCBs), polychlorinated dibenzo-p-dioxins and furans (PCDD/Fs), total petroleum hydrocarbons (TPHs), and volatile organic compounds (VOCs). These chemicals enter the environment through leaks, spills, improper waste disposal, and atmospheric deposition through wet or dry precipitation (Canosa & Bertucci, 2023; Kumar et al., 2023). Microbial pollution originates from domestic sewage systems, farmyard sewerage, and breaches in safe water systems. Effluents from chemical factories also introduce a variety of complex mixture of organic, inorganic, and synthetic pollutants (Levin et al., 2024).

2.1 Heavy metals

Heavy metals constitute a group of metals and metalloids that possess a density higher than 5 gcm^{-3} (Briffa et al., 2020; Maria Raffa et al., 2021). Some of these metals serve as micronutrients essential for cellular processes, including cobalt, Cu, Cr, iron (Fe), lithium (Li), manganese (Mn), nickel (Ni), and zn. Nevertheless, metals such as As, cCd, hexavalent Cr, and Pb have no known biological function and exhibit extreme toxicity even at low concentrations. The persistence, bioaccumulation, and biomagnification tendencies of non-essential metals cause them to accumulate in soil, water, and sediments, thereby disrupting biological processes and accumulating in the food chain.

2.2 Pesticides

Pesticides encompass various chemical compounds such as insecticides, herbicides, rodenticides, fungicides, and larvicides intended to eliminate, deter, or control pests. Their usage surged in the 20th century following the introduction of synthetic chemical pesticides like dichlorodiphenyltrichloroethane (DDT), aldrin, dieldrin, 2-methyl-4-chlorophenoxyacetic acid (MCPA), and 2,4-dichlorophenoxyacetic acid (2,4-D). Following World War II, the agrochemical industry underwent substantial global expansion, driven by population growth and technological advancements. Pesticides also found applications in aquaculture, horticulture, household settings, and disease control.

The intensification of pesticide use has led to widespread soil contamination in agricultural soils. Global consumption increased from 1.55 to 2.63 kg ha^{-1} between 1990 and 2018, with fungicides and bactericides constituting the largest fractions. As the agricultural sector continues to develop, reliance on pesticides remains a significant issue despite regulatory efforts. Residues persist in soils, negatively impacting microbial communities and human health through exposure via water, air, and direct contact (Sharma et al., 2019). The toxicity of pesticides toward microorganisms and humans depends on molecular structure, dosage, and exposure duration. Contaminated water extraction for irrigation can promote the further spread of pollutants (Levin et al., 2024).

Bioremediation represents an economically viable and environmentally friendly approach, buoyed by recent scientific and technological progress (Azubuike et al., 2016). Exploiting microorganisms capable of degrading pesticides into less harmful substances primarily CO_2, water, and mineral salts-the strategy encompasses diverse techniques tailored to contamination profiles. In situ methods include natural attenuation, bioaugmentation, biostimulation, bioventing, and biosparging, while ex-situ and on-site options encompass biopiles, bioreactors, composting, and land farming. Operational parameters modulate effectiveness, dictating the selection and deployment of methods (Alori et al., 2022).

Chemical pesticides rank among the most important environmental contaminants, constituting an unequivocal risk to ecological balance. The problem is particularly acute in developing countries due to inadequate infrastructure, limited funding, and ineffective information dissemination (Zhou et al., 2025). Areas subjected to intensive or improper pesticide application risk contaminating food crops, water, air, and soils. Common removal methodologies often suffer from elevated costs, dependence on specialized equipment, high power consumption, the formation of toxic by-products, and a lack of efficiency in fully degrading complex molecules. Hence, adsorption emerges as a particularly suitable technique due to simplicity, cost-effectiveness, high efficiency, and tolerance to toxic substances. Agricultural wastes such as watermelon peels and used tea leaves represent low-cost, non-conventional adsorbents that are readily available and easy to dispose of after treatment. Investigations into their application for pesticide removal from soil and washing water harness batch-mode operations, with potential to convert saturated residues into rodenticides. Agricultural waste sorbents like watermelon peels and tea leaves exhibit reduced selectivity in multi-solute conditions due to competitive adsorption, where ions compete for limited binding sites (e.g., functional groups), lowering efficiency for target pollutants by 20%–50% at higher concentrations. To address this, sequential or selective modification (e.g., acid/alkali treatment) enhances specificity. Pollutant-laden spent adsorbents are safely managed via regeneration (acid/alkaline desorption, 70%–90% recovery), pyrolysis (300–700°C for biochar), or safe landfilling/composting after stabilization to prevent leaching (Tan et al., 2015).

2.3 Industrial chemicals

Industrial chemicals constitute a complex group of compounds with diverse physicochemical properties widely used in modern industry, agriculture, and pharmaceuticals. Chemicals such as chlorinated hydrocarbons (CHCs), organic solvents, polychlorinated biphenyls (PCBs), hydrocarbons, and industrial oils have become widespread environmental contaminants (Gavrilescu et al., 2015). They are generally characterized by low biodegradability and varying degrees of toxicity. Their widespread use in a variety of industrial and commercial activities results in a diffuse release into the environment, with potential impacts on ecosystems. Contamination occurs as a result of faulty land disposal practices, leakage or rupture of containers or pipes, poor maintenance of industrial plants, and accidental discharges or operational releases (Vågen et al., 2016).

Annually, about 4.5 million tons of chemicals are discharged into the soil as a result of industrial activities (United Nations Environment Program (UNEP, 2019). Despite the wide range of chemicals produced and global spread of industrial pollution, organic compounds such as polycyclic aromatic hydrocarbons (PAHs), chlorinated hydrocarbons (CHCs), polychlorinated biphenyls (PCBs), petroleum products, and pesticides receive the most scrutiny. Of particular concern are chlorinated aliphatics (CAs), which are globally widespread pollutants due to their large-scale production and extensive use as metal degreasers, paint and varnish removers, and adhesives. Trichloroethene (TCE) and tetrachloroethene (PCE) are two widely spread CA pollutants. They are toxic and carcinogenic to humans, persisting in soil and groundwater, where they can be stored in various forms, including dense non-aqueous-phase liquids (DNAPLs), as seen in PCE and TCE contamination (Rossi et al., 2022).

2.4 Microbial contaminants

Microbial contaminants constitute a critical category of soil pollutants, emerging when infectious agents, such as bacteria, parasites, viruses, or vectors, are introduced into soil that is depleted of oxygen and nutrients (under anaerobic or nutrient-depleted conditions) (Pande et al., 2022). These contaminants, which may persist in the environment for extended durations, have the potential to travel via water or air and disrupt natural soil processes by altering the activities of soil microorganisms key agents in the organic matter decomposition, pollutant detoxification, and nutrient cycling (Karaś et al., 2021). Soil microorganisms also affect the sorption and bioavailability of many contaminants, thereby modulating the extent of pollution. The input of organic waste products substantially contributes to microbial contamination. Animal wastes are a major source of microbial contamination due to the presence of pathogens (Gavrilescu, 2021). Urban development and farming operations further exacerbate contamination by acidifying valuable soils and deactivating microbes critical for various soil processes and pathogen denitrification. Although some soil microorganisms can facilitate the degradation and immobilization of pollutants with techniques such as bioremediation employing microbial activity to address organic and inorganic contaminants the introduction of foreign microorganisms remains a contentious strategy due to concerns about environmental unpredictability and safety. Microbial contaminants are notably pervasive in environments characterized by inadequate sanitary infrastructure, underscoring the necessity for ongoing research to mitigate associated risks to contour, feed, and human wellbeing (Pande et al., 2022).

3. Impacts of soil pollution

Soil pollution disrupts ecosystems, adversely affects plant growth, and deteriorates soil, water, and air quality. Mitigating the impact of soil pollution on humans and the environment remains a significant challenge. Inorganic compounds, including As, Pb, Hg, Cd, Ni, selenium (Se), (F^-), and Cr, are among the most hazardous contaminants (Roy et al., 2022). Similarly, illegal pesticides, shrimp waste, domestic sewage, and industrial waste materials are common pollutants. Soil pollution can also introduce microbial pathogens and parasites.

Heavy metals and organic compounds often accumulate in raw food materials, fish, and the groundwater used for drinking. Numerous environmental pathways facilitate the transmission of polluted soil to humans, such as agricultural soil, landfills, and fertilizers. Furthermore, hazardous constituents discharged from industrial and mining activities contaminate groundwater, streams, lakes, surface soils, and sediments. Soil pollution endangers approximately three billion people worldwide, and a significant proportion of crop soils, forage, and agricultural lands are contaminated with toxic metals, posing a serious threat to food and energy security (UNEP, 2021). Globally, soil pollution is primarily caused by rapid industrialization, economic development, waste discharge, extensive mining activities, improper agricultural practices, excessive pesticide use, and other anthropogenic activities.

3.1 On human health

Soil heavily contaminated by human activities has been reported to exhibit a sustained release of contaminants into the urban atmosphere, both via dust and aerosols, as well as through subsurface groundwater transport (Rodríguez-Espinosa et al., 2021). This matter makes humans vulnerable to suffering from adverse health effects due to exposure to a series of trace toxic elements. Pollution in urban environments remains a significant challenge due to the ongoing process of urban growth and un-favorable population pressure, despite the clear impact of such contamination on human health.

3.2 On ecosystems

The increasing clarity regarding the impacts of soil pollution on ecosystems, together with escalating atmospheric greenhouse gas concentrations, has intensified public attention on the cascading effects of contamination originating from the soil, as well as on appropriate remediation methods (Orcen, 2017). Atmospheric emissions, wastewater irrigation, and crustal dust contribute substantially to the circulation of contaminants within terrestrial environments. The deposition of pollutants onto forest floors reduces soil nutrient content. These contaminant pools then leach into subsoils and groundwater, entering streams and, eventually, oceans. Once in water, chemicals may concentrate within aquatic biota or remain suspended in the water, allowing them to reach further areas. Ecosystems that provide vital functions and services consequently suffer negative effects at diverse biological scales, reaching from intracellular and individual levels up to communities and entire ecosystems. Ecosystem properties and processes under stress encompass primary production, water and nutrient cycling, soil respiration and degradation, vegetation dynamics, and species composition and diversity, all of which underpin the critical ecosystem services upon which humans depend.

3.3 On agriculture

Increasing land degradation combined with rising productivity demandsimpose considerable pressures on agricultural lands. A considerable body of recent literature advises the development and adoption of innovative decision-support tools that enable a systemic approach to stimulating sustainable behavior in managing farmland and conserving agricultural landscapes. Implemented during the agri-cycle timeline, these tools should serve as a benchmark for soil monitoring, optimize decision-making on innovative management practices, promote soil quality assessments and forecasting, and empower awareness-raising activities. Environmental sustainability in agriculture entails not only achieving specified targets related to land quality but also embracing a proper combination of multifunctionality, competitiveness, resource efficiency, climate change mitigation, and knowledge transfer. Any direct guidelines for managing contaminated soil should be based on a coherent assessment conducted for the entire agri cycle using well documented and experimentally proven methods (Kupincts & Tiutiunnyk, 2017).

4. Regulatory frameworks and policies

Soil pollution is regulated at the international, national, and local levels by a wide range of legislation addressing waste, pollutants, pesticides, and other related issues. Considerable effort is devoted to preventing and mitigating risks. At the national level, pollution prevention (quantified as a rate of pollutant input per unit area of land) and quality guidelines (expressed in terms of both concentrations and loads) help to set priorities and drive action programs. Countries have complied through a variety of measures derived from common principles, including: - Prioritizing sites based on risk, considering viable options and selecting the most sustainable course of action; - Biasing the selection of remedial options in favor of the equally efficient alternative with the lowest consumption of energy, materials and resources, which produces the least waste, releases fewer and less hazardous emissions or is least disruptive to land use; - Developing appropriate partnerships to identify, work toward and maintain sustainable solutions to address both within-site contamination and longer-term diffuse emission sources; - Establishing effective monitoring frameworks able to detect adverse changes in the quality of land, water and ecosystems in advance of significant damage and which support appropriate management responses; and - Considering long-term sources of soil pollution and the distribution of impacts on particular communities. National guidance compels effective and demonstrable progress toward a sub-set of sustainable development objectives, but at the local level, nevertheless, the essential role of the involved stakeholders is still often overlooked (Pollard et al., 2004).

4.1 International treaties

The Revised World Soil Charter 2014 recognizes that soils form a key reservoir of biodiversity. They support soil functions, including maintaining atmospheric and climatic equilibrium, providing a medium for plant growth and biomass production, filtering and buffering, and transforming nutrients and substances, thereby regulating water and air quality. Various international treaties highlight government obligations related to land and environmental management, including the Convention on Climate Change (1992), the United Nations Convention on the Law of the Sea (1982) (UNCLOS), and the Convention on Long-Range Transboundary Air Pollution (1979). The potential benefits of a global soil regime, which would outline the rights and duties of states and safeguards regarding soil, should be considered alongside the challenges of negotiating further soil-related international agreements. A global soil regime is best established under a formal legal instrument to stimulate effectiveness; support developing countries, which often lack the research and information necessary to make informed policy decisions; facilitate the sharing of information and technology related to the ecological condition of land; encourage capacity building, education, and public participation; commit parties to adopt comprehensive national or sub-national legislation for sustainable soil use, supported by institutional systems and legal frameworks; support development of technical guidelines and methodologies for appropriate legislative and institutional arrangements; and provide a mechanism linking existing international instruments and institutions, with guidance on its scope, form, and interpretation available from the International Union for Conservation of Nature and Natural Resources (IUCN) (Boer & Hannam, 2003).

4.2 National regulations

In the USSR, the Soil Code of 1968 established the initial legal framework for soil conservation, followed by the Forest Code of 1977 and the Law on Environmental Protection of 1982. Complementary regulations encompassed Laws on the Protection of Natural Resources (1972), Water (1979), and Wildlife (1983). In Georgia, soil contamination is addressed by the Law on Environmental Protection (1996) and the Law on Ambient Air Protection (1996), which regulate the release of air pollutants, the establishment of protected natural areas, and environmental conservation (Kupinets & Tiutiunnyk, 2017). The Ukrainian Government's Decree No. 832 (Aug 25, 1998) delineates a system of ecological norms, standards, and limits, including Maximum Contaminant Limits (MCL) for soils and soils around protected areas, as well as Excess Degrees of Contamination (EDC) for complex risk evaluation to water, soil, plants, and humans. Since fertilizer Cd/ha (~0.1–1 g) is less than ambient soil Cd/ha (~960 kg), fertilizer Cd inputs contribute very little to soil Cd burdens (~1 ng/kg soil increase per application); if leaching were to be ignored, it would take approximately 1800 years to double soil Cd levels (Kupinets & Tiutiunnyk, 2017). Policies should prioritize soil pH management (>6.5) and liming to minimize Cd bioavailability via reduced solubilization in addition to low-Cd phosphate fertilizers (<20 mg/kg P_2O_2) only if site-specific mass-balance confirms risk, in addition to generic MCL compliance (Briffa et al., 2020; Roy et al., 2022). This increases the effectiveness of remediation by focusing on the underlying causes acidulation and ion effects over input limits (Kupinets & Tiutiunnyk, 2017). Soil contamination remains one of the most difficult environmental problems and the control of soil pollution is frequently overlooked in state programs and policies. Both the USA and the EU have adopted comparable legislation to control contaminated site investigations and clean-up procedures. Moreover, numerous environmental treaties signed worldwide serve to prevent soil pollution and promote corrective measures to restore degraded sites.

4.3 Local initiatives

Local initiatives that enhance crop productivity address specific local problems, including conflict resolution, afforestation, and water management. Land users, often farmers, as well as local government and non-governmental organisations, shareresponsibility for these activities. Understanding the processes and feedback associated with land degradation enables intervention that addresses the root causes rather than the symptoms. Factors influencing land management and restoration include the limited options for intensifying production, high economic risk, poor market conditions, and unstable institutions, such as land tenure and water users' groups. Successful interventions emphasize strong community participation, situated within the context of local needs and demands. Incentive mechanisms which may include targeted subsidies, access to capital, resolution of land tenure issues, and reward policies encourage farmers to make investments that yield both private and social benefits. Removing barriers and creating new ones can shift the focus from reactive rehabilitation to proactive maintenance of soil and land health, as well as ecosystem health.

5. Assessment of soil pollution

The assessment of soil pollution is the indispensable preliminary step required to effectively reduce the dispersion of contaminants, particularly in agricultural settings. This assessment is crucial for the early detection of pollution and the identification of its sources. It enables the accurate quantification of contamination levels, thereby guiding the design of appropriate prevention and management strategies. Successful examination requires systematic soil sampling, combined with a suitable analytical procedure, that provides reliable data on the nature, concentration, and bioavailability of pollutants in various environmental compartments (Swartjes, 2011).

Soil pollution constitutes the largest repository of waste worldwide (Roy et al., 2022). Notwithstanding the crucial role of soil protection outlined in Directive 2004/35/CE, the European Union still lacks an overarching soil legislation framework, despite several initiatives. The absence of a regulatory regime increases the complexity of pollution assessment, which remains necessary from both environmental and human health perspectives.

Several international organizations have encouraged state members to take initiative in this field. The Ministry of Environment of the People's Republic of China implemented a programme called "Action Plan for Soil Pollution Prevention and Control" to prevent and control soil pollution and contaminated soil sites, to achieve sound soil quality that will be considered a vital environmental factor and efficiently protect people's health. The United States EPA developed a review process framework to select appropriate risk-based methodologies for estimating sources and types of contaminants released. It also measures the potential adverse impacts on human health and the environment. In addition, the EPA proposed the "referential soil screening value" and the "soil screening level" to support site assessment and restoration, respectively. The Canadian Environmental Quality Guidelines define soil quality criteria correspondingly (Fagnano, 2018).

5.1 Soil sampling techniques

Soil pollutants are typically present at extremely high concentrations in hotspots, necessitating rapid remediation to mitigate environmental risk and prevent the expansion of polluted areas. Different soil sampling techniques are commonly used, such as discrete, composite, and grid sampling (Carter & Gregorich, 2007; Fagnano, 2018). Discrete sampling, which involves sampling at individual, defined locations, is often used when pollution sources are point-specific or where spatial variability is high. However, the presence of soil heterogeneity may affect data interpretation, necessitating the collection of many samples distributed over a very large area. Composite sampling is more appropriate for obtaining an average concentration over a wide area; it consists of mixing the individual samples collected according to a predetermined criterion, usually randomly or systematically, into a single sample for laboratory analysis. Finally, grid sampling is a combination of discrete and composite sampling where samples are collected from nodes, squares, or triangles and submitted either independently or composited by block.

5.2 Analytical methods

Soil pollution is defined as the contamination of soil by human-made chemicals or other alterations (physical changes) that can cause a change in the soil environment and ecosystem, resulting in various problems for human health. It typically occurs as a result of the presence of xenobiotic (manufactured) chemicals or other alteration such as radioactive contamination (Münzel et al., 2023; Rai et al., 2023). Pollutants often leak into the underlying groundwater. Pollutants present in soil can also enter the food chain through plants growing in the affected soil, animals feeding on these plants, and, as a result, pollute the food chain, harming consumers. The most common anthropogenic soil pollutants include inorganic contaminants, such as heavy metals like Cr, Pb, and As; organic pollutants like polycyclic aromatic hydrocarbons (PAHs), polychlorinated biphenyls (PCBs), and polyaromatic ethers; xenobiotic agrochemicals; e-wastes; and high concentrations of both organic and inorganic salts (Roy et al., 2022). Several analytical techniques are being utilized for the detection of soil pollutants from the contaminated soil matrix (Cr; Pb; As; polycyclic aromatic hydrocarbons (PAHs); polychlorinated biphenyls (PCBs); polyaromatic ethers; xenobiotic agrochemicals; e-wastes; organic salts (high concentration); inorganic salts (high concentration) (Liu et al., 2023; Xia et al., 2024). Physical, chemical, and biological analytical methods are available for determining pollutants. An analytical method stands on the long history of many divisions of the analytical chemistry discipline. It employs the principles of an analytical technique, which encompasses concepts such as measurement, detection, concentration, and separation (Bansod et al., 2017). In addition to spectroscopy, sensors and chromatographic techniques are widely used for separation through the application of instruments.

Tiered analytical workflow linking remote sensing, ultra-trace aerosol metal sensing, and ground-based laboratory analyses. In addition to conventional laboratory analytics, an operationally efficient soil-pollution investigation can be organized as a tiered assessment strategy that combines (i) landscape-scale satellite screening, (ii) rapid ultra-trace aerosol metal detection for pathway/source indication, and (iii) confirmatory ground sampling with laboratory analysis for regulatory-grade quantification and risk assessment.

Tier 1—Sentinel-2 time-series indices for screening and prioritization. Sentinel-2 satellite image time series can be processed into spectral indices and anomaly maps to provide repeated, broad-area screening that highlights parcels or zones requiring follow-up, rather than serving as a stand-alone diagnostic of contamination (Vågen et al., 2016; Xia et al., 2024). For example, index-based monitoring approaches have been used to track field-scale changes associated with exogenous organic matter applications using indices such as EOMI and standard vegetation indices (NDVI, EVI), and machine-learning models have been trained for parcel-level detection (reported F1-scores up to 0.85), demonstrating the feasibility of scalable, repeatable screening from Sentinel-2 data streams (Xia et al., 2024). The Tier 1 output is a ranked set of "priority polygons" (persistent deviations from baseline, abrupt changes, or spatial patterns consistent with known transport pathways) that directly informs where and when to intensify field measurements (Vågen et al., 2016).

Tier 2—Ultra-trace metal aerosol probing (FIFS) to resolve atmospheric contribution and guide sampling geometry. Where atmospheric deposition, fugitive emissions, or resuspension are plausible contributors to soil metal loading, Filament-Induced Fluorescence Spectroscopy (FIFS) provides a rapid, remote approach for screening metal-bearing aerosols and identifying short-lived plumes or micro-hotspots that may be missed by periodic soil sampling (Bansod et al., 2017).

Tier 3—Ground sampling plus laboratory analysis (confirmation, speciation/bioavailability, and compliance). Tier 3 applies established soil sampling designs (discrete, composite, grid) to the Tier 1–2 prioritized areas to quantify total concentrations and, where relevant, bioavailable fractions and co-contaminants using appropriate analytical chemistry workflows (Carter & Gregorich, 2007). This tier provides the evidentiary basis for risk assessment, source apportionment follow-up, remediation selection, and post-remediation verification, while the tiered design reduces cost by increasing sampling density only where screening indicates elevated likelihood or consequence (Fagnano, 2018; Swartjes, 2011).

5.3 Risk assessment protocols

Risk assessment protocols are implemented to analyze and evaluate soil pollutants for effective remediation and management (Fagnano, 2018). The assessment involves estimating soil quality and health, which mainly depends on various environmental conditions coupled with human activities. The subsoil tension value, for example, plays an important role in nitrogen movement within soil profiles due to soil-water interactions. The state-of-the-art protocol, called ECORISK, is an integrated agronomic approach to characterizing contaminated soils and assessing risk according to European law (Vanni et al., 2016). The economic and social feasibility of remediation and decontamination actions has also been taken into consideration. An agricultural risk index is generally obtained based on specific risk models, as per the regulatory framework of the Land Protection Act.

6. Bioremediation techniques

Soil pollution originates from many sources. Heavy metals such as Pb, Cd, Hg, Cr, Zn, Ni, and Cu enter the environment through industries such as tannery and electroplating, as well as through refinery waste, traffic emissions, coking, and coal power plants (Roy et al., 2022). The increased use of pesticides, often entering the soil via spraying or direct application, also contributes to pollution. The same is true of phenolic compounds, which are primarily associated with the dying and paper-making industries. Inorganic chemicals such as sodium chloride (NaCl) and sodium sulfide (Na_2S), widely used in textile industries for dyeing, and sulfur dioxide (SO_2) emitted by power plants significantly affect the soil. Microbiological contamination arises from sewage, effluents, animal carcasses, hospital waste, and heavy metals. Fig. 15.1 summarizes an integrated remediation workflow (site characterization → pre-treatment → active treatment → post-treatment monitoring) and explicitly links (i) dominant contaminant classes (metals/metalloids vs. organics), (ii) principal removal pathways (degradation/transformations vs. immobilization), and (iii) field constraints that typically govern performance (moisture, temperature, pH, oxygen/redox, and soil texture/organic matter) (Azubuike et al., 2016; Biswas, 2015; Pande et al., 2022).

Waste generated by various sectors is often buried directly in the soil, creating bans on such areas and degrading the state's natural capital. Remediation of polluted soil is therefore of prime importance. Different methods have been proposed to mitigate soil pollution with heavy metals and polycyclic aromatic hydrocarbons (PAHs) in aqueous systems. Among available techniques, bioremediation is considered eco-friendly, economical, the least disruptive to the landscape, and the safest for workers and the public. Bioremediation utilizes the natural ability of certain organisms such as microbes or plants that degrade, transform, or accumulate contaminants (Biswas, 2015). Microorganisms use different mechanisms for contaminant removal. Microbial remediation of heavy metals involves fungi, bacteria, actinomycetes, and algae, with strategies including bioaccumulation, biosorption, biomineralization, biotransformation, biomethylation, and biodegradation processes (Pande et al., 2022).

Decision criteria for selecting in situ versus ex situ biological treatment: Prefer in situ when contamination is shallow to moderate, the formation is permeable enough to distribute oxygen/electron donors/acceptors, and contaminants are predominantly biodegradable organics with manageable heterogeneity; in situ options include natural attenuation, biostimulation, bioaugmentation, bioventing, and biosparging (Alori et al., 2022; Azubuike et al., 2016). Prefer ex situ/on-site (excavation + biopiles/composting/landfarming) or engineered bioreactors when contamination is deep, access is constrained, soils are low-permeability (delivery limitations), source zones/hotspots require rapid mass removal, or the plume is mixed/complex (e.g., organics with metals) requiring tighter control of moisture, aeration, nutrients, residence time, and treatment sequencing (Biswas, 2015;

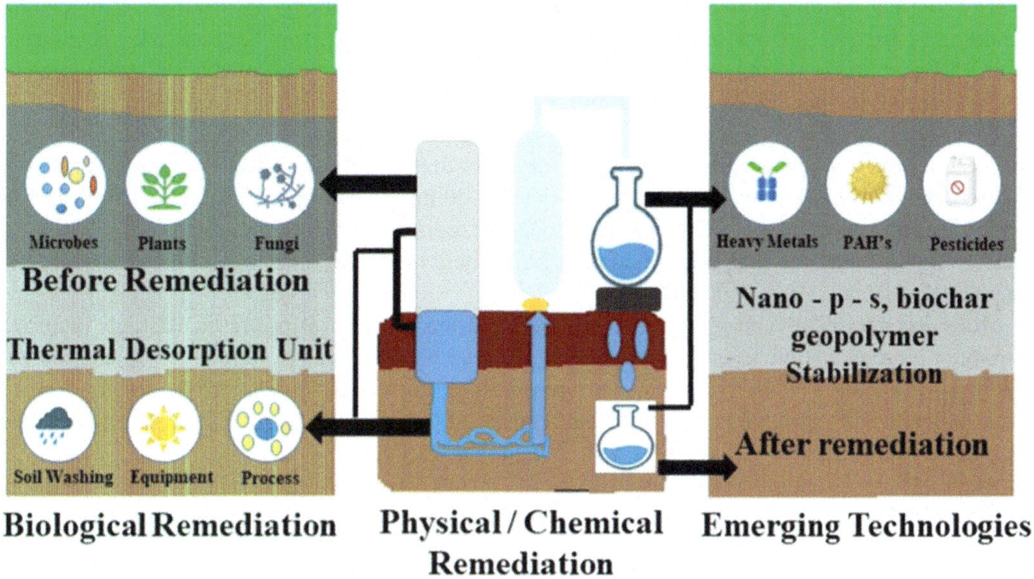

FIGURE 15.1

Integrated approaches for remediation of contaminated soils across the project lifecycle (pre-treatment, treatment, and post-treatment). Biological options (microbial, plant, fungal) primarily enable degradation/transformation of organic contaminants and/or biologically mediated immobilization; physico-chemical and nano-enabled stabilization options are commonly used to accelerate removal of organics or immobilize metals, especially where field constraints (moisture, temperature, pH, oxygen/redox, and heterogeneity) limit biological activity.

Pande et al., 2022). For mixed contaminant profiles, use a staged approach: biologically degrade organics where feasible, while managing metals primarily via immobilization/stabilization and controlling bioavailability; hybrid trains (e.g., excavation of hotspots + in situ polishing) are often most practical (Wuana & Okieimen, 2011).

6.1 Microbial bioremediation

Bioremediation, an environmentally friendly, non-invasive, and cost-effective method, has gained prominence in studies focused on soil pollution control. Indigenous microbes that utilize hydrocarbon contaminants often exist in polluted environments; specific populations can be increased by adjusting environmental factors (Biswas, 2015). Effective bioremediation requires demonstrating the presence of an active microbial population capable of degrading the target contaminants. Hydrocarbon-utilizing bacteria are usually present if the site contains petroleum hydrocarbons and has water content, pH, and porosity within reasonable limits (Hamouda et al., 2023). Thus, initial monitoring consists of chemical analyses to identify pollutants and microbiological techniques to quantify microbial populations. Detection of hydrocarbon-utilizing microorganisms is a routine, rapid task; enrichment cultures can reveal important degraders even when pure culture isolation is challenging. Biological remediation technologies harness natural organisms to degrade or immobilize soil contaminants sustainably. Table 15.1 compares microbial bioremediation, phytoremediation, mycoremediation, algal remediation, and bacterial consortia across target pollutants, mechanisms, advantages, limitations.

To ensure that Table 15.1 and Fig. 15.1 align, selection of a biological option should be made by matching (i) the dominant contaminant class (organics that can be degraded vs. metals that largely require immobilization or valence change), (ii) the dominant mechanism(s) expected under site conditions, and (iii) the main field constraints (moisture regime, temperature seasonality, and pH/redox controls on bioavailability and microbial/plant/fungal activity) (Azubuike et al., 2016; Biswas, 2015; Pande et al., 2022).

Molecular microbial-ecology methods that do not depend on cultivation are increasingly employed to monitor bioremediation. Changes in the microbial community can be examined in conjunction with analytical techniques such as gas chromatography, high-performance liquid chromatography, and tests using radiolabelled substrates to confirm biodegradation or mineralization. These approaches have been successfully applied in laboratory and field studies.

Algae species including Zooglea sp., Phormidium valderium, and Chlorella vulgaris can remove heavy metals such as Co, Ni, Cd, Cu, Hg, Zn, Au, U, Pb, and As from polluted environments (Mani & Kumar, 2014). Higher plants including Helianthus annuus, Brassica juncea, and Solanum nigrum efficiently remediate Cu, Pb, Cr, Ni, Zn, Cd, Ag, Se, Hg, Ti, and chlorinated aliphatic (Muthusaravanan et al., 2018; Rizwan et al., 2016). Phytoremediation offers an attractive alternative to physical and chemical methods, as it improves soil health and ecosystem quality while remaining economically viable. Enhancing understanding of the physiological and biochemical responses of microorganisms and plants, as well as insights into their genomes, expands their capacity to degrade pollutants and contributes to the restoration of contaminated land and water sites (Zhang et al., 2024).

6.2 Phytoremediation

Vegetation-based remediation processes termed phytotechnologies can remediate, contain, or prevent contaminants in soil, sediments, and groundwater; enhance soil quality by adding nutrients, porosity, and organic matter. Phytotechnologies provide an economical, in situ approach that reduces the amount of contaminated waste requiring disposal (Pilon-Smith, 2005). Typical soil contaminants include organic chemicals, such as petroleum hydrocarbons and chlorinated solvents, as well as metals, which are also relevant in older sites. Only organic substances can be transformed through phytotechnologies mainly phytodegradation and rhizodegradation. Inorganic contaminants, such as metals, are either phytoextracted or immobilized through stabilization. Various phytotechnology processes exist, including phytodegradation, rhizodegradation, phytovolatilization, phytometabolism, phytoextraction, phytohydraulics, phytostabilization, and rhizofiltration. The latter is commonly applied in water and sewage management, constructed wetlands and stormwater filters (Muthusaravanan et al., 2018).

Recent field-scale demonstrations further support the practicality and scalability of these green technologies. For example, large-scale phytoremediation systems using emergent and aquatic macrophytes have been deployed to remove metals such as chromium and mercury, pharmaceuticals, and other organics from contaminated waters and adjacent soils, highlighting the feasibility of long-term

Table 15.1 Comparison of biological remediation technologies.

Technology	Primary target pollutants (soil)	Core mechanisms (use consistent verbs)	Field constraints/operating windows (moisture–temperature–pH and key controls)	Advantages	Limitations (make the constraint explicit)	Key references
Microbial bioremediation (native or augmented bacteria/actinomycetes)	Petroleum hydrocarbons; PAHs; some pesticide residues; metals/metalloids mainly via immobilization/valence change rather than "destruction"	Biodegradation/mineralization of organics; cometabolism; biosurfactant-assisted mobilization; biotransformation and biomineralization; biosorption/bioaccumulation (mainly for metals)	Needs bioavailable substrate, nutrients and electron acceptors; performance drops under very low moisture (limited mass transfer) or waterlogging (oxygen limitation) and outside near-neutral pH where many degraders are less active; temperature controls kinetics	In situ, comparatively low-cost and minimally disruptive	Slow at low temperature/dry soils; inhibited by extreme pH/salinity/toxicity; metals cannot be degraded—often only immobilized or transformed	Pande et al., 2022; Biswas, 2015; Hamouda et al., 2023; Chikere et al., 2011
Phytoremediation (plants + rhizosphere)	Metals/metalloids in topsoil (uptake or stabilization); petroleum hydrocarbons/solvents and some pesticides via rhizodegradation/phytodegradation	Phytoextraction; phytostabilization; rhizodegradation (root-stimulated microbial degradation); phytodegradation; phytovolatilization (limited cases)	Root-zone limited (typically shallow treatment depth); requires plant establishment—thus constrained by soil moisture regime, seasonal temperature, and pH that controls metal bioavailability and phytotoxicity; drought/waterlogging reduces growth and rhizosphere activity	Low-input, improves soil structure/organic matter; high public acceptance	Long timeframes; not suitable for deep/hotspot contamination; biomass handling may be needed when metals are phytoextracted	Muthusaravanan et al., 2018; Pilon-Smith, 2005
Mycoremediation (fungi; often ligninolytic)	Recalcitrant organics (PAHs, some pesticides); some chlorinated organics; metals mainly via sorption/immobilization	Extracellular oxidative enzymes (laccases, peroxidases) for polymeric/recalcitrant organics; biosorption to mycelium; co-metabolic transformation	Strongly constrained by moisture and temperature (mycelial growth and enzyme activity), oxygen availability, and pH affecting enzyme stability; requires suitable substrate/support matrix and competition control in field	Broad enzyme repertoire; effective for difficult-to-degrade organics	Field scale-up can be difficult; sensitive to drying/over-wetting and suboptimal temperature; needs cultivation/logistics	Harms et al., 2011

Continued

Table 15.1 Comparison of biological remediation technologies.—cont'd

Technology	Primary target pollutants (soil)	Core mechanisms (use consistent verbs)	Field constraints/operating windows (moisture–temperature–pH and key controls)	Advantages	Limitations (make the constraint explicit)	Key references
Algal remediation (microalgae/cyanobacteria; usually ex situ or engineered in situ niches)	Metals/metalloids (biosorption/uptake); nutrient co-removal in some systems	Biosorption to cell surfaces; bioaccumulation; precipitation/immobilization via pH shifts in microenvironments	Requires sufficient moisture/water phase and light (if photosynthetic); pH and ionic strength strongly affect metal binding and biomass stability; biomass harvesting/disposal is a practical constraint	High binding capacity; rapid biomass generation under suitable conditions	Limited full-field soil application; harvesting and safe disposal/regeneration of metal-laden biomass required	Mani & Kumar, 2014
Bacterial consortia/engineered mixed cultures (often with amendments)	Complex petroleum mixtures; mixed organics; some metal(loid) transformations (e.g., Cr(VI) reduction) and co-contaminated sites	Synergistic biodegradation (division of labor); amendment-assisted metabolism (co-substrates, nutrients); redox-driven transformations	Requires control of competing electron acceptors/donors and moisture–oxygen balance; temperature and pH shifts can destabilize community composition and pathway expression; needs monitoring to maintain function	Higher robustness and broader substrate range than single strains	Requires consortium design/optimization and field monitoring; may fail if site conditions fluctuate (moisture/pH/temperature)	Chikere et al., 2011; Ha mouda et al., 2023

operation under real site hydrological and geochemical constraints (Muthusaravanan et al., 2018; Rizwan et al., 2016). Similarly, an in situ coupled bioventing–biosparging system integrated with phytoextraction has been successfully applied at a semi-saturated construction site co-contaminated with petroleum hydrocarbons and cement-derived heavy metals, where oxygen-enriched air injection in the deep vadose and saturated zones, combined with metal-accumulating plants in the shallow soil, achieved substantial reductions in BTEX, PAHs and heavy metal concentrations while avoiding excavation and off-site treatment (Pilon-Smith, 2005). Together, these studies provide robust case-based evidence that integrated phytoremediation and enhanced in situ bioremediation can deliver cost-effective restoration of complex, mixed-contaminant soils at field scale (Muthusaravanan et al., 2018; Sharma et al., 2019).

Phytoextraction involves plants absorbing and translocating metals to harvestable shoots. Phytostabilization uses plants to form a layer of roots and/or shoots that stabilize contaminated soil, whereas rhizofiltration employs plants to clean aquatic environments. Phytodegradation describes the uptake, metabolism, and subsequent degradation of organic contaminants. Some plants extract volatile metals from soil and release these into the atmosphere through volatilization (Rizwan et al., 2016). Species selected for phytoremediation should exhibit fast growth, high biomass yields, substantial metal tolerance, hyperaccumulative capacity, ease of cultivation, and straightforward harvestability (Zahangir et al., 2022). Such plants undertake detoxification through mechanisms that include exclusion, excretion, accumulation, antioxidant activity, and chelation via metallothionein and phytochelatin pathways. Microbial activity further enhances phytoremediation by promoting contaminant removal, particularly with heavy metals, and by supporting plant growth in contaminated habitats. Relevant microbial groups encompass phosphate-solubilizing bacteria, endophytes, mycorrhizal fungi, and plant growth-promoting rhizobacteria (Sharma et al., 2019).

6.3 Mycoremediation

Mycoremediation employs fungal species to degrade, transform, or immobilize environmental contaminants, making it a promising approach to restoring polluted soils (Harms et al., 2011). Both micro- and macrofungi saprotrophic and biotrophic basidiomycetes in particular—participate in the enzymatic degradation of diverse contaminants (Shourie & Vijayalakshmi, 2022). The fungi produce a wide array of extracellular enzymes, including laccases, manganese peroxidases, lignin peroxidases, other peroxidases, tyrosinases, and hydrolases, which catalyze the transformation and detoxification of several major classes of soil pollutants. Intracellular degradation is also driven by enzymes such as cytochrome P450 monooxygenases and glutathione transferases (Dinakarkumar et al., 2024). Physico-chemical parameters govern biodegradation efficiency; hence, processes can be stimulated through biostimulation or enhanced via bioaugmentation to achieve optimal pollutant removal (Tan et al., 2015).

Laboratory-scale microcosm studies constitute a valuable tool for evaluating the practicability of selected biodegradation operations. Mycoremediation has the potential to become a crucial technique in soil conservation and restoration, providing a straightforward, versatile, and cost-effective approach for detoxifying polluted sites and preserving soil quality. The application of fungal technologies is expected to gain increasing significance in environmental remediation in the coming years (Mohamadhasani & Rahimi, 2022). Table 15.2 presents a quantitative performance metrics for soil remediation classes are summarized below, drawing from established field studies and reviews.

Table 15.2 Quantitative benchmarks for soil remediation technologies: Efficiency, timeframe, and cost (field ranges).

Remediation class	Removal efficiency (typical/Best-case)	Treatment duration	Indicative cost (US $/ton soil)	Representative field case
Microbial	50%–80%/>95%	Months to 2 years	50-200	Bacterial consortia on petroleum/Cr (VI) sites (Hamouda et al., 2023)
Phyto-	40%–70%/90%	1–10 years	5-40	Hyperaccumulators (e.g., brassica juncea) on mine/agricultural soils (Muthusaravanan et al., 2018)
Myco-	60%–85%/>95%	3–12 months	30-150	Fungal enzymes (laccases) on PAHs/pesticides sites (Harms et al., 2011)
Adsorption	70%–90%/98%	Weeks to months	50-250	Biochar/agricultural wastes on pesticides/heavy metals (Beesley et al., 2011)
Stabilization	80%–95% leachability reduction/>99%	Days to weeks (initial); years monitoring	50-150	Cement/lime binders on heavy metals

7. Physical and chemical remediation methods

A wide array of technologies and strategies have been deployed globally to address heavy metal contamination in soils and sediments, a persistent hazard in many regions. Conventional remediation methods encompass physical and chemical treatments, such as soil washing, electrokinetics, thermal desorption, and vitrification, as well as chemical immobilization using activated carbon or limestone. Excavation and landfilling of contaminated soil represented complete yet costly solutions. Surface capping and encapsulation provided rapid yet temporary containment of pollutants. Electrokinetics enabled removal of ionized metal contaminants through electrically induced adsorption. Chemical amendments sought to precipitate or sequester heavy metals, delivering quick but transitory results (Wuana & Okieimen, 2011). Several innovative technologies also demonstrated promising potential for restoring polluted soils. Geo-polymer remediation approaches have yielded solidification and stabilization of heavy metal contaminants, demonstrating their applicability across diverse materials. Recycling methods generated reusable products from contaminated sediments through a three-stage cleaning process. Multi-walled carbon nanotubes effectively adsorb a range of heavy metals, however, their application is constrained by cost and potential ecotoxicity. Nanomaterials, including zero-valent iron particles, induced concurrent oxidation and reduction of soil pollutants. Electrothermal techniques, which synergistically utilize asymmetrical alternating current and electrochemical generation of oxidants, were employed to eliminate multiple pollutants in soil.

7.1 Soil washing

Soil washing is a well-established off-site remediation technique for treating soils contaminated by toxic metals (Dermont et al., 2008). The process involves washing the soil surface and pores with a chemical solution that transfers metals from the soil to the solution, thereby facilitating their removal. This method provides rapid and highly efficient metal extraction, significantly reducing associated ecological risks. In addition, it is relatively simple to implement and cost-effective compared to many other remediation options (Nkonya et al., 2016).

The type of soil and contaminants present greatly influence performance, as well as the nature of the washing agents employed. Various solutions have been used, including inorganic acids, organic acids, surfactants, and chelating agents. Some possess drawbacks, such as altering soil properties and potential toxicity (Dermont et al., 2008). More recently, natural chelating agents have garnered considerable interest due to their high metal binding capacity, along with biocompatibility and biodegradability.

Although chitosan fulfills the first two criteria, its application to soil remediation is limited by poor water solubility at neutral pH (Pande et al., 2022). Typically, the soil is excavated and may be subjected to pretreatment steps such as the removal of oversized debris and consolidation into discrete clods. The material is then agitated with fluids to transfer contaminants into the aqueous phase or to concentrate pollutants within a much reduced soil volume.

Size separation screening is often employed to enhance the latter effect. The subsequent handling strategy normally integrates soil washing with complementary methods such as incineration, thermal desorption, bioremediation, or landfilling. The system architecture commonly comprises five main functional units dedicated respectively to pretreatment, contaminant separation, coarse-grained treatment, fine-grained treatment, water purification, and residuals management. One influential parameter is the soil particle size distribution, as higher contents of sand and gravel generally improve the removal performance.

Persistent surficial contamination of these fractions can nevertheless occur because adhesion and compaction promote residual entrapment of fine solids. Abrasive scouring or scrubbing can fully restore separation efficiency; in some cases, the addition of chemicals proves beneficial. Prior to recycling, residual contaminants are extracted from the fluids employed for separation. Mechanical agitation ensures intimate contact between soil and washing fluid thereby supporting efficient contaminant transfer.

Physical and chemical methods provide rapid, high-efficiency contaminant removal for diverse soil pollution scenarios. Table 15.3 evaluates soil washing, thermal desorption, chemical stabilization, electrokinetics, and vitrification based on target contaminants, processes, efficiency, costs.

7.2 Thermal desorption

Thermal desorption constitutes a mature and widely employed soil remediation technology that rapidly heats contaminated medium to volatilize organic pollutants, subsequently capturing them in the resulting gas stream (Khan et al., 2004). During thermal desorption, contaminated soil is heated to

Table 15.3 Physical and chemical remediation technologies: Performance and cost analysis.

Technology	Target contaminants	Process description	Efficiency	Cost factors	Citation technologies-for-soil-pollution-reduction-and-restoration-final.docx
Soil washing	Heavy metals, PAHs, pesticides	Chemical extraction + physical separation	70%–95% metal removal	Moderate (100-300 $/ton), soil type dependent	(Dermont et al., 2008)
Thermal desorption	VOCs, SVOCs, oil, PCBs, pesticides	Heating (90–600°C) to volatilize organics	>95% for volatiles, soil texture dependent	High energy costs (200-500 $/ton)	(Khan et al., 2004)
Chemical stabilization/ solidification	Heavy metals, inorganic wastes	Binders (cement, lime) immobilize contaminants	>90% leachability reduction	Low-moderate (50-150$/ton)	(Falciglia et al., 2012)
Electrokinetics	Heavy metals, charged organics	Electrically induced migration	60%–90% for ionic species	High electricity costs, slow	(Wuana and Okieimen, 2011)
Vitrification	Heavy metals, radionuclides	High-temperature glass formation	>99% immobilization	Very high capital costs	(Gill et al., 2020)

drive off organic contaminants without destroying the soil or its organic carbon content. Thermal desorption methods encompass a broad temperature range, typically from 90 °C to 560°C. Heating mechanisms include conductive heating via thermal conduction, convective heating through hot gas injection, and radiative heating. Higher treatment temperatures facilitate the efficient degradation of resistant pollutants, such as dioxins and furans (Canosa & Bertucci, 2023; Kumar et al., 2023).

Thermal desorption systems can be categorized into three types: direct, indirect, and contact heating. Direct-heating systems rely on the injection of hot gases during gas-solid contact, whereas indirect-heating systems do not require such interaction. Contact-heating systems employ a heated medium that contacts the material without direct exposure between the bed and the heat source (Khan et al., 2004).

In the presence of oxygen, thermal treatment processes can be further classified as thermal desorption or thermal oxidation. Thermal desorption, also known as "low-temperature thermal decontamination," operates within a temperature range of 90–600 °C. In contrast, thermal oxidation, or "high-temperature thermal decontamination," occurs at elevated temperatures where organic compounds undergo oxidation into carbon dioxide, water, and light hydrocarbons.

Thermal desorption is highly effective for removing a wide spectrum of organic contaminants, including volatile and semi-volatile compounds such as solvents, oil, gasoline, paint, creosote, PCBs, pesticides, and other hazardous pollutants. However, the technology is generally not considered suitable for soils contaminated with heavy metals, PCBs, or coal tar. The efficacy of lower-temperature thermal desorption depends on the type of soil texture; loamy sand achieves maximum efficiency with a 15-minute exposure, while loamy soil exhibits reduced effectiveness even at 5-minute treatment durations.

7.3 Chemical stabilization

Chemical stabilization has been recognized for centuries as an effective method to enhance soil strength and increase resistance to water-induced softening (Reddi & Inyang, 2000). The technique achieves this by binding soil particles, water-resistant them, or employing a combination of both mechanisms. Heavy metal-contaminated soils tend to exhibit increased erodibility, with specimens polluted with lead found to be more susceptible than their uncontaminated counterparts. Consequently, soil stabilization serves as a cost-effective, long-term remediation approach wherein pollutants such as Pb are either immobilized in a stable, insoluble form or encapsulated within a cementitious matrix (Pham et al., 2024).

Solidification/stabilization (S/S) is similarly regarded as an affordable remediation technology for contaminated sites and hazardous waste treatment (Chen et al., 2022). Traditionally, S/S applications have been limited to inorganic wastes due to their resistance to thermal and biological degradation, coupled with the need to retard their mobility in aqueous environments. Increasing attention is now focused on extending S/S to hazardous organic wastes, particularly those arising from processes such as the Synthol method in the oil industry. Thermally based techniques tend merely to reconfigure these materials rather than eliminate them, and established bioprocessing methods are unsuitable for the vast majority of organic sludges (Luhar et al., 2023). Organics are therefore fixed at source employing either organic or inorganic binders. Synthetic products including asphalt, modified clays, and thermoplastics such as polypropylene have been introduced as binders. Although effective when carefully selected and applied, they require specialized equipment and controlled conditions that raise costs and complicate implementation (Nkonya et al., 2016).

Laboratory evaluations demonstrate that admixtures comprising Portland cement, bentonite, and water can be optimally blended to achieve solidification/stabilization (S/S) of contaminated soils (Sainju et al., 2003). Analyses using protocols such as the Synthetic Precipitation Leaching Procedure (SPLP) and the Toxicity Characteristic Leaching Procedure (TCLP) establish reductions in contaminant leachability. Air emission potentials assessed via TO-3 testing during treatment further inform operational parameters. The chosen additives serve both to physically bind pollutants physically thereby reducing their mobility and to induce chemical transformations rendering hazardous constituents less reactive or toxic. Substitution of Portland cement with organo-clay binders provides comparable performance, broadening the palette of remediation options.

8. Innovative technologies for soil restoration

Innovative restoration technologies demonstrate considerable potentialin rapidly and effectively removing persistent soil pollutants. Nanotechnologies advance conventional soil remediation approaches by interacting with contaminants at a molecular scale, enabling degradation or transformation through high surface reactivity. However, challenges remain in achieving widespread application (Sadegh et al., 2016). Biochar consistently improves soil structure, water and nutrient-holding capacity, pH, biological activity and microbial biomass (Tan et al., 2015). Geo-polymer technologies restore structural integrity, immobilize contaminants through encapsulation or fixation, and enhance ancient building materials, thereby bridging the gap between construction and environmental science (Wuana & Okieimen, 2011). Beyond proof-of-concept, geopolymer remediation needs process standardization particularly for activator solution preparation because undefined batching/equilibration practices undermine reproducibility and scale-up; recent thermochemical modeling explicitly motivates SOP development for ready-to-use, stable activators (Pham et al., 2024). For environmental deployment, durability and contaminant-retention should be reported using coupled mechanical integrity and scenario-based leachability metrics under realistic stressors (e.g., wet–dry cycling, temperature variability, carbonation/acidic infiltration), since monolithic solidified/stabilized matrices often leach by diffusion and performance is scenario-dependent (Chen et al., 2022; Luhar et al., 2023).

8.1 Nanotechnology applications

Nanotechnology is a branch of science and engineering that focuses on the synthesis and development of materials on the atomic, and molecular, and nanoscale. Many materials exhibit a drastic change in toughness, appearance, conductivity, and sensitivity between the macro- and nanoscales (Table 15.4). Nanomaterials possess a high surface-area-to-volume ratio, which confers greater biogeological and chemical reactivity. Applications extend to various fields, including environment, agriculture, and medicine. In particular, the removal of pollutants through processes such as surface assimilation, absorption, catalysis, oxidation, screening, and filtering is widely investigated. Nanomaterials have been utilized for soil and groundwater remediation at hazardous waste sites contaminated with chlorinated solvents, heavy metals, and oil spills (Crane & Scott, 2012).

Silver nanoparticles (AgNPs) are among the most widely commercialized nanomaterials worldwide, leading to increased environmental and biological release. Given their widespread use in various industries, textiles, and everyday products, extensive research has been devoted to understanding their associated toxicological consequences. Long-term accumulation of AgNPs alters several cellular pathways, induces oxidative stress, and triggers DNA damage. As a consequence, the stimulation of phytoremediation the use of plants to remove, uptake, or degrade metal and organic compounds from soil, water, or air has attracted renewed attention. Such strategies include rhizosphere biodegradation, phytoextraction, phytovolatilization, and phytodegradation. Phytoremediation is a natural, cost-effective, and publicly approved technique providing an outlet for sewage sludge, industrial, and household wastes. It is regarded as a sustainable, green, and solar-driven method with reduced disturbance to the environment. Beginning in the 1980s, it has been used to clean heavy metals and radionuclides, oil spills, pesticides, and chlorinated solvents at contaminated sites (Das et al., 2022).

A general trend toward greener, more sustainable, and smarter nanoremediation solutions has become evident. The global deployment of safe and more eco-friendly formulations has the potential to revolutionize the remediation sector. Eco safety achieved through an eco-design approach should become a priority feature. Ecotoxicity testing should be more ecologically grounded and include

Table 15.4 Emerging technologies for soil restoration: Opportunities and challenges.

Technology	Target contaminants	Key features	Field applications	Challenges	Citation technologies-for-soil-pollution-reduction-and-restoration
Biochar application	Heavy metals, pesticides, organics	High CEC, surface area, soil amendment	Agricultural soils, mine sites	Feedstock variability, long-term effects	(Beesley et al., 2011)
Nanotechnology (nZVI)	Chlorinated solvents, Cr (VI), metals	High reactivity, nanoscale particles	Groundwater plumes, industrial sites	Toxicity concerns, aggregation	(Crane & Scott, 2012)
Geopolymer stabilization	Heavy metals, industrial wastes	Alkali-activated binders, low CO2	Contaminated sediments, urban soils	Material optimization needed	(Davidovits, 2015)
Nano-biochar	Heavy metals, PFAS, pharmaceuticals	Functionalized biochar nanoparticles	Emerging contaminants sites	Scalability, cost	(Tan et al. (2015)
Electrothermal remediation	Mixed organics/ metals	AC current + electrochemical oxidation	Complex industrial sites	Energy intensive, electrode fouling	(Gill et al., 2020)

realistic environmental scenarios. From the perspective of environmental sustainability, life cycle assessment (LCA) serves as a valuable tool for implementing a life cycle thinking strategy when selecting the most sustainable solution. However, its full implementation remains constrained by methodological issues that are expected to progress in parallel with ecotoxicological characterization. LCA can support eco-design at the early stages of nanomaterial development and guide the transition from laboratory-scale research to full-scale applications (Keller et al., 2013). Innovative technologies like biochar, nanotechnology, and geopolymers address persistent pollutants with enhanced precision and sustainability.

8.2 Biochar utilization

Biochar is a carbon-rich, highly porous substance produced through pyrolysis or gasification at temperatures ranging from 300 to 1000 °C, with limited or oxygen free condition, using feedstocks such as wood, manure, and various crop residues (Joseph, 2015). It attracts considerable attention as an environmentally friendly medium for the remediation and regeneration of polluted soil, due to its ability to modify soil physicochemical properties and aid in the remediation of contaminated soils (Beesley et al., 2011). Biochar contains diverse array of surface functional groups and presents a high specific surface area, combined with a high cation exchange capacity (CEC), which significantly facilitates the adsorption and retention of toxic metals. Moreover, biochar directly enhances soil physical and chemical properties, thereby improving microbial activity and soil fertility. An emerging trend involves the application of nanotechnology to enhance biochar properties, producing functionalized biochars and nano-biochar to improve remediation capacities and specificity (Tan et al., 2015).

8.3 Geopolymer technologies

Geopolymer technologies provide a novel approach to soil remediation, combining soil stabilization with environmental protection. Geopolymers are inorganic polymers formed by the alkali activation of aluminosilicate-rich precursors such as fly ash and ground granulated blast furnace slag (GGBFS). These materials react under highly alkaline conditions to form three-dimensional aluminosilicate polymeric networks that can bind soil particles, thereby enhancing soil strength and reducing plasticity. Geopolymer-based soil stabilizers exhibit high compressive strength and greater adhesion compared to traditional Portland cement, making them suitable for reinforcing soils that have degraded. Unlike conventional techniques, geopolymers enable simultaneous soil mixing and waste immobilization, thereby reducing the demand for natural raw materials and carbon emissions (Pham et al., 2024).

9. Sustainable agricultural practices

Optimizing sustainable soil management through conservation agriculture holds great potential for safeguarding soil health and achieving a food security agenda that benefits all producers, adhering to internationally accepted standards. The key to effective crop rotation is to grow individual crops, as this prevents the build-up of pathogens and pests that often occur when a single species is continuously cropped in the same place. Making crop rotation a valuable strategy for soil restoration or improvement. Organic farming has the potential to restore and maintain soil health; however, farming practices must be adapted to local conditions, and fertility management must be more intensive to achieve optimal yields. Even in long-term trials, yields reach 80%–90% of those achieved by conventional farming methods, but overall, the quality of the food and the environmental benefits usually outweigh the yield disadvantage. Hence, in addition to technical measures, well-established rural advisory services and government support and policies are necessary to conserve soils and contribute to resilient agricultural landscapes. By reversing degradation, healthy soils can continue to support the livelihoods of approximately 1.2 billion smallholder farmers and sustain the essential functions of the Earth system, particularly the soil-carbon pool.

9.1 Crop rotation

Crop rotation is one of the oldest known sustainable agricultural cultivation methods, alongside recreational grazing. It is generally considered the keystone agricultural practice for sustainable agriculture and is a well-established technique to reduce the buildup of pathogens that often occurs when a single species is continuously cropped (Sharma et al., 2019). Crop rotation also increases general biodiversity within the rotation, which tends to enhance soil biodiversity by providing more diverse food sources and habitats. This biodiversity is often linked to healthy, functional soils. It can also provide temporal heterogeneity that supports greater diversity. Furthermore, crop rotation is known to improve physical and chemical soil conditions (Briffa et al., 2020). Different crops have varying effects on physicochemical soil properties, but crop rotation generally provides benefits over continuous cropping. This may be related to the lack of continuous harvest of one particular nutrient pool. Including winter rye in crop rotations is advantageous, as the ground cover it provides can reduce soil erosion while simultaneously allowing for more sustainable soil management. The cultivation of winter rye has a peculiar effect on soil properties and biodiversity. Soil temperature is closely related to the release of CO_2 into the environment; as temperature increases, the release of CO_2 intensifies. Earthworms play a crucial role in soil structure and fertility, enhancing the availability and release of C and N by stimulating the microbial mineralization of organic matter. Crop rotation schemes generally result in yield increases of 4.5%–6.6% compared to continuous cropping of the main crop, better performance under harsh conditions, and an increase in microbial diversity, enzymatic activities, and soil respiration (Shourie & Vijayalakshmi, 2022). Certain crop combinations can reduce yields, so not all rotations are beneficial, requiring careful planning. Key considerations when designing a crop rotation include crop diversity, variation in root depths, restoration of soil fertility, suppression of weeds and pests, and prevention of erosion. Assessing changes in soil microbiota with next-generation sequencing (NGS) techniques can provide indicators of soil health and guide the selection of effective crop rotation schemes. Testing various crop combinations and analyzing the response of the soil microbial community, especially through short-term experiments, helps predict the most effective rotations. Microbial diversity assessments, such as 16S rDNA sequencing, are useful for evaluating soil responses to different rotations and supporting large-scale crop selection (Chen et al., 2022).

9.2 Organic farming

Maintenance of soil quality is critical in regions like Europe, where extensive moldboard plowing elevates erosion risk and threatens long-term soil fertility. Organic farming, emphasizing low external inputs, crop rotations, legumes, green manures, and organic fertilization, constitutes a viable approach to reconcile sustainable agricultural production with societal demands for food security while reducing environmental impact (Reganold & Wachter, 2016).

The application of crop rotations and conservation tillage plays a major role in reducing soil pollution strategies. Crop rotations and cover crops increase plant diversity, thereby enhancing the number and diversity of nutrients and stimulating the levels of soil organic matter. Conservation tillage reduces soil disturbance, thereby lowering the risk of soil erosion and soil organic matter mineralization. It contributes to improved water infiltration and soil-water relations, and hence, can protect against the leaching of chemical pollutants. Combined with cover crops and vegetable rotations, conservation tillage also increases soil organic matter levels and reduces nutrient losses. Organic farming is widely recognized as a resilient agricultural practice that can provide a positive land-use option to strengthen ecosystem services, especially in addressing climate change and biodiversity loss (Seufert et al., 2012).

9.3 Conservation tillage

Sustainable agricultural practices, including crop rotation, organic farming, and conservation tillage, offer effective strategies for mitigating soil degradation and promoting land restoration (Sainju et al., 2003). Crop rotation involves the successive cultivation of different crops, allowing them to serve as soil cover and forage during various months, thereby protecting the soil from erosion and enhancing its fertility (Li, Huang, et al., 2020a, Li, Li et al., 2020b). Organic farming eliminates mineral fertilizers and synthetic chemicals, relying on organic residues, crop residues, and biofertilizers to maintain biodiversity and replenish soil nutrients, while avoiding soil acidification and salinity associated with chemical inputs (Vågen et al., 2016). Conservation tillage encompasses methods that limit soil disturbance and maintain residue cover, such as direct drilling, mulch tillage, and ridge tillage. The core principles include less intensive loosening without turning the furrow, and retaining crop residues on the soil surface—practices recognized as environmental protection strategies. Conservation tillage improves soil physical and chemical properties by reducing disturbance and increasing cover, enhancing soil moisture storage and organic matter, and ultimately promoting soil health. However, continuous no-till may cause soil hardening and increased bulk density, which can constrain root growth and nutrient absorption (Lampurlanés & Cantero-Martínez, 2003). Integrating soil rotation measures can mitigate these effects and sustain productivity (Chen et al., 2022). The adoption of conservation tillage contributes to reduced soil erosion, enhanced soil carbon and nitrogen storage, and minimized environmental contamination, while potentially mitigating greenhouse gas emissions and maintaining crop yields comparable to or superior to those of conventional tillage (Sainju et al., 2003).

10. Community involvement and education

The widespread occurrence of contaminated land imposes substantial risks to the environment, society and the economy. Integrating the diverse perspectives of private landowners, relevant social sectors and various government institutions is essential in determining suitable soil pollution management options. The public must therefore become aware of the potential detrimental impacts of polluted land and actively participate in restoring soil health. Public access to relevant information about the state of soil quality and land restoration efforts enables the adoption of new technologies and forms of land use, facilitates cooperation, and supports stakeholder involvement (Bampa et al., 2019).

Soil is among the crucial natural resources and stakeholders can undertake a wide range of positive actions toward sustainable land use. Climatic conditions, limited knowledge about soil quality and recognized pollutants, land tenure status, and community-based energy consumption all influence the willingness and capacity of local communities to participate in soil restoration activities. Better information flow and government spatial planning are, nonetheless, important prerequisites. Awareness-raising and education campaigns on soil pollution and degraded land, along with standby community technical assistance, can generate a better understanding of the extent of the damage and instil a sense of responsibility. The dissemination of scientific information for industrial planning and more sustainable local practices may also prove valuable. Enhancing training, education, and skill development among authorities and community members ensures the long-term viability of soil restoration. The effective marketing of organically grown and free-range products, constructed around the positive social, environmental, and economic impacts of sustainable land management, deepens community support.

10.1 Public awareness campaigns

Any public awareness campaign must be designed and implemented with a coherentawareness strategy, with its particular aims and measurable objectives clearly defined in advance. Much can be learned from social marketing campaigns in related fields, such as water pollution and water conservation. Public awareness campaigns can be used as a stand-alone communication tool or to support and complement broader initiatives (O'Brien, 2005).

'Public awareness' campaigns aim at increasing people's knowledge and awareness about soil pollution. Typically, this involves the adoption of several communication tools that work in combination and complement each other. Typical objectives include: providing information about soil pollution with particular emphasis on those aspects which are of interest to the target groups; raising awareness of the negative environmental and health consequences of soil pollution; enhancing knowledge of potential sources and types of soil pollution; developing a better understanding of the authorities' work to reduce soil pollution; teaching the public how to recognize and avoid polluted areas; heightening the public's ability to detect polluting activities; addressing rumors and misunderstandings which emerge on particular soil pollution issues; lobbing policy makers to create adequate institutional arrangements; and raising the political profile of the soil pollution issue.

10.2 Educational programs

Educational programs play a vital role in reducing soil pollution and actively engaging in community efforts to restore soil and promote sustainable land use. Low-cost programs have emerged as effective and practical tools for mitigating erosion and remain indispensable to successful soil management (Shelton et al., 1989). In eastern Nebraska, two subsidized programs were developed and implemented to address soil erosion in targeted regions. Traditional extension methods, along with other approaches, were extensively employed. Over the course of 1 year, more than 81,000 m of terraces were constructed, resulting in an annual soil erosion reduction approximately 170,000 tons (Shelton et al., 1989). These projects demonstrated that educational programs focused on soil conservation can substantially reduce erosion. Soil erosion and sedimentation continue to represent major water-quality concerns in Nebraska and the Midwest, particularly in areas with steep slopes and erodible soils, where some fields experience erosion rates exceeding 225 t ha^{-1} y^{-1} (Shelton et al., 1989).

10.3 Stakeholder engagement

Land and Ecosystems researchers argue that community participation and responsiveness to local demands are fundamental to sustainable land management. Incentive mechanisms such as subsidies and reward systems motivate farmers to make investments that benefit both themselves and society. Overcoming institutional barriers including uncertain land tenure, a lack of community-level information, and short-term compensation schemes can promote a more proactive approach to soil and land health. To support decision-making, tools, maps, and information have been developed to target interventions aimed at restoring productive and resilient landscapes. The Soil Best Bets Compendium provides a selection of practices designed to maintain or enhance soil fertility under various conditions, complemented by a catalog of exclosure management options that deliver both ecological and economic benefits. Sequestering soil organic carbon constitutes an additional avenue to restore soil health, enhance fertility, reduce erosion and leaching, and mitigate climate change. Globally, soils store approximately five gigatons of carbon annually (Georgiou et al., 2022).

Drawing on multi-criteria decision analysis, Bampa et al. investigate stakeholders' knowledge and future needs regarding soil quality, the prioritization of soil functions, and land management at regional and European levels (Bampa et al., 2019). Perceptions of soil quality differ across phases, with terminology shaped by regional usage and disciplinary background. Although stakeholders demonstrate a good understanding of soil functioning, notable gaps remain. High technical, activity and policy needs emerge in areas such as financial incentives, credible information on sustainable management practices, locally relevant advice, discussion groups for farmers, training programs, funding for applied research and monitoring, and enhanced inclusion of soil science within education. Faced with the challenge of implementing agricultural systems that are both productive and sustainable, stakeholders emphasize the necessity of interdisciplinary cooperation and early involvement through participatory methods that foster ownership, trust, and reciprocity. Effective soil management requires the explicit prioritization of soil functions to secure sustainable land use. These findings provide a relevant starting point for developing strategies to improve the supply and demand of soil functions through informed decision support.

11. Case studies of successful soil restoration

Several restoration approaches have been employed successfully to remediate contaminated soils. In urban areas, remediation of former industrial sites and heavily contaminated soils has been conducted using various techniques including soil excavation and replacement, soil washing/flushing, solidification/stabilization, thermal treatments, chemical oxidation, phytoremediation, and bioremediation (Khalid et al., 2017). In regions affected by past mining activities, appropriate remediation techniques have been applied to reduce the risks posed by heavy metals (Li et al., 2014). Heavy-metal-contaminated agricultural soils have been restored through various methods, including soil amendment, physicochemical treatments, agricultural practices, phytoremediation, microbial remediation, and phyto-microbial remediation (Mahar et al., 2016). Wetland-soil restoration for environmental protection and sustainable development has been achieved through hydrological restoration and ecological restoration.

11.1 Urban soil remediation

Urban soils are highly heterogeneous because their physicochemical composition and biological characteristics depend on their source (natural or anthropogenic) and subsequent disturbance (Puskás & Farsang, 2009). Their structure may still be visible, a good indicator of fertility. These soils are generally stored for a short time on-site or ex situ. They can be reused without remediation for non-sensitive uses, such as parks, swales, or embankments, or can undergo rapid physical/chemical treatment with an aging process to recover their natural properties (De Kimpe & Morel, 2000). When contaminated, urban soils can be remediated and reused on-site following international standards and regulations. If soil cannot be treated on-site, it must undergo chemical stabilization or extraction before being sent to landfill. Chemical stabilization (solidification/stabilization) involves adding reagents to contaminated soils to immobilize pollutants and prevent environmental release. This technique significantly reduces pollutant leachability and is cost-effective due to its rapid treatment and avoidance of cross-contamination between uncontaminated soil and treated soil (Shen et al., 2019). Extraction methods include soil washing and soil flushing. Soil washing, used only ex situ, is a separation process (particle size, gravity, electrokinetic separation) divided into three steps: physical separation to remove coarse, less contaminated particles; chemical extraction to transfer pollutants from soil to a washing solution; and pollutant treatment and washing solution recycling (Dermont et al., 2008). Soil flushing is an in-situ technique in which the soil is flushed with water or solvents to mobilize and recover pollutants, minimizing disturbance to the contaminated site. Air sparging accelerates soil flushing kinetics by enhancing penetration, mixing and aeration of injected fluids (Reddy et al., 1995). Integrated soil washing and soil flushing technologies combine the advantages of each and can be selected according to site geological heterogeneity after arid maturity evaluation.

11.2 Agricultural land restoration

Soils exhibit considerable heterogeneity in their agricultural potential owing to a wide range of physical soil properties, such as mineralogy, texture and structure. Agricultural land restoration reverses degradation processes in field cropping areas, often by amending or removing contaminants from soils, sediments, or vegetation. Soil health is essential for sustainable agricultural productivity. Land degradation adversely affects crop yields, water quality, ecological function and, ultimately, human wellbeing, with large portions of the annual costs falling on smallholder farmers and poor communities (Seufert et al., 2012). Land degradation costs about USD 300 billion annually, with the most severely affected major farming regions located in South Asia and sub-Saharan Africa, where soil and crop nutrient depletion are severe, with the annual financial cost of degraded lands in these two regions amounting to approximately USD 68 billion equivalent to a reduction in the agricultural GDP of about 3% (Nkonya et al., 2016). Costs vary considerably, depending on the severity and type of land degradation [that is, salinization, soil fertility loss, erosion and/or pesticide pollution], being estimated between USD 20 and USD 4000 per hectare for establishment of restoration activities (Nkonya et al., 2016). Recommended practices include: targeting soil-management challenges to local biophysical and socio-economic conditions, aided by risk-assessment, soil and nutrient-balance and land-management pathway assessment tools; encouraging farmer investments in soil health through payments, access to finance and approaches to the recovery and recycling of crop nutrients; accentuating the role of a circular economy in the nutrient cycle, and its livelihood development opportunities, through the recycling of nutrients from waste into cost-competitive fertilizer; and efforts to measure progress in improving agricultural soil health, including through improved remote sensing and earth observation by satellite technology (Vågen et al., 2016).

11.3 Wetland restoration projects

Wetlands provide essential ecological, biological, and socioeconomic services, including water purification, flow regulation, subsidence control, soil stabilization, erosion control, nutrient recycling, retention and absorption of metals, and flood control. They provide critical habitat for migratory waterfowl and resident aquatic species, offering esthetic and recreational opportunities (Tomes, 2013). Construction activities, the disposal of municipal waste, and over-extraction of groundwater have degraded many geomorphic and hydrological characteristics of wetlands, resulting in the complete loss of some wetlands (Zedler & Kercher, 2005). Accordingly, wetland rehabilitation and restoration programs have emerged globally to enhance further and sustain wetland-dependent resources.

Wetland restoration is a mitigation option considered during the Natural Resource Damage Assessment (NRDA) process. It is often the preferred compensation method because it can potentially return the injured resource to baseline and restore important services lost due to injury. Large areas of wetlands are being restored in response to NPDES Multi-Sector General Permit (MSGP) stormwater management requirements and other regulatory needs. Although more than 130 hydrogeomorphic type C wetlands have been U.S. Army Corps of Engineers (USACE) jurisdictional determinations, only 20%–25% of these have on-site and off-site wetland restoration opportunities in the vicinity of the site. Wetland restoration projects have become an increasingly popular groundwater NRDA compensation strategy. Because these projects are non-groundwater-based, the connection between restoration and injury may be more difficult to establish.

Wetland restoration is one form of natural resource restoration considered as part of the NRDA process. Other types of natural resource restoration include the restoration, replacement, or acquisition of groundwater, surface water, or biological resources. Restoration may be more appropriate when addressing interim losses because it can begin immediately at the onset of injury, especially in cases where groundwater-impaired natural resources cannot be restored until groundwater returns to baseline concentrations. Furthermore, because restoration is considered a preferred remedy, restoration-focused projects are strongly favored when allocating compensatory damages for natural resource injuries. In general, restoration activities must return injured natural resources to their baseline, and a direct nexus between injury and restoration must be established before using restoration projects for compensation. Establishing the connection between injury and restoration remains one of the greatest challenges that adjudicators face when considering wetland restoration projects as compensation for groundwater injuries.

12. Future directions in soil pollution research

Further research is needed on contaminant sources and tailoring remediation to specific soil types and conditions (Li, Huang, et al., 2020, Li, Li, et al., 2020). New procedures are emerging for industrial chemical contamination, while microbial community bioremediation, which has recently attracted interest, requires further investigation for many soil contaminants. Emerging pollutants, such as flame retardants, surfactants, PFAS, antibiotics, pharmaceuticals, and nanoparticles, are receiving growing attention. Updated knowledge on environmental behavior and improved analytical methods are needed for soil assessment and remediation (Yang et al., 2024). The selection of soil remediation technologies can be based on an optimization procedure that combines environmental impacts with economic and social considerations. The future will also address the effects of global warming and climate change as key issues that will influence the behavior and fate of contaminants in soil and the environment (Roy et al., 2022; Yang et al., 2024).

12.1 Emerging contaminants

Emerging contaminants such as per- and polyfluoroalkyl substances (PFASs), pharmaceuticals, personal care products, endocrine-disrupting chemicals, nanoparticles, and microplastics have attracted increasing public and scientific attention as anthropogenic pollutants in soil (Fagnano, 2018; Li, Huang, et al., 2020, Li, Li et al., 2020). These pollutants enter the soil environment through various sources, including illicit dumping, land application of biosolids, direct and indirect discharges of industrial waste, landfill leachate, and atmospheric deposition. PFASs, which comprise numerous compounds containing at least one per- or polyfluorinated carbon moiety, are widely used in industrial applications and consumer products due to their unique physicochemical properties, including chemical and thermal stability (Yang et al., 2024). Among PFASs, pentadecafluorooctanoate acid (PFOA) and perfluorooctanoate sulfonate (PFOS) are the most commonly detected in water and soil samples. Pharmaceuticals, personal care products, and endocrine-disrupting chemicals are emerging classes of soil contaminants that were previously of concern in aquatic environments but are now recognized to accumulate appreciably in terrestrial ecosystems, often at concentrations that may be ecotoxicological (Ali et al., 2019). Among ecotoxicological endpoints studied, cell viability and growth inhibition are the most sensitive to these emerging contaminants. The advent of nanotechnologies has led to the development of novel engineered nanomaterials (NMs), whose increased production also raises the risk of release into soil and water environments. Microplastics smaller than 5 mm have also emerged as serious contaminants in water, sediments, and soil, posing increasing threats to organisms at the base of the food chain and, consequently, to the entire ecosystem.

12.2 Climate change impacts

Climate change has become a prominent issue, recognized as an urgent global problem that human society must address promptly and effectively if future generations are to enjoy a healthy and sustainable environment (Yang et al., 2024). Its consequences are diverse, harmful, and largely unpredictable. Many effects are already observable in parts of the world, implying more severe consequences in the near future. Thus, significant efforts are primarily aimed at mitigating rather than preventing climate change impacts. Numerous international agreements and actions highlight the need for continued commitment at all governance levels. Morocco, for instance, has integrated climate change variables and parameters into its national economic development strategies, encompassing water, agricultural production, public health, forestry, environmental protection, and rural development.

Restoring degraded lands is an urgent requirement worldwide and would provide significant benefits (Yang et al., 2024). Best practices include: regulating and incentivizing land management to encourage the adoption of sustainable land-use practices; stimulating local supply chains and employment creation focused on restoration efforts such as agroforestry and forestry; building on natural capital with integrated adaptation and mitigation objectives; and promoting a circular economy that recycles nutrients and other materials to create closed-loop agricultural systems.

12.3 Technological innovations

Technological innovations for removing soil pollutants constitute a significant demand worldwide, and a variety of approaches to efficient and sustainable soil restoration have been proposed. Conventional physicochemical approaches include soil extraction and landfilling of hazardous components, chemical immobilization, and thermal treatment. Newer approaches demonstrate potential for combining efficiency and sustainability, including bioremediation, nanotechnology, biochar, and geopolymer technology (Tan et al., 2015).

Bioremediation techniques exploit the advantages of microbial and phytoremediation. Microbial remediation applies to both organic and inorganic soil contaminants, utilizing the degradation and transformation activities of a wide variety of microorganisms, most notably fungi and bacteria. In phytoremediation, metals are removed from soil by tolerant plant species, such as Populus, Salix, and Brassicaceae, which accumulate contaminants in their roots, shoots, and leaves. Mycoremediation involves the use of fungi to degrade soil contaminants through the secretion of various extracellular enzymes that break down complex molecules and convert them into less toxic metabolites (Mohamadhasani & Rahimi, 2022). The combined use of these techniques has enabled the removal of Pb, Cr (VI), and petroleum hydrocarbons through sophisticated systems incorporating microalgal and bacterial consortia using organic amendments.

Ethical approval

Not applicable.

Consent of participate

Not applicable.

Consent to publish

Not applicable.

Funding statement

No funds or grants were received for this study.

Data availability

Data available within the article.

Statement of competing interest

There is no conflict of interest between the authors regarding this paper.

Acknowledgments

We extend our gratitude to all the universities involved in this study for their support in facilitating my research.

References

Ali, H., Khan, E., & Ilahi, I. (2019). Environmental chemistry and ecotoxicology of hazardous heavy metals: Environmental persistence, toxicity, and bioaccumulation. *Journal of Chemistry, 2019*(1), Article 6730305.

Alori, E. T., Gabasawa, A. I., Elenwo, C. E., & Agbeyegbe, O. O. (2022). Bioremediation techniques as affected by limiting factors in soil environment. *Frontiers in Soil Science, 2*, Article 937186.

Azubuike, C. C., Chikere, C. B., & Okpokwasili, G. C. (2016). Bioremediation techniques–Classification based on site of application: Principles, advantages, limitations and prospects. *World Journal of Microbiology and Biotechnology, 32*(11), 180.

Bampa, F., O'Sullivan, L., Madena, K., Sandén, T., Spiegel, H., Henriksen, C. B., … Debeljak, M. (2019). Harvesting European knowledge on soil functions and land management using multi-criteria decision analysis. *Soil Use and Management, 35*(1), 6–20.

Bansod, B., Kumar, T., Thakur, R., Rana, S., & Singh, I. (2017). A review on various electrochemical techniques for heavy metal ions detection with different sensing platforms. *Biosensors and Bioelectronics, 94*, 443–455.

Beesley, L., Moreno-Jiménez, E., Gomez-Eyles, J. L., Harris, E., Robinson, B., & Sizmur, T. (2011). A review of biochars' potential role in the remediation, revegetation and restoration of contaminated soils. *Environmental pollution, 159*(12), 3269–3282.

Biswas, K. (2015). *Biological agents of bioremediation: A concise review.*

Boer, B., & Hannam, I. (2003). Legal aspects of sustainable soils: International and national. *Review of European, Comparative & International Environmental Law., 12*, 149.

Briffa, J., Sinagra, E., & Blundell, R. (2020). Heavy metal pollution in the environment and their toxicological effects on humans. *Heliyon, 6*(9).

Cachada, A., Rocha-Santos, T., & Duarte, A. C. (2018). Soil and pollution: An introduction to the main issues. In *Soil pollution* (pp. 1–28). academic press.

Canosa, L. F., & Bertucci, J. I. (2023). The effect of environmental stressors on growth in fish and its endocrine control. *Frontiers in Endocrinology, 14*, Article 1109461.

Carter, M. R., & Gregorich, E. G. (2007). *Soil sampling and methods of analysis.* CRC press.

Chen, S., Yao, F., Mi, G., Wang, L., Wu, H., & Wang, Y. (2022). Crop rotation increases root biomass and promotes the correlation of soil dissolved carbon with the microbial community in the rhizosphere. *Frontiers in Bioengineering and Biotechnology, 10*, Article 1081647.

Chikere, C. B., Okpokwasili, G. C., & Chikere, B. O. (2011). Monitoring of microbial hydrocarbon remediation in the soil. *3 Biotech, 1*(3), 117–138.

Crane, R. A., & Scott, T. B. (2012). Nanoscale zero-valent iron: Future prospects for an emerging water treatment technology. *Journal of hazardous materials, 211*, 112–125.

Das, S., Das, S., & Ghangrekar, M. M. (2022). Efficacious bioremediation of heavy metals and radionuclides from wastewater employing aquatic macro-and microphytes. *Journal of Basic Microbiology, 62*(3–4), 260–278.

Davidovits, J. (2015). False values on CO_2 emission for geopolymer cement/concrete published in scientific papers. *Technical Paper, 24*, 1–9.

De Kimpe, C. R., & Morel, J. L. (2000). Urban soil management: A growing concern. *Soil Science, 165*(1), 31–40.

Dermont, G., Bergeron, M., Mercier, G., & Richer-Laflèche, M. (2008). Soil washing for metal removal: A review of physical/chemical technologies and field applications. *Journal of hazardous materials, 152*(1), 1–31.

Dinakarkumar, Y., Ramakrishnan, G., Gujjula, K. R., Vasu, V., Balamurugan, P., & Murali, G. (2024). Fungal bioremediation: An overview of the mechanisms, applications and future perspectives. *Environmental Chemistry and Ecotoxicology, 6*, 293–302.

Fagnano, M. (2018). Definition of a site as contaminated: Problems related to agricultural soils. *Italian Journal of Agronomy, 13*, 1–5.

Falciglia, P. P., Cannata, S., Romano, S., & Vagliasindi, F. (2012). Assessment of mechanical resistance,??-radiation shielding and leachate??-radiation of stabilised/solidified radionuclides polluted soils: Preliminary results. *Chemical Engineering Transactions, 28*, 127–132.

Gavrilescu, M. (2021). Water, soil, and plants interactions in a threatened environment. *Water, 13*(19), 2746.

Gavrilescu, M., Demnerová, K., Aamand, J., Agathos, S., & Fava, F. (2015). Emerging pollutants in the environment: Present and future challenges in biomonitoring, ecological risks and bioremediation. *New Biotechnology, 32*(1), 147–156.

Georgiou, K., Jackson, R. B., Vindušková, O., Abramoff, R. Z., Ahlström, A., Feng, W., ... Torn, M. S. (2022). Global stocks and capacity of mineral-associated soil organic carbon. *Nature communications, 13*(1), 3797.

Gill, S., Alshankiti, A., Shahid, S. A., & Rodriguez, J. P. (2020). Amending soil health to improve productivity of alternate crops in marginal sandy soils of the UAE. In *Emerging research in alternative crops* (pp. 93–123). Cham: Springer International Publishing.

Hamouda, R. A., Alhumairi, A. M., & Saddiq, A. A. (2023). Simultaneous bioremediation of petroleum hydrocarbons and production of biofuels by the micro-green alga, cyanobacteria, and its consortium. *Heliyon, 9*(6).

Harms, H., Schlosser, D., & Wick, L. Y. (2011). Untapped potential: Exploiting fungi in bioremediation of hazardous chemicals. *Nature Reviews Microbiology, 9*(3), 177–192.

Joseph, S. (2015). *Biochar for environmental management: Science, technology and implementation.* Routledge.

Karaś, M. A., Wdowiak-Wróbel, S., & Sokołowski, W. (2021). Selection of endophytic strains for enhanced bacteria-assisted phytoremediation of organic pollutants posing a public health hazard. *International Journal of Molecular Sciences, 22*(17), 9557.

Keller, A. A., McFerran, S., Lazareva, A., & Suh, S. (2013). Global life cycle releases of engineered nanomaterials. *Journal of Nanoparticle Research, 15*(6), 1692.

Khalid, S., Shahid, M., Niazi, N. K., Murtaza, B., Bibi, I., & Dumat, C. (2017). A comparison of technologies for remediation of heavy metal contaminated soils. *Journal of Geochemical Exploration, 182*, 247–268.

Khan, F. I., Husain, T., & Hejazi, R. (2004). An overview and analysis of site remediation technologies. *Journal of Environmental Management, 71*(2), 95–122.

Kumar, P., Babita, M., Kailasam, M., Muralidhar, M., Hussain, T., Behera, A., & Jithendran, K. P. (2023). Effect of changing environmental factors on reproductive cycle and endocrinology of fishes. *Outlook of Climate Change and Fish Nutrition*, 377–396.

Kupinets, L., & Tiutiunnyk, H. (2017). Quality of soil in the system of agro economic natural use supplying. *Economics Ecology Socium, 1*(1), 60–70.

Lampurlanés, J., & Cantero-Martínez, C. (2003). Soil bulk density and penetration resistance under different tillage and crop management systems and their relationship with barley root growth. *Agronomy journal, 95*(3), 526–536.

Levin, R., Villanueva, C. M., Beene, D., Cradock, A. L., Donat-Vargas, C., Lewis, J., ... Deziel, N. C. (2024). US drinking water quality: Exposure risk profiles for seven legacy and emerging contaminants. *Journal of exposure science & environmental epidemiology, 34*(1), 3–22.

Li, Z., Ma, Z., van der Kuijp, T. J., Yuan, Z., & Huang, L. (2014). A review of soil heavy metal pollution from mines in China: Pollution and health risk assessment. *Science of the Total Environment, 468*, 843–853.

Li, D., Huang, G., Zhang, G., & Wang, J. (2020a). Driving factors of total carbon emissions from the construction industry in Jiangsu Province, China. *Journal of Cleaner Production, 276*, Article 123179.

Li, Y., Li, Z., Cui, S., & Zhang, Q. (2020b). Trade-off between soil pH, bulk density and other soil physical properties under global no-tillage agriculture. *Geoderma, 361*, Article 114099.

Liu, J., Kang, H., Tao, W., Li, H., He, D., Ma, L., ... Li, X. (2023). A spatial distribution–Principal component analysis (SD-PCA) model to assess pollution of heavy metals in soil. *Science of the Total Environment, 859*, Article 160112.

Luhar, I., Luhar, S., Abdullah, M. M. A. B., Sandu, A. V., Vizureanu, P., Razak, R. A., ... Imjai, T. (2023). Solidification/stabilization technology for radioactive wastes using cement: An appraisal. *Materials, 16*(3), 954.

Münzel, T., Hahad, O., Daiber, A., & Landrigan, P. J. (2023). Soil and water pollution and human health: What should cardiologists worry about? *Cardiovascular Research, 119*(2), 440–449.

Mahar, A., Wang, P., Ali, A., Awasthi, M. K., Lahori, A. H., Wang, Q., ... Zhang, Z. (2016). Challenges and opportunities in the phytoremediation of heavy metals contaminated soils: A review. *Ecotoxicology and Environmental Safety, 126*, 111–121.

Mani, D., & Kumar, C. (2014). Biotechnological advances in bioremediation of heavy metals contaminated ecosystems: An overview with special reference to phytoremediation. *International Journal of Environmental Science and Technology, 11*(3), 843–872.

Maria Raffa, C., Chiampo, F., & Shanthakumar, S. (2021). Remediation of metal/metalloid-polluted soils: A short review. *Applied sciences, 11*(9), 4134.

Mohamadhasani, F., & Rahimi, M. (2022). Growth response and mycoremediation of heavy metals by fungus pleurotus sp. *Scientific Reports, 12*(1), Article 19947.

Muthusaravanan, S., Sivarajasekar, N., Vivek, J. S., Paramasivan, T., Naushad, M., Prakashmaran, J., ... Al-Duaij, O. K. (2018). Phytoremediation of heavy metals: Mechanisms, methods and enhancements. *Environmental Chemistry Letters, 16*(4), 1339–1359.

Nkonya, E., Mirzabaev, A., & Von Braun, J. (2016). *Economics of land degradation and improvement–a global assessment for sustainable development* (p. 686). Springer Nature.

O'Brien, K. (2005). Making a splash: The clean water campaign from grassroots to mass media. In *Proceedings of the 2005 Georgia Water Resources Conference*.

Orcen, N. (2017). Stomatal parameters and growth responses of nicotiana and atriplex to Cd, Pb and Cd–Pb-contaminated soil. *Fresenius Environmental Bulletin, 26*, 6340–6345.

Pande, V., Pandey, S. C., Sati, D., Bhatt, P., & Samant, M. (2022). Microbial interventions in bioremediation of heavy metal contaminants in agroecosystem. *Frontiers in Microbiology, 13*, Article 824084.

Pham, T. H., Acharya, P., Bachina, S., Osterloh, K., & Nguyen, K. D. (2024). Deep-learning framework for optimal selection of soil sampling sites. *Computers and Electronics in Agriculture, 217*, Article 108650.

Pilon-Smith, E. (2005). *Phytoremediation: Annual review plant biology.*

Pollard, S. J., Brookes, A., Earl, N., Lowe, J., Kearney, T., & Nathanail, C. P. (2004). Integrating decision tools for the sustainable management of land contamination. *Science of the Total Environment, 325*(1–3), 15–28.

Puskás, I., & Farsang, A. (2009). Diagnostic indicators for characterizing urban soils of Szeged, Hungary. *Geoderma, 148*(3–4), 267–281.

Rai, M., Pant, G., Pant, K., Aloo, B. N., Kumar, G., Singh, H. B., & Tripathi, V. (2023). Microplastic pollution in terrestrial ecosystems and its interaction with other soil pollutants: A potential threat to soil ecosystem sustainability. *Resources, 12*(6), 67.

Reddi, L., & Inyang, H. I. (2000). *Geoenvironmental engineering: Principles and applications*. CRC Press.

Reddy, K. R., Kosgi, S., & Zhou, J. (1995). A review of in-situ air sparging for the remediation of VOC-contaminated saturated soils and groundwater. *Hazardous Waste and Hazardous Materials, 12*(2), 97–118.

Reganold, J. P., & Wachter, J. M. (2016). Organic agriculture in the twenty-first century. *Nature Plants, 2*(2), 1–8.

Rizwan, M., Ali, S., Abbas, T., Zia-ur-Rehman, M., Hannan, F., Keller, C., ... Ok, Y. S. (2016). Cadmium minimization in wheat: A critical review. *Ecotoxicology and Environmental Safety, 130*, 43–53.

Rodríguez-Espinosa, T., Navarro-Pedreño, J., Gómez-Lucas, I., Jordán-Vidal, M. M., Bech-Borras, J., & Zorpas, A. A. (2021). Urban areas, human health and technosols for the green deal. *Environmental Geochemistry and Health, 43*(12), 5065–5086.

Rodríguez-Eugenio, N., McLaughlin, M., & Pennock, D. (2018). *Soil pollution: A hidden reality.*

Rossi, M. M., Matturro, B., Amanat, N., Rossetti, S., & Petrangeli Papini, M. (2022). Coupled adsorption and biodegradation of trichloroethylene on biochar from pine wood wastes: A combined approach for a sustainable bioremediation strategy. *Microorganisms, 10*(1), 101.

Roy, S., Gupta, S. K., Prakash, J., Habib, G., & Kumar, P. (2022). A global perspective of the current state of heavy metal contamination in road dust. *Environmental Science and Pollution Research, 29*(22), 33230–33251.

Sadegh, H., Shahryari, G. R., Masjedi, A., Mahmoodi, Z., & Kazemi, M. (2016). *A review on carbon nanotubes adsorbents for the removal of pollutants from aqueous solutions.*

Sainju, U. M., Whitehead, W. F., & Singh, B. P. (2003). Agricultural management practices to sustain crop yields and improve soil and environmental qualities. *The Scientific World Journal, 3*(1), 768–789.

Seufert, V., Ramankutty, N., & Foley, J. A. (2012). Comparing the yields of organic and conventional agriculture. *Nature, 485*(7397), 229–232.

Sharma, A., Kumar, V., Shahzad, B., Tanveer, M., Sidhu, G. P. S., Handa, N., … Thukral, A. K. (2019). Worldwide pesticide usage and its impacts on ecosystem. *SN Applied Sciences to Discover Applied Sciences, 1*(11), 1446. https://link.springer.com/article/10.1007/s42452-019-1485-1.

Shelton, D. P., Dickey, E. C., Jasa, P. J., & Biere, D. A. (1989). *Enhancing the adoption of soil conservation practices with targeted educational programs.*

Shen, Z., Jin, F., O'Connor, D., & Hou, D. (2019). *Solidification/stabilization for soil remediation: An old technology with new vitality.*

Shourie, A., & Vijayalakshmi, U. (2022). Fungal diversity and its role in mycoremediation. *Geomicrobiology Journal, 39*(3–5), 426–444.

Swartjes, F. A. (Ed.). (2011). *Dealing with contaminated sites: From theory towards practical application.* Springer Science & Business Media.

Tan, X., Liu, Y., Zeng, G., Wang, X., Hu, X., Gu, Y., & Yang, Z. (2015). Application of biochar for the removal of pollutants from aqueous solutions. *Chemosphere, 125*, 70–85.

Tomes, A. L. (2013). *Evaluating wetland restoration performance through a case study on the*. University of Maine Campus.

UNEP, U. (2019). Global chemicals outlook II-From legacies to innovative solutions: Implementing the 2030 agenda for sustainable development.

UNEP, F. (2021). Global assessment of soil pollution: Summary for policymakers. Food and Agriculture Organization of the United Nations.

Vågen, T. G., Winowiecki, L. A., Tondoh, J. E., Desta, L. T., & Gumbricht, T. (2016). Mapping of soil properties and land degradation risk in africa using MODIS reflectance. *Geoderma, 263*, 216–225.

Vanni, F., Scaini, F., & Beccaloni, E. (2016). Agricultural areas in potentially contaminated sites: Characterization, risk, management. *Annali dell'Istituto Superiore di Sanità, 52*(4), 500–504.

Wuana, R. A., & Okieimen, F. E. (2011). Heavy metals in contaminated soils: A review of sources, chemistry, risks and best available strategies for remediation. *International Scholarly Research Notices, 2011*(1), Article 402647.

Xia, F., Zhao, Z., Niu, X., & Wang, Z. (2024). Integrated pollution analysis, pollution area identification and source apportionment of heavy metal contamination in agricultural soil. *Journal of Hazardous Materials, 465*, Article 133215.

Yang, Y., Tilman, D., Jin, Z., Smith, P., Barrett, C. B., Zhu, Y. G., … Zhuang, M. (2024). Climate change exacerbates the environmental impacts of agriculture. *Science, 385*(6713), Article eadn3747.

Zahangir, M. M., Rahman, M. L., & Ando, H. (2022). Anomalous temperature interdicts the reproductive activity in fish: Neuroendocrine mechanisms of reproductive function in response to water temperature. *Frontiers in Physiology, 13*, Article 902257.

Zedler, J. B., & Kercher, S. (2005). Wetland resources: Status, trends, ecosystem services, and restorability. *Annual Review of Environment and Resources, 30*(1), 39–74.

Zhang, D., Lei, Y., Wang, C., Lan, S., Li, X., & Xie, Y. (2024). Responses of composition and metabolism of microbial communities during the remediation of black and odorous water using bioaugmentation and aeration. *Environmental Research, 243*, Article 117895.

Zhou, W., Li, M., & Achal, V. (2025). A comprehensive review on environmental and human health impacts of chemical pesticide usage. *Emerging Contaminants, 11*(1), Article 100410.

PART

Soil and food security

V

CHAPTER

Impacts of chemical soil degradation: Pesticides, heavy metals on food insecurity and soil health

16

Maryam Zahedifar

Department of Range and Watershed Management (Nature Engineering), Faculty of Agriculture, Fasa University, Fasa, Iran

1. Introduction

One of the major challenges of the present century is to pay special attention to agriculture in order to provide human food and solve related problems and dilemmas. Many human and natural factors threaten the sustainability of agriculture. One of these factors is soil degradation. Soil degradation refers to the reduction of land's productive capacity, soil fertility, and biodiversity. This phenomenon is caused by factors such as erosion, pollution, drought, and widespread fires, and affects the ecological functions of ecosystems. Soil degradation takes various forms, such as physical, chemical, and biological degradation (Fig. 16.1). Wind and water erosion and loss of soil structure are considered a type of physical degradation. Chemical degradation involves the accumulation of toxic chemicals and chemical processes that affect important chemical properties of the soil (Logan, 1990). Changes in the chemical properties of soils have adverse effects on soil fertility directly or indirectly. One of these changes is the accumulation of toxic chemicals that not only disrupt the vital activities of the soil but also disturb the availability of nutrients, their absorption, and mobility (Suraj et al., 2001).

In recent years, soil pollution, which is considered one of the main sources of chemical soil degradation, has been the result of human activities such as urban waste disposal, energy production, and transportation. Of course, the cause of chemical soil degradation will vary depending on the human activities of the residents of the area. The loss of soil organic matter and its biodiversity is a type of biological soil degradation. Maintaining soil health is the permanent capacity of the soil to produce and function as a vital living system in the ecosystem, in order to maintain biological productivity, maintain water and air quality, and improve plant, animal, and human health (Doran et al., 1996). Maintaining soil health is the path to achieving human health and well-being because soil is the main site of food and biomass production, carbon storage and sequestration, biodiversity habitat, food cycle, and more. The reasons for the decline in soil health can be drought, land degradation, industrial agriculture, chemicals and pollution, diet, and nutrition. Over one-third of the global population resides in regions facing water scarcity. With land degradation, soil loses its ability to retain water, which in turn leads to the loss of

The Soil Nexus. https://doi.org/10.1016/B978-0-443-45286-4.00007-9

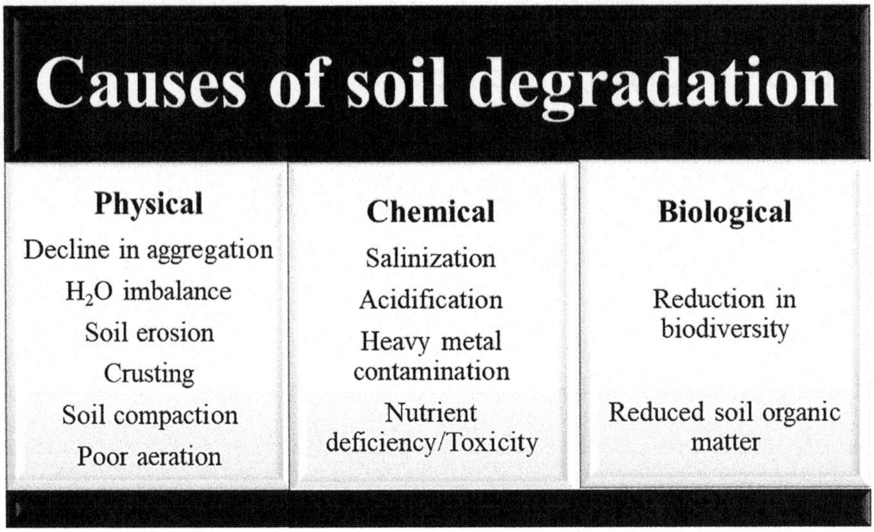

FIGURE 16.1

Causes of soil degradation.

vegetation and erosion. The use of heavy machinery, plowing, monoculture systems, excessive use of pesticides and fertilizers, declines soil quality, pollutes water resources, loses biodiversity and emit greenhouse gases. Industrial activities, improper waste management, and unsustainable agricultural practices add inappropriate chemical compounds and heavy metals to the soil. Excessive use of fertilizers upsets the balance of nutrients, and pesticides harm the activity of beneficial soil organisms. The accumulation of heavy metals disrupts microbial activity and nutrient absorption. As the tendency toward monoculture agriculture and the consumption of crops such as wheat, corn, and rice are widespread among people, nutrient depletion of the soil, reduction of organic matter, and ultimately soil erosion and compaction occur. Diets based on animal meat led to overgrazing and soil degradation.

Soil degradation affects food security, the development of sustainability, and environmental health. In order to prevent this dangerous and worrying phenomenon, it is essential to understand the factors that cause it and to understand how to control them.

2. Chemical soil degradation

2.1 Definition and types of chemical soil degradation

Chemical soil degradation is the second most prevalent type of soil degradation, following erosion, and poses a significant threat to soil health. Chemically degraded soils contain large amounts of toxic chemicals that disrupt vital soil processes. Chemical soil degradation results from soil contamination due to the accumulation of chemicals. This process includes diffuse pollution and spatial pollution (Kavvadias, 2014). Diffuse pollution includes substances such as atmospheric sediments, acidifying and eutrophicating composites or potentially destructive chemicals, sedimentation of pollutants from running waters or eroded soils, pesticides, sewage sludge, and fertilizers that may contain heavy metals. Some soil properties affect diffuse pollution. It is difficult to predict the fate of chemical compounds added to soil. Chemicals introduced into the soil may be adsorbed by soil micelles or leached with water, causing groundwater contamination. Given the behavior of these chemicals in soil, it is very difficult to determine their fate in the soil. As diffuse chemical degradation begins, some soil functions are disrupted. The most important of these functions are the buffering, filtering, and transformation capacities of the soil. As these functions are lost or disrupted, the soil loses its capacity to remove harmful chemicals or reduce their impact on crop growth and yield. Spatial pollution typically impacts regions with high urban density, a long history of heavy industry, or proximity to military installations. The primary pollutants in these areas include heavy metals, organic contaminants like chlorinated hydrocarbons, and mineral oils. The result of the presence of these pollutants is a threat to plant, human, and environmental health.

2.2 Causes of chemical degradation

2.2.1 Agricultural activities

2.2.1.1 Soil salinity and sodicity

The accumulation of water-soluble salts on or near the soil surface is defined as salinity (Kavvadias, 2014). If the amount of exchangeable sodium is high compared to other cations, the soil is called sodic infertile. The entry of soluble salts into the soil can be through irrigation water, weathering of minerals, fertilizers, transport of salts from the sea, or lateral movement of groundwater from saline areas (Hedge et al., 2011). Three main mechanisms of salinization include: salt accumulation, salt leaching, and wind-driven deposition (West et al., 1994). Soil salinization is very common in areas where evaporation and transpiration exceed precipitation. The use of non-conventional water sources such as wastewater for irrigation exacerbates the problem of soil salinization.

2.2.1.2 Chemical fertilizers

These fertilizers contain phosphate, nitrate, ammonium and potassium salts. Some of them also contain heavy metals such as mercury (Hg), cadmium (Cd), hydrogen (H), lead (Pb), copper (Cu) and nickel (Ni). According to Savci (2012), adding chemical fertilizers destroys the structure of the soil. Adding these fertilizers can affect the balance of nutrients in the soil by changing the soil pH of the soil and preventing them from being absorbed by the plant.

2.2.1.3 Pesticides

The toxicity created in the soil following the long-term use of pesticides should be considered. The addition of pesticides to the soil can occur through the spread of the poison during spraying operations on the aerial parts of plants and then washing off from the foliage, release from treated granules or seeds. The stability and persistence of pesticides in the soil lead to substantial permanent changes in the soil's microbial population. Fungicides are toxic to fungi and actinomycetes in the soil by changing the structure of the microbial community. Nitrifying bacteria are also very sensitive to pesticides (Liebich et al., 2003). Research results showed that the use of organochlorine pesticides destroyed symbiotic nitrogen fixation and reduced crop yield. The pesticides pentachlorophenol, DDT and methyl parathion disrupted the function of soil symbiotic bacteria that were active in alfalfa, pea and soybean plants. Some other pesticides had a negative effect on the activity of mycorrhizal fungi that improve the uptake of nutrients by the plant (Chiocchio et al., 2000).

2.2.2 Urbanization and industrial activities

Urbanization and the increasing concentration of population in cities cause significant changes in the ecosystem. Urban construction, industrial activities and land use change significantly degrade the soil. When an asphalt road is built, water can practically not penetrate the ground, resulting in increased surface runoff, flooding and reduced groundwater recharge. The natural cycle of water, nutrients and carbon is disrupted and the ability of the soil to support plant life is severely reduced. Construction activities compact the soil and reduce its porosity and ability to drain, aerate and penetrate roots. Topsoil, which is usually rich in organic matter and fertile, is buried under urban buildings. Industrial activities, greenhouse gas emissions and improper waste disposal are other factors that cause soil degradation. Soil acidification is caused by a decrease in pH or acid deposition. Acid deposition is caused by the deposition of exhaust gases from vehicles such as sulfur dioxide, power plants, other industrial processes, and natural biogeochemical cycles on the soil surface through precipitation and dry deposition. Soil acidification can reduce the availability of some nutrients. This occurs through binding to soil particles or complexation, or leaching and removal from the surface layers of the soil (European Commission, 2013). The impact of the aforementioned physical and chemical degradations in the soil creates unfavorable conditions for the life of soil microorganisms and reduces their abundance and biodiversity. These organisms play an important role in the decomposition of organic matter, the cycling of nutrients, and the maintenance of soil structure, so their reduction will weaken the functioning of the soil as a healthy and efficient ecosystem.

3. Pesticides and their effects

3.1 Types of pesticides

Pesticides are chemical or biological compounds used to control or eliminate pests that harm agricultural products, animals, or humans (Yasir et al., 2025). Their use is very important in reducing pests, increasing crop yields, reducing livestock diseases, and controlling plant disease vectors (Rani et al., 2021). Pesticides are of various types based on chemical classes, functional groups, mechanisms of action, and toxicity (Kumar et al., 2021). They are classified into two categories based on chemical structure: organic and inorganic. Pesticides are also classified based on their target pests, such as insecticides, fungicides, and herbicides, which are used to control insects, fungi, and weeds, respectively (Minif et al., 2011) (Fig. 16.2).

The use of pesticides provides food security by increasing agricultural productivity on the one hand, and can be a serious threat to soil and environmental health on the other.

3.2 Effect of pesticide application in soil

3.2.1 Microbial communities

Soil microorganisms play an important role in basic soil functions such as nutrient cycling, plant growth enhancement, and decomposition of organic and inorganic compounds. In some cases, they increased the growth of microorganisms, in other cases they inhibited their growth, or sometimes they had no effect on their activity (Abdelfattah et al., 2018) (Fig. 16.3).

3.2.2 Decomposition of soil organic matter

Humic soil organic matter is more stable and resistant to microbial degradation than non-humic soil organic matter because it contains more stable free radicals. This may facilitate the interaction of pesticides with organic matter. Non-humic soil organic matter, however, is less stable and is more

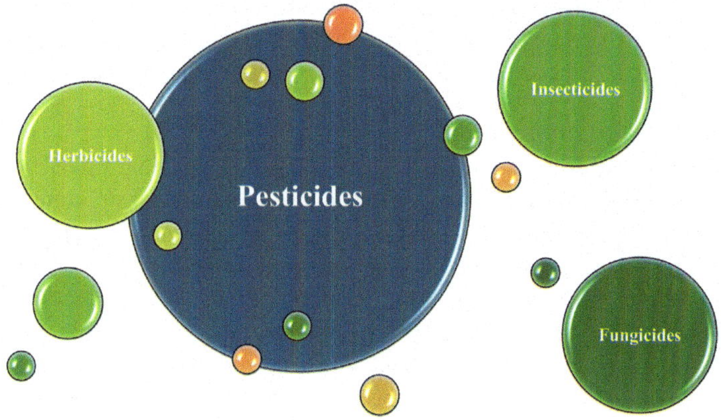

FIGURE 16.2

Types of pesticides.

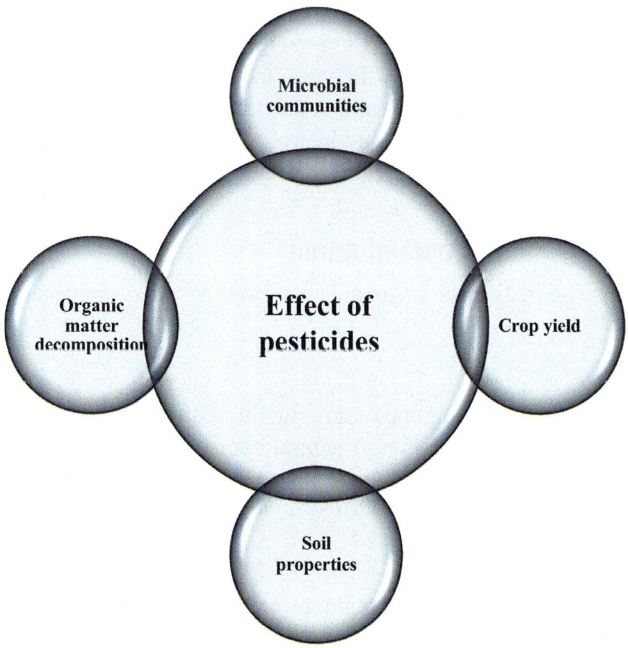

FIGURE 16.3

Effect of pesticide application in soils.

readily biodegradable (Williams et al., 2006). Soluble organic matter can affect pesticide mobility by reducing or increasing adsorption (Huang & Lee, 2001). Some pesticides increase carbon mineralization, which affects the decomposition of soil organic matter (Rasool et al., 2022).

3.2.3 Soil properties and crop yield

The capacity and mechanism of pesticide uptake in soils with different physicochemical properties vary. In some cases, residual pesticide compounds affect soil texture and structure, blocking soil pores, reducing porosity, and water infiltration. Pesticide degradation products may react with clay minerals and organic matter, reducing the number of exchange sites available for essential nutrients such as calcium (Ca), magnesium (Mg), potassium (K), and ammonium (NH_4^+) (Bending et al., 2007). This reduces cation exchange capacity and consequently reduces soil fertility. Excessive use of pesticides disrupts plant physiology, affects crop yield, and sometimes causes phytotoxicity and reduced chlorophyll. Excessive use of pesticides contaminates the soil. In addition, it may harm populations of non-target species, including beneficial species. Once these substances enter the environment, it is virtually impossible to remove them. They can remain in the soil for weeks to years with high toxicity (Wang et al., 2017). In some cases, they enter the soil or groundwater through surface or subsurface runoff.

4. Heavy metals in soil

Heavy metals are elements with high atomic weight and a density of at least 5 g cm^{-2}, the accumulation of which in the soil leads to a decrease in soil quality, and in agricultural production, environmental pollution and ultimately a threat to human health (Zhang et al., 2019). Hg, Pb, Cd, and arsenic (As) are among the most dangerous of these metals. Heavy metals are usually stable in the ecosystem and do not decompose and bioaccumulate in the food chain (Acharya, 2024). Skin diseases, cancer, liver and kidney failure, behavioral and neurological problems are samples of the destructive effects of heavy metal accumulation in the human body.

4.1 Sources of heavy metal contamination

In general, there are two sources of heavy metals entering the ecosystem: natural and anthropogenic (Fig. 16.4).

4.1.1 Natural

Heavy metals can be released following the weathering of bedrock and accumulate in nature. Igneous and sedimentary rocks are the most common natural sources of heavy metals (Deng et al., 2022). Erosion, leaching, volcanic eruptions, biological processes and surface winds lead to the release of these metals into the soil (Muradoglu et al., 2015).

4.1.2 Anthropogenic

Sources of heavy metals entering the environment are those that occur with human intervention. Human activities such as smelting and mining, burning fossil fuels for energy, disposal of municipal waste, use of fertilizers, pesticides and irrigation with sewage, increase the entry of heavy metals into the soil (Muradoglu et al., 2015).

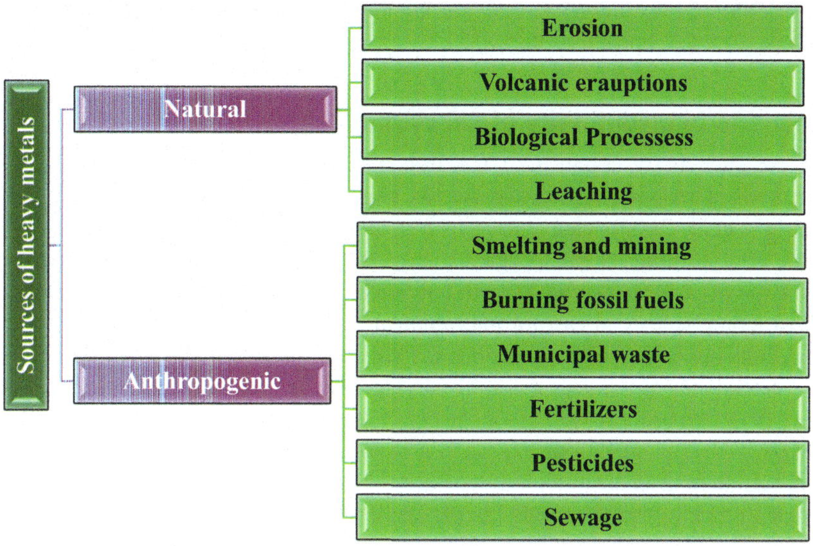

FIGURE 16.4

Sources of heavy metals in soils.

4.1.2.1 Industrial

Heavy metals enter the environment as a result of increased human activities, such as industrial developments, especially in areas that are rapidly industrializing (Jin et al., 2019). One of these metals is lead, which accumulates in the environment as a result of the combustion of fossil fuels, paints, burning of industrial waste, explosives, industrial dust and smoke, pesticides, production of lead-acid batteries and urban runoff (Obeng-Gyasi, 2019). Iron and steel production, industrial dust, metal plating, mining and coal combustion are some of the ways in which Ni enters the environment (Shrestha et al., 2021). Textile, rubber, photography, leather and mining industries cause chromium to enter the environment (Sharma et al., 2021). Industrial dust, burning of urban waste, explosives, fluorescent lamps, rubber and plastic are sources of mercury entry. Pharmaceuticals, sewage, pesticides, phosphate fertilizers and steel and iron production cause arsenic to enter the ecosystem (Bhardwaj et al., 2023).

4.1.2.2 Agriculture

The agricultural sector is a potential source of fertilizers, pesticides, animal waste, and sewage. The soil surface is a place for heavy metals to accumulate, and plants can absorb them through their roots and vessels along with water (Xiao et al., 2017). Organic and inorganic fertilizers (chemicals) contain heavy metals. Arsenic is found in both organic and inorganic forms in soil. Biofertilizers, lime, and phosphate fertilizers are the most common inorganic fertilizers responsible for the release of heavy metals into the soil (Fan et al., 2018). Pesticides can be naturally or synthetically produced. Some of them contain hazardous compounds. Long-term use of fungicides has caused soil degradation and plant toxicity to Cu, reducing crop production potential (Khanam et al., 2020). The use of agricultural solid wastes, including biosolids and farmyard manure, increases heavy metals such as Ni, Cd, zinc (Zn), chromium (Cr), Cu, and Pb in the soil.

4.2 Toxicity and bioavailability of heavy metals

Some metal ions such as Cr, cobalt (Co), Cu, iron (Fe), Ni, manganese (Mn) and Zn play important roles in biochemical reactions, for example, as components of various enzymes or for osmotic regulation (Bruins et al., 2000). Many heavy metals such as gold (Au), Hg, Pb and silver (Ag) have no biological role and are toxic to microorganisms. In large quantities, these ions form complex compounds of unknown nature inside the cell and cause toxic effects, compromising physiological function. Toxic compounds can disrupt microbial metabolism by either displacing essential metals from their binding sites or through interactions with ligands (Bruins et al., 2000). For instance, Hg, Cd, and Ag have a tendency to bind to the sulfhydryl groups of key enzymes, thereby inhibiting their activity (Nies, 1999). In order for metal ions to affect the function of microorganisms, they must enter their cells. Many heavy metal cations have similar structures, even the structures of oxyanions such as chromate and sulfate or arsenate and phosphate are similar. This causes the replacement of toxic metal ions with physiologically essential ions in enzymes and inactivates them. High levels of metal concentration damage cell membranes and DNA structure, disrupting cellular function (Bruins et al., 2000). Bioavailability refers to the portion of the total metal that is either available or can be taken up by microorganisms from the surrounding environment under specific conditions and at a given time. It is the bioavailability, not the total metal concentration, that affects the overall physiological and toxic effects of a metal (Traina & Laperche, 1999). Various physicochemical processes in soil control the behavior and fate of metals. In the aqueous phase, the chemical form of a metal determines its biological availability and chemical reactivity (adsorption/desorption, precipitation/dissolution) with respect to other components of a system. Several factors affect the availability of heavy metals. Binding components and physicochemical properties of the soil. Some compounds can complex and precipitate metals, for example, phosphate easily dissociates metals and the insoluble metal phosphate species formed decrease the metals bioavailability (Nakamura & Sawada, 2000). The physicochemical properties of the soil affect the mobility, bioavailability and toxicity of metals (McLean & Bledsoe, 1992). Metals are distributed between soil components in several ways such as ion exchange, adsorption, precipitation, complex formation or their presence in the minerals structure. The amount of soil organic matter has a great influence on the cation exchange capacity, buffering capacity and retention and absorption of heavy metals. The mobility and bioavailability of metals in organic soils are lower than in mineral soils (Balasoiu et al., 2001). Other properties such as the length of incubation time and soil moisture also affect the bioavailability of metals (Tom-Petersen et al., 2004). Another factor affecting the availability of heavy metals is the solubility and oxidation-reduction potential of the soil. Usually, at high pH, metals form insoluble complexes with phosphates and inorganic carbonates, while at low pH, free ionic species or soluble organic forms of metals are present (Rensing & Maier, 2003). At acidic pH, there are more protons to saturate the metal binding sites, so a large portion of the phosphate is protonated and the possibility of forming insoluble precipitates with phosphate is reduced (Huges & Poole, 1991). Therefore, in many cases, a decrease in soil pH causes metal toxicity, and an increase in it causes a decrease in metal toxicity. However, in some cases, an increase in metal toxicity with an increase in soil pH has also been reported. One possibility is that microbial cells are able to absorb more metal ions under high pH conditions (Sandrin & Maier, 2002). Additionally, the functional groups linked to the microbial membrane become protonated in acidic conditions, which decreases the electrostatic attraction between metal cations and the membrane. Another potential explanation is that under acidic conditions, metals are removed from the cell more effectively by efflux pumps that are powered by the proton motive force (Sandrin & Maier, 2002). At high pH, there is also a possibility of forming species such as hydroxy-metals, which are more toxic (Ivanov et al., 1997). Another factor affecting the bioavailability of metals is the oxidation-reduction potential. In anaerobic environments and reducing conditions (negative Eh), there is a possibility of metal precipitation with environmental components. In oxidizing conditions (positive Eh), metals are likely to exist in free ionic form and have greater solubility in water.

4.3 Impacts on soil fertility and crop production

The presence of heavy metals in soil solution and their downward leaching through the soil profile is significantly influenced by their concentration, physicochemical properties, and the type of soil (Shaheen & Iqbal, 2018). Heavy metals can adsorb onto the surfaces of soil particles and may precipitate in various forms (Moral et al., 2005). As a result, these processes, along with the availability and toxicity of heavy metals in soil for plants, are contingent upon environmental conditions. Natural resources are polluted by human activities and the addition of heavy metals, which reduces plant performance because it disrupts the absorption and transport of nutrients in plants (Mengel & Kirkby, 2001; Tanhan et al., 2007). According to the findings of Ghasemi et al. (2009), excessive absorption of nickel disrupts the absorption and transport of copper and iron within the plant.

Moosavi et al. (2015) showed that Ni had different impacts on canola micronutrient contents. Liu et al. (2003) reported that the interaction between Cd and essential nutrients such as K, Ca, Mg, Ni, Fe, Mn, Cu, and Zn can lead to nutrient deficiencies, ultimately reducing plant growth and yield due to the similarity in their trans-membrane carriers (Sarwar et al., 2010). Li et al. (2016) noted that Cd can affect the absorption, transport, and metabolism of nutrients in plants. Additionally, exposure to Cd can result in several consequences, including changes in plasma membrane permeability, alterations in protein composition, variations in enzyme activity, and interference with nutrient uptake. Ye et al., (2002) highlighted that soils contaminated with heavy metals typically have low organic matter content and low nutrient levels, leading to deficiencies in elements such as nitrogen (N), phosphorus (P), and Zahedifar et al. (2019) found that the application of 40 mg Cd kg^{-1} decreased the concentrations of Zn, Cu, and Mn in basil shoots while increasing the concentration of Fe. Moreover, the transfer factor of the studied heavy metals was significantly influenced by Cd and K treatments. Overall, the uptake of Mn, Zn, and Cu in basil shoots was negatively correlated with Cd levels, likely due to competition between Cd and these metals for root uptake.

4.4 Impacts on human health

Currently, exposure to heavy metals poses a significant threat to human health, primarily due to the frequent introduction of these metals into the soil as a result of rapid urbanization, intensive agricultural practices, and industrial activities. The recognition of soil as a critical reservoir for heavy metals, owing to its high retention capacity, has led to an increased discharge of heavy metals into the environment. Consequently, the role of soil in environmental health, particularly regarding human health, is becoming increasingly important (Zhang et al., 2019).

The primary sources of heavy metals in soils include parent rocks and various human activities, such as the application of organic and inorganic amendments, disposal of industrial waste, and sewage sludge. In addition to natural sources, factors like over-fertilization, wastewater irrigation, compost use, pesticide application, substandard fertilizers, and industrial activities contribute to the contamination of agricultural soils with heavy metals (Keshavarzi & Kumar, 2019; Khelfaoui et al., 2020). In arid and semi-arid regions, where high-quality irrigation water is scarce, some farmers resort to using sewage sludge alone or in combination with water to irrigate their fields. Wastewater often contains elevated concentrations of heavy metals, leading to significant infiltration of these metals into the soil. Thus, the long-term effects of irrigating with wastewater on heavy metal levels in soils under vegetable cultivation are a major concern (Shaheen & Iqbal, 2018). In soils, heavy metals can adhere to soil particles, migrate downward through the soil profile, and ultimately reach groundwater. They may also accumulate in the soil or be released into the soil solution, where they can be absorbed by plant roots (Keshavarz et al., 2021; Moosavi et al., 2018; Zahedifar & Moosavi, 2020a). Ultimately, heavy metals can enter the food chain of animals and humans. The mobility and bioavailability of heavy metals, and consequently their transfer into the food chain, are primarily influenced by soil properties such as pH, organic matter content, and texture (Zahedifar & Moosavi, 2017; Cao et al., 2019; Zahedifar et al., 2019; Xu et al., 2022). Enjavinejad et al. (2024) were evaluated several ecological and health risk indices, based on ingestion, inhalation, and dermal exposure pathways for both adults and children to evaluate soil quality and the risks associated with multi-element contamination (Fig. 16.5).

According to calculated indices, soil quality was unsuitable. Based on their results, soils were contaminated for several heavy metals especially for children. The main exposure routes were ingestion, inhalation, and dermal exposure, with ingestion being the most significant contributor to health risks (Enjavinejad et al., 2024).

5. Food insecurity

5.1 Definition and dimensions of food insecurity

Food security is the assured access to sufficient amounts of safe and affordable food. In a situation of food insecurity, there is limited or uncertain access to adequate food. The six pillars of food security are availability, access, utilization, stability, agency and sustainability (Food and Agriculture Organization, 2012) (Fig. 16.6).

- **Food Availability**: The presence of enough quantity/quality of food in the supply chain (through domestic production or imports) to satisfy dietary and cultural requirements.
- **Food Access**: The financial ability to obtain food for a nutritious diet without compromising other essential needs.
- **Food Utilization**: The capacity to securely prepare meals that fulfill nutritional requirements, supported by access to adequate storage, cooking equipment, and clean water.
- **Food Stability**: The assurance of food security in the face of unexpected shocks (such as climate events, economic crises, or conflicts) or persistent stresses (like drought).

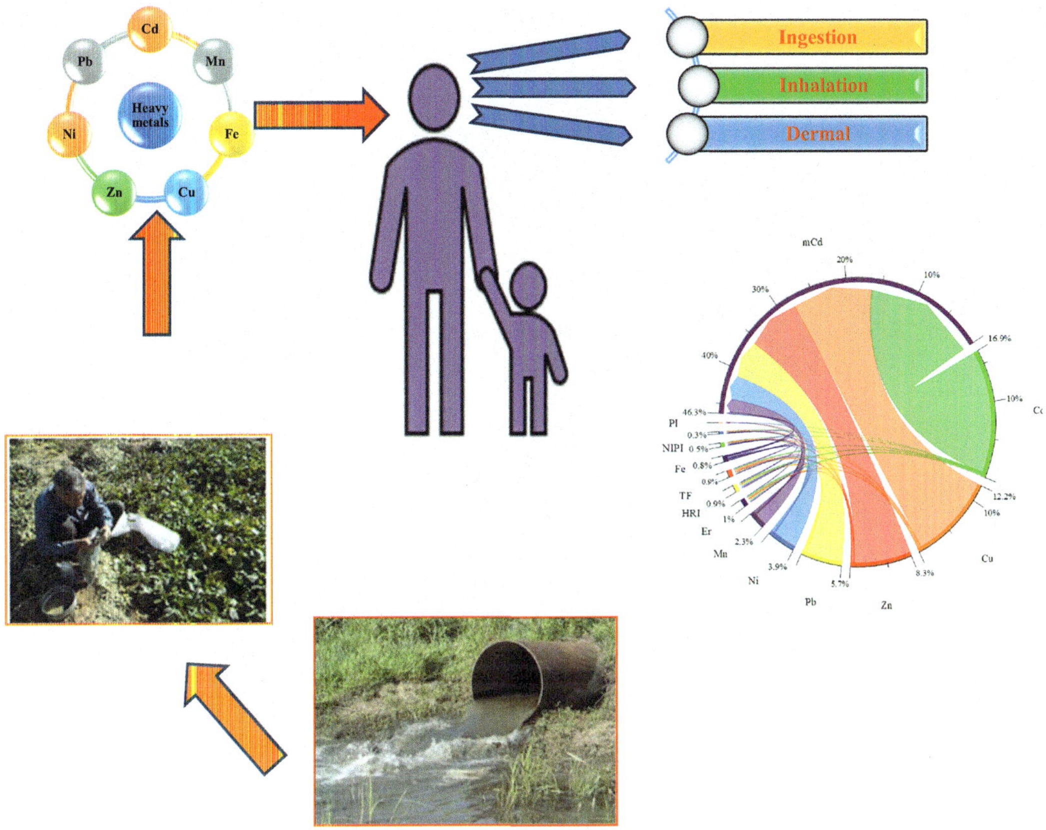

FIGURE 16.5

Impacts of heavy metals on adults and children health.

- **Food Agency**: The ability to make independent choices regarding food consumption and to engage in policy decisions that influence food systems.
- **Food Sustainability**: Meeting the food needs of the current generation without jeopardizing the food needs of future generations, while contributing to the long-term regeneration of natural, social, and economic systems (Bhanye et al., 2024).

5.2 Relationship between soil health and food security

Soil health refers to the soil's ongoing ability to function as a dynamic, living ecosystem that supports plants, animals, and humans. Consequently, achieving long-term food security is impossible without healthy soils. Plants rely on soil for essential nutrients and water. Healthy soil not only yields food for

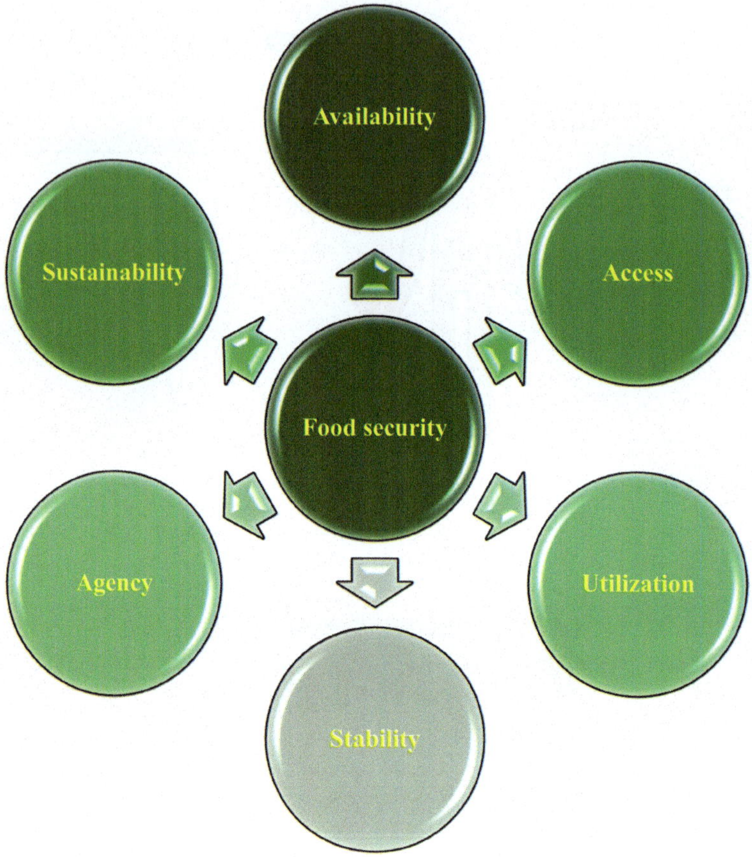

FIGURE 16.6

Dimension of food security.

humans but also provides vital ecosystem services such as water purification and nutrient cycling. Some indicators of soil health are shown in Fig. 16.7.

However, factors like urbanization, intensive agriculture, deforestation, overgrazing, and industrial pollution can lead to soil degradation. When soil becomes degraded, it loses its capacity to serve effectively as a source of food, water, and production. As soil quality declines, food security is jeopardized. Therefore, sustainable soil management is essential for maintaining soil fertility and ensuring ongoing food production.

Employing sustainable agricultural practices is a key strategy for enhancing both food security and soil health. Some effective methods include:

I. **Crop Rotation:** This practice improves soil fertility and reduces pest buildup.
II. **Cover Cropping:** Planting cover crops protects the soil from erosion and enhances soil structure.

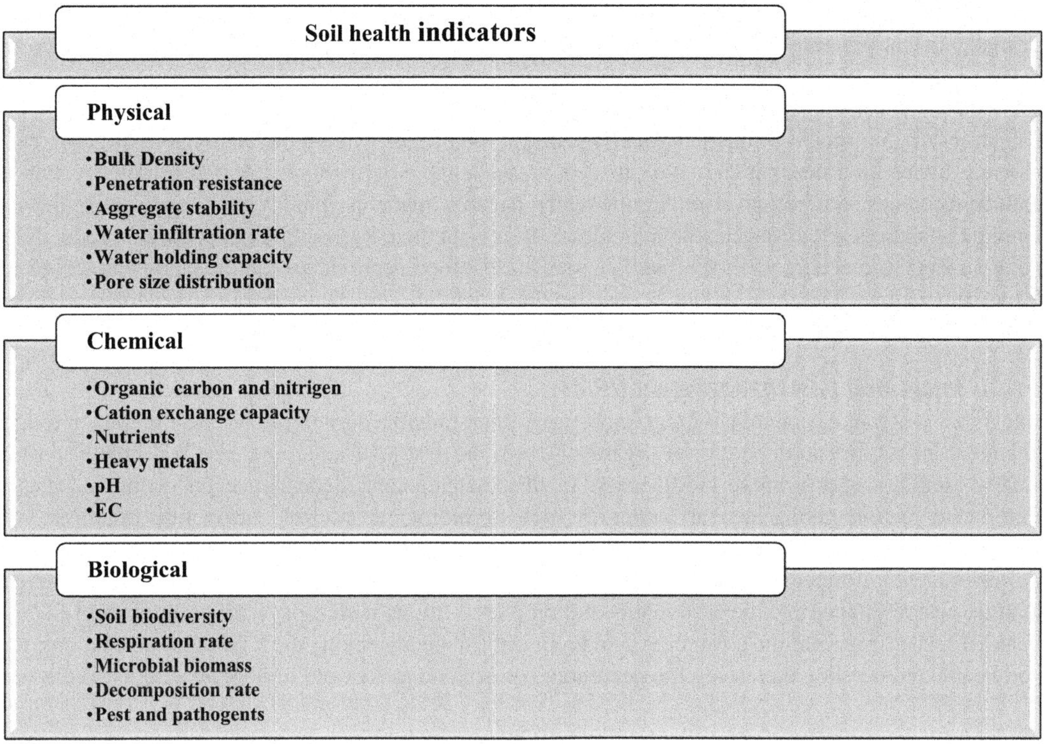

FIGURE 16.7

Soil health indicators.

III. **Agroforestry:** Integrating trees and shrubs into agricultural systems boosts soil health and preserves biodiversity.
IV. **Minimizing Chemical Use:** Reducing the application of chemical fertilizers and pesticides is crucial to prevent soil degradation.
V. **Selecting Appropriate Plant Species:** It is vital to cultivate plants that are well-suited to local environmental conditions.

Overall, soil health and food security are intricately connected. Without healthy soil, we face significant challenges in sustaining crop yields and food quality. As highlighted, soil degradation threatens our ability to produce nutritious food and impacts the lives of millions globally.

6. Mitigation strategies

6.1 Sustainable agricultural practices

Today, feeding the world's population and simultaneously protecting the environment from various types of pollution is one of the most fundamental concerns in human societies. To achieve a balance between the environment and agricultural production while meeting human food needs, priority is given to the use of methods that preserve biodiversity, ecosystem health, and climate resilience. In order to reduce the negative impacts of agriculture on the environment and increase food security, integrated pest management strategies, implementation of ecological farming principles, education on soil conservation, and habitat restoration should be utilized. Meeting current food needs without threatening or worrying about meeting future food needs is the goal pursued by sustainable agriculture. Long-term conservation of natural resources, promotion of environmentally friendly methods, biodiversity, and soil health are among the approaches to sustainable agriculture. In this method, by combining economic productivity, environmental protection, and social welfare, the world's food is provided, and the land is preserved for future generations. The main pillars of sustainable agriculture are shown in Fig. 16.8. Integrated pest management and organic farming are ways to achieve sustainable agriculture.

6.1.1 Integrated pest management (IPM)

Integrated pest management is of great importance in sustainable agriculture because it aims to reduce chemical inputs, maintain beneficial organisms, prevent pest resistance, increase soil health, protect water resources, and promote biodiversity. In this management, dependence on chemical pests is minimized by integrating several strategies such as biological control, agronomic practices, and mechanical methods. Reducing the use of chemical pesticides helps reduce environmental pollution, protect beneficial organisms, and maintain ecosystem health. Continuous use of chemical pesticides leads to pest resistance and ineffectiveness of their effects, along with ecosystem pollution and adverse health effects. Crop rotation, cover crops, and reduced tillage are among the IPM methods that improve soil health and fertility. Increasing the percentage of organic matter and biological activity in the soil,

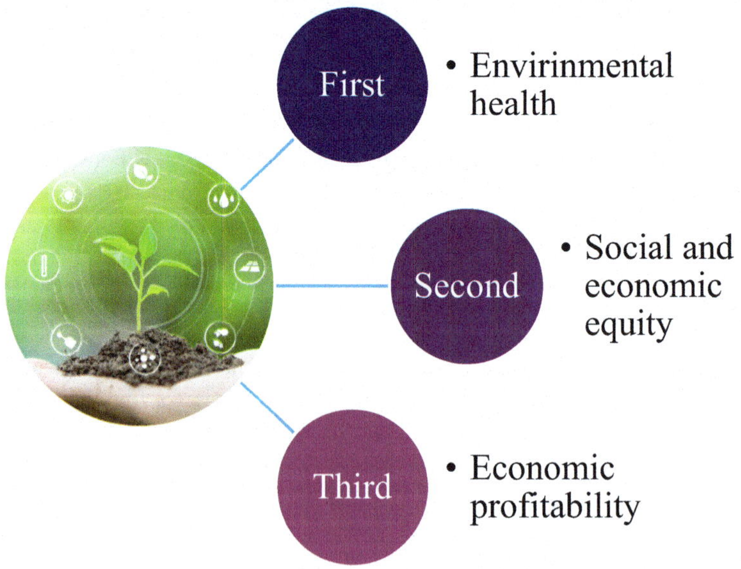

FIGURE 16.8

The pillars of sustainable agriculture.

controlling erosion, sustainable management and nutrient cycling, reducing runoff, surface and groundwater pollution are among the beneficial results of using IPM. In addition to the above, reducing input costs, increasing product yield, and achieving desired economic returns are also appropriate consequences of this management method (Ilieva et al., 2025).

6.1.2 Organic farming

It is a type of agriculture that uses organic compounds and biological elements to support the health of the soil, ecosystem, and human. It aims to create a self-sufficient agricultural system in harmony with nature that relies on biological processes, biodiversity, and cycles that are tailored to local conditions, rather than using harmful inputs (Gugalia, 2021). Some of the methods of organic agriculture are shown in Fig. 16.9.

Organic agriculture is based on four principles:

- ➢ *Principle of Health,* to sustain and enhance the well-being of the land, plants, animals, humans, and the global ecosystem.
- ➢ *Principle of Care,* to uphold the welfare and well-being of both current and future generations, it is essential to practice organic farming responsibly and with care, ensuring the health of the environment.
- ➢ *Principle of Ecology,* Organic farming should consistently align with and support natural systems and life cycles, aiming to cooperate with, emulate, and contribute to their long-term sustainability.

FIGURE 16.9

Methods of organic farming.

➢ *Principle of Fairness,* Organic farming must be founded on relationships that promote fair treatment and equity toward the environment as a whole (Gairola & Rawat, 2025).

6.2 Remediation techniques for contaminated soil

Contaminated soil can significantly damage ecosystems. Direct contact with contaminated soil or its vapors is dangerous to humans and animals. The use of certain fertilizers, improper disposal of waste and leakage of their leachate into the surrounding environment, and the release of heavy metals from mines are some of the ways in which soils become contaminated. Soil remediation methods offer ways to return contaminated soil to its natural, pollution-free state. Contaminated soil remediation methods can be carried out in three ways: in situ (on site), ex situ, and off-site treatment. In situ and ex-site methods use on-site remediation technologies such as physical, chemical, bioremediation, and thermal treatment on site. In situ methods, at a lower cost, apply remediation methods directly to the soil, without removing them from their site. Ex-site remediation is done by carefully excavating the contaminated soil and then using appropriate treatment methods. This prevents the spread of contamination (Santos et al., 2025). In the off-site soil remediation method, the waste is directed to an off-site treatment facility.

6.2.1 Physical and chemical remediation technologies

The main physical and chemical remediation technologies include solidification/stabilization (S/S), soil vapor extraction (SVE), chemical oxidation, soil washing, and electrokinetic remediation (Muhammad et al., 2018):

I. *Solidification/Stabilization (S/S)* combines contaminated media with reagents to alter physical properties (solidification) and reduce leachability/toxicity (stabilization). It can be performed in-situ (in place) or ex-situ (excavated). These methods are effective for low-level volatile organic compounds (VOCs). It requires pre-treatment for high contamination levels.
II. *Soil Vapor Extraction (SVE)* removes VOCs from the vadose zone by collecting vapors through perforated pipes. It is less effective in fine-grained, moist soils.
III. *Chemical Oxidation* involves redox reactions to convert hazardous contaminants into less toxic forms. Ozone, hydrogen peroxide, and permanganates are used as oxidants. This method can be in-situ or ex-situ, with considerations for soil conditions and contaminant types.
IV. *Soil Washing* is a physical/chemical process that separates contaminants from soil using liquids and mechanical processes. It is effective for coarse soils; less effective for fine soils where contaminants bind tightly to particles. This process involves excavation, screening, and scrubbing with solvents.
V. *Electrokinetic Remediation* applies low-intensity direct current to move contaminants through soil, enhancing the removal of heavy metals and organics. It involves electro-osmosis and electromigration, affecting soil pH and contaminant adsorption.

6.2.2 Biological remediation techniques

Biological methods utilize microorganisms and plants to treat contaminated media (Muhammad et al., 2022; Omokhagbor et al., 2020):

I. *Bioremediation:* Uses microorganisms to biodegrade organic contaminants, effective in both aerobic and anaerobic conditions.

II. *Phytoremediation:* Employs plants to detoxify contaminants through various mechanisms (e.g., uptake, degradation).
III. *Bioventing:* Supplies oxygen to indigenous microorganisms to enhance degradation in the vadose zone.

6.2.3 Soil amendments

Heavy metal pollution is a significant global issue that poses serious threats to both the environment and human health. The addition of organic and inorganic amendments can enhance the physical, chemical, and biological properties of soil by increasing soil organic matter and reducing the availability of heavy metals through mechanisms such as complexation, precipitation, oxidation-reduction reactions, immobilization, and adsorption. Organic amendments specifically help decrease metal mobility by transforming the speciation of heavy metals from more available forms to less available ones (Zahedifar, 2017). Common organic amendments include biosolids, wood chips, various composts, animal manures, sewage sludge, and wood ash. These amendments are relatively cost-effective and can aid in restoring the vegetation cover of organic soils, ultimately improving soil quality by altering its properties. One notable organic amendment is biochar, which features a porous structure with a high specific surface area and reactive surfaces, enabling it to effectively bind heavy metals. The interactions between biochar and heavy metals involve several mechanisms, including cation exchange associated with biochar, surface adsorption through functional groups, complex formation with hydroxyl groups on mineral oxides and other surface deposits, and physical adsorption (Gavili et al., 2019). Research by Zahedifar and Moosavi (2020b) examined the impact of three levels of biochar (0%, 2%, and 4%) derived from sugarcane bagasse, two levels of Cd, on Cd adsorption in different soil types: saline, sodic, saline-sodic, and normal soils. Their findings indicated that biochar increases the pH of the soil solution, enhancing the negative surface charges of the soil, which in turn promotes cadmium adsorption by soil particles. They recommended applying 2% biochar as the optimal level for mitigating cadmium pollution. In addition to organic amendments, inorganic options such as lime and gypsum can also influence the availability and mobility of heavy metals in the soil (Lwin et al., 2018).

7. Conclusion

In summary, addressing the challenges posed by soil degradation is crucial for ensuring sustainable agriculture and food security in the 21st century. Soil degradation, driven by various human and natural factors, significantly threatens soil fertility, biodiversity, and the overall health of ecosystems. The consequences of this degradation manifest in multiple forms, including physical, chemical, and biological deterioration, which not only compromise agricultural productivity but also jeopardize environmental health. As the global population continues to rise, the demand for food increases, making it imperative to adopt sustainable agricultural practices that prioritize soil health. Strategies such as crop rotation, cover cropping, and organic farming can enhance soil fertility and resilience while mitigating the adverse effects of chemical inputs and urbanization. Additionally, effective soil remediation techniques are essential for restoring contaminated soils, thereby safeguarding human health and the environment. The intricate relationship between soil health and food security underscores the necessity for a holistic approach to land management. By prioritizing the maintenance and restoration of soil health, we can secure food production for future generations and contribute to the sustainability of our ecosystems. It is essential for policymakers, farmers, and communities to collaborate in implementing these practices and to raise awareness about the importance of healthy soils in achieving global food security and environmental sustainability. Only through collective action can we combat soil degradation and ensure a resilient agricultural system that meets the needs of both people and the planet.

References

Abdelfattah, A., Malacrinò, A., Wisniewski, M., Cacciola, S. O., & Schena, L. (2018). Metabarcoding: A powerful tool to investigate microbial communities and shape future plant protection strategies. *Biological Control, 120*, 1–10.

Acharya, S. (2024). Heavy metal contamination in food: Sources, impact, and remedy. In *Food Safety and Quality in the Global South* (pp. 233–261). Singapore: Springer Nature.

Balasoiu, C. F., Zagury, G. J., & Deschênes, L. (2001). Partitioning and speciation of chromium, copper, and arsenic in CCA-contaminated soils: Influence of soil composition. *Science of the Total Environment, 280*, 239–250.

Bending, G. D., Rodríguez-Cruz, M. S., & Lincoln, S. D. (2007). Fungicide impacts on microbial communities in soils with contrasting management histories. *Chemosphere, 69*, 82–88.

Bhanye, J. I., Matooane, L., Matamanda, A., & Bhanye, A. S. (2024). Food and housing insecurity: Addressing the dual burden of health risks. *Developments in Environmental Science, 15*, 477–502. https://doi.org/10.1016/B978-0-443-21948-1.00023-6

Bhardwaj, L. K., Kumar, D., & Kumar, A. (2023). Phytoremediation potential of ocimum sanctum: A sustainable approach for remediation of heavy metals. *Preprints*. https://doi.org/10.20944/preprints202308.0593.v1

Bruins, M. R., Kapil, S., & Oehme, F. W. (2000). Microbial resistance to metals in the environment. *Ecotoxicology and Environmental Safety, 45*, 198–207.

Cao, X., Wang, X., Tong, W., Gurajala, H. K., Lu, M., Hamid, Y., et al. (2019). Distribution, availability and translocation of heavy metals in soil-oilseed rape (Brassica napus L.) system related to soil properties. *Environmental Pollution, 252*, 733–741.

Chiocchio, V., Venedikian, N., Martinez, A. E., Menendez, A., Ocampo, J. A., & Godeas, A. (2000). Effect of the fungicide benomyl on spore glomus mosseae. *International Microbiology, 3*, 173–175.

Deng, H., Tu, Y., Wang, H., et al. (2022). Environmental behavior, human health effect, and pollution control of heavy metals (loid)s toward full life cycle processes. *Eco-Environmental Health, 4*.

Doran, J. W., Sarrantonio, M., & Liebig, M. A. (1996). Soil health and sustainability. *Advances in Agronomy, 56*, 1–54.

EC (European Commission). (2013). Soil contamination: Impacts on human health. *Science for Environment Policy*. Retrieved November 19, 2014, from www.ec.europa.eu/environment/integration/research.

Enjavinejad, S. M., Zahedifar, M., Moosavi, A. A., & Khosravani, P. (2024). Integrated application of multiple indicators and geographic information system-based approaches for comprehensive assessment of environmental impacts of toxic metals-contaminated agricultural soils and vegetables. *Science of the Total Environment, 926*, Article 171747. https://doi.org/10.1016/j.scitotenv.2024.171747

Fan, Y., Li, Y., Li, H., & Cheng, F. (2018). Evaluating heavy metal accumulation and potential risks in soil-plant systems applied with magnesium slag-based fertilizer. *Chemosphere, 197*, 382–388.

Food and Agriculture Organization. (2012). *Report committee on world food security. CFS 2012/39 final report*. Rome: FAO. http://www.fao.org/3/a-mf115e.pdf.

Gairola, A., & Rawat, A. (2025). Organic farming. In *Modern agriculture: Exploring current trends (Chapter 12)*.

Gavili, E., Moosavi, A. A., & Zahedifar, M. (2019). Integrated effects of cattle manure-derived biochar and soil moisture conditions on soil chemical characteristics and soybean yield. *Archives of Agronomy and Soil Science, 65*(17), 1758–1774. https://doi.org/10.1080/03650340.2019.1576864

Ghasemi, R., Ghaderian, S. M., & Kramer, U. (2009). Interference of nickel with copper and iron homeostasis contributes to metal toxicity symptoms in the nickel hyperaccumulator plant Alyssum inflatum. *New Phytologist, 184*, 566–580.

Gugalia, G. (2021). A sustainable agriculture: Organic farming: A review. *Bhartiya Krishi Anusandhan Patrika, 36*(3), 204–207. https://doi.org/10.18805/BKAP307

Hedge, R., Natarajan, A., Naidu, L. G. K., & Dipak, S. (2011). Soil degradation, soil erosion issues in agriculture. In *Tech*.

Huang, X., & Lee, L. S. (2001). Effects of dissolved organic matter from animal waste effluent on chlorpyrifos sorption by soils. *Journal of Environmental Quality, 30*, 1258–1265.

Huges, M. N., & Poole, R. K. (1991). Metal speciation and microbial growth-the hard (and soft) facts. *General Microbiology, 137*, 725–734.

Ilieva, T., Karova, A., & Ivanova, M. (2025). Sustainable agriculture through integrated pest management: Strategies for effective implementation. In E. G. Nathanail, N. Gavanas, & E. Adamos (Eds.), *Climate crisis and resilient transportation systems (Lecture Notes in Intelligent Transportation and Infrastructure* (p. 38). Springer. https://doi.org/10.1007/978-3-031-82818-8_38

Ivanov, A. Y., Fomchenkov, V. M., Khasanova, L. A., & Gavryushkin, A. V. (1997). Toxic effect of hydroxylated heavy metal ions on the plasma membrane of bacterial cells. *Microbiology, 66*, 490–495.

Jin, Y., O'Connor, D., Ok, Y. S., Tsang, D. C. W., Liu, A., & Hou, D. (2019). Assessment of sources of heavy metals in soil and dust at children's playgrounds in beijing using GIS and multivariate statistical analysis. *Environmental International, 124*, 320–328. https://doi.org/10.1016/j.envint.2019.01.024

Kavvadias, V. (2014). Soil degradation. In *Soil Science Institute of Athens, National Agricultural Research Foundation*. Retrieved November 20, 2014, from www.prosodol.gr/sites/.

Keshavarz, S., Ghasemi-Fasaei, R., Ronaghi, A., et al. (2021). Innovative assisted phytoremediation of multi-elements contaminated soil by ryegrass: An electro-bio-chemical approach. *Journal of Soils and Sediments, 21*, 2604–2618. https://doi.org/10.1007/s11368-021-02961-8

Keshavarzi, A., & Kumar, V. (2019). Ecological risk assessment and source apportionment of heavy metal contamination in agricultural soils of Northeastern Iran. *International Journal of Environmental Health Research, 29*(5), 544–560.

Khanam, Z., Gupta, S., & Verma, A. (2020). Endophytic fungi-based biosensors for environmental contaminants: A perspective. *South African Journal of Botany*. https://doi.org/10.1016/j.sajb.2020.08.007

Khelfaoui, M., Medjram, M. S., Kabir, A., Zouied, D., Mehri, K., Chikha, O., & Trabelsi, M. A. (2020). Chemical and mineralogical characterization of weathering products in mine wastes, soil, and sediment from the abandoned Pb/Zn mine in Skikda, Algeria. *Environmental Earth Sciences, 79*, 1–15.

Kumar, M., Yadav, A. N., Saxena, R., Paul, D., & Tomar, R. S. (2021). Biodiversity of pesticides degrading microbial communities and their environmental impact. *Biocatalysis and Agricultural Biotechnology, 31*, Article 101883.

Li, X., Zhou, Q., Sun, X., & Ren, W. (2016). Effects of cadmium on uptake and translocation of nutrient elements in different welsh onion (Allium fistulosum L.) cultivars. *Food Chemistry, 194*, 101–110.

Liebich, J., Schäffer, A., & Burauel, P. (2003). Structural and functional approach to studying pesticide side effects on specific soil functions. *Environmental Toxicology and Chemistry, 22*(4), 784–790.

Liu, J., Li, K., Xu, J., Liang, J., Lu, X., Yang, J., & Zhu, Q. (2003). Interaction of cadmium and five mineral nutrients for uptake and accumulation in different rice cultivars and genotypes. *Field Crops Research, 83*(3), 271–281.

Logan, T. J. (1990). In R. Lal, & B. A. Stewart (Eds.), *Advances in soil science, 11*. New York, NY: Springer. https://doi.org/10.1007/978-1-4612-3322-0_6.

Lwin, C. S., Seo, B. H., Kim, H. U., Owens, G., & Kim, K. R. (2018). Application of soil amendments to contaminated soils for heavy metal immobilization and improved soil quality—A critical review. *Soil Science & Plant Nutrition, 64*(2), 156–167. https://doi.org/10.1080/00380768.2018.1440938

McLean, J. E., & Bledsoe, B. E. (1992). Behaviour of metals in soils. In *Technical research document, EPA/540/S-92018*. Ada, Oklahoma, USA: Environmental Research Laboratory.

Mengel, K., & Kirkby, E. A. (2001). *Principles of plant nutrition* (5th ed.). Kluwer Academic Publishers.

Minif, W., Hassine, A. I. H., Bouaziz, A., Bartegi, A., Thomas, O., & Roig, B. (2011). Effect of endocrine disruptor pesticides: A review. *International Journal of Environmental Research and Public Health, 8*, 2265–2303.

Moosavi, A. A., Mansouri, S., & Zahedifar, M. (2015). Effect of soil water stress and nickel application on micronutrient status of canola grown on two calcareous soils. *Plant Production Science, 18*(3), 377–387.

Moosavi, A. A., Zahedifar, M., & Mansouri, S. (2018). The uptake and partitioning of nickel and some nutrient elements in canola grown in two differently textured soils as influenced by nickel and soil moisture conditions. *International Journal of Environmental Studies, 77*(1), 30–47.

Moral, R., Gilkes, R., & Jordan, M. M. (2005). Distribution of heavy metals in calcareous and non-calcareous soils in Spain. *Water, Air, and Soil Pollution, 162*(1–4), 127–142. https://doi.org/10.1007/s11270-005-5997-5

Muhammad, H. M., Sharifah, S. S. S., Ummi, M. B., & Zaleha, M. I. (2022). Remediation technology inventions for soil and groundwater contamination. *Medical Health, 17*, 13–30.

Muhammad, H. M., Shettima, I., Eberemu, A. O., Kundiri, A., & Muhammed, A. (2018). *Remediation techniques for contaminated soils: A review.* Seminar Series, 9(2), 53–61.

Muradoglu, F., Gundogdu, S., Ercisli, T., Encu, F., Balta, H. Z. E., Jaafar, M., & Zia-Ui-Haq. (2015). Cadmium toxicity affects chlorophyll II and b content, antioxidant enzyme activities, and mineral nutrient accumulation in strawberry. *Biological Research, 48*, 1–7.

Nakamura, Y., & Sawada, T. (2000). Biodegradation of phenol in the presence of heavy metals. *Chemistry and Technology of Biotechnology, 75*, 137–142.

Nies, D. H. (1999). Microbial heavy metal resistance. *Applied and Environmental Microbiology, 51*, 730–750.

Obeng-Gyasi, E. (2019). Sources of lead exposure in various countries. *Reviews on Environmental Health, 34*(1), 25–34.

Omokhagbor, A.,G., Tawari Fufeyin, P., Eruke Okoro, S., & Ehinomen, I. (2020). Bioremediation, biostimulation and bioaugmentation: A review. *International Journal of Environmental Bioremediation and Biodegradation, 3*, 28–39.

Rani, L., Thapa, K., Kanojia, N., Sharma, N., Singh, S., Grewal, A. S., Srivastav, A. L., & Kaushal, J. (2021). An extensive review on the consequences of chemical pesticides on human health and environment. *Journal of Cleaner Production, 283*, Article 124657.

Rasool, S., Rasool, T., & Gani, K. M. (2022). A review of interactions of pesticides within various interfaces of intrinsic and organic residue amended soil environment. *Chemical Engineering Journal Advances, 11*, Article 100301.

Rensing, C., & Maier, R. M. (2003). Issues underlying use of biosensors to measure metal bioavailability. *Ecotoxicology and Environmental Safety, 56*, 140–147.

Sandrin, T., & Maier, R. M. (2002). Effect of pH on cadmium toxicity, speciation, and accumulation during naphthalene biodegradation. *Environmental Toxicology and Chemistry, 21*, 2075–2081.

Santos, M., Rebola, S., & Evtuguin, D. V. (2025). Soil remediation: Current approaches and emerging bio-based trends. *Soil Systems, 9*(2), 35. https://doi.org/10.3390/soilsystems9020035

Sarwar, N., Saifullah, S., Malhi, S. S., Zia, M. H., Naeem, A., Bibi, S., & Farid, G. (2010). Role of mineral nutrition in minimizing cadmium accumulation by plants. *Journal of the Science of Food and Agriculture, 90* (6), 925–937.

Savci, S. (2012). An agricultural pollutant: Chemical fertilizer. *International Journal of Environmental Science and Development, 3*(1), 73–80.

Shaheen, A., & Iqbal, J. (2018). Spatial distribution and mobility assessment of carcinogenic heavy metals in soil profiles using geostatistics and random forest, Boruta algorithm. *Sustainability, 10*(3), 799. https://doi.org/10.3390/su10030799

Sharma, P., Pandey, A. K., Udayan, A., & Kumar, S. (2021). Role of microbial community and metal-binding proteins in phytoremediation of heavy metals from industrial wastewater. *Bioresource Technology, 326*, Article 124750. https://doi.org/10.1016/j.biortech.2021.124750

Shrestha, R., Ban, S., Devkkota, S., Sharma, S., Joshi, R., Tiwari, P., Kim, H. Y., & Joshi, K. (2021). Technological trends in heavy metals removal from industrial wastewater: A review. *Journal of Environmental Chemical Engineering, 9*, Article 105688.

Suraj, B., Reddy, C. P., & Manda, V. (2001). Scientific management and development of saline land. Addressing the challenges of land degradation: An overview and Indian perspective. In *Proceedings of the National Seminar on Land Resource Management for Food and Environmental Security* (pp. 73–83). Soil Conservation Society of India.

Tanhan, P., Kruatrachue, M., Pokethitiy, P., & Chaiyarat, R. (2007). Uptake and accumulation of cadmium, lead and zinc by Siam weed [Chromolaena odorata (L.) King & Robinson]. *Chemosphere, 68*, 323–329. https://doi.org/10.1016/j.chemosphere.2006.12.064

Tom-Petersen, A., Hansen, H. C. B., & Nybroe, O. (2004). Time and moisture effects on total and bioavailable copper in soil water extracts. *Journal of Environmental Quality, 33*, 505–512.

Traina, S. J., & Laperche, V. (1999). Contaminant bioavailability in soils, sediments, and aquatic environments. *Proceedings of the National Academy of Sciences of the United States of America, 96*, 3365–3371.

Wang, C., Zhang, Q., Wang, F., & Liang, W. (2017). Toxicological effects of dimethomorph on soil enzymatic activity and soil earthworm (Eisenia fetida). *Chemosphere, 169*, 316–323.

West, N. E., Stark, J. M., Johnson, D. W., Abrams, M. M., Wright, J. R., Heggem, D., & Peck, S. (1994). Effects of climate change on the edaphic features of arid and semi-arid lands of western North America. *Arid Soil Research and Rehabilitation, 8*, 307–351.

Williams, C. F., Letey, J., & Farmer, W. J. (2006). Estimating the potential for facilitated transport of napropamide by dissolved organic matter. *Soil Science Society of America Journal, 70*, 24–30.

Xiao, R., Wang, S., Li, R., Wang, J. J., & Zhang, Z. (2017). Soil heavy metal contamination and health risks associated with artisanal gold mining in Tongguan, Shaanxi, China. *Ecotoxicology and Environmental Safety, 141*, 17–24. https://doi.org/10.1016/j.ecoenv.2017.03.002

Xu, D., Shen, Z., Dou, C., Dou, Z., Li, Y., Gao, Y., & Sun, Q. (2022). Effects of soil properties on heavy metal bioavailability and accumulation in crop grains under different farmland use patterns. *Scientific Reports, 12*(1), 9211.

Yasir, M., Hossain, A., & Pratap-Singh, A. (2025). Pesticide degradation: Impacts on soil fertility and nutrient cycling. *Environments, 12*, 272. https://doi.org/10.3390/environments12080272

Ye, Z. H., Shu, W. S., Zhang, Z. Q., Lan, C. Y., & Wong, M. H. (2002). Evaluation of major constraints to revegetation of lead/zinc mine tailings using bioassay techniques. *Chemosphere, 47*, 1103–1111.

Zahedifar, M. (2017). Sequential extraction of zinc in the soils of different land use types as influenced by wheat straw derived biochar. *Journal of Geochemical Exploration, 182*, 22–31. https://doi.org/10.1016/j.gexplo.2017.08.007

Zahedifar, M., & Moosavi, A. A. (2017). Modeling desorption kinetics of the native and applied zinc in biochar-amended calcareous soils of different land uses. *Environmental Earth Sciences, 76*, 567.

Zahedifar, M., & Moosavi, A. A. (2020a). Adsorptive removal of nickel from different soil orders as a function of physicochemical properties. *Protection of Metals and Physical Chemistry of Surfaces, 56*, 54–62.

Zahedifar, M., & Moosavi, A. A. (2020b). Assessing cadmium availability of contaminated saline-sodic soils as influenced by biochar using the adsorption isotherm models. *Archives of Agronomy and Soil Science, 66*(12), 1735–1752. https://doi.org/10.1080/03650340.2020.1779231

Zahedifar, M., Moosavi, A. A., Zarei, Z., Shafigh, M., & Karimian, F. (2019). Heavy metals content and distribution in basil (Ocimum basilicum L.) as influenced by cadmium and different potassium sources. *International Journal of Phytoremediation, 21*(5), 435–447. https://doi.org/10.1080/15226514.2018.1537253

Zhang, Y., Hou, D., O'Connor, D., Shen, Z., Shi, P., Ok, Y. S., Tsang, C. D. W., Wen, Y., & Luo, M. (2019). Lead contamination in Chinese surface soils: Source identification, spatial-temporal distribution and associated health risks. *Critical Reviews in Environmental Science and Technology, 49*, 1386–1423.

CHAPTER

Saline soils and the challenges of food production

17

Yousef Hasheminejhad and Najmeh Mazloom

Khorasan Razavi Agricultural and Natural Resources Research and Education Center, Agricultural Research, Education and Extension Organization (AREEO), Mashhad, Iran

1. Introduction

Salinity is one of the most pervasive forms of land degradation affecting agriculture in arid and semi-arid regions. In these climates, high evaporation rates, limited rainfall, and shallow groundwater tables naturally promote the accumulation of salts in the soil profile. At the same time, the expansion of irrigated agriculture has intensified the development of secondary salinization. As a result, saline and sodic soils now represent a major constraint to global food production, limiting plant growth through osmotic stress, ion toxicity, soil structural deterioration, and reduced water infiltration.

The global extent of salt-affected soils highlights the seriousness of the problem: more than 424 million hectares of agricultural topsoil and 833 million hectares of subsoil are negatively impacted by salinity or sodicity (Basak et al., 2022, pp. 107–129). These conditions are particularly critical in countries where agricultural productivity depends heavily on irrigation. In such environments, the chemical equilibrium between soil and irrigation water governs the long-term evolution of soil salinity and sodicity. Thus, understanding the composition of irrigation water and its interactions with soil properties is essential for developing sustainable management strategies.

Iran, located in the arid belt of the Middle East, is among the countries where both natural and human-induced salinization are widespread. Nearly 50% of irrigated lands exhibit some degree of secondary salinity, driven by factors such as low-quality irrigation water, insufficient leaching, inadequate drainage, and the use of groundwater resources with rising salinity (Moameni, 2011). In addition to these longstanding challenges, new agricultural and environmental developments such as coastal aquaculture, the use of halophytes for forage and food, the recycling of saline drainage water, and the expansion of desalination technologies require updated frameworks for assessing salinity and designing suitable management strategies.

Existing classification systems for evaluating soil and water salinity—including the USDA Handbook No. 60, the Wilcox diagram (Wilcox, 1955), and the FAO guidelines have served the scientific community for decades. However, many were developed under conditions that differ from those currently encountered in Iran and other arid regions. Emerging issues such as magnesium-rich waters, expanding saline groundwater exploitation, drainage water reuse, and environmental applications of saline resources further necessitate a refined, context-specific classification system. This

The Soil Nexus. https://doi.org/10.1016/B978-0-443-45286-4.00011-0

chapter therefore, provides an overview of saline soils and the challenges they pose for food production. It synthesizes global knowledge on the formation and impacts of saline and sodic soils, reviews the principles governing soil-water chemical equilibrium under irrigation, and introduces management strategies tailored to both water-scarce and salt-affected environments. Drawing on Iran's extensive practical experience—ranging from the use of halophytes to novel applications of saline drainage water—the chapter presents an integrated framework for sustainable food production under saline conditions.

A large portion of Iran's agricultural soils, especially irrigated lands, are saline. About 6.8 million hectares of the country's agricultural land are saline to varying degrees (Moameni, 2011). Irrigation water resources are also becoming increasingly saline over time. Improper use of these saline water resources in saline or non-saline soils leads to the expansion of soil salinity. The continuity and sustainability of agricultural production using these saline water and soil resources depend on making appropriate decisions and changing resource-use management practices. It is obvious that management strategies for resource utilization are not the same for all salinity levels and water-soil quality conditions. Therefore, saline water and soil resources must first be classified into groups that, as far as possible, require similar management practices, and then a specific management strategy should be adopted for each group or class.

Various classifications have been presented for evaluating water and soil salinity, including the Wilcox salinity and alkalinity classification diagram (Wilcox, 1955), the U.S. Department of Agriculture Handbook No. 60 (Richards, 1954), FAO Irrigation and Drainage Paper No. 29 (Ayers & Westcot, 1985), and even classifications proposed for Iran and India (Abedi et al., 2002). All these classifications are based on assumptions and conditions that make their use feasible for a given time period and location. Some of these conditions and assumptions may have changed over time or may apply only to specific locations (Asadi et al., 2019), which can limit their applicability under current agricultural conditions in Iran. Therefore, an updated classification tailored to the country's new conditions is essential.

The emergence of certain new issues and aspects related to the use of saline water and soil resources further highlights the need to revise existing classifications. For example, the expansion of magnesium-rich water and soil resources reported from Iran, the Central Asian countries, and Pakistan (Qadir et al., 2018); the increasing use of desalination systems in agriculture (Rahimi et al., 2021); newly proposed uses for saline waters and drainage water recycling in halophyte-based agriculture (Khorsandi et al., 2020); and environmental and non-agricultural applications of saline water and soil resources (Salehi, 2020) are among the issues that have received attention in recent years. In this chapter, we attempt to present a comprehensive classification for the agricultural applications of saline water and soil resources, based on the country's current conditions.

The quality of irrigation water varies among regions, countries, and locations depending on how groundwater is extracted and used, the intensity of rainfall, and consequently the amount of aquifer recharge. The use of groundwater for agriculture in hot and arid countries may lead to increased groundwater salinity, which in turn restricts the range of crops that can be grown. Therefore, it is essential to determine irrigation water suitability. The concentration and composition of dissolved salts in water define its irrigation quality. Four criteria have been proposed for evaluating water quality: water salinity (EC), sodium hazard (SAR), residual sodium carbonate (RSC), and ion toxicity. In irrigated regions, soil salinity and dissolved ions are in equilibrium with the corresponding components in the irrigation water. In this publication, the four criteria for evaluating irrigation water quality are also used to assess soil quality, and corresponding calculation methods based on irrigation water are introduced as needed.

2. Materials and methods

Sustainable agricultural management requires the accurate characterization and classification of water and soil salinity, particularly in arid and semi-arid regions. This section discusses the fundamental scientific criteria employed in this domain, covering various parameters such as electrical conductivity (EC), specific ion concentrations and their ratios and the role of pH. A critical aspect is the quantitative relationship between irrigation water salinity and the resultant soil salinity alongside the effects of key cations like sodium, potassium, and magnesium on soil physical properties, specifically flocculation and dispersion dynamics.

Furthermore, this section transitions from fundamental metrics to applied regional assessments and management strategies. It systematically addresses specific challenges associated with calcareous and high-pH environments, examining the current status of water resources and irrigated soils in Iran against conventional classification standards. The subsequent sections propose an alternative, specialized classification for saline agricultural water resource and outline a comprehensive suite of management technologies designed for the effective use and mitigation of saline water and soil impacts.

2.1 Common criteria for classifying water and soil salinity

The most widely used criterion for classifying irrigation water salinity is the Wilcox diagram. Based on this system, water resources with salinity below 250 $\mu mhos/cm$ (1000 $\mu mhos/cm$ equals one dS/m) are classified as low salinity (C1); between 250 and 750 as medium salinity (C2); between 750 and 2250 as high salinity (C3); and above 2250 micromhos per centimeter as very high salinity (C4) (Wilcox, 1955). This classification is based on the relationship between irrigation water salinity and EC of the soil saturation extract, developed from more than 1300 surface and groundwater samples collected from the western United States. According to Wilcox's findings at that time, more than half of the irrigation water sources examined had EC values below 750 $\mu mhos/cm$, and only 10% had salinity levels above 2250 $\mu mhos/cm$.

In the same publication, Wilcox correctly identifies soil salinity as a function of irrigation water salinity and leaching fraction, and based on USDA Handbook No. 60 (Richards, 1954), who presents the following classification for the EC of the soil saturation extract Table 17.1.

Although research activities in the decades following the publication of this guideline table have expanded the list of tolerant, sensitive, and semi-sensitive agricultural and horticultural crops, an important point in this classification is that there is virtually no relationship between irrigation water salinity and soil salinity.

2.2 Converting irrigation water salinity to soil salinity

Ayers and Westcot (1985) showed that under long-term (steady-state) conditions, and assuming a water-uptake pattern of 40–30-20–10, the following relationship exists between irrigation water salinity and soil salinity (Ayers & Westcot, 1985). The 40-30–20-10 water-uptake pattern assumes 40,

Table 17.1 Classification of soil saturated extract's electrical conductivity (Richards, 1954).

Electrical conductivity of the saturation extract (dS/m)	Corresponding plant response
0–2	Salinity effects are mostly negligible.
2–4	Yields of more sensitive crops decrease; such as avocado, citrus, strawberry, peach, apricot, almond, plum, celery, and radish.
4–8	Yields of many crops decline. More sensitive crops in this group include cauliflower, cabbage, broccoli, tomato, rye, wheat, oat, alfalfa, Sudan grass, dallisgrass, clover, perennial ryegrass, and sweet clover.
8–16	Only salt-tolerant crops produce satisfactory yields. This group includes date palm, Astragalus spp., kale, beet, barley, many wheatgrass species, Rhodes grass, Bermuda grass, some cotton varieties, and sugar beet.
> 16	Satisfactory yields are obtained only from certain native rangeland species.

30, 20 and 10% of total water requirement to be uptake from sequential quarters of the root zone. 40% from the first quarter, to 10% for the last quarter. Although this is a rule of thumb for simulating solute transport but it has a comparable precision with more complicated methods and models (Hoffman & Van Genuchten, 1983).

$$EC_e \text{ (dS/m)} = X.\ EC_{iw} \text{ (dS/m)} \tag{17.1}$$

where EC_e and EC_{iw} are the electrical conductivity of the soil saturation extract and the irrigation water, respectively, and X is the concentration factor, which depends on the leaching fraction. They also provided Table 17.2 for obtaining different values of X based on the leaching fraction.

Table 17.2 Calculation of concentration factor from leaching fraction.

Leaching fraction (%)	Required water (percentage of water requirement)	Concentration factor (x)
5	105.3	3.2
10	111.1	2.1
15	117.6	1.6
20	125	1.3
25	133.3	1.2
30	142.9	1
40	166.7	0.9
50	200	0.8
60	250	0.7
70	333.3	0.6
80	500	0.6

Based on Table 17.2, the concentration factor at a leaching fraction between 15% and 20% is estimated to be about 1.5. Accordingly, under these leaching fractions which reflected typical irrigation management practices in the United States at that time, the EC of the soil saturation extract in the long term and under steady-state conditions was approximately 1.5 times the salinity of the irrigation water ($EC_e = 1.5\ EC_{iw}$).

A closer examination of the Wilcox classification, particularly when applying the concentration factor of 1.5 historically prevalent in the United States, suggests this system tends to be overly conservative. For instance, water designated as "saline" under that scheme would typically lead to soil EC values below 4 dS/m which is only problematic for the most sensitive crops. Subsequent revisions to this framework (Abedi et al., 2002; Fipps, 2003), have significantly raised the threshold for water deemed unsuitable for agriculture, setting it at 3 and 6 dS/m, respectively.

In contrast, the Indian classification system incorporates a broader set of variables beyond just irrigation water salinity, factoring in rainfall amount, crop salt tolerance, and soil clay content (Minhas, 1996). As illustrated in Table 17.3, this classification demonstrates that the acceptable salinity range for irrigation water expands when rainfall is higher and soil texture is coarser (i.e., lower clay content), especially for crops tolerant to salinity. A notable example is that, in areas with more than 500 mm of rainfall and coarse-textured soils, irrigation water with salinity up to 12.5 dS/m can be used for salt-tolerant crops.

Although this table is much more realistic than the Wilcox diagram, we will show in later sections that there are differences between agricultural conditions in Iran and India that limit the applicability of this table for Iran.

Perhaps in recognition of some of these differences, a new classification for Iran's saline waters was proposed in 1998 (Abedi et al., 2002). In this classification as the Neyrizi Classification (Table 17.4) the effect of the leaching fraction on soil salinity is taken into account. Based on irrigation water distribution efficiency in typical irrigation systems used in Iran, a leaching fraction of about 20% was

Table 17.3 The Indian guideline for irrigation water salinity (Minhas, 1996).

Soil texture (Clay percentage)	Crop tolerance	Rainfall (mm) and irrigation water salinity (dS/m)		
		Less than 350	350–550	More than 550
Fine (> 30%)	Sensitive	1	1	1.5
	Semi-tolerant	1.5	2	3
	Tolerant	2	3	4.5
Relatively fine (20% to 30%)	Sensitive	1.5	2	2.5
	Semi-tolerant	2	3	4.5
	Tolerant	4	6	8
Relatively coarse (10% to 20%)	Sensitive	2	2.5	3
	Semi-tolerant	4	6	8
	Tolerant	6	8	10
Coarse (Less than 10%)	Sensitive	–	3	3
	Semi-tolerant	6	7.5	9
	Tolerant	8	10	12.5

Table 17.4 Classification of saline water for agricultural use under a 20% leaching fraction (Abedi et al., 2002).

Irrigation water salinity	Effects on crop production
Low salinity EC < 4 dS/m	Most crops will produce satisfactory yields. Some salt-sensitive crops such as maize lose around 30% of yield, and very sensitive crops such as beans may lose about 50%.
Slightly saline 4 < EC < 8 dS/m	Salt-tolerant crops such as sugar beet, cotton, and barley give acceptable yields. Semi-tolerant crops such as wheat lose about 30% of yield, and alfalfa loses about 50%.
Moderately saline 8 < EC < 12 dS/m	Salt-tolerant crops lose about 30% of their yield, and semi-tolerant crops lose more than half.
Highly saline EC > 12 dS/m	Suitable for growing halophytic (salt-loving) plants. Even salt-tolerant crops lose more than 50% of their yield.

assumed. Then, referring to Table 17.2, it was assumed that under these conditions, the average salinity of the soil saturation extract in the root zone would be approximately 1.3 times the salinity of the irrigation water. Using this estimated average root-zone soil salinity, the suitability of water for irrigating various crops was determined.

- With an increase in the leaching fraction above 20%—due to higher irrigation amounts or lighter-textured soils—crop yields will increase beyond the ranges presented in the table, and the use of saline water becomes more justifiable.

2.3 Sodium

Excess sodium on soil exchange sites causes soil aggregates to disperse. As a result, fine soil particles block the pore spaces between aggregates and reduce the soil's infiltration rate. Therefore, the hazard associated with excess sodium in soil is important. To evaluate the sodium hazard in soil, the Exchangeable Sodium Percentage (ESP) index is used:

$$\text{ESP }(\%) = \frac{Na_{ex}\left(\text{cmol}_c/_{Kg}\right)}{\text{CEC}\left(\text{cmol}_c/_{Kg}\right)} \times 100 \tag{17.2}$$

where Na_{ex} is the exchangeable sodium and CEC is the cation exchange capacity, both expressed in centimoles of positive charge per kilogram of soil. Since measuring both exchangeable sodium and total cation exchange capacity involves various sources of error, efforts were made about 70 years ago to estimate this value using the Sodium Adsorption Ratio (SAR) of the irrigation water.

$$\text{SAR} = \frac{Na^+}{\sqrt{\dfrac{Ca^{2+} + Mg^{2+}}{2}}} \tag{17.3}$$

where Na^+, Ca^{2+}, and Mg^{2+} are the concentrations of sodium, calcium, and magnesium ions in the irrigation water, respectively, expressed in milliequivalents per liter.

$$\mathrm{ESP} = \frac{100(-0.0126 + 0.01475SAR)}{1 + (-0.0126 + 0.01475SAR)} \tag{17.4}$$

Using the empirical Eq. (17.4), which is derived from the concept of the Gapon coefficient for cation exchange between monovalent and divalent ions, the ESP of the soil can be estimated from SAR of the irrigation water with which the soil has reached equilibrium. Thus, although the hazard of excess sodium is actually determined by its effect on the soil, irrigation water quality criteria are commonly used for its evaluation. Although the derived relationship was empirical, its basis lies in the stoichiometric principles of cation exchange, which today are accepted as fundamental concepts of cation exchange.

Irrigation waters with high SAR levels tend to accumulate high amounts of sodium over time, which in turn can adversely affect soil permeability due to soil particle dispersion. Higher SAR levels can lead to crusting of the soil surface, poor seedling emergence, and inadequate aeration.

Similar to salinity classification, Wilcox (1955) proposed a classification for different SAR levels. According to this system, SAR values between 0 and 10 fall into the low category (S1), values between 10 and 18 into the medium category (S2), values between 18 and 26 into the high category (S3), and values above 26 into the very high category (S4).

He also combined the irrigation water electrical conductivity classes (C1–C4) with the sodium hazard classes (S1–S4) to introduce 16 different groups for classifying irrigation water salinity and sodium hazard. Fig. 17.1 shows this classification, commonly known as the Wilcox diagram.

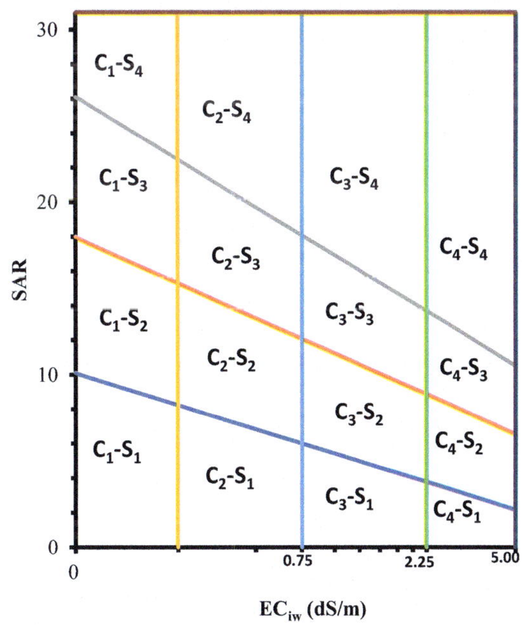

FIGURE 17.1

Wilcox diagram for classification of salinity and sodicity hazards (Wilcox, 1955).

According to this diagram, salinity levels impose constraints on the permissible sodium content of irrigation waters. For example, if a water source has a salinity of up to 0.25 dS/m and falls into class C1, its SAR must be around nine or lower to fall into the S1 category. However, if the same water has a salinity of about 3 dS/m and falls into class C4, its SAR must be at most about four to remain in category S1. Later findings, however, contradict this conclusion.

Rengasamy and Olsson (1991) proposed a classification for sodic soils in Australia based on SAR, the pH of a 1:5 soil-to-water suspension, and EC. These properties were selected because they have the most significant detrimental effects on soil structure in sodic soils. Due to difficulties in measuring ESP, these researchers used the SAR of the 1:5 soil extract to define sodic soils (which is thermodynamically related to ESP). The ESP of the soil is approximately twice the SAR of the 1:5 soil-to-water extract.

Increasing the salinity of the soil solution has a positive effect on soil structure. At higher soil solution salinities, the thickness of the double diffuse layer (DDL) decreases, allowing attractive forces between clay particles to dominate, which leads to the stabilization of soil structure (Quirk & Schofield, 1955). It is important to note that the expansion and contraction of the DDL without particle separation even if separation occurs at the angstrom scale constitutes a form of swelling and contraction (Bennett et al., 2019). Consequently, the effect of salinity on soil structure is a major factor determining soil pore geometry, which governs hydraulic conductivity. For this reason, the hazard of excess sodium in soil should be evaluated together with the effect of salinity.

Oster et al. provided a guideline table for interpreting the hazard of using waters with different salinities and SAR levels on the risk of reduced water infiltration in soils (Oster et al., 2016). This guideline is based on the principle that with increasing irrigation water salinity, the range of SAR values that can cause a reduction in soil infiltration rate also increases (Table 17.5; Oster et al., 2016).

Fig. 17.2 shows a diagram known as the Suarez diagram, which is used to assess the sodium hazard at different salinity levels (Ayers & Westcot, 1985). An important point is that although it is used to evaluate the risk of reduced water infiltration in soil, its input parameters are the salinity and SAR of the irrigation water. This approach is based on the assumption of equilibrium among the irrigation water, the soil solution, and the soil exchange phase.

According to Fig. 17.2, at a salinity level of 2 dS/m, the SAR of the irrigation water must exceed 21 for its use to pose a risk of severe reduction in soil infiltration rate, whereas at a salinity of 4 dS/m, this value must exceed 39. For ease of decision-making, California State University, Davis introduced the

Table 17.5 Integrative guideline for the effects of EC and SAR on infiltration rate (Oster et al., 2016).

	SAR effects in relation with EC (dS/m)		
SAR $(mmol_c/L)^{0.5}$	**Low**	**Moderate to high**	**High**
0–3	More than 0.7	0–0.72	Less than 0.2
3–6	More than 1.2	1–2.3	Less than 0.3
6–12	More than 1.9	1–8.5	Less than 0.5
12–20	More than 2.9	2.1–9.3	Less than 1.3
20–40	More than 5	2–5.9	Less than 2.9

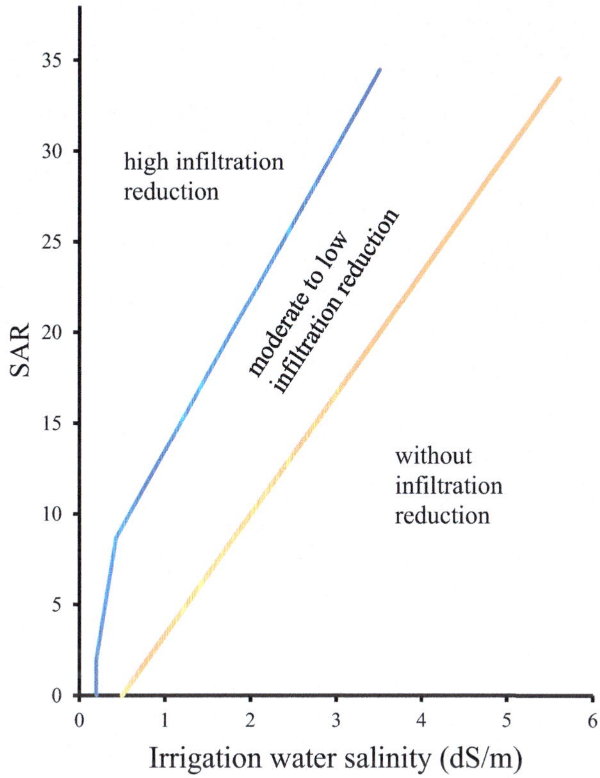

FIGURE 17.2

Suarez diagram for the sodium hazard at different salinity levels (Ayers & Westcot, 1985).

"five-times rule" (Marino et al., 2021). Accordingly, at any given salinity level, if the SAR of the irrigation water exceeds five times its EC, using that water will pose a risk of reduced soil infiltration. Similarly, the concentration of sodium in solution should be more than five times that of calcium; or in the case of soil, the exchangeable sodium concentration should exceed five times the exchangeable calcium concentration.

2.4 Concentration and ratio of specific ions

Sodium hazard assessment criteria have been in use since the 1980s and are based on evaluating salinity (EC) and sodium (SAR). In these criteria, the positive effects of calcium and magnesium are considered against the negative effect of sodium. However, recent research provides substantial evidence of negative impacts of both potassium and magnesium, in addition to sodium, on soil physical properties (Qadir et al., 2021).

2.4.1 Calcium-to-magnesium ratio

Soil salinization under irrigation and the decline in irrigation water quality due to salinityespecially sodium salts—are well-known phenomena. Nevertheless, evidence from irrigated regions worldwide shows higher magnesium concentrations in both water and soils compared to several decades ago

(Qadir & De Pauw, 2007). Specific examples include the Amu Darya and Syr Darya basins in Uzbekistan and Kazakhstan, the Aral Sea Basin in Central Asia, the Central Iranian Plateau, the Indus Basin in Pakistan, the Yellow River Basin in China, the Ganges–Indus Basin in India, the Euphrates Basin in Syria and Iraq, the San Joaquin Valley in California, and the Murray-Darling Basin in Australia (Qadir et al., 2018). These are all regions critical to national and regional food security and economies, and thus their management cannot be neglected. The impacts of irrigating with high-magnesium water include gradual deterioration of soil structure, reduced soil permeability, and significant reductions in crop yield.

Magnesium is a crucial element for plant growth, being an essential nutrient with key roles in chlorophyll formation and as an enzymatic cofactor. However, under certain conditions, it can be toxic to plants grown in high-magnesium soils. For instance, soils containing serpentine, which naturally have high magnesium, reduce plant growth. Similarly, hydrated magnesium silicates with very high Mg:Ca ratios (8 to over 4000), and serpentine-derived clay minerals, can lead to the formation of problematic soils.

Although it has been established that magnesium ions are less effective than calcium ions in particle flocculationa point discussed further below—it is generally accepted that in most agricultural soils and irrigation waters, calcium concentrations exceed magnesium concentrations.

Because calcium and magnesium concentrations in soils are usually sufficiently high, ignoring their ratio typically does not cause significant problems. Research results support this notion. However, the concept of the Ca:Mg ratio persists, despite decades of research suggesting otherwise. This concept originated from the work of Bear et al. in the 1940s (Bear et al., 1951). In their experiments, they found no significant difference in alfalfa response to pH adjustment in an acidic soil by changing the Ca:Mg ratio. Other studies at the time also indicated that this ratio was not particularly important. Many similar experiments showed no effect on plant production (McLean et al., 1983).

Nevertheless, concern has always existed that magnesium could damage soil structure and cause magnesium-induced sodicity. Laboratory studies also indicate that magnesium can be detrimental under certain conditions (Shainberg & Letey, 1984). This phenomenon is observed even in more recent research. For example, in a 2002 laboratory study (Dontsova & Norton, 2002), as shown in the Figure below, increasing calcium concentration in the suspension enhanced flocculation, causing the suspension to settle and clarify. This effect is likely due to the relatively large hydrated radius of magnesium ions, which is larger than that of calcium or even sodium ions.

However, under non-laboratory conditions, these negative effects generally do not manifest because calcium is usually present at higher concentrations than magnesium. In fact, the presence of lime in the soil can largely neutralize the potentially negative effects of magnesium.

Various factors cause soil colloids to preferentially adsorb specific cations. For example, cations with higher valence are typically preferred over monovalent cations. In addition to, these general preferences, certain colloids exhibit an exceptional affinity for specific cations. A notable example is the high exchangeable magnesium in vermiculite. The hydrated radius of magnesium ions matches the interlayer spacing of partially degraded vermiculite so well that magnesium is preferentially adsorbed over a wide range of cation concentrations. low Mg concentrations, vermiculite prefers Ca over Mg (Peterson et al., 1965). This is expected because hydrated magnesium ions are larger than hydrated calcium ions. However, once magnesium exceeds a sufficient threshold (over ~40% of the total cations in solution), adsorption shifts strongly toward magnesium.

2.4.2 Effect of potassium and magnesium on flocculation/dispersion

Similar to SAR, the U.S. Salinity Laboratory proposed a Potassium Adsorption Ratio (PAR), in which potassium concentration replaces sodium in Eq. (17.3). However, likely because sodium concentrations in irrigation water are several times higher than potassium, and exchangeable potassium has negligible effects on soil physical properties, no irrigation water quality guidelines based on PAR have been developed.

In both SAR and PAR, it was assumed that magnesium has a positive effect on soil permeability, and thus its effect, along with that of the divalent cation calcium, is included in the denominator of Eq. (17.3). Bresler et al. noted that under typical conditions, where calcium concentration is 2–5 times higher than magnesium, the negative effects of magnesium are likely masked by calcium (Bresler et al., 2012). Additionally, Eq. (17.3) does not account for the dissolution or precipitation of soil lime, which affects equilibrium calcium concentration; this issue is discussed in detail in the section on modified SAR.

High potassium and magnesium concentrations are common in recycled wastewater, which is increasingly used for irrigation. There are also numerous examples where high K and Mg concentrations negatively affected soil permeability, water quality, and crop growth (Oster et al., 2016; Qadir et al., 2018; Zhang et al., 2025).

The four cationscalcium, magnesium, potassium, and sodiumare common in irrigation water, soil solution, and soil exchange sites. Their concentrations and relative distribution are determined by both natural and anthropogenic factors. Research beginning in the 1950s, and continuing over subsequent decades, has shown negative effects not only of sodium but also of potassium (Liang et al., 2021) and magnesium (Zhu et al., 2019) on soil physical properties worldwide. These negative effects depend on factors such as clay mineralogy and the relative abundance of potassium and magnesium in soil exchange sites and irrigation water compared to sodium and calcium.

Soil aggregates consist of clay, silt, sand, and organic matter, bound together by various chemical forces. The overall surface charge of soil aggregates arises from the combined contributions of mineral particles and organic matter and therefore differs from the charge behavior of individual components. Importantly, the surface charge of both clay minerals and soil organic matter is pH-dependent. As soil pH increases, dissociation of functional groups (e.g., carboxyl and phenolic groups in organic matter) enhances negative charge, whereas at pH values below the zero point of charge (ZPC), protonation leads to the development of positive charge. When the net surface charge is negative, it is predominantly balanced by exchangeable cations, while under positive charge conditions, anions become the dominant counter-ions.

Many salt-affected soils have a net negative charge, balanced primarily by Ca^{2+}, Mg^{2+}, K^{+}, and Na^{+}, which link soil particles via covalent and/or ionic bonds. Polar water molecules interact with these ionic bonds, leading to hydration reactions dependent on the ionic index (Marchuk & Rengasamy, 2011). The ionic index is a geochemical molecular-scale parameter indicating the relative affinity of cations to bond (ionically or covalently) with charged particles. Rengasamy and Sumner defined cation flocculation capacity as follows (Rengasamy & Sumner, 1998):

$$\text{Flocculation Power} = 100(I_z/I_{z+1})^2 Z^3 \qquad (17.5)$$

In this formulation, I_z and I_z+1 are the ionization potentials of a cation with valence Z and the cations with valence Z+1, respectively. Using this approach, the flocculation capacity of sodium, potassium, magnesium, and calcium cations was calculated to be 1, 1.8, 27, and 45, respectively.

Following the proposal of this hypothesis (Rengasamy, 2002), the dispersion potential of these cations relative to calcium was determined. The values are: calcium = 1, magnesium = 1.7, potassium = 25, and sodium = 45. While exchangeable cations in dry soils participate directly in the dispersion of clay particles, cations in the soil solution counteract this dispersion process due to their flocculation potential.

A new concept called "net dispersive charge," which determines the amount of clay dispersed from a soil according to the following relationship (Rengasamy, 2016):

$$\text{Net Dispersive Charge} = \text{Dispersive Charge} - \text{Flocculating Charge} \tag{17.6}$$

The dispersive charge of a soil is calculated by multiplying the concentration of exchangeable cations ($\text{cmol}_c Kg^{-1}$)measured at a specific soil pHby the dispersion potential weighting factor of each cation:

$$\text{Dispersive Charge}\left(\text{cmol}_c Kg^{-1}\right) = (Ca) + 1.7(Mg) + 25(K) + 45(Na) \tag{17.7}$$

Similarly, the flocculation charge is calculated by multiplying the concentration of cations (cmol_c/kg, based on soil) in a soil–water dispersed suspension by the flocculation potential weighting factor of the corresponding cation:

$$\text{Flucculating charge}\left(\text{cmol}_c Kg^{-1}\right) = 45(Ca) + 27(Mg) + 1.8(K) + (Na) \tag{17.8}$$

For the effect of ions on the dispersion or flocculation of clay particles, there is a threshold at each salinity level. e.g., in defining SAR, a threshold of 13 is considered for a salinity of 2 dS/m. Similarly, for dispersive charge, there is a dispersion threshold. When the dispersive charge exceeds this threshold, clay particles separate and form a diffuse double layer. The effects of cation flocculation occur only when a diffuse double layer has been established.

At low flocculation charges, soil aggregates break into smaller aggregates because larger aggregates are fragmented into smaller ones; this process is not influenced by the electrolyte concentration (i.e., salinity) of the soil solution.

2.4.3 Cation ratio of structural stability

To evaluate the negative effects of potassium and magnesium cations on soil physical properties, Rengasamy and Marchuk introduced an index as an alternative to SAR (Rengasamy & Marchuk, 2011). This index, called the Cation Ratio of Structural Stability (CROSS), is based on the assumption that the inverse of the critical flocculation concentration (CFC) for a clay saturated with a specific cation is a quantitative measure of that cation's flocculation power.

Using the CFC values for four soils saturated with sodium, potassium, magnesium, and calcium cations, they were able to determine the relative flocculation strength of each cation by dividing the mean CFC of that cation by the mean CFC of sodium, which was taken as the reference.

Compared to sodium's flocculation power, which was set to 1, the average flocculation powers of potassium, magnesium, and calcium for the four soils studied were 1.8, 27, and 45, respectively.

By combining SAR and PAR, they introduced the CROSS index, in which the relative flocculation power ratios for potassium and magnesium are used as weighting factors. The potassium concentration factor was calculated as the ratio of sodium flocculation power to potassium flocculation power ($1 \div 1.8 = 0.56$), and the magnesium concentration factor as the ratio of magnesium flocculation power to calcium flocculation power ($27 \div 45 = 0.60$).

Thus, the CROSS index was defined as follows:

$$\text{CROSS} = \frac{\text{Na} + 0.56\text{K}}{\sqrt{\frac{\text{Ca} + 0.60\text{Mg}}{2}}} \tag{17.9}$$

Rengasamy and Marchuk (2011) suggested that by accounting for the dispersive effect of potassium relative to sodium and the reduced flocculation effect of magnesium compared to calcium, the CROSS index should provide a more accurate prediction of the risk of reduced water infiltration in soils than the SAR criterion.

However, as noted by Jayawardane et al., the CROSS index is not based on exchange isotherms and therefore cannot predict the intensity of surface adsorption of sodium and potassium (Jayawardane et al., 2011). Nevertheless, it can be used to predict the effect of clay particle dispersion on soil stability and hydraulic properties, depending on the relative amounts of the four cations in the soil solution at equilibrium.

The underlying assumption in applying Eq. (17.9) is that the theoretically derived values for cation flocculation strength accurately reflect their effect on soil structural stability and hydraulic properties. Jayawardane et al. (2011) noted that the two numerical factors used in CROSS are soil-dependent and proposed that, to generalize the equation, these factors be replaced with variable parameters, a and b, representing the concentration coefficients for potassium and magnesium, respectively.

Specific values for the coefficients a and b were calculated for 17 soils whose clay fractions had been isolated and chemically treated with sodium, potassium, magnesium, and calcium by (Zhu et al., 2019). The soils differed in clay mineralogy, clay type, pH, and EC.

The mean critical flocculation concentration (CFC) in mmol_c/L for the four cations in the clays was determined as follows:

$$\text{Na } (28.38) > \text{K } (15.24) > \text{Mg } (1.29) > \text{Ca } (0.85)$$

At the same time, the CFC values varied considerably among soils, particularly for sodium-saturated clays, which ranged from 5.66 to 98.1 mmol_c/L with a standard deviation of 26.9 mmol_c/L. Furthermore, clays dominated by montmorillonite required lower electrolyte concentrations to flocculate compared to illite-dominated clays.

Based on this, the numerical coefficient for magnesium in the CROSS calculation in this study was determined as 0.677, ranging from 0.488 to 0.855 with a standard deviation of 0.097. This average value for the magnesium coefficient aligns well with the mean value of 0.6 used by Rengasamy and Sumner (Rengasamy & Sumner, 1998).

Zhu et al. (2019) determined the numerical coefficient for potassium in CROSS to range from 0.309 to 0.777, with a mean of 0.605 and a standard deviation of 0.140, which also aligns with the concentration factor used in Eq. (17.9).

Smith et al. (2015) proposed that the coefficients a and b in the CROSS calculation should be based on the electrolyte concentration that causes clay dispersion, not the concentration that causes clay flocculation, since clay dispersion is the process that leads to reduced water infiltration in soils (Smith et al., 2015). Accordingly, they calculated a and b by optimizing the linear relationship between CROSS and a series of threshold electrolyte concentration (TEC) data, which resulted in a 15% reduction in saturated hydraulic conductivity of a sodic soil in the Riverina region of Australia (Arienzo et al., 2012). Because reductions in hydraulic conductivity are linked to clay particle contraction and dispersion, the electrolyte concentration at which this occurs is lower than the concentration required for clay flocculation (Dang et al., 2018).

The resulting values of a and b were 0.335 ± 0.038 and 0.0758 ± 0.012, respectively, yielding an R^2 of 0.95 for the linear relationship between CROSS and TEC. The optimized CROSS relationship derived by Smith et al. (2015), referred to as $CROSS_{opt}$, is as follows (Smith et al., 2015):

$$CROSS_{opt} = \frac{Na + 0.335K}{\sqrt{\frac{Ca + 0.0758Mg}{2}}} \quad (17.10)$$

The dispersive power of potassium is only one-third that of sodium. This value is also consistent with the coefficient *a* reported by Zhu et al. (2019) based on CFC data, where their value of 0.335 was not statistically far from the lower end of the coefficient *a* range, which was 0.309. However, the coefficient *b* determined by Smith et al. (2015) does not align with the range reported by Zhu et al. (2019), in which the coefficient *b* was never below 0.488.

The coefficient *b* reported by Smith et al. (2015) implies that the magnesium concentration required to prevent clay particle dispersion must be 13 times higher than the calcium concentration under the same conditions. In terms of clay flocculation, the relative effect of magnesium compared to calcium is about half of the value reported by Zhu et al. (2019), as their study indicated that the magnesium concentration must be 14 times higher than calcium to achieve flocculation. Therefore, Eqs. (17.9) and (17.10) provide different interpretations of the CROSS concept.

Regardless of how the coefficients *a* and *b* are determined, Qadir et al. (2021) plotted the risk of reduced soil permeability as a linear function of CROSS versus salinity (Fig. 17.3). Based on Fig. 17.3, water or soil samples are classified into two categories: no risk of reduced infiltration and severe risk of reduced water infiltration. The values along the guideline in Fig. 17.3 are based on the average threshold for reduced infiltration, and the error bars indicate variability around the mean CROSS value at each point (Qadir et al., 2021).

Variability in soil water infiltration depends on factors such as soil texture, clay mineralogy, organic matter content, gypsum content, soil depth, presence of compacted layers in the subsoil, type

FIGURE 17.3

Graphical illustration of Qadir et al. (2021) for interpretation of CROSS in relation to EC for infiltration rate reduction risk.

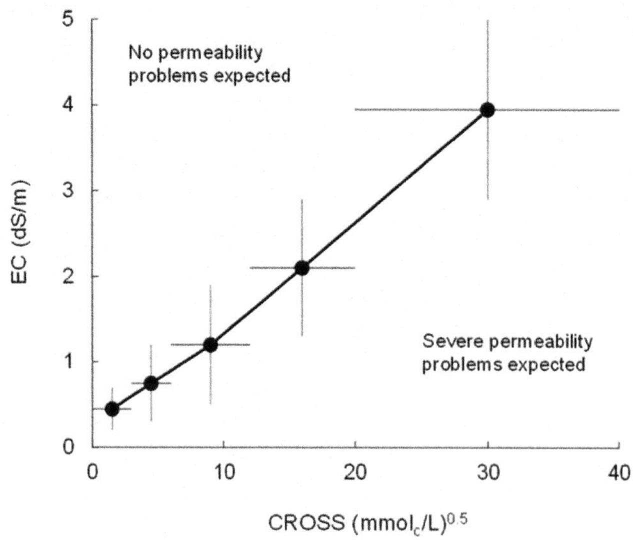

of crops, depth and quality of groundwater, irrigation method and scheduling, drainage system and its efficiency, rainfall pattern, and overall climatic conditions.

The authors switched the vertical and horizontal axes in Fig. 17.3 so that, similar to Fig. 17.2, the no-risk zone appears at the bottom of the graph and the severe-risk zone at the top. Based on the error bars in Fig. 17.3, they also delineated a moderate-risk range.

According to Fig. 17.4, samples of water or soil that fall below the green line are considered free of risk for reduced infiltration, while samples above the red line face a high risk of reduced infiltration. Samples located between the two lines are classified as having a moderate risk. The moderate infiltration problem region in Fig. 17.4 corresponds to the original guideline points in Fig. 17.3.

2.5 Alkalinity

Soils with a pH greater than 7 are considered alkaline. However, soils with a pH between 7 and 8 generally do not impose significant limitations on plant growth. Soils with pH above 8 are typically referred to as "alkaline soils" (Rengasamy et al., 2022). As pH increases beyond 8, nutrient deficiencies and ionic toxicities become more severe.

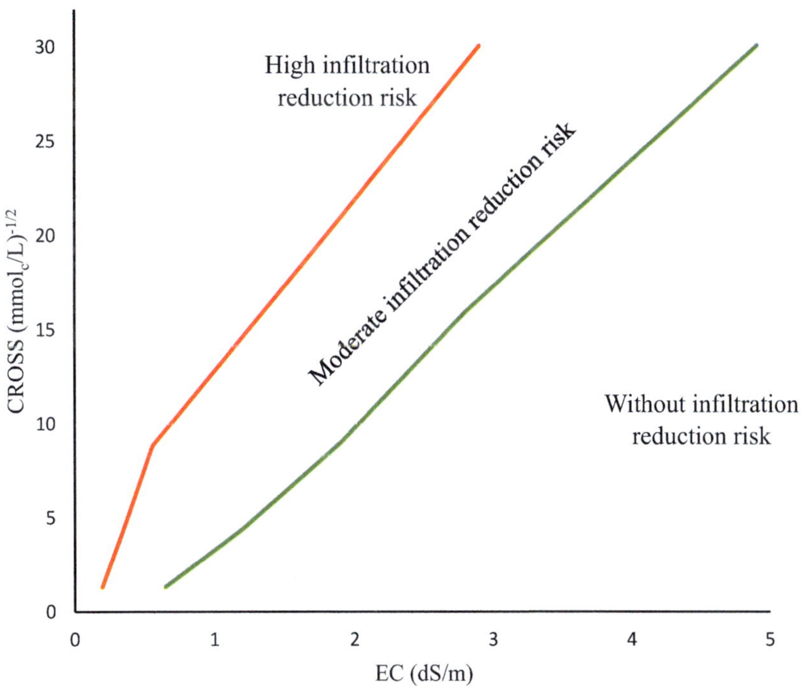

FIGURE 17.4

Modified graphical illustration for interpretation of CROSS in relation to EC for infiltration rate reduction risk. *Modified from Qadir, M., Sposito, G., Smith, C. J., & Oster, J. D. (2021). Reassessing irrigation water quality guidelines for sodicity hazard. Agricultural Water Management, 255, 107054.*

Alkalinity of the soil solution at pH above 7, without considering organic species, can be defined as follows (Sposito, 2008):

$$\text{Alkalinity} = \left[HCO_3^-\right] + 2\left[CO_3^{2-}\right] + \left[H_2PO_4^-\right] + 2\left[HPO_4^{2-}\right] + 3\left[PO_4^{3-}\right] + \left[B(OH)_4^-\right] + \left[OH^-\right] - \left[H^+\right] \quad (17.11)$$

Carbonate Alkalinity: In soils with pH greater than 8, carbonate ($CO_3{}^{2-}$) and bicarbonate ($HCO_3{}^-$) ions are predominant. These ions together constitute what is called "carbonate alkalinity."

$$\text{Carbonate Alkalinity} = \left[HCO_3^-\right] + 2\left[CO_3^{2-}\right] \quad (17.12)$$

3. Result and discussion

This section combines the findings presented across various analyses, starting with the specific challenges posed by soil chemistry, moving through the national water and soil status, detailing the proposed classification, exploring management strategies, and concluding with overall recommendations.

3.1 Contextual findings: Soil chemistry and water status

3.1.1 Calcareous soils and High-pH environments

Calcareous soils, containing lime (calcite, $CaCO_3$) in either powdered or nodular form, exhibit a buffered pH between 8 and 8.5, regardless of their lime content. Due to the low solubility of lime, the soil pH is largely controlled by the buffering capacity of calcite. Highly alkaline environments ($pH > 9$) occur in wide areas of Western Asia, Eastern Europe, Southern South America, and Southern Australia, and their prevalence is increasing with irrigation expansion in dry valleys, such as the Ganges in India or the Nile in Egypt (Jobbágy et al., 2017).

High-pH conditions develop when the concentration of carbonate and bicarbonate ions in the soil solution exceeds the sum of calcium and magnesium ions. In such soils, the solution is often dominated by sodium ions. Consequently, when sodium-induced salinity increases in calcareous soils, sodium reacts with calcite, forming sodium carbonate and bicarbonate salts, which also leads to an increase in exchangeable sodium. It is important to note, however, that high pH (alkalinity) results from elevated carbonate and bicarbonate concentrations rather than high ESP or soluble sodium (Tavakkoli et al., 2015).

Accumulation of chloride and sulfate salts in the lower soil layers can reduce soil pH. Generally, in salt-affected soils, a negative correlation between soil pH and salinity (EC) is observed (Al-Busaidi & Cookson, 2003). Rengasamy et al. (2022) observed that measuring pH in highly alkaline soils ($pH > 9$) using a 0.01 M $CaCl_2$ solution consistently yields pH values around 8.2, indicating that carbonate precipitation by calcium in the $CaCl_2$ solution increases the total calcium and magnesium concentration above that of carbonate and bicarbonate.

Interestingly, although NaCl does not precipitate carbonates nor increase Ca and Mg concentrations, adding increasing NaCl concentrations ($EC = 0.5$–6.2 dS/m) to an alkaline soil gradually lowered the pH from 9.7 to 8.9 (Rengasamy et al., 2022).

3.1.2 Specific limitations in alkaline soils (pH 8.5–10)

The specific limitations in alkaline soils largely depend on the soil pH. For example, bicarbonate toxicity tends to be predominant in the pH range of eight to nine, while iron deficiency can become problematic within the same range (Fig. 17.5). Conversely, iron toxicity has been observed in flooded soils in India at pH values above 9 (Setter et al., 2009). Similarly, both boron toxicity and deficiency are pH-dependent. At pH levels above 9, the concentration of $B(OH)_4^-$ species increases, which, together with limited leaching caused by clay dispersion, can lead to boron accumulation and toxicity (Barrett-Lennard et al., 2016).

Recent studies indicate that when soil pH rises above 9, the formation of non-ionic aluminate species may make aluminum toxicity a potential concern (Brautigan et al., 2012).

As pH increases, monovalent cations, particularly sodium, increase in the soil solution, while the concentrations of divalent cations such as calcium and magnesium decrease. This leads to enhanced clay dispersion and degradation of soil structure. Consequently, in the subsoil, temporary waterlogging and reducing conditions may develop, in which interactions between pH and redox potential (pE) can induce micronutrient toxicities (e.g., Mn, Fe, Al, and B) due to their elevated concentrations in the soil solution or the inability of plant roots to exclude these ions (Setter et al., 2009).

Although most soils have relatively high Ca:Mg ratios, conditions such as solutions where calcium carbonate precipitates, residual salts from ancient marine deposits, or magnesium-rich micas from vermiculite degradation can favor magnesium over calcium. In such cases, vermiculite becomes a strong sink for magnesium and remains nearly saturated with Mg even in the presence of high concentrations of calcium or monovalent cations. To displace magnesium from natural vermiculite, the mineral must be washed several times with high concentrations of replacing cations, reducing Mg to below the preferential adsorption threshold.

In magnesium-rich water (Ca:Mg < 1) or magnesium-rich soil (Ca:Mg in soil solution < 1), the potential hazard of sodium may increase slightly. In other words, at a Ca:Mg ratio below 1, a given SAR value may have a somewhat greater effect. Research shows that at a given irrigation water SAR, using water with a Ca:Mg ratio below 1 produces higher than normal ESP in the soil (Rahman & Rowell, 1979).

However, a concern is that crop performance in magnesium-rich soils, or in soils irrigated with high-magnesium water, is sometimes reported to be lower even if infiltration problems are not evident. This may result from calcium deficiency caused by high exchangeable magnesium levels in the soil. Some research indicates that yields of barley, wheat, maize, and sugar beet decline when the Ca:Mg ratio in soil solution falls below 1. While the role of calcium in crops is not fully understood, it appears that calcium mitigates potential toxic effects of other ions (Na, Mg) in the root zone. When the Ca:Mg ratio approaches or falls below 1, calcium uptake and translocation from soil solution to aerial parts of the crop may be reduced due to magnesium antagonism or competition for binding sites. This can lead to calcium deficiency, even if the water or soil solution has relatively high calcium concentrations compared to conditions with a higher Ca:Mg ratio. Although not conclusively confirmed, irrigation water with a similar Ca:Mg ratio (<1) is expected to have a similar effect if no readily available calcium source exists in the soil.

3.1.3 Status of water resources in Iran based on common classifications

In this section, the status of water resources is evaluated based on 135 chemical analyses of irrigation water samples collected from pistachio orchards in northern Ardakan, Yazd province (Hasheminejhad, 2019). These water resources range from low to very high salinity and are primarily used for pistachio irrigation.

Fig. 17.5 shows the distribution of salinity (EC) and SAR of these irrigation water sources on the Wilcox diagram. As mentioned in the introduction to the Wilcox diagram, this classification is much

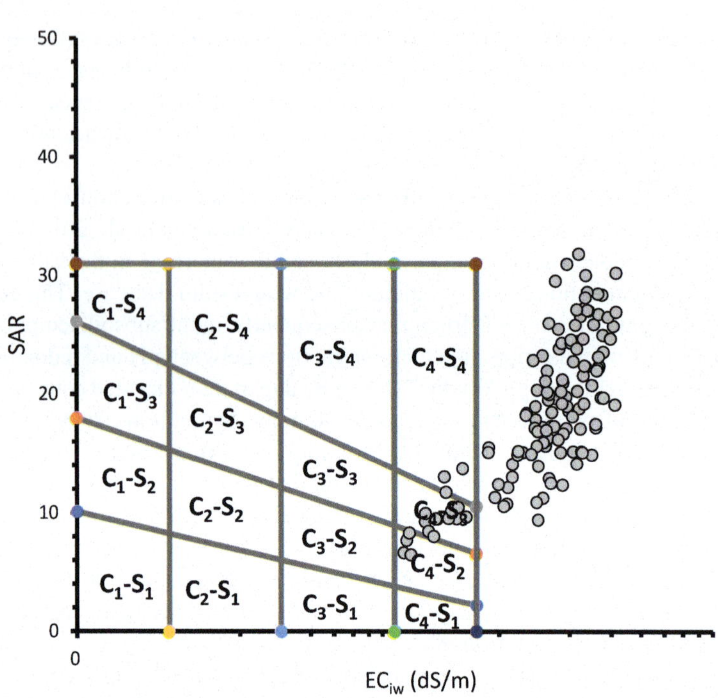

FIGURE 17.5

Distribution of irrigation water EC and SAR in Wilcox diagram for 135 samples.

more restrictive compared to the salinity levels of irrigation water currently used in Iran. All 135 water samples examined in this study fall into class 4, and a substantial portion exceed this class. Regarding sodium hazard, most of the water resources examined lie outside the Wilcox classification range.

Based on the Suarez diagram (Fig. 17.6), all the water resources examined fall within the safe range with no risk of reduced soil permeability. This indicates that, although the irrigation waters exhibit relatively high values of SAR, their high EC ensures that the coagulating (flocculating) effect of cations exceeds the dispersive effect. Consequently, adverse impacts of exchangeable sodium on soil physical properties are not expected.

Fig. 17.7 shows the distribution of the examined water samples in terms of the magnesium-to-calcium (Mg/Ca) ratio. Most of the samples have an Mg/Ca ratio greater than 1, while a smaller proportion of the samples have an Mg/Ca ratio exceeding 2, and only one sample out of the total has an Mg/Ca ratio greater than 3.

Fig. 17.8 illustrates the status of the examined water samples based on the CROSS index across three ranges: no risk, moderate risk, and high risk. According to this Figure, all water samples fall within the no-risk range.

3.1.4 Status of irrigated soils in conventional classifications

To assess the status of soils under different classification schemes, 17 representative soils corresponding to the 135 irrigation water sources examined previously were sampled to a depth of 90 cm, and their relevant properties were measured in the laboratory.

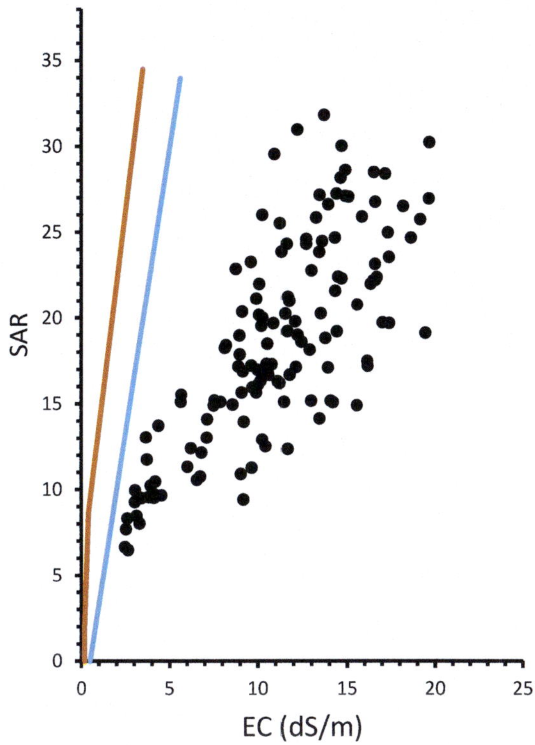

FIGURE 17.6

Distribution of EC and SAR in 135 water samples. The points which are below blue line have no infiltration reduction problem.

In Fig. 17.9, lines representing each leaching fraction were drawn based on the relationships between irrigation water salinity and soil salinity at different leaching fractions (Table 17.2). According to the referenced table, a linear relationship is assumed between irrigation water salinity and mean soil salinity for each leaching fraction. Consequently, a guideline line can be plotted at each irrigation water salinity, representing the mean soil salinity for a given leaching fraction. Points located below each leaching fraction line indicate effective leaching fractions higher than the line above.

Accordingly, only one case shows a leaching fraction below 10%, which—considering the irrigation water salinity (less than 4 dS/m) and the salinity tolerance of pistachio (approximately 12 dS/m in soil)—keeps the soil salinity below the threshold under long-term (steady-state) conditions. However, with increasing irrigation water salinity, farmers apply higher leaching fractions. For example, at an irrigation water salinity of approximately 19 dS/m, the leaching fraction is estimated to exceed 50%, maintaining the soil salinity around 13 dS/m.

At the same salinity level, in another field where a leaching fraction of about 25% was applied, the average soil salinity exceeded 22 dS/m, which would naturally lead to a severe decline in pistachio yield and may even halt tree growth. In general, the average soil salinity levels were maintained mostly within the range of 4–16 dS/m.

In Fig. 17.10, the status of the 17 soil samples, together with their corresponding irrigation water samples, is evaluated using the Suarez diagram to assess the risk of reduced soil infiltration rates. Here as well, all samples fall within the no-risk category.

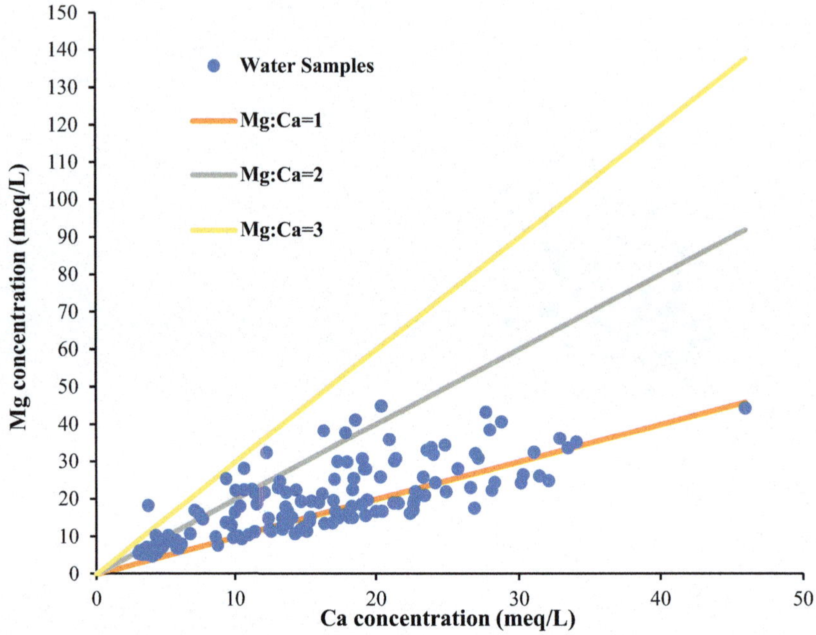

FIGURE 17.7

The distribution of Ca and Mg concentrations in 135 water samples around the 1:1, 2:1 and 3:1 of Mg:Ca lines.

FIGURE 17.8

The distribution of EC and CROSS in 135 water samples. The points which are below blue line have no infiltration reduction problem.

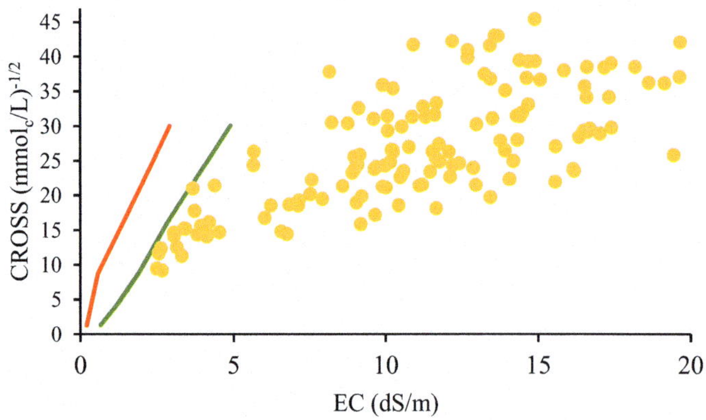

With respect to the magnesium-to-calcium ratio, the values observed in the soil were generally lower than those in the corresponding irrigation water. As shown in Fig. 17.11, only in samples 4 and 6 and 13 was the Mg/Ca ratio in the soil higher than that of the irrigation water, a finding that requires further investigation.

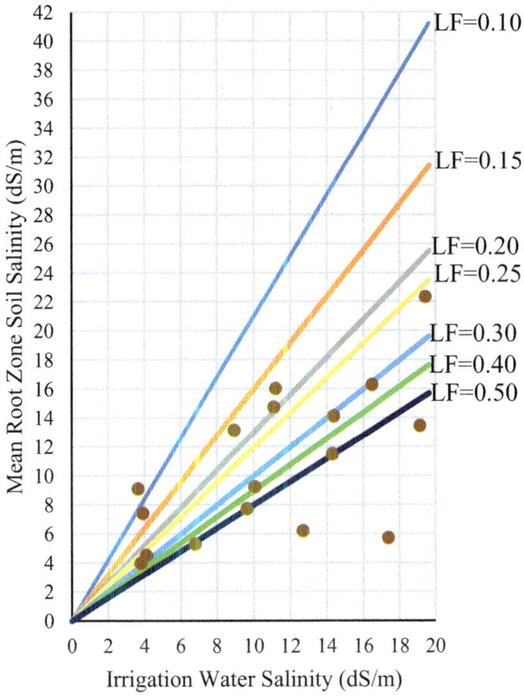

FIGURE 17.9

Distribution of irrigation water and mean root zone soil salinity of 17 sampling points in comparison with different LF levels.

3.2 Proposed framework and management strategies

3.2.1 Classification of saline water resources in agriculture

A proposed classification system for saline irrigation waters and for soils irrigated with such water resources, tailored to the conditions in Iran, is presented in Table 17.6. This scheme considers, for each level of irrigation water salinity, three groups of soil, management, and drainage conditions:

Group A: Light-textured soils with a leaching fraction greater than 30% and naturally well-drained conditions—under which the average root-zone soil salinity in steady-state conditions equals or is less than that of the irrigation water. This group also includes situations where rainfall contributes significantly to crop water requirements or when high-quality irrigation water is available during sensitive growth stages.

Group B: Medium-textured soils with a leaching fraction ranging from 15% to 30% and naturally well-drained conditions, where the average root-zone soil salinity under steady-state conditions is between 1 and 1.5 times the salinity of the irrigation water.

Group C: Heavy-textured soils with a leaching fraction below 15% and/or imperfect drainage conditions, where the average root-zone soil salinity under steady-state conditions exceeds 1.5 times the salinity of the irrigation water.

In this classification, crops are divided into six tolerance groups:

- *Very sensitive:* citrus (*Citrus* spp)
- *Sensitive:* bean (*Phaseolus vulgaris*), maize (*Zea mays*)
- *Moderately sensitive:* tomato (*Solanum lycopersicum*)

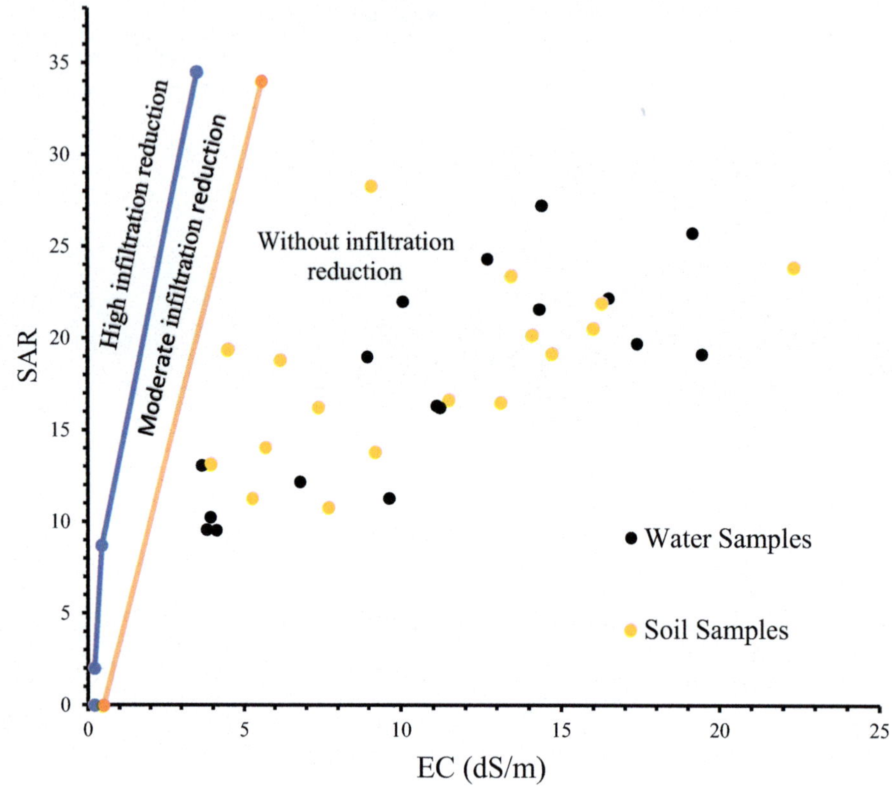

FIGURE 17.10

Distribution of EC and SAR in 17 water and their corresponding soil samples in Suarez diagram for identification on infiltration problem.

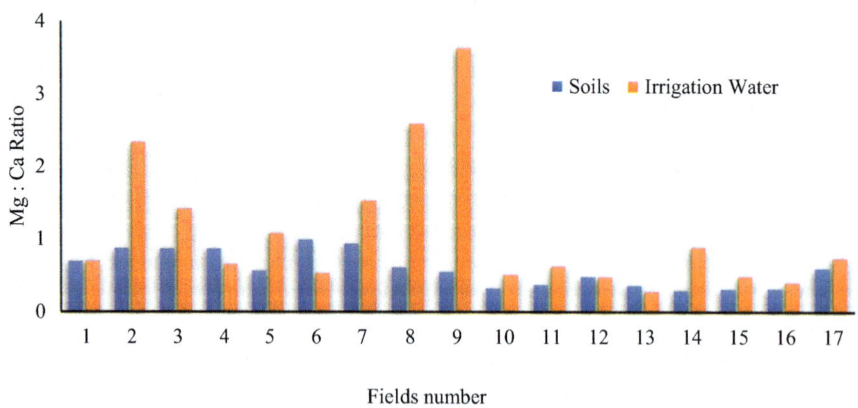

FIGURE 17.11

Magnesium to calcium ratio in 17 water and their corresponding soil samples.

Table 17.6 Proposed classification of saline irrigation waters and soils irrigated with these water resources under Iranian conditions.

Irrigation water salinity (dS/m)	Group	Very sensitive	Sensitive	Moderately sensitive	Moderately tolerant	Tolerant	Highly tolerant
< 2	A	1	1	1	1	1	1
< 2	B	2	1	1	1	1	1
< 2	C	3	2	1	1	1	1
2–4	A	2	1	1	1	1	1
2–4	B	3	2	1	1	1	1
2–4	C	4	3	2	1	1	1
4–6	A	3	2	2	1	1	1
4–6	B	4	3	3	2	1	1
4–6	C	5	4	4	3	2	1
6–8	A	5	4	4	3	2	1
6–8	B	5	5	4	4	3	2
6–8	C	5	5	5	5	4	2
8–10	A	5	5	4	4	3	2
8–10	B	5	5	5	4	4	2
8–10	C	5	5	5	5	5	2
10–12	A	5	5	5	4	4	2
10–12	B	5	5	5	5	4	3
10–12	C	5	5	5	5	5	3
12–14	A	5	5	5	5	4	3
12–14	B	5	5	5	5	5	3
12–14	C	5	5	5	5	5	4
14–16	A	5	5	5	5	5	3
14–16	B	5	5	5	5	5	4
14–16	C	5	5	5	5	5	4
> 16	A	5	5	5	5	5	5
> 16	B	5	5	5	5	5	5
> 16	C	5	5	5	5	5	5

- *Moderately tolerant:* wheat (*Triticum aestivum*)
- *Tolerant:* sugar beet (*Beta vulgaris*), cotton (*Gossypium* spp)
- *Highly tolerant:* pistachio (*Pistacia vera*)

Crop responses to various levels of salinity, as shown in Table 17.6, are categorized into five classes:

1. **Suitable:** Crop production under these conditions faces no specific limitations.
2. **Moderately suitable:** Crop production is feasible but requires attention to controlling salt accumulation in the root zone.

3. **Moderately unsuitable:** Managing soil salinity requires specific interventions such as reducing evaporation or implementing precise irrigation practices.
4. **Unsuitable:** Crop production is at the threshold of feasibility. If salinity is carefully managed with appropriate practices, conditions may improve to level 3 or level 2.
5. **Completely unsuitable:** Crop production is economically unviable, or the quantity of water needed for salinity control is unjustifiable. In such cases, cultivation should shift toward more salt-tolerant species.

3.2.2 Use and management of saline water and soil in agriculture

In recent years, the salinization of groundwater resources and soils has been reported in all provinces of the country. The degree of its severity, however, varies significantly from one province to another. As a result, under such conditions, farmers have had to either switch to growing more salt-tolerant crops or even abandon lands degraded by salinity. On the other hand, in central parts of Iran, saline water resources have long been used for crop production. Very well-known examples include the cultivation of high-value pistachios in Yazd and Kerman provinces. In addition, several crops have traditionally been produced from saline soils affected by high groundwater tables. For example, Khuzestan is one of the highest-ranking provinces in irrigated wheat production, and Golestan ranks among the top in irrigated cotton production. While the irrigation water in these areas is not highly saline, agricultural lands face very high levels of soil salinity due to a lack of adequate drainage.

On the whole, different regions of the country have gained abundant practical experience in using saline water and soil resources. It is very instructive to study these experiences and draw from them to guide salinity management in areas that have just started to face this problem.

Technologies for the reclamation and use of saline soil and water resources can be divided into the following three categories: Biological technologies, Chemical technologies, Physical technologies.

The following sections present a detailed discussion of these technologies.

3.3 Technological and practical management solutions

3.3.1 Biological technologies

Biological technologies are based on the ecological and biological capability of saline environments, which can develop the effective use of soil and water in saline conditions. Among them, these technologies cover a lot of methods, and we will only introduce some key examples here.

3.3.1.1 Development of salt-tolerant cultivars

Salt tolerance is a genetic quantitative trait controlled by multiple genes (Shannon & Grieve, 1998). It represents an adaptive characteristic of a plant to continue growing and producing despite the fact that it is being subjected to elevated levels of soluble salt in either its root zone or on its leaves. Despite existing volumes of studies on breeding salt-tolerant varieties, success in creating plant commercially viable plant for cultivation has been restricted. According to Ranjbar and Salehi, only a small number of salt-tolerant wheat varieties have been released in most of the agricultural areas, including one variety in Egypt, two varieties in Pakistan and one variety in India (Ranjbar & Salehi, 2014).

Several genotypes of cereals have been introduced in Iran and identified as salt-tolerant, albeit differing significantly from each other with respect to growth potential. Anagholi et al. evaluated eight wheat genotypes (Akbari, Sistan, Arg, Ofogh, Roshan, Morvarid, Kuhdasht and Falat) in the presence or absence of saline medium (Anagholi & Galeshi, 2017). They observed that Kuhdasht had the minimum reduction of grain yield from salt stress ranks, indicating a relatively acceptable tolerance, though it is not among officially declared salt-tolerant cultivars.

An extensive assessment by Ranjbar and Salehi (2014), including a decade of field trials with 475 wheat genotypes from national and international origins compared to several commercial cultivars (Arg, Neishabour, Chamran, Shiroudi, Pishtaz, Shiraz, Marvdasht, Falat Bahar Yavarus [durum] Debira, Ghods, Roshan and Mahouti) in saline fields or Khuzestan-Yazd-Golestan, showed equivalent results: salt-tolerant cultivars did not necessarily perform better than the rest. The authors stressed that breeding should focus on salt-tolerant classifications not only based on face value, but also on the interaction with locally adapted genotypes showing high yield potential.

Factors such as climate, soil characteristics, management of irrigation, development stage of the plant and soil fertility influence salt tolerance. However, the variation among cultivars of a specific crop in their salinity thresholds is relatively minor. For example, found salinity thresholds of 6.06, 5.27 and 4.00 dS/m for the wheat varieties Falat, Kuhdasht and Ofogh, respectively. Therefore, salt-tolerant varieties cannot change the range of salty water or land that can be used for economic crop production. In this case, all wheat varieties shown here are still categorized as moderately salt-tolerant, and the cultivation of wheat with highly saline water is not a sustainable practice from an economic aspect. The same is true for a wide range of other crops. So sensational announcements—claims like growing rice or potatoes with seawater, for example—must not be associated with plant genetics. The real trick, however, lies in reducing root-zone salinity via rainfall or alternating irrigation or mixed water sources.

The relative nature of salt tolerance is another critical aspect. Salt tolerance would be misinterpreted if expressed in relative yield only, since an 80% relative yield of potato cannot be equated with 60% relative wheat. Thus, practical management decisions must therefore consider absolute yield and ultimately the economic value of the product.

Differences between cultivars of the same species can be small, but inter-species and inter-generic differences can be large, so this information is useful to choose crops appropriate for saline areas. For instance, the genus *Gossypium* contains four cultivated species of cotton. Although over 95% of global cotton production comes from *G. hirsutum*, genetic diversity in salt tolerance exists, and indigenous Iranian cottons (*G. herbaceum*) exhibit higher tolerance. Hybrids from interspecific crosses may also combine high yield with improved salt tolerance.

3.3.1.2 Bioremediation

Bioremediation is the breakdown, conversion, decrease in concentration, and/or elimination of contaminants in the environment by organisms (plants and microorganisms) to a level regarded as safe or below the limit in the environment. Using the term salinity, the contaminants are soluble salts, the environment refers to the root zone, and the organism is plants, but the ions in salts cannot be degraded biologically, and no plant has the capacity to remove the high amount of salts introduced into the environment every year by irrigation. The amount of salts removed by plant uptake is low compared to the amounts introduced through irrigation.

The salt balance can be described in detail this way: Irrigated and well-drained fields receive salts in the irrigation water and release them in the drainage. The total dissolved solids (TDS in mg/L or kg/m^3) in water can be determined by multiplying the electrical conductivity (EC in dS/m) by a factor of 0.7 (Richards, 1954). For example, an EC value of 1 dS/m represents a salt content of 0.7 kg/m^3 in the water to be irrigated. Assuming an irrigation need of 10,000 m^3 per ha per year, the salt added to the soil per ha per year will be 7000 kg per ha per year for each dS/m increase in water salinity value.

The plant uptake of salts is, however, limited. For field crops, the maximum contribution of salts to the total above-ground biomass is estimated to be up to 3%; in obligatory halophytes, the value is up to 13%. The mean increase in salt and dry biomass removed by harvesting in the case of 10 tons/ha biomass will be approximately 300 kg/ha of salt per year. By contrast, the application of water with an electrical conductivity of 4 dS/m will add a total of approximately 28 tons/ha per year to the soil salt burden. Even in the case of the most tolerant species of the genus *Salicornia,* with a reported production of approximately 2 tons/ha of above-ground biomass and 260 kg/ha salt per 1 ton/ha biomass, the water needed to satisfy plant demands (estimated to be 20,000 m^3/ha) will add 56 tons/ha per 1 year to the soil. Thus, salt input via irrigation far exceeds salt removal via biomass harvest. Reports of long-term salinity reduction under halophyte cultivation apply only to non-irrigated systems where rainfall meets plant water requirements.

Bioremediation in irrigated ecosystems, in fact, has little effect in decreasing the soil salinity, but its effectiveness in decreasing the sodicity of the soil is substantially high. For sodic and saline-sodic soils, a source of calcium in the form of a calcium-rich chemical has to be added to replace the exchangeable sodium ions on the exchangeable sites. This has been practiced through the application of chemical amendments in the soil (Qadir et al., 2001). Gypsum application in sodic and saline-sodic soils has been practiced widely because the dissolution of the applied materials leads to the release of calcium ions in the soil, thus improving the structure and permeability of the soil (Rengasamy & Olsson, 1991). The application of sulfuric acid to the soil, in fact, has been noted to be effective in the dissolution of calcite and the release of calcium ions in the soil, and in some studies, its effectiveness is higher than the application of gypsum in reducing ESP (Abdul Qadir et al., 2022). The sulfuric acid has also been found to be effective in the infiltration of water in sodic soils (Daba, 2025). However, monitoring and documenting changes in soil physical properties after reclamation are often overlooked. Chemical amendments can also be costly and inaccessible for some farmers.

Studies conducted in the past decades demonstrated the effectiveness of salt-tolerant and moderately salt-tolerant plants such as *Leptochloa fusca* (L.), *Sesbania acuelata*, and *Medicago sativa* in decreasing the values of salinity and sodicity (Qadir et al., 1996). The plants enhance the partial pressure of CO_2 in the rhizosphere and decrease the pH of the soil solution, thereby increasing the dissolution of calcite (Qadir & De Pauw, 2007; Tetteh et al., 2025).

By conducting a pot experiment, the results of Mazloom and Hasheminejhad indicated the effectiveness of both chemical methods (gypsum and sulfuric acid) and biological methods (growing *Sesbania acuelata*, *Cyanodon dactylon*, and *Rubia tinctorum* (L.)) in improving the properties of sodic soil (Mazloom & Hasheminejhad, 2025). The results demonstrated that the application of *Rubia tinctorum* was much more effective than the chemical methods in improving the physical and chemical properties of the soil. For instance, Fig. 17.12 reveals its effectiveness in improving the mean weight diameter of the aggregate. Yazd farmers also plant the species *Rubia tinctorum* before and in conjunction with pistachio plants in saline-sodic soils. The plant has a deep root system, and its susceptibility to rainfall infiltration leads to an increase in the efficiency of leaching and the downward movement of salts in the soil. The plant also provides a source of natural dye through its root fibers.

3.3.2 Chemical technologies

Chemical methods of saline-sodic soil reclamation are well known globally, but in Iran, they are sometimes mistakenly referred to as "salinity reduction agents." The fact is, however, salinity

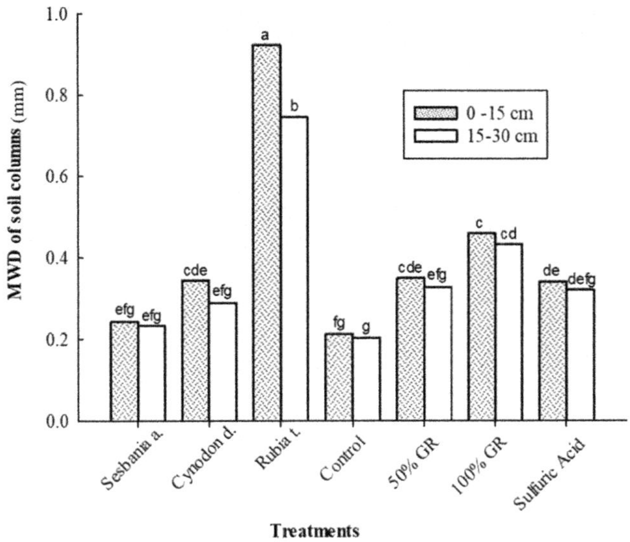

FIGURE 17.12

The effects of chemical and biological amendments on the mean weight diameter of aggregates in a sodic soi (Mazloom & Hasheminejhad, 2025).

reclamation is not a chemical process and can only be achieved by preventing salt entry into the root zone or by leaching salts below it. Fertilizers and bio-stimulators that enhance plant growth in the presence of high levels of salts are referred to as "anti-salinity agents," but an anti-salinity agent is a chemical that reduces saline levels in the root zone. The next section will discuss the application of chemical amendments in saline-sodic soil reclamation in Iran.

3.3.2.1 Mineral gypsum

The chemical treatment of sodic soils is mostly performed by mineral gypsum in Iran. Gypsum deposits with varying levels of purity are geographically widely distributed in the central Iranian plateau, and their important deposits are believed to be in the Semnan gypsum mines. Gypsum soils are also found in plenty, especially in non-irrigated lands. Easy access to these resources, low transportation costs, the lack of need for processing before agricultural use, and their safety during handling have led manufacturers of construction gypsum to supply part of their raw product as "agricultural gypsum."

Even though results from irrigated soils in the Yazd Province and other samples from other areas proved that the majority of the irrigated soils in Iran are not at risk of reduced water infiltration, the application of the gypsum soil treatment is quite popular even in saline soils and in the absence of the need for their treatment. Moreover, the results suggest that the enhancement in plant growth derives mostly from the nutritional and competition functions of calcium, and less from the real ameliorative effects of gypsum.

The effect of concentrations of gypsum solution in the process of sodic soil reclamation in the Doroodzan-Marvdasht Plain in Fars Province was investigated in a field test conducted by Rasouli and Kiani Pouya (2010) and the results of chemical analyses of soil solution ions compared to those predicted in the WatSuit model (Fig. 17.13) (Rasouli & Kiani Pouya, 2010). Fig. 17.13 shows a moderate risk of reduced soil infiltration. The visual appearance of the soil confirmed this issue. Application of 3 meq/L gypsum balanced this ratio.

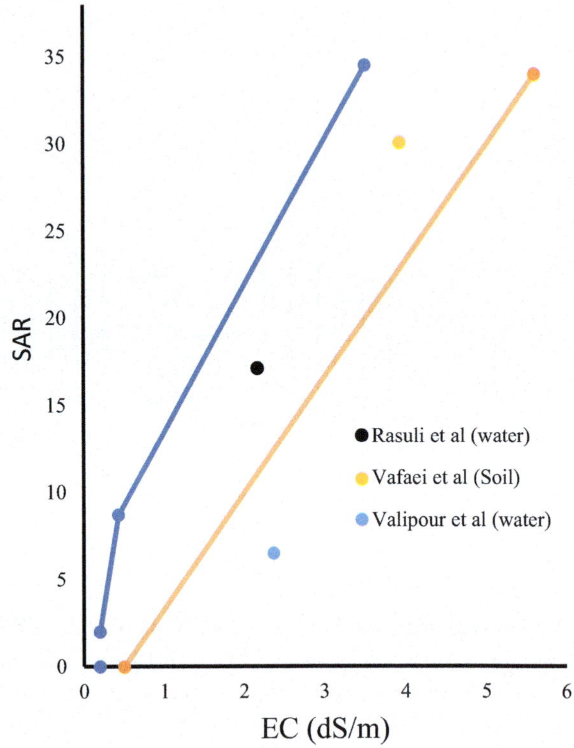

FIGURE 17.13

Classification of sodium hazard in the studied experiments.

Vafaei et al. (2019) studied the effect of mineral gypsum and organic amendments in the reclamation of sodic soil (Vafaei et al., 2019). Though the quality of the irrigation water was unreported, the involved soil was of moderate infiltration risk in the Suarez graph (Fig. 17.13). The results indicated that longer contact time between soil and amendments improved the soil chemical properties. The efficiency was higher in the treatment involving the full dose of gypsum, in the absence of organic amendments.

The effect of the application of mineral gypsum on the leaching of salts in highly saline soil with an electrical conductivity of saturated extract ranging from 70 to 99 dS/m and highly sodic soil with SAR ranging from 67 to 113 ($mmol_c$/L) was investigated in another study conducted by Valipour & Sakhaeirad, 2011. The irrigation water in this treatment was non-hazardous in the Suarez chart (Fig. 17.13). Soil analysis conducted after the reclamation process indicated that the application of gypsum in this treatment didn't affect the leaching of salts in the soil.

The above three examples illustrate the importance of the quality of irrigation water in making the decision to use a soil conditioner. In the study by Rasouli and Kiani Pouya (2010), the soil and water were sodic but nonsaline, and the use of gypsum resulted in the improvement of the soil properties and wheat yield. The soil in the study (Vafaei et al., 2019) was very sodic in the first season, and the application of gypsum worked in the long term. The application of gypsum in the study by Valipour and Sakhaeirad proved to be unproductive because the irrigation water had no sodium hazard and was able to leach the soil naturally (Valipour & Sakhaeirad, 2011). In summary, irrigation water should be classified in the moderate to severe category regarding the sodium hazard in order to merit the application of chemical amendments to the soil.

3.3.2.2 Acids

In calcareous soils or in irrigation waters containing high bicarbonate levels, acids are commonly used to mitigate the negative effects of bicarbonates on nutrient availability. These amendments are particularly common in greenhouse production systems. Based on a study conducted by Motallebifard et al., the treatment of sulfuric acid to decrease pH and subsequently lower the bicarbonate concentrations in the irrigation solution, containing 4 meq/L bicarbonate, to a pH of 6 significantly resulted in an increase in the diameter of flower heads and leaf chlorophyll index at the 5% significance level in carnations (*Dianthus caryophyllus*) (Motallebifard et al., 2008). The treatment also significantly resulted in an increase in the height of flower stems, thickness, and life span by 8.5%, 13%, and 34%, respectively, at the 1% significance level. The treatment resulted in a significant decrease in the flexibility of flower stems and the concentrations of leaf Zn and B at the 5% significance level, while the increase in leaf P content was significant at the 1% significance level. Irrigation water with high bicarbonate concentrations, despite proper fertilization, results in the reduced quality of cut flowers, and thus the bicarbonate levels and pH values of the irrigation water should be reduced.

The effect of various levels of irrigation water acidification on the growth of two geranium varieties was studied by Sharifiasl et al. (2012). The results showed acidification significantly increased the number of flowers per inflorescence, number of open flowers, chlorophyll content, and leaf Mg concentration at the 1% significance level, but had no significant effect on peduncle length. The interaction of cultivar × acidification significantly affected stem length, leaf size, shoot fresh and dry weight, number of inflorescences, and leaf Fe concentration. The results revealed that the reduced pH of the irrigation water to 5.5 enhanced the uptake and quality attributes of the two varieties.

The employment of mineral acids to rectify the effects of alkaline and bicarbonate-rich water/soils has proved well-founded in practice. Other than the application of mineral acids, organic acids began to be employed to promote plant growth in saline conditions, primarily because of facilitated nutrient transport and availability and less because of genuine alleviative properties. Among the examples are the favorable effects exerted by fulvic and citric acids in pistachio plants (Mirabzadeh Ardakani et al., 2021), humic acids in wheat plants (Rousta & Enayati, 2019), and humic acids in cotton plants (Esmaili et al., 2016).

3.3.2.3 Elemental sulfur and acid-forming fertilizers

The application of mineral acids in soil and water restoration works involves a number of technical issues. Consequently, sulfur and acid-producing fertilizers are gaining popularity. The sulfur in its natural forms has an oxidation state that ranges from −2 to +6. The reduced form of sulfur is the sulfide ion (HS^-), while the sulfate ion (SO_4^{2-}) is the most oxidized form. The sulfur element (S^0) is formed through the oxidation of the hydrogen sulfide ion and is the most stable form of sulfur. The form in which sulfur is available to plants is the sulfate ion; thus, the sulfur element has to be oxidized to the sulfate form through microbial action to be applied either for its action in reducing the pH in the soil or for its nutritional value (Yasin, 2024).

Over 150 years ago, it was demonstrated that the sulfur to sulfate conversion is a microbial activity. The first recorded agricultural application of sulfur was made in 1877 in the United States of America, in the state of South Carolina, by Charles F. Panknin, who recommended the mixture of 95% bone meal and ground phosphate rocks with 5% sulfur to make the phosphorus soluble through the sulfur oxidation and the production of sulfuric acid. Though he knew the sulfur oxidation resulted in the production of sulfuric acid, he didn't know that the process was microbial in nature (Lucheta & Lambais, 2012). Advances in the field of microbiology allowed numerous findings to be made in the area of sulfur-catabolizing microbes.

Sulfur-oxidizing bacteria function significantly in the ecosystem in terms of nutrient cycling processes. The presence of sulfur-oxidizing bacteria has been reported in agricultural and forest soils in numerous countries. Though chemotrophic microorganisms are the main sulfur oxidizers, a variety of other heterotrophic microorganisms such as fungi, bacteria, and actinomycetes also perform the function of sulfur oxidation (El-Tarabily et al., 2006). Moreover, the process of inoculating sulfur-oxidizing microorganisms into the soil, seeds, and fertilizers has gained considerable importance to improve the availability of nutrients in calcareous and alkaline soils. El-Tarabily et al. (2006) also examined the effectiveness of four bacterial isolates in a calcareous soil in the UAE, and their results proved that the application of *Paracoccus versutus* and elemental sulfur significantly reduced the pH, enhanced the soil sulfate content, and resulted in higher N, S, Fe, Mn, and Zn uptake in maze plants.

Apart from the enhanced availability of plant nutrients, the sulfur's capacity to produce an acidifying effect has been explored for reclaiming sodic soils. The basic assumption in the process is that the sulfuric acid formed due to sulfur oxidation will dissolve an amount of calcium carbonate in the soil and thereby increase the soluble calcium ions, resulting in the displacement of sodium ions to improve the soil structure. Ghaneei Motlagh et al. (2010) used a combination of gypsum, acid, and elemental sulfur in a saline-sodic soil (Ghaneei Motlagh et al., 2010). Despite the fact that the soil was not in the infiltration risk zone in the Suarez chart, the application of both gypsum and sulfuric acid reduced the soluble calcium and soluble sodium in the soil, but the effect of the latter was not statistically significant. Moreover, irrigation water had no sodium hazard, and the irrigation alone was enough to avert sodication. There were no significant changes in the soil salinity in this treatment, either. Another form of sulfur oxidation is performed through the combustion process. For this, a sulfur-burning equipment has been designed in the Iranian Soil and Water Research Institute (SWRI), Science and Technology Park.

One of the misunderstandings commonly perpetuated by some farmers and agricultural scientists is that any fertilizer that contains a sulfate form will result in the acidification process in the soil environment. But based on the information presented in the earlier sections, the process of producing an acid environment in the soil results from the conversion of sulfur from the reduced compounds to the oxidized form. For instance, in the case of the potassium sulfate fertilizer, the sulfur within the fertilizer is in the form of a sulfate and there is no conversion process taking place. The application of the fertilizer to the soil environment, therefore, should not be expected to decrease the pH levels in the soil environment. The case of ammonium sulfate is somewhat different, because the conversion of ammonium to nitrate (nitrification)—which is a microbial process—releases protons and acidifies the environment.

3.3.3 Hydraulic technologies

3.3.3.1 Leaching

Salinity of irrigation water is the principal cause of salinization in irrigated lands for large areas of the central plateau of Iran. For the many such lands that do not suffer from drainage problems, salts must be leached out of the root zone by proper leaching in order to prevent accumulation at levels that reduce crop yield. Additional salts are carried away by the downward, natural drainage of excess water out of the root zone. The soil salinity level not causing a reduction of crop yield depends on the type of crop and is itself influenced by irrigation water salinity and the leaching fraction.

The ratio of the volume—or depth—of drainage water exiting the bottom of the root zone to the total depth of irrigation water applied is termed the leaching fraction. This ratio may also be estimated from the ratio of irrigation water salinity to drainage water salinity, based on: no plant uptake of salts, no dissolution or precipitation processes, and steady-state equilibrium conditions. Whether or not these conditions are actually met, when natural drainage is occurring, the drainage water is not accessible, hence cannot be directly sampled or measured for salinity. In consequence, to manage root-zone salinity by leaching, three steps must be carried out:

1. Estimating the existing leaching fraction
2. Calculating the required leaching fraction
3. Comparing the present leaching fraction with the amount required

The minimum leaching fraction that maintains the salt balance and keeps soil salinity within a crop's tolerance is called the required leaching fraction (Chhabra, 2022, pp. 487–544; Richards, 1954). If the actual leaching fraction is less than this amount, salts will be accumulating and yield will eventually be reduced; if it is greater, deep percolation losses will be unnecessarily excessive. Therefore, the optimal situation is to maintain the leaching fraction close to the required amount.

3.3.3.1.1 Leaching fraction estimation. For soil salinity prediction, irrigation water salinity and drained water salinity can be used in combination with water and salt balance computations. However, the problem is that in the absence of artificial drainage systems—or when drains are not placed at the bottom of the root zone—the drainage water salinity cannot be measured directly. Other than computational methods, some tools can collect free water infiltrating through the soil profile.

The wetting front detector is a device designed to sample soil solution that is moving freely downward, following irrigation. Installed at various depths, it signals the arrival of water at each depth, from which the wetting front movement can be estimated. The wetting front is positioned between detectors—one above that collects water and one below that does not (Fig. 17.14). With this modest tool, the depth to which water has infiltrated can be determined for varying irrigation management programs, and the salinity of the soil solution can be used to determine whether leaching is occurring adequately.

Hasheminejhad (2011), using this device in pistachio orchards of northern Ardakan, estimated leaching fractions and used them in steady-state models for soil salinity prediction. Results showed that this tool can estimate both leaching fraction and soil salinity. In pistachio orchards, it is not easy to identify the exact depth of the active water-absorbing roots. Therefore, steady-state salinity models, which depend on root uptake distribution, need knowledge of active root depth.

3.3.3.1.2 Calculation of required leaching fraction. Several relationships are available to estimate the required leaching fraction. One common formula is:

$$LR = \frac{EC_{iw}}{5\mathrm{EC}_t - EC_{iw}} \tag{17.13}$$

where:

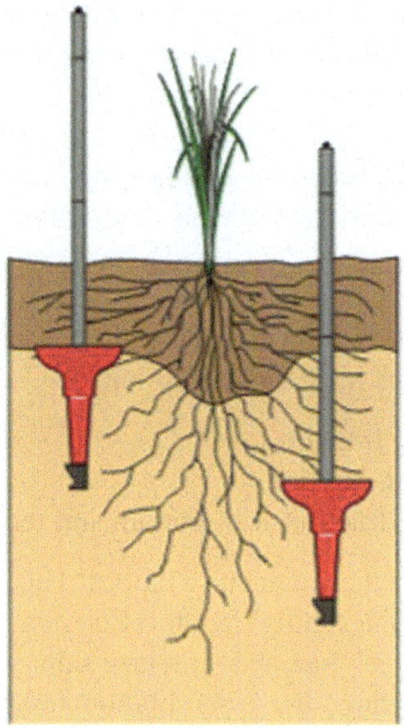

FIGURE 17.14

Schematic illustration of wetting front detector (WFD).

LR = required leaching fraction
EC_{iw} = electrical conductivity of irrigation water (dS/m)
EC_t = salinity threshold of the crop (dS/m)

Because salinity tolerance varies among crops, the required leaching fraction also varies for the same irrigation water salinity. Fig. 17.15 shows simplified results of Eq. (17.13) for several common salt-affected crops. Where the required leaching fraction exceeds 50%, excessive water losses occur, and thus salt-tolerant crops should be selected. Fortunately, a wide range in tolerance levels exists among agricultural and halophytic species, which enables the selection of suitable crops for a wide range of salinity and climatic conditions.

3.3.3.2 Drainage

Proper physical conditions of the root zone are crucial for plant growth. Drainage systems control the water and salt balance in the root zone by removing salts and preventing salt accumulation in the soil (Soltani et al., 2018). Conventional drainage systems promote better conditions for cultivation on lands troubled by salinity and waterlogging, but have disadvantages related to installation and maintenance costs, and environmental hazards at the discharge point (Akram et al., 2008; Parsinejad & Akram, 2018).

Agricultural drainage criteria define the maximum allowable water table level which prevents yield loss because of waterlogging. Where the actual water table is higher than this, artificial drainage or improvement of existing natural systems is required. Where the water table is kept lower than is required, it means that the system has been overdesigned. Technical criteria, aimed at cost reduction, and environmental and economic criteria, aimed at impact minimization and profit maximization, respectively, can be used besides agricultural criteria.

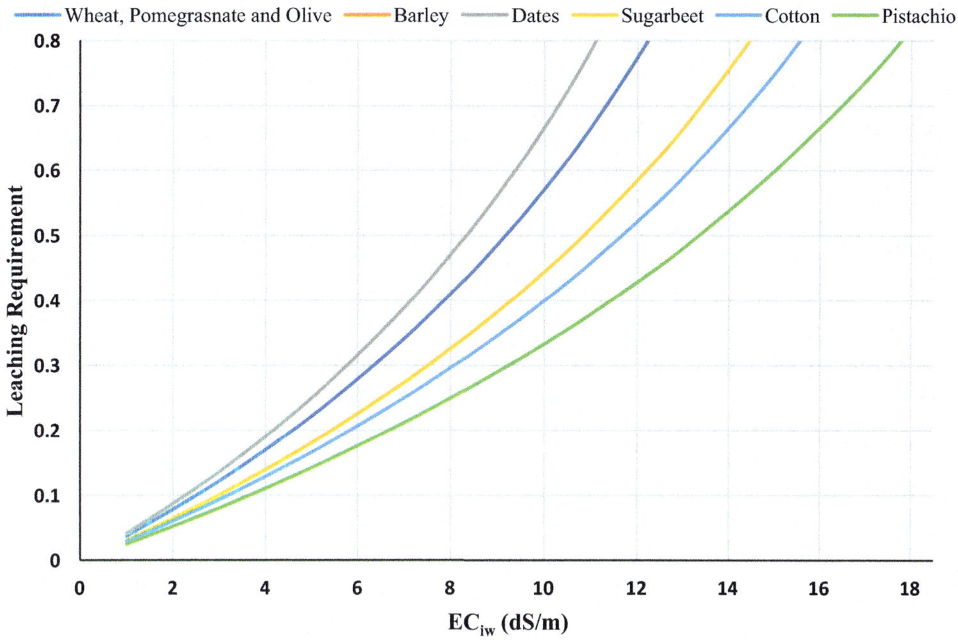

FIGURE 17.15

Estimation of the leaching requirement for several common crops under saline conditions using irrigation water salinity (Hasheminejhad & Besharat, 2017).

Correct assessment of drainage criteria involves:

- Knowledge of various drainage systems
- A proper index of waterlogging
- Clear description of the agriculture objectives
- Information about the relationship between the index and objectives

3.3.3.2.1 Types of agricultural drainage systems. Agricultural drainage systems facilitate the movement of water off the land so that farming benefits from reduced water table levels. These systems may be designed to improve water movement on the soil surface or below it. Thus, they are classified as surface drainage and subsurface drainage. Each type includes an internal field drainage system (lowering the water table within the field) and an external collector drain conveying the water to an outlet.

Surface drainage is employed when waterlogging appears on the soil surface, while subsurface drainage is employed when waterlogging occurs within the soil. In several cases, subsurface drainage is utilized to reduce surface waterlogging and vice versa, although this is not normally recommended; however, combined systems may be logical in specific situations.

Agricultural drainage systems do not necessarily create peak drainage flows, although in surface systems, after rainfall events, this may occur. Ideally, subsurface drainage should produce a relatively steady discharge, not sudden fluctuations. An exception is rice cultivation, where controlled drainage is used—discharge allowed only during specific periods such as the harvesting season.

This definition excludes drainage systems for cities, sports fields, and other non-agricultural uses and it also excludes natural drainage and levees.

3.3.3.2.2 Agricultural drainage systems analysis. Agricultural drainage aims at land improvement for agricultural purposes, increasing crop yield, making cultivation of more valuable crops possible, allowing multiple cropping, and decreasing production costs in waterlogged lands. These objectives are the result of direct and indirect effects.

Direct effects include:

- Reduced average soil water content—drier and less waterlogged soil
- Water removal from the soil-water system
- Less salt accumulation because of leaching

Direct effects lead to a variety of indirect effects, depending on climate, soil, crops, and socio-economic management. Such indirect effects can be positive or negative.

3.3.3.2.3 Positive indirect effects

- Increased aeration
- Soil structure improvement
- Increased nitrogen availability
- Increased yields and varieties of crops
- Easier field operations
- Earlier planting
- Reduced peak discharge by temporary storage of water in the soil

3.3.3.2.4 Negative indirect impacts

- Organic matter decomposition
- Soil subsidence
- Acidification of Potential Acid Sulfate Soils
- Increased risk of drought
- Ecological damage

Changes due to drainage also affect weeds, pests, and diseases—sometimes positively, sometimes negatively. It allows the removal of harmful salts and provides drainage water for reuse, but often involves environmental impacts downstream. Many indirect effects balance each other, for example, an improved soil structure may increase the infiltration and subsurface drainage while reducing runoff.

Thus, agricultural drainage requires a conceptual and systematic framework for evaluation. Monitoring would be continuously required to bring about further refinement in the criteria and identify malfunctioning systems or adverse impacts after initial improvements.

3.3.3.2.5 Irrigation, soil salinity, and subsurface drainage. In arid and semi-arid areas, subsurface drainage is usually installed in waterlogged irrigated land as a method to reduce or eliminate soil salinity. Salt balance is driven primarily by the water balance, particularly it's amount. Under adequate irrigation, the impact of drainage on salinity is influenced chiefly by the salt removal through drainage water. Therefore, the function of drainage for controlling salinity rests more on the aspect of discharge than on the lowering of the water table.

Properly designed irrigation systems maintain the water table at depths sufficient to control salinity. Even a deep water table cannot prevent salinity if the irrigation design is poor. For example, studies in the Nile Delta (Abdel-Dayem & Ritzema, 1990; Oosterbaan & Abu Senna, 1989) showed that maintaining the seasonal average water table at 1–1.2 m is sufficient; deeper tables can harm irrigation efficiency. Rao et al. (1990) found that during monsoon-season drainage, the water table should not be deeper than 0.8 m to allow soil desalinization (Rao et al., 1990).

Often, the depth for salinity control is associated with upward capillary flow. Shallow water tables, in the absence of irrigation or rainfall, contribute to upward seepage. However, in irrigated/drained systems, upward flow stops and net flow is downward. Thus, van Hoorn (1979) observed that "discussions on deep drainage to reduce capillary rise apply mainly where they are not relevant." (van Hoorn, 1979). In semi-arid regions that have well-defined wet and dry seasons, drainage is usually confined to the wet seasons. Salt disposal during these months is adequate to maintain the annual salt balance, even if partial salinization occurs during the dry months. Irrigation with moderately saline waters (EC $\approx$ 10 dS/m) during the dry seasons does not give any yield loss if sufficient leaching takes place in rainy seasons (Derbew et al., 2026; Sharma et al., 1990).

Advantages of using drainage water for irrigation during the dry season and disposing salts during the wet season:

- Prevent environmental discharge of saline drainage water into rivers during low-flow dry months
- Preserve limited supplies of freshwater
- Use high-flow rivers comfortably to dispose of saline drainage during wet months.

Rao et al. (1992) indeed report successful experiments where the drainage was completely stopped during dry seasons, allowing crops to benefit from capillary rise in addition to conserving irrigation water (Rao et al., 1992).

3.3.3.2.6 Saline drainage water reuse. Saline-sodic water (EC 4–30 dS/m; SAR 13–40 $mmol_c/L$) is suitable for irrigation of salt-tolerant crops, if root-zone salinity is kept under control and soil permeability is maintained. Reusing drainage water for salt-tolerant crops increases the salt concentration while decreasing the volume of water used, thereby reducing environmental impact on aquifers and downstream areas. It is assumed that percolated water below the root zone is captured by the drainage system and delivered to subsequent crops. Depending on land slope and hydrology, some drainage water may move downslope, increasing cumulative volume; the opposite occurs where no clay layer exists to restrict deep percolation.

Another consideration is the time taken for quasi-steady-state drainage water composition; this is often decades. Reuse requires that drainage water be available when crops need irrigation. In the western San Joaquin Valley, peak drainage occurs in winter when crop demand is low. Thus, storage would be needed—either through controlled drainage (Ayars, 2002) or reservoirs (Oster & Grattan, 2002). Other needs would include choosing crops tolerant to salt and boron (Maas & Grattan, 1999) and having adequate leaching. Rainfall after saline-sodic irrigation can reduce surface salinity sufficiently to result in crusting, reduced infiltration, and poor soil tilth. These effects can be ameliorated with the use of amendments like gypsum. Saline drainage water reuse for salt-tolerant perennial grasses relegates soil degradation to specific sites and makes control manageable (Oster et al., 1999).

Sequential reuse of drainage water can be continued until salinity increases up to 20–30 dS/m levels that can be tolerated by halophytes, such as *Distichlis spicata*, bermudagrass, *Salicornia bigelovii*, various *Atriplex* species, and tamarisk. In California, sequential reuse began with eucalyptus (*Eucalyptus camaldulensis*), and ended with halophytes like *Salicornia* (Cervinka, 1994, pp. 237–250). A field study conducted from 1994 to 1998 indicated that irrigation with water having an EC of 8–10 dS/m and SAR of 25–30 $mmol_c/L$ reduced the eucalyptus water use and growth significantly at a soil salinity of 20–22 dS/m (Oster et al., 1999). Eucalyptus is moderately sensitive in laboratory studies, the threshold being around 6 dS/m. Evapotranspiration was reduced by 53% at 15 dS/m. Strong boron–salinity interactions were noticed: boron affected growth at low salinity but not above 22 dS/m (Grattan et al., 1996)

Between 1994 and 1997, the yield performance of the "Kerman" pistachio cultivar grafted onto various rootstocks and irrigated with saline drainage water ranging from 0.5 to 8 dS/m was evaluated (Ferguson et al., 2002). No significant differences in pistachio yield appeared among salinity levels or among the different rootstocks. Nevertheless, 1 year of irrigation with water of 12 dS/m in 1997 caused a severe yield reduction (46%) in the hybrid rootstock UCB1, whereas the reduction in other rootstocks ranged from 5% to 11%.

Agricultural drainage water disposal with high salinity, trace elements, and other contaminants can have negative impacts on both terrestrial and aquatic ecosystems (Díaz et al., 2013). For instance, the mid-1980s concentration of agricultural drainage waters from the western San Joaquin Valley caused selenium bioaccumulation and toxicity in fish and migratory birds (Ohlendorf et al., 1986; Saiki & Lowe, 1987). Thus, there is a necessity for developing cost-effective and efficient techniques for agricultural drainage water and other saline effluent reuse to reduce the volume of this type of water, minimize the risks of salt, trace elements, and nutrient accumulation in receiving water bodies, and ultimately guarantee sustainability in irrigated agriculture, as stated by (Oster & Grattan, 2002; Qadir & Oster, 2004).

The concept of reuse of drainage water—also referred to as Integrated Farm Drainage Management—implies successive use of drainage waters for irrigation purposes. In this practice, drainage water is used first for irrigation of salt-sensitive commercial crops like lettuce, broccoli, tomato, and alfalfa; afterward for moderately salt-tolerant forage crops; and lastly for crops like *Salicornia* and *Atriplex* before final disposal to evaporation ponds (Jacobsen & Basinal, 2004). Although a considerable amount of information exists regarding forage production and water use under saline irrigation (Benes et al., 2012; Diaz & Grattan, 2009; Grattan et al., 2004), much less is known about halophyte cultivation.

Halophytes are plants that can complete their life cycle at a minimum concentration of 200 mM sodium chloride, a level common in their natural habitat (Flowers et al., 1986). Thus, they can tolerate salt concentrations that are lethal to most plant species, which are glycophytes (Flowers & Colmer, 2008). While most plants lack the genetic capacity to tolerate very high salinity, halophytes are endowed with specialized biological mechanisms that reduce the entry of salt into the plant or into the cytoplasm and thus enable growth in saline soils (Munns, 2005; Song et al., 2009). Because of this ability to grow well in saline soils, halophytes have become promising crops for saline environments, with a plethora of potential products, such as animal forage (El Shaer, 2010; Masters et al., 2007), edible products for human consumption (Nanduri et al., 2019; Ventura et al., 2011; Weber et al., 2007), industrial raw materials (Reddy et al., 2008), renewable green energy (Abideen et al., 2011), and plant cover restoration or bioremediation of saline landscapes (Glenn et al., 2009; Manousaki & Kalogerakis, 2011).

However, large-scale crop production from halophytes irrigated with hypersaline waters, either seawater or similar salinity, remains at the exploratory stage as of the writing of Rozema and Flowers (Rozema & Flowers, 2008). Some consider that halophytes inherently manifest only average growth rates, water use efficiency, or forage quality and hence do not have much potential as new crops (O'Connell et al., 2006). Others insist that systematic domestication of salt-loving crops is a strategic option that should be pursued (Rozema & Flowers, 2008).

3.3.3.3 Reducing evaporation losses

Saline water irrigation inevitably adds salt to the soil. Evaporation is an unproductive part of irrigation loss, and it increases the demand on irrigation. Thus, a reduction in evaporative losses reduces the amount of salt added to the soil by irrigation. There are a number of ways to reduce evaporation, and two main approaches will be described below.

3.3.3.3.1 Capillary rise reduction. Capillary rise is the upward movement of water, under unsaturated conditions, from deeper soil layers toward the surface, where it evaporates into the atmosphere. Capillary rise is greater for soils with finer pores. Unsaturated hydraulic conductivity depends on soil moisture content and pore size: with decreasing soil moisture, hydraulic conductivity decreases exponentially to near zero under air-dry conditions.

In fine-textured soils with small pores, capillary rise can occur over very long periods, furnishing water to the soil surface where it evaporates and leaves salts behind. In addition to wasting water, this greatly reduces leaching efficiency. Using coarse-textured surface mulches offers three advantages:

a) Coarse materials have larger pores, which disrupt capillary flow when rising water reaches the mulch layer and no longer allows for evaporation.
b) The mulch dries rapidly and its unsaturated hydraulic conductivity drops sharply, thereby limiting evaporation.
c) These combined effects act to conserve more water in the root zone, reduce irrigation demand, and improve leaching efficiency by inhibiting salt accumulation near the soil surface.

Therefore, mulches that are designed to reduce capillary rise have been adopted as a practical management strategy under saline conditions.

3.3.3.3.2 Reducing wetted surface area. Improving leaching efficiency depends more on increasing irrigation depth than on applied volume. Greater irrigation depth pushes the salt front deeper into the soil profile, away from the root zone. Reducing the wetted area decreases evaporative and capillary rise losses, while focusing the irrigation volume on a smaller area reduces total water use.

Localized irrigation systems, including drip and bubbler irrigation, by nature, encourage deeper wetting and better leaching. For surface irrigation, there are approaches to decrease wetted area and increase irrigation depth. For example, alternate-furrow irrigation in furrow-irrigated systems allows for both advantages.

In orchard surface strip irrigation, strip width reduction to match canopy width increases irrigation depth and leaching. Research in pressurized tape systems that continue from the earlier work is directed at reducing wetted surface area and increasing irrigation depth for saline conditions. Preliminary results for wheat show reduced tape use, decreased soil salinity within the wetted zone, and even a yield increase per unit area.

For row crops under surface irrigation, alternate-row irrigation decreases the wetted area and increases infiltration depth while enhancing salt leaching. For non-row crops such as cereals, under tape irrigation, certain studies are taking place to introduce unplanted zones to increase soil water infiltration depth.

3.3.4 Using highly saline water resources

Due to the limitations in allocation of fresh water to the agriculture sector regarding to the competition of industrial and urban sectors, more attention is paid to the use of highly saline water resources. These resources may include drain water, desalination plans' bine and even sea water. In some cases, the environmental concerns may lead to the use of saline water resources for irrigation. The examples may be the fixation of dust storms hotspots and prevention of eutrophication in coastal areas due to the efflux comes from shrimp production plans at the sea shore. In this section we discuss about the capabilities of saline farming (biosaline agriculture and/or haloculture) (Khorsandi et al., 2020) to response a part of these concerns with examples from pilot plans executed in Iran.

In extensive areas of the Central Plateau of Iran, the use of saline water for agricultural production has long been a common practice. In addition, in regions such as northeastern and southwestern Iran, particularly in Golestan and Khuzestan provinces, saline soils have developed despite the presence of freshwater resources, mainly due to inadequate drainage conditions. Through the implementation of artificial drainage systems, these lands have also been brought into agricultural production. Field crops such as wheat, barley, cotton, and sugar beet, as well as horticultural crops including pomegranate, pistachio, olive, and date palm, are cultivated in salt-affected soils of Iran (Cheraghi & Karimi, 2016) depending on climatic conditions and their relative tolerance to salinity, generally up to soil salinity levels of about 15 dS/m. In the case of pistachio, the use of saline irrigation water with salinity levels of up to approximately 20 dS/m has also been observed; however, production under such conditions is marginal and becomes feasible only through the application of specific management practices and very high irrigation volumes.

Several issues have raised the possibility and, in some cases, the necessity of using more saline and highly saline water resources for crop production. It is evident that the crops suitable for such high salinity levels are facultative or obligative halophytes. The first issue concerns the reuse of agricultural drainage water. This issue is of particular importance in Khuzestan Province, both in terms of increasing overall agricultural production and reducing the volume of drainage water and the associated challenges of its disposal. In recent years, Khuzestan Province has also been severely affected by dust storms. Biological stabilization of dust storm hotspots requires new water resources; however, given the current freshwater scarcity, allocating freshwater resources for environmental purposes is not feasible. The use of drainage water for agricultural production, while contributing to the stabilization of these hotspots, enables the economic utilization of saline water resources with the participation of local communities. Based on this approach, in 2015 the idea of multiple reuse of drainage water generated downstream of the Karkheh River was proposed for the production of aquatic organisms, forage, edible seeds, and woody crops, while simultaneously stabilizing a supercritical dust storm hotspot in southwestern Iran. This 50-ha pilot project represents a successful example of multipurpose and repeated use of highly saline water resources (Hasheminejhad, 2011).

The second issue relates to the development of aquaculture using seawater resources along the northern and southern coasts of Iran. Currently, large areas—particularly along the southern coasts of Iran—are dedicated to shrimp farming. Aquaculture activities in these units require the discharge of effluent drainage water, which is rich in phosphorus and nitrogen compounds. Direct discharge of this effluent into the sea causes eutrophication problems in marine environments. Conversely, irrigation of plants with this return water can significantly reduce nutrient concentrations and largely mitigate disposal challenges. Given the salinity of seawater, particularly along the southern coasts of Iran in the Persian Gulf and the Gulf of Oman, Salicornia is virtually the only plant species suitable for cultivation under these conditions. A similar pilot project using seawater for irrigation of Salicornia was implemented along the coastal areas of Bushehr Province, Iran.

Field results determining the relationships between yield and water consumption of Salicornia in the Bushehr region indicated that irrigation volume plays a highly critical and decisive role in biomass production. Comparisons showed that increasing applied irrigation water up to 15,000 m^3/ha had a significant positive effect on Salicornia dry matter yield, increasing it to approximately 4600 kg/ha. Furthermore, based on calculations for the Bushehr region and considering irrigation efficiency and leaching requirements, the required volume of saline water for Salicornia cultivation was estimated to range between 19,000 and 27,000 m^3/ha, with an average of about 22,000 m^3/ha. It is anticipated that, with timely and sufficient water supply (through shorter irrigation intervals), yields exceeding 5 t/ha could also be achieved. Under seawater irrigation conditions, Salicornia water use efficiency (WUE) ranged from 0.15 to 0.44 kg/m^3. Field evaluations of leaching fraction indicated that values varied between 43% and 49% across different irrigation events, with an average of approximately 46% (Ranjbar et al., 2021)

The third issue concerns forage production in the country and its dependence on freshwater resources. Despite the persistent annual deficit in forage production from croplands and rangelands—particularly in light of the national demand for red meat production—almost all conventional forage crops exhibit relatively low tolerance to salinity. Therefore, to exploit saline and highly saline water resources for forage production, halophytic plants must be considered.

It is believed that part of the livestock forage requirement can be met by cultivating halophyte species using saline water and salt-affected soils. Accordingly, it is necessary to introduce forage halophyte species under farmers' field conditions. To evaluate the quantitative and qualitative performance of forage halophytes, a comparative study was conducted on kochia (Kochia scoparia), quinoa (Chenopodium quinoa), Salicornia (Salicornia sinus-persica), panicum (Panicum antidotale), and camelthorn (Alhagi persarum), in comparison with forage beet (genotype 052). The study was carried out under saline conditions in two provinces: Yazd (Ardakan) and Razavi Khorasan (Sabzevar).

Based on measurements conducted in Ardakan, Salicornia and kochia produced the highest dry forage yields, with 18.89 and 17.07 t/ha, respectively, compared to forage beet, which produced 10.39 t/ha. Panicum produced approximately 7.00 t/ha of dry forage. Under Sabzevar conditions, dry forage yields of kochia, quinoa, and camelthorn were 7.82, 4.58, and 4.00 t/ha, respectively. The highest and lowest ash contents of forage, at 51.28% and 5.30%, were observed for Salicornia and beet root, respectively. Ash content of kochia ranged from 9.38% to 21.65% depending on location, while values for panicum, quinoa, camelthorn, and beet leaves were 10.61%, 23.94%, 12.11%, and 28.38%, respectively. In Ardakan, the highest crude protein content (13.00%) was observed in kochia and beet leaves. In the same region, crude protein contents of Salicornia, panicum, and beet root were 10.18%, 6.62%, and 7.74%, respectively. In Sabzevar, crude protein contents for quinoa, kochia, and camelthorn were 9.09%, 3.46%, and 7.72%, respectively.

Based on the results of this study, halophytic plants are capable of producing considerable yields under conditions where economic production of conventional crops is not feasible. However, despite their relatively high protein content, complete inclusion of these forages in livestock diets is not recommended due to certain anti-nutritional characteristics, such as high ash content. Therefore, complementary feeding trials are required to determine appropriate inclusion rates of these forages in livestock rations under practical conditions. Nevertheless, panicum—with its relatively low ash content and acceptable protein level—can be considered a promising forage species for inclusion in livestock diets (Ranjbar et al., 2022).

4. Conclusions and recommendations

Analysis of Iran's agricultural water and soil resources in relation to empirical evidence suggests that under the current conditions, salinity is the most dominant problem in salt-affected soils, whereas sodicity is less often a problem, except in a few places. In assessing the hazard of decreased soil infiltration, CROSS and the criteria where calcium is equal to five times electrical conductivity or five times sodium concentration may be used. In heavy-textured soils and for crops sensitive to crown rot, soil amendments should be used even when conditions appear borderline.

As irrigation of salt-tolerant crops with drainage water, wastewater, desalination brines, or seawater becomes more widespread, emerging problems -higher magnesium-to-calcium ratios, increased soil pH, and higher bicarbonate concentrations-will be expected and will require special management practices.

With regards to the state of Iran's water and soil resources, and based on available guidelines, a practical decision table has been proposed for using saline water and soils in agriculture by incorporating irrigation water salinity, crop salt tolerance, leaching requirements, and soil texture.

Among the various technologies available, excluding hydraulic methods (leaching and drainage), there is no direct method for soil salinity reduction. Only sodic or alkaline soils can be subjected to chemical and biological amendments and these must be used according to soil physico-chemical characteristics.

A deeper understanding of the soil–water–plant–atmosphere continuum will provide details necessary for the formulation of good management practices for not only sustainable crop production in saline environments but also the prevention of salt accumulation in the root zone. Several such strategies relevant to Iranian agriculture are highlighted in this publication.

Declaration of generative AI and AI-assisted technologies in the manuscript preparation process

The authors used ChatGPT (OpenAI) to assist with the translation of selected sections of this chapter. All AI-generated translations were carefully reviewed, edited, and verified by the authors, who take full responsibility for the final content.

References

Abdel-Dayem, S., & Ritzema, H. P. (1990). Verification of drainage design criteria in the Nile Delta, Egypt. *Irrigation and Drainage Systems, 4*, 117–131.

Abdul Qadir, A., Murtaza, G., Zia-ur-Rehman, M., & Waraich, E. A. (2022). Application of gypsum or sulfuric acid improves physiological traits and nutritional status of rice in calcareous saline-sodic soils. *Journal of Soil Science and Plant Nutrition, 22*(2), 1846–1858.

Abedi, M. J., Neyrizi, S., Ebrahimi Birang, N., Maherani, M., Mehrdadi, N., Khaledi, H., & Cheraghi, S. A. M. (2002). Using saline water in sustainable agriculture. *Iranian Committee of Irrigation and Drainage.*

Abideen, Z., Ansari, R., & Khan, M. A. (2011). Halophytes: Potential source of ligno-cellulosic biomass for ethanol production. *Biomass and Bioenergy, 35*, 1818–1822.

Akram, M., Kashkouli, H. A., & Pazira, E. (2008). Sensitive variables controlling salinity and water table in a bio-drainage system. *Irrigation and Drainage Systems, 22*, 271–285.

Al-Busaidi, A. S., & Cookson, P. (2003). Salinity–pH relationships in calcareous soils. *Journal of Agricultural and Marine Sciences, 8*(1), 41–46.

Anagholi, A., & Galeshi, S. (2017). Comparison of salinity response in tolerant wheat cultivars with introduced cultivars for non-saline condition. *Journal of Crop Production, 10*(1), 203–226. https://doi.org/10.22069/ejcp.2017.8474.1641

Arienzo, M., Christen, E. W., Jayawardane, N. S., & Quayle, W. C. (2012). The relative effects of sodium and potassium on soil hydraulic conductivity and implications for winery wastewater management. *Geoderma, 173*, 303–310.

Asadi, E., Isazadeh, M., Samadianfard, S., Firuz Ramli, M., Mosavi, A., Nabipour, N., Shamshirband, E. H., & Chau, K. W. (2019). Groundwater quality assessment for sustainable drinking and irrigation. *Sustainability, 12*, 177.

Ayars, J. E. (2002). Field crop production in areas with saline soil and shallow saline ground water. *Journal of Crop Production, 2.*

Ayers, R. S., & Westcot, D. W. (1985). *Water quality for agriculture.* Rome: Food and Agriculture Organization of the United Nations.

Barrett-Lennard, E. G., Anderson, G. C., Holmes, K. W., & Sinnott, A. (2016). High soil sodicity and alkalinity cause transient salinity in South-Western Australia. *Soil Research, 54*(4), 407–417.

Basak, N., Rai, A. K., Barman, A., Mandal, S., Sundha, P., Bedwal, S., Kumar, S., Yadav, R. K., & Sharma, P. C. (2022). *Salt affected soils: Global perspectives soil health and environmental sustainability: Application of geospatial technology.* Springer.

Bear, F. E., Prince, A., & Toth, S. (1951). *Magnesium in plants and soils.*

Benes, S. E., Adhikari, D. D., Grattan, S. R., & Snyder, R. L. (2012). Evapotranspiration potential of forages irrigated with saline-sodic drainage water. *Agricultural Water Management, 105*, 1–7.

Bennett, J. M., Marchuk, A., Marchuk, S., & Raine, S. R. (2019). Towards predicting the soil-specific threshold electrolyte concentration of soil as a reduction in saturated hydraulic conductivity: The role of clay net negative charge. *Geoderma, 337*, 122–131.

Brautigan, D. J., Rengasamy, P., & Chittleborough, D. J. (2012). Aluminium speciation and phytotoxicity in alkaline soils. *Plant and Soil, 360*, 187–196.

Bresler, E., McNeal, B. L., & Carter, D. L. (2012). *Saline and sodic soils: Principles-dynamics-modeling: Springer science & business media.*

Cervinka, V. (1994). Agroforestry farming system for the management of selenium and salt on irrigated farmland. In *Selenium in the environment.*

Cheraghi, S. A. M., & Karimi, M. (2016). The relation between irrigation water salinity and soil salinity. *Water Management in Agriculture, 3*(1), 1.

Chhabra, R. (2022). *Irrigation and salinity control salt-affected soils and marginal waters: Global perspectives and sustainable management.* Springer.

Daba, A. W. (2025). Rehabilitation of soil salinity and sodicity using diverse amendments and plants: A critical review. *Discover Environment, 3*(1), 53.

Dang, A., Bennett, J. M., Marchuk, A., Marchuk, S., Biggs, A. J. W., & Raine, S. R. (2018). Validating laboratory assessment of threshold electrolyte concentration for fields irrigated with marginal quality saline-sodic water. *Agricultural Water Management, 205*, 21–29.

Derbew, Z., Abate, S., Ali, H., Tefera, S., & Assen, S. (2026). Review on impacts of soil salinity and sodicity on crop yield and management option. *International Journal, 11*(1), 1–12.

Díaz, F. J., Benes, S. E., & Grattan, S. R. (2013). Field performance of halophytic species under irrigation with saline drainage water in the San Joaquin Valley of California. *Agricultural Water Management, 118*, 59–69.

Diaz, F. J., & Grattan, S. (2009). Performance of tall wheatgrass (Thinopyrum ponticum, cv.'Jose') irrigated with saline-high boron drainage water: Implications on ruminant mineral nutrition. *Agriculture, Ecosystems & Environment, 131*(3–4), 128–136.

Dontsova, K. M., & Norton, L. D. (2002). Clay dispersion, infiltration, and erosion as influenced by exchangeable Ca and Mg. *Soil Science, 167*(3), 184–193.

El Shaer, H. M. (2010). Halophytes and salt-tolerant plants as potential forage for ruminants in the near East region. *Small Ruminant Research, 91*, 3–12.

El-Tarabily, K. A., Soaud, A. A., Saleh, M. E., & Matsumoto, S. (2006). Isolation and characterisation of sulfur-oxidising bacteria from calcareous sandy soils and their effects on nutrient uptake and growth of maize. *Australian Journal of Agricultural Research, 57*, 101–111.

Esmaili, S., Tadayon, M. R., Tadayyon, A., & Rafiee Alhossainy, M. (2016). Response of some quantitative and qualitative traits of cotton cultivars to foliar application of humic acid in saline soil. *Journal of Oil Plant Peoduction, 3*(1), 1–14.

Ferguson, L., Poss, J. A., Grattan, S. R., Grieve, C. M., Wang, D., Wilson, C., Donovan, T. J., & Chao, C. T. (2002). Pistachio rootstocks influence scion growth and ion relations under salinity and boron stress. *Journal of the American Society for Horticultural Science, 127*, 194–199.

Fipps, G. (2003). *Irrigation water quality standards and salinity management strategies.*

Flowers, T. J., & Colmer, T. D. (2008). Salinity tolerance in halophytes. *New Phytologist, 179*, 945–963.

Flowers, T. J., Hajibagheri, M. A., & Clipson, N. J. W. (1986). Halophytes. *The Quarterly Review of Biology, 61*, 313–337.

Ghaneei Motlagh, G. R., Pashaei Aval, A., Khormali, F., & Mosaedi, A. (2010). Investigating effect of some amendments on soil chemical properties in a saline-sodic soil. *Management Researches, 23*(1), 24–31.

Glenn, E. P., McKeon, C., Gerhart, V., Nagler, P. L., Jordan, F., & Artiola, J. (2009). Deficit irrigation of a landscape halophyte for reuse of saline wastewater in a desert city. *Landscape and Urban Planning, 89*, 57–64.

Grattan, S., Grieve, C., Poss, J., Robinson, P., Suarez, D., & Benes, S. (2004). Evaluation of salt-tolerant forages for sequential water reuse systems: III. Potential implications for ruminant mineral nutrition. *Agricultural Water Management, 70*(2), 137–150.

Grattan, S., Shannon, M., Grieve, C., Poss, J., Suarez, D., & Leland, F. (1996). Interactive effects of salinity and boron on the performance and water use of eucalyptus. *Paper presented at the II International Symposium on Irrigation of Horticultural Crops, 449.*

Hasheminejhad, Y. (2011). Irrigation management under saline conditions using wetting front detector (WFD). *Iranian Journal of Soil Research, 24*(3), 265–272.

Hasheminejhad, Y. (2019). *Establishment of a monitoring network for pistachio orchards in North Ardakan based on water quality and remote sensing data.*

Hasheminejhad, Y., & Besharat, N. (2017). *Anti-salt or anti-sodium? A guideline for leaching and gypsum requirements calculations* (p. 24) (Extension paper No. 51908).

Hoffman, G. J., & Van Genuchten, M. T. (1983). Soil properties and efficient water use: water management for salinity control. *Limitations to Efficient Water Use in Crop Production, 1*, 73–85.

Jacobsen, T., & Basinal, L. (2004). *Managing agricultural irrigation drainage water: A landowner's manual.*

Jayawardane, N. S., Christen, E. W., Arienzo, M., & Quayle, W. C. (2011). Evaluation of the effects of cation combinations on soil hydraulic conductivity. *Soil Research, 49*, 56–64.

Jobbágy, E. G., Tóth, T., Nosetto, M. D., & Earman, S. (2017). On the fundamental causes of high environmental alkalinity (pH$\geq$ 9): An assessment of its drivers and global distribution. *Land Degradation & Development, 28*(7), 1973–1981.

Khorsandi, F., Siadati, S. M. H., & Rastegary, J. (2020). Haloengineering as a vital component of sustainable development in salt-affected ecosystems. *Environmental Development, 35*, Article 100545.

Liang, X., Rengasamy, P., Smernik, R., & Mosley, L. M. (2021). Does the high potassium content in recycled winery wastewater used for irrigation pose risks to soil structural stability? *Agricultural Water Management, 243*, Article 106422.

Lucheta, A. R., & Lambais, M. R. (2012). Sulfur in agriculture. *Revista Brasileira de Ciência do Solo.*

Maas, E. V., & Grattan, S. R. (1999). Crop yields as affected by salinity. *Agricultural Drainage, 38*, 55–108.

Manousaki, E., & Kalogerakis, N. (2011). Halophytes present new opportunities in phytoremediation of heavy metals and saline soils. *Industrial and Engineering Chemistry Research, 50*, 656–660.

Marchuk, A., & Rengasamy, P. (2011). Clay behaviour in suspension is related to the ionicity of clay–cation bonds. *Applied Clay Science, 53*, 754–759.

Marino, G., Zaccaria, D., Lagos, L. O., Souto, C., Kent, E. R., Grattan, S. R., Shapiro, K., Sanden, B. L., & Snyder, R. L. (2021). Effects of salinity and sodicity on the seasonal dynamics of actual evapotranspiration and surface energy balance components in mature micro-irrigated pistachio orchards. *Irrigation Science, 39*(1), 23–43.

Masters, D. G., Benes, S. E., & Norman, H. C. (2007). Biosaline agriculture for forage and livestock production. *Agriculture, Ecosystems and Environment, 119*, 234–248.

Mazloom, N., & Hasheminejhad, Y. (2025). Phytoremediation of saline-sodic soils in arid regions using Rubia tinctorum L.: Effects on soil chemical and physical restoration. *Eurasian Soil Science, 58*(11), 159.

McLean, E. O., Hartwig, R. C., Eckert, D. J., & Triplett, G. B. (1983). Basic cation saturation ratios as a basis for fertilizing and liming agronomic crops. II. Field studies. *Agronomy Journal, 75*(4), 635–639.

Minhas, P. S. (1996). Saline water management for irrigation in India. *Agricultural Water Management, 30*, 1–24.

Mirabzadeh Ardakani, M., Jafari, A., Javanshah, A., & Shirmardi, M. (2021). Improving pistachio yield in saline soil with the application of organic acids. *Journal of Pistachio Science and Technology, 6*(11), 69–84.

Moameni, A. (2011). Geographical distribution and salinity levels of soil resources of Iran. *Iranian Journal of Soil Research, 24*(3), 203–215.

Motallebifard, R., Malakouti, M. J., & Kafi, M. (2008). Effect of acidifying irrigation water on the yield and quality of carnations. *Iranian Journal of Soil Research, 22*(1), 103–111. https://doi.org/10.22092/ijsr.2008.126990

Munns, R. (2005). Genes and salt tolerance: Bringing them together. *New Phytologist, 167*, 645–663.

Nanduri, K. R., Hirich, A., Salehi, M., Saadat, S., & Jacobsen, S. E. (2019). *Quinoa: A new crop for harsh environments* sabkha ecosystems. Springer.

O'Connell, M., Young, J., & Kingwell, R. (2006). The economic value of saltland pastures in a mixed farming system in Western Australia. *Agricultural Systems, 89*(2–3), 371–389.

Ohlendorf, H. M., Hoffman, D. J., Saiki, M. K., & Aldrich, T. W. (1986). Embryonic mortality and abnormalities of aquatic birds: Apparent impacts of selenium from irrigation drainwater. *Science of the Total Environment, 52*, 49–63.

Oosterbaan, R. J., & Abu Senna, M. (1989). *Using SaltMod to predict drainage and salinity in the Nile Delta.*

Oster, J., & Grattan, S. (2002). Drainage water reuse. *Irrigation and Drainage Systems, 16*(4), 297–310.

Oster, J., Shainberg, I., & Abrol, I. (1999). Reclamation of salt-affected soils. *Agricultural drainage, 38*, 659–691.

Oster, J., Sposito, G., & Smith, C. (2016). Accounting for potassium and magnesium in irrigation water quality assessment. *California Agriculture, 70*, 71–76.

Parsinejad, M., & Akram, M. (2018). A fresh look at drainage for agriculture. *Irrigation and Drainage, 67*, 8–16.

Peterson, F., Rhoades, J., Arca, M., & Coleman, N. (1965). Selective adsorption of magnesium ions by vermiculite. *Soil Science Society of America Journal, 29*(3), 327–328.

Qadir, M., & De Pauw, E. (2007). *Emerging challenges addressing the characterization and mapping of salt-induced land degradation.*

Qadir, M., & Oster, J. (2004). Crop and irrigation management strategies for saline-sodic soils and waters aimed at environmentally sustainable agriculture. *Science of the Total Environment, 323*(1–3), 1–19.

Qadir, M., Qureshi, R. H., Ahmad, N., & Ilyas, M. (1996). Salt-tolerant forage cultivation on a saline-sodic field for biomass production and soil reclamation. *Land Degradation and Development, 7*, 11–18.

Qadir, M., Schubert, S., Ghafoor, A., & Murtaza, G. (2001). Amelioration strategies for sodic soils: A review. *Land Degradation and Development, 12*, 357–386.

Qadir, M., Schubert, S., Oster, J. D., Sposito, G., Minhas, P. S., Cheraghi, S. A. M., Murtaza, G., Mirzabaev, A., & Saqib, M. (2018). High-magnesium waters and soils: Emerging environmental and food security constraints. *Science of the Total Environment, 642*, 1108–1117.

Qadir, M., Sposito, G., Smith, C. J., & Oster, J. D. (2021). Reassessing irrigation water quality guidelines for sodicity hazard. *Agricultural Water Management, 255*, Article 107054.

Quirk, J. P., & Schofield, R. K. (1955). The effect of electrolyte concentration on soil permeability. *Journal of Soil Science, 6*(2), 163–178. https://bsssjournals.onlinelibrary.wiley.com/doi/abs/10.1111/j.1365-2389.1955.tb00841.x.

Rahimi, B., Afzali, M., Farhadi, F., & Alamolhoda, A. (2021). Reverse osmosis desalination for irrigation in a pistachio orchard. *Desalination, 516*, Article 115236.

Rahman, W., & Rowell, D. (1979). The influence of magnesium in saline and sodic soils: A specific effect or a problem of cation exchange? *Journal of Soil Science, 30*(3), 535–546.

Ranjbar, G., Dehghani, F., AlaEddin, A., Soltani Gerdfaramarzi, & Keshtkar, S. (2021). Yield evaluation of some Salicornia species and ecotypes irrigated with seawater and groundwater. *Journal of Water Research in Agriculture, 35*(2), 187–199.

Ranjbar, G. H., Hasheminejhad, Y., Babaei Zarch, M. J., Mazinani, Kh., Fotohi, M. R., & Rivandi, A. (2022). Comparison of Performance and nutritional value of some marine species in Ardakan and Sabzevar regions. *Salt Farming, 21*, 9–17.

Ranjbar, G., & Salehi, M. (2014). *Wheat cultivars: Salt-tolerant or adapted to the salinity of each region?.*

Rao, K. V. G. K., Kumbhare, P. S., Kamra, S. K., & Oosterbaan, R. J. (1990). *Reclamation of waterlogged saline alluvial soils in India by sub-surface drainage.*

Rao, K. V. G. K., Sharma, D. P., & Oosterbaan, R. J. (1992). *Subirrigation by groundwater management with controlled subsurface drainage in semi-arid areas.*

Rasouli, F., & Kiani Pouya, A. (2010). Use of WATSUIT model for predicting the chemical composition of the soil solution reclaimed with mined gypsum. *Iranian Journal of Soil Research, 24*(2), 193–202.

Reddy, M. P., Shah, M. T., & Patolia, J. S. (2008). Salvadora persica, a potential species for industrial oil production in semiarid saline and alkali soils. *Industrial Crops and Products, 28*, 273–278.

Rengasamy, P. (2002). *Clay dispersion Soil physical measurement and interpretation for land evaluation: CSIRO Publishing.*

Rengasamy, P. (2016). *Soil salinization Oxford research encyclopedia of environmental science.*

Rengasamy, P., de Lacerda, C. F., & Gheyi, H. R. (2022). *Salinity, sodicity and alkalinity subsoil constraints for crop production*. Springer.

Rengasamy, P., & Marchuk, A. (2011). Cation ratio of soil structural stability (CROSS). *Soil Research, 49*, 280–285.

Rengasamy, P., & Olsson, K. A. (1991). Sodicity and soil structure. *Soil Research, 29*, 935–952.

Rengasamy, P., & Sumner, M. (1998). *Processes involved in sodic behavior sodic soil: Distribution, management and environmental consequences*. Oxford university press.

Richards, L. A. (1954). *Diagnosis and improvement of saline and alkali soils*. US department of agriculture.

Rousta, m. j., & Enayati, k. (2019). The effects of humic acid application on yield and yield components of wheat and some chemical properties of a saline-sodic soil. *Journal of Soil Management and Sustainable Production, 8*(4), 95–109. https://doi.org/10.22069/ejsms.2019.14096.1778

Rozema, J., & Flowers, T. (2008). Crops for a salinized world. *Science*, 1478–1480.

Saiki, M. K., & Lowe, T. P. (1987). Selenium in aquatic organisms from subsurface agricultural drainage water. *Archives of Environmental Contamination and Toxicology, 16*, 657–670.

Salehi, M. (2020). Hypersaline water for alternative crop irrigation. In *Iran emerging research in alternative crops.* Springer.

Setter, T. L., Waters, I., Sharma, S. K., Singh, K. N., Kulshreshtha, N., Yaduvanshi, N. P. S., Ram, P. C., Singh, B. N., Rane, J., & McDonald, G. (2009). Review of wheat improvement for waterlogging tolerance. *Annals of Botany, 103*, 221–235.

Shainberg, I., & Letey, J. (1984). Response of soils to sodic and saline conditions. *Hilgardia, 52*(2), 1–57.

Shannon, M., & Grieve, C. (1998). Tolerance of vegetable crops to salinity. *Scientia horticulturae, 78*(1–4), 5–38.

Sharifiasl, R., Shojaeiyan, A., Seyedi, M., & Giti, A. (2012). Effects of acidifying irrigation water on quality and quantity of two pelargonium varieties. *Journal of Horticultural Science, 26*(2).

Sharma, D. P., Singh, K. N., Rao, K. V. G. K., & Kumbhare, P. S. (1990). *Reuse of saline drainage water for irrigation in a sandy loam soil.*

Smith, C. J., Oster, J. D., & Sposito, G. (2015). Potassium and magnesium in irrigation water quality assessment. *Agricultural Water Management, 157*, 59–64.

Soltani, M., Rahimikhoob, A., Sotoodehnia, A., Mendicino, G., Akram, M., & Senatore, A. (2018). Numerical evaluation of the effects of increasing ratio of cropped to uncropped width on dry drainage efficiency in salty soils. *Irrigation and Drainage, 67*, 91–100.

Song, J., Chen, M., Feng, G., Jia, Y., Wang, B., & Zhang, F. (2009). Effect of salinity on growth, ion accumulation and osmotic adjustment of two populations of suaeda salsa. *Plant and Soil, 314*, 133–141.

Sposito, G. (2008). *The chemistry of soils.* Oxford university press.

Tavakkoli, E., Rengasamy, P., Smith, E., & McDonald, G. K. (2015). The effect of cation–anion interactions on soil pH and solubility of organic carbon. *European Journal of Soil Science, 66*, 1054–1062.

Tetteh, K., Guggenberger, G., Sauheitl, L., & Zamanian, K. (2025). Modeling calcite's sensitivity to biogenic CO2 production: A pathway to soil CO2 efflux partitioning. *Environmental Science & Technology, 59*(44), 23869–23882.

Vafaei, M., Golchin, A., & Shafiei, S. (2019). Effect of incubation times, gypsum, organic carbon, and leaching on some properties of a sodic soil. *Iranian Journal of Soil Research, 33*(1), 101–113.

Valipour, M., & Sakhaeirad, H. (2011). Appraisal use of gypsum for reclamation of saline and sodic soils in Southern Khuzestan. *Journal of Water Sciences and Engineering, 1*(1), 95–104.

van Hoorn, J. W. (1979). *Effect of capillary flow on salinization and the concept of critical depth for determining drain depth.*

Ventura, Y., Wuddineh, W. A., Myrzabayeva, M., Alikulov, Z., Khozin-Goldberg, I., Shpigel, M., Samocha, T. M., & Sagi, M. (2011). Effect of seawater concentration on productivity and nutritional value of salicornia and sarcocornia. *Scientia Horticulturae, 128*, 189–196.

Weber, D. J., Ansari, R., Gul, B., & Khan, M. A. (2007). Potential of halophytes as source of edible oil. *Journal of Arid Environments, 68*, 315–321.

Wilcox, L. (1955). *Classification and use of irrigation waters.* US department of agriculture.

Yasin, S. (2024). Soil sulfur dynamics and their role in plant growth and development. *Jurnal Agronomi Tanaman Tropika (Juatika), 6*(3), 850–868.

Zhang, B., Hu, X., Yang, Y., Deng, X., Li, B., Gong, X., Xiang, X., Cai, X., & Liu, T. (2025). Comprehensive evaluation of groundwater quality in population-dense and extensive agricultural regions and study on its relationship with agricultural production and human activities. *Environmental Geochemistry and Health, 47*(3), 62.

Zhu, Y., Ali, A., Dang, A., Wandel, A. P., & Bennett, J. M. (2019). Re-examining the flocculating power of sodium, potassium, magnesium and calcium for a broad range of soils. *Geoderma, 352*, 422–428.

PART

VI

The future of soil science: Prospects and challenges

CHAPTER

Soil and biodiversity: A complex underground network

18

Narges Abdar[1] **and Mehdi Zarei**[1, 2]

[1]*Department of Soil Science, College of Agriculture, Shiraz University, Shiraz, Iran;* [2]*Department of Agriculture and Natural Resources, Higher Education Center of Eghlid, Eghlid, Iran*

1. Introduction

The multiplicity of soil biota communities is important for soil wellness and differentiation. A large continuum of organisms partake in important functions of the soil. Soil biota are classified into various groups according to their ecological function. Microflora (Bacteria, Actinobacteria, Fungi, and Algae) and micro fauna (Nematoda, protozoa, and Rorifera) groups comprise chemical engineers, like the decomposers and transformers and also macrofauna group comprise ecosystem engineers, like the earthworms and termites (Trivedi et al., 2018). Hence soil biota are fundamental for human existence on planet Earth. These include the sustenance of agricultural and animal husbandry. They also include the supervision of greenhouse gases and climate change. Moreover, they constitute nutrients cycling and the synthesis of Antimicrobials (Morgado et al., 2018).

Soil biodiversity encounters threats from activities of human, such as land-use change, pollution, intensive agricultural practices, and change of climate (Tibbett et al., 2020). The decline of soil organisms can lead to a decline in soil quality, productivity of agriculture, and resilience of ecosystem. In general, biodiversity of soil is an essential part of ecosystems that supports critical ecosystem functions for sustainable agriculture and ensuring food security (Dervash et al., 2024). Therefore, soil biodiversity conservation, such as sustainable land management techniques, organic farming, and reforestation, is essential. The purpose aim of this chapter is to review strategies for conserving and enhancing soil biodiversity to maintain soil health despite environmental challenges.

2. Materials and methods

2.1 Soil sampling

Studies have shown that the soil sampling stage is important for measuring soil biodiversity. The reason for this can be examined in several ways. Soil biota vary significantly in size, from microscopic bacteria to larger insects and earthworms. It is advisable to collect samples from areas of land that have not been disturbed. The efficiency of soil sampling is influenced by a combination of physical and chemical soil properties (such as texture, pH, EC, OC, N, P, and CEC), as well as site-specific factors (encompassing elevation, slope, direction, annual rainfall, average temperature, humidity, and

The Soil Nexus. https://doi.org/10.1016/B978-0-443-45286-4.00015-8

vegetation cover). Typically, plots measuring 3 × 3 m must be evenly spaced across the field at the sampling location. Subsequent to the elimination of plant residues, five soil samples (0–20 cm) were gathered from each section using an auger. Soils are mixed and a sample is separated for each location. Immediately after collection, samples of soil are to be transported to the laboratory in an insulated receptacle, on ice at 4°C. Samples are typically stored at a temperature of -20°C for analysis DNA extraction. It is recommended that sampling be accomplished upon the termination of the rainy season and under stable conditions. Therefore, the sampling system employed in any ecological survey must be meticulously designed as a fundamental component of the entire study (González et al., 2020). Counting all soil biota at unit area of soil is nearly impossible; therefore, populations must be estimated based on the numbers obtained from small soil samples. Population of soil biota can then be estimated at a specific depth of 25–30 cm, provided samples number collected are sufficient (González et al., 2020).

2.2 Extracting soil fauna

Two methods are used to separate soil fauna: direct methods, such as washing, hydrocarbon adhesion, or flotation, which physically remove the fauna, and indirect methods, including dry funnel methods and wet funnel methods, which employ a trapping device to capture the fauna after triggering their migration through environmental changes. Thus, combinations of extraction methods are used to thoroughly sample soil organisms, containing microfauna, mesofauna, and macrofauna (Geisen et al., 2019).

Berlies described the dry funnel apparatus in the year 1905. The design of dry funnels includes a mesh for placing the sample, and the surrounding warm water stream creates a temperature gradient that encourages arthropods to move into the collection vial. Subsequently, Tullgren (1918) made a modification to Berlies by incorporating an electric lamp as the heat source to facilitate the drying of the sample material and the collection of arthropods from it (Fig. 18.1a). Wet funnels depend on mechanisms that Baermann first developed in 1917. These funnels are effective for isolating hydrophilic organisms. This method is advised for the extraction of fauna that are highly sensitive to desiccation, and is used to isolate nematodes and rotifers, anchytrids and nematode larvae from soil samples (Fig. 18.1b). The Baermann apparatus includes a funnel that is affixed to a clamped rubber tube. Samples of soil are contained in bags of cloth that are firmly closed and affixed to an upper holding rod. Water is introduced into the funnel, making certain that the sample is entirely submerged in water, and left undisturbed for no less than 24 h. Nematode larvae move from the sample into the water and subsequently descend to the bottom. The lower section is subsequently retrieved for the purpose of counting and identifying soil fauna (Demelash et al., 2016). These sampling methods were selected as the most commonly used methods based on the size of the soil fauna and allow researchers with limited experience in soil ecology to use them (Eisenhauer et al., 2022). Recently, pitfall traps are not used as the primary method, but rather as an auxiliary method for collecting soil biota (Jia et al., 2023).

Pitfall traps are based on the movement of soil fauna and serve the purpose of collecting active fauna that fall into the trap (Fig. 18.1c). This trap may include a vial that is embedded in the ground and meticulously aligned with the surface to avoid any protruding edges that could obstruct the movement of organisms. Pitfall traps generally incorporate a cone-shaped device that serves to stop fallen organisms from escaping, as well as a ceiling to protect the trap from being inundated by rain. Each trap is filled with ethanol 96% and has bait placed inside (Soto-Adames & Taylor, 2010). The effectiveness

FIGURE 18.1

Methods of extracting soil fauna a) Dry funnel berlies-tollgren b) Wet funnels baermann c) Pitfall traps.

of pitfall traps varies due to a variety of factors, including location, type, and traps size, and sampling timing (Brown & Matthews, 2016). Gonzalez et al. (2020) reported that glass vials are more impressive than metal and plastic because there are virtually no escape routes in glass vials. Furthermore, the dimensions of organisms can influence their ability to evade capture, as larger organisms tend to escape from smaller traps more effectively, while smaller organisms are more adept at escaping from larger traps. As a result, there have been calls for standardization of the use of these traps, such as the activity level of the organisms (Brown & Matthews, 2016).

Berlese-Tullgren funnels and pits were compared in terms of efficiency by Tuf (2015) for assessing *Chilopoda* diversity, with the result being that more *Chilopoda* were collected using funnels than using pits, however some ecological groups, like large *lithobiomorphs*, were only found in pits (Tuf, 2015). Research has shown that the Berlese-Tullgren funnel has high efficiency for extracting arthropods from soil samples. Some families, such as *microarthropods*, *Chilopoda*, *Coleoptera*, and *Pselaphinae*, were collected only using the Berlese-Tullgren funnel. Dry Funnels Compared to pit traps, funnels can be used to collect smaller, slower-moving fuana from soil and substrate samples. Furthermore, less active soil fauna or specific life stages of a species (such as pupae) are less frequently encountered compared to active adult insects. Emergence traps are employed to capture insect larvae from diptera, beetles, and hymenoptera. These traps are provided with photoreceptors. They make use of the geotropism of negative and phototropism of positive observed in the majority of insects (González et al., 2020).

In genetic studies and selective breeding, marking or tagging methods are especially important, given that identification of individual is necessary (Arce et al., 2003). Identification of visual is frequently challenging. The use of fluorescent markers is emerging as the premier technique for tagging aquatic fauna and insects (Doupé et al., 2003) but are underused in the study of invertebrates of soil (González et al., 2020). Conducting short-term studies using fluorescent markers can provide valuable insights into the biology, demography, and basic behavior of earthworms (González et al., 2020).

Observational studies regarding earthworms have emphasized their activities. Rhea-Fournier & González (2017) indicated that the burrows created by soil-dwelling earthworms extend to the surface of the soil, offering visual proof that enables researchers to collect samples of the burrow soil (Gorres et al., 2001). On the other hand, researchers can compare drillosphere and non-drillosphere soils by sampling the castings themselves (Hong et al., 2011) or the underlying soil (Bohlen et al., 2004). In general, soil species are difficult to identify due to their difference size and high biodiversity. Therefore, no method can efficiently extract all soil fauna from soil (Eisenhauer et al., 2017; González et al., 2020; Stork, 2018).

2.3 Rhizobia bacteria

Leguminous species that can nodulate are known in advance. Remove the roots of herbaceous plants from the soil, making sure that the nodes are not separated from the roots. Carefully remove the soil from the roots. This action can be effectively aided by placing the roots in water and letting the soil come loose. Then, the roots, in conjunction with the nodules, are placed in a sampling bag and can be kept in the refrigerator overnight. Nodules can be kept for a duration exceeding 1 year at -20°C in vials that contain about glycerol 50%. The nodules are immersed in 96% ethanol for 5–10 s and then in disinfectant for 3–4 min. Rinse at least 5 times in distilled water. Utilize a sterile glass rod to crush each nodule in a test tube, add sterile water to generate a suspension, and subsequently transfer a drop to YMA plates. Streak the drop of suspension across the agar surface to gradually dilute the suspension. The plates are incubated inverted at the desired temperature. Individual, completely isolated colonies are picked and replated onto new plates to obtain pure cultures. For short-term storage, pure culture isolates are transferred to YMA agar slants with screw caps. For longer storage periods, suspend rhizobium cells in a YMB medium containing 25% glycerol and store at -80°C. The most probable number (MPN) technique for plant contamination has been used for many years to enumerate rhizobia in soils. This method is based on four core principles: (1) a rhizobium cell will form nodules when introduced into its host in a nitrogen-free environment, (2) the formation of nodules proves the presence of infectious rhizobia, (3) the test's accuracy is confirmed by the lack of nodules in uninoculated plants, and (4) the failure of inoculated plants to form nodules indicates the absence of infectious rhizobia. Soil samples are serially diluted to enumerate rhizobia. The dilution ratio can vary from 10–4 to 10–9, depending on the expected cell concentration in the soil sample; consequently, more significant dilutions are required for soils with a larger abundance of rhizobia (Prévost & Antoun, 2007).

2.4 Microbial biomass carbon

This technique is derived from the research conducted by Amato and Ladd (1988). The first step is to determine water-holding capacity. Approximately 50 g of soil is loosely packed into a large strainer funnel, which has been lined with glass wool and placed in a beaker; 100 mL of water is then added, and it is allowed to sit overnight. The moist soil and funnel are weighed, then dried in an oven at 105°C for 24 h, and weighed again. Remove the dry soil from the funnel and record its weight separately 20 g of moist soil are placed in small beakers, and the water retention ratio is adjusted to 40%. One sample is treated with chloroform and subsequently extracted using a KCl solution, while the other sample, serving as a control, is extracted with KCl solution without any exposure to chloroform. Position the fumed samples within a spacious vacuum desiccator that holds a beaker containing approximately 50 mL of chloroform. Chloroform is added and then evacuated using a desiccator until it begins to boil. Following this, the desiccator is closed, and the sample is permitted to sit at room temperature for 24 h. Soil samples used as controls do not undergo exposure to chloroform. Each soil sample is then shaken with KCl after a period of 24 h. Following this, the sample undergoes centrifugation at a speed of 2500 rpm for 5 min. Take out 10 mL of the supernatant and move the supernatant through a pre-filter into a 10 mL vial. This filtered solution may subsequently be preserved until analysis.

2.5 Mycorrhizal arbuscular fungus

The extraction of mycorrhizal spores is conducted following the methodology established by Boyno et al. (2023). The spores contained in 100 g of soil are removed from the soil via wet sieving, which involves a series of sieves (spanning from 25 to 400 mesh). The content from the final sieve was comprehensively blended with a 55% sucrose solution and centrifuged at 1500 rpm for 7 min. Through a 400 mesh sieve and thoroughly washed with water to remove sucrose (Boyno et al., 2023). The spores that have been collected are analyzed under a stereomicroscope in accordance with the technique described by Boyno et al. (2023). The newly harvested roots were washed with water and preserved in a 50% alcohol solution; subsequently clarified using 2.5% KOH at 90 °C for 10–30 min; and then rinsed with water. Roots with pigments undergo a process involving H2O2 for 10–30 min, then are rinsed, acidified with 1% HCl at room temperature for 10–30 min, and stained with trypan blue in acidic glycerol. The root colonization is established by inspecting arbuscules, vesicles and hyphae (Merryweather & Fitter, 1991).

3. Result and discussion

3.1 The diversity of soil life: From microbes to macrofauna

Soil organisms are classically organized by body size, which correlates with their ecological roles.

3.1.1 Microfauna and microflora (<0.1 mm)

This group includes microflora (bacteria, archaea, fungi and microalgae) and micro fauna (nematodes, protozoa and rotifers). Bacteria and fungi are responsible for over 90% of the chemical transformations in soil, including decomposition and nutrient cycling. They releases nutrients in plant-available forms, a process known as mineralization. In contrast, microfauna regulate microbial populations by feeding on bacteria and fungi (Nadarajah, 2018).

3.1.1.1 Bacteria

Bacteria are single-celled organisms, too small to be seen without powerful microscopes, and that they represent the highest biomass of all microorganisms in the soil, particularly in the plant root zone known as the rhizosphere.

Soil bacteria are considered some of the most numerous and various organisms participating in essential processes such as element cycling, fixation-N_2, and the decomposition of organic matter (Lladó et al., 2017). Species of nitrogen-fixing bacteria, including *Rhizobium*, *Bradyrhizobium*, , *Azospirillum* and *Azotobacter*, can convert N_2 to NH_3 through a process termed fixation-N_2 (Lindström & Mousavi, 2020) (Fig. 18.2). These bacteria establish symbiotic associations with leguminous plants, supplying them with fixed nitrogen. Nitrifying bacteria, including *Nitrosomonas* and *Nitrobacter*, transform NH_4^+ into NO_3^- throughout the nitrification process (Grzyb et al., 2021) (Fig. 18.2). The denitrification process involves the conversion of nitrate (NO_3^-) back into nitrogen gas (N_2), which is then released into the atmosphere (Grzyb et al., 2021). Examples of bacteria that perform nitrification include species such as *Pseudomonas* and *Bacillus*. Ammonifying bacteria break down organic nitrogen compounds, such as proteins, into ammonium (NH_4^+), a form that plants can take up (Lamb et al., 2014) (Fig. 18.2). A number of bacterial types, like *Bacillus* and *Clostridium* species, contribute to ammonia production. *Cellulolytic* bacteria, including *Cellulomonas*, *Bacillus*, and *Clostridium* species, break down cellulose in the walls of plant cells into simpler sugars. This procedure is crucial for the breakdown of plant debris and organic material in the soil (Datta, 2024). Phosphate-solubilizing bacteria, which comprise species like *Bacillus*, *Pseudomonas*, and *Rhizobium*, can transform soil insoluble phosphorus into available phosphorus for plants (Tian et al., 2021). Sulfur-oxidizing bacteria are responsible for the oxidation of elemental sulfur or sulfide to sulfate, in contrast to sulfur-reducing bacteria, which reduce SO_4^{2-} back to SO_3^{2-} (Gonsior et al., 2018). Examples bacteria of oxidizing and

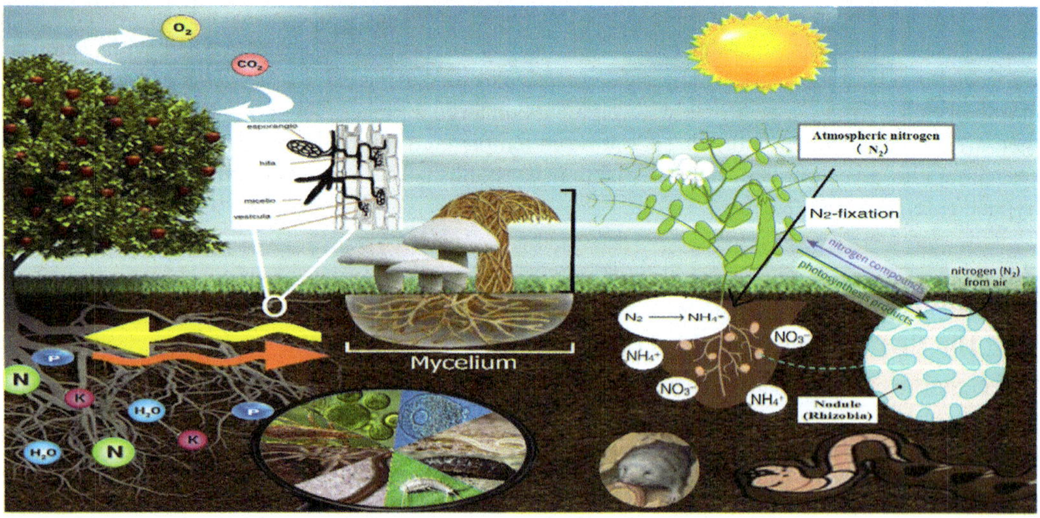

FIGURE 18.2

The connection between soil and plants, as well as the function of soil organisms in the elemental cycle, is examined.

reducing of sulfur include *Thiobacillus* and *Desulfovibrio*, respectively. Cyanobacteria, including *Anabaena* and *Nostoc* species, and play a significant role in N_2 fixation, particularly in paddy. Cyanobacteria are used as inoculants in rice fields, which increase soil fertility and improve rice yield. Cyanobacteria are able to produce oxygen due to their ability to photosynthesize and provide the plant with the elements it needs by decomposing organic matter. The symbiosis of *Azolla* and *Anabaena* is crucial for N_2 fixation in rice fields, specifically for the process of N_2 fixation (Battistuzzi et al., 2023). These represent a selection of the varied types of bacteria present in soil ecosystems. All types of bacteria are crucial for cycling of nutrient, decomposition, and fertility of soil.

3.1.1.2 Archaea

Several researchers have documented archeal communities found in soil by 16S rRNA studies (Mukhtar et al., 2017). These research efforts have identified that the most common archaea found in soil are classified in the phylum *Thaumarchaeota* (Eme et al., 2017). *Euryarchaeota* are methanogens in flooded soils and are anaerobic and oligotrophic (Angel et al., 2012). The Archaea, methanogenic and change CO_2 into CH_4. The leading methanogens identified in soil are classified under the genera *Methanosaeta*, *Methanosarcina*, *Methanocella*. *Methanosarcina* and *Methanosaeta* can effectively reduce acetate to produce methane (Yadav et al., 2019).

3.1.1.3 Actinomycetes

Actinomycetes are filamentous bacteria, they are usually G+, aerobic; we discuss them separately here because of their importance in soil. They are known to play an important role in the synthesis of bioactive compounds and have been utilized by the pharmaceutical industry and other sectors in the formulation of enzymes and antibiotics (Terkina et al., 2006). Furthermore, genus like *Streptomyces* sp., *Actinomyces*, *Micromonospora*, *Frankia*, and *Nocardia* have been frequently cited as belonging to the actinomycetes phylum, which have a crucial role in enhancing soil fertility and managing diseases. Actinomycetes contribute to the production of compost by decomposing organic matter. They mineralize soil phosphorus by producing the enzyme phosphatase (Richardson et al., 2009). These organisms' compounds, including antimicrobials and siderophores, have been discovered to increase plant growth while also constraining the spread of soil pathogens (Franco-Correa et al., 2010).

3.1.1.4 Algae

Soil algae are organisms of eukaryotic that possess chlorophyll, are phototrophic (light-loving), and utilize light as their energy source. Soil algae predominantly fall into the categories of yellow-green algae (*Heterokontophyta: Xanthophyceae*), green algae (*Chlorophyta*, *Streptophyta*), diatoms (*Heterokontophyta*: *Bacillariophyceae*). Green algae and cyanobacteria (classified among bacteria) engage in symbiosis with some ascomycetes. The result of these symbiotic is referred to as lichen Lichens represent a symbiotic partnership between algae and fungi. The fungus, by covering the rock substrate and breaking it down, absorbs minerals and makes them available to the algae. The algae produce carbohydrates through photosynthesis and provide energy to the fungus (Crouzet et al., 2019). Algae are instrumental in promoting soil health and fertility in a variety of ways: They sequester carbon through photosynthesis and maintain soil organic carbon content (Jassey et al., 2022). They assist in stabilizing soil particles and preventing erosion by creating a filaments that connect particles of soil together (Crouzet et al., 2019). Algae can engage with various soil microbes, including fungi and bacteria, thereby affecting cycling of nutrients and the management of microbial communities (Solomon et al., 2023). Although algae are not as dominant as bacteria and fungi in soil ecosystems, they still have serious functions in processes of ecosystem, cycling of nutrient, and health of soil.

3.1.1.5 Fungi

Currently, there are seven phyla of soil fungi, including *Ascomycota, Basidiomycota, Blastocladiomycota, Neocalymestigomycota, Glomeromycota, Chytridiomycota*, and *Microsporidia* (Dervash et al., 2024). Ascomycota is one of the most common fungal phyla, characterized by its symbiotic interactions with algae and cyanobacteria. Mutualistic relationships between ascomycetes and algae (dibuspores) or cyanobacteria (trispores) lead to the development of lichens. The fungus, functioning as a saprophyte, decomposes organic matter via a variety of enzymes. Given that these fungi acquire their sustenance from algae, the presence of complete fuel facilitates their production of more enzymes than that of other saprotrophs (Yadav et al., 2019).

3.1.1.6 Protozoa

Microorganisms consisting of a single cell that derive their nourishment from other organisms, like bacteria, present in the soil. Protozoa exist in the uppermost 15 cm of the soil surface. They inhabit aquatic environments and navigate through layers of water and water-saturated pores in the soil. Protozoa are categorized into two main groups: flagellates, which use flagella to propel themselves and mainly rely on bacteria for sustenance; and ciliates, the smallest of the protozoa, which employ cilia for movement and have a diet comprising amebae, flagellates, and bacteria. Many protozoa undergo a life cycle that features an active stage, termed the trophozoite, alongside a resting stage, called the cyst, where the cell develops a thick protective coating. During the cyst stage, many species can endure extreme environmental conditions and survive for extended periods until the environment becomes more suitable (Laban et al., 2018; Nielsen et al., 2015).

3.1.1.7 Nematodes

Nematodes, also known as roundworms, are microscopic, non-fragment worms with narrow ends, belonging to the phylum Nemata. While often recognized for the few species that cause infectious diseases in plants and animals, the vast majority play crucial and beneficial roles in soil. Physically, they are very small, allowing them to inhabit the pores within soil. As aquatic, they require adequate soil moisture to move. Despite their simple, worm-like form, they are complex fauna, possessing a well-developed central nervous system, a complete digestive system, and a reproductive system, leading to their classification as the most primitive fauna (Romero et al., 2023).

A key aspect of their ecological importance lies in their highly diverse feeding habits, which place them at several trophic levels within the soil food web. They are categorized into five broad dietary groups. The first four are free-living in the soil: bacterivores that consume bacteria; fungivores that puncture and feed on fungal cells; omnivores that eat a variety of microflora and microfauna, including plant roots; and predators that consume other nematodes and protozoa. The fifth group, root-feeders, are parasites of plants that live on or inside roots and are not considered free-living (Potapov et al., 2022).

These feeding activities drive critical soil processes. Nematodes are major agents of nutrient mineralization, particularly nitrogen. By ingesting bacteria and fungi, they convert the nutrients locked within these microbes into inorganic, such as NH_4^+ and PO_4^-, which are then excreted as waste and become readily available for plant uptake. This process increases nitrogen mineralization significantly, by between 8% and 19%. Furthermore, when plant-grazing nematodes feed on roots, they stimulate the plant to release root exudates. This, in turn, fuels the soil microbial, creating a "priming effect" that accelerates the decomposition of organic matter (Potapov et al., 2022).

The soil food web is interconnected, and the activity of nematodes is part of a larger regulatory system. For instance, predatory nematodes help regulate nitrogen mineralization by preying on the microbivorous (bacteria- and fungus-eating) nematodes, controlling their populations and thus modulating the flow of nutrients from the bottom to the top of the trophic chain. Due to their abundance and diverse functions, nematodes are integral to fundamental processes like soil organic matter decomposition and the maintenance of plant health, making them ecologically significant organisms in both natural and managed ecosystems (Potapov et al., 2022; Romero et al., 2023).

3.1.2 Mesofauna (0.one to two mm) - the litter transformers

Includes mites, springtails (Collembola), and enchytraeids. They live in air-filled pores and are primarily herbivores, bacterivores, or fungivores.

They feed on decaying organic matter, fungi, and bacteria. By fragmenting litter, they enhance the surface area for microbial attack and contribute to nutrient cycling and soil aeration (as Mites (Acari), Springtails (Collembola) and Tardigrades (Water Bears)).

3.1.3 Macrofauna (2–20 mm) - the ecosystem engineers

Includes earthworms, ants, beetles, and spiders. Many are "ecosystem engineers" that significantly modify the soil habitat.

3.1.3.1 Earthworms

Ecologically, earthworms are classified into three distinct groups, each with specific traits and functions:

- Epigeic: earthworms are small (0.5–5 cm), red, and live in the surface organic litter. They are not strong burrowers, can tolerate acidic soil, and their primary role is consuming litter.
- Endogeic: earthworms are medium-sized (1–8 cm), have no pigment, and live in the upper mineral soil. They are good burrowers that create horizontal tunnels, feed on humus and dead roots, and are sensitive to low pH (below 5).
- Anecic: Earthworms are sizable (frequently exceeding 5 cm), reddish-brown in color, and are proficient burrowers that form deep vertical tunnels. They reside in the mineral soil but surface to feed on litter, avoiding both waterlogged and dry soils. The activity of these earthworms profoundly affects soil structure and chemistry. Their most direct physical effect is "burrowing", which creates a network of tunnels. This dramatically increases soil porosity by 20%–30%, which enhances aeration, improves water circulation, and creates pathways for other soil organisms to move.

Chemically, their most significant contribution comes from producing "worm casts" (vermicasts), which are the waste products expelled after digestion. These casts are not simply waste; they are a highly enriched soil amendment. The process of digestion mixes organic matter with mineral particles and mucous, and the resulting casts are rich in nutrients like nitrogen, calcium, magnesium, and potassium in a form that plants can easily use.

Furthermore, the casts have a enhanced "microbial ecosystem". The earthworm's gut acts as an incubator, increasing the number and activity of bacteria, fungi, and actinomycetes. These microbes thrive in the casts due to the ideal conditions of moisture, surface area, and nutrient concentration, and they play a crucial role in further breaking down organic matter through humification and mineralization.

A critical behavior that enhances soil fertility is the "mixing of soil layers" via feeding and casting. Anecic earthworms, which live deep in the soil, bring their mineral-rich casts to the surface. In contrast, endogeic earthworms, which live near the surface, deposit their organic-rich casts deeper in the soil. This constant translocation of material creates a more homogeneous and fertile soil profile.

Earthworms, along with other soil fauna like molluscs, secrete cutaneous mucous. This mucous acts as a natural glue, cementing soil particles together into stable aggregates. This improved soil structure makes the ground more resistant to erosive forces like wind and water (Fig. 18.2).

Finally, earthworms contribute to the "nitrogen cycle" beyond their casts. They excrete nitrogen directly in plant-available forms like ammonia and urea. When they die, their bodies, which are composed of 72% protein, decompose and return a significant amount of nitrogen back to the soil (Fig. 18.2).

3.2 Geographic distribution of biodiversity for soil organisms

The composition of soil life is not uniform but varies significantly with climate and location. Among protists, Apicomplexa, Cercozoa, Ciliophora, and Dinophyceae are the most abundant, with their distributions closely tied to moisture; Dinophyceae flourish in dry soils, while Ciliophora and Apicomplexa prefer moist soils, and Cercozoa are found in a broader range of environments. Most soils contain a large number of multicellular nematodes, which comprise a substantial portion of the faunal biomass in deserts; their community composition is also influenced by climate, with plant-parasitic nematodes generally found in warm, humid environments and bacterial feeders thriving in colder, drier areas. Fauna consisting of larger organisms exhibit distinct patterns, with mites and springtails being relatively common in forests but reaching high levels of abundance. In contrast, earthworms and enchytraeids typically dominate faunal biomass due to their size, as enchytraeids are abundant in boreal forests and tundra, and earthworms are prevalent in temperate grasslands and deciduous forests. Social insects like ants are common from temperate to tropical zones, and termites, though less widespread, are abundant in subtropical and tropical ecosystems, with their numbers increasing toward the equator. While the tropics are biodiversity hotspots, the acidic soils there can make them "deserts" for soil bacteria, which prefer a neutral pH. Consequently, bacterial diversity is often higher in grassland and desert soils than in tropical rainforests. Similarly, soil nematode diversity appears to peak in temperate zones, and microbial diversity follows different patterns along elevation gradients compared to larger organisms.

Climatic conditions are overarching determinant of the entire system. They control plant growth, which defines the amount and quality of organic matter entering the soil. This, in turn, dictates which ecosystem engineers dominate. For instance, in cold or dry deserts, earthworms are absent, and engineers like ants and termites have a relatively small impact, leaving microbial and fungal-based food webs in control. Different ecosystems foster distinct soil communities. Forests produce low-quality litter, favoring fungal and arthropod-dominated decomposition with few earthworms. In contrast, grasslands support more earthworms, shifting the system to bacterial-dominated decomposition. Generally, converting natural vegetation to managed agricultural systems leads to a decline in larger soil fauna communities.

3.3 The underground network: Interactions and symbiosis

Soil Microbial Hotspots and Hot Moments. Based on the complex and heterogeneous nature of the soil biosphere, it is evident that biological activity is not uniformly distributed but is concentrated in specific microsites known as microbial hotspots. These hotspots, including the rhizosphere, detritusphere, and biopores, are characterized by an accelerated rate of biochemical processes, higher microbial respiration, and greater enzyme activity compared to the bulk soil. The fundamental driver of hotspot formation is the localized input of labile organic carbon, which provides energy for microbial growth and metabolism. The dynamic nature of these inputs also creates "hot moments," short sequences of events where process rates are significantly accelerated. The duration and intensity of these hotspots are determined by the quantity and quality of the substrate and the rate at which it is utilized by the microbial community.

The rhizosphere, the soil directly influenced by plant roots, is a primary biotic hotspot. Plants stimulate microbial activity through a continuous flow of rhizodeposits, including root exudates, mucigels, and sloughed-off cells. This creates a chemically, physically, and biologically distinct zone where carbon acts as a primary currency, fueling a dense and active microbial community. In contrast, the detritusphere, the zone surrounding decaying plant litter and dead roots, represents another primary biotic hotspot. Here, the decomposition of organic material provides a temporally concentrated, longer-lasting carbon source. This often results in an even greater biomass of active microorganisms compared to the rhizosphere, partly due to the absence of direct competition with living roots for nutrients.

Secondary hotspots include biopores, which are macropores created by the activity of roots or soil invertebrates like earthworms. These structures are crucial hotspots, particularly in subsoils, as they improve physical conditions for air and water circulation and become enriched with organic materials. The interaction between plants and soil invertebrates, such as earthworms, further amplifies hotspot activity through multiple mechanisms, including increased nutrient release, improved soil structure, enhanced mycorrhizal colonization, and the direct control of pests.

A critical feature of all these hotspots is not only an increase in total microbial biomass—often two to three times greater than bulk soil—but also a significant shift in the physiological state of the community. While a large portion of the total soil microbial biomass may be dormant, hotspots support a disproportionately high share of active microorganisms. This is exemplified by reports of hydrolytic enzyme activity being 3–5 times higher in the rhizosphere and N_2O emissions being 2–9 times more intensive in the detritusphere. The community structure itself is also shaped by the locality and type of substrate, with microbial abundance and diversity changing with distance from the root or residue.

These intensified processes have profound ecological significance. Hotspots are central to the decomposition of organic matter, the mineralization of nutrients, and the transformation of carbon. They govern microbial immobilization of nitrogen, consumption of oxygen, and the use of other electron acceptors. The management of these zones, including the symbiotic relationships with mycorrhizal and N_2-fixing bacteria, is therefore critical for agroecosystem productivity and nutrient cycling. Ultimately, the spatial and temporal heterogeneity created by microbial hotspots is a fundamental property of soil, driving the efficient cycling of energy and nutrients that sustains terrestrial ecosystems.

Symbiosis and Plant Nutrition. The sustainability of terrestrial ecosystems is fundamentally underpinned by two key mutualistic associations: the fixation of atmospheric nitrogen by prokaryotes and the enhanced nutrient acquisition facilitated by mycorrhizal fungi. While nitrogen is the element most required by plants, its availability is often limited, making biological nitrogen fixation a critical process. This function is exclusively performed by certain bacteria and archaea, which can exist freely or in symbiosis with hosts, most notably with legumes via rhizobia. The economic and ecological significance of this relationship is profound, as evidenced in agricultural systems like soybean fields, where bacterial inoculation entirely replaces synthetic fertilizers, showcasing a successful application of a biological process for sustainable production. The discovery of extensive diversity within rhizobial species, facilitated by molecular techniques, highlights the need for precise pairings between specific bacterial strains and plant hosts to ensure effective nitrogen fixation.

Concurrently, phosphorus frequently acts as the primary growth-limiting nutrient, particularly in tropical ecosystems, due to its propensity to form insoluble compounds and its poor soil mobility. This limitation is overcome by mycorrhizal, which form symbiotic relationships with the vast majority of plant species. Through their extensive hyphal networks, these fungi explore a greater soil volume than roots alone, mobilizing and transporting phosphorus and other immobile nutrients to their host plants. The two most prevalent types, arbuscular (AM) and ectomycorrhizal (EM) fungi, exhibit distinct biogeographical patterns, with AM fungi being more common in species-rich tropical ecosystems and EM fungi dominating in colder boreal and temperate forests. Beyond nutrition, these fungi enhance host plant resilience to abiotic and biotic stresses, including drought and pathogens.

A particularly significant ecological extension of the mycorrhizal symbiosis is the formation of common mycorrhizal networks (CMNs), which interconnect multiple plants of the same or different species. This connectivity transforms plant communities from collections of individuals into integrated networks, facilitating a range of interplant interactions. Research indicates that CMNs can improve seedling establishment, influence community composition, induce defense responses, and enable the transfer of nutrients and informational signals between connected plants. The mechanisms governing resource allocation within these networks are theorized through two main frameworks. The source-sink theory suggests a passive flow of resources along concentration gradients, which could promote a more equitable distribution. In contrast, the biological market theory posits a more dynamic, reciprocal exchange where both plant and fungal partners adjust their resource allocation based on the "value" of the resources received from their partner. This theory suggests that larger plants, capable of supplying more carbon to the fungi, may receive a greater share of limiting nutrients like nitrogen and phosphorus, thereby potentially gaining a competitive advantage and influencing plant competition and community dynamics.

In conclusion, the synergistic and individual contributions of N_2-fixing prokaryotes and mycorrhizal are indispensable for nutrient cycling and plant productivity. The rhizobia-legume symbiosis provides a critical input of available nitrogen, while mycorrhizal associations, particularly through the formation of CMNs, drastically improve phosphorus acquisition and create complex belowground communication and resource-transfer systems. Understanding the principles governing these mutualisms, including the biological market dynamics within CMNs, is crucial for advancing sustainable agricultural practices and for comprehending the fundamental processes that maintain the structure and function of natural ecosystems worldwide.

3.4 Soil biodiversity and ecosystem services

Soil is a dynamic, living ecosystem that serves as the cornerstone of terrestrial life. Its functionality is not merely a product of its chemical and physical properties but is fundamentally driven by the immense diversity of organisms within it—from microbes and microfauna to earthworms and plant roots. The interactions among these organisms and their habitat underpin a vast array of ecosystem functions, which in turn provide critical services that sustain human societies and the biosphere. The loss of this biodiversity disrupts the flows of energy, water, and nutrients, thereby impairing the ecosystem services upon which life depends.

Plants absorb atmospheric carbon (CO_2) through photosynthesis. This carbon is transferred to the soil system through root exudates and substrate sediment. This organic matter is the main source of energy for soil heterotrophs and forms a complex food web. Through their activities, soil organisms perform essential functions such as decomposition, nutrient cycling, and soil structure formation.

Supporting and Provisioning Services. Supporting services, such as soil formation and habitat provision, are the foundation for all other services. Organisms are involved in all stages of soil development, from weathering rocks to creating stable soil aggregates. Plants, earthworms, macroarthropods, and fungi are key "ecosystem engineers" that structure the soil. Roots penetrate and bind soil, while earthworms and other invertebrates create burrows and pores through their movement and feeding. This activity directly determines the soil's physical architecture—its porosity, aeration, and water infiltration capacity—which in turn creates the habitat for smaller organisms.

This engineered habitat is crucial for provisioning services, most notably food and fiber production. Decomposition, a key function carried out by a consortium of organisms, releases essential nutrients like N and P from organic matter, making them available for plant uptake. This process, known as mineralization, is a core component of soil fertility. Reductions in soil biodiversity consistently slow decomposition rates, directly impacting crop yields. Furthermore, a biodiverse soil community supports services like:

Biocontrol: Suppressing agricultural pests and diseases, potentially reducing the need for agrochemicals.

Biotechnology: Serving as a source of pharmaceutical products like penicillin and tetracyclines.

Erosion Control: Stabilizing soil and preventing the loss of fertile land.

Regulating Services: Maintaining Global Balances. Soil biodiversity is a master regulator of global biogeochemical cycles.

Carbon Cycling and Climate Regulation: Soil is a major carbon sink. The decomposition of plant litter by soil organisms is a double-edged process. While respiration by bacteria, fungi, and fauna returns CO_2 to the atmosphere, soil biota also help stabilize organic carbon by promoting its incorporation into soil aggregates. The delicate balance between mineralization (releasing nutrients) and humification (preserving organic matter) is regulated by the diversity of both the decomposer community and the organic inputs.

Nitrogen and Phosphorus Cycling: Microbes are the primary agents in the nitrogen cycle, acting as fixers, nitrifiers, and denitrifiers. They transform nitrogen into plant-available forms (ammonium, nitrate), but an imbalance can lead to environmental problems like water pollution. Similarly, while soil contains abundant phosphorus, much of it is in insoluble forms. Specific fungi and bacteria solubilize this P, making it accessible to plants. The efficiency of these cycles is entirely dependent on the health of the soil microbial.

Water Quality Regulation: By regulating decomposition and nutrient transformation, soil organisms prevent the leaching of excess nutrients (e.g., nitrates, phosphates) into groundwater and surface waters, where they can cause eutrophication and algal blooms.

Cultural Services: The Human Connection. Soils also provide cultural services that are deeply intertwined with human history and identity. These include:

✓ The cultural heritage of farming and ranching communities.
✓ The use of clay soils for pottery, cookware, and jewelry.
✓ Soils as a source of pigments for art and dyes.
✓ Soils as archeological repositories, preserving evidence of past civilizations.

Biological indicator. The intricate web of soil life is highly vulnerable to disturbance. Soil contamination, for example from heavy metals or pesticides, can severely impact ecosystem engineers like earthworms, reducing their burrowing activity and leading to soil compaction and degraded structure. This demonstrates how biodiversity loss directly impairs ecosystem functions.

Because soil organisms are highly sensitive to environmental changes and are inextricably linked to their habitat, they serve as excellent "bioindicators". The structure of invertebrate communities (e.g., nematodes, collembolans, mites) contains valuable information about overall soil health. Monitoring changes in their diversity, species richness, and functional groups provides a holistic assessment of soil quality that complements traditional physicochemical analyses (Yang et al., 2022).

3.5 Threats to soil biodiversity

An Integrated Discussion on Soil Integrity, Biodiversity, and Function Human activities, the most important driving force behind the global soil degradation process, direct threaten soil integrity, biodiversity, and the critical ecosystem services it provides. Soil Integrity—The capacity of a soil to carry out all its normal ecosystem functions under the dominant influence from natural soil-forming factors—is being severely eroded by a convergence of human pressures. The biggest threats to soil biodiversity and integrity come from land use change, agricultural intensification, climate change, pollution, and urbanization.

Land Use Change. The most dramatic alterations occur during land-use conversion. Studies showed that deforestation caused a major reduction in soil organic carbon. Importantly, the loss of soil carbon was particularly severe after forest alteration to cropland, which decreased by 48%. The most important reason for the loss of soil organic carbon is due to the reduction in carbon input from plant remains (Qu et al., 2024). There was also a decreasing trend in total soil nitrogen after forest conversion (23%), which led to a decrease in the carbon-to-nitrogen ratio. Native forest conversion led to a more substantial decrease in soil carbon reserves than in soil nitrogen levels, a change possibly associated with nitrogen fertilizer introduction and reduced ligninized carbon in litter due to the shift from woody vegetation to grasses following deforestation. Conversion of native forests to managed systems also enhanced pH. Meanwhile, soil phosphorus increases with deforestation, mainly due to the intensive use of phosphorus fertilizers in managed systems. Soil composition variations, coupled with decreasing soil organic carbon levels and rising fertilizer use, cause fluctuations in the carbon to nitrogen and phosphorus ratio, leading to elevated soil pH levels and a significant increase in bacterial diversity. Deforestation, on the other hand, had a less negative impact on fungal richness than other land uses. This can be attributed mainly to livestock grazing and manure produced by livestock, which can expand the functional niche of fungi. This increased fungal richness is not inherently positive for ecosystem function, as it may lead to a higher diversity of pathogens rather than beneficial species (Van Gestel et al., 2022). Agricultural intervention has also led to a homogenization of soil microbial. The uniformity of communities within managed ecosystems is probably due to the decline of native microorganisms (Trivedi et al., 2018). Research indicates that managed ecosystems harbor a higher percentage of bacterial groups including *Bacteroidetes*, *Gemmatimonadetes*, *Firmicutes*, *Chloroflexi*, and *Nitrospira*, whereas the prevalence of *Proteobacteria*, Acidobacteria, and Verrucomicrobia is decreased. The post-deforestation fungal community supported a greater proportion of Ascomycota at the expense of a reduced abundance of Basidiomycota. Thus, agricultural intensification generally leads to smaller, less diverse soil communities compared to natural ecosystems like forests or grasslands (Qu et al., 2024).

Erosion. Accelerated by deforestation, tillage, and removal of perennial vegetation, leading to the loss of organic matter-rich topsoil where most biological activity occurs; it disrupts nutrient cycles, causing nutrient imbalances through fertilizer application; and it depletes soil organic carbon stocks, directly destroying habitats and relocating soil organisms. Also, an estimated 1% of topsoil is lost annually to erosion (Qu et al., 2024).

Climate Chang. Climate change and how it affects soil biota pose a unique challenge. Results showed that in warmer and wetter ecosystems, climate has a greater negative impact on fungal diversity than on bacteria, accelerating the loss of native fungal species (Zhang et al., 2023). In contrast, the survival of bacteria that are more resistant to higher temperatures and environmental disturbances is more likely to dominate ecosystems in warmer and wetter regions (Qu et al., 2024). This suggests that climate can greatly influence soil biodiversity and functions. Predicting the future composition of soil fauna remains difficult. Wang et al. (2021) reported that soil respiration and microbial growth have decreased with increasing temperatures in the Arctic, boreal, temperate, and tropical biomes. Therefore, deforestation and increased greenhouse gases can synergistically affect soil biodiversity with climate change.

Soil pollution. Pollutants such as heavy metals from sewage sludge or industrial waste, or pesticides from agricultural activities, impose major stresses on soil organisms. The response to these pollutants is not uniform, but strongly depends on the properties of the pollutant, the physicochemical characteristics (e.g., texture, pH, OC) (Jia et al., 2023). Also, pollutants change the structure of the soil microbial community and some species may decline, while more resistant species increase in population and become the dominant species in the soil. On the other hand, the indiscriminate use of chemical fertilizers increases soil salinity. Jia et al. (2023) reported that soil salinity reduces microbial diversity and activity through ion toxicity and osmotic stress. Although increasing soil organic matter can reduce salinity stress by diluting ions, producing protective compounds, and improving soil structure, its effectiveness is limited in highly saline soils. In addition, salinity drives communities toward bacterial dominance, as bacteria adapt more quickly and utilize the volatile organic matter common in agricultural ecosystems, while rare microbial species are particularly vulnerable to changes in salinity and soil organic matter. In soils with low salinity (<2 ds/m), soil organic matter increases enzyme activity, supporting microbial diversity by providing energy (Jia et al., 2023). While, under high salinity conditions, stress disrupts microbial metabolism, reducing organic matter decomposition and nutrient cycling, it also forces microbes to direct their energy toward survival rather than other ecological functions, thus suggesting that adding external OM – such as compost or manure – can better alleviate salinity stress and restore soil function. Romero et al. (2023) reported that one of the main factors altering soil microbial communities is nitrogen fertilization, which increases Proteobacteria and pathogenic fungi, while decreasing oligotrophic groups. Fertilization also stimulates some enzymatic activities but reduces soil respiration (Jia et al., 2023).

Urbanization. Instead of modifying and slowly expanding existing ecosystems as with agriculture or forestry, however, urbanization often amounts to a near-total replacement of the natural system with built-in environment. It's a fundamental shift from a biologically driven system to a physically engineered one.

Habitat Destruction, Fragmentation, and Soil Sealing are the three direct mechanisms of impact:

Habitat Destruction: Ore soil and the plant communities it supports are removed in order to make way for buildings, roads and other infrastructure.

Fragmentation: Remaining patches of green space (except for parks, gardens) are like isolated "islands." This obstructs the movement of soil organisms between patches. It reduces genetic diversity, making local populations more vulnerable to extinction.

Soil Sealing: A very direct impact. Soil surfaces receive impervious coverings such as asphalt or concrete. So,lackss gas exchange. At the same time, it deprives soil organisms of oxygen and traps CO_2. It also prevents water infiltration. There is a "hydrological drought." Nutrient cycling is stifled. Moreover, energy input is blocked. Leaf litter and other forms of organic matter can't get into the soil food web. Soil sealing is essentially a form of eco-asphyxiation and emaciation.

Invasive Species. The introduction of non-native soil organisms (e.g., European earthworms in North America) or plants can disrupt decomposition processes, nutrient cycles, and outcompete native species, leading to fundamental shifts in soil community structure and function.

The connection between the composition of soil communities and their functioning is a central theme in soil ecology, governed by several competing hypotheses. The "redundant species" hypothesis posits that high functional redundancy within incredibly diverse soil communities provides an "insurance" effect. Under this view, moderate species loss may not immediately impair ecosystem processes like decomposition or nutrient cycling, as other species can compensate. This has led to a historical belief that functional measurements are less sensitive indicators of disturbance than structural ones.

However, this view is disputed by the "rivet" theory, which claims that each species constitutes a unique, non-redundant role and any such role lost leads only to one less thing. Evidence along contamination gradients has shown that functional indicators such as soil respiration may sometimes be more sensitive than structural measures. Another idea, the "idiosyncratic" hypothesis, argues we can't predict the relationship because a few "keystone" species or functional groups disproportionately affect it. Their loss-for example, specific ecosystem engineers like earthworms that stabilize soil against erosion or major predators that limit pest outbreaks-can have effects extending to other parts of the food web far beyond what is lost with other species. The application of pesticides, which can wipe out natural enemy populations and lead to subsequent pest outbreaks, is a classic case of this phenomenon.

This intricate interplay means that although changes in soil community structure do not necessarily lead directly to functional loss, they should still be seen as early warnings of possible ecosystem deterioration. The point at which compensation is no longer possible and the function of the ecosystem collapses, is a critical turning point that must be avoided through careful management practices.

3.6 Conservation and restoration of soil biodiversity

Human soil management practices are a major factor in maintaining soil microbial, which in turn improve soil ecosystem function. With the exception of forests, which are natural ecosystems, managed agricultural systems cause profound and often detrimental changes in soil biodiversity, with significant environmental consequences. Despite these threats, there are strategies to conserve and restore soil biodiversity. These strategies are interconnected and have synergistic effects.

Vegetation. Plants are the primary engineers of the soil environment. Agroecosystems that mimic the original vegetation (e.g., agroforestry in forest zones, modified perennial grasslands in savannas) consistently support higher soil biodiversity than annual monocultures. Plants influence the soil biota through the quantity, quality, and diversity of their organic inputs (litter, root exudates). Legumes, for instance, have a dual positive effect by fixing nitrogen and improving soil structure via enhanced earthworm activity, the selection of plant cultivars that fail to engage in beneficial symbioses can degrade soil biological function.

Rehabilitating degraded lands, such as through reforestation, can rebuild soil carbon and restore microbial biomass and activity. Soil organisms can even be deployed directly in bioremediation to detoxify pollutants. Protecting intact natural ecosystems, particularly forests which host the highest soil biodiversity, remains the most effective strategy for preserving soil integrity and its associated services.

Tillage. The tillage system generates an impact in varying degrees, going from the strongly detrimental conventional plowing down to the moderate minimum-tillage as well as no-till systems. The negative relationship between the intensity of tillage and the diversity and abundance of soil faunal groups (from earthworms to microarthropods) is well documented. Conservation tillage facilitates transition to fungal-dominated decomposition pathways and more complex food webs similar to those present in undisturbed ecosystems. These include no-tillage, conservation tillage with agronomic sowing, agroforestry and the use of perennial crops system, which improve soil biodiversity and raise the pool of carbon contents; such techniques could also be used to slash external inputs by fully exploiting soil natural functions.

Tillage represents a classic form of high-intensity disturbance that keeps soil in an early successional condition. It kills soil organisms, disrupts fungal hyphal networks and earthworm homes, and depletes available food resources to support decomposers on the way toward mold domination of a system over development of complex, bacteria dominated systems. The comparison regarding the microbiological properties of no-till and plowed systems is relatively unanimous for microbial biomass, ammonia-oxidizing bacteria diversity, and microarthropods (mites and collembolan) abundance in several studies. This provides evidence that, despite strong impacts of restored complexity on the community structure of microorganisms, soil ecosystem processes can be reconstituted with diminished disturbance by tillage.

Fertilization and Organic Inputs. The source of nutrients critically shapes the soil community. Chemical fertilizers can lead to "starvation" of the decomposer community by reducing organic residue inputs, thereby Disrupted food webs. In contrast, organic amendments (e.g., manure, compost, cover crops) provide a diverse and continuous food source, stimulating biological activity across trophic levels. This leads to increased microbial biomass, enhanced abundance of litter-dwelling invertebrates, and improved soil structure through the action of ecosystem engineers. The positive effects of organic and low-input farming systems on soil species richness and abundance are a direct result of this principle. The quality of organic matter matters, with different types of compost yielding vastly different outcomes for invertebrate communities.

Inoculation. The practice of inoculating soils with beneficial organisms (e.g., rhizobia, mycorrhizal fungi, earthworms) is a promising tool for agro-ecological manipulation. Successful cases, such as earthworm inoculation in pastures, show dramatic improvements in soil structure and crop quality. The results showed that the utilization of biofertilizers significantly altered the structure of the soil microbial, which in turn accelerated the rate of plant residue decomposition. Biofertilizer inoculants successfully competed with the native microbiome and modulated it to favor a more active decomposer community. Biofertilizers can therefore fundamentally influence nutrient cycling and soil fertility through microbially-directed manipulation of decomposition processes (Baldi et al., 2021). The utilization of biofertilizers, especially those containing Trichoderma spp., resulted in a measurable increase in total soil microbial biomass. Nevertheless, inoculation is not a silver bullet; it is most effective when integrated with management practices that create a generally favorable and complex soil habitat (Eisenhauer et al., 2022).

4. Conclusion and future perspectives

The biodiversity of soil is a crucial foundation for global sustainability, being integral to the regulation of climate, food security, and the vitality of terrestrial and aquatic ecosystems. The vast array of soil organisms, within micro-habitats or "hotspots" such as the rhizosphere and detritusphere, evidently serves as the driving force behind effective nutrient cycling and the stability of ecosystems. Management practices that enhance organic matter inputs, minimize soil disturbance, and sustain diverse plant cover are essential for nurturing complex and resilient soil communities.

Looking forward, some immediate priorities emerge in order to translate this knowledge into effective action. First, we have a call for more focussed research to develop novel and practical indicators of soil biological health that can readily be used in field trials alongside management practices to conserve or exploit this resource. Furthermore, this knowledge needs to be further assimilated into the prospective agroecosystems in a timely manner particularly in severely degraded areas which the loss of soil biodiversity is one of the major causes of food insecurity and poverty.

Instead, the future is contingent upon our collective commitment to responsible stewardship. The deepening of the plow layer and land use transformation are important signs showing the significant anthropogenic effect on soil. Addressing this requires coordinated, global actions for conservation and promotion of soil biodiversity. Far from just another crop pushing exercise, it is a sine qua non of long-term management for the Earth's renewable resources, and essential to maintaining the resilience and productivity of our ecosystems for future generations.

Declaration of generative AI and AI-assisted technologies in the manuscript preparation process

During the preparation of this work, used DeepSeek to improve the readability and language of the manuscript. After using this tool, the authors reviewed and edited the content as needed and take(s) full responsibility for the final content of the publication.

References

Amato, M., & Ladd, J.N. (1988). Assay for microbial biomass based on ninhydrin-reactive nitrogen in extracts of fumigated soils. *Soil Biology and Biochemistry, 20* (1), 107–114.

Angel, S., Parent, J., & Civco, D.L. (2012). The fragmentation of urban landscapes: global evidence of a key attribute of the spatial structure of cities, 1990–2000. *Environment and Urbanization, 24*(1), 249–283.

Arce, S. M., Argue, B. J., Thompson, D. A., et al. (2003). Evaluation of a fluorescent, alphanumeric tagging system for penaeid shrimp and its application in selective breeding programs. *Aquaculture, 228*, 267–278. https://doi.org/10.1016/S0044-8486(03)00267-9

Baldi, E., Gioacchini, P., Montecchio, D., Mocali, S., Antonielli, L., Masoero, G., & Toselli, M. (2021). Effect of biofertilizers application on soil biodiversity and litter degradation in a commercial apricot orchard. *Agronomy, 11*(6), 1–18.

Battistuzzi, M., Cocola, L., Claudi, R., Pozzer, A. C., Segalla, A., Simionato, D., Morosinotto, T., Poletto, L., & La Rocca, N. (2023). Oxygenic photosynthetic responses of cyanobacteria exposed under an M-dwarf starlight simulator: Implications for exoplanet's habitability. *Frontiers in Plant Science, 14*, Article 1070359.

Bohlen, P. J., Pelletier, D. M., Groffman, P. M., et al. (2004). Influence of earthworm invasion on redistribution and retention of soil carbon and nitrogen in northern temperate forests. *Ecosystems, 7*, 13–27. https://doi.org/10.1007/s10021-003-0127-y

Boyno, G., Demir, S., Rezaee Danesh, Y., Durak, E. D., Çevik, R., Farda, B., … Pellegrini, M. (2023). A new technique for the extraction of arbuscular mycorrhizae fungal spores from rhizosphere. *Journal of Fungi, 9*(8), 845.

Brown, G. R., & Matthews, I. M. (2016). A review of extensive variation in the design of pitfall traps and a proposal for a standard pitfall trap design for monitoring ground-active arthropod biodiversity. *Ecology and Evolution, 6*, 3953–3964. https://doi.org/10.1002/ece3.2176

Crouzet, O., Consentino, L., Pétraud, J.P., Marrauld, C., Aguer, J.P., Bureau, S., ... & Bérard, A. (2019). Soil photosynthetic microbial communities mediate aggregate stability: influence of cropping systems and herbicide use in an agricultural soil. *Frontiers in Microbiology, 10*, 1319.

Datta, R. (2024). Enzymatic degradation of cellulose in soil: A review. *Heliyon, 10*, Article e24022. https://doi.org/10.1016/j.heliyon.2024.e24022

Demelash, K., Abebaw, M., Negash, A., et al. (2016). A review on diagnostic techniques in veterinary helminthology. *Natural Sciences, 14*, 109–118. https://doi.org/10.7537/marsnsj140716.15

Dervash, M. A., Yousuf, A., Bhat, M. A., & Ozturk, M. (2024). Introduction to soil organisms: Deciphering the life beneath our feet. In *Soil organisms: Deciphering the life beneath our feet* (pp. 1–3). Cham: Springer Nature Switzerland.

Doupé, R. G., Partridge, G. J., & Lymbery, A. J. (2003). Visible implant fluorescent elastomer tags as pedigree markers for applied aquaculture: An evaluation using black bream *Acanthopagrus butcheri*. *Aquaculture Research, 34*, 681–683. https://doi.org/10.1046/j.1365-2109.2003.00860.x

Eisenhauer, N., Antunes, P. M., Bennett, A. E., et al. (2017). Priorities for research in soil ecology. *Pedobiologia, 63*, 1–7. https://doi.org/10.1016/j.pedobi.2017.05.003

Eisenhauer, N., Bender, S. F., Calderón-Sanou, I., de Vries, F. T., Lembrechts, J. J., Thuiller, W., & Potapov, A. (2022). Frontiers in soil ecology—Insights from the World Biodiversity Forum 2022. *Journal of Sustainable Agriculture and Environment, 1*(4), 245–261.

Eme, L., Spang, A., Lombard, J., Stairs, C. W., & Ettema, T. J. (2017). Archaea and the origin of eukaryotes. *Nature Reviews Microbiology, 15*(12), 711–723.

Franco-Correa, M., Quintana, A., Duque, C., Suarez, C., Rodríguez, M.X., & Barea, J.M. (2010). Evaluation of actinomycete strains for key traits related with plant growth promotion and mycorrhiza helping activities. *Applied Soil Ecology, 45*(3), 209–217.

Geisen, S., Wall, D. H., & van der Putten, W. H. (2019). Challenges and opportunities for soil biodiversity in the anthropocene. *Current Biology, 29*(19), R1036–R1044.

Gonsior, M., Hertkorn, N., & Hinman, N. (2018). Yellowstone hot springs are organic chemodiversity hot spots. *Scientific Reports, 8*, Article 14155. https://doi.org/10.1038/s41598- 018- 32593- x

González, G., Barberena-Arias, M. F., Huang, W., & Ospina-Sánchez, C. M. (2020). Sampling methods for soil and litter fauna. In *Measuring arthropod biodiversity: A handbook of sampling methods* (pp. 495–522). Cham: Springer International Publishing.

Gorres, J. H., Savin, M. C., & Amador, J. A. (2001). Soil micropore structure and carbon mineralization in burrows and casts of an anecic earthworm (*Lumbricus terrestris*). *Soil Biology and Biochemistry, 33*, 1881–1887. https://doi.org/10.1016/S0038-0717(01)00068-2

Grzyb, A., Wolna-Maruwka, A., & Niewiadomska, A. (2021). The significance of microbial transformation of nitrogen compounds in the light of integrated crop management. *Agronomy, 7*, 1415. https://doi.org/10.3390/agronomy11071415

Hong, N. H., Rumpel, C., Henry, des T. T., et al. (2011). How do earthworms influence organic matter quantity and quality in tropical soils? *Soil Biology and Biochemistry, 43*, 223–230. https://doi.org/10.1016/j.soilbio.2010.09.033

Jassey, V.E., Walcker, R., Kardol, P., Geisen, S., Heger, T., Lamentowicz, M., ... & Lara, E. (2022). Contribution of soil algae to the global carbon cycle. *New Phytologist, 234*(1), 64–76.

Jia, J., Zhang, J., Li, Y., Koziol, L., Podzikowski, L., Delgado-Baquerizo, M., & Zhang, J. (2023). Relationships between soil biodiversity and multifunctionality in croplands depend on salinity and organic matter. *Geoderma, 429*, Article 116273.

Laban, P., Metternicht, G., & Davies, J. (2018). *Soil biodiversity and soil organic carbon: Keeping drylands alive* (Vol 10). Gland, Switzerland: IUCN.

Lamb, J. A., Fernandez, F. G., & Kaiser, D. E. (2014). *Understanding nitrogen in soils*. Minneapolis: University of Minnesota.

Lindström, K., & Mousavi, S. A. (2020). Effectiveness of nitrogen fixation in rhizobia. *Microbial Biotechnology, 13*, 1314–1335. https://doi.org/10.1111/1751- 7915.13517

Lladó, S., López-Mondéjar, R., & Baldrian, P. (2017). Forest soil bacteria: Diversity, involvement in ecosystem processes, and response to global change. *Microbiology and Molecular Biology Reviews, 81*. https://doi.org/10.1128/mmbr.00063- 16. e00063–e00016.

Merryweather, J.W., & Fitter, A.H. (1991). A modified method for elucidating the structure of the fungal partner in a vesicular-arbuscular mycorrhiza. *Mycological Research, 95*(12), 1435–1437.

Morgado, R. G., Loureiro, S., & González-Alcaraz, M. N. (2018). Changes in soil ecosystem structure and functions due to soil contamination. In *Soil pollution* (pp. 59–87). Academic Press.

Mukhtar, S., Zaheer, A, Aiysha, D., Malik, K. A., & Mehnaz, S. (2017). Actinomycetes: A source of industrially important enzymes. *Journal of Proteomics & Bioinformatics, 10*, 316–319. https://doi.org/10.4172/0974-276X.1000456

Nadarajah, K. K. (2018). In silico identification of plant-derived secondary metabolites in defense. In *In silico approach for sustainable agriculture* (pp. 275–293). Singapore: Springer.

Nielsen, U. N., Wall, D. H., & Six, J. (2015). Soil biodiversity and the environment. *Annual Review of Environment and Resources, 40*, 63–90.

Potapov, A. M., Sun, X., Briones, M. J., Brown, G. G., Cameron, E. K., Cortet, J., … Wall, D. H. (2022). Global monitoring of soil animal communities using a common methodology. *bioRxiv, 2022–01*.

Prévost, D., & Antoun, H. (2007). Root nodule bacteria and symbiotic nitrogen fixation. *Soil sampling and methods of analysis*, 379–398.

Qu, X., Li, X., Bardgett, R. D., Kuzyakov, Y., Revillini, D., Sonne, C., & Delgado-Baquerizo, M. (2024). Deforestation impacts soil biodiversity and ecosystem services worldwide. *Proceedings of the National Academy of Sciences, 121*(13), Article e2318475121.

Rhea-Fournier, D., & González, G. (2017). Methodological considerations in the study of earthworms in forest ecosystems. In S. Chakravarty, & G. Shukla (Eds.), *Forest ecology and conservation*. Rijeka: InTech.

Richardson, A.E., Barea, J.M., McNeill, A.M., & Prigent-Combaret, C. (2009). Acquisition of phosphorus and nitrogen in the rhizosphere and plant growth promotion by microorganisms.

Romero, F., Hilfiker, S., Edlinger, A., Held, A., Hartman, K., Labouyrie, M., & van der Heijden, M. G. (2023). Soil microbial biodiversity promotes crop productivity and agro-ecosystem functioning in experimental microcosms. *Science of The Total Environment, 885*, Article 163683.

Solomon, W., Mutum, L., Janda, T., & Molnár, Z. (2023). Potential benefit of microalgae and their interaction with bacteria to sustainable crop production. *Plant Growth Regulation, 101*, 53–65. https://doi.org/10.1007/s10725- 023- 01019- 8

Soto-Adames, F. N., & Taylor, S. J. (2010). Status assessment survey for springtails (collembola) in Illinois caves: The Salem Plateau. *Illinois Natural History Survey Technical Report, 13*, 1–76.

Stork, N. E. (2018). How many species of insects and other terrestrial arthropods are there on earth? *Annual Review of Entomology, 63*, 31–45. https://doi.org/10.1146/annurev-ento-020117-043348

Terkina, I.A., Parfenova, V.V., & Ahn, T.S. (2006). Antagonistic activity of actinomycetes of Lake Baikal. *Applied Biochemistry and Microbiology, 42*(2), 173–176.

Tian, J., Ge, F., Zhang, D., Deng, S., & Liu, X. (2021). Roles of phosphate solubilizing microorganisms from managing soil phosphorus deficiency to mediating biogeochemical P cycle. *Biology, 10*, 158. https://doi.org/10.3390/biology10020158

Tibbett, M., Fraser, T. D., & Duddigan, S. (2020). Identifying potential threats to soil biodiversity. *PeerJ, 8*, Article e9271.

Trivedi, P., Singh, B. P., & Singh, B. K. (2018). Soil carbon: Introduction, importance, status, threat, and mitigation. In *Soil carbon storage* (pp. 1–28). Academic Press.

Tuf, I. H. (2015). Different collecting methods reveal different ecological groups of centipedes (chilopoda). *Fortschritte der Zoologie, 32*, 345–350. https://doi.org/10.1590/S1984-46702015000500003

Tullgren, A. (1918). *Zur morphologie und systematik der Hemipteren I.* Almquist & Wiksells.

Van Gestel, C. A., Mommer, L., Montanarella, L., Pieper, S., Coulson, M., Toschki, A., & Römbke, J. (2022). Soil biodiversity: State-of-the-art and possible implementation in chemical risk assessment. *Integrated Environmental Assessment and Management, 17*(3), 541–551.

Wang, C., Morrissey, E. M., Mau, R. L., Hayer, M., Piñeiro, J., Mack, M. C., & Hungate, B. A. (2021). The temperature sensitivity of soil: Microbial biodiversity, growth, and carbon mineralization. *The ISME journal, 15*(9), 2738–2747.

Yadav, R. K., Yadav, M. R., Mahala, D. M., Kumar, R., Kumar, D., Yadav, N., & Yadav, S. (2019). Soil microbial hotspots and hot moments: Management vis-a-vis soil biodiversity. In *Plant growth promoting rhizobacteria for agricultural sustainability: From theory to practices* (pp. 189–202). Singapore: Springer Singapore.

Yang, G., Ryo, M., Roy, J., Lammel, D. R., Ballhausen, M. B., Jing, X., & Rillig, M. C. (2022). Multiple anthropogenic pressures eliminate the effects of soil microbial diversity on ecosystem functions in experimental microcosms. *Nature Communications, 13*(1), 4260.

Zhang, M., Delgado-Baquerizo, M., Li, G., Isbell, F., Wang, Y., Hautier, Y., & Wang, L. (2023). Experimental impacts of grazing on grassland biodiversity and function are explained by aridity. *Nature Communications, 14*(1), 5040.

CHAPTER

19 Climate change impacts on soil futures

Arun Karnwal[1], Neha Saklani[1], Abdel Rahman Mohammad Said Al-Tawaha[2], Abdel Razzaq Al-Tawaha[3], Natalia Nesterova[4], Gaurav Kumar[5], Anupam Kumar[6] and Deepak Kumar[7]

[1]*Department of Microbiology, Graphic Era (Deemed to be University), Dehradun, Uttarakhand, India;* [2]*Department of Biological Sciences, Al Hussein Bin Talal University, Ma'an, Jordan;* [3]*National Agricultural Research Center (NARC), Baq'a, Jordan;* [4]*National University of Life and Environmental Sciences of Ukraine, Kyiv, Ukraine;* [5]*Amity Institute of Microbial Technology, Amity University Rajasthan, Jaipur, Rajasthan, India;* [6]*Department of Biotechnology, Faculty of Applied, Pure and Health Sciences, Parul University, Quepem, Goa, India;* [7]*Department of Chemistry, Manipal University Jaipur, Jaipur, Rajasthan, India*

1. Methods

Literature Search and Selection Criteria. A comprehensive literature search was conducted to synthesize evidence on climate change impacts on soil properties, health, and management strategies. Databases included Google Scholar, PubMed, Scopus, Web of Science, and Elsevier/Springer platforms (aligned with the authors' publishing expertise) (Panagos et al., 2015; Burke et al., 2017; Smith et al., 2019).

Search Parameters. Keywords encompassed "climate change AND soil health," "soil degradation climate impacts," "soil organic carbon warming," "soil erosion precipitation," "nutrient cycling temperature," "soil microbiome climate," and region-specific terms like "tropical soil climate" or "arid soil erosion," combined with Boolean operators (AND/OR/NOT). The date range covered 2000–25 to capture recent empirical studies, models (e.g., CMIP5/6), and projections, prioritizing high-impact journals (IF > 4) and peer-reviewed articles (Eyring et al., 2016; O'Neill et al., 2016).

Inclusion and Exclusion Criteria. Inclusion required empirical data (field experiments, time-series, meta-analyses), modeling outputs, or case studies directly linking climate variables (temperature rise, precipitation shifts, extremes) to soil properties (moisture, erosion, SOC, nutrients, microbes). Excluded were non-peer-reviewed sources, pre-2000 studies without longitudinal relevance, and unrelated topics (e.g., non-soil climate effects). Approximately 500 initial records yielded ~150 key references after duplicate removal and title/abstract screening (Moher et al., 2009).

Handling Conflicting Evidence. Conflicting findings (e.g., SOC increase vs. loss under warming) were weighed by study quality (sample size, methodology rigor, regional replication), recency, and consensus from meta-analyses (e.g., IPCC-aligned reports). Preference given to multi-biome studies resolving discrepancies via biome-specific factors like moisture offsets or erosion rates; unresolved conflicts noted in discussions (e.g., boreal vs. tropical Rs trends) (Crowther et al., 2016; Van Gestel et al., 2018).

The Soil Nexus. https://doi.org/10.1016/B978-0-443-45286-4.00001-8

2. Introduction to climate change and soil health

Climate change has emerged as one of the most pressing issues of the 21st century, exerting profound socio-economic consequences and posing significant threats to global food security (Ullah et al., 2021). While numerous studies have documented the effects of climate change on water, energy, agriculture, and land use, comparatively little attention has been given to soil quality and soil health, fundamental elements for food production and long-term ecosystem resilience (Lal et al., 2011). This chapter aims to systematically synthesize the direct and indirect impacts of climate change on soil properties, processes, and health across diverse biomes, while evaluating adaptive strategies like regenerative agriculture and soil monitoring technologies. Its primary contribution fills a critical literature gap by integrating quantitative empirical evidence—from time-series data and model projections—with region-specific case studies, providing actionable insights for policy-aligned soil management to safeguard food security and ecosystem resilience under future scenarios. Soils are complex systems comprising minerals, organic matter, and living organisms. The mineral fraction includes sand, silt, and clay particles that form the soil skeleton, providing the physical matrix for plant growth and governing essential functions such as water- and nutrient-holding capacity, porosity, stability, and aeration. Organic matter encompasses the living and decomposing remains of plants and animals, acting as a platform for chemical and biological reactions. It serves as a source of essential nutrients and energy, binds soil particles, regulates water-holding capacity, pH, and aeration, and promotes microbial diversity. Soil microorganisms, including bacteria, fungi, archaea, and microarthropods, decompose organic matter into simpler compounds, releasing the nutrients required by plants (Gelybó et al., 2018; Jeyanthi & Kanimozhi, 2018).

Climate change affects several soil properties, resulting in degradation across various environments. Observations and models demonstrate that shifts in temperature and variability in precipitation regulate moisture supply, significantly influencing biochemical and physical properties of soils and the functioning of associated ecosystems (Bannari & Al-Ali, 2020). In semi-arid regions, climate change may exacerbate soil salinity, whereas in humid environments, it may lead to increased drying and enhanced erosion (Corwin, 2021). The dynamics of soil organic carbon and organic matter in soil aggregates are closely linked to these erosion processes, with changes in particle fractions causing measurable differences in soil composition and fertility. Soil acidification, another phenomenon driven by elevated temperatures and atmospheric carbon dioxide concentrations, may be mitigated by extended periods of drought. Projected regional climate characteristics suggest potential alterations in the stability and productivity of agricultural soils, which could compromise food security and threaten dams and reservoirs. Soil-carbon sequestration emerges as a promising strategy for both mitigating climate change and enhancing food security, although land-use patterns modulate the associated risk of erosion (Lal et al., 2011). Paleosols represent a valuable archive of past-paleoclimatic data, providing insights into climate-soil interactions. Extreme weather events further threaten soil stability and productivity. Addressing these multifaceted impacts necessitates the development of a coordinated scientific and technical research strategy to sustain soil health and associated ecosystem services in a changing climate.

Soil constitutes the outermost layer of the Earth's crust. Soils form through the interaction of bedrock material, minerals, organic compounds, and air and water under the influence of the surrounding climatic zone (Gelybó et al., 2018). The weathering of larger mineral material gives rise to the smaller mineral fraction of the soil. Soil development is a long-term process that occurs over thousands of years, as mineral weathering in temperate regions takes place at approximately 30 mm per 1000 years. The mineral fraction of the soil includes clay, silt, and sand and is a carrier of essential cationic nutrients such as Ca+2, Mg+2, and K+ as well as microelements such as Fe, Mn, Cu, and Zn (Fay et al., 2015). Soils with coarser textures have lower moisture retention, lower microbial activity, and decreased nutrient cycling and retention. Mineral particles are also the source of these preferred micronutrients, Si and Mn, for plants. Clay minerals, such as illite and vermiculite, which are considered inherent components of soil, contribute to resource availability through processes like nutrient sorption and desorption. Clay material also controls the supply of important trace elements, directly impacts enzymatic activity and plant growth, aids in preventing moisture loss, and facilitates the reduction of nutrient losses. The chemical composition of soil minerals can also be altered by soil warming, as the increase in weathering rates can lead to accelerated nutrient release (Heckman & Rasmussen, 2018).

Soil organic matter content represents a key determinant of heat transfer capabilities and, when combined with other factors, contributes to an understanding of how soils respond to climate change. Climate change can influence the soil water balance by altering the equilibrium between precipitation and potential evapotranspiration, while temperature fluctuations also lead to the redistribution of soil moisture. These modifications to fundamental hydrological variables have the potential to modify biogeochemical processes within soils. Percolating water can infiltrate deeper into porous soils during the wet season, subsequently replenishing deeper groundwater tables, increasing vertical runoff, and evoking faster erosion if nutrients are not retained. These processes can be magnified by warming temperatures, especially in regions where movement away from ice-covered states towards seasonal water flux is prominent (Zhu et al., 2019).

Furthermore, heightened frequencies of heavy precipitation can exacerbate erosive processes across much of Europe. Soil microorganisms play a critical role in maintaining soil function and fertility. These microorganisms can not only sustain themselves without re-inoculation but also produce essential plant growth hormones, decompose relatively resistant organic material, such as lignin, and suppress soil-born plant diseases (Priyadarshini et al., 2025).

2.1 Mineral components

Steady-state weathering of minerals constitutes the basis for soil formation, supplying critically important nutrients such as potassium (K), calcium (Ca), and magnesium (Mg). Among the four major soil constituents, mineral components, organic matter, soil microorganisms, and soil fauna, mineral components form the largest fraction (Gelybó et al., 2018). Soil formation progresses gradually through the chemodeterioration of primary minerals; consequently, the mineral composition reflects the bedrock type. The mineral composition controls the soil's physicochemical properties and undergoes continuous changes during soil development. Weathering rates increase with rising temperature and precipitation. Elevated rainfall events (>25–45 mm single storms) and warming (+1–3°C) increase leaching rates under low evapotranspiration deficits in semiarid environments (mean annual precipitation 250–500 mm, temperatures 15–25°C), dissolving fine clays (<2 μm) and reducing surface horizon clay content by 5%–15% while accumulating them in subsoils via illuviation (Heckman & Rasmussen, 2018). When percolating water exceeds 300 mm per year during wet phases, this process intensifies in soils with smectitic clays and moderate base saturation, encouraging the formation of argillic horizons.

2.2 Organic matter

The mineral material gives the physical structure of soil, while organic matter (SOM) makes it a living substrate and decomposes slowly via microbes and meso-fauna to persist and accumulate (Melillo et al., 2017). Nutrient cycling drives fertility through its interaction with minerals. Different fractions have different properties: labile (sugars, amino acids; residence time days–months), intermediate plant-derived materials such as lignin-rich compounds that last between 5 to 33 years and stable mineral-associated organic matter (MAOM), which lasts for centuries. This information comes from field studies. In Swiss sub-alpine grasslands at altitudes ranging from 1005 to 3000 meters above sea level under long-term management, particulate organic matter (POM) turned over within a period ranging between 5to 20 years while MAOM has been known to persist for more than 100 years and constitutes 60% – 80 %of total SOC pools as reported by Budge et al. (2011). Ramped thermal analysis helps separate labile fractions RT 150–350°C MRT weeks-years from recalcitrant RT $\geq$ 450°C MRT decades showing turnover based on activation energy (Sanderman & Grandy, 2020). The resistance of lignin to oxidation enriches MAOM by 20%–50% in Fe-rich tropical soils under redox fluctuations, as demonstrated by meadow microcosm incubations where retention was higher for lignin-C at 48% compared to bulk litter-C, which was only 31% Huang et al. (2019). Climatic factors—temperature, moisture, oxygen—control the transformation of SOM from plant inputs to minerals. Incubations over a range of elevations indicated that Q_{10} increased with moisture; decomposition was twice as fast in moist soils compared with dry ones; protection by topsoil aggregates reduced Q_{10} by 24%–37%, with microbes being dominant in labile pools (Huang et al., 2019). These climate-sensitive interactions control the SOM quantity, quality, and resilience of carbon cycling.

2.3 Soil microorganisms

The soil is a complex ecosystem, which hosts an estimated 25% of the planet's biodiversity (Guerra et al., 2021). Bacteria and fungi are the dominant components of the soil microbiome, conducting essential functions worldwide, including the decomposition of organic matter, nutrient cycling, and pathogenesis, which impact critical ecosystem processes and human well-being. Although research over recent decades has promoted a mechanistic understanding of their ecological preferences and biogeography, it is only recently that global investigations have been combined with mechanistic insights to generate global and national atlases for the soil microbiome.

3. Impacts of climate change on soil properties

Soil degradation stands among the primary climate change challenges. Accelerated temperature increases, altered rainfall patterns, and intensified erosion amplify soil deterioration (Trnka et al., 2013). The soil climate, integrating soil temperature and hydric regimes, defines a region's potential soil quality and, by extension, the site-specific suitability for sustainable agricultural production. Climate change forecasts indicate that an adverse soil climate will dominate throughout much of the central plains of the United States and Central Europe during summer. The likelihood of extreme soil climate conditions in the central plains may increase from 30% to 40% in the current climate to 60%–70% by 2070. Similarly, in Central Europe, extreme soil climate conditions during the summer could become the prevailing average after 2050. The frequency of extremely dry soil conditions is also expected to increase substantially, heightening the risk of soil erosion and reducing overall productivity. The importance of soil carbon cannot be overstated. Increased soil respiration positively correlates directly with temperature rises, which diminishes soil organic carbon content and, consequently, hampers carbon fixation (Gelybó et al., 2018). Climate warming influences soil-forming factors by affecting mineral weathering, plant growth, and microbial activity, potentially transforming soil organic carbon along with total carbon and nitrogen stocks into an emissions source, thereby intensifying the greenhouse effect.

3.1 Soil temperature changes

Climate change can significantly impact soil properties through changes in soil temperature, moisture, acidity, organic carbon concentration, and microbial activity; these impacts show notable regional variation (Gelybó et al., 2018). Although soil temperature is influenced largely by human-induced climate forces and atmospheric interactions, the soil–atmosphere relationship itself remains incompletely understood. Available global reconstructions have traditionally relied on models or sparse in situ measurements.

Air temperature and soil temperature should not be considered directly comparable, especially within the coldest and cold biomes. While seasonal air and soil temperatures exhibit a strong correlation with no significant spatial variations, the mean annual air temperature is a poor approximation of the mean annual soil temperature due to additional factors that mediate the signal from air to soil. Furthermore, biome-specific differences exist in mean annual temperature offset, with the largest discrepancies and the lowest spatial variation observed in boreal and polar regions.

The incident energy penetrating the soil is partially offset by biotic factors (e.g., snow cover and canopy interception), which, in turn, are influenced by the soil temperature itself. This generates complex feedback loops with the capacity to alter soil temperature regimes in response to climate warming. The spatial variation of this offset is therefore strongly coupled with multiple processes, such as plant phenology, ice cover, and soil moisture. CMIP5 model ensembles project that deep soils (0.5–3.2 m depth) in boreal and tundra regions will warm 1.5–3.5 times faster than near-surface air temperatures by 2100 under the RCP8.5 scenario, with global mean soil warming reaching 4.5 ± 1.1°C at 1 m depth compared to lower air warming rates (Trnka et al., 2013). This counterintuitive rapid deep soil warming reconciles with insulating effects (e.g., snow cover and organic layers causing offsets up to -8.7°C in tundra) because phase lags and thermal diffusion in permafrost lead to amplified responses at depth once insulation diminishes under sustained warming, as offsets shrink and heat penetrates faster in models. Biome-specific offsets currently buffer shallow warming (soil lags air by 20%–50%), but projections indicate disproportionate vulnerability in cold biomes as these effects weaken (Maclean & Early, 2023). Biome-specific offsets ($\Delta T_{surf} =$ soil-air) buffer shallow warming and reduce spatial variability by up to 50% in cold regions. They average -3.9°C globally, but they can reach -8.7°C in tundra and -6.3°C in boreal forests (De Lombaerde et al., 2022). These biggest offsets, which are caused by organic layers and snow insulation, slow down air warming rates in cold biomes (soil lags 20%–50% behind), having a significant impact on subterranean processes like carbon decomposition and root growth. These results indicate that cold soils will warm at a slower rate than the macroclimate to which species are currently adapted as offsets shrink. Consequently, boreal and polar systems, which have received little scientific attention in the context of future soil warming, could be disproportionately vulnerable to climate change. In contrast, soil temperatures in hot and dry biomes could become increasingly warmer than those experienced by the air, a tendency amplified at higher (air) temperatures. Changes in precipitation and soil moisture have a significant impact on the relationship between air and soil temperatures, with reductions in precipitation potentially causing soil temperatures to warm more rapidly than the macroclimate.

Dedicated observations and global datasets of recorded soil temperatures are imperative for advancing knowledge in this field. Soil temperature data also enable an evaluation of the mechanisms shaping the relationships between air and soil temperatures, an analysis currently hampered by the coarse spatial resolution of existing datasets. An enhanced understanding of these mechanisms would aid in the study of belowground ecological processes influenced by soil temperature, which remain understudied due to the inadequate testing of the common assumption that equates soil and air temperature. The addition of soil temperature data, alongside air temperature and precipitation, would facilitate investigations of soil-climate warming scenarios under future socio-economic pathways, capturing the complex interplay among changing variables and ecosystem responses.

3.2 Moisture variability

Shifting rainfall patterns, including variations in the timing and intensity of precipitation, are already evident in many regions and are projected to intensify during this century, posing significant risks to soil health. In temperate zones, increased rainfall extremes will elevate the likelihood of soil-erosion events, while non-monsoonal deserts are expected to experience enhanced water deficits (Trnka et al., 2013). Moreover, the cyclical nature of wet and dry phases has amplified in certain continental interiors, potentially affecting the stability and availability of soil moisture reserves. These trends exacerbate concerns regarding land degradation and jeopardize soil productivity, emphasizing the need for effective strategies to counteract the loss of ecosystem services and biodiversity. Fig. 19.1 illustrates climate change effects on soil systems, showing elevated temperatures accelerating soil respiration (CO_2 emissions), shifting precipitation patterns causing drought/flood cycles, and extreme events driving erosion and wildfires. The figure depicts soil profile degradation (organic matter loss, nutrient leaching, microbial decline) with regional vulnerability mapping for tropical (accelerated nutrient cycling), temperate (soil drying), and arid zones (enhanced erosion).

3.3 Soil erosion and degradation

Soil erosion is one of the main threats driving soil degradation globally, affecting crop yields, soil biota, biogeochemical cycles, and human nutrition (Guerra et al., 2020). It also impacts soil-water interactions, organic matter content, surface soil organic carbon stocks, atmospheric dust input, and can even alter nutrient stoichiometry in arbuscular mycorrhizal fungi in severely affected systems. The increasing frequency and intensity of precipitation, especially during the monsoon, systematically contribute to soil degradation, leading to the loss of vegetative cover and topsoil from both cultivated and barren lands. Most of the world's soil resources are in a fair, poor, or very poor condition. The principal threats to soil functioning at the global scale are erosion, loss of soil organic carbon, and nutrient imbalance (Montanarella et al., 2016). Climate change is one of the drivers affecting land use, land cover change, and management, and therefore further contributes to the alteration of the functions

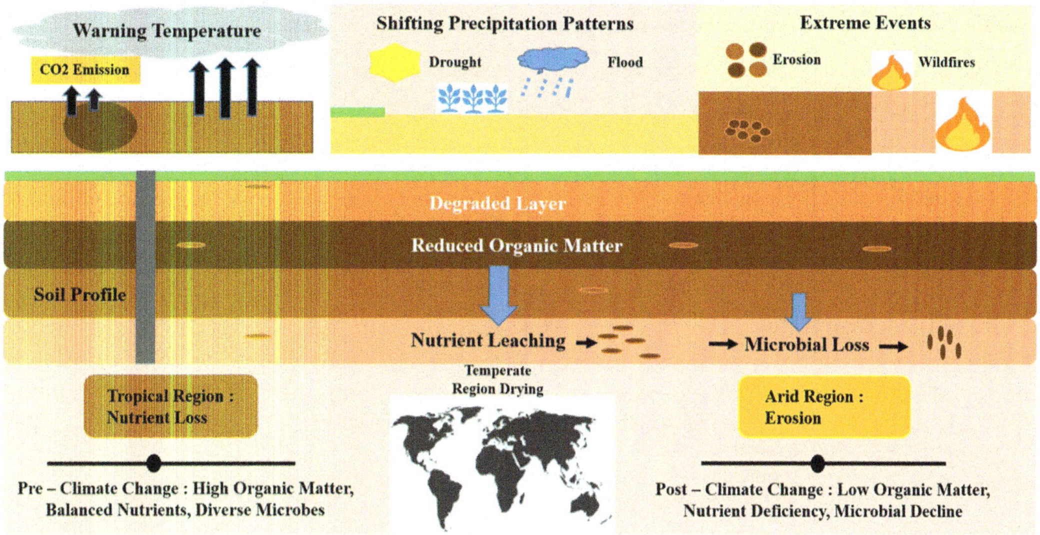

FIGURE 19.1

Climate change impacts on soil properties and regional vulnerabilities.

that soil can provide. Changes in precipitation regimes directly impact water availability, evaporation, groundwater recharge, and runoff. The global soil erosion rate is already 20–100 times higher than the soil formation rate, rendering soil erosion an unsustainable process that contributes to annual economic losses in the tens of billions of U.S. dollars. Rising intensities and frequencies of rainfall events under climate change will intensify the extent and severity of soil erosion, thus accelerating soil degradation (Guerra et al., 2020). Table 19.1 summarizes the primary effects of climate change on soil temperature, moisture, organic carbon, and other essential properties across global regions. It highlights biome-specific vulnerabilities and quantifies risks like erosion rates 20–100 times higher than soil formation, drawing from the document's analysis of regional forecasts.

4. Soil fertility and nutrient dynamics

Soil fertility cycles essential biochemical elements and compounds vital for living organisms. These nutrients, nitrogen (N), phosphorus (P), potassium (K), sulfur (S), calcium (Ca), and magnesium (Mg), derive primarily from rock weathering and organic matter cycling (Al-Shammary et al., 2024). Their availability hinges on an equilibrium between losses and cycling rates, which climate change significantly disrupts. Extreme weather events, such as droughts, heatwaves, floods, and wildfires, also affect nutrient dynamics and the emissions of climate-relevant trace gases (Montanarella et al., 2016).

4.1 Essential nutrients for plant growth

Soil fertility refers to the ability of soil to supply essential nutrients for the growth of a wide range of plants. Essential nutrients required by plants are carbon (C), hydrogen (H), oxygen (O), nitrogen (N), phosphorus (P), potassium (K), calcium (Ca), magnesium (Mg), sulfur (S), iron (Fe), manganese (Mn),

Table 19.1 Key impacts of climate change on critical soil properties.

Soil property	Climate change impact	Regional effects	Key references
Soil temperature	Accelerated warming; biome-specific offsets (slower in cold biomes)	Boreal/polar vulnerable; hot/dry warmer than air	Gelybó et al. (2018)
Soil moisture	Shifting rainfall; extremes and deficits	Temperate: Erosion risk; deserts: Water deficits	Trnka et al. (2013)
Soil organic carbon	Increased respiration reduces content; emissions source	Central U.S./Europe: Adverse summer conditions by 2070 (60%–70% extreme soil climate risk)	Gelybó et al. (2018)
Erosion risk	20-100x higher than formation rate (global erosion ~35.9 Pg yr^{-1}; formation ~0.01–2.45 t ha^{-1} yr^{-1})	Global: Tens of billions USD losses annually	Guerra et al. (2020)
Nutrient cycling	Disrupted by extremes, leaching	Tropical: accelerated losses; semi-arid: degradation	Montanarella et al. (2016)
Microbial activity	Altered composition; reduced by drought	Global atlases emerging	Guerra et al. (2021)

copper (Cu), zinc (Zn), molybdenum (Mo), chlorine (Cl) and boron (B). Irrigated and fertilized cropping systems are highly dependent on the supply of only the major nutrients N, P and K during the entire crop growth. Most soils contain sufficient quantities of calcium, magnesium, iron, and other micronutrients for crop growth. Deficiencies of calcium, magnesium, and sulfur are reported only in highly weathered tropical soils that are highly acidic. However, continuous removal of grains and straw causes a deficiency of calcium and magnesium in dryland and irrigated cultivation.

Carbon and oxygen are supplied through air, and hydrogen is supplied through water. Nitrogen, phosphorus, and potassium are primary nutrients that should be available in sufficient quantities for the completion of the plant life cycle. Calcium, magnesium, and sulfur are secondary nutrients and do not need to be supplied to crops as fertilizers. Micronutrients such as iron, manganese, copper, zinc, molybdenum, chlorine and boron are required in very small quantities. Elevated atmospheric CO2 generally enhances plant growth and photosynthesis, often improving water-use efficiency. In simultaneous exposure to other climate change factors, the interaction is variable and strongly species-specific. However, overall, it can be predicted that climate change will reduce grain quality in relation to mineral nutrient concentrations and grain nitrogen concentration. Owing to specific interactions with the mechanisms of plant C/N metabolism and mineral uptake and transport, Fe and Zn uptake are also frequently reduced under elevated CO2.

4.2 Nutrient cycling in changing climates

Nutrient cycling underpins the productivity of both natural and managed ecosystems, yet climate change threatens the availability of soil nutrients and the global food supply that relies on these stocks.

Most soils have about 1000–2000 kg N ha^{-1} as the main nutrient in surplus of crop needs, but only 1%–5% of this N is plant-available at any one time, with variation due to soil type and climate. In temperate soils high in clay, mineralizable N is on average 20–50 kg/ha year (1%–3%), while sandy tropical soils display lower fractions (~0.5%–2%) because of quick leaching and mineralization under heat. Climate change worsens this susceptibility by changing microbial processes; therefore, it reveals a big danger to the global food system: a large stockpile of non-renewable nutrients mostly out of farmers' control against increasing supply uncertainties once thought impossible (Al-Shammary et al., 2024). Maintaining nutrient availability during increasingly extreme weather and variable drought/flood cycles is a key requirement for ensuring a sustainable agricultural supply. Nutrient cycling also links natural and managed resources. If either is lost, the other becomes a transient source; a key reason why the resilience of more-productive agricultural systems depends heavily on the long-term stability of lower-fertility natural systems.

4.3 Impact of extreme weather events

Extreme weather events can imperil resilient farming systems. During the 2011 drought, high-input maize farms in South Africa experienced yield losses of 32%–78%, whereas hybrid maize with optimum inputs in the United States suffered losses of 25%–33% (Brown & Funk, 2008). Lower rainfall in Malawi and Zambia over the past 2 decades has contributed to a decline in maize yield of up to 30%, exacerbating existing problems in food production (Brown & Funk, 2008). Shifting seasonal distributions of precipitation, combined with an increasing frequency and severity of drought and flooding events, exacerbate vulnerabilities to water erosion and nutrient depletion on arable land, thereby diminishing farm productivity and reducing the potential for intensification (Borrelli et al., 2020). The same holds for tropical forests. Under severe drought conditions, Asian forests typically exhibit a decline in productivity. Increased fire, wind storms, and the recruitment of drier forest types are elements of a shortening fire rotation cycle linked to tropical drought and associated with shifting annual precipitation patterns.

5. Case studies: Soil health in different regions

Numerous regions, including the Mediterranean Basin, Central Europe, the central United States, and semi-arid and tropical zones, provide case-study examples of changing soil dynamics under climate stress. The Mediterranean Basin exemplifies how warmer temperatures, shifting precipitation patterns, and recurrent droughts are degrading organic carbon content, nutrients, and clay in soils, which contrasts with many other areas of Western Europe (Trnka et al., 2013). In Central Europe and the central United States, rising temperatures and reduced summer precipitation are also degrading soil properties and increasing climatological dryness, drought frequency, and the likelihood of heatwaves. In semi-arid areas, water-deficit conditions have led to increased soil erosion, accelerated loss of organic matter, and a reduction in available nutrients (Borrelli et al., 2020). In tropical areas, sustained precipitation and temperature increases have accelerated nutrient cycling, leading to significant losses from the soil and a subsequent decline in the overall pool of available nutrients (Cleveland & Townsend, 2006).

5.1 Tropical regions

Climatic fluctuations and their consequences pose a formidable challenge to the sustainability of tropical ecosystems and the survival of human populations that are closely dependent on a narrow set of animals, crops, and trees. Climate stressors, including drought, heat stress, altered rainfall patterns, and increased pest and disease outbreaks, can substantially impact the productivity of both commodity and staple food crops. The resulting changes in overall agricultural productivity not only bear direct implications for human nutrition but also indirectly influence households that rely on agricultural systems as sources of income and employment (Brown & Funk, 2008). Africa, with its rain-fed agricultural traditions, already faces threats to regional monsoon patterns stemming from ENSO phenomena and modifications to oceanic currents (Brown & Funk, 2008). The low levels of agricultural productivity characteristic of many food-insecure African countries present a significant opportunity to transform these systems through improved seeds, fertilizers, land-use strategies, and governance structures, intending to attain food security for all (Brown & Funk, 2008).

5.2 Temperate regions

The decrease in precipitation by the end of the 21st century is expected to substantially impact soil productivity in temperate regions (Trnka et al., 2013). Variations in temperature and rainfall influence plant physiological processes, reducing yields by approximately 5%–20% in much of central Europe (Brown & Funk, 2008). Predictions suggest that in much of Europe, warmer temperatures, reduced winter precipitation, and increased evapotranspiration will accelerate soil drying and nutrient removal by up to 50% by 2080. Soils of northern and central Europe will be affected by warmer temperatures, reduced winter precipitation, and accelerated evapotranspiration (Trnka et al., 2013).

5.3 Arid and semi-arid regions

Soil-climate analysis indicates that the soil climate of arid and semi-arid areas will change substantially by reducing cold and moist conditions toward drier and hotter conditions (Trnka et al., 2013). Changes to near-surface soil moisture are more marked than those to temperature or precipitation, as it significantly impacts soil genesis by increasing the probability of dry years intensifying soil erosion and decreasing productivity (Borrelli et al., 2020). Global models predict a +17.3% increase in soil erosion rates in semi-arid zones due to climate change, with rangelands being susceptible up to 4.25 t ha^{-1} yr^{-1} under RCP2.6 by 2070. In Morocco's High Atlas semi-arid watersheds, 10%–15% declines in rainfall could increase sediment yields by 5%–10%, thereby increasing nutrient losses (e.g., N/P via runoff) (Chaikin et al., 2022). These changes present major challenges for adaptation strategies while increasing pressure on the resource base. The projections for the Sahel suggest very large increases in drought events and rainfall variability under climate scenarios, which threaten further degradation of fertility and structure in conventional agriculture systems (Vicente-Serrano et al., 2022). Food security is threatened in eastern and southern Africa, where there are already major problems of access to food (Brown & Funk, 2008), but transforming agricultural systems through better seed, fertilizer, land use, and governance could be a pathway to resilience (Brown & Funk, 2008).

Table 19.2 presents case studies from tropical, temperate, arid, and other regions, detailing climate stressors, soil degradation impacts, and productivity consequences such as 30%–78% maize yield losses. It illustrates how precipitation shifts and warming exacerbate erosion and nutrient loss differently across ecosystems. These examples demonstrate the global scope of soil challenges, emphasizing food security threats in Africa and Europe while identifying adaptation opportunities through regional management.

6. Agricultural practices and soil management

Agricultural practices and soil management that prioritize soil conservation and regenerative agriculture are crucial pathways for enhancing ecosystem resilience and food security in a changing climate. The types of agricultural practices and management are crucial determinants of maintaining soil carbon and soil structure, as well as soil fertility and the flows of other ecosystem services associated with soil (Loboguerrero et al., 2019). Conventional agricultural strategies designed to increase crop production are typically aimed at reducing the soil's susceptibility to erosion, thereby maintaining long-term agricultural productivity. Conservation agriculture involves maintaining a permanent organic soil cover, minimizing soil disturbance, and diversifying crops through rotations and cover crops. Crop residues contribute to soil organic carbon, and the roots of rotations and cover crops affect the soil microbial biodiversity. Fertilisers play an important role in enhancing the rate of incorporation of crop residue into soil organic matter and so contribute to its formation. Conventional

Table 19.2 Regional vulnerabilities of soil health to climate stress.

Region	Key climate stressors	Soil impacts	Productivity consequences	References
Tropical	Heat, altered rain, pests	Accelerated nutrient cycling/ losses	30% maize yield decline (Malawi/ Zambia); up to 32%–41% losses	Trnka et al. (2013)
Temperate (Central Europe)	Reduced precipitation, heatwaves	Soil drying, nutrient removal (up to 50% by 2080)	5%–20% yield reductions	Trnka et al. (2013)
Arid/Semi-arid (Sahel)	Drought variability, hotter climates	Erosion intensification (17.3% increase projected); fertility loss	Food security threats	Brown and Funk (2008)
Mediterranean Basin	Recurrent droughts, warming	Degraded organic C, nutrients, clay	Contrasts Western Europe's stability	Trnka et al. (2013)
Central U.S. Plains	Extreme soil climate	60%–70% risk by 2070; dry conditions, erosion	Reduced suitability	Trnka et al. (2013)
Southern Africa	2011 drought	32%–78% maize losses	Exacerbated food insecurity	Brown and Funk (2008)

agriculture is effective in many situations, but the administration of appropriate nutrients needs to be holistic, and conserving soil and soil water also needs to be considered, as they encourage healthy root growth and the development of soil microbial biodiversity. The success of green revolution-type strategies for increasing productivity will be limited where irrigation water, soil fertility, and soil and water conservation measures are unavailable or limited.

6.1 Conventional versus sustainable practices

Agricultural practices play a critical role in determining soil health status at regional to global scales. The majority of agricultural systems worldwide rely on conventional tillage methods, resulting in well-known negative effects, such as increased erosion and degradation of soil quality in rain-fed farmland. These effects occur in both managed and unmanaged areas. Coupled with the expansion of human populations, current agricultural techniques will not sustain the necessary global food supply in the long term. Sustainable agricultural practices have the potential to counterbalance some of these negative impacts and produce positive effects on soil quality (Basch et al., 2012).

Agricultural practices have a profound influence on soil quality by altering its physical, chemical, and biological properties. Conventional systems with intensive tillage and monocultures tend to degrade soil quality, whereas agroforestry and sustainable practices enhance it by reducing erosion rates, improving physical properties and nutrient cycling, and increasing beneficial biota such as microorganisms (Basch et al., 2012). Conservation Agriculture, a global initiative consisting of no-till seeding, maintaining soil mulch, and crop diversification, has been successfully implemented worldwide for over 15 years. It has substantially improved soil health status and fertility levels while also supporting carbon sequestration, nutrient and water use efficiency, and profitability. Shifting from the business-as-usual model of intensive tillage to this agroecological approach is essential for sustaining both human well-being and economic welfare by safeguarding soil quality and fertility.

6.2 Soil conservation techniques

Soil conservation practices can maintain and improve soil health, particularly soil fertility, under a changing climate for sustainable food production and ecosystem services (Blanchy et al., 2023). Conventional soil management tends to degrade soil health because tillage erodes soil organic carbon and disrupts soil structure (Lal, 2015). Overfertilization can lead to acidification, nutrient imbalance, and environmental pollution (Guo et al., 2010). The removal of crop biomass and residues decreases organic matter input, resulting in erosion and nutrient loss (Turmel et al., 2015). Conservation agriculture can protect soil health and facilitate climate change adaptation through minimal disturbance, soil-covering crops, and crop rotation (Hobbs et al., 2008). Minimum tillage avoids organic matter decomposition and soil erosion (Montgomery, 2007). Soil-covering crops prevent eolian and water erosion, while also enhancing soil biological activity (Kaye & Quemada, 2017). Crop rotation controls pests and consequently reduces agrochemical use (Dias et al., 2015). Nutrients removed by harvests can be compensated for by organic amendments, such as manure, compost, and biochar, as well as fertilizers (Lehmann et al., 2011). The addition of organic matter improves soil structure and increases bioturbation by earthworms, both of which contribute to enhanced water retention and increased nutrient storage (Blanchy et al., 2023). Regenerative agriculture is also important for soil conservation and improvement because this system combines several conservation practices, accounting for environmental, social, and economic sustainability (Schreefel et al., 2020). It can potentially sequester 2 billion tons of soil carbon annually, primarily in conventional agricultural lands, by improving organic matter, nutrient availability, and biological functioning (Zomer et al., 2017). No-till with cover crops improved nutrient retention by 15%–25%, stabilized yields 10%–40% above conventional baselines, and increased soil carbon by 0.2–0.6 Mg C ha^{-1} yr^{-1} in Midwest U.S. corn-soy systems during drought conditions. These methods improved N retention by 20%–40% and increased yield by 5%–15% in semi-arid Australian wheat fields, accumulating 0.4–1.2 Mg C ha^{-1} yr^{-1} (Farmaha et al., 2022; Jacobos et al., 2022). These estimates, such as the potential for regenerative agriculture to sequester up to 2 Gt C yr^{-1} globally and yield stabilization ranging from 10% to 40%, are derived from meta-analyses and long-term field trials (Giller et al., 2021; Schreefel et al., 2020; Zomer et al., 2017). The sequestration figure assumes widespread adoption (e.g., >50% of cropland) of practices like no-till, cover cropping, and organic amendments on arable soils, primarily in temperate and tropical climates, over short-to medium-term horizons (10–50 years); however, saturation limits long-term potential, with feasible rates often 0.2–1.2 Mg C ha^{-1} yr^{-1} and high uncertainty (±30%–50%) due to soil type variability and management intensity (Schreefel et al., 2020; Smith et al., 2019; Zomer et al., 2017). Yield stabilization benefits are most pronounced in rainfed systems under drought-prone conditions (e.g., semi-arid to temperate regions), relying on medium-high intensity practices like cover crops and reduced tillage, with gains emerging after 3–7 year transitions but varying 5%–40% based on crop (e.g., maize/soy) and extremes; conventional systems may drop to 30%–70% of baseline yields in droughts, while regenerative ones retain 95%+ (Brown & Funk, 2008; Hobbs et al., 2008; Kaye & Quemada, 2017). Table 19.3 compares conventional versus sustainable practices like conservation agriculture, minimum tillage, and regenerative systems, outlining mechanisms, climate benefits, and challenges such as equipment costs. It quantifies potential carbon sequestration at 2 billion tons annually from organic amendments.

6.3 Regenerative agriculture

Regenerative Agriculture prioritizes revitalizing degraded and depleted soils as a high-value ecosystem service while producing nutritious food. The approach aims to rebuild soil organic matter and biodiversity, thereby restoring soil health and contributing to both carbon sequestration and climate resilience. Practices such as cover cropping, no- or reduced tillage, planned grazing, crop rotation, composting, and the avoidance of synthetic agrochemicals underpin this philosophy (Giller et al., 2021). These methods mostly predate the formal concept of Regenerative Agriculture, indicating long-standing recognition of their benefits.

In the shorter term, the yield benefits of increased soil organic matter (SOM) under Regenerative Agriculture are inconclusive (Oldfield et al., 2019). Greater SOM generally confers resistance to abiotic stressors primarily through improvements in soil physical properties, whose influence on crop performance varies (Williams et al., 2016). Both direct and indirect climate benefits accrue from increased SOM, yet the net impact and extent of such benefits remain uncertain (Powlson et al., 2011).

Regenerative Agriculture has been proposed as a pathway toward "zero carbon farming" or as a means of offsetting greenhouse gas (GHG) emissions with negative emissions elsewhere (Minasny et al., 2017). While certain systems can sequester carbon and decrease emissions relative to Alternative Peasant Agriculture, quantitative evaluations focusing on climate mitigation are limited. For instance, in Midwest U.S. corn-soy rotations, cover cropping and no-till boosted soil organic matter (SOM) by

Table 19.3 Comparative effectiveness of soil conservation practices.

Practice	Mechanism/ Benefit	Climate adaptation role	Limitations/ Challenges	References
Minimum tillage	Preserves structure/organic matter	Avoids decomposition losses	Equipment costs	Montgomery (2007)
Cover crops	Prevents erosion, enhances biology	Water retention, nutrient storage (15%–25% better retention)	Competition with cash crops	Kaye and Quemada (2017)
Crop rotation	Pest control, biodiversity	Reduces agrochemicals	Planning complexity	Dias et al. (2015)
Organic amendments (biochar/manure)	Improves structure, bioturbation	C Sequestration (~2 Gt yr^{-1} potential globally)	Nutrient loss risks	Lehmann et al. (2011)
Regenerative agriculture	Combines practices for SOM/ biodiversity	Climate resilience; 0.2–1.2 Mg C ha^{-1} yr^{-1} gain	Short-term yield uncertainty	Giller et al. (2021)
Precision nutrient Mgmt.	Holistic fertilizer use	Counters overfertilization/ acidification	Data/technology needs	Guo et al. (2010)
Agroforestry/No-till	Erosion reduction, nutrient cycling	Enhances properties; 10%–40% yield stability	Initial transition costs	Schreefel et al. (2020)

0.2–0.6 Mg C ha^{-1} yr^{-1}, cutting N_2O emissions 30%–50% via enhanced denitrification while improving N retention by 15%–25% (Rising et al., 2022; Zhou et al., 2022). In Australian wheat systems, these practices accumulated 0.4–1.2 Mg C ha^{-1} annually alongside 20%–40% GHG reductions, underscoring nutrient synergies (Singh et al., 2024). A study of a 6-year maize–wheat rotation in North–Central China found that soil carbon levels tracked increases in crop yields, accumulating with enhanced net primary production. Nonetheless, the increasing footprint of N fertilizer and irrigation somewhat counteracted the associated climate advantages (Zhang et al., 2013).

Conventional agriculture in a developed region, such as the United Kingdom, exhibits potential for Operational Reform to deliver climate benefits; however, GHG emissions sourced from agricultural practices, including fertilizer application, livestock, and land-use change, typically exceed the maximum carbon sequestered by soil (Powlson et al., 2014). Reformative and Conservative Agriculture, which largely exclude GHG-intense practices and external nutrient inputs such as mineral fertilizers, embodies principles aligned with Regenerative Agriculture. Organic fertilizers like manure, whilst capable of elevating SOM and yield, are subject to nutrient losses that may partly offset mitigation gains (Smith et al., 2019). The elemental composition of SOM is confined within a relatively narrow range of C: N:P:S ratios (Kirkby et al., 2013). In situations where nutrients are limiting, such as nitrogen, phosphorus, or sulfur, added carbon may not be stabilized and instead is released as CO_2 (Richardson et al., 2014). The advantage of sequestered nutrients within soil reservoirs as opposed to their immediate availability for crop uptake continues to be a relevant consideration.

The ambition to reverse biodiversity decline constitutes a fundamental objective of Regenerative Agriculture; yet, bibliometric and systematic reviews reveal its limited emphasis within the discourse (Giller et al., 2021; Schreefel et al., 2020). While fostering plant diversity and minimizing pesticide use are frequently cited, the promotion of integrated pest and disease management, crop rotations, cover crops, intercropping, and strip cropping, common elements of Reformative Agriculture, receives comparatively less mention.

6.4 Trade-offs and constraints of conservation and regenerative practices

While conservation agriculture (CA) and regenerative practices offer substantial benefits for soil health and climate resilience, they involve notable trade-offs and constraints that must be managed contextually. Key challenges include limited residue availability in low-biomass systems, elevated pest and disease pressures under no-till, potential increases in N_2O emissions, and higher upfront equipment costs (Pittelkow et al., 2015):

- **Residue availability in low-biomass systems**: In smallholder or dryland farming with yields below 4 t/ha, insufficient crop residues hinder mulch coverage, compromising erosion control and moisture retention; residues often compete with livestock feed needs, exacerbating shortages in mixed crop-livestock systems (Naudin et al., 2015).
- **Pest/disease pressure under no-till**: Surface residues can harbor pathogens (e.g., Fusarium, Septoria) and pests, increasing disease incidence in high-residue environments, particularly during wet conditions; initial years require intensified herbicide/insecticide use and rotations to mitigate (Nicolopoulou-Stamati et al., 2016).
- **N_2O emissions**: No-till with residues may elevate N_2O by 6%–26% short-term due to denitrification in wet soils or N immobilization, though long-term adoption (with rotations/cover crops) reduces emissions by up to 26% in dry climates; high N inputs amplify risks (Abdalla et al., 2016).
- **Equipment costs**: Specialized no-till drills/planters cost $5000-$150,000, posing barriers for smallholders; transition requires investment in precision tools, offset partially by fuel/labor savings long-term (Friedrich et al., 2012).

These practices underperform in cold/dry/acidic/compacted/low-SOM soils (deep tillage superior), humid high-N silt–clay loams (higher N_2O), early adoption phases (yield dips from N immobilization), and residue-poor systems without supplements (Pittelkow et al., 2015). Success demands site-specific adaptations like subsurface nutrient placement, integrated pest management, and hybrid conventional-regenerative approaches (Pittelkow et al., 2015).

7. The role of technology in soil monitoring

Monitoring soil health, the foundation of ecosystem resilience and food security, has been revolutionized by the application of digital technologies (Behrens et al., 2018). Advances in remote sensing enable large-scale assessments of soil moisture, erosion risk, and vegetation health (Mulder et al., 2011). In-situ technologies, such as ion-selective field effect transistors, electrochemical and electromagnetic sensors, and ground-penetrating radar, allow for continuous tracking of nutrient dynamics

(Obade & Gaya, 2021). Remote sensing platforms can monitor evapotranspiration, soil moisture deficiency, runoff, and rainfall at regional scales, thereby informing models for drought risk assessment and crop yield evaluations (Mulla, 2013). Forecasting systems in agriculture depend critically on data produced and aggregated by these remote and in situ monitoring technologies. Integrating digital soil physics with plant modeling activities allows the evaluation and continuous monitoring of indicator properties at the pedon and crop profile levels (Vereecken et al., 2016). Together, these technologies form powerful tools that strengthen analytical and soil management capabilities, thereby enhancing the resilience of agricultural systems to climate change.

The rapid pace of digital technological advances necessitates a corresponding transformation in educational curricula linked to soil physics and associated disciplines. Today's graduates require a new spectrum of skills ranging from computer programming to plant physiology and agronomy in addition to traditional soil physics expertise (Field et al., 2011). Innovative laboratory and field practicals, delivered through online learning platforms, are essential for cultivating graduates with the requisite hands-on experience in state-of-the-art technology systems and applications. To operationalize "soil monitoring technologies" at national scale, we envision an EO/ML pipeline integrated as a modular service layer inside the national soil information system (NSIS), rather than as a one-off map product (Mulla, 2013; Obade & Gaya, 2021). The NSIS would store (i) versioned EO-derived covariates and model artifacts, (ii) model outputs as 3D + T layers (space × depth × time) for priority indicators (e.g., SOC, N, P, K, pH), and (iii) pixel-level uncertainty plus "validity domain" flags derived from spatial cross-validation to explicitly manage spatial generalization limits during deployment (Smith et al., 2019; Vereecken et al., 2016). Because EO signals and management effects are time-varying, each prediction would be linked to an acquisition window (e.g., composited bare-soil periods and seasonal summaries) and would be continuously re-calibrated using time-stamped ground truth from a tiered observation network: routine laboratory sampling at sentinel sites, dense low-cost sensor clusters for high-frequency dynamics, and targeted campaigns triggered where the model indicates out-of-domain conditions or rising uncertainty (Obade & Gaya, 2021). This creates a closed loop—EO/ML provides scalable coverage, while NSIS governance ensures traceable updates (model version, training data period, sensor calibration metadata) and prevents over-interpretation by requiring that policy-facing outputs be accompanied by uncertainty thresholds and minimum ground-truth density criteria for reporting at each administrative level (Montanarella et al., 2016; Smith et al., 2019).

7.1 Remote sensing technologies

Climate change, through rising temperatures and erratic precipitation, is impacting soil fertility, carbon stocks, and ecosystem diversity on a global scale (Obade & Gaya, 2021). Soils maintain essential ecosystems that fix, filter, transform, and provide nutrients. They also act as the foundation for a diverse array of life-support systems. Assessing soil quality is crucial for monitoring ecosystem health; however, the significant variation in soil properties poses a challenge to the use of soil quality indices in evaluating the effects of external factors, such as land use (Bunemann et al., 2018).

Remote sensing technologies can contribute to digital soil information with broad spatial and temporal coverage. Capabilities are available to rapidly collect high-resolution data over large areas (Adams et al., 1993; Smith et al., 2019). These technologies have applications for resource management, characterization, conservation, and agricultural planning (Mulder et al., 2011). They also provide the basis for mapping small-scale spatial variability in soil information, which is required for precision agriculture (Mulla, 2013). This can inform the delineation of functional land management zones, defined as regions where activities are managed to meet specific functions, such as food production, water purification, or carbon storage (Schulte et al., 2014).

Status and changes in soil properties that control ecosystem functions are needed to monitor the impacts of climate change and functional land management. Digital soil mapping and geostatistical techniques can estimate variability in several key properties, including soil resilience to erosion, soil carbon sequestration potential, and soil pH, which influences nutrient availability (McBratney et al., 2003). Projected climate change over the next 50 years is likely to alter the spatial distribution of soil properties and the land's ability to deliver food and ecosystem services (Smith et al., 2019). Information on the changes in soil properties that control the land's ability to deliver functionality can support declarations under international frameworks, such as the United Nations Framework Convention on Climate Change (Minasny et al., 2017). Functional land management offers a pathways approach to regional-to-global governance, balancing land use through activities such as reduced tillage, afforestation, and nutrient application, all of which can be informed by the support of remote sensing technologies (Schulte et al., 2014).

7.2 Soil sensors and data collection

Many environmental challenges underscore the crucial role of soil ecosystems in sustaining life. Soil is fundamental for provisioning services such as food, fiber, and pharmaceuticals, for regulating services that stabilize the climate and its cycles, and for cultural services associated with natural and human landscapes. Rapidly expanding human demands, particularly from agricultural intensification and changes in agricultural land use, have altered the composition of many soils worldwide (Amundson et al., 2015). Most such domesticated soils suffer biodiversity losses and physical degradation, and are often unable to provide many of their original qualities (Kibblewhite et al., 2008). Monitoring soil health is a vital step toward tackling key global issues related to food, climate and human health. Complementary to traditional laboratory and field studies, sensor networks have emerged as a promising and scalable approach to observe soil in the field (Obade & Gaya, 2021). Work to date has primarily focused on sensor deployment, aiming for guaranteed coverage by planning the optimal locations for sensor placement. Scenario 1: Sensing an extended area requires many sensors to cover a large terrain, which significantly increases the deployment cost (Akyildiz et al., 2002). Scenario 2: Such coverage is typically offered as rigid and limited sensing services that cannot be extended or adapted to different soil properties (Othman & Shazali, 2012). Scenario 3: Continuous soil monitoring is needed, implying the need to redeploy new sensors frequently to cover any uncovered or emerging subarea (Dong et al., 2013). Planning-based deployment methods often require detailed user requirements on the properties to sense, as well as the spatial and temporal resolutions. In many real-world cases, agricultural specialists gradually determine optimal soil monitoring parameters, rendering prior specifications incomplete amid heterogeneous terrains and evolving needs (Trnka et al., 2013). This is addressed by low-cost capacitive soil moisture sensors, which allow for scalable, adaptive deployment; units range in price from \$1 to \$2, while high-end models, such as 5TM, cost about \$350. Cosmic ray neutron sensors (CRNS) monitor fields with a radius of up to 240 m; active learning frameworks with mobile robots reduce deployment costs by optimizing paths in vast areas and accurately predicting moisture despite variability. After calibration, adaptability shines: inexpensive models perform well in sandy loam (RMSE 0.03–0.29 cm^3/cm^3), but they need to be adjusted for clayey soils, where factory calibrations overestimate; improvements such as 16-bit ADCs increase precision across textures (SD drops to 0.04%–0.14%) (Wu et al., 2023). These metrics support practical, thorough monitoring strategies.

7.3 Predictive modeling for soil health

Predictive modeling represents a new frontier that sheds light on the shifting dynamics of soil health, fertility, and resilience. Technological advances underlie the development of numerical soil models, parsing processes from a governing set of soil-related equations (Trnka et al., 2013). Such an approach, exemplified by the Crop Environment Resource Synthesis (CERES) suite, allows for the simulation of critical soil properties fundamental to ecosystem productivity, including soil water content, nutrient fluxes, and humus dynamics. These breakthroughs enable the generation of novel indices for assessing soil quality and erosion potential, as well as numerical guidelines for sustainable fertilizer use and quantitative maps of plant-available macro- and micronutrients at regional or global scales (Jones et al., 2003). Thermal energy and soil moisture influence the rates of important microbiological processes (Pietikäinen et al., 2005).

7.4 Spatially explicit erosion and risk modeling

To enhance predictions for extreme events like intense rainfall and erosion under climate change, spatially explicit models such as RUSLE2015 provide map-based insights into long-term average annual soil loss from runoff (Panagos et al., 2015). RUSLE2015, calibrated with high-resolution (100m) pan-European datasets on rainfall erosivity, soil erodibility, cover-management, topography, and support practices, estimates current erosion rates (e.g., >5 t ha^{-1} yr^{-1} on 12.7% of EU arable lands) and can incorporate climate projections for intensified rainfall by 2050.

Projections using Field Scale Soil Erosion Model (FSSLM) under CMIP6 scenarios forecast U.S. soil erosion increases of 8%–21% by 2050 due to shifting rainfall extremes and land use changes, shifting hotspots eastward (Borrelli et al., 2020). Integrated approaches coupling erosion with landslide models (e.g., LANDPLANER's hydrological-erosion-stability framework) quantify uncertainties, revealing morphometry-vegetation-soil interactions and risks like 17.3% erosion rises in semi-arid zones (Panagos et al., 2015).

These models generate GIS-compatible risk maps with uncertainty ranges (e.g., via ensemble CMIP6 inputs), enabling targeted conservation in vulnerable regions like Mediterranean Basin or Central Europe, as highlighted in chapter case studies (Panagos et al., 2015; Borrelli et al., 2020; Trnka et al., 2013). Future RUSLE-FSSLM applications could extend to chapter biomes, linking to remote sensing (Section 7.1) for dynamic monitoring.

8. Policy implications and global responses

The policy landscape surrounding soil science, climate change, and food security influences responses from international to community levels, often directly shaping strategies to address soil-related threats. Categories of practice encompass international treaties, governance and regulation, national and regional strategy, local policy and government, as well as community and indigenous policies and engagement (Montanarella et al., 2016). Regardless of scale, policies articulate priorities that influence the allocation of incentives, investments, and resources for the development of practice.

Internationally, treaties establish agreements around protocols and sustainable management that influence actions across regions. Organizations such as the FAO or Global Soil Partnership maintain global frameworks for policy coordination, including the development of advisory guidance and technical resources (Food and Agriculture Organization (FAO, 2015). Elements such as governance, sustainable land management, food and energy security, regional program coordination, and intergovernmental coordination facilitate integration across regions. Governance and regulation further support favorable practice development, promoting use through land development codes, nutrient licenses, waste management directives, and water efficiency regulations. In the UK, the Committee on Climate Change recommends including soils within strategies to achieve net-zero emissions (Committee on Climate Change, 2019). In the USA, the National Resources Inventory collects data on the health of soil and related land resources to inform strategies for erosion control, soil conservation, groundwater availability, and land use development (U.S. Department of Agriculture (USDA, 2020).

8.1 International agreements on climate change

Current projections indicate that global average temperatures will reach 2°C above pre-industrial levels within the next 2 decades (Zhai et al., 2018). This trend is expected to drive significant hydrological changes worldwide, including intensified droughts in semi-arid and Mediterranean regions, as well as increased flooding in temperate areas with sufficiently wet soils (Loboguerrero et al., 2018). Even when countries commit to the goal of keeping global temperatures below 2°C, there will be inevitable overshoot and substantially higher temperature increases before reductions become effective. In its third assessment report, the Intergovernmental Panel on Climate Change (IPCC) concluded that most land carbon models indicate an increase in terrestrial carbon storage under scenarios of elevated atmospheric CO2. However, a few indicate a loss, particularly under temperature rise.

8.2 National policies for soil health

National policies for soil health form a critical component of government frameworks addressing land systems and agricultural production. The Soil Health Policy of India (2017) connects soil health to broader policies spanning agriculture, industry, energy, environment, and health. Similarly, Nigeria's Policy on Sustainable Soil Management (2012) integrates soil health with the national food security objective. Effective policy alignment, bridging biophysical and social sciences, is key to exploiting synergies across land and agricultural systems (Montanarella et al., 2016). The regulations in China reflect this very well. The "Action Plan for Organic-Substitute-Chemical-Fertilizer" incentivized organic alternatives to the overuse of chemical fertilizers. It has successfully reduced chemical fertilizer application by 4.1%–7.2% without any compromise on yields, showing a possible path forward for sustainable intensification (Liu & Wu, 2022). In the same way, India's Soil Health Card (SHC) scheme brought scientific assessment together with extension services and empowered farmers with site-specific nutrient data; as a result, SHC adoption has increased crop yields by 5%–6% while lowering fertilizer costs by 8%–10%. This effectively harmonizes economic viability with soil stewardship. In the U.S., policy-supported adoption of soil health management systems—like no-till and cover cropping—across varied climates has produced annual net income gains of $2–1257 per acre (Karki Nepal et al., 2025; Madhuri et al., 2024). These practices have also reduced greenhouse gas emissions by up to 560% and nutrient runoff nearly by 96% which shows how aligned regulatory frameworks drive measurable improvements in both ecosystem resilience as well as agricultural productivity. Achieving integrated monitoring of Earth's resources involves overcoming existing limitations in coverage, data continuity, data harmonization, and an integrated approach to Earth surface genealogy through international efforts and capacity building. Equipping citizens to participate actively in soil health management through improved policies and tools contributes to an inclusive, sustainable monitoring framework (Chander et al., 2018).

8.3 Community-led initiatives

Community-led approaches to soil health have the potential to restore degraded ecosystems and sustain agricultural productivity. Local innovations help communities manage their agro-ecosystems effectively and can provide fertile ground for scaling up rehabilitation initiatives. Proper management of domestic and agricultural waste, along with efficient resource recovery, supports a circular economy by returning nutrients to the agricultural cycle (Drechsel et al., 2015). Co-composting converts materials such as municipal solid waste, excreta, bio-solids, and manure into nutrient-rich fertilizer. A review of 200 case studies across Asia, Africa, and Latin America identifies promising business models for safe reuse of human waste and assesses the supportive role of policies such as Ghana's fertilizer subsidy program and the approval of Fortifer compost for commercial use (Otoo & Drechsel, 2018). Public-private partnerships and non-governmental organizations promote initiatives such as co-composting plants to produce Fortifer and the dissemination of vermicompost to smallholders, aiming to improve soil fertility and crop production. A land health surveillance framework based on scientific principles can inform landscape management and enable the development of preventive strategies. The 4 Pour 1000 Initiative, launched at COP 21, aims to mitigate climate change by sequestering four‰ of carbon in soils annually, focusing on levels of key risk factors for land degradation (Minasny et al., 2017).

9. Future directions for research and innovation

Innovation and discovery continue to drive soil science toward new horizons (Karlen et al., 2003). Climates and soils are shifting on a planetary scale, posing complex new challenges. International initiatives, such as the Global Soil Partnership and the Global Soil Information System, encourage interdisciplinary collaboration, recognizing the crucial role that soil science will play in addressing these challenges (Montanarella et al., 2016). Leveraging satellite and drone observations, crowd-sourcing, and soil sensors has become increasingly important, requiring a concerted effort to translate advances in soil monitoring into solutions at various scales (Loboguerrero et al., 2018; Mulla, 2013). Remotely sensed data and predictive-model outputs, designed to enable informed management decisions, must also be integrated into production systems. Soil scientists, who aim to elucidate the cellular and molecular mechanisms behind soil degradation, need to remain aware of those contraindications. Without sustained and synergistic collaborative efforts, innovations risk failing to reach those communities where their adoption might prove most effective.

9.1 Emerging trends in soil science

The interdependence of soils, plants, animals, and microorganisms creates a resilient web of life that forms the basis for both human well-being and ecosystem resilience (Obade & Gaya, 2021). However, this vital support system is in danger of being disrupted by changes in soil properties brought on by tropical cyclones, rising temperatures, severe drought, wildfires, and ash deposition. According to Trenberth et al. (2014), warming increases the intensity and duration of soil drought, reducing vital biota for plant growth and nutrient cycling. Variations in soil health brought on by climate change are expected to last into the 21st century. These changes frequently appear out of the blue, having detrimental effects in some areas and beneficial ones in others, with dire repercussions for ecosystem resilience and food security (Kraskura et al., 2023). Empirical data from worldwide time-series analyses further support this spatial heterogeneity in soil responses to climate stressors. For example, a high-resolution (1 × 1 km) dataset of annual soil respiration (Rs), a crucial indicator of soil health, from 2000 to 2014 showed different biome-specific trends: Rs increased significantly due to warming in tropical regions like the Amazon and Congo Basins (+3.66% relative change), decreased due to drought and vegetation shifts in temperate zones like the central U.S. and northeast China (−1.53%), and displayed mixed boreal gains (+6.80%) associated with thawing and moisture dynamics (Chen et al., 2022; Dettleff et al., 2022). At finer scales, a 60-ha tropical montane rainforest plot's geostatistical analysis revealed a moderate spatial dependence of Rs (nugget-to-sill ratio 68.1%, patch size ~112 m), which was negatively correlated with slope but positively correlated with soil phosphorus and organic carbon at higher elevations, underscoring micro-scale topographic influences on nutrient responses and drought (Helmer et al., 2023). In addition to these findings, model projections under CMIP6 scenarios that use geographically weighted logistic regressions forecast sharp regional differences in drought-fire risks—higher in southwest China compared to wetter northwest regions through 2099—further amplifying heterogeneous effects on soil biota and resilience (Lv et al., 2025). In the end, this kind of spatial variability can either mitigate or exacerbate risks like drought and wildfires, resulting in both localized advantages (like increased boreal carbon turnover) and disadvantages (like temperate Rs declines).

9.2 Interdisciplinary approaches

Addressing climate change demands collaboration across disciplines and scales. Projects focused on extreme events and sectoral impacts call for broader integration of climate dialog. The scope of climate research expands to consider larger spatial areas and extended time periods, examining regional damages and adaptation measures. Interdisciplinary analyses combine science with political economy, revealing the roles of various actors in climate responses (Tanner & Allouche, 2011). Bridging gaps between climate effects, adaptation/disaster relief, and sustainable development requires cumulative and iterative scientific efforts. Enhanced socio-political understanding is crucial for integrating climate issues with development and disaster reduction efforts (Schipper & Pelling, 2006).

9.3 The role of citizen science

The DSPT framework introduced by Koiter et al. defines the digital soil map DpTi as the point estimate of soil property p at location i, with p representing a particular soil-sensing technique, and Ti referring to the type of the associated soil property (e.g., measured, predictedEmpirical (E), predictedVirtual (V)) (Koiter et al., 2013). Each element estimates the property at a specific location, and the framework supports integrating data from different sources when their measurements support the digitization of a common property. Although the initial focus was empirical and virtual data, the framework has been extended to empirical, numerical, and virtual data combined. Platforms like CROPMAN analyze weather data, soil information, and crop requirements to generate recommendations (Gijsman et al., 2002). The FAO AgroClimate Tool forecasts the impacts of climate change on smallholder farming systems, and projects such as AFarCloud leverage machine learning for enhanced agricultural insights. Furthermore, digitally acquired soil data, whether empirical or virtual, span a range of soil characteristics to hydrological, ecological, plant-biophysical, and biochemical variables and can be related to various soil ecosystem services.

Given the confluence of digital technologies, emergent AI methods for soil property prediction, and an increasing number of open soil datasets, "digital soil demographic" scholars should consider citizen science as a potential approach to help resolve the Soil Quality Index conundrum (SQI-9). The democratization of the science-politics interface promotes a more interactive policy-making process, diminishing the strict boundaries between design and implementation. At the same time, the expertise of research institutions remains highly sought after, and civic science increasingly forms one among several valuable inputs in policy decisions (Okner, 2015).

10. Conclusion

Climate change has emerged as one of the defining challenges of the 21st century. Its wide-ranging consequences affect every country, disrupting national economies and threatening lives worldwide. Soil is an integral part of terrestrial ecosystems, and its health is pivotal in determining the resilience of ecosystems and their capacity to provide essentials for the survival and growth of living beings. Soil health is therefore fundamental to food and nutritional security and is vital in addressing the challenges posed by climate variation. Soil serves as a fundamental principal component of the terrestrial ecosystem. Its health is central to the capacity of ecosystems to provide essential support for the survival and growth of living beings, underpinning food and nutritional security. The World Overview of Conservation Approaches and Technology (WOCAT) defines Sustainable Land Management as a non-degrading or improving system of land use and management, integrating facets of soils, vegetation, and water, which helps reduce the destruction and degradation of land resources, thereby improving production while conserving resources. Unsustainable land management, poor agricultural practices, and recurring droughts have severely contributed to land degradation in many arid and semi-arid regions. The gross primary production of terrestrial plants, which influences biomass accumulation in soil, increases with rising atmospheric CO2 concentrations and temperature on a global scale. However, soil decomposer activities, corona activity, organic matter decomposition, and mineralization rates also increase with temperature. Hence, under increasing atmospheric CO2 concentrations and ambient temperature, it remains uncertain whether soil carbon runoff from terrestrial environments to aquatic systems will increase or decrease, indicating a complexity in soil carbon dynamics under changing climatic conditions.

The soil microbiota, comprising bacteria and fungi, plays a crucial role in the health and productivity of soils. The physical and chemical properties of the soil, as well as the climate, constitute some of the key parameters influencing the composition and spatial distribution of the microbiota. Carbon, for example, is the dominant energy and nutrient source utilized by soil microorganisms; thus, the heterotrophic activity of the soil microbiota, which has consequences for the viability of terrestrial ecosystems and atmospheric CO2 concentrations, is regulated by the availability of carbon compounds within the soil matrix. Decomposers convert organic compounds within the soil into various fractions of soil organic matter. The Organic Matter in Soil (OMIS) characterizes the soil organic carbon pools into eight fractions and provides specific decomposition and mineralization rates for these fractions.

Acknowledgments

We extend our gratitude to all the universities involved in this study for their support in facilitating my research.

Data availability

Data available within the article.

References

Abdalla, K., Chivenge, P., Ciais, P., & Chaplot, V. (2016). No-tillage lessens soil CO2 emissions the most under arid and sandy soil conditions: Results from a meta-analysis. *Biogeosciences, 13*(12), 3619–3633.

Adams, J. B., Smith, M. O., & Gillespie, A. R. (1993). Imaging spectroscopy: Interpretation based on spectral mixture analysis. In *Remote geochemical analysis: Elemental and mineralogical composition, 7* (pp. 145–166). Cambridge University Press.

Akyildiz, I. F., Su, W., Sankarasubramaniam, Y., & Cayirci, E. (2002). Wireless sensor networks: A survey. *Computer Networks, 38*(4), 393–422.

Al-Shammary, A. A. G., Al-Shihmani, L. S. S., Fernández-Gálvez, J., & Caballero-Calvo, A. (2024). Optimizing sustainable agriculture: A comprehensive review of agronomic practices and their impacts on soil attributes. *Journal of Environmental Management, 364*, Article 121487.

Amundson, R., Berhe, A. A., Hopmans, J. W., Olson, C., Sztein, A. E., & Sparks, D. L. (2015). Soil and human security in the 21st century. *Science, 348*(6235), Article 1261071.

Bannari, A., & Al-Ali, Z. M. (2020). Assessing climate change impact on soil salinity dynamics between 1987–2017 in arid landscape using Landsat TM, ETM+ and OLI data. *Remote Sensing, 12*(17), 2794.

Basch, G., Kassam, A., Friedrich, T., Santos, F. L., Gubiani, P. I., Calegari, A., … Dos Santos, D. R. (2012). *Sustainable soil water management systems.*

Behrens, T., Schmidt, K., Viscarra Rossel, R. A., Gries, P., Scholten, T., & MacMillan, R. A. (2018). Spatial modelling with euclidean distance fields and machine learning. *European Journal of Soil Science, 69*(5), 757–770.

Blanchy, G., Bragato, G., Di Bene, C., Jarvis, N., Larsbo, M., Meurer, K., & Garré, S. (2023). Soil and crop management practices and the water regulation functions of soils: A qualitative synthesis of meta-analyses relevant to European agriculture. *Soil, 9*(1), 1–20.

Borrelli, P., Robinson, D. A., Panagos, P., Lugato, E., Yang, J. E., Alewell, C., … Ballabio, C. (2020). Land use and climate change impacts on global soil erosion by water (2015-2070). *Proceedings of the National Academy of Sciences, 117*(36), 21994–22001.

Brown, M. E., & Funk, C. C. (2008). Food security under climate change. *Science, 319*(5863), 580–581.

Budge, K., Leifeld, J., Hiltbrunner, E., & Fuhrer, J. (2011). Alpine grassland soils contain large proportion of labile carbon but indicate long turnover times. *Biogeosciences, 8*(7), 1911–1923.

Bunemann, E. K., Bongiorno, G., Bai, Z., Creamer, R. E., De Deyn, G. B., de Goede, R., … Brussaard, L. (2018). Soil quality–A critical review. *Soil Biology and Biochemistry, 120*, 105–125.

Burke, E. J., Chadburn, S. E., & Ekici, A. (2017). A vertical representation of soil carbon in the JULES land surface scheme (vn4. 3_permafrost) with a focus on permafrost regions. *Geoscientific Model Development, 10* (2), 959–975.

Chaikin, S., Dubiner, S., & Belmaker, J. (2022). Cold-water species deepen to escape warm water temperatures. *Global Ecology and Biogeography, 31*(1), 75–88.

Chander, G., Wani, S. P., Gopalakrishnan, S., Mahapatra, A., Chaudhury, S., Pawar, C. S., … Rao, A. K. (2018). Microbial consortium culture and vermi-composting technologies for recycling on-farm wastes and food production. *International Journal of Recycling of Organic Waste in Agriculture, 7*(2), 99–108.

Chen, H., Hu, Q., Kong, L., Rong, H., & Bi, B. (2022). Effects of temperature on the growth performance, biochemical indexes and growth and development-related genes expression of juvenile hybrid sturgeon (Acipenser baerii♀× Acipenser schrenckii♂). *Water, 14*(15), 2368.

Cleveland, C. C., & Townsend, A. R. (2006). Nutrient additions to a tropical rain forest drive substantial soil carbon dioxide losses to the atmosphere. *Proceedings of the National Academy of Sciences, 103*(27), 10316–10321.

Committee on Climate Change. (2019). *Net zero: The UK's contribution to stopping global warming.*

Corwin, D. L. (2021). Climate change impacts on soil salinity in agricultural areas. *European Journal of Soil Science, 72*(2), 842–862.

Crowther, T. W., Todd-Brown, K. E., Rowe, C. W., Wieder, W. R., Carey, J. C., Machmuller, M. B., … Bradford, M. A. (2016). Quantifying global soil carbon losses in response to warming. *Nature, 540*(7631), 104–108.

De Lombaerde, E., Vangansbeke, P., Lenoir, J., Van Meerbeek, K., Lembrechts, J., Rodríguez-Sánchez, F., … De Frenne, P. (2022). Maintaining forest cover to enhance temperature buffering under future climate change. *Science of the Total Environment, 810*. Article 151338.

Dettleff, P., Zuloaga, R., Fuentes, M., Gonzalez, P., Aedo, J., Estrada, J. M., … Valdés, J. A. (2022). High-temperature stress effect on the red cusk-eel (Geypterus chilensis) liver: Transcriptional modulation and oxidative stress damage. *Biology, 11*(7), 990.

Dias, T., Dukes, A., & Antunes, P. M. (2015). Accounting for soil biotic effects on soil health and crop productivity in the design of crop rotations. *Journal of the Science of Food and Agriculture, 95*(3), 447–454.

Dong, X., Vuran, M. C., & Irmak, S. (2013). Autonomous precision agriculture through integration of wireless underground sensor networks with center pivot irrigation systems. *Ad Hoc Networks, 11*(7), 1975–1987.

Drechsel, P., Heffer, P., Magen, H., Mikkelsen, R., & Wichelns, D. (2015). *Managing water and fertilizer for sustainable agricultural intensification.* Paris, France: International Fertilizer Industry Association (IFA) Colombo. Sri Lanka: International Water Management Institute Georgia, USA: International Plant Nutrition Institute (IPNI) Horgen, Switzerland: International Potash Institute (IPI).

Eyring, V., Bony, S., Meehl, G. A., Senior, C. A., Stevens, B., Stouffer, R. J., & Taylor, K. E. (2016). Overview of the coupled model intercomparison project phase 6 (CMIP6) experimental design and organization. *Geoscientific Model Development, 9*(5), 1937–1958.

Fay, P. A., Prober, S. M., Harpole, W. S., Knops, J. M., Bakker, J. D., Borer, E. T., … Yang, L. H. (2015). Grassland productivity limited by multiple nutrients. *Nature Plants, 1*(7), 1–5.

Field, D. J., Koppi, A. J., Jarrett, L. E., Abbott, L. K., Cattle, S. R., Grant, C. D., … Weatherley, A. J. (2011). Soil science teaching principles. *Geoderma, 167*, 9–14.

Food and Agriculture Organization [FAO]. (2015). *Status of the World's soil resources.* Rome: FAO.

Friedrich, T., Derpsch, R., & Kassam, A. (2012). Overview of the global spread of conservation agriculture. *Field actions science reports. The Journal of Field Actions,* (Special Issue 6).

Gelybó, G., Tóth, E., Farkas, C., Horel, A., Kása, I., & Bakacsi, Z. (2018). Potential impacts of climate change on soil properties. *Agrokémia És Talajtan, 67*(1), 121–141.

Gijsman, A. J., Hoogenboom, G., Parton, W. J., & Kerridge, P. C. (2002). Modifying DSSAT crop models for low-input agricultural systems using a soil organic matter–residue module from CENTURY. *Agronomy Journal, 94* (3), 462–474.

Giller, K. E., Hijbeek, R., Andersson, J. A., & Sumberg, J. (2021). Regenerative agriculture: An agronomic perspective. *Outlook on Agriculture, 50*(1), 13–25.

Guerra, C. A., Delgado-Baquerizo, M., Duarte, E., Marigliano, O., Görgen, C., Maestre, F. T., & Eisenhauer, N. (2021). Global projections of the soil microbiome in the anthropocene. *Global Ecology and Biogeography, 30* (5), 987–999.

Guerra, C. A., Rosa, I. M., Valentini, E., Wolf, F., Filipponi, F., Karger, D. N., … Eisenhauer, N. (2020). Global vulnerability of soil ecosystems to erosion. *Landscape Ecology, 35*(4), 823–842.

Guo, J. H., Liu, X. J., Zhang, Y., Shen, J. L., Han, W. X., Zhang, W. F., … Zhang, F. S. (2010). Significant acidification in major Chinese croplands. *Science, 327*(5968), 1008–1010.

Heckman, K., & Rasmussen, C. (2018). Role of mineralogy and climate in the soil carbon cycle. In *Developments in soil science, 35* (pp. 93–110). Elsevier.

Helmer, E. H., Kay, S., Marcano-Vega, H., Powers, J. S., Wood, T. E., Zhu, X., … Ruzycki, T. S. (2023). Multiscale predictors of small tree survival across a heterogeneous tropical landscape. *PLoS One, 18*(3), Article e0280322.

Hobbs, P. R., Sayre, K., & Gupta, R. (2008). The role of conservation agriculture in sustainable agriculture. *Philosophical Transactions of the Royal Society B: Biological Sciences, 363*(1491), 543–555.

Huang, W., Hammel, K. E., Hao, J., Thompson, A., Timokhin, V. I., & Hall, S. J. (2019). Enrichment of lignin-derived carbon in mineral-associated soil organic matter. *Environmental Science & Technology, 53*(13), 7522–7531.

Jeyanthi, V., & Kanimozhi, S. (2018). Plant growth promoting rhizobacteria (PGPR)-prospective and mechanisms: A review. *Journal of Pure and Applied Microbiology, 12*(2), 733–749.

Jones, J. W., Hoogenboom, G., Porter, C. H., Boote, K. J., Batchelor, W. D., Hunt, L. A., … Ritchie, J. T. (2003). The DSSAT cropping system model. *European Journal of Agronomy, 18*(3–4), 235–265.

Karki Nepal, A., Choudhary, D., & Pandit, N. R. (2025). Farmers' response to soil health card recommendations: input usage and crop yield. *Nutrient Cycling in Agroecosystems, 131*(3), 473–486.

Karlen, D. L., Ditzler, C. A., & Andrews, S. S. (2003). Soil quality: Why and how? *Geoderma, 114*(3–4), 145–156.

Kaye, J. P., & Quemada, M. (2017). Using cover crops to mitigate and adapt to climate change. A review. *Agronomy for Sustainable Development, 37*, 1–17.

Kibblewhite, M. G., Ritz, K., & Swift, M. J. (2008). Soil health in agricultural systems. *Philosophical Transactions of the Royal Society B: Biological Sciences, 363*(1492), 685–701.

Kirkby, C. A., Richardson, A. E., Wade, L. J., Batten, G. D., Blanchard, C., & Kirkegaard, J. A. (2013). Carbon-nutrient stoichiometry to increase soil carbon sequestration. *Soil Biology and Biochemistry, 60*, 77–86.

Koiter, A. J., Owens, P. N., Petticrew, E. L., & Lobb, D. A. (2013). The behavioural characteristics of sediment properties and their implications for sediment fingerprinting as an approach for identifying sediment sources in river basins. *Earth-Science Reviews, 125*, 24–42.

Kraskura, K., Hardison, E. A., & Eliason, E. J. (2023). Body size and temperature affect metabolic and cardiac thermal tolerance in fish. *Scientific Reports, 13*(1), Article 17900.

Lal, R. (2015). Restoring soil quality to mitigate soil degradation. *Sustainability, 7*(5), 5875–5895.

Lal, R., Delgado, J. A., Groffman, P. M., Millar, N., Dell, C., & Rotz, A. (2011). Management to mitigate and adapt to climate change. *Journal of Soil and Water Conservation, 66*(4), 276–285.

Lehmann, J., Rillig, M. C., Thies, J., Masiello, C. A., Hockaday, W. C., & Crowley, D. (2011). Biochar effects on soil biota–a review. *Soil Biology and Biochemistry, 43*(9), 1812–1836.

Liu, T., & Wu, G. (2022). Does agricultural cooperative membership help reduce the overuse of chemical fertilizers and pesticides? Evidence from rural China. *Environmental Science and Pollution Research, 29*(5), 7972–7983.

Loboguerrero, A. M., Campbell, B. M., Cooper, P. J., Hansen, J. W., Rosenstock, T., & Wollenberg, E. (2019). Food and earth systems: Priorities for climate change adaptation and mitigation for agriculture and food systems. *Sustainability, 11*(5), 1372.

Loboguerrero, A. M., Boshell, F., León, G., Martinez-Baron, D., Giraldo, D., Mejía, L. R., … & Cock, J. (2018). Bridging the gap between climate science and farmers in Colombia. Climate Risk Management, 22, 67-81.

Lv, Q., Chen, Z., Wu, C., Peñuelas, J., Fan, L., Su, Y., … Wang, Q. (2025). Increasing severity of large-scale fires prolongs recovery time of forests globally since 2001. *Nature Ecology & Evolution*, 1–13.

Maclean, I. M., & Early, R. (2023). Macroclimate data overestimate range shifts of plants in response to climate change. *Nature Climate Change, 13*(5), 484–490.

Madhuri, C. R., Naik, A., Kumar, A., Padhy, C., & Ray, S. (2024). Perception and adoption of soil health card by paddy growers of North Coastal Andhra Pradesh. *Indian Journal of Extension Education. 60*(4), 24–29.

McBratney, A. B., Mendonça Santos, M. L., & Minasny, B. (2003). On digital soil mapping. *Geoderma, 117*(1–2), 3–52.

Melillo, J. M., Frey, S. D., DeAngelis, K. M., Werner, W. J., Bernard, M. J., Bowles, F. P., … Grandy, A. S. (2017). Long-term pattern and magnitude of soil carbon feedback to the climate system in a warming world. *Science, 358*(6359), 101–105.

Minasny, B., Malone, B. P., McBratney, A. B., Angers, D. A., Arrouays, D., Chambers, A., … Winowiecki, L. (2017). Soil carbon 4 per mille. *Geoderma, 292*, 59–86.

Moher, D., Liberati, A., Tetzlaff, J., & Altman, D. G. (2009). Preferred reporting items for systematic reviews and meta-analyses: The PRISMA statement. *Bmj, 339.*

Montanarella, L., Pennock, D. J., McKenzie, N., Badraoui, M., Chude, V., Baptista, I., … Vargas, R. (2016). World's soils are under threat. *Soil, 2*(1), 79–82.

Montgomery, D. R. (2007). Soil erosion and agricultural sustainability. *Proceedings of the National Academy of Sciences, 104*(33), 13268–13272.

Mulder, V. L., De Bruin, S., Schaepman, M. E., & Mayr, T. R. (2011). The use of remote sensing in soil and terrain mapping—A review. *Geoderma, 162*(1–2), 1–19.

Mulla, D. J. (2013). Twenty five years of remote sensing in precision agriculture: Key advances and remaining knowledge gaps. *Biosystems Engineering, 114*(4), 358–371.

Naudin, K., Husson, O., Scopel, E., Auzoux, S., Giner, S., & Giller, K. E. (2015). PRACT (Prototyping Rotation and Association with Cover crop and no Till)–a tool for designing conservation agriculture systems. *European Journal of Agronomy, 69*, 21–31.

Nicolopoulou-Stamati, P., Maipas, S., Kotampasi, C., Stamatis, P., & Hens, L. (2016). Chemical pesticides and human health: The urgent need for a new concept in agriculture. *Frontiers in Public Health, 4*, 148.

Obade, V. D. P., & Gaya, C. (2021). Digital technology dilemma: On unlocking the soil quality index conundrum. *Bioresources and Bioprocessing, 8*(1), 6.

Okner, T. (2015). *The role of research institutions in civic science & sustainable development diplomacy.*

Oldfield, E. E., Bradford, M. A., & Wood, S. A. (2019). Global meta-analysis of the relationship between soil organic matter and crop yields. *Soil, 5*(1), 15–32.

O'Neill, B. C., Tebaldi, C., Van Vuuren, D. P., Eyring, V., Friedlingstein, P., Hurtt, G., … Sanderson, B. M. (2016). The scenario model intercomparison project (ScenarioMIP) for CMIP6. *Geoscientific Model Development, 9* (9), 3461–3482.

Othman, M. F., & Shazali, K. (2012). Wireless sensor network applications: A study in environment monitoring system. *Procedia Engineering, 41*, 1204–1210.

Otoo, M., & Drechsel, P. (Eds.). (2018). *Resource recovery from waste: Business models for energy, nutrient and water reuse in low-and middle-income countries*. Routledge.

Panagos, P., Borrelli, P., Poesen, J., Ballabio, C., Lugato, E., Meusburger, K., … Alewell, C. (2015). The new assessment of soil loss by water erosion in Europe. *Environmental Science & Policy, 54*, 438–447.

Pietikäinen, J., Pettersson, M., & Båth, E. (2005). Comparison of temperature effects on soil respiration and bacterial and fungal growth rates. *FEMS Microbiology Ecology, 52*(1), 49–58.

Pittelkow, C. M., Liang, X., Linquist, B. A., Van Groenigen, K. J., Lee, J., Lundy, M. E., … Van Kessel, C. (2015). Productivity limits and potentials of the principles of conservation agriculture. *Nature, 517*(7534), 365–368.

Powlson, D. S., Stirling, C. M., Jat, M. L., Gerard, B. G., Palm, C. A., Sanchez, P. A., & Cassman, K. G. (2014). Limited potential of no-till agriculture for climate change mitigation. *Nature Climate Change, 4*(8), 678–683.

Powlson, D. S., Whitmore, A. P., & Goulding, K. W. (2011). Soil carbon sequestration to mitigate climate change: A critical re-examination to identify the true and the false. *European Journal of Soil Science, 62*(1), 42–55.

Priyadarshini, C., Lal, R., Yuan, P., Liu, W., Adhikari, A., Bhandari, S., & Xia, Y. (2025). Plant disease suppressiveness enhancement via soil health management. *Biology, 14*(8), 924.

Richardson, A. E., Kirkby, C. A., Banerjee, S., & Kirkegaard, J. A. (2014). The inorganic nutrient cost of building soil carbon. *Carbon Management, 5*(3), 265–268.

Rising, J., Tedesco, M., Piontek, F., & Stainforth, D. A. (2022). The missing risks of climate change. *Nature, 610* (7933), 643–651.

Sanderman, J., & Grandy, A. S. (2020). Ramped thermal analysis for isolating biologically meaningful soil organic matter fractions with distinct residence times. *Soil, 6*(1), 131–144.

Schipper, L., & Pelling, M. (2006). Disaster risk, climate change and international development: Scope for, and challenges to, integration. *Disasters, 30*(1), 19–38.

Schreefel, L., Schulte, R. P. O., De Boer, I. J. M., Schrijver, A. P., & Van Zanten, H. H. E. (2020). Regenerative agriculture–the soil is the base. *Global Food Security, 26*, Article 100404.

Schulte, R. P., Creamer, R. E., Donnellan, T., Farrelly, N., Fealy, R., O'Donoghue, C., & O'hUallachain, D. (2014). Functional land management: A framework for managing soil-based ecosystem services for the sustainable intensification of agriculture. *Environmental Science & Policy, 38*, 45–58.

Singh, S., Kiran, B. R., & Mohan, S. V. (2024). Carbon farming: A circular framework to augment CO 2 sinks and to combat climate change. *Environmental Science: Advances, 3*(4), 522–542.

Smith, P., Soussana, J. F., Angers, D., Schipper, L., Chenu, C., Rasse, D. P., … Klumpp, K. (2019). How to measure, report and verify soil carbon change to realize the potential of soil carbon sequestration for atmospheric greenhouse gas removal. *Global Change Biology, 26*(1), 219–241.

Tanner, T., & Allouche, J. (2011). Towards a new political economy of climate change and development. *IDS Bulletin, 42*(3), 1–14.

Trenberth, K. E., Dai, A., van der Schrier, G., Jones, P. D., Barichivich, J., Briffa, K. R., & Sheffield, J. (2014). Global warming and changes in drought. *Nature Climate Change, 4*(1), 17–22.

Trnka, M., Kersebaum, K. C., Eitzinger, J., Hayes, M., Hlavinka, P., Svoboda, M., … Žalud, Z. (2013). Consequences of climate change for the soil climate in Central Europe and the Central Great Plains of the United States. *Climatic Change, 120*(1), 405–418.

Turmel, M. S., Speratti, A., Baudron, F., Verhulst, N., & Govaerts, B. (2015). Crop residue management and soil health: A systems analysis. *Agricultural Systems, 134*, 6–16.

Ullah, A., Bano, A., & Khan, N. (2021). Climate change and salinity effects on crops and chemical communication between plants and plant growth-promoting microorganisms under stress. *Frontiers in Sustainable Food Systems, 5*, Article 618092.

U.S. Department of Agriculture [USDA]. (2020). *Summary report: 2017 national resources inventory.* Natural Resources Conservation Service.

Van Gestel, N., Shi, Z., Van Groenigen, K. J., Osenberg, C. W., Andresen, L. C., Dukes, J. S., … Hungate, B. A. (2018). Predicting soil carbon loss with warming. *Nature, 554*(7693), E4–E5.

Vereecken, H., Schnepf, A., Hopmans, J. W., Javaux, M., Or, D., Roose, T., … Young, I. M. (2016). Modeling soil processes: Review, key challenges, and new perspectives. *Vadose Zone Journal, 15*(5), Article vzj2015-09.

Vicente-Serrano, S. M., Peña-Angulo, D., Beguería, S., Domínguez-Castro, F., Tomás-Burguera, M., Noguera, I., … El Kenawy, A. (2022). Global drought trends and future projections. *Philosophical Transactions of the Royal Society A, 380*(2238), Article 20210285.

Williams, A., Hunter, M. C., Kammerer, M., Kane, D. A., Jordan, N. R., Mortensen, D. A., … Davis, A. S. (2016). Soil water holding capacity mitigates downside risk and volatility in U.S. rainfed maize: Time to invest in soil organic matter? *PLoS ONE, 11*(8), Article e0160974.

Wu, N., Wen, H., Xu, P., Chen, J., Xue, M., Li, J., … Li, H. (2023). PPAR signaling maintains metabolic homeostasis under hypothermia in freshwater drum (Aplodinotus grunniens). *Metabolites, 13*(1), 102.

Zhai, P., Pörtner, H. O., Roberts, D., Skea, J., Shukla, P. R., Pirani, A., & Matthews, R. (2018). Global Warming of 1.5° C. An IPCC Special Report on the impacts of global warming of 1.5° C above pre-industrial levels and related global greenhouse gas emission pathways, in the context of strengthening the global response to the threat of climate change. *Sustainable Development, and Efforts to Eradicate Poverty, 32*.

Zhang, W. F., Dou, Z. X., He, P., Ju, X. T., Powlson, D., Chadwick, D., … Zhang, F. S. (2013). New technologies reduce greenhouse gas emissions from nitrogenous fertilizer in China. *Proceedings of the National Academy of Sciences, 110*(21), 8375–8380.

Zhou, J., Wang, L., Zhong, X., Yao, T., Qi, J., Wang, Y., & Xue, Y. (2022). Quantifying the major drivers for the expanding lakes in the interior Tibetan Plateau. *Science Bulletin, 67*(5), 474–478.

Zhu, D., Ciais, P., Krinner, G., Maignan, F., Jornet Puig, A., & Hugelius, G. (2019). Controls of soil organic matter on soil thermal dynamics in the northern high latitudes. *Nature Communications, 10*(1), 3172.

Zomer, R. J., Bossio, D. A., Sommer, R., & Verchot, L. V. (2017). Global sequestration potential of increased organic carbon in cropland soils. *Scientific Reports, 7*(1), Article 15554.

CHAPTER

Emerging technologies for soil data collection 20

Arun Karnwal[1], Abdel Rahman Mohammad Said Al-Tawaha[2], Abdel Razzaq Al-Tawaha[3], Natalia Nesterova[4], Gaurav Kumar[5], Neha Saklani[1], Anupam Kumar[6] and Deepak Kumar[7]

[1]Department of Microbiology, Graphic Era (Deemed to be University), Dehradun, Uttarakhand, India; [2]Department of Biological Sciences, Al Hussein Bin Talal University, Ma'an, Jordan; [3]National Agricultural Research Center (NARC), Baq'a, Jordan; [4]National University of Life and Environmental Sciences of Ukraine, Kyiv, Ukraine; [5]Amity Institute of Microbial Technology, Amity University Rajasthan, Jaipur, Rajasthan, India; [6]Department of Biotechnology, Faculty of Applied, Pure and Health Sciences, Parul University, Quepem, Goa, India; [7]Department of Chemistry, Manipal University Jaipur, Jaipur, Rajasthan, India

1. Introduction to soil data collection

Soil is an indispensable resource that plays a crucial role in agricultural production and the overall well-being of human populations on Earth (Lal, 2015). It serves as a vital interface for the complex interactions that occur among the geosphere, hydrosphere, biosphere, and atmosphere. Due to its integral role in these interactions, the importance of soil data has grown significantly, particularly for assessing weather conditions and understanding the broader environmental state. Traditional techniques for acquiring soil data have included methods such as augers, tubes, pits, trenches, and drive points, all of which typically require physical probing of the soil's spatial domain (Carter & Gregorich, 2007). These conventional methods enable researchers and agronomists to analyze soil samples, providing valuable insights into hydraulic and chemical parameters. Such parameters are essential as they provide valuable insights into soil quality, fertility, and overall health, which, in turn, are crucial for effective agricultural management and environmental conservation. However, despite their utility, these traditional soil sampling methods have several drawbacks. They are often labor-intensive and require a significant amount of resources, which can limit their practicality and scalability, especially in vast agricultural areas (de Paul Obade & Gaya, 2021). These methods also face limitations regarding sampling resolution and overall accuracy, which can result in incomplete or misleading data, ultimately affecting land management decisions and agricultural practices. As a response to these challenges, emerging technologies are beginning to reshape the landscape of soil data collection and analysis. Innovations such as remote sensing, soil moisture sensors, and advanced imaging techniques are being employed to enhance data collection procedures (Wu, Yang et al., 2023). These technologies can significantly enhance the efficiency of soil sampling, enabling the survey of larger areas more quickly and with greater accuracy than traditional methods.

The Soil Nexus. https://doi.org/10.1016/B978-0-443-45286-4.00020-1

Furthermore, the integration of data analytics and machine learning can help in the interpretation of the vast amounts of data generated by these new techniques, allowing for more informed decision-making. The adoption of these advanced technologies not only streamlines the data collection process but also enhances our understanding of soil dynamics in relation to environmental factors. This is essential for developing sustainable agricultural practices and for addressing pressing global challenges such as food security and climate change. Overall, the shift toward incorporating technology in soil data acquisition reflects a significant evolution in our approach to understanding and managing this critical resource, ultimately benefiting agriculture, environmental management, and human well-being on Earth (de Paul Obade & Gaya, 2021; Peets et al., 2012).

2. Overview of emerging technologies

Emerging technologies offer alternative methods for capturing, studying, and modeling soil data. For these purposes, geographic information systems (GIS) and the application of precision agriculture are among the most prominent areas of modern technological innovation. Since the advent of computing, efforts have been devoted to the development of soil data collection (SDC) techniques and instruments. In recent decades, technologies such as remote sensors, the Internet of Things (IoT), and unmanned aerial vehicles (UAVs) have stimulated soil research worldwide (Peets et al., 2012). Sensors for on-the-go data collection on soil and crops are essential for the successful implementation of precision agriculture. IoT architectures underpin multiple-sensor monitoring systems that centralize data for diagnosis. A number of such systems for soil, water, crop, and weather monitoring have been designed and prototyped. The development of complete frameworks for custom monitoring systems opens up new possibilities for the farming sector. The aerial perspective of UAVs complements data from ground sensors and remote sensing by providing high spatial and temporal resolution (Wu, Zhang et al., 2023).

2.1 Definition of emerging technologies

Emerging technologies refer to innovative methods that leverage novel digital data sources to enhance soil data acquisition beyond traditional means. Consequently, remote sensing (electromagnetic observation of surface conditions), the Internet of Things (IoT; networked sensors for real-time monitoring), and unmanned aerial vehicles (UAVs, or drones; aerial platforms for high-resolution surveys) constitute the primary emerging approaches for effective soil data collection (de Paul Obade & Gaya, 2021). Their integration improves accuracy, efficiency, and decision-making in soil science (Peets et al., 2012; Wu, Zhang et al., 2023).

2.2 Importance in agriculture

Soil data and information form the essential foundation for acquiring precise knowledge, effective planning, systematic action, and rigorous follow-up in various agricultural and environmental activities. A diverse array of remote sensing technologies, including but not limited to LANDSAT, hyperspectral sensors, synthetic aperture radar (SAR), and light detection and ranging (LIDAR), provide a rich variety of useful soil-related information at different levels of detail and accuracy (Wu, Zhang et al., 2023). These advanced technologies enable one to thoroughly assess soil type and mineralogical composition, soil moisture content, organic matter levels, salinity, and many other key soil properties of interest. With the demand for food continuing to rise due to the ever-growing global population, reliable soil data are becoming increasingly paramount in enhancing crop productivity and optimizing the effective application of limited resources (Lal, 2015). Continuous monitoring of soil data over vast spaces and extended timeframes facilitates a significantly better understanding of the complex soil environment and plays a critical role in helping both agricultural and environmental sectors make informed decisions regarding land management practices. In this crucial regard, satellite and airborne sensors have been extensively investigated and utilized in soil surveys and assessments, demonstrating their capacity to deliver vital inputs and insights toward achieving sustainable development goals (Peets et al., 2012; Postolache et al., 2022). Fig. 20.1 illustrates the hierarchical integration of soil data collection technologies, showing satellite remote sensing providing broad-scale monitoring, UAV platforms delivering high-resolution aerial surveys, and IoT ground sensors enabling continuous point-specific measurements. Central soil profile highlights key monitored parameters (moisture, pH, nutrients, organic matter) with data flow arrows converging toward cloud-based analytics platforms, demonstrating the complementary nature of these systems for comprehensive soil health assessment.

3. Remote sensing technologies

Remote sensing utilizes a range of electromagnetic sensors to effectively observe, scan, or monitor various objects and conditions on the Earth's surface. This monitoring process generally covers

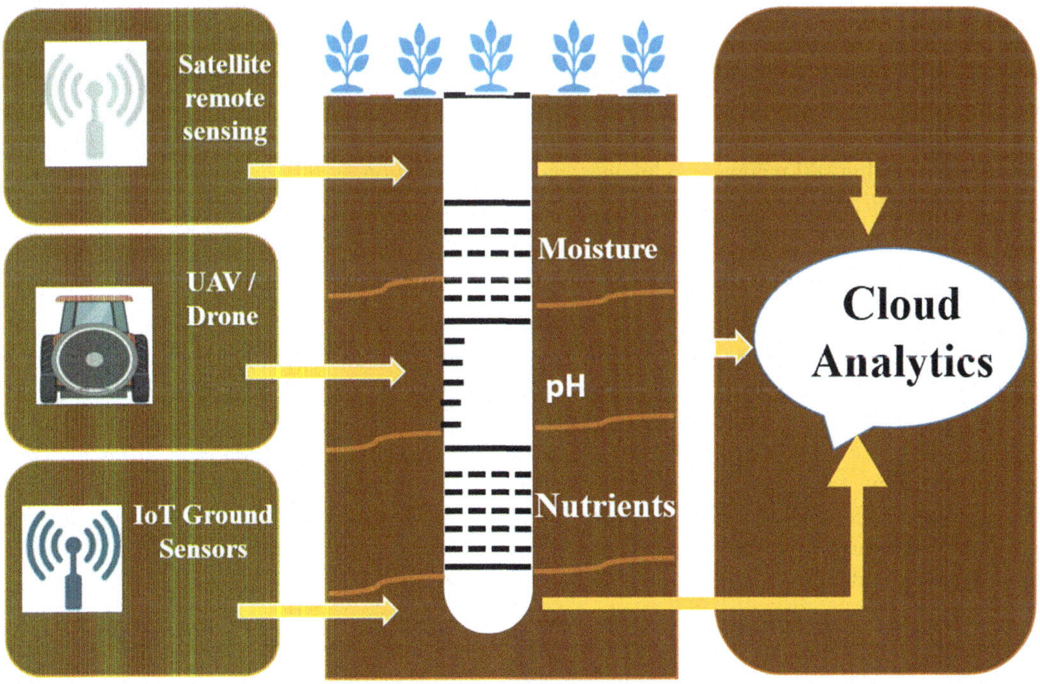

FIGURE 20.1

Integrated framework of emerging soil data collection technologies. A conceptual diagram illustrating satellite remote sensing, UAV platforms, and IoT sensor integration for comprehensive soil monitoring.

significantly larger areas compared to other technologies that may be utilized. Remote sensing systems can be categorized into two main types: active systems, which generate their own electromagnetic signals, and passive systems, which measure existing electromagnetic signals, such as sunlight. One common application of active remote sensing systems in agriculture is the use of LiDAR technology, which offers precise measurements of physical characteristics. In contrast, passive remote sensing systems often incorporate multispectral and hyperspectral imagery. Hyperspectral imagery is akin to multispectral imagery; however, it collects data across a much broader range, capturing data from many more and narrower bands throughout the electromagnetic spectrum. This typically spans from the visible spectrum through to thermal infrared wavelengths. In a general sense, remote sensing technologies enhance the ability to develop repeatable, consistent, and continuous monitoring methodologies that provide crucial information about soils, crops, and additional features of the landscape. As a result, these technologies present significant promise for integration into agricultural decision-support systems, which can guide farmers in making informed decisions based on reliable data. Remote sensing techniques offer a range of frameworks for effectively mapping and monitoring various soil processes. For instance, hyperspectral imagery captures data from many contiguous bands, generating immense interest in soil analysis due to its unparalleled vertical and horizontal resolution. Through remote sensing, a wide range of critical soil characteristics can be estimated, including texture, pH levels, carbon content, moisture content, porosity, salinity, mineral content, and temperature. Key climatic and biological factors in the soil, such as soil temperature, the presence of plant residues, available nitrogen, and the content of organic matter, play vital roles in influencing seed germination, plant growth, and overall crop yield. Consequently, collecting and integrating environmental information into cropping models is of pivotal importance. The use of remote sensing displays remarkable potential in consistently monitoring various soil and crop characteristics, which are essential processes that affect both environmental sustainability and human safety. An emerging trend in geo-education within the soil sciences is anticipated to focus on enhancing the utility and effectiveness of existing soil mapping techniques and the information available regarding land resources, thus contributing to improved agricultural practices and resource management (Pinter et al., 2003).

3.1 Types of remote sensing

Remote sensing technologies offer a promising alternative for collecting soil data, enabling the observation and measurement of soil parameters without direct contact. These methods include satellite-based and hyperspectral remote sensing (Pinter et al., 2003). Satellite remote sensing utilizes various spectral bands—optical, thermal, microwave, and radar—to assess soil characteristics. The selection of spectral bands depends on the specific soil information required, and different spatial resolutions allow for targeted observation of soil parameters. The advantages of satellite remote sensing include extensive spatial coverage, long-term archive availability, and cost-effective data acquisition. However, limitations arise from factors such as canopy cover, terrain shading, and soil moisture variability, which can affect data quality.

Hyperspectral remote sensing captures data across hundreds of narrow spectral bands, offering detailed reflectance measurements down to sub-pixel scales. This approach enhances soil analysis by providing comprehensive spectral information that supports the quantification of various soil properties. Although stations equipped with hyperspectral sensors may be costly, the technology's precision and depth of data contribute to its increasing adoption in soil monitoring.

The use of remote sensing technologies for soil analysis has advanced alongside the integration of complimentary tools such as IoT and UAVs. These methods collectively address the demand for affordable, non-invasive solutions that can capture soil data at diverse spatial scales and index data to specific geographic locations. By overcoming challenges related to traditional sampling methods—such as expense, scale limitations, and invasive procedures—remote sensing and related emerging technologies contribute to the ongoing transformation of soil data collection strategies.

3.2 Applications in soil analysis

Emerging technologies employed in soil data collection encompass remote sensing, IoT devices, and UAVs. Precisely, remote sensing offers satellite, hyperspectral, thermal, visual, and multispectral data acquisition for soils. IoT devices in this category include soil sensors and their associated network connectivity. Meanwhile, unmanned aerial vehicles facilitate the acquisition of soil data from fixed-wing, multi-rotor, or helicopter platforms (Landeta-Escamilla et al., 2019).

Optical and spectroscopic techniques hold a prominent position among emerging technologies used for soil assessment, enabling the detection of physicochemical and biological information through spectral analysis of field-specific surfaces. Alternatively, sensors affixed to mobile platforms or vehicles facilitate the in situ estimation of soil properties by acquiring reflectance or radiance spectra. Reflectance spectroscopy caters to temporal monitoring, incorporates spatial context, and tends to be non-invasive; conversely, spectroscopic inverse modeling (SIM) excels in optimally characterizing soil composition. The suitability of soil mapping, laboratory, or field-scale applications influences the selection among these techniques (Nawar & Mouazen, 2019).

Remote sensing technologies enable the derivation of several soil-related parameters, including soil moisture, bulk density, temperature, transpiration, organic carbon, and salinity levels. Nevertheless, the frequent presence of clouds constitutes a limitation within the geographical context of Mexico.

IoT devices actively monitor soil and crop development by transmitting data remotely for analytical purposes. These devices encompass various commercially available options that measure parameters such as temperature, moisture, conductivity, pressure, salinity, relative humidity, and pH. Network connectivity is enabled through Wi-Fi (WLAN), cellular networks, LoRa, Sigfox, or ZigBee systems. Data received from sensors undergo delivery to centralized management systems capable of processing and executing predefined actions upon data reception.

UAVs play an increasingly significant role in soil data collection activities. They acquire a comprehensive range of optical and spectroscopic data types from vantage points unavailable to other aerial platforms. Besides their primary function of soil data acquisition, UAVs also operate as communication extenders, enhancing IoT transmitters beyond the scope of traditional Internet coverage. Mandated by regulatory frameworks stipulating specific operational zones and distances, UAV UAS operate effectively within urbanized protected areas, expanding their applicative reach.

Remote sensing technologies revolutionize soil data collection by providing non-invasive, large-scale monitoring across diverse spectral bands and platforms. Table 20.1 systematically compares multispectral, hyperspectral, thermal, LiDAR, and radar systems, detailing measured soil parameters, spatial resolutions, advantages, and limitations, enabling researchers to select optimal technologies for specific soil monitoring objectives.

Table 20.1 Remote sensing technologies for comprehensive soil parameter assessment.

Technology	Soil parameters measured	Platforms	Spatial resolution	Advantages	Limitations	Citations
Multispectral imaging	Moisture, organic matter, texture	Satellite (LANDSAT), UAV	10–30m (satellite), 5–10 cm (UAV)	Wide coverage, cost-effective	Cloud interference, surface only	Pinter et al. (2003); Wu, Zhang et al. (2023)
Hyperspectral sensing	pH, carbon content, minerals, salinity	Airborne, ground-based	1–5m	High spectral detail, mineral detection	High cost, data processing intensive	Gomez et al. (2008); Nawar and Mouazen (2019)
Thermal infrared	Soil temperature, moisture	Satellite, UAV	30–100m	Non-contact temperature mapping	Affected by vegetation cover	Babaeian et al., (2019)
LiDAR (light detection)	Bulk density, topography, porosity	Airborne UAV	0.5–2m	3D soil surface mapping	Expensive, limited penetration	Shi et al., (2016)
Synthetic aperture radar (SAR)	Moisture, texture (wet conditions)	Satellite	1–20m	All-weather, penetrates clouds	Complex data interpretation	Wang et al., (2023)
Visible-NIR spectroscopy	Organic carbon, nutrients	Ground sensors, proximal	1–10 cm	Real-time, non-destructive	Calibration required	Viscarra Rossel et al., (2011)
Microwave remote sensing	Moisture, salinity	Satellite	25–50m	Deep penetration, dynamic monitoring	Lower resolution	Robinson et al., (2008)
Radar (active)	Surface roughness, erosion	UAV, satellite	1–5m	Day/night operation	Speckle noise	d'Oleire-Oltmanns et al., (2012)
VNIR field spectroscopy	Clay content, iron oxides	Portable spectrometers	Field-scale	High accuracy for minerals	Weather dependent	Landeta-Escamilla et al., (2019)
Thermal imaging	Evapotranspiration, heat flux	UAV thermal cameras	5–20 cm	Identifies stress zones	Atmospheric interference	Maes and Steppe (2019)

3.3 Advantages and limitations

Each technology offers distinct advantages and limitations for acquiring information about soil quality and its spatial distribution across arable and rangeland environments. The choice among them depends on factors such as operational range and scale, repeat frequency, site accessibility, spatial resolution goals, data quantities and formats, incident angle, and weather conditions (Adamchuk et al., 2004). All three—remote sensing, the Internet of Things (IoT), and unmanned aerial vehicles (UAV)—enable the acquisition of digital soil data over extensive areas with relatively high spatial detail. However, the deployment of remote sensors does not require personnel to be on-site; data can be obtained without field visits by using pre-existing satellite or airborne acquisitions for locations within the receiving range. The broad availability of such imagery facilitates the examination of areas that are not easily accessible by land vehicles or individuals. Satellite platforms typically provide information from the visible through the middle and thermal infrared spectral intervals, while many commercial airborne systems extend further into the far infrared. In contrast, regional (field-scale) surveys typically require the deployment of in situ equipment as part of IoT or UAV operations. Such local engagements yield more intensive spatial information.

4. Internet of things (IoT) in soil monitoring

The Internet of Things (IoT) represents a strategic approach to soil data collection, comprising a wide range of devices that gather and transmit information on soil conditions and quality. The resulting data sets capture essential soil properties under diverse geographical and environmental conditions. The selection of an appropriate soil monitoring system depends on the specific applications and soil parameters of interest. Soil moisture data, a critical component of multiple soil-probing technologies, is gathered using a variety of sensors, including piezometers, tensiometers, block sensors, resistance sensors, and capacitance sensors, as illustrated by Wu, Zhang et al. (2023). Soil salinity sensors complement these devices by enabling the detection of ion concentrations in the soil. In a comprehensive IoT-enabled monitoring framework, IoT devices survey soil conditions and transmit the collected data to edge devices deployed within the monitoring region. These edge devices process, analyze, and ultimately upload the processed data to cloud platforms. In the cloud environment, soil data are stored and provided for access by users and applications. A variety of technical challenges associated with communication, routing, and data-processing delay management for soil-probing IoT multisensor data are thereby introduced, as discussed by Chamara et al., (2022).

4.1 IoT devices for soil data collection

The detection and collection of soil data, vital for precision agriculture and environmental monitoring, have traditionally relied on remote sensing technologies and manual sampling methods. These methodologies often produce data that are delayed, coarse-grained, or spatially limited, necessitating the emergence of alternative approaches capable of offering near-real-time information at high spatial resolutions.

Recent years have witnessed the deployment of a wide array of Internet of Things (IoT) devices designed to capture soil-related variables, including temperature, moisture, pH value, salinity, micronutrients, and organic compounds. IoT sensors can be situated in fixed positions to provide continuous, point-specific data or mounted on moving platforms, including robots, Unmanned Aerial Vehicles (UAVs), or vehicles, to perform detailed surveys over broader areas (Wu, Zhang et al., 2023). The choice of IoT technology hinges on application-specific demands, balancing factors such as coverage, energy consumption, durability, and cost before proceeding to the sensor installation phase. The abundance of data generated by IoT deployments challenges existing open-source platforms, which often fall short of providing high-capacity cloud storage and possess limited analytical capabilities. Consequently, the successful implementation of IoT-based soil monitoring has largely depended on customized solutions that integrate both hardware infrastructures and software systems tailored to manage and effectively exploit large data volumes (Talavera et al., 2017).

4.2 Data transmission and management

One of the advantages of the Internet of Things (IoT) is the combination of simple and cheap sensors into complex networks that capture a wide range of measurements and then make this data available to a wide base of data users to support decision-making (Wu, Zhang et al., 2023). They installed a soil data perception module in the farmland, deployed a gateway and base station that supports wireless functions, detected soil data through the perception module, uploaded the soil data to the base station via the gateway, and finally synthesized and uploaded the data to the host. The host comprises data servers, application servers, and web servers. The application server hosts the data analytic services, which include historical data analysis, soil quality anomaly detection, and soil quality prediction. The web server manages the interaction between the smart farming application and the other farm management systems.

4.3 Case studies of IoT applications

Internet of Things (IoT) technologies represent a rapidly evolving avenue for harvesting precise and dynamic soil data to enhance agricultural productivity. Diverse sensors—such as static, vision, thermal, proximity, electrochemical, and microwave—enable the collection of granular details on moisture, temperature, nitrogen, and pH levels, among other pertinent parameters in the soil–plant nexus. Typically, IoT systems deployed at the field level comprise several soil-monitoring units linked to a central gateway (Wu, Zhang et al., 2023). These devices convert analog or digital sensor signals into electrical currents for relaying soil metrics, including temperature, moisture, and pH values. Upon aggregation, the gateway transmits this information over mobile cellular networks to stakeholder platforms. Raw data—once collated—undergo categorization, storage, and exploration within a centralized database, facilitating a seamless conduit between on-site measurements and regulatory entities. Incorporating the Efficient Data Allocation Algorithms (EDAA) framework further optimizes data transmission efficiency and prolongs the operational lifespan of IoT nodes (Sheng et al., 2013). People, on the other hand, can interface with these platforms through common browsers and phones to access real-time, localized soil conditions and forecasts, potentially forestalling adverse events.

5. Unmanned aerial vehicles (UAV) for soil assessment

Emerging methods for collecting soil data include the use of portable scanners, remote sensing, Internet of Things (IoT) devices, and unmanned aerial vehicles (UAVs). UAVs, categorized into fixed-wing, single-rotor, and multi-rotor platforms, offer unique capabilities for large-area surveillance and close-profiling tasks. Leaf-area index, surface spectral reflectance, surface temperature, soil color, soil surface microtopography, soil moisture, and weed count can be mapped via multispectral sensors, thermal sensors, laser profilers, ground-penetrating radar, and hyperspectral sensors mounted on UAVs. Additionally, UAV data can be integrated with ground-based information to enhance monitoring strategies for pest infestations and canopy development (d'Oleire-Oltmanns et al., 2012). Soil data acquisition is fundamental to agricultural and environmental management. Emerging technologies have therefore converged on automated and digital data-gathering systems.

5.1 Types of UAVs used in soil studies

Several types of unmanned aerial vehicles (UAVs) are utilized in soil data-collection activities, varying in design, size, and capabilities (Shi et al., 2016). Different UAVs can be selected to suit the specific requirements of a given mission, depending on factors such as the extent of coverage and the desired level of detail in the information.

The principal categories of UAVs include rotary-wing and fixed-wing platforms. Rotary-wing UAVs are equipped with multiple propellers that generate the necessary thrust. Such devices can take off and land vertically, hover close to features of interest, and carry out detailed inspections or measurements. However, their relative complexity reduces endurance and limits the area that can be covered in a single operation.

Fixed-wing UAVs feature one or more rigid lifting surfaces and generate lift via forward flight. These aircraft typically require a runway, catapult, or launcher for deployment, but they are generally capable of higher cruise speeds and longer flight endurance than rotary-wing systems. Fixed-wing UAVs are therefore better suited for complex, extensive surveys, but may require specially designed sensors to operate effectively at the increased relative speed.

Currently, rotary- and fixed-wing UAVs are the dominant categories in soil-related data collection activities (Sona et al., 2016). Several subtypes of UAVs exist, each with specific roles in acquisition. The platforms can carry instruments and payloads, including cameras and sensors suitable for surveys. Digital images and spectral scans enable users to generate orthorectified or photogrammetric products, whereas light detection and ranging (LiDAR) instruments provide topographic and volumetric information.

5.2 UAV-measured soil parameters

UAVs equipped with multispectral, hyperspectral, thermal, LiDAR, and GPR sensors map surface and near-surface soil properties at centimeter resolutions, bridging gaps between satellite and ground data. Key parameters include moisture (via NDMI from multispectral), organic matter and texture (hyperspectral reflectance), temperature (thermal IR), microtopography/porosity (LiDAR), and salinity/color (RGB/multispectral). These enable erosion risk assessment, compaction mapping, and fertility zoning (Shi et al., 2016; d'Oleire-Oltmanns et al., 2012). Table 20.2 presents the UAV sensor capabilities compared by measured parameters, resolution, advantages, limitations, and citations, highlighting their role in high-resolution soil mapping.

5.3 Integration examples

UAV-IoT fusion: Drones relay IoT ground data via onboard gateways, creating hybrid networks for real-time validation (e.g., UAV hyperspectral calibrates IoT moisture sensors; Platel et al., 2025).

Table 20.2 UAV technologies for soil parameter assessment.

Technology	Soil parameters measured	Platforms	Spatial resolution (cm)	Advantages	Limitations	Citation
Multispectral UAV	Moisture, texture, organic matter, salinity	Multi-rotor	5–10	High-res NDVI/ NDMI, affordable	Vegetation interference	Shi et al., (2016)
Hyperspectral UAV	Minerals, carbon content, pH proxies	Fixed/multi-rotor	2–5	Detailed spectra for composition	Data volume, processing	Sona et al., (2016)
Thermal UAV	Temperature, moisture flux	Multi-rotor	10–20	Stress/flux detection	Atmospheric effects	Maes and Steppe (2019)
LiDAR UAV	Bulk density, topography, porosity	Fixed-wing	1–5	3D structure mapping	Cost, no subsurface	Igwenagu et al., (2025)
GPR UAV	Subsurface moisture, compaction	Specialized drone	10–50	Penetration to 1m	Speed limits, soil type	d'Oleire-Oltmanns et al., (2012)

UAV-satellite synergy: High-res UAV fills cloud gaps in Sentinel/Landsat imagery for moisture mapping (Wu, Zhang et al., 2023). Sensor fusion platforms: Multi-payload UAVs (LiDAR + thermal) integrate with GIS for 4D soil models in erosion studies (Igwenagu et al., 2025). These enhance precision by 20%–30% in variable fields.

6. Comparative analysis of technologies

Emerging technologies for soil data collection include remote sensing, Internet of Things (IoT), and unmanned aerial vehicles (UAVs). The three technologies augment traditional methods such as Automated Soil Sampling (ASS) and Proximal Soil Sensing (PSS) (Viscarra Rossel et al., 2011). ASS generates high-quality measurements but is expensive and time-consuming. PSS achieves high spatial and temporal resolutions with relatively low operational costs, yet suffers from a lack of accuracy and sensibility (Viscarra Rossel et al., 2011).

Remote sensing encompasses both passive and active systems, as well as ground-based and aerial systems. Each subset offers different solutions to problems related to spatial scale, temporal resolution, and spectral resolution (Mulla, 2013). Ultimately, remote sensing enables the capture of spatial and temporal variabilities over a wide area with high temporal resolution. The key factors restricting the use of remote sensing technologies are weather sensitivity, time integration, costs, and ease of access (Mulla, 2013). Furthermore, remote sensing reflects the fact that only surface information is captured, contrary to the interest in volume.

The IoT enables real-time data acquisition of various soil parameters through the deployment of a large number of sensors. The technology creates an infrastructure for developing advanced applications and decision support tools relevant to numerous domains, including agriculture, precision irrigation, and soil conservation (Villa-Henriksen et al., 2020). The massive increase in volume of soil data leads to remote storage and analysis in a cloud computing environment. The methodology is suitable for both homogeneous and highly heterogeneous monitoring fields and is robust to environmental stimuli (Villa-Henriksen et al., 2020).

UAVs are a promising platform for soil data acquisition, with four major changes arising in soil data capture. The first is the type of soil data collected. They offer easy access to characterize surface data and potentially sub-soil parameters. Secondly, the spatial and temporal scales of soil parameter measurements change. Thirdly, the methods of soil data collection have evolved to include active and passive remote sensing techniques, utilizing novel sensors. Finally, the complementarity of UAVs with remote sensing, IoT, and satellite systems constitutes a central development in soil data capture. Table 20.3 presents key IoT devices and UAV types with their sensors, measured parameters, data transmission methods, applications, and challenges, illustrating their complementary roles in comprehensive soil health assessment systems.

6.1 Cost-effectiveness

Precision agriculture involves regularly collecting detailed data on physico-chemical, biological, and meteorological soil characteristics to improve agricultural production (Reynolds et al., 2019). Soil data collection has seen significant technological advances; however, many techniques are often expensive due to initial investment, maintenance, and calibration requirements. Cost-effectiveness emerges as a

Table 20.3 IoT and UAV platforms: Real-time soil monitoring capabilities.

Technology	Key sensors	Measured parameters	Data transmission	Applications	Challenges	Citation
IoT soil moisture sensors	Capacitance, TDR	Moisture, temperature	LoRa, ZigBee, NB-IoT	Precision irrigation	Battery life, connectivity	Wu, Zhang et al. (2023); Talavera et al. (2017)
Electrochemical IoT probes	Ion-selective electrodes	pH, EC, nutrients (N,P,K)	WiFi, Cellular	Nutrient management	Sensor drift, calibration	Postolache et al., (2022)
Multi-sensor IoT nodes	Temperature, humidity, pressure	Soil climate, salinity	GPRS, SigFox	Smart farming	Data overload	Chamara et al., (2022)
Fixed-wing UAV	Multispectral cameras	Moisture, NDVI, texture	Onboard storage	Large area mapping	Requires runway	Shi et al. (2016); Sona et al. (2016)
Multi-rotor UAV (drone)	Hyperspectral, LiDAR	Organic matter, topography	Real-time telemetry	High-resolution surveys	Limited flight time (20–30min)	Maes and Steppe (2019)
Single-rotor UAV	Thermal, RGB cameras	Temperature, erosion mapping	4G/5G downlink	Precision agriculture	High cost	d'Oleire-Oltmanns et al., (2012)
IoT gateway systems	Multiple sensor fusion	Comprehensive soil health	Cloud platforms	Farm management	Cybersecurity	Sheng et al. (2013)
UAV-LiDAR integration	Laser scanners	Soil microtopography	RTK GPS	Erosion monitoring	Vegetation interference	Igwenagu et al. (2025)
NB-IoT networks	Low-power sensors	Long-term moisture, salinity	Narrowband IoT	Remote areas	Low bandwidth	Weizhong et al. (2019)
Sensor fusion UAV-IoT	Combined payloads	All parameters	Hybrid networks	Complete monitoring	Data integration complexity	Platel et al. (2025)

key consideration when developing and community-adopting large-scale ground observation technologies.

An open hardware design strategy employed commercially available electronics and sensors to produce soil moisture data loggers (Reynolds et al., 2019). Usability increased through GPRS data transmission, which allows results to be collated off-site. Small workstations provide plot-level crop growth traits and micro-environment variables. Integrating multiple sensor types into single-board computers creates scalable, multi-point in-field networks to support decisions such as crop management (Fisher & Gould, 2012). Spatial variability modeling in greenhouse platforms reduces sensor count. Sensor networks advance breeding through "envirotyping," which identifies environmental scenarios that affect yield and the target population, facilitating the development of resilient genotypes (Xu, 2016). However, cost calculations for field and greenhouse phenotyping depend on assumptions, simplifications, and factors such as the device's market availability and intended application (Reynolds et al., 2019).

Agricultural geophysics utilizes non-invasive sensing to characterize soil variability and more accurately quantify soil-plant management relationships (Allred et al., 2008). Geospatial information on soil characteristics provides insights into optimal cultivation approaches and supports yield estimation. Attributes such as salt content, bulk density, minerals, pH, and temperature are evaluated by measuring the field to generate a land profile (Corwin & Lesch, 2005). Aproximated spatial variability helps understand water flow, identify irregular spots, and determine the requirement for additional nutrients or irrigation. Soil moisture is estimated mostly in situ with moisture meters, time-domain reflectometers, and electromagnetic induction (EMI) devices (Robinson et al., 2008). These instruments deliver data only from discrete points, which can translate into less efficient tactics when applied over larger fields. Geospatial measurement of soil electrical conductivity obtains continuous or sparse measurements across the field. Unmanned Ground Vehicles (UGVs) and aerial vehicles, such as UAVs, perform data collection, enabling on-the-fly decision-making for field management. A collaborative approach that integrates ground and aerial robots can yield more detailed and broader inspections. Several methods introduced by the literature exploit robots to collect soil moisture measurements.

6.2 Data accuracy and reliability

Advanced soil data collection technologies have shown significant potential for enhancing the accuracy of collected data, both in terms of spatial location and measurement precision (Babacian et al., 2019). Unprecedented volumes of spatial and temporal data on climate, soils, and nutrients can now be acquired by sophisticated digital technologies, increasing the range and complexity of information available to the adoption of digital agriculture. Soil moisture varies dynamically in response to infiltration, evaporation, and redistribution within the soil profile. Reliable capture and analysis of such dynamic variation is essential for complementary crop modeling. Processing of temporal and spatial moisture information with other data layers, such as crop health and soil chemical constituents, refines understanding of nutrient and water stress (Seneviratne et al., 2010). Initial developments of novel data acquisition methods have contributed to an expectation among farmers and stakeholders that continuous monitoring over a range of parameters can be provided. Constant monitoring of soil, crop, weather, and ambient data on a range of configurable parameters in text, image, and video formats for any field or plot anywhere on the globe is considered achievable.

6.3 User accessibility

User accessibility is crucial for encouraging widespread adoption of emerging soil data collection technologies while minimizing the associated training, labor, maintenance, and operational costs. Devices that require extensive calibration or specialized local knowledge pose significant obstacles to operational implementation (Peets et al., 2012). Emerging technology represents a fundamental shift like soil data acquisition, transforming it from a manual, inference-based process into a faster and more scalable data capture method. However, this change introduces an operational challenge in the acquisition and analysis of large volumes of heterogeneous data collected at high spatial resolutions with high temporal sampling frequencies. Instrumentation that enables scalability in terms of data acquisition or station establishment consequently produces large volumes of heterogeneous information that require expertise and specialized platforms to collect, process, and analyze. A platform capable of managing multiple sensors simultaneously deployed to track, process, and store data facilitates real-time decision-making, eliminates the need for bespoke data-management schemes, and reduces the level of specialist knowledge required to deploy data-acquisition instrumentation (Piccoli et al., 2022).

6.4 Practical case studies

Real-world examples show how IoT, UAVs, and remote sensing improve soil sampling accuracy in farms by supplying high-resolution, real-time data that lowers the variability and human error associated with traditional grid sampling.

6.4.1 IoT sensors

Variable-rate irrigation and fertilization are made possible by wireless soil probes (such as capacitance and TDR types) that track moisture, pH, nutrients, and salinity at various depths. LoRaWAN sensors used precise dosing based on real-time data to reduce water use by 40% in an almond farm in California, and a citrus orchard study optimized inputs to increase yields. LTE-M/NB-IoT networks are used for continuous monitoring across thousands of hectares in Scandinavian projects, increasing accuracy over intermittent manual testing.

6.4.2 UAVs and drones

Soil moisture, texture, and topography are mapped at cm-scale resolution by multirotor drones equipped with multispectral, thermal, and LiDAR sensors (80%–90% accuracy). Throughout the season, the STaPLE Drone autonomously gathers physical soil and leaf samples, allowing for frequent nutrient analysis to identify deficiencies early and avoid overfertilization. Vodafone's New Zealand system uses GPS-equipped drones to map fertilizer, identify waste, and refine applications in precision farming.

6.4.3 Remote sensing

Targeted sampling is guided by maps of soil variability for moisture, organic matter, and pH created by remote sensing satellites (like LANDSAT) and hyperspectral imagery. In US farms (more than 50 million acres), GIS integration evaluates suitability, with input zoning supported by nutrient estimates derived from the NDVI.

7. Challenges in soil data collection

Several challenges, including complexity, cost, and reliability, limit the widespread adoption of emerging technologies for soil data collection.

Soil data underpins decision-making across agricultural, ecological, and hydrological domains, yet precise and timely data acquisition remains a challenge. Traditional laboratory-based methods such as chromatography and spectrometry provide high accuracy and specificity but are resource-intensive and inflexible. Emerging techniques—particularly remote sensing, the Internet of Things (IoT), and unmanned aerial vehicles (UAVs)—offer pathways to higher data granularity and enhanced decision support.

Remote sensing utilizes the spectral responses of the soil surface to assess various variables, including temperature, moisture, organic carbon, texture, structure, nutrient availability, pH, and salinity. Satellite-based systems offer broad spatial coverage, while hyperspectral imaging enables the detection of mineralogical and organic components; however, both face cost and data processing hurdles, and satellite measurements are hindered by cloud cover and environmental variability (Wang et al., 2023). IoT devices—ranging from electrochemical sensors to embedded systems—facilitate real-time monitoring of individual soil attributes, often transmitting data wirelessly to centralized platforms. The resulting information flows involve multivariate data, often sampled irregularly and containing missing values, thus posing significant interpretative and management challenges (Postolache et al., 2022). Similarly, UAVs capture high-resolution soil information through remote sensing or proximal soil interaction; their integration with IoT and satellite systems further enriches data collection capabilities. Comprehensive comparative studies highlight disparities in equipment costs, data accuracy, and accessibility among the three technologies, informing strategic selection that aligns with specific application requirements.

7.1 Technical challenges

Emerging technologies widely applied for collecting soil data include remote sensing, IoT systems, and unmanned aerial vehicles (UAVs). These technologies are each useful for supporting the establishment of soil-monitoring regimes by farmers or other land-users. Requirements for each regime vary, and the proper selection of an appropriate set of technologies may involve considering numerous factors, including budget, desired data accuracy, and terrain accessibility.

Technical challenges across the sensing sector introduce several issues to implementation. The widespread use of IoT technologies raises concerns about both data privacy and the ability of existing regional regulatory frameworks to adapt to omnipresent sensors throughout managed areas. Complementary regulatory gaps exist alongside a shortage of experienced personnel capable of preserving and maintaining the technological infrastructures on which these instruments depend. Global emergence of application-level communication solutions and cloud-based data-processing platforms that support this body of technologies in soil monitoring nevertheless enables frameworks that combine methods in parallel. Large-scale collection of aggregated information from multiple sources allows soil scientists to characterize a wider breadth of soil conditions and more accurately determine the effects particular environmental factors have on their evolution (Peets et al., 2012; Wang et al., 2023).

7.2 Data privacy and security concerns

Environmental data must be public and openly shared to contribute to the collective pool of knowledge (de Paul Obade & Gaya, 2021). However, discussion of data collection, utilization, and distribution at regional, national, and international levels has renewed interest in the associated privacy challenges (De Capitani di Vimercati et al., 2007). The demand for prompt information to combat global issues, such as climate change and peacekeeping, will increase, requiring further expansion of environmental monitoring (EM) capabilities. These capabilities utilize instruments to collect data from controlled environmental phenomena, synthesize the data into information via electronic methods, and then distribute it to end users. EM systems can be centralized, distributed, or remote sensing. Centralized systems transfer all raw acquired data to a single central unit where data fusion occurs, and a single signal is generated and then disseminated. Distributed systems perform decentralized computation of local EM processes, which are then shared among nodes for global evaluation. Remote sensing systems gather information remotely without any physical contact between the instrument and monitored scene or object. Challenges facing the IoT and sensor networks have been outlined. However, concerns relating to privacy and security are most pronounced when dynamic systems, such as IoT, which involve human interaction in some manner, are considered. With the high growth projected for IoT, the issue should be tackled on a much larger scale than was necessary for WSNs. The discussion around drone safety should also form the basis of IoT privacy, because many of the exemplary comments, prevalent concerns, and many issues raised are quite similar. When adapted to Internet legislation, the ramifications of proper implementation become extensive, as IoT encompasses many of the concerns found within both the Internet and drone legislation. For a country to develop an effective and appropriate legislative framework, it must initially assess the specific requirements of its economy, identify potential threats and risks, and adapt accordingly to suit its unique setting.

7.3 Regulatory and compliance issues

Technologies underpin monitoring systems, but privacy and regulatory concerns remain (de Paul Obade & Gaya, 2021). For example, using proprietary hardware and software could limit interoperability and dictate recourse in the event of disputes (Kaur et al., 2022). The use of unmanned ground systems and aerial vehicles by military, law enforcement, and border patrol has already prompted legislation covering treaties, liability, data security, privacy issues, and more (Dutta et al., 2023). Laws and regulations vary by state and municipality; for example, the use of GPS tracking systems to monitor employees in California requires express written authorization (Arias v. Intermex Wire Transfer, 2019). Meanwhile, differing Occupational Health and Safety Administration requirements and regulations can will impede their dissemination across the supply chain. Hence, design considerations may include robust encryption, oversampling, deduplication, and other techniques to protect data and enable verifiability (de Paul Obade & Gaya, 2021; Wu, Zhang et al., 2023).

Initiatives like the Open Geospatial Consortium (OGC) SensorThings API offer standards, but adoption lags. Diverse IoT sensors, UAV payloads, and remote sensing platforms generate heterogeneous formats (e.g., CSV vs. GeoTIFF, varying spectral resolutions), making fusion and analysis more difficult. Multivendor networks are hampered by poor interoperability between proprietary systems (LoRaWAN vs. NB-IoT), and calibration varies by manufacturer (e.g., capacitance probes require site-specific soil corrections). Precision agriculture will be able to scale smoothly by addressing these through open-source calibration tools and unified protocols (such as ISO 19115 for metadata).

8. Future trends in soil data collection technologies

Data acquisition using soil sensors is playing an increasingly important role in sustainability and environmental protection. Trends in soil data collection technologies include the continuous reuse of sensors, low-power design, AI-based sensors, and integration (de Paul Obade & Gaya, 2021). The reuse of the sensor and low-power design can reduce waste materials and labor costs associated with data collection, respectively (Katharria et al., 2024). AI-based sensing helps reduce the reading interval, providing a more continuous and rapid response for soil monitoring. Finally, integrating the sensor enables fixed installation and integration with autonomous robots and UAVs, allowing for continuous, long-term remote operation without human interference (Igwenagu et al., 2025).

The merger of these recent technological advances with artificial intelligence/machine learning (AI/ML) paves the way for more advanced sensing solutions for soil data acquisition and monitoring. There is a growing need for soil-quality information worldwide, which supports impending food safety and security policies, as well as environmental protection efforts (de Paul Obade & Gaya, 2021). Real-time and continuous data analysis becomes a major focus of the growing soil-monitoring infrastructure. Precision agriculture now demands an advanced characterization of soil properties, particularly the spatial variability of several constituents related to soil quality, crop development, and water circulation (Faqir et al., 2024).

Measurement of soil constituents in real-time and continuous modes is still in its early stages of research, despite numerous analyses that can be carried out on-site. The use of ion-selective electrodes (ISEs), optical sensors, and absorption techniques, among others, offers a solution (Zhang et al., 2024). Integrating AI/ML and big data in the near future will help resolve problems related to crop cultivation in extreme environments and challenges to soil quality resulting from management, pedoclimatic conditions, and global changes that influence both human and environmental health (Katharria et al., 2024).

8.1 Advancements in sensor technology

Emerging sensor technologies for soil data collection have transformed the farming industry by providing real-time information during the growing season. Although the integration of various sensors provides a plethora of useful data, reliability and repeatability are challenging (Dhillon et al., 2010). Current applications include soil moisture monitoring and soil compaction mapping (Hardie, 2020). Early studies encourage the mapping of additional variables, such as soil organic matter, nitrogen, and pH, to optimize fertilizer applications (Igwenagu et al., 2025). Sensory data can be processed for mapping or integrated with decision support systems, but the sequence of sensor data acquisition influences the quality presented to the farmer. Numerous sensors exist, each estimating a spectrum of soil characteristics. The field that has received considerable attention is soil compaction because current approaches are cumbersome and time-consuming. Traditionally, soil compaction is measured with tools such as penetrometers and analysis of soil water content (Davies et al., 2025).

Recent advancements focus on sensor technologies for continuous measurement of soil properties in space. Sensors based on soil water content measurement (time domain reflectometry, neutron probe, frequency domain reflectometry, dielectric constant) require further development, whereas sensors for soil organic matter (visible near- and mid-infrared, remote sensing, laser-induced fluorescence, X-ray fluorescence) show greater promise (Adamchuk & Rossel, 2010). Sensor fusion, which combines multiple sensor outputs to derive complementary information, offers a path to improved accuracy but is currently at an initial stage and would benefit from enhanced laboratory testing (Platel et al., 2025). The integration of various sensors onto a single platform is a key goal, as different sensors can create data-antagonistic effects if not properly sequenced and combined. The current state-of-the-art includes direct soil measurement, sensor fusion, and platform integration (Dhillon et al., 2010).

8.2 Integration of AI and machine learning

Artificial Intelligence and Machine Learning (AI/ML) have generated significant interest within the agriculture sector since 2017 (de Paul Obade & Gaya, 2021). The adoption of sophisticated algorithms into new processes enables the effective management of complex agricultural systems that were previously difficult to control (Katharria et al., 2024). Researchers are continuously improving machine-learning models to enhance accuracy by incorporating more parameters, and creating frameworks and guidelines to determine the most suitable machine-learning methods for specific agricultural tasks, such as soil monitoring (Faqir et al., 2024). The accessible application of AI/ML extends beyond large agricultural enterprises to include smaller operations through the use of unmanned systems and readily available computational resources. The recent increase in remote sensing data and the availability of high-performance computing resources have opened up new avenues for AI/ML applications in agriculture. Researchers have developed technological approaches to describe, measure, predict, classify, and map soil properties using available soil data, frequently utilizing extended datasets and empirical equations. Alternative predictive modeling techniques, notably those based on AI/ML, have been proposed to improve performance and predictive capabilities (Deshmukh & Deshpande, 2025). A variety of models, such as regressions, cluster analyses, classification procedures, and neural networks, may be employed to capture complex data patterns and derive rules applicable to soil prediction. Emphasis must be placed on collecting ample data and rigorously evaluating model performance to avoid overfitting and other modeling pitfalls (Katharria et al., 2024).

8.3 Potential for real-time data analysis

Real-time analysis of data collected by the emerging technologies discussed above would offer valuable information to farmers, allowing new possibilities beyond the employment of strong indicators such as vegetation indices (NDVI, NDWI) as a proxy for above-ground vegetation or soil moisture (de Paul Obade & Gaya, 2021). Remote-sensing sensors may record visible, near-infrared and thermal-infrared solar or temperature radiation, which contains a wealth of information about soil and vegetation (Nițu et al., 2025). Industrial-standard data-analysis procedures are available to process these data carefully. The vegetation indices are fast but rather simplistic methods, and a new generation of spectral-index analyses is emerging; these remain a relevant topic for further investigation (McCormick et al., 2025).

9. Case studies and real-world applications

Emerging technologies for collecting soil data are finding practical applications in smart agriculture worldwide. Examples include remote sensing, the Internet of Things (IoT) and unmanned aerial vehicles (UAV). With confidence drawn from these cases, others can replicate or adapt procedures and operations while anticipating and responding proactively to challenges and limitations."Methods and procedures for automatic collection and management of data acquired from on-the-go sensors with application to on-the-go soil sensors" (Peets et al., 2012) can be used to analyze the potentials developed in general procedures for on-the-go data acquisition from soil sensors. Methods and procedures used for data management relative to a farm managing information system (FMIS) build data management capability into collection "Internet-of-Things-Based Multiple-Sensor Monitoring System for Soil Information Diagnosis Using a Smartphone" (Wu, Zhang et al., 2023). illustrates IoT device versatility and application."Jirapond Muangprathub designed a control system utilizing node sensors to collect data on agricultural practices. A smartphone and a web system managed data, while a LINE API service delivered notifications to users," demonstrating device integration and information transparency. "Shadi Atalla proposed a new classification method for agricultural Internet of Things (IoT) based on structural and service features. Fixed and mobile scenarios relevant to IoT deployment were studied when evaluating the performance of six data forwarding schemes in 6LoWPAN networks (Atalla et al., (2023). New classification methods and performance metrics support optimisation of network deployment for specific applications. "Huan Juan developed a Narrowband Internet of Things (NB-IoT) based water quality monitoring system for aquaculture ponds. In this system, the water quality sensor functional blocks sent data collected on-site to the cloud platform in real-time, before being transmitted over long distances. Huan et al., (2020). "Shola Usharani designed a system using multiple sensors and IoT hardware to monitor agricultural fields, enabling remote observation to prevent losses caused by an animal invasion." (Usharani et al., 2023). Monitoring capability can be extended from soil-companion variables into complementary conditions of the wider environment. "Yang Weizhong developed an NB-IoT-based soil moisture monitoring system for real-time measurement" (Weizhong et al., 2019). The collected data have advantages and disadvantages, but a combination of remote sensing, IoT and UAV techniques can create a comprehensive picture. "Matteo Francia built fine-grained two- and three-dimensional soil moisture profiles using a sensor grid, demonstrating that three sensors placed appropriately can obtain soil moisture profiles that effectively describe the soil moisture distribution." (Zhuo et al., 2020).

9.1 Successful implementations

Successful implementations demonstrate the critical role of emerging technologies in soil science, as evidenced by field and laboratory trials. Various types of on-the-go sensors installed on tractors, mounted on unmanned aerial vehicles (UAVs), or integrated into sensor networks successfully deliver essential information for the preparation and operational phases of agronomic management. Methodologies are also developed for the automatic collection and management of soil and crop data in relation to a farm management information system (Peets et al., 2012). Sensors for on-the-go collection of soil and crop data have become essential for successful implementation of precision agriculture.

Emerging technologies such as remote sensing, IoT, and UAVs provide innovative means for data acquisition. Remote sensing includes satellite, aerial (Micasense), and hyperspectral imaging (Venμs). IoT applications use soil moisture sensors, weather stations, atmospheric sensors for soil nutrient estimation, heat flux sensors, and multi-specification integrated devices with GPS and communication modules that transmit information autonomously. UAVs encompass a range of multi-environment types for terrestrial surveys, capable of carrying various sensors, including LiDAR, RGB, and multispectral cameras, as well as satellite imagery. A comprehensive understanding of the structure and characteristics of big data is essential to meet the accelerating demand for precise data relevant to soil science, which underpins the effective management of soil resources and facilitates sustainable development of the involved region.

9.2 Lessons learned

Emerging technologies promise to alleviate many of the challenges in soil data collection. However, failures often characterize the use of new technologies in agriculture, delivering valuable lessons. Wi-Fi, for instance, does not prove to be a suitable communication mean in outdoor deployments. Radio technologies resonate well with large outdoor domains. Data transmission requires configuration of a software infrastructure, often involving third parties, and data management is an equally demanding activity. The sensor's lifetime is short, and calibrations must be repeated regularly. These technologies improve the collection, transmission, and storage of data, fostering cross-comparisons and repeatability while vehicles and hands take care of themselves. However, the number of sensors remains limited to tens of points across a whole farm and the resolution of the picture that emerges does not always compare to the level of detail required by soil science (Peets et al., 2012). Doubling the number of instruments increases the cost and complexity of the setup.

10. Conclusion

The potential of emerging technologies to revolutionize soil data collection processes has been extensively highlighted in the preceding sections of this discussion. The remarkable capabilities of remote sensing, the innovative Internet of Things (IoT), and the evolving field of Unmanned Aerial Vehicles (UAVs) enable the generation of timely, spatially explicit, and high-resolution soil information, which is truly impressive and has the potential to fundamentally transform agricultural decision-making. However, it is crucial to recognize that several significant challenges related to data quality, system integration, associated costs, and various social considerations must be effectively addressed. Only then can we ensure the successful deployment of these exciting technologies. As these innovative technologies continue to advance and new innovations emerge, we can expect progress toward establishing practical, accessible, and intelligent soil data collection systems to accelerate, further enriching the realm of precision agriculture and ultimately benefiting farmers and the broader agricultural community (Peets et al., 2012; Wu, Zhang et al., 2023).

References

Adamchuk, V. I., Hummel, J. W., Morgan, M. T., & Upadhyaya, S. K. (2004). On-the-go soil sensors for precision agriculture. *Computers and Electronics in Agriculture, 44*(1), 71–91.

Adamchuk, V. I., & Rossel, R. V. (2010). Development of on-the-go proximal soil sensor systems. *Proximal Soil Sensing*, 15–28.

Allred, B., Daniels, J. J., & Ehsani, M. R. (2008). *Handbook of agricultural geophysics*. CRC Press.

Atalla, S., Tarapiah, S., Gawanmeh, A., Daradkeh, M., Mukhtar, H., Himeur, Y., … Daadoo, M. (2023). IoT-enabled precision agriculture: Developing an ecosystem for optimized crop management. *Information, 14* (4), 205.

Babaeian, E., Sadeghi, M., Jones, S. B., Montzka, C., Vereecken, H., & Tuller, M. (2019). Ground, proximal, and satellite remote sensing of soil moisture. *Reviews of Geophysics, 57*(2), 530–616.

Carter, M. R., & Gregorich, E. G. (Eds.). (2007). *Soil sampling and methods of analysis* (2nd ed.). CRC Press.

Chamara, N., Islam, M. D., Bai, G., Shi, Y., & Ge, Y. (2022). Ag-IoT for crop and environment monitoring: Past, present, and future. *Agricultural Systems, 203*, Article 103497.

Corwin, D. L., & Lesch, S. M. (2005). Apparent soil electrical conductivity measurements in agriculture. *Computers and Electronics in Agriculture, 46*(1–3), 11–43.

d'Oleire-Oltmanns, S., Marzolff, I., Peter, K. D., & Ries, J. B. (2012). Unmanned aerial vehicle (UAV) for monitoring soil erosion in Morocco. *Remote Sensing, 4*(11), 3390–3416.

Davies, M. A., Davidson, S. J., Deane, P. J., Filicetti, A., Ketcheson, S., Korsah, P., … Strack, M. (2025). A data compilation and synthesis of the impacts of seismic surveys on surface soil properties in boreal Alberta, Canada. *Canadian Journal of Forest Research, 55*, 1–18.

De Capitani di Vimercati, S., & Samarati, P. (December 2007). Data privacy–problems and solutions. In *International conference on information systems security* (pp. 180–192). Berlin, Heidelberg: Springer Berlin Heidelberg.

de Paul Obade, V., & Gaya, C. O. (2021). Digital technology dilemma: On unlocking the soil quality index conundrum. *Bioresources and Bioprocessing, 8*. Article 8.

Deshmukh, H., & Deshpande, H. (2025). Advances in soil health monitoring: A review of digital and sensor-based tools. *International Journal of Plant & Soil Science, 37*(11), 72–89.

Dhillon, R. S., l Adamchuk, V., Holland, K. H., & Hempleman, C. R. (2010). Development of an integrated on-the-go sensing system for soil properties. In *2010 Pittsburgh, Pennsylvania, June 20-June 23, 2010* (p. 1). American Society of Agricultural and Biological Engineers.

Dutta, S., Bhattacharya, S., Sadhukhan, R., Sengupta, S., Chatterjee, R., Roy Choudhury, M., & Das, S. (2023). Transforming agriculture and securing food using digital technologies: Past evolution, present practices and challenges, and future perspectives. *Present Practices and Challenges, and Future Perspectives*.

Faqir, Y., Qayoom, A., Erasmus, E., Schutte-Smith, M., & Visser, H. G. (2024). A review on the application of advanced soil and plant sensors in the agriculture sector. *Computers and Electronics in Agriculture, 226*, Article 109385.

Fisher, D. K., & Gould, P. J. (2012). Open-source hardware is a low-cost alternative for scientific instrumentation and research. *Modern Instrumentation, 1*(2), 8–20.

Gomez, C., Viscarra Rossel, R. A., & McBratney, A. B. (2008). Soil organic carbon prediction by hyperspectral remote sensing and field vis-NIR spectroscopy: An Australian case study. *Geoderma, 146*(3–4), 403–411.

Hardie, M. (2020). Review of novel and emerging proximal soil moisture sensors for use in agriculture. *Sensors, 20*(23), 6934.

Huan, J., Li, H., Wu, F., & Cao, W. (2020). Design of water quality monitoring system for aquaculture ponds based on NB-IoT. *Aquacultural Engineering, 90*, Article 102088.

Igwenagu, U. T. I., Debnath, R., Ahmed, A. A., & Alam, M. J. B. (2025). An integrated approach for Earth infrastructure monitoring using UAV and ERI: A systematic review. *Drones, 9*(3), 225.

Katharria, A., Pant, M., Velásquez, J. D., Snášel, V., Rajwar, K., & Deep, K. (2025). Information fusion in smart agriculture: Machine learning applications and future research directions. *Information Fusion*, 104040.

Kaur, J., Hazrati Fard, S. M., Amiri-Zarandi, M., & Dara, R. (2022). Protecting farmers' data privacy and confidentiality: Recommendations and considerations. *Frontiers in Sustainable Food Systems, 6*, Article 903230.

Lal, R. (2015). Soil carbon sequestration and aggregation by cover cropping. *Journal of Soil and Water Conservation, 70*(6), 329–339.

Landeta-Escamilla, O., Sandoval-Gonzalez, O., Martínez-Sibaja, A., Flores-Cuautle, J. D. J. A., Posada-Gómez, R., & Alvarado-Lassman, A. (2019). Intelligent spectroscopy system used for physicochemical variables estimation in sugar cane soils. *Sensors, 19*(2), 240.

Maes, W. H., & Steppe, K. (2019). Perspectives for remote sensing with unmanned aerial vehicles in precision agriculture. *Trends in Plant Science, 24*(2), 152–164.

McCormick, R., Thenkabail, P. S., Aneece, I., Teluguntla, P., Oliphant, A. J., & Foley, D. (2025). Artificial neural network multi-layer perceptron models to classify California's crops using harmonized landsat sentinel (HLS) data. *Photogrammetric Engineering & Remote Sensing, 91*(2), 91–100.

Mulla, D. J. (2013). Twenty five years of remote sensing in precision agriculture: Key advances and remaining knowledge gaps. *Biosystems Engineering, 114*(4), 358–371.

Nawar, S., & Mouazen, A. M. (2019). On-line vis-NIR spectroscopy prediction of soil organic carbon using machine learning. *Soil and Tillage Research, 190*, 120–127.

Nițu, A., Florea, C., Ivanovici, M., & Racoviteanu, A. (2025). NDVI and beyond: Vegetation indices as features for crop recognition and segmentation in hyperspectral data. *Sensors (Basel, Switzerland), 25*(12), 3817.

Peets, S., Mouazen, A. M., Blackburn, K., Kuang, B., & Wiebensohn, J. (2012). Methods and procedures for automatic collection and management of data acquired from on-the-go sensors with application to on-the-go soil sensors. *Computers and Electronics in Agriculture, 81*, 104–112.

Piccoli, F., Locatelli, S. G., Schettini, R., & Napoletano, P. (2023). An open-source platform for GIS data management and analytics. Sensors, *23*(8), 3788.

Pinter, P. J., Jr., Hatfield, J. L., Schepers, J. S., Barnes, E. M., Moran, M. S., Daughtry, C. S. T., & Upchurch, D. R. (2003). Remote sensing for crop management. *Photogrammetric Engineering & Remote Sensing, 69*(6), 647–664.

Platel, A., Sandino, J., Shaw, J., Bollard, B., & Gonzalez, F. (2025). Advancing sparse vegetation monitoring in the Arctic and Antarctic: A review of satellite and UAV remote sensing, machine learning, and sensor fusion. *Remote Sensing, 17*(9), 1513.

Postolache, S., Sebastião, P., Viegas, V., Postolache, O., & Cercas, F. (2022). IoT-based systems for soil nutrients assessment in horticulture. *Sensors, 23*(1), 403.

Reynolds, D., Baret, F., Welcker, C., Bostrom, A., Ball, J., Cellini, F., … Tardieu, F. (2019). What is cost-effective phenotyping? Optimizing costs for different scenarios. *Plant Science, 282*, 14–22.

Robinson, D. A., Campbell, C. S., Hopmans, J. W., Hornbuckle, B. K., Jones, S. B., Knight, R., … Wendroth, O. (2008). Soil moisture measurement for ecological and hydrological watershed-scale observatories: A review. *Vadose Zone Journal, 7*(1), 358–389.

Seneviratne, S. I., Corti, T., Davin, E. L., Hirschi, M., Jaeger, E. B., Lehner, I., … Teuling, A. J. (2010). Investigating soil moisture–climate interactions in a changing climate: A review. *Earth-Science Reviews, 99*(3–4), 125–161.

Sheng, Z., Yang, S., Yu, Y., Vasilakos, A., Mccann, J., & Leung, K. (2013). A survey on the IETF protocol suite for the internet of things: Standards, challenges, and opportunities. *IEEE Wireless Communications, 20*(6), 91–98.

Shi, Y., Thomasson, J. A., Murray, S. C., Pugh, N. A., Rooney, W. L., Shafian, S., … Yang, C. (2016). Unmanned aerial vehicles for high-throughput phenotyping and agronomic research. *PLoS One, 11*(7), Article e0159781.

Sona, G., Passoni, D., Pinto, L., Pagliari, D., Masseroni, D., Ortuani, B., & Facchi, A. (2016). UAV multispectral survey to map soil and crop for precision farming applications. *The International Archives of the Photogrammetry, Remote Sensing and Spatial Information Sciences, 41*, 1023–1029.

Talavera, J. M., Tobón, L. E., Gómez, J. A., Culman, M. A., Aranda, J. M., Parra, D. T., ... Garzeta, L. E. (2017). Review of IoT applications in agro-industrial and environmental fields. *Computers and Electronics in Agriculture, 142*, 283–297.

Usharani, S., Rajarajeswari, S., Kishore, D. S., & Depuru, S. (May 2023). IoT based animal trespass identification and prevention system for smart agriculture. In *2023 7th international conference on intelligent computing and control systems (ICICCS)* (pp. 983–990). IEEE.

Villa-Henriksen, A., Edwards, G. T., Pesonen, L. A., Green, O., & Sørensen, C. A. G. (2020). Internet of things in arable farming: Implementation, applications, challenges and potential. *Biosystems Engineering, 191*, 60–84.

Viscarra Rossel, R. A., Adamchuk, V. I., Sudduth, K. A., McKenzie, N. J., & Lobsey, C. (2011). Proximal soil sensing: An effective approach for soil measurements in space and time. *Advances in Agronomy, 113*, 243–291.

Wang, J., Zhen, J., Hu, W., Chen, S., Lizaga, I., Zeraatpisheh, M., & Yang, X. (2023). Remote sensing of soil degradation: Progress and perspective. *International Soil and Water Conservation Research, 11*(3), 429–454.

Weizhong, Y., Wang, Y., Yao, Y., & Jingbo, S. (2019). Soil moisture monitoring system based on narrow band internet of things. *Transactions of the Chinese Society for Agricultural Machinery, 50*(Supp), 243–247.

Wu, Y., Yang, Z., & Liu, Y. (2023). Internet-of-things-based multiple-sensor monitoring system for soil information diagnosis using a smartphone. *Micromachines, 14*(7), 1395.

Wu, B., Zhang, M., Zeng, H., Tian, F., Potgieter, A. B., Qin, X., & Yan, N. (2023). Challenges and opportunities in remote sensing-based crop monitoring: A review. *National Science Review, 10*(4), Article nwac290.

Xu, Y. (2016). Envirotyping for deciphering environmental impacts on crop plants. *Theoretical and Applied Genetics, 129*(4), 653–673.

Zhang, X., Feng, G., & Sun, X. (2024). Advanced technologies of soil moisture monitoring in precision agriculture: A review. *Journal of Agriculture and Food Research, 18*, Article 101473.

Zhuo, L., Dai, Q., Zhao, B., & Han, D. (2020). Soil moisture sensor network design for hydrological applications. *Hydrology and Earth System Sciences, 24*(5), 2577–2591.

Index

'*Note:* Page numbers followed by "f" indicate figures and "t" indicate tables.'

Printed and bound by CPI Group (UK) Ltd, Croydon, CR0 4YY
14/09/2026
02216410-0017